COLLOIDS IN BIOTECHNOLOGY

SURFACTANT SCIENCE SERIES

FOUNDING EDITOR

MARTIN J. SCHICK
1918–1998

SERIES EDITOR

ARTHUR T. HUBBARD
Santa Barbara Science Project
Santa Barbara, California

COLLOIDS IN BIOTECHNOLOGY

Edited by

Monzer Fanun
Al-Quds University
East Jerusalem, Palestine

CRC Press
Taylor & Francis Group
Boca Raton London New York

CRC Press is an imprint of the
Taylor & Francis Group, an **informa** business

CRC Press
Taylor & Francis Group
6000 Broken Sound Parkway NW, Suite 300
Boca Raton, FL 33487-2742

First issued in paperback 2017

ISBN-13: 978-1-4398-3080-2 (hbk)
ISBN-13: 978-1-138-11519-4 (pbk)

Library of Congress Cataloging-in-Publication Data

Colloids in biotechnology / Monzer Fanun.
    p. cm. -- (Surfactant science series)
   Includes bibliographical references and index.
   ISBN 978-1-4398-3080-2 (hardcover : alk. paper)
   1. Colloids--Biotechnology. I. Fanun, Monzer. II. Title. III. Series.

R857.C66.C655 2010
541'.345--dc22
                                                           2010027818

Visit the Taylor & Francis Web site at
http://www.taylorandfrancis.com

and the CRC Press Web site at
http://www.crcpress.com

Contents

Preface

Colloids have been used for a long time, since the invention of ink, but only recently have scientists been able to decode some of their unique properties. Colloids are heterogeneous mixtures made of tiny particles or droplets that appear visually as homogeneous solutions, in the sense that they are actually a mixture of two phases but look as if they consist of only one phase. They have shown great potential for use in a wide variety of applications, and the design and fabrication of such systems have consequently attracted considerable interest in recent years. Many innovations related to biological activity such as drug delivery systems and medical imaging are based on colloid and interface science. Colloids are utilized as biofunctional products, or serve as cell-like man-made compartments in which biological activity takes place. Many of the systems currently under development, particularly those for bio-applications, draw their inspiration from naturally occurring structures. That is, by mimicking the supramolecular architecture of structures found in nature, one can prepare complex materials capable of highly sophisticated functions. Colloidal particles with the possibility to bind or encapsulate active substances such as drugs or agents are a prerequisite for advanced applications of colloidal materials as carriers. The gap between the extensive research and translation into commercially viable options needs to be bridged to find more new applications in the fields of biomedicine and biotechnology. This book describes experimental and theoretical developments in the field of biotechnological applications of colloids over the past decade. We have tried to strike a reasonable balance between theory and experiment, between principles and applications, and between molecular and physical approaches to the subject. We discuss new types of biosurfactants; mixtures of surfactants; and peptides, proteins, and polyelectrolytes. The formation and properties of magnetic colloids and their applications in chemical biology and medicine are also described. The current progress in the design of self-assembled materials for biotechnology is highlighted. The formation of nanofibers and the use of sol-gel technology in biotechnology are also covered.

Chapter 1 by Lindman covers the recent advances in the study of amphiphilic biopolymers with respect to their bulk self-assembly and surface-modifying ability. Chapter 2 by Shibata et al. summarizes the current topics on the surface property, phase behavior, and orientational change of synthetic pulmonary surfactant mixtures with a newly designed 18-mer amphiphilic α-helical peptide consisting of 13 hydrophobic and 5 hydrophilic amino acid residues at the interface. Self-assembling oligopeptides represent a new class of natural or bio-inspired surfactants, which can form a wide variety of supramolecular self-organized structures ranging from micelles and vesicles to tubules and gels. Chapter 3 by Déjugnat and Zemb provides an overview of supramolecular self-organized structures that are formed by amphiphilic oligopeptides mainly in water but also, in some cases, in organic media. The adsorption of polymers and biopolymers on the surface and interface plays an important role in different areas of natural sciences, including materials science, colloid science, and biophysics. Chapter 4 by Jachimska covers the bulk characteristic of polyelectrolytes and proteins. The bulk characterization of the molecules was used for the interpretation of the molecule conformation obtained by adsorption on the surface. Visualization of the conformation of individual molecules on the adsorption surface clearly reveals the flexible nature of the macromolecule (change of the conformation from fully stretched to a bent, worm-like shape).

Lysine amino acid–based surfactants have been synthesized in a variety of structural forms. Because of the possibility of linking the hydrophobic group through the carboxylic group and the two amino groups, this amino acid offers a wide range of possible structures for surfactant formation. The ionic nature of surfactants with lysine in the head group depends on the pH and the specific structure of the surfactant. Chapter 5 by Pons et al. deals with the synthetic, self-aggregation, and biological properties of lysine amino acid. Protein production is an important step for biotechnology and functional proteomics. Chapter 6 by Wang and He reviews the recent developments in cell-free systems and their proteomic applications. Milk contains many components that are valuable from a nutritional point of view but that can also find many other alternative applications. Due to their specific structural and functional properties, milk proteins are excellent emulsifying, stabilizing, and adhesive agents. Chapter 7 by Kalicka et al. deals with the historical, current, and potential nonfood applications of milk proteins. The demand for an efficient delivery system for DNA-based vaccines/therapeutics has increased vastly in response to rapid advances in the use of plasmid DNA molecules as nonviral vectors for vaccination and gene therapy applications. However, several limitations such as ineffective cellular uptake and intracellular delivery and degradation of pDNA need to be overcome. Colloidal delivery systems are noted for their flexibility in tailoring the internal structure and surface of the colloids in order to achieve improved delivery modes and therapeutic delivery profiles. Chapter 8 by Ho et al. discusses the role of these colloidal systems in improving and advancing DNA genetic therapeutics. The benefits and drawbacks of using the colloidal system in alternative routes of administration are also discussed. In addition, the features of colloidal delivery systems that provide the pharmaceutical industry with a highly flexible tool to solve formulation and clinical problems are also addressed. The biological events in cells are caused by specific interactions among proteins, nucleotides, compounds, drugs, and so on. The identification of unknown interactions among these molecules is important for chemical biology and drug discovery. Chapter 9 by Iizumi et al. reviews the development of high-performance affinity beads, which enable the identification of unknown target molecules—something that will contribute to the advancement of chemical biology and to the development of medicine.

Nanoparticles can be defined as particles of any material having dimensions of 100 nm or less with novel properties that distinguish them from bulk material due to size and surface effects. The use of nanoparticles for any application is strongly dependent on their physicochemical characteristics and interactions with surface modifiers. Chapters 10 and 11 deal with colloidal nanoparticles and their use in biotechnology. Extracellular biological synthesis of nanoparticles is an emerging branch of science that interconnects nanotechnology and biotechnology. Increasing awareness toward green chemistry and other biological processes has led to the development of eco-friendly approaches for the synthesis of nanoparticles. Therefore, there is growing concern among scientists to develop environment-friendly and sustainable methods. Since the synthesis of nanoparticles of different compositions, sizes, shapes, and controlled dispersity is an important aspect of nanotechnology, new cost-effective procedures are being developed. In recent years, plant-mediated biological synthesis of nanoparticles has gained importance due to its simplicity and eco-friendliness. Chapter 10 by Narayanan and Sakthivel describes the recent developments on the extracellular biological synthesis of nanoparticles and their applications. Regulatory peptide receptor proteins are over-expressed in numerous human cancer cells. These receptors have been successfully used as molecular targets for peptides to localize cancer tumors. Gold nanoparticles (AuNPs) undergo a plasmon resonance with light incidence and are resistant to oxidation. Both properties are relevant for possible applications in optical biodetection, cellular imaging, and photo-thermal therapeutic medicine. Nanoconjugates used in the diagnosis or therapy must be nontoxic, biocompatible, and stable in biological media with high selectivity for biological targets. Recent studies have demonstrated that conjugating AuNPs to peptides produces biocompatible and stable multifunctional systems with target-specific molecular recognition. Chapter 11 by Ferro-Flores et al. reviews the strategies of design, synthesis, physicochemical characterization, and molecular recognition assessment for AuNPs conjugated to peptides as potential nanopharmaceuticals are discussed. An overview of the therapeutic and diagnostic possibilities of these nanoconjugates is also included.

The term "bioadhesion" refers to any bond formed between two biological surfaces, or a bond between a biological and a synthetic surface. In the case of bioadhesive drug delivery systems, this term is typically used to describe the adhesion between polymers, either synthetic or natural and soft tissue (i.e., gastrointestinal mucosa). In general, bioadhesion is an all-inclusive term to describe adhesive interactions with any biological or biologically derived substance, "mucoadhesion" is used only when describing a bond involving mucus or a mucosal surface, and "cytoadhesion" is the cell-specific bioadhesion. Patel deals with bioadhesive microspheres tailoring, preparations, and applications in Chapter 12. Microencapsulation is a defined technology, involving the coating or entrapment of a core material into capsules in the size range between a few micrometers and a few millimeters. A prevalent objective of microencapsulation is to protect the core material from degradation by reducing its reactivity to the outside environment. This is mainly achieved by control of the mass transfer between the core and surrounding ambient by using the shell or matrix material as a physical barrier. Therefore, microencapsulation can serve as a technological method to protect living probiotic cells against adverse external conditions when introduced for the creation of functional foods. Chapter 13 by Heidebach et al. discusses the future approaches toward encapsulation processes that selectively combine the advantages of different encapsulation techniques. Promising strategies that include the application of water-insoluble microcapsules from spray-drying, based on proteins or polysaccharides, are also reviewed.

The engineering of emulsion structures has attracted the interest of many research laboratories over the last few years due to increasing industrial demand for colloidal matrices that could be used as bio-microreactors in biotechnology, vehicles for delivering and controlled release of lipophilic (or hydrophilic) active compounds in pharmaceutical and cosmetic formulations, tunable systems for controlling texture, stability and rheological properties of food products, and many other well-defined and profitable applications. Chapters 14 through 18 deal with the formulation and use of emulsions, multiple emulsions, and microemulsions for biotechnological applications. In Chapter 14, Macierzanka and Szeląg have shown an example of how an overall structure of emulsion with desired functional properties can be controlled by tailoring the emulsifier composition. Multiple emulsions are complex polydispersed systems and are also called "emulsion of emulsions" or, more recently, "emulsion liquid membranes." Multiple emulsions have shown promise in the pharmaceutical and cosmetic industry, and in food technology and separation sciences. Recently, many interesting fields of applications, like vaccine formulations, enzyme immobilization, etc., have been found. Multiple emulsions also serve as a substitute for blood. These systems also have some other advantages, such as the protection of the entrapped substances in the inner aqueous phase, and modulate their release rate to improve the efficacy of the product and the incorporation of several active compounds in the different compartments. The encapsulation effect of these systems is of primary importance as a prolonged release system. Chapter 15 by Ozer and Kantarci reviews the recent biomedical and biotechnological applications of multiple emulsions. Microencapsulation by solvent evaporation is widely applied in pharmaceutical investigations in order to control drug release. Forming polymer nano- or microspheres can degrade and release the encapsulated drug slowly with a specific release profile. The sustained drug release has clinical benefits: it reduces the dosing frequency; is more convenient and acceptable for patients; and targets drugs to specific locations, resulting in higher efficiency. The appropriate choice of the encapsulation method to achieve efficient drug incorporation depends on the hydrophilicity or the hydrophobicity of the drug. Multiple emulsion solvent evaporation methods have been used primarily in pharmaceutical investigations for entrapping water-soluble compounds into water-insoluble polymers in order to synthesize nano- or microparticles suitable for sustaining the release of active agents. The multiple emulsion technique most generally applied is the water-in-oil-in-water system. Numerous hydrophilic peptides and proteins have been encapsulated by the double emulsion solvent evaporation method. Chapter 16 by Feczkó discusses the multiple emulsion–solvent evaporation technique and its application to synthesize pharmaceutical agents incorporated in protecting polymer. Recent years have witnessed an increase in the use of microemulsions in biotechnological applications, in general, with the establishment of microemulsions as an inevitable means for protein-based reactions, in particular. The reason underlying the marked increase in popularity and need of microemulsions for protein-based materials is mainly the enhanced solubilization of proteins in nonpolar and lipophilic solvents and the retention of their activity in microemulsions, which is otherwise negligible. Chapter 17 by Misra et al. reviews the recent applications of microemulsions in biotechnology. Fanun, in Chapter 18, reviews the formulation and applications of biocompatible microemulsions.

Nanofibers are ultrafine solid fibers of very small diameters, although fibers of hundreds of nanometers in diameter are also included among nanofibers. Nanofibers have a large surface area per unit mass and a very small pore size. The surface functionality of nanofibers can be controlled and tailored flexibly, and they possess superior mechanical performance like stiffness and tensile strength. Electrospinning seems to be the only suitable candidate for the large-scale production of one-by-one continuous nanofibers from various polymers or biopolymers. Nanofibers prepared from biopolymers or biocompatible synthetic polymers are very attractive for medical and cosmetic applications. Especially, biopolymer nanofibers can be used in regenerative medicine as wound dressings or three-dimensional scaffolds (cell supports for tissue engineering). Chapter 19 by Pekař et al. reviews the hyaluronan-based nanofibers.

Sol-gel chemistry is a versatile approach for fabricating biocompatible materials and has attracted considerable interest among researchers for various bio-applications. Ambient sol-gel processing makes it possible to utilize sol-gel matrices for the entrapment of various types of biological materials from proteins to whole cells. Although conventional sol-gel bioencapsulation has demonstrated the functional activity of encapsulated biomolecules to some extent, it cannot lead to biosensor development. This is because of the local environment of the entrapped biomolecules that affect both the stability and the functionality of the entrapped species, as well as the interaction with analyte molecules. Recently, sol-gel processing has come up with new modifications including a new class of biocompatible precursors, organically modified silanes, and other additives, as well as tuning the level and nature of the additive within the sol-gel-derived matrix to optimize the activity and stability of the entrapped biomolecules. In addition, sol-gel chemistry in combination with other techniques like microarray technology, molecular imprinting technique, and nanotechnology can provide a better approach for various biomedical, environmental, defense, and biotechnological applications. Chapter 20 by Gupta and Kumar reviews sol-gel technology with a specific focus on bio-applications of sol-gel materials. This chapter describes the significance of molecular imprinting in the sol-gel matrix for such biomolecules for which no recognizing biological receptors are available, and highlights the importance of the sol-gel process for preparing bioactive materials and sol-gel-derived nanocomposites for biosensing and immunosensing applications.

Finally, this book will be of immense help, not only for those involved in research and development over a range of different technologies dealing with colloids in biotechnology, but also for scientists studying the field of colloid and surface science. It summarizes recent research in the field of colloids in biotechnology, therefore avoiding the need to search through stacks of journals for critical information. This unique book presents all the facts that may be required by the researcher in the laboratory or in the classroom, and can be used as an effective guide for planning future research. An international community of colloid scientists has come together to create this book. It should complement other existing books on biocolloids very well, which, in general, take a more traditional approach and provide a systematic review of the fundamental (pure) aspects. An important feature of this book is that the authors of each chapter have been given the freedom to present, as they see fit, the spectrum of the relevant science, from pure to applied, in their particular topic. Author will have views on, and approaches to, a specific topic, molded by their experience. This book covers the recent advances in the formulation and characterization of the properties of colloids for biotechnological applications; it also covers new colloidal systems used for biotechnological applications that include nanoparticles, emulsions, and others.

I would like to thank all the contributing authors for sacrificing their precious time despite their demanding schedules. In total, 56 individuals from 17 countries contributed to this book. All of them are specialists in their specific areas. Any errors or omissions that remain are entirely my responsibility. I am especially grateful to all the reviewers for their valuable comments. I am also thankful to Barbara Glunn of Taylor & Francis for all her help during this project.

Monzer Fanun
Al-Quds University
East Jerusalem, Palestine

Editor

Monzer Fanun is a professor in surface and colloid science and the head of the colloids and surfaces research laboratory at Al-Quds University, East Jerusalem, Palestine. He is the editor of the books *Microemulsions: Properties and Applications* and *Colloids in Drug Delivery* and the author or coauthor of more than 50 professional papers. He is a member of the European Colloid and Interface Society and a fellow of the Palestinian Academy for Science and Technology. In 2003, he received his PhD in applied chemistry from the Casali Institute of Applied Chemistry, a part of the Institute of Chemistry at the Hebrew University of Jerusalem, Jerusalem, Israel. His research focuses on colloidal systems for health-care products and surfactant-based alternatives to organic solvents.

Contributors

Helena Bilerová
CPN spol. s.r.o.
Dolní Dobrouč, Czech Republic

Michael K. Danquah
Bio Engineering Laboratory
Department of Chemical Engineering
Monash University
Clayton, Victoria, Australia

Christophe Déjugnat
Laboratory of Molecular and Chemical- and
    Photochemical Reactivity
Paul Sabatier University
Toulouse, France

Monzer Fanun
Faculty of Science and Technology
Colloids and Surfaces Research Laboratory
Al-Quds University
East Jerusalem, Palestine

Tivadar Feczkó
Chemical Research Center
Institute of Materials and Environmental
    Chemistry
Hungarian Academy of Sciences

and

Research Institute of Chemical and Process
    Engineering
University of Pannonia
Veszprém, Hungary

and

Faculty of Mechanical Engineering
Institute for Engineering Materials and Design
University of Maribor
Maribor, Slovenia

Guillermina Ferro-Flores
Department of Radioactive Materials
Instituto Nacional de Investigaciones Nucleares
Estado de México, Mexico

Kiruba Florence
Pharmacy Department
Faculty of Technology and Engineering
The Maharaja Sayajirao University of Baroda
Vadodara, Gujarat, India

Petra Först
Z I E L Research Center for Nutrition and Food
    Science
Institute for Food Process Engineering and
    Dairy Technology
Technische Universität München
Weihenstephan, Germany

Tadeusz Grega
Faculty of Food Technology
Department of Animal Production
University of Agriculture
Krakow, Poland

Radha Gupta
Department of Biological Sciences and
    Bioengineering
Indian Institute of Technology
Kanpur, Uttar Pradesh, India

Claudia E. Gutierrez-Wing
Department of Materials
Instituto Nacional de Investigaciones Nucleares
Estado de México, Mexico

Hiroshi Handa
Graduate School of Bioscience and
    Biotechnology
Tokyo Institute of Technology
Yokohama, Japan

and

Integrated Research Institute
Tokyo Institute of Technology
Yokohama, Japan

Mamoru Hatakeyama
Integrated Research Institute
Tokyo Institute of Technology
Yokohama, Japan

Mingyue He
The Babraham Institute
Cambridge, United Kingdom

Thomas Heidebach
Z I E L Research Center for Nutrition and
    Food Science
Institute for Food Process Engineering and
    Dairy Technology
Technische Universität München
Weihenstephan, Germany

Jenny Ho
Bio Engineering Laboratory
Department of Chemical Engineering
Monash University
Clayton, Victoria, Australia

Yi Huang
Department of Chemical Engineering
Monash University
Clayton, Victoria, Australia

Yosuke Iizumi
Graduate School of Bioscience and
    Biotechnology
Tokyo Institute of Technology
Yokohama, Japan

María Rosa Infante
Institut de Química Avançada de Catalunya
Consejo Superior de Investigaciones Científicas
Barcelona, Spain

Barbara Jachimska
Institute of Catalysis and Surface Chemistry
Polish Academy of Sciences
Cracow, Poland

Yasuaki Kabe
Department of Biochemistry and Integrative
    Medical Biology
School of Medicine
Keio University
Tokyo, Japan

Dorota Kalicka
Faculty of Food Technology
Department of Animal Production
University of Agriculture
Krakow, Poland

Gulten Kantarci
Faculty of Pharmacy
Department of Pharmaceutical Biotechnology
Ege University
Izmir, Turkey

Ulrich Kulozik
Z I E L Research Center for Nutrition and Food
    Science
Institute for Food Process Engineering and
    Dairy Technology
Technische Universität München
Weihenstephan, Germany

Ashok Kumar
Department of Biological Sciences and
    Bioengineering
Indian Institute of Technology
Kanpur, Uttar Pradesh, India

Manisha Lalan
Pharmacy Department
Faculty of Technology and Engineering
The Maharaja Sayajirao University of Baroda
Vadodara, Gujarat, India

Sannamu Lee
Faculty of Science
Department of Chemistry
Fukuoka University
Fukuoka, Japan

and

Faculty of Pharmaceutical Sciences
Department of Biophysical Chemistry
Nagasaki International University
Nagasaki, Japan

Elena Leeb
Z I E L Research Center for Nutrition and Food
    Science
Institute for Food Process Engineering and
    Dairy Technology
Technische Universität München
Weihenstephan, Germany

Björn Lindman
Physical Chemistry
Centre for Chemistry and Chemical
    Engineering
Lund University
Lund, Sweden

and

Department of Chemistry
University of Coimbra
Coimbra, Portugal

Shan Liu
Bio Engineering Laboratory
Department of Chemical Engineering
Monash University
Clayton, Victoria, Australia

Adam Macierzanka
Food Structure and Health Programme
Institute of Food Research
Norwich, United Kingdom

and

Chemical Faculty
Department of Fats and Detergents Technology
Gdansk University of Technology
Gdansk, Poland

Ambikanandan Misra
Pharmacy Department
Faculty of Technology and Engineering
The Maharaja Sayajirao University of Baroda
Vadodara, Gujarat, India

María del Carmen Morán
Chemistry Department
Coimbra University
Coimbra, Portugal

Consuelo Arteaga de Murphy
Department of Nuclear Medicine
Instituto Nacional de Ciencias Médicas y
    Nutrición Salvador Zubirán
Tlalpan, Mexico

Dorota Najgebauer-Lejko
Faculty of Food Technology
Department of Animal Production
University of Agriculture
Krakow, Poland

Hiromichi Nakahara
Faculty of Pharmaceutical Sciences
Department of Biophysical Chemistry
Nagasaki International University
Nagasaki, Japan

K. Badri Narayanan
Department of Biotechnology
Pondicherry University
Puducherry, India

Blanca E. Ocampo-García
Department of Radioactive Materials
Instituto Nacional de Investigaciones

and

Faculty of Medicine
Universidad Autónoma del Estado de México
Estado de México, Mexico

Ozgen Ozer
Faculty of Pharmacy
Department of Pharmaceutical Technology
Ege University
Izmir, Turkey

Jayvadan K. Patel
Department of Pharmaceutical Technology
Nootan Pharmacy College
Visnagar, Gujarat, India

Miloslav Pekař
Faculty of Chemistry
Institute of Physical and Applied Chemistry
Brno University of Technology
Purkyňova, Czech Republic

Lourdes Pérez
Institut de Química Avançada de Catalunya
Consejo Superior de Investigaciones Científicas
Barcelona, Spain

Aurora Pinazo
Institut de Química Avançada de Catalunya
Consejo Superior de Investigaciones Científicas
Barcelona, Spain

Ramon Pons
Institut de Química Avançada de Catalunya
Consejo Superior de Investigaciones Científicas
Barcelona, Spain

Flor de María Ramírez
Department of Chemistry
Instituto Nacional de Investigaciones Nucleares
Mexico, Japan

Satoshi Sakamoto
Graduate School of Bioscience and
    Biotechnology
Tokyo Institute of Technology
Yokohama, Japan

N. Sakthivel
Department of Biotechnology
Pondicherry University
Puducherry, India

Clara L. Santos-Cuevas
Department of Radioactive Materials
Instituto Nacional de Investigaciones
    Nucleares

and

Faculty of Medicine
Universidad Autónoma del Estado de México
Estado de México, Mexico

Tapan Shah
Pharmacy Department
Faculty of Technology and Engineering
The Maharaja Sayajirao University of Baroda
Baroda, Gujarat, India

Osamu Shibata
Faculty of Pharmaceutical Sciences
Department of Biophysical Chemistry
Nagasaki International University
Nagasaki, Japan

Halina Szeląg
Chemical Faculty
Department of Fats and Detergents Technology
Gdansk University of Technology
Gdansk, Poland

Vladimír Velebný
CPN spol s.r.o.
Dolní Dobrouč, Czech Republic

Mingrong Wang
Center for Protein Therapeutics
Sichuan Industrial Institute of Antibiotics
Sichuan, People's Republic of China

Thomas Zemb
Institute of Separative Chemistry
Commission for Atomic Energy and
    Alternative Energies, Marcoule Center
Bagnols-sur-Cèze, France

1 Amphiphilic Biopolymers: DNA and Cellulose

Björn Lindman

CONTENTS

1.1 INTRODUCTION

Amphiphilic compounds, that is, those that have distinct hydrophilic and lipophilic parts, are used in most branches of industry and are ubiquitous in biological systems. They range from low molecular weight substances (like surfactants and lipids) to macromolecules (comprising synthetic graft and block copolymers) and biomacromolecules (like proteins, lipopolysaccharides, and nucleic acids) [1–4]. Amphiphilic molecules are those that have an affinity for two different types of environments. Amphiphilic molecules self-organize both in bulk solution and at interfaces. Low-molecular-weight amphiphilic compounds, mainly constituted of surfactants and polar lipids, have been thoroughly investigated extensively and are well understood both with respect to their bulk self-assembly and surface-modifying ability. The elaborate research efforts have been stimulated by numerous applications, ranging from soil removal to pharmaceutical and other formulations, as well as the biological implications—cell membranes being a prime example. The broad use of surfactants is, however, not without problems; in recent years, there has been an increased focus on the environmental and toxic effects displayed, in particular, by cationic surfactants. However, a remedy to this can often be found in designing surfactants, based on constituents derived from nature, for example, amino acids. Such work is quite recent and has been pioneered by Maria Rosa Infantes and her group in Barcelona, one of the contributors in this book. The study of high molecular weight amphiphilic molecules is of much more recent. There are two reasons for this. First, synthetic amphiphilic polymers, illustrated by block and graft copolymers, have been synthesized to any important extent only recently. Second, the recognition of amphiphilicity of biomacromolecules has been very limited and, in our view, the significance of hydrophobic interactions in biology has not been taken into consideration, but rather given a cold shoulder. While proteins, in which the secondary structure is determined by a balance between hydrophilic and hydrophobic interactions, and lipopolysaccharides are obvious examples of amphiphilic biological macromolecules, there are many cases where the role of amphiphilicity is not properly considered. As a means of illustration of these aspects, one can take into consideration two typical examples: DNA and cellulose. As indicated below, the double helix structure of DNA owes its stability to hydrophobic interactions, and this aspect is also applicable to the insolubility of cellulose in water. Often, the association of DNA and cellulose is

1

discussed in terms of hydrogen bonding. However, it is our contention that hydrogen bonding is, in general, not the driving force for association in the presence of excess of water; water itself has a very strong hydrogen bonding ability. DNA and cellulose are, of course, among the biopolymers that receive the largest attention because of their biological roles, and because of their broad biotechnological and industrial applications. We will here emphasize that they are both amphiphilic. As indicated earlier, this concept has not been given sufficient attention, with consequences both for a general understanding and, in particular, in steering the technological developments in wrong directions. We will here briefly consider these issues.

1.2   SOME ASPECTS OF DNA SELF-ASSEMBLY

The association of two DNA strands into the double helix is driven by the hydrophobic interactions between the bases. Polar interactions, associated with the phosphate and carbohydrate groups, counteract the association. Hydrogen bonding and specific packing of the bases control the details of the double helix structure [5]. The electrostatic interactions of DNA have been analyzed in detail [5]. The hydrophobic interactions have been much less discussed; in particular, the balance between the polar and nonpolar interactions have a deep impact into how DNA interacts with cosolutes, including electrolytes, nonpolar molecules, surfactants, lipids, and macromolecules, as well as with interfaces [6]. Here, we will briefly comment on the amphiphilic nature of DNA and its consequences for the solution behavior—we have recently reviewed these aspects in more detail [6]. DNA is clearly different from both block and graft copolymers, but closer to the graft copolymer concept, with hydrophobic grafts on a hydrophilic backbone. However, the segregation between hydrophilic and lipophilic parts is less pronounced in DNA, and the force opposing self-assembly is stronger due to a high charge density and a large persistence length. While the detailed structure of the double helix has been extensively investigated, it should be noted that the balance between the hydrophobic force driving self-assembly and the opposing force is very subtle. Two consequences are as follows: First, the stability of the double helix (ds-DNA) is critically dependent on the concentration of the electrolyte. In the absence of the electrolyte, the opposing force dominates, and dissociation occurs. Small amounts of electrolyte, or essentially any cationic cosolute, overcome the electrostatic repulsion and stabilize ds-DNA. Second, if the driving force is changed, for example, by changing the base composition, there is a significant change in the stability of the double helix.

Manifestations of the hydrophobic interactions include the following:

- *Solubilization of hydrophobic molecules* [7–15]. This concept can be illustrated by the so-called intercalating agents. Ethidium bromide (EtBr) is a well-known fluorescent dye commonly used to study the interaction between DNA and cosolutes due to its displacement when other molecules bind to DNA. Other dyes binding to DNA are not soluble in water. A recent work has focused on the role of the ligand hydrophobicity on DNA binding and it was found, not surprisingly, that the most hydrophobic compounds have a higher binding affinity to DNA. In this case, however, the ligands did not interact with DNA by intercalation, but by hydrophobic interactions with the surface of the DNA, that is, the pockets of the groves. This type of interaction is common for some fluorescent dyes, such as DAPI, and in protein–DNA interactions.

- *Adsorption on hydrophobic surfaces.* It was observed by ellipsometry that, whereas both ds- and ss-DNA molecules adsorb on hydrophobic surfaces, ss-DNA adsorbs, in general, more preferentially than ds-DNA [16,17]. In addition, whereas ds-DNA molecules form a very thick and diffuse layer on the surface, the ss-DNA molecules adsorb in a thin layer of about 20 Å, indicating that the molecules are parallel to the surface [17]. This is naturally due to the larger hydrophobicity of the ss-DNA, as each base serves as an attachment point to the surface, which overcomes the entropy loss of the adsorption; ss-DNA is much more flexible than ds-DNA. In fact, the bases were shown to have different adsorption properties depending on their hydrophobicity. The purine bases, more hydrophobic due to the two aromatic rings, present a larger adsorption than the pyrimidine bases [18,19].

- *Effects of hydrophobic cosolutes on DNA melting.* The interactions between DNA and alkyltrimethylammonium bromide salts with short hydrophobic chains and the influence of the chain length on the melting have been previously studied [20]. It was observed that the melting temperature of DNA decreases with the increase of the hydrophobic group in a linear fashion up to the pentyl substitution. Short-chain alcohols showed the same behavior. The melting temperature of DNA was found to decrease in water/methanol solutions [21]. Furthermore, the midpoint of the solvent denaturation decreased in the order of methanol, ethanol, and propanol; that is, the secondary structure stability was lowered as the length of the aliphatic chain was increased [21].

- *Differences in interactions of cationic surfactants between ss- and ds-DNA.* One other indication that points to the importance of the hydrophobic moieties of DNA on the interaction with cosolutes is the difference in interactions of ss- and ds-DNA with cationic surfactants. It was observed that the precipitation behavior for DNA—dodecyltrimethylammonium bromide ($C_{12}TAB$) is different when DNA is in the denatured or the double-helix conformation [22]. In this case, the DNA conformation was controlled by the temperature. The fact that $C_{12}TAB$ interacts preferentially with ss-DNA for low concentrations of surfactant signifies that the melting temperature of DNA shifts with lower temperature [22]. Other illustrations on the role of hydrophobic interactions in DNA self-assembly [6] relate to DNA–protein interactions [23–25], dependence of DNA melting on base sequence, and preparation of DNA chemical and physical gels. As regards the chemical gels [26], chemically cross-linked ss- and ds-DNA interact differently with cationic surfactants. With reference to the physical gels [27], it is notable that DNA enables them to be formed in combination with hydrophobically modified cationic polymers.

1.3   DISSOLUTION OF CELLULOSE

Cellulose is the most abundant natural polymer and has numerous applications. Some applications involve dissolution of cellulose, but these have different limitations. Cellulose is difficult to dissolve. It is insoluble in water and in typical organic solvents, but soluble in a few classes of solvents, which, according to current understanding, have no clear common properties. On the other hand, glucose is highly soluble in water (for example 9% at 25°C and 56% at 90°C), but insoluble in nonpolar solvents. Also glucose derivatives, like alkylpolyglucosides (APGs), made less polar by introduction of alkyl chains, may be highly soluble in water. The aqueous insolubility of cellulose contrasts to some other nonionic polysaccharides, like dextran, another polyglucose. Cellulose is quite a polar molecule, with several –OH groups, and thus has a good hydrogen-bonding ability. The insolubility in nonpolar organic solvents, therefore, poses no problem of understanding. On the other hand, the aqueous insolubility is more difficult to understand and has created a lot of interest and concomitant research. The scientific literature on the topic is huge and broadly spread. The opinions are very scattered. However, there seems to be a consensus among leaders in the field that the insolubility of cellulose is due to its ability to form intra- and intermolecular hydrogen bonds; some authors refer, in addition, to the crystallinity of cellulose—a matter questioned by others. However, an analysis of the literature together with common knowledge of intermolecular interactions and solution physical chemistry demonstrates that completely different factors are behind the cellulose solubility characteristics. This analysis is expected to lay a better ground for understanding the mechanism behind cellulose solubility in novel solvents, like ionic liquids, and, in particular, provide a better roadmap for further selection of solvents for cellulose.

The driving force for dissolution, or miscibility in general, is almost always the entropy of mixing, and not favorable interactions. For dissolution, the change in free energy needs to be negative. The free-energy change is the sum of two terms, one normally positive energetic (enthalpy), which depends on intermolecular interactions, and one entropic, which is negative. The reason for low polymer solubility is that the number of molecules, and thus the entropic term, becomes smaller. These considerations refer to the translational entropy of mixing. There is also another term for polymers, related to the conformational freedom. For a polymer that increases its conformational freedom on going into solution, dissolution is more favorable than for a polymer that cannot change conformation. Flexible polymers are, therefore, more soluble than stiff ones. Applying this to cellulose, the fully equatorial conformation of β-linked glucopyranose residues stabilizes the chair structure, minimizing its flexibility (for example, relative to the slightly more flexible α-linked glucopyranose residues in amylose). The stiffness of cellulose thus directly, due to a low configurational entropy in solution, contributes to the lowering of the solubility. There is also another effect: A polymer with hydrophilic and hydrophobic parts will adjust its conformation in water to reduce the contacts between hydrophobic parts and water. For a stiff polymer, such conformational changes are hindered. This effect would contribute to a low solubility of cellulose in water. Much of the current literature considers that the insolubility of cellulose is a result of the hydrogen bond systems. The reason for cellulose solubility in ionic liquids and other solvents is that they "break" these hydrogen bonds; for original work and excellent reviews on cellulose dissolution in ionic liquids see the works in [28–38]. Some authors also refer to the crystallinity of cellulose as a contributing cause for its insolubility. Indeed, solubility is a balance between the energy of solid and solution states. For a given solution state, we expect the solubility to decrease as the stability of the solid state, "crystallinity," increases. Water is a very strong hydrogen-bonded liquid. Breaking these hydrogen bonds increases the energy. Therefore, solutes, which are unable to form hydrogen bonds, but decrease the number of hydrogen bonds of water, tend to have low solubility. Therefore, the general teaching is that compounds capable of significant hydrogen bonding should be soluble in water. This lesson seems to be forgotten in the cellulose field, while the understanding is different in other fields dealing with polysaccharides. In a textbook on *Food Chemistry* [39], the solubility of polysaccharides in water is entirely discussed in terms of hydrogen bonding with water as a promoter of solubility. The behavior of a complex system is due to a balance between different intermolecular interactions—in this case, hydrogen bonding, van der Waals, and hydrophobic interactions, which all need to be considered. In the cellulose field, it is striking that the discussion focuses on hydrogen bonding as driving cellulose association and insolubility in water. How can this issue of hydrogen bonding mechanism of cellulose insolubility be tested and compared with hydrophobic interactions? There are many examples. For instance, dextran, which should have a similar capacity for hydrogen bonding as cellulose, is soluble in water. Furthermore, cellulose derivatives, like methyl cellulose and hydroxyethyl cellulose, may be highly soluble in water even if they have a high, often as high as cellulose itself, capacity for intermolecular hydrogen bonding. Another example is the one mentioned above, that of glucose. If intermolecular hydrogen bonding should be very important, glucose should show a strong tendency for self-association and phase separation, which is not observed. As said, solubility has to be considered in the light of the balance between different interactions; and focusing on hydrogen bonding for aqueous solubility of carbohydrates, we have to take into account not only water–carbohydrate interactions, but also water–water and carbohydrate–carbohydrate hydrogen bonding. Regarding a hydrogen-bonding mechanism, it is striking that all these interactions are very similar in magnitude, ca. 5 kcal/mol (Gunnar Karlström, personal communication). Then, in the presence of excess water, clearly cellulose should be highly soluble if hydrogen bonding is the sole interaction. Entropy would here play the largest role but we also have to consider the fact that water–cellulose contacts promote a larger number of hydrogen bonds (water O- carbohydrate –OH contacts) (Gunnar Karlström, personal communication). Insolubility due to hydrogen bonding would only occur if the carbohydrate–carbohydrate hydrogen-bonding would be very much stronger, which is clearly not the case.

1.3.1 Cellulose Is Amphiphilic

Many polymers are amphiphilic, that is, contain both polar and nonpolar groups/segments/sides. Amphiphilic self-assembly is well known in the surfactant and lipid field, as well as for block and graft copolymers, but seems to be rather neglected for homopolymers. It should be noted that for high molecular weight polymers, even a slight amphiphilicity may play a very important role for properties like solubility. An example is poly(ethylene glycol), which can induce surfactant self-assembly, another is ethylene oxide–propylene oxide block copolymers showing a strong self-assembly even at molecular weights of a few thousands. What is the evidence for amphiphilicity of cellulose or of its constituent glucose rings? An interesting example is cyclodextrins, which have a high aqueous solubility at the same time as they can incorporate in their interior very nonpolar molecules. This demonstrates that a chain of glucose rings can have sides of very different polarity. Another demonstration is single helix amylose, which behaves similarly to cyclodextrins by possessing a relatively hydrophobic inner surface. Therefore, hydrophobic molecules like hydrophobic lipids and aroma molecules can be found in these hydrophobic "pockets." A similar conclusion has been drawn from the structure of cellulose crystals. Thus, it is argued that, due to intra- and intermolecular hydrogen bonding, there is a formation of rather flat ribbons, with sides that differ markedly in their polarity. The hydrophobic sides would have a large tendency to stick to each other in an aqueous environment, contributing to a low solubility. Although, on the whole, cellulose is not more or less polar than other polysaccharides, the distinction between polar and less polar regions would strongly affect its solubility behavior.

1.4 CONCLUSIONS

It is argued that DNA and cellulose are biopolymers that have a significantly amphiphilic character, and that hydrophobic interactions are important for their association behavior and solubility. Hydrogen bonding is important in dictating the structures formed but is not expected to give a net driving force for association in aqueous systems. It is clear that mastering the dissolution of cellulose better has deep implications, not least for industrial developments. Preliminary accounts on the amphiphilic properties of cellulose were given at recent conferences [40]. Much emphasis has been placed on ionic liquids, but there are other interesting directions that may turn out to be more important [41]. It is found that hydrophobic interactions are often overlooked for biopolymers; the considerations presented are expected to apply for many other biopolymers.

ACKNOWLEDGMENTS

The work on DNA has been performed in collaboration with several colleagues and coworkers in Lund and Coimbra; in this context, I would like to especially mention Rita Dias and Maria Miguel. My interest in cellulose was stimulated by Lars Stigsson as well as by contacts with colleagues within two European research networks: the European Polysaccharide Network of Excellence (EPNOE) and an Initial Training Network (ITN) "Shaping and Transformation in the Engineering of Polysaccharides" (STEP). I am grateful to Thomas Heinze, Patrick Navard, Karin Stana-Kleinschek, John Mitchell, and others for fruitful discussions. Furthermore, I would also like to acknowledge the discussions with Gunnar Karlström, Håkan Wennerström, and Lennart Piculell on intermolecular interactions and polymer solubility.

REFERENCES

1. D.F. Evans and H. Wennerström, *The Colloidal Domain: Where Physics, Chemistry, Biology, and Technology Meet.* Wiley-VCH, New York, 1999.
2. P. Alexandridis and B. Lindman (Eds.), *Amphiphilic Block Copolymers. Self-Assembly and Applications.* Elsevier, Amsterdam, the Netherlands, 2000.

3. P. Alexandridis, U. Olsson, and B. Lindman, A record nine different phases (four cubic, two hexagonal, and one lamellar lyotropic liquid crystalline and two micellar solutions) in a ternary isothermal system of an amphiphilic block copolymer and selective solvents (water and oil), *Langmuir*, 14, 2627–2638, 1998.

4. K. Holmberg, B. Jönsson, B. Kronberg, and B. Lindman (Eds.), *Surfactants and Polymers in Aqueous Solution*, 2nd edn., Wiley, London, U.K., 2002.

5. R.S. Dias and B. Lindman (Eds.), *DNA Interactions with Polymers and Surfactants*. Wiley, Hoboken, NJ, 2008.

6. R.S. Dias, M. Miguel, and B. Lindman, DNA as an amphiphilic polymer. In *DNA Interactions with Polymers and Surfactants*. Dias, R.S. and Lindman, B. (Eds.), Wiley, Hoboken, NJ, 2008, pp. 367–378.

7. B. Gaugain, J. Barbet, N. Capelle, B.P. Roques, J.B. Lepecq, and M. Lebret, DNA bifunctional intercalators. 2. Fluorescence properties and DNA binding interaction of an ethidium homodimer and an acridine ethidium heterodimer, *Biochemistry*, 17, 5078–5088, 1978.

8. H.P. Spielmann, D.E. Wemmer, and J.P. Jacobsen, Solution structure of a DNA complex with the fluorescent bis-intercalator TOTO determined by NMR-spectroscopy, *Biochemistry*, 34, 8542–8553, 1995.

9. H.S. Rye and A.N. Glazer, Interaction of dimeric intercalating dyes with single-stranded-DNA, *Nucleic Acids Res.*, 23, 1215–1222, 1995.

10. M. Howegrant and S.J. Lippard, Binding of platinum(Ii) intercalation reagents to deoxyribonucleic-acid: Dependence on base-pair composition, nature of the intercalator, and ionic-strength, *Biochemistry*, 18, 5762–5769, 1979.

11. A.D. Richards and A. Rodger, Synthetic metallomolecules as agents for the control of DNA structure, *Chem. Soc. Rev.*, 36, 471–483, 2007.

12. V. Brabec and O. Novakova, DNA binding mode of ruthenium complexes and relationship to tumor cell toxicity, *Drug Resist. Updates*, 9, 111–122, 2006.

13. I. Kostova, Ruthenium complexes as anticancer agents, *Curr. Medicinal Chem.*, 13, 1085–1107, 2006.

14. P.U. Maheswari, V. Rajendiran, M. Palaniandavar, R. Thomas, and G.U. Kulkarni, Mixed ligand ruthenium(II) complexes of 5,6-dimethyl-1,10-phenanthroline: The role of ligand hydrophobicity on DNA binding of the complexes, *Inorg. Chim. Acta*, 359, 4601–4612, 2006.

15. J. Kapuscinski, Dapi: A DNA-specific fluorescent-probe, *Biotech. Histochem.*, 70, 220–233, 1995.

16. K. Eskilsson, C. Leal, B. Lindman, M. Miguel, and T. Nylander, DNA-surfactant complexes at solid surfaces, *Langmuir*, 17, 1666–1669, 2001.

17. M. Cárdenas, A. Braem, T. Nylander, and B. Lindman, DNA compaction at hydrophobic surfaces induced by a cationic amphiphile, *Langmuir*, 19, 7712–7718, 2003.

18. S. Sowerby, C.A. Cohn, W.M. Heckl, and N.G. Holm, Differential adsorption of nucleic acid bases: Relevance to the origin of life, *Proc. Natl. Sci. USA*, 98, 820–822, 2001.

19. A.-M. Chiorcea Paquim, T.S. Oretskaya, and A.M. Oliveira Brett, Adsorption of synthetic homo- and hetero-oligodeoxynucleotides onto highly oriented pyrolytic graphite: Atomic force microscopy characterisation, *Biophys. Chem.*, 121, 131–141, 2006.

20. J.M. Orosz and J.G. Wetmur, DNA melting temperatures and renaturation rates in concentrated alkylammonium salt-solutions, *Biopolymers*, 16, 1183–1199, 1977.

21. E.P. Geiduschek and T.T. Herskovits, Nonaqueous solutions of DNA. Reversible and irreversible denaturation in methanol, *Archiv. Biochem. Biophys.*, 95, 114–129, 1961.

22. M. Rosa, R. Dias, M.D. Miguel, and B. Lindman, DNA–cationic surfactant interactions are different for double- and single-stranded DNA, *Biomacromolecules*, 6, 2164–2171, 2005.

23. T. Härd and T. Lundbäck, Thermodynamics of sequence-specific protein–DNA interactions, *Biophys. Chem.*, 62, 121–139, 1996.

24. L. Jen-Jacobson, L.E. Engler, and L.A. Jacobson, Structural and thermodynamic strategies for site-specific DNA binding proteins, *Structure*, 8, 1015–1023, 2000.

25. M. West and V.G. Wilson, Hydrophobic residue contributions to sequence-specific DNA binding by the bovine papillomavirus helicase E1, *Virology*, 296, 52–61, 2002.

26. D. Costa, M.G. Miguel, and B. Lindman, Responsive polymer gels: Double-stranded versus single-stranded DNA, *J. Phys. Chem. B*, 111, 10886–10896, 2007.

27. D. Costa, S. dos Santos, F.E. Antunes, M.G. Miguel, and B. Lindman, Some novel aspects of DNA physical and chemical gels, *Arkivoc*, 4, 161–172, 2006.

28. P. Walden, Molecular weights and electrical conductivity of several fused salts, *Bull. Acad. Imper. Sci.*, 1800, 1914.

29. C. Graenacher, U.S. Patent 1,943,176, 1934.

30. R.P. Swatloski, R.D. Rogers, J.D. Holbrey, WO Patent 03,029,329, 2003.

31. R.P. Swatloski, S.K. Spear, J.D. Holbrey, and R.D. Rogers, Dissolution of cellulose with ionic liquids, *J. Am. Chem. Soc.*, 124, 4974–4975, 2002.

32. J. Wu, J. Zhang, H. Zhang, J. He, Q. Ren, and M. Guo, *Biomacromolecules*, 5, 266–268, 2004.

33. O.A. El Seoud, A. Koschella, L.C. Fidale, S. Dorn, and T. Heinze, Applications of ionic liquids in carbohydrate chemistry: A window of opportunities, *Biomacromolecules*, 8, 2629–2647, 2007.

34. T. Liebert and T. Heinze, Interaction of ionic liquids with polysaccharides 5. Solvents and reaction media for the modification of cellulose, *Bioresources*, 3(2), 576–601, 2008.

35. T. Heinze, S. Dorn, M. Schöbitz, T. Liebert, S. Köhler, and F. Meister, Interactions of ionic liquids with polysaccharides: 2: Cellulose, *Macromol. Symp.*, 262, 8–22, 2008.

36. L. Feng and Z.I. Chen, Research progress of dissolution and functional modification of cellulose in ionic liquids. *J. Molec. Liquids*, 142, 1–5, 2008.

37. S.D. Zhu, Y.X. Wu, Q.M. Chen, C.W. Wang, S.W. Jin, Y.G. Ding, and G. Wu, Dissolution of cellulose with ionic liquids and its application: A mini-review. *Green Chem.*, 8, 325–327, 2006.

38. Y. Fukaya, A. Sugimoto, and H. Ohno, Superior solubility of polysaccharides in low viscosity, polar, and halogen-free, 1,3-dialkylimidazolium formats, *Biomacromolecules*, 7, 3295–3297, 2006.

39. J. N. BeMiller and L. Whistler, In *Food Chemistry*. Owen R. F. (Eds.), Marcel Dekker, New York, 1996, Chap. 4, p. 157.

40. B. Lindman, *Conferences of the Initial Training Network (ITN) "Shaping and Transformation in the Engineering of Polysaccharides"* (STEP) in Nottingham, U.K., March 2009; Jena, Germany, September 2009.

41. L. Stigsson, B. Lindman, 61/272080, U.S. Patent application, 2010.

2 Mode of Interaction of Artificial Pulmonary Surfactant with an Amphiphilic α-Helical Peptide at the Air/Water Interface

Osamu Shibata, Hiromichi Nakahara, and Sannamu Lee

CONTENTS

2.1 INTRODUCTION

Langmuir monolayers at the air/water interface induce experimentally simple, convenient, and useful model systems for biophysical studies in cell biology and physiology. In fact, Langmuir monolayer behavior of a mixture of proteins and phospholipids is converted to pulmonary surfactant (PS) behavior *in vivo* in order to treat respiratory distress syndrome (RDS), acute respiratory distress syndrome (ARDS), etc., in the clinical field. In recent times, some researchers have clarified the correlations from *in vitro* to *in vivo* behavior (Goerke 1998, Lipp et al. 1997).

A pulmonary surfactant is a complex mixture of multiple lipids (~90 wt%) and four surfactant proteins (SP-A, -B, -C, and -D, ~10 wt%) (Krüger et al. 2002, Veldhuizen et al. 1998, Yu and Possmayer 2003). It aids significant functions in the alveoli such as the defense from exotic viruses and bacteria and the prevention of alveolar collapse during respiration. The material consists mainly of phosphatidylcholines (especially, dipalmitoylphosphatidylcholine (DPPC), ~50 wt%) and smaller but significant amounts of phosphatidylglycerol (PG), palmitic acid (PA), and four proteins (surfactant proteins: SP-A, -B, -C, and -D) (Krüger et al. 2002, Postle et al. 2001, Veldhuizen et al. 1998, Yu and Possmayer 2003). DPPC is not common in cell membranes and has been recognized as an important surfactant component soon after the discovery that surface active agents are present in the lungs (Brown 1964). DPPC, as is known to form tightly packed and solid surface films, has properties of lowering surface tension near zero in the multilayer state during π–A (surface pressure–molecular area) compression–expansion cycling. The disaturated phospholipids in a PS are thought to become enriched in the surface films at the expense of fluid components during compression to low interfacial area or high surface concentration. However, the resultant rigidity of the condensed films at physiological conditions causes it to adsorb slowly from the alveolar fluid and to respread poorly from multiple materials (Fleming and Keough 1988). The unsaturated and anionic PG is thought to help DPPC molecules adsorb and respread rapidly despite the fact that it does not lower surface tension as effectively as DPPC. In addition, the potential for specific molecular biophysical associations between anionic phospholipids and one or more of the surfactant apoproteins has also been widely suggested (Baatz et al. 1990, Chang et al. 1998, Johansson et al. 1991). PA is present in a relatively low amount in comparison to DPPC, but it is used as a very important additive for the proper functioning of both natural and synthetic PS replacement formulations. The addition of ~10 wt% PA to natural PS extracts obtained from animal sources has induced a significant improvement in their properties both *in vitro* and *in vivo* (Cockshutt et al. 1991, Gorree et al. 1991). Furthermore, two relatively small hydrophobic surfactant proteins, SP-B and SP-C, promote the spreading and adsorption of DPPC from close-packed states and vesicular aggregates. The molecular basis remains obscure, although there are many experimental evidences for these actions (Chang et al. 1998, Oosterlaken-Dijksterhuis et al. 1991). This is thought to be due to large molecular structures of SP-B and SP-C compared with lipids in pulmonary surfactants. Therefore, SP-B and SP-C analogous peptides such as KL_4, rSP-C, and Hel 13-5 have been investigated to clarify the pulmonary functions (Grigoriev et al. 2003, Ma et al. 1998, Nakahara et al. 2008b).

A deficiency of PS has been shown to cause neonatal respiratory distress syndrome (NRDS) in premature infants (Avery and Mead 1959). In a related distress for adults, an acquired inactivation of PS is likely to set up adult respiratory distress syndrome (ARDS). So far, a common treatment of NRDS is the surfactant replacement therapy in which exogenous surfactant preparations are administered to NRDS infants. These medicines can be roughly divided into three types. The first one is the natural-based source obtained from animals, which is mostly used in clinical cases. For example, Curosurf (Chiesi Pharmaceutici, Parma, Italy), Survanta (Ross Laboratories, Columbus, Ohio), and Surfacten (Surfactant TA; Mitsubishi Pharma Corporation, Osaka, Japan) are analogue to the native pulmonary surfactant, and they have been clinically used in several countries. Although the preparations above are dramatically effective to the patients, they involve the risk of animal infections such as bovine spongiform encephalopathy (BSE), potential viral contamination, and inherent immunity. Other drawbacks include a high-cost purification procedure and the difficulty of producing batch-to-batch uniformity. The second medicine is a synthetic type made from synthetic surfactants without proteins. It had been previously studied as a substitute for natural types, and then some medicines (Exosurf, ALEC, etc.) were developed (Kirkness et al. 2003a,b). Although these medicines can be used at a lower cost, they do not have a sufficient beneficial effect because they contain no proteins. Therefore, the preparations fully made of synthetic surfactants with proteins (or peptides) would be hoped for in the clinical surfactant replacement therapy for NRDS. Recently, in addition, a basic and clinical research on an expanded application of these preparations to the ARDS has been actively carried out (Notter 2000). As a most effective lipid replacement, Tanaka and coworkers (Tanaka et al. 1986) have originally developed a DPPC/PG/PA (=68:22:9, by weight) mixture. The composition of this mixture mimics that of the lipids existing in the amniotic liquid. It has been reported that the surface properties of this mixture are enhanced by adding a small amount of the lipid-binding protein (Tanaka et al. 1986). Therefore, the DPPC/PG/PA mixture has been examined by many researchers to develop new pulmonary preparations and to clarify pulmonary functions (Gustafsson et al. 1996, 2000, Ma et al. 1998). Recently, Surfaxin, a synthetic PS preparation with KL_4 peptide (consisting of novel 21-amino acid residues) has been approved for clinical use (Cai et al. 2003, Ma et al. 1998, Revak et al. 1996). The preparation is composed of the DPPC/PG/PA (=68:22:9, by weight) mixture. Although a few medicines containing new protein analogues have been developed (Ma et al. 1998, Veldhuizen et al. 2000), no replacement surfactant has been found comparable to the complete native surfactant with SP-B and -C. Therefore, the development of such synthetic-type medicine is strongly desired.

Recently, the authors synthesized a series of 18-mer amphiphilic α-helical peptides made up of hydrophobic and hydrophilic residues of a ratio of 5:13, 7:11, 9:9, 11:7, and 13:5 (abbreviated as Hel 5-13, Hel 7-11, Hel 9-9, Hel 11-7, and Hel 13-5, respectively), which are the shorter peptides by three residues than KL_4 (Figure 2.1) (Kiyota et al. 1996).

When these peptides take α-helical structures, the hydrophobic part and the hydrophilic part are completely separated in the α-helical structure (Figure 2.2). It has been suggested that the most hydrophobic Hel 13-5 of the five residues interacts with specific phospholipid mixtures to adopt long nanotubular structures (Kitamura et al. 1999). In addition, the interaction could induce the formation of membrane structures resembling cellular organelles such as the Golgi apparatus (Furuya et al. 2003, Lee et al. 2001). The structure bears a strong resemblance to the tubular myelin structure made of the PS secreted from type II pneumocytes. Therefore, it is expected to mimic the biophysical functions of SP-B and to be safe for RDS patients. Furthermore, monolayer experiments using a modified Wilhelmy surface balance showed that a mixture of Hel 13-5 and phospholipids quickly spread and adsorbed comparably with Surfacten (Lee et al. 2004, Yukitake et al. 2008), which is a modified natural bovine PS and is used well for RDS patients in Japan.

Herein, the authors have targeted an arc light specifically on newly designing PS preparations and elucidating the interfacial biophysical behavior and mechanism of SP-B during respiration using the model peptide (Hel 13-5). The interfacial behavior of spread monolayers for the DPPC/Hel 13-5, DPPC/PG/Hel 13-5, DPPC/PA/Hel 13-5, and DPPC/PG/PA/Hel 13-5 systems based on the DPPC/PG/PA (=68:22:9, by weight) lipid mixture and Surfacten has been investigated employing

Number of residue	10	18
Hel 5-13:	K K L K K L K K K W	K K L K K K L K
Hel 7-11:	K K L K K L L K K W	K K L L K K L K
Hel 9-9:	K L L K K L L K L W	K K L L K K L K
Hel 11-7:	K L L K L L L K L W	K K L L K L L K
Hel 13-5:	K L L K L L L K L W	L K L L K L L L

FIGURE 2.1 Amino acid sequences of five synthetic peptides (Hel 5-13, Hel 7-11, Hel 9-9, Hel 11-7, and Hel 13-5).

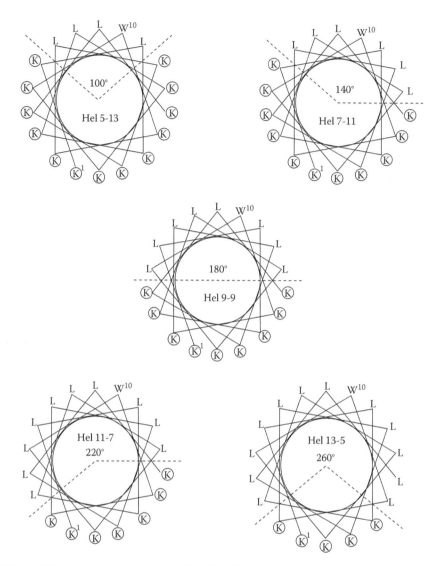

FIGURE 2.2 Helical wheel representations of Hel 5-13, Hel 7-11, Hel 9-9, Hel 11-7, and Hel 13-5. The abbreviations with and without open cycles denote hydrophilic and hydrophobic amino acids, respectively.

Langmuir (π–A and ΔV–A) isotherms, infrared reflection–absorption spectrometry (IRRAS), fluorescence microscopy (FM), and atomic force microscopy (AFM).

2.2 MODEL PEPTIDE (HEL 13-5) PROPERTIES

2.2.1 Peptide Structures

The amphiphilic α-helical peptide (Hel 13-5) consists of 13 hydrophobic residues (12 Leu and 1 Trp) and 5 hydrophilic residues (5 Lys) (Figures 2.1 and 2.2). The hydrophobic–hydrophilic balance (HHB) was estimated both theoretically from the calculated hydrophobicity value (or the magnitude of hydrophobic faces) and experimentally from the retention times in reverse phase high-performance liquid chromatography (RP-HPLC). For this modulation (hydrophobicity of Hel 13-5 is 0.07) (Kiyota et al. 1996), it is supposed that Hel 13-5 mimics SP-B of the surface-staying amphipathic helical structure rather than SP-C of the membrane-spanning helical structure at the alveolar air/liquid interface. In addition, when it forms an ideal α-helical structure, it has the amphiphilic structure with a 260° hydrophobic sector region as shown in the helical wheel representation (Figure 2.2). This means that the hydrophobic part and the hydrophilic one are completely separated in the α-helical structure. KL_4 peptides have been used as an element in synthetic-type RDS medicines (or Surfaxin) to mimic the native SP-B (Cai et al. 2003, Gustafsson et al. 1996, Ma et al. 1998), whereas rSP-C peptides (based on recombinant SP-C) are in clinical trials as that in Venticute (Amrein et al. 1997, Grigoriev et al. 2003, Krol et al. 2000, Nahmen et al. 1997). Although rSP-C forms a membrane-spanning α-helix across the air/water interface, Hel 13-5 and KL_4 take surface-staying amphiphilic α-helical structures at the interface. Notice that the α-helical representations for KL_4 and rSP-C reveal a disheveled configuration of hydrophobic amino acids and hydrophilic ones.

2.2.2 ADSORPTION ISOTHERMS

Surface pressure (π)–time (t) isotherms of pure DPPC and the representative DPPC/Hel 13-5 monolayer ($X_{\text{Hel 13-5}} = 0.1$) formed by adsorption from their vesicle solutions are shown in Figure 2.3. It indicated a slow adsorption of DPPC molecules from a 0.15 M NaCl subphase as reported previously (Serrano et al. 2005). For the DPPC/Hel 13-5 ($X_{\text{Hel 13-5}} = 0.1$) system, on the contrary, the π–t isotherm had a shoulder at ~15 min and reached the equilibrium surface pressure of ~37 mN m^{-1}. SP-B accelerates the adsorption rate of vesicular phospholipids from the subphase to the air/water interface (Oosterlaken-Dijksterhuis et al. 1991). Similar to native SP-B, Hel 13-5 of our mimicking peptide also facilitated the interfacial adsorption of DPPC toward the air/water interface (Nakahara et al. 2006b).

2.2.3 TEMPERATURE DEPENDENCE

Figure 2.4 shows surface pressure (π)–molecular area (A) and surface potential (ΔV)–A isotherms of Hel 13-5 monolayers. For the π–A isotherms, Hel 13-5 had no kink point and was more expanded, indicating that the monolayer of Hel 13-5 was in the expanded state over the whole range of surface pressures and temperatures studied. Hel 13-5 formed the expanded (disordered) film up to ~42, ~40, and ~39 mN m^{-1} at 298.2, 303.2, and 310.2 K, respectively. However, the extrapolated area of Hel 13-5 was ~2.6 nm^2 irrespective of temperature. This suggests that Hel 13-5 forms stable disordered films under biological temperature ranges and takes on the low solubility into the aqueous subphase.

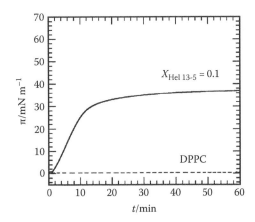

FIGURE 2.3 π–t isotherms of pure DPPC and the representative DPPC/Hel 13-5 system ($X_{\text{Hel 13-5}} = 0.1$) adsorbing from their vesicular solutions in 0.15 M NaCl at 298.2 ± 1.0 K. (From Nakahara, H. et al., *Langmuir*, 22, 1182, 2006b. With permission.)

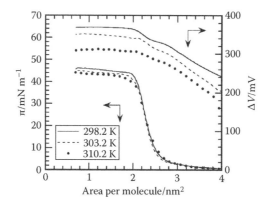

FIGURE 2.4 π–A and ΔV–A isotherms of the pure Hel 13-5 monolayer on a 0.02 M Tris buffer solution (pH 7.4) with 0.13 M NaCl at 298.2, 303.2, and 310.2 K.

However, the low solubility does not have a relation to the exclusion of Hel 13-5 from surface monolayers into the subphase upon compression. The value of the extrapolated area reflects the molecular conformation of peptides. The value of ~2.6 nm² supports an α–helical structure rather than β-sheet, random, and other structures. Further investigation of its conformation at the interface was referred to in the previous paper (Nakahara et al. 2006b). The surface potential (ΔV) of Hel 13-5 always showed a positive variation under compression (Figure 2.4). The surface potential of Hel 13-5 monotonically increased up to ~380, ~350, and ~310 mV on compression at 298.2, 303.2, and 310.2 K, respectively. The reduction of the ΔV values reflects an aggravation of molecular orientation due to an increase in molecular motion.

2.2.4 FLUORESCENCE MICROSCOPY

Fluorescence images of the phase state in the Hel 13-5 films are shown in Figure 2.5. It is widely accepted that the fluorescent probe is selectively dissolved in the disordered phase, not dissolving in the ordered phase (Lösche and Möhwald 1984). Therefore, ordered domains can be visualized as dark domains in the disordered/ordered coexistence region. The Hel 13-5 films indicated the bright homogeneous images from the low surface pressure to the collapse one. It provides evidence that Hel 13-5 forms the disordered film miscible with the FM probe independent of surface pressure. The results also correspond to those of the π–A isotherms for Hel 13-5 (Figure 2.4).

2.2.5 ATOMIC FORCE MICROSCOPY

AFM for this study provided both topographic and phase contrast images. The topographic image reflects the sample topography, whereas the phase contrast image, which originates from

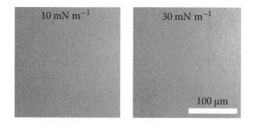

FIGURE 2.5 Representative FM images of pure Hel 13-5 spread on a 0.02 M Tris buffer with 0.13 M NaCl at 298.2 K. The monolayers contain 1 mol% fluorescent probe (R18). The scale bar in the lower right represents 100 μm. (From Nakahara, H. et al., *Langmuir*, 22, 1182, 2006b. With permission.)

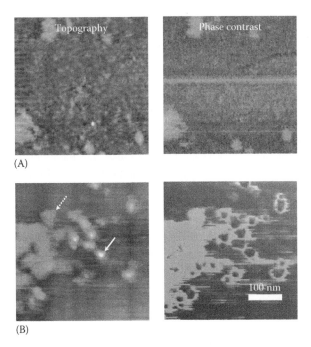

(A)

(B)

FIGURE 2.6 AFM images of the Hel 13-5 monolayer in a tapping mode: topography and corresponding phase contrast images transferred onto mica at (A) 35 and (B) 45 mN m^{-1} at the scan area of 400×400 nm. The solid and dashed arrows indicate the domain with and without the bright midpoint, respectively. (From Nakahara, H. et al., *Langmuir*, 22, 1182, 2006b. With permission.)

the energy loss of the oscillating AFM tip, shows the chemical structures of heterogeneous samples. Figure 2.6 shows AFM images of a pure Hel 13-5 monolayer transferred onto mica substrates below and beyond the collapse pressure (~42 mN m^{-1}). AFM images of Hel 13-5 at below 35 mN m^{-1} were homogeneous (not shown). At 35 mN m^{-1}, on the other hand, some bright domains appeared in the topographic image (Figure 2.6A), which are slightly higher than the surrounding regions by ~0.2 nm. The small protrusions suggest the existence of intermediate states toward monolayer collapses because the height of the protrusions is much smaller than the diameter of α-helical Hel 13-5 (~1 nm) estimated by a computer simulation (CS ChemOffice Ultra 5.0). After the collapse of Hel 13-5 at 45 mN m^{-1}, many protrusions are observed over the whole range as shown in Figure 2.6B. They are located in a line, implying that the collapse of Hel 13-5 promotes another collapse beside it. Zasadzinski and coworkers (Ding et al. 2001) reported that above the plateau pressure, many protrusions appeared in the natural system of Survanta, which was a commercial RDS medicine containing a natural bovine pulmonary surfactant. The resultant patches were also observed in the DPPC/POPG (palmitoyloleoylphosphatidylglycerol) films containing low amounts of protein analogues (Diemel et al. 2002), indicating that they were induced by the squeeze-out of fluid compositions (one of the PS functions). In the topographic image (Figure 2.6B), the protrusions (bright) and monolayers (dark) of single-species Hel 13-5 molecules coexist. The height of these protrusions is ~2.0–3.0 nm, suggesting that Hel 13-5 molecules are excluded from the monolayer after plateau regions on π–A isotherms (Figure 2.4), and then a three-dimensional folding made of two or three Hel 13-5 molecules is formed. Notice that two kinds of domains are observed: the domain with the bright midpoint (indicated by an arrow) and that without it (indicated by a dashed arrow). The former shows a central higher protrusion, as shown in the corresponding phase contrast image. These protrusions are found to be disk-like in shape with a typical diameter of ~33 nm, corresponding to aggregations containing ~340 molecules of Hel 13-5.

2.3 TWO-COMPONENT MODEL SYSTEM (DPPC/HEL 13-5)

2.3.1 LANGMUIR ISOTHERMS

The π–A and ΔV–A isotherms of DPPC containing Hel 13-5 have been studied to assess the effect of Hel 13-5 as a model pulmonary surfactant protein. In Figure 2.7, π–A isotherms exhibit two small plateau regions (or two clear kink points) at the surface pressures of ~12–19 and ~42 mN m^{-1} as indicated by the arrows for $X_{\text{Hel 13-5}} = 0.3$ (curve e). The first plateau at lower surface pressures and the second one at higher surface pressures indicate the first-order disordered/ordered transition and the monolayer collapse of Hel 13-5, respectively. On increasing the mole fraction of Hel 13-5, the first transition points of all the π–A isotherms increased from ~12 to 19 mN m^{-1} and became more and more unclear. The second kink point in the π–A isotherm appeared at ~42 mN m^{-1} (or the collapse pressure of pure Hel 13-5). It is quite notable that the second kink points appeared at the same surface pressure (~42 mN m^{-1}), independent of their constituents. The DPPC monolayer formed a stable film up to ~55 mN m^{-1}, and the monolayer for different compositions of the DPPC/Hel 13-5 systems ($0 < X_{\text{Hel 13-5}} \leq 0.3$) remained stable up to ~55 mN m^{-1} despite differences in the ratio of Hel 13-5, indicating that Hel 13-5 stabilizes DPPC films, similar to SP-B functions (Piknova et al. 2001, Taneva and Keough 1994). This behavior also confirms that Hel 13-5 weakly interacts with rigid compositions of DPPC at high surface pressures, and then only the Hel 13-5 component is squeezed out of the DPPC/Hel 13-5 system at ~42 mN m^{-1}. The squeeze-out phenomena are managed to be gradually advanced upon further compression, and then only the DPPC component locates at the air/water interface at ~55 mN m^{-1}. Beyond 42 mN m^{-1}, the ΔV–A isotherms of $0.05 \leq X_{\text{Hel 13-5}} \leq 0.3$ gradually increases, whereas the ΔV of the other mole fractions shows nearly constant value. Judging from the above phenomena, if the surface pressure goes beyond the collapse pressure, the ΔV–A isotherms become almost parallel to the area axis. Namely, the rising of ΔV beyond the second kink point means that the packing state or orientation of the molecule is still but slightly developing. At least, these results support the squeeze-out phenomenon of Hel 13-5 from the monolayer surface. However, the value of ΔV became almost similar (~500 mV) at ~55 mN m^{-1}, indicating that Hel 13-5 molecules do not squeeze out completely from the monolayer.

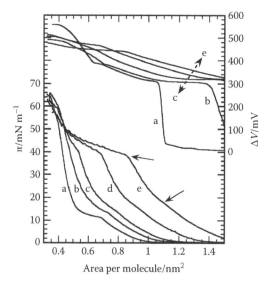

FIGURE 2.7 π–A and ΔV–A isotherms of the DPPC/Hel 13-5 mixtures on a 0.02 M Tris buffer solution (pH 7.4) with 0.13 M NaCl at 298.2 K; (a) DPPC, (b) $X_{\text{Hel 13-5}} = 0.05$, (c) 0.1, (d) 0.2, and (e) 0.3. The arrows indicate two clear kink points for $X_{\text{Hel 13-5}} = 0.3$.

2.3.2 INFRARED REFLECTION–ABSORPTION SPECTROSCOPY MEASUREMENTS

The reflectance–absorbance (RA) intensity is equal to $-\log (R/R_0)$, where R and R_0 are the reflectivities of the monolayer-covered and pure buffer solution surfaces, respectively. The original IRRAS data were plotted as RA intensity versus wavenumber as shown in Figure 2.8, where the spectra of spread DPPC monolayers prior to film compression and at the compressed state were presented. Both the spectra had specific peaks of an asymmetric methylene stretching vibration (v_a-CH$_2$) at ~2920 cm^{-1} and of a symmetric methylene stretching vibration (v_s-CH$_2$) at ~2850 cm^{-1}. The v_a-CH$_2$ peak with larger absolute values and better sensitivities is easy to probe the change in surface DPPC concentrations upon compression compared with the v_s-CH$_2$ peak, although both the peaks can reflect it (Yin and Chang 2006). Thus, the v_a-CH$_2$ RA intensity and its wavenumber were utilized to perform the IRRAS analysis (Nakahara et al. 2009). The repeated compression–expansion behavior of pure DPPC monolayers at the air/water interface is shown as a plot of the v_a-CH$_2$ RA intensity against relative trough area (Figure 2.9). The representative data of the first and the third cycles are presented, and upper abscissa is labeled by the calculated area per DPPC molecule in the figure. All of the v_a-CH$_2$ RA intensity data obtained in the present study included an uncertainty of 0.0002 (Dluhy 1986, Dluhy and Cornell 1985). The absorption bands of monolayers at the interface are negative, and the basis for the negative absorbance in the monolayer spectra has been explored (Dluhy 1986, Dluhy and Cornell 1985). As shown in Figure 2.9A, the initial absolute RA value of the v_a-CH$_2$ band for a DPPC monolayer at ~0.86 nm^2 molecule^{-1} was 0.0011. Upon compression, the absolute RA value increased gradually and reached 0.0032 at ~0.45 nm^2 molecule^{-1} (before the collapse area). This is expected because a higher surface concentration of molecules always results in a stronger RA response. During the expansion stages, the absolute RA intensity coincided with that under film compression due to the insufficient compression of the monolayer. The monolayer compression up to the DPPC collapse state induced a loss of free DPPC molecules during the cycle (Yin and Chang 2006). Thus, the analogy of the v_a-CH$_2$ value between the first and the third cycle indicates no loss of the DPPC molecules. As for Figures 2.9 and 2.10 as described later, note that a spread amount of DPPC was kept constant in both the pure DPPC and the DPPC/Hel 13-5 mixture ($X_{\text{Hel 13-5}} = 0.1$) to perform the quantitative comparison between their RA intensities.

The corresponding v_a-CH$_2$ wavenumber versus the relative area for DPPC monolayers is plotted in Figure 2.9B. At the beginning of the first compression, the maximum in wavenumber

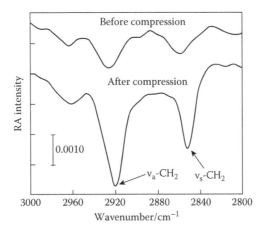

FIGURE 2.8 IRRAS spectra of the methylene stretching band region (2800–3000 cm^{-1}) for pure DPPC monolayers spread on a NaH$_2$PO$_4$/Na$_2$HPO$_4$ buffer solution (pH 7.0) at ~303.2 K. The representative spectra before ($A = ~0.86$ nm^2) and after compression ($A = ~0.45$ nm^2) are presented. The antisymmetric methylene stretching vibration (v_a-CH$_2$) at ~2920 cm^{-1} is indicated by an arrow. The bar reflects the intensity in the reflectance–absorbance unit. (From Nakahara, H. et al., *Colloids Surf. B*, 68, 61, 2009. With permission.)

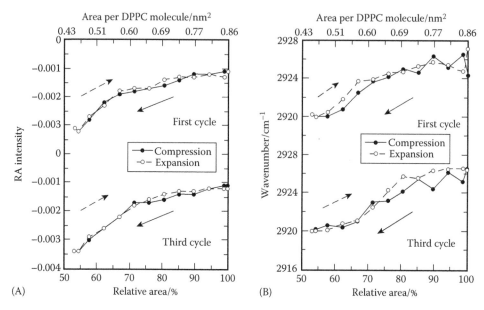

FIGURE 2.9 Consecutive (A) ν_a-CH_2 RA intensity–relative area and (B) ν_a-CH_2 wavenumber–relative area curves of a DPPC monolayer with an initial area per molecule of $0.86\,nm^2$. The relative area represents the ratio of the actual interfacial area during the compression–expansion cycle. The upper abscissa is labeled by the calculated area per DPPC molecule. (From Nakahara, H. et al., *Colloids Surf. B*, 68, 61, 2009. With permission.)

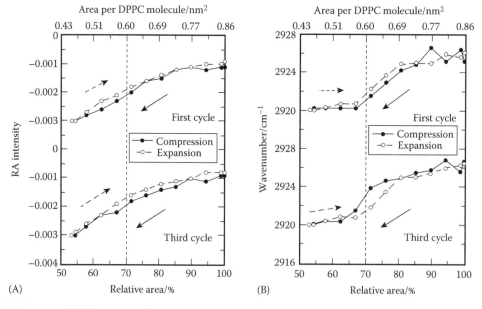

FIGURE 2.10 Consecutive (A) ν_a-CH_2 RA intensity–relative area and (B) ν_a-CH_2 wavenumber–relative area curves of a binary DPPC/Hel 13-5 monolayer at $X_{Hel\ 13\text{-}5}=0.1$ with an initial area per DPPC molecule of $0.86\,nm^2$. The onset area of the squeeze-out phenomenon is indicated by dashed lines ($0.60\,nm^2$). The upper abscissa is labeled by the calculated area per DPPC molecule. (From Nakahara, H. et al., *Colloids Surf. B*, 68, 61, 2009. With permission.)

was ~2926 cm^{-1}. The wavenumber decreased with shifting to a small area upon compression, and a value of 2920 cm^{-1} was found at the end of the first compression stage. It has been reported that the wavenumber corresponding to the minimum in the v_a-CH$_2$ band is sensitive to the packing, the orientation, and the conformational change of the molecular acyl chains in the monolayer state, and a lower wavenumber is a characteristic of the highly ordered conformation (or the all-*trans* conformation) of the aliphatic chains (Gericke and Hühnerfuss 1993, Mendelsohn et al. 1995, Mitchell and Dluhy 1988, Wen et al. 2000). Therefore, the decreasing behavior demonstrated that the DPPC molecules became well oriented on compression. In the first expansion, the wavenumber increased with an increase in relative area to accord with those in the corresponding compression. The same behavior was also observed in the third cycle, suggesting that the variation in the orientation of DPPC acyl chains during repeated compression and expansion processes occurred reversibly.

For the binary DPPC/Hel 13-5 monolayer at $X_{\text{Hel 13-5}} = 0.1$, the hysteresis curves of v_a-CH$_2$ RA intensity with the relative area are plotted in Figure 2.10A. The figure contains dashed lines at ~0.60 nm^2 molecule^{-1}, where Hel 13-5 states are squeezed out of surface monolayers (Nakahara et al. 2006b, 2005b). Similar to pure DPPC, the initial absolute RA value of the binary monolayer was 0.0011 due to the same surface amount as the pure DPPC system. Upon compression, the absolute RA value increased gradually and reached 0.0030 at ~0.45 nm^2 molecule^{-1}, where the monolayers were in the close-packed state. This value was the same as that in the DPPC system at ~0.45 nm^2 molecule^{-1}. That is, no loss of DPPC upon compression was observed at the air/liquid interface. This result supported that the surface was refined to DPPC monolayers and almost all Hel 13-5 molecules were squeezed out into the subphase. These plots in the first and the third cycles showed the same behavior. It indicates that Hel 13-5 makes DPPC molecules in the solid monolayer state respread to the interface repeatedly.

For hysteresis curves of the corresponding v_a-CH$_2$ wavenumber with the relative area (Figure 2.10B), on the other hand, an interesting behavior was observed in the squeeze-out regions. Upon the first compression, the wavenumber decreased gradually as the same in a pure DPPC system. Beyond the onset of the squeeze-out indicated by dashed lines, however, the wavenumber kept the value of ~2920 cm^{-1} regardless of shifting to small areas. This demonstrated that the orientation for hydrophobic chains of DPPC became in the most packed state soon after the emergence of the Hel 13-5 squeeze-out action and the packed orientation continued up to the collapse state of the binary monolayer at ~0.45 nm^2 molecule^{-1}. During this surface refinement, the Hel 13-5 was gradually excluded to the subphase, with orientation and conformation for the DPPC aliphatic chains sustained. For the third cycle, similar behavior was observed. It supports the fact that the reversible change in the orientation and the good respreading ability of the binary monolayer exist (Nakahara et al. 2005b, 2006b).

2.3.3 FM Images

Through the FM observation, a spreading solution of the surfactants was prepared as a mixed solution doped with 1 mol% of 3,6-bis(diethylamino)-9-(2-octadecyloxycarbonyl) phenyl chloride (R18) as a fluorescence probe. It is widely accepted that fluorescent probes are selectively dissolved in LE (disordered) phases and not in LC (ordered) phases. Therefore, LC (ordered) domains can be visualized as dark domains in the LE/LC (disordered/ordered) coexistence region. A series of FM images for the DPPC/Hel 13-5 system at 15 and 20 mN m^{-1} is presented in Figure 2.11, where all of them show the disordered/ordered coexistence states; that is, the dark domains reflect the ordered phase of DPPC, meanwhile the bright regions reflect the disordered phases of DPPC and Hel 13-5. The ordered domains of all molar fractions grow in size with increasing surface pressure from 15 to 20 mN m^{-1}. When the films were compressed further, they formed dark homogeneous images consisting of almost the ordered phase up to the collapse pressure except for $X_{\text{Hel 13-5}} = 0.025$ (data not shown). Note that the addition of a small amount of Hel 13-5 to DPPC induced the "moth-eaten" aggregation (shown by an arrow) of the LC domains of DPPC. This moth-eaten aggregation

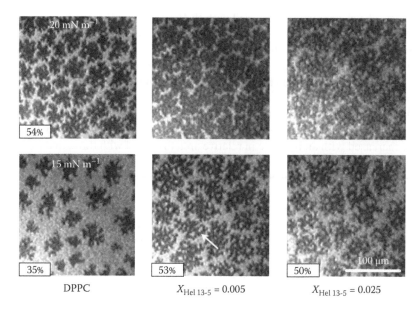

FIGURE 2.11 FM images of the DPPC/Hel 13-5 mixture system at 298.2 K at 15 and 20 mN m^{-1} for pure DPPC, $X_{Hel\,13-5}$=0.005, and $X_{Hel\,13-5}$=0.025. In the coexistence phase, percentage refers to the ordered domains in the micrograph. The monolayers contain 1 mol% fluorescent probe (R18). The arrow shows the "moth-eaten" aggregation of LC domains made of pure DPPC. The scale bar in the lower right represents 100 μm. (From Nakahara, H. et al., *Langmuir*, 22, 1182, 2006b. With permission.)

occurred only in the specific case where a small amount of Hel 13-5 coexists with DPPC. In addition, its aggregation enlarged and expanded the regions of each LC domain. Therefore, the disordered phases of Hel 13-5 penetrated into the DPPC ordered domains and promoted their nucleation.

2.3.4 AFM IMAGES

AFM images (500 × 500 nm) of an LB film of DPPC films containing small amounts of Hel 13-5 ($X_{Hel\,13-5}$=0.1) transferred onto mica at 35, 45, and 55 mN m^{-1} are shown in Figure 2.12. Pure DPPC monolayers provided a homogeneous AFM image in the previous paper (Nakahara et al. 2005a). In contrast, the monolayer containing Hel 13-5 ($X_{Hel\,13-5}$=0.1) at 35 mN m^{-1} shows two different phases as shown in Figure 2.12A. Because nontilted DPPC molecules at the interface are ~2.5 nm in height, bright phases represent the DPPC monolayer and dark ones indicate the Hel 13-5 monolayer in the topography. The height difference of these phases is ~1.0 nm, demonstrating that experimental values agree considerably well with theoretical ones. The corresponding phase contrast image shows more clearly the morphological character than the topography due to the difference in the surface physicochemical property between DPPC and Hel 13-5. The behavior of these AFM images agrees with that of FM images in the way to disperse the DPPC ordered domains by adding a small amount of Hel 13-5. On the other hand, three different phases appear at 45 mN m^{-1}, where Hel 13-5 is squeezed out of binary DPPC/Hel 13-5 monolayers (Figure 2.12B). The brightest lobes (indicated by an arrow) by the squeezed-out particles appear. Interestingly, most of the Hel 13-5 protrusions sit on DPPC monolayers. The height of protrusions from the DPPC monolayer is ~1.5 nm, which corresponds to that of one and a half α-helical Hel 13-5 molecules (~1 nm). The corresponding phase contrast image (Figure 2.12B) clearly shows morphological changes between respective monolayers. Upon further compression to 55 mN m^{-1}, the topography image (Figure 2.12C) indicates that DPPC regions and the protrusions increase in number, whereas Hel 13-5 dark regions decrease in comparison with those in Figure 2.12A and B. This indicates that the ratio of DPPC-occupied areas increases

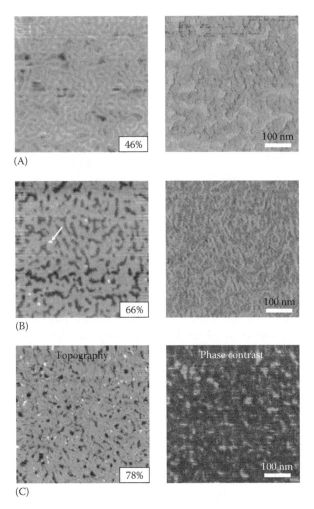

FIGURE 2.12 AFM images of the DPPC monolayer containing moderately low amounts of Hel 13-5 (at $X_{\text{Hel 13-5}} = 0.1$) in a tapping mode at the scan area of $500 \times 500\,\text{nm}$: topography and corresponding phase contrast image transferred onto mica at (A) 35, (B) 45, and (C) $55\,\text{mN m}^{-1}$. The ratio (%) of occupied areas by DPPC monolayers to whole images is shown in each topographic image. The arrow shows the brightest lobes by the squeezed-out particles. (From Nakahara, H. et al., *Langmuir*, 22, 1182, 2006b. With permission.)

as the surface pressure increases from 35 to $55\,\text{mN m}^{-1}$. That is, the content of ordered domain is increased by compression. The percentage of ordered domain becomes 46%, 66%, and 78% in turn. This result demonstrates that only Hel 13-5 is squeezed out of binary monolayers beyond the plateau regions on the π–A isotherms and that the squeezed-out Hel 13-5 molecules occupy 3-D surface-associated reservoirs. As a result, the surface is refined to a DPPC monolayer. There is no difference in the height and size of the protrusions between 45 and $55\,\text{mN m}^{-1}$. The diameter of the disk-like protrusions is found to be ~5.6 nm, corresponding to 10 molecules of Hel 13-5. Note that all of the protrusions locate on DPPC monolayers, whereas the Hel 13-5 molecules that are not still squeezed out exist as a monolayer regardless of the high surface pressure ($55\,\text{mN m}^{-1}$), indicating that all Hel 13-5 molecules are not completely squeezed out of the binary system. This supports the result from the quantitative analysis based on π–A isotherms. In the phase contrast image (Figure 2.12C), the protrusions become indistinct.

Unfortunately, the individual AFM image does not reveal directly whether the protrusions are located toward the air or the subphase. In general, the two possibilities can be described in

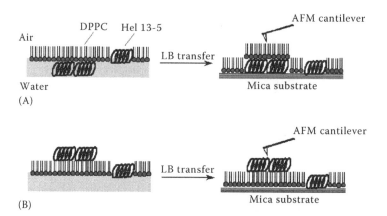

SCHEME 2.1 Two possible orientations of the protrusions (or surface-associated nanoparticles that adhere to the monolayer) across the air/water interface and the corresponding molecular structures induced by Langmuir–Blodgett transfer to a mica substrate. The illustrations are based on the assumptions that (A) the protrusions locate in the subphase and that (B) they form to the air. (From Nakahara, H. et al., *Langmuir*, 22, 1182, 2006b. With permission.)

Scheme 2.1. When these results are analyzed by combining the topography and the phase contrast image, the problem may be resolved. First, assuming that the protrusions form to the air (Scheme 2.1B), the surface orientation would be translated on mica throughout the process of an LB transfer as it is (Amrein et al. 1997). At that time, the AFM tip would scan the protrusions consisting of Hel 13-5 molecules on DPPC monolayers. Because the DPPC monolayers are different from Hel 13-5 in terms of a surface physicochemical property, the protrusions (Hel 13-5) should possess a different color from the surrounding DPPC monolayers in the phase contrast image. Second, assuming that the protrusions locate in the subphase (Scheme 2.1A), however, they would be locked between the DPPC monolayer and the mica support upon an LB transfer (Amrein et al. 1997). The resultant LB films would make the protrusions covered by DPPC monolayers accessible to the AFM tip. That is, the AFM tip would come to scan the protrusions of the hydrophobic chains of the DPPC monolayer located on Hel 13-5 molecules. In this case, the protrusions (DPPC monolayers) should possess a color similar to that of surrounding DPPC monolayers in the phase contrast image. This is supported by the fact that the protrusions in the phase contrast image disappear in the corresponding phase contrast image (Figure 2.12C). To confirm a contrast change of protrusions based upon the topography and the phase contrast image in more detail, the expanded AFM images at 55 mN m⁻¹ are shown in Figure 2.12. These results are consistent with those reported by others (Takamoto et al. 2001), who concluded that the disordered and fluid LE monolayers folded toward the aqueous subphase for native protein-containing films (Scheme 2.1A) (Krol et al. 2000, Nakahara et al. 2006b).

2.3.5 HYSTERESIS ISOTHERMS

The cycling experiments of the compression–expansion isotherm (known as hysteresis curves) of the DPPC/Hel 13-5 ($X_{Hel\ 13-5} = 0.1$) system are shown in Figure 2.13. The films were compressed up to the surface pressure of 55 mN m⁻¹ and then expanded to the starting area. In general, surfactants would be continuously exhausted at the interface during the compression and expansion cyclic processes. However, these cycles did not lead to loss of materials from the surface. The first kink point became unclear by the cyclic processes from the first to the fifth cycle in Figure 2.13. The second kink point, however, remained clear despite repeated cycling processes, indicating that Hel 13-5 was desorbed from the interface. Although the first cycling π–A isotherm almost coincided with the fifth cycling one at low surface pressures, the π–A isotherms shifted to smaller

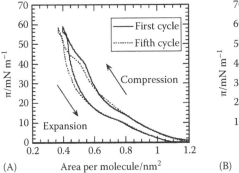
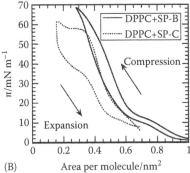

FIGURE 2.13 (A) Cyclic compression and expansion isotherms (or hysteresis curves) of the DPPC/Hel 13-5 ($X_{Hel\ 13-5}$ = 0.1) mixture on a 0.02 M Tris buffer solution (pH 7.4) with 0.13 M NaCl at 298.2 K. The compression and expansion cycle was repeated five times at a compression rate of 0.1-0.2 nm^2 molecule^{-1} min^{-1}. (From Nakahara, H. et al., *Langmuir*, 22, 1182, 2006b. With permission.) (B) Hysteresis curves of mixtures of DPPC + 0.5 mol% SP-B and DPPC + 2.0 mol% SP-C. (From Wüstneck, N. et al., *Colloids Surf. B*, 21, 200, 2001; Wüstneck, R. et al., *Langmuir*, 18, 1127, 2002. With permission.)

areas in the repeated cycling processes at high surface pressures. Figure 2.13 indicates that the easy respreading behavior indicates the ability for desorbed molecules to reenter into the interface and works for a highly reproducible hysteresis loop. These results demonstrate that Hel 13-5 can accelerate the spreading of DPPC and induce the good respreading. Furthermore, these hysteresis curves nicely resemble those of the DPPC/SP-B and DPPC/SP-C mixtures reported previously (Wüstneck et al. 2001, 2002).

2.4 THREE-COMPONENT MODEL SYSTEMS (DPPC/ PG/HEL 13-5 AND DPPC/PA/HEL 13-5)

2.4.1 Langmuir Isotherms

Typical isotherms for the fixed composition of DPPC/egg-phosphatidylglycerol (PG) (68/22, by weight) and DPPC/PA (90/9, by weight) systems with Hel 13-5 peptides have been investigated to assess each role of the PG and PA components in the ternary systems. For this purpose, the π–A and ΔV–A isotherms of the three component monolayers were measured at various Hel 13-5 molar fractions ($X_{Hel\ 13-5}$) in Figures 2.14 and 2.15. For the DPPC/PG/Hel 13-5 system, as shown in Figure 2.14, the π–A isotherms exhibit two plateau regions at ~18–20 and ~42 mN m^{-1}; they are indicated by arrows at representative $X_{Hel\ 13-5}$ = 0.1 (curve d). All of the π–A isotherms have the first plateaus at similar surface pressures (transition pressures), and the kink points become unclear with increasing $X_{Hel\ 13-5}$. The second kink point at $X_{Hel\ 13-5}$ = 0.1 appears at ~42 mN m^{-1} (or the collapse pressure of Hel 13-5), and the second plateau range expands as the amount of Hel 13-5 increases. The DPPC/PG monolayer forms a stable film up to ~45 mN m^{-1}. On the other hand, the ternary DPPC/PG/Hel 13-5 monolayers ($0 < X_{Hel\ 13-5} \leq 0.1$) can retain stable states up to ~55 mN m^{-1}. This is the first evidence that Hel 13-5 attractively interacts with the fluid component (PG), and the two fluid components begin to be squeezed out of the ternary mixtures at ~42 mN m^{-1}. Then the degree of squeezing-out is gradually enhanced upon further compression, and the interface is ultimately refined to DPPC-rich monolayers at ~55 mN m^{-1} because the collapse pressure of DPPC monolayers is ~55 mN m^{-1} (Figure 2.7). At ~42 mN m^{-1}, the ΔV–A isotherm at $X_{Hel\ 13-5}$ = 0.1 shows a kink point corresponding to the second plateau on its π–A isotherm, as indicated by an arrow, and then the ΔV value gradually rises upon further compression, indicating that monolayer orientations slightly change due to the exclusion of fluid materials. Finally, it becomes constant above 55 mN m^{-1}. This also supports the squeeze-out phenomenon of Hel 13-5.

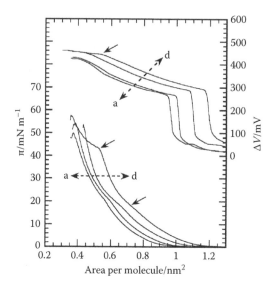

FIGURE 2.14 π–A and ΔV–A isotherms of the ternary DPPC/PG/Hel 13-5 (fixed DPPC/PG ratio) mixtures on a 0.02 M Tris buffer solution (pH 8.4) with 0.13 M NaCl at 298.2 K; (a) DPPC/PG, (b) $X_{Hel\ 13-5}=0.01$, (c) 0.05, and (d) 0.1. Two kink points are indicated by arrows on the π–A isotherm of representative $X_{Hel\ 13-5}=0.1$. The kink point corresponding to the second plateau on its π–A isotherm is also shown by an arrow on the corresponding ΔV–A isotherm.

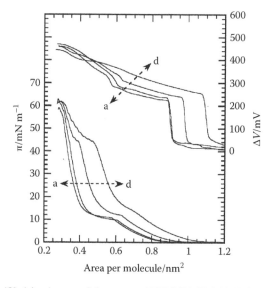

FIGURE 2.15 π–A and ΔV–A isotherms of the ternary DPPC/PA/Hel 13-5 (fixed DPPC/PA ratio) mixtures on a 0.02 M Tris buffer solution (pH 8.4) with 0.13 M NaCl at 298.2 K; (a) DPPC/PA, (b) $X_{Hel\ 13-5}=0.01$, (c) 0.05, and (d) 0.1.

For the DPPC/PA/Hel 13-5 system shown in Figure 2.15, π–A isotherms also have two plateau regions at ~10–14 mN m^{-1} and ~42 mN m^{-1}. The π–A isotherm at $X_{Hel\ 13-5}=0.005$ has only the first plateau (Nakahara et al. 2006a). On the other hand, two plateaus are observed in the π–A isotherms of $0.01<X_{Hel\ 13-5}<0.1$. Their transition pressures of the first plateau change from ~11 to ~14 mN m^{-1} and become unclear with increasing $X_{Hel\ 13-5}$, which is the same as the DPPC/PG/Hel 13-5 system. The second kink points in $0.01<X_{Hel\ 13-5}<0.1$ also appear at ~42 mN m^{-1}, and the second plateau range elongates as the amount of Hel 13-5 increases. This behavior is the same as that of the DPPC/PG/Hel 13-5 system. However, the DPPC/PA/Hel 13-5 system displays a jagged second plateau, revealing that the Hel 13-5 species cannot be easily squeezed out of the ternary monolayers because of the absence of fluid components such as PG. In contrast to the collapse behavior of the DPPC/PG/Hel 13-5 system, the DPPC/PA monolayer forms a stable film up to $\sim$60 mN m^{-1}, and all of the DPPC/PA/Hel 13-5 monolayers ($0 < X_{Hel\ 13-5} < 0.1$) also remain stable up to $\sim$60 mN m^{-1} despite adding Hel 13-5 to the DPPC/PA monolayer. This also reveals that Hel 13-5 interacts with rigid components (DPPC and undissociated PA) even at high surface pressures, and Hel 13-5 with dissociated PA is gradually squeezed out of the ternary monolayers above $\sim$42 mN m^{-1}. That is, the squeeze-out manages to gradually start upon further compression, and the interface becomes rich in the DPPC/PA monolayer at $\sim$60 mN m^{-1}. The more detailed mechanism of PA molecules during the squeeze-out motion is mentioned in the next section. At more than 42 mN m^{-1}, the ΔV–A isotherms within $0.01 < X_{Hel\ 13-5} < 0.1$ indicate the same behavior as that of the above-mentioned DPPC/PG/Hel 13-5 system.

2.4.2 IRRAS MEASUREMENTS

Shown in Figure 2.16A are the v_a-CH$_2$ RA intensities versus the relative area for the lipid mixture of DPPC + 10 wt% PA at the first and the third cycles. To carry out the quantitative comparison between the DPPC/PA mixtures with and without Hel 13-5, a spread amount of DPPC/PA was fixed in between Figures 2.16 and 2.17. Thus, the upper abscissa is labeled by the calculated area per DPPC/PA molecule. For the DPPC/PA monolayer, the initial absolute RA value of the v_a-CH$_2$ band at $\sim$0.65 nm^2/molecule was 0.0017. Upon compression, the absolute RA value increased gradually and then reached 0.0050 at $\sim$0.35 nm^2 molecule^{-1}, where the monolayer was in a close-packed state (see Figure 2.15). When the films were expanded, the absolute RA intensity decreased to reach the initial value. In addition, this behavior was reproducible from the first to the third cycles. Thus, it is also suggested that no quantitative loss of DPPC and PA molecules from the surface during compression–expansion cycles occurs in the monolayer. Figure 2.16B presents the plots of the

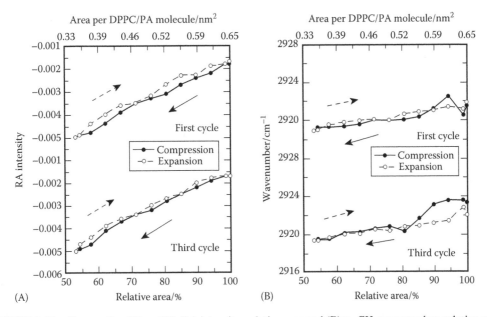

FIGURE 2.16 Consecutive (A) v_a-CH$_2$ RA intensity–relative area and (B) v_a-CH$_2$ wavenumber–relative area curves of DPPC + 10 wt% PA with an initial area per molecule of 0.65 nm^2. The upper abscissa is labeled by the calculated area per DPPC/PA molecule. (From Nakahara, H. et al., *Colloids Surf. B*, 68, 61, 2009. With permission.)

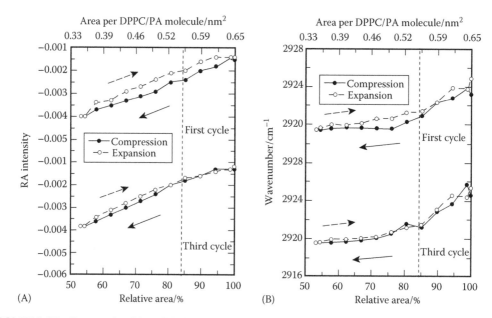

FIGURE 2.17 Consecutive (A) v_a-CH$_2$ RA intensity–relative area and (B) v_a-CH$_2$ wavenumber–relative area curves of a ternary DPPC/PA/Hel 13-5 monolayer at $X_{Hel\,13-5}$ = 0.1 with an initial area per DPPC/PA molecule of 0.65 nm^2. The onset area of the squeeze-out phenomenon is indicated by dashed lines (0.55 nm^2). The upper abscissa is labeled by the calculated area per DPPC/PA molecule. (From Nakahara, H. et al., *Colloids Surf. B,* 68, 61, 2009. With permission.)

corresponding v_a-CH$_2$ wavenumber with the relative area for the DPPC/PA monolayer. Under the first compression, the v_a-CH$_2$ wavenumber slightly decreased from 2922 to 2919 cm^{-1}. The slight reduction results from the lower compressibility of the DPPC/PA monolayer compared with the DPPC monolayer. During the first expansion, the wavenumber increased with an increase in the relative area to accord with those under the corresponding compression. The good reproducibility between the first and the third cycles indicates the reversible orientational variation in acyl chains for DPPC and PA during repeated cycles.

The hysteresis curves of v_a-CH$_2$ RA intensity with the relative area for the ternary DPPC/PA/Hel 13-5 monolayer at $X_{Hel\,13-5}$ = 0.1 are shown in Figure 2.17A. The dashed lines mean the molecular area of ~0.55 nm^2 molecule^{-1}, where the squeeze-out action of Hel 13-5 states occurred. The absolute RA values increased from 0.0015 to 0.0040 with decreasing relative area. It is quite interesting that at the close-packed state of monolayers, the value of 0.0040 was smaller compared with that for the DPPC/PA mixture (difference: 0.0010). Furthermore, the trend was reproducible from the first to the third cycles. If only the peptides are squeezed out of surface monolayers, the absolute RA value should reach ~0.0050 due to the surface refinement to binary DPPC/PA mixtures. Consequently, it is suggested that Hel 13-5 is excluded together with parts of dissociated PA molecules from the surface to form the surface-associated aggregates. This action is also reversible, and the dissociated PA molecules utilized into the aggregates can steadily respread to the surface. The corresponding v_a-CH$_2$ wavenumber versus relative area for the DPPC/PA/Hel 13-5 mixture ($X_{Hel\,13-5}$ = 0.1) is plotted in Figure 2.17B. In analogy with the DPPC/PA mixture, the wavenumber of DPPC/PA/Hel 13-5 slightly decreased from 2923 to 2919 cm^{-1} on the first compression. However, the reduction mode is different in the vicinity of the squeeze-out area indicated by dashed lines. That is, in the squeeze-out region, the wavenumber showed almost the same value of ~2920 cm^{-1} despite the film compression. This means that the aliphatic chain orientation of DPPC/PA is kept in the close-packed state. Under the situation, the squeeze-out motion of Hel 13-5 together with dissociated PA molecules occurs, and the surface is gradually refined to the monolayer of DPPC and undissociated PA molecules. These motions across the surface are reversible without loss of materials. The events are also observed in the binary DPPC/Hel 13-5 system. Therefore, it is demonstrated that the close-packed orientation of acyl chains at the surface triggers an ejected movement of the proteins or peptides (with positive charges) together with negatively charged molecules or LE components (Nakahara et al. 2009).

2.4.3 FM IMAGES

An FM observation has been carried out to understand the morphological effects of Hel 13-5 on binary DPPC/PA monolayers. Figure 2.18 shows FM images of ternary DPPC/PA/Hel 13-5 monolayers at $X_{\text{Hel 13-5}} = 0.01$. Two-phase coexistence states similar to the resultant images in the DPPC/PA monolayer system are also clearly observed (Nakahara et al. 2006a). They indicate a homogeneous disordered phase below the transition pressure (like FM images in Figure 2.5) and disordered/ordered coexistence phases at 15–25 mN m⁻¹. The micrograph at 35 mN m⁻¹ shows dark homogeneous images. Once the micrographs became dark homogeneously, the dark images remained, as is the case in the DPPC/PA system. Generally, aggregations of fluorescent probes induce a self-quenching, and the probe cannot emit fluorescence. From the image at 45 and 55 mN m⁻¹, the micrographs are released from quenching, and the contrast becomes gradually clearer. It is clearly shown that this specific behavior reflects the squeeze-out phenomenon of Hel 13-5 on lateral compression. In addition, the micrographs at 15 and 55 mN m⁻¹ have the same ordered domains in size. However, smaller disordered regions are observed in the FM image at 55 mN m⁻¹; ordered domains in the image at 55 mN m⁻¹ are more packed, supporting the squeeze-out of Hel 13-5.

A direct imaging with FM clearly showed that the morphological behavior of the synthetic Hel 13-5 peptide and the PA molecule in the DPPC monolayer were nonideal and led to the novel morphological evidence of squeeze-out events of Hel 13-5. Aggregated fluorescent probes upon lateral compression ultimately induce self-quenching phenomena, resulting from the high probe concentration at the surface. Fluorescent probes form stable LE (disordered) films and selectively dissolve in disordered lipids. Therefore, the fluorescent quenching sensitively depends on the surface quantities of such lipids. Thus, employing this quenching phenomenon can confirm the occurrence of squeeze-out phenomena from the interface. There are a few papers that interpret this behavior by applying

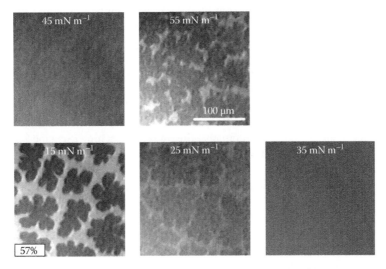

FIGURE 2.18 FM micrographs of the DPPC/PA/Hel 13-5 mixture system at $X_{\text{Hel 13-5}} = 0.01$ from $\pi = 15$–55 mN m⁻¹. In the coexistence phase, the percentage refers to ordered domains in the micrograph. The monolayers contain 1 mol% of fluorescent probe (R18). The scale bar in the lower right represents 100 μm. (From Nakahara, H. et al., *Langmuir*, 22, 5792, 2006a. With permission.)

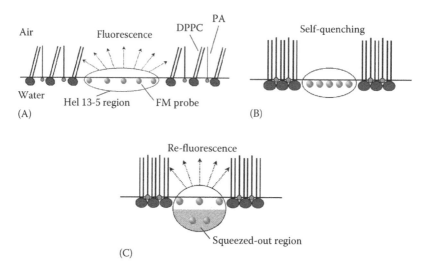

SCHEME 2.2 New concept describing the refluorescent phenomenon through self-quenching of FM probes led to by the squeeze-out of Hel 13-5 peptides from monolayers at (A) low surface pressure (<25 mN m^{-1}), (B) middle surface pressure (from 25 to 42 mN m^{-1}), and (C) high surface pressure (>42 mN m^{-1}). (From Nakahara, H. et al., *Langmuir*, 22, 5792, 2006a. With permission.)

fluorescent probe quenching. The papers reported that in the DPPC/dipalmitoylphosphatidylglyc-erol (DPPG)/rSP-C system clear contrast FM images of the disorder/order coexistence phase at low surface pressure were converted to unclear ones upon compression, and refluorescing images that were the same as those at low surface pressure were later observed beyond the plateau regions at high surface pressure, showing that the fluorescent probes as well as rSP-C were excluded from the interface (Amrein et al. 1997, Bourdos et al. 2000). This behavior surely confirms the occurrence of squeeze-out phenomena. However, mechanisms for the restoration of the contrast of fluorescent probes during the exclusion of rSP-C peptides cannot be clarified. In this system, the molar fraction change of Hel 13-5 peptides ($X_{\text{Hel 13-5}}$) in a fixed DPPC/PA ratio or the variation of surface quantities of disordered films, in which the fluorescent probes can preferentially dissolve, provided a sharp insight into an elucidation of the mechanism for the contrast changes corresponding to squeeze-out behavior (Figure 2.18). In addition, the comparison between DPPC/Hel 13-5 (Nakahara et al. 2005b, 2006b) and DPPC/PA/Hel 13-5 (Nakahara et al. 2006a) systems shows that the addition of PA into the DPPC/Hel 13-5 system effectually induces the squeeze-out behavior of Hel 13-5 because such refluorescent observations are not observed in the former system.

First, Scheme 2.2 shows a proposed model based on our results for the refluorescent phenomenon. FM probes emit fluorescence in a proper surface concentration (Scheme 2.2A). Further compres-sion leads to a higher surface concentration, resulting in self-quenching (Scheme 2.2B), although it depends sensitively on the number of disordered films at the surface. Assuming that there are small surface quantities of Hel 13-5 ($X_{\text{Hel 13-5}} = 0.01$ in Figure 2.18) at the interface, the FM images indicate complete self-quenching. On the other hand, more surface quantities ($X_{\text{Hel 13-5}} = 0.05$ (Nakahara et al. 2006a)) induce incomplete self-quenching because of the existence of abundant disordered regions. In general, further compression causes dark homogeneous patterns in FM images. However, the squeeze-out process for Hel 13-5 peptides again recovers the coexistent images (Scheme 2.2C). The compression beyond the Hel 13-5 collapse pressure provides the Hel 13-5 monolayer with a driving force to squeeze out from the interface, and then a portion of the Hel 13-5 molecules is excluded into the subphase with some fluorescent probes (LE film), which is supported by AFM measurements in a previous study (Nakahara et al. 2006b). Consequently, this specific behavior also demonstrates the squeeze-out phenomenon of our synthetic peptide Hel 13-5 at the same time.

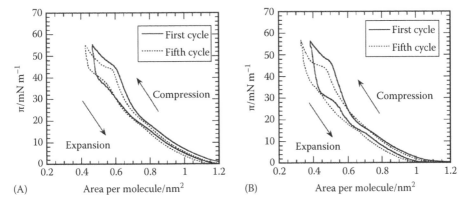

FIGURE 2.19 Cyclic compression and expansion isotherms of the (A) DPPC/PG/Hel 13-5 and (B) DPPC/PA/Hel 13-5 mixtures at $X_{Hel\ 13-5} = 0.1$ on a 0.02 M Tris buffer solution (pH 8.4) with 0.13 M NaCl at 298.2 K. The compression and expansion cycle was performed five times at the compression rate of ~0.1 nm² molecule⁻¹ min⁻¹.

2.4.4 HYSTERESIS ISOTHERMS

The repeated compression–expansion cycling π–A and ΔV–A isotherms of ternary (A) DPPC/PG/Hel 13-5 and (B) DPPC/PA/Hel 13-5 systems at $X_{Hel\ 13-5} = 0.1$ are shown for the first and the fifth rounds in Figure 2.19. These two systems were compressed up to a surface pressure of ~55 mN m⁻¹ and then expanded to the respective starting area. In Figure 2.19A, the second kink point clearly appears despite the repeated cycling processes, indicating that Hel 13-5 is reversibly desorbed into the subphase in the π–A isotherms. In Figure 2.19B, for π–A isotherms, the second kink shoulder for repeated hysteresis curves clearly appears without becoming unclear as it does in Figure 2.19A. The surface pressure decreases sharply like a solid film in the expansion process. Consequently, Figure 2.19 indicates good respreading behavior from the first to the fifth rounds, showing the ability of the desorbed molecules to come back into the interface. These results demonstrate that small amounts of Hel 13-5 peptides can accelerate the good respreading of rigid components (DPPC and PA).

2.5 COMPARISON OF THE NEW MULTICOMPONENT MODEL SYSTEM WITH SURFACTEN

2.5.1 LANGMUIR ISOTHERMS

The isotherms for the fixed composition of DPPC/PG/PA (=68/22/9, by weight) with Hel 13-5 peptides have been investigated to decide an optimum adding amount of Hel 13-5 to the DPPC/PG/PA mixture and to verify whether our artificial synthetic preparations have effective pulmonary functions by *in vitro* examinations. The π–A and ΔV–A isotherms of the multicomponent monolayers (DPPC/PG/PA/Hel 13-5) have been measured at various Hel 13-5 mole fractions ($X_{Hel\ 13-5}$) (Figure 2.20). All of the π–A isotherms have the transition pressures (π^{eq}), and one of them is indicated by a straight-line arrow ($X_{Hel\ 13-5} = 0.1$, curve d). The π^{eq} slightly increases with increasing $X_{Hel\ 13-5}$. The increase means that lipid monolayers are difficult to transform from LE to LC states due to the addition of Hel 13-5. This behavior also occurred in the DPPC/PA/Hel 13-5 system (Nakahara et al. 2006a). At ~42 mN m⁻¹, where Hel 13-5 starts to change from monolayer states to other states, the π–A isotherms exhibit plateau regions typically indicated by dashed arrows (curves c and d). These plateaus on the π–A isotherms correspond to the squeeze-out of Hel 13-5 from surface monolayers. That was previously supported by AFM measurements (Nakahara et al. 2006b). The plateau regions in the present system are almost parallel to the abscissa of area compared with those in the previous systems. In particular, the slope of the plateau region at $X_{Hel\ 13-5} = 0.1$ (curve d) is kept

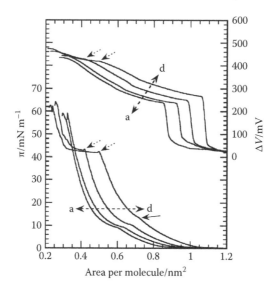

FIGURE 2.20 π–A and ΔV–A isotherms of the DPPC/PG/PA/Hel 13-5 preparations with the fixed DPPC/PG/PA ratio on a 0.02 M Tris buffer solution (pH 7.4) with 0.13 M NaCl at 298.2 K; (a) DPPC/PG/PA, (b) $X_{Hel\ 13-5} = 0.01$, (c) 0.05, and (d) 0.1. The transition pressure for $X_{Hel\ 13-5} = 0.1$ is indicated by a straight-line arrow. The plateau regions corresponding to the squeeze-out are shown by dashed arrows on the π–A isotherms of $X_{Hel\ 13-5} = 0.05$ and 0.1.

nearly zero on the lateral compression from 0.5 down to 0.3 nm^2, meaning the small slope. The slope difference between the present mixtures and the previous ternary systems shows that the present preparations generate a critical effect on the squeeze-out phenomenon. The smaller slope implies that the squeeze-out of Hel 13-5 with PG preferentially occurs at the expense of the improvement in orientation and packing of the DPPC/PA surface monolayers in the plateau region during the compression process. On further compression, the surface pressures rapidly increase up to a monolayer collapse, resulting in the production of close-packed monolayers (DPPC/PA) at the interface. As for the monolayer collapse, the DPPC/PG/PA/Hel 13-5 and DPPC/PA/Hel 13-5 systems indicate similar π^c values at $X_{Hel\ 13-5} = 0.05$ and 0.1. This also supports that the mixed DPPC/PA monolayers dominate the air/liquid interface at higher surface pressures in the plateau region (Nakahara et al. 2006a, 2008b).

For the surface potentials, the ΔV–A isotherms successively change with $X_{Hel\ 13-5}$ at low surface pressures. However, the inclination of curves c and d at the onset of the squeeze-out starts to vary and then is kept nearly zero (indicated by dashed arrows). These ΔV results support the dominant exclusion of Hel 13-5 with PG from surface monolayers (at the plateau region) at the expense of the improvement in the orientation of the binary DPPC/PA monolayer. On further compression, all of the ΔV–A isotherms converge to ~460 mV, suggesting that the close-packed DPPC/PA monolayers are formed in the high surface pressure region. These ΔV results also support the observations for the corresponding π–A isotherms. These isotherms demonstrate that the preparations in the present study take on better pulmonary functions (the squeeze-out action, high collapse pressure, and so on) and require a smaller amount of Hel 13-5 for wider plateau length on the π–A isotherm than the previous ternary systems.

2.5.2 FM Images

Shown in Figure 2.21 are the selected FM micrographs of the DPPC/PG/PA/Hel 13-5 preparations. In $X_{Hel\ 13-5} = 0.01$ (Figure 2.21A), where there is no plateau region corresponding to the squeeze-out on the π–A isotherm, two-phase coexistence states similar to the resultant images in

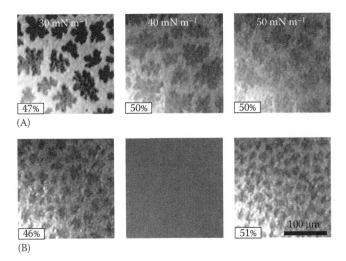

FIGURE 2.21 FM micrographs of the DPPC/PG/PA/Hel 13-5 preparations at (A) $X_{\text{Hel }13\text{-}5}=0.01$ and (B) 0.05 on a 0.02 M Tris buffer solution (pH 7.4) with 0.13 M NaCl at 298.2 K. In the coexistence phases, the percentage (%) refers to ordered domains in the micrograph. The monolayers contain 1 mol% of fluorescent probe (R18). The scale bar in the lower right represents 100 μm. (From Nakahara, H. et al., *Langmuir*, 24, 3370, 2008b. With permission.)

the lipid mixture are observed (Nakahara et al. 2008b). The FM image at 30 mN m^{-1} indicates a clear contrast between ordered domains and disordered ones. With an increase in surface pressure, however, the images become slightly unclear due to the concentration quenching of FM probes. Moreover, the ratio (%) of ordered domains does not vary in spite of an increase in surface pressure, showing the formation of phase-separated films at high surface pressures. In $X_{\text{Hel }13\text{-}5}=0.05$ (Figure 2.21B), at which the π–A isotherm has the plateau region, FM images indicate a different behavior from those of $X_{\text{Hel }13\text{-}5}=0.01$. Although a completely dark image appears at 40 mN m^{-1} due to the quenching, the image at 50 mN m^{-1} is released from it and gets better in contrast. The recovery of FM contrast results from the event that Hel 13-5 is squeezed out together with fluid components (PG and dissociated PA) and FM probes in the plateau region, the surface concentration of FM probes decreasing. This behavior was previously observed in the DPPC/PA/Hel 13-5 systems as a "refluorescence" phenomenon, providing a novel morphological evidence of the squeeze-out events of Hel 13-5 (Nakahara et al. 2006a).

2.5.3 HYSTERESIS ISOTHERMS

Repeated cycling π–A and ΔV–A isotherms of the selected DPPC/PG/PA/Hel 13-5 preparations for the first and the fifth rounds are shown in Figure 2.22. These monolayers are compressed up to a surface pressure of ~62 mN m^{-1} and then expanded to the respective starting molecular areas. Both of the π–A isotherms have large hysteresis area and indicate good reproducibility during cycling. After reaching the maximum surface pressure on compression, the surface pressure rapidly decreases by ~20 mN m^{-1} by a small expansion, suggesting that the squeezed-out Hel 13-5 instantly reenters into the surface and then disperses the close-packed monolayers (DPPC/PA) during the expansion. The ability of Hel 13-5 to respread with PG in the present preparations was found improved in terms of the rapid reentry to the surface in comparison with that in the previous systems (Nakahara et al. 2005b, 2006b). Furthermore, reproducible plateaus at 42 mN m^{-1} demonstrate that the exclusion and the successive inclusion of Hel 13-5 are a reversible process during cycling. Consequently, these actions during the compression–expansion cycling demonstrate that our preparations have sufficient pulmonary functions in the basic *in vitro* level (Notter 2000).

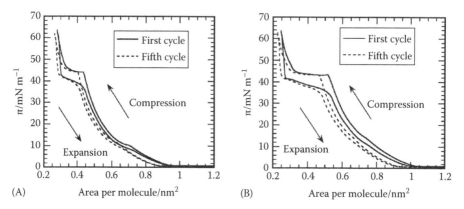

FIGURE 2.22 Cyclic compression and expansion isotherms of the DPPC/PG/PA/Hel 13-5 preparations at (A) $X_{Hel\,13-5}=0.05$ and (B) 0.1 on a 0.02 M Tris buffer solution (pH 7.4) with 0.13 M NaCl at 298.2 K. The compression–expansion cycle was performed five times at the compression rate of ~0.17 nm^2 molecule^{-1} min^{-1}.

2.5.4 Comparison to Surfacten (Surfactant TA)

The cyclic compression and expansion isotherms for Surfacten (a modified natural bovine pulmonary surfactant used well for RDS patients in Japan) are shown in Figure 2.23A. The π–A (or trough area) isotherms have a large hysteresis area and indicate good reproducibility during the cyclic process. After reaching the maximum surface pressure on compression, the surface pressure results in a rapid reduction of ~15 mN m^{-1} by a small expansion. The isotherms of the first cycle for $X_{Hel\,13-5}=0.05$ and 0.1, and Surfacten are superimposed in Figure 2.23B. The hysteresis area on the π–A isotherms for $X_{Hel\,13-5}=0.1$ is the largest of the three preparations by an integral of area, implying that the preparation ($X_{Hel\,13-5}=0.1$) has a better capability for the respreading than Surfacten. In addition, the analogous trend in the loop area and the profile between $X_{Hel\,13-5}=0.05$ and Surfacten is observed for π–A isotherms. These results demonstrate that when Hel 13-5 is mixed with the DPPC/PG/PA lipid mixture, it has similar capabilities to SP-B and SP-C such as rapid respreading and adsorption through the interface.

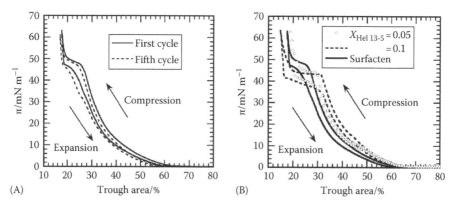

FIGURE 2.23 (A) Cyclic compression and expansion isotherms of Surfacten on a 0.02 M Tris buffer solution (pH 7.4) with 0.13 M NaCl at 298.2 K. (B) Hysteresis curves at the first round for the DPPC/PG/PA/Hel 13-5 preparations at $X_{Hel\,13-5}=0.05$ and 0.1, and for Surfacten. The compression–expansion cycle was performed five times at the compression rate of ~90 cm^2 min^{-1}. Herein, area (abscissa axis) represents the relative area (%) to the initial surface area prior to compression.

2.6 SUMMARY

The present chapter has highlighted the interfacial behavior of synthetic pulmonary surfactants containing a synthetic peptide (Hel 13-5)—a mimic peptide of the human surfactant protein B (SP-B)—during lateral compression and expansion. Hel 13-5 itself is an amphiphilic peptide and can form an α-helical structure at the air/water interface. DPPC is a main phospholipid component in pulmonary surfactants and contributes to lowering surface tension during exhalation. DPPC alone has some defects in pulmonary functions such as rapid respreading, adsorption, and large hysteresis. However, a small-amount addition of Hel 13-5 to DPPC leads to improvement in the adsorption of DPPC. Moreover, the addition results in the enlargement of a hysteresis loop of reversible π–A curves. These results support the possibility of Hel 13-5 as a substitute for SP-B (Nakahara et al. 2005b, 2006b). In addition, the IRRAS measurements particularly focusing on v_a–CH_2 peaks suggested that the acyl chain orientation of DPPC becomes in the close-packed state soon after the occurrence of the squeeze-out behavior of Hel 13-5, and the packed orientation is retained up to the collapse state of monolayers (Nakahara et al. 2009).

Both PG and PA components prevent the DPPC monolayer from collapsing above the squeeze-out motion of Hel 13-5. PG is thought to selectively and electrostatically interact with cationic Hel 13-5 during the squeeze-out process, and then 3-D aggregates containing Hel 13-5 and PG can be formed just below the surface monolayer (Alonso et al. 2004, Diemel et al. 2002, Ding et al. 2003, Nakahara et al. 2006b, 2008a). This formation contributes to the stability of the DPPC monolayers. On the other hand, the PA components have a complex role during the squeeze-out motion. The undissociated PA molecules are useful for the accomplishment of close-packed orientation of DPPC aliphatic chains at the surface, whereas the dissociated PA components contribute to the squeeze-out motion of Hel 13-5 to form 3-D aggregates associated with surface monolayers (Nakahara et al. 2009).

Finally, various kinds of measurements for surface chemistry (Langmuir isotherms, fluorescent images, temperature dependence, and hysteresis curves) have been made for the comparison between the DPPC/PG/PA/Hel 13-5 preparations and Surfacten, which is commercially used for NRDS patients in Japan (Nakahara et al. 2008b). These results reveal that the preparations are comparable to Surfacten in terms of the surface activity and the interfacial behavior. The present work on pulmonary surfactant substitutes containing a synthetic peptide will provide a deep insight into the clarification of functional mechanisms of pulmonary surfactants near the air/alveolar liquid interface and the introduction of "tailoring" surfactant preparations to various states of surfactant deficiency, insufficiency, and inactivation.

ABBREVIATIONS

A	Mean molecular area (nm^2 $molecule^{-1}$)
ΔV	Surface potential (mV)
$\mu_\perp$	Surface dipole moment (mD)
π	Surface pressure (mN m^{-1})
π^c	Collapse pressure (mN m^{-1})
π^{eq}	Phase transition pressure (mN m^{-1})
R	Reflectivity of the monolayer-covered surface
R_0	Reflectivity of the subphase (pure buffer solution) surface
t	Time (min)
X	Mole fraction
AFM	Atomic force microscopy
ARDS	Acute (adult) respiratory distress syndrome
BSE	Bovine spongiform encephalopathy

3-D	Three-dimensional
DPPC	Dipalmitoylphosphatidylcholine
DPPG	Dipalmitoylphosphatidylglycerol
FM	Fluorescence microscopy
HHB	Hydrophobic–hydrophilic balance
IRRAS	Infrared reflection–absorption spectroscopy
LB	Langmuir–Blodgett film
LC	Liquid-condensed
LE	Liquid-expanded
Lys (K)	Lysine
Leu (L)	Leucine
NRDS	Neonatal respiratory distress syndrome
PA	Palmitic acid (Hexadecanoic acid)
PG	Egg-phosphatidylglycerol
POPG	Palmitoyloleoylphosphatidylglycerol
PS	Pulmonary surfactant
R18	3,6-bis(diethylamino)-9-(2-octadecyloxycarbonyl) phenyl chloride
RA	Reflection–absorption
RDS	Respiratory distress syndrome
RP-HPLC	Reverse phase high-performance liquid chromatography
SP	Surfactant (associated) protein
Trp (W)	Tryptophan
v_s-CH$_2$	Symmetric methylene stretching vibration
v_a-CH$_2$	Asymmetric methylene stretching vibration

ACKNOWLEDGMENTS

This work was supported by a Grant-in-Aid for Scientific Research 20500414 from the Japan Society for the Promotion of Science (JSPS). This work was also supported by a Grant-in-Aid for Young Scientists (B) 22710106 from JSPS (H.N.).

REFERENCES

Alonso, C., Alig, T., Yoon, J. et al. 2004. More than a monolayer: Relating lung surfactant structure and mechanics to composition. *Biophys. J.* 87:4188–4202.

Amrein, M., von Nahmen, A., and Sieber, M. 1997. A scanning force- and fluorescence light microscopy study of the structure and function of a model pulmonary surfactant. *Eur. Biophys. J.* 26:349–357.

Avery, M.E. and Mead, J. 1959. Surface properties in relation to atelectasis and hyaline membrane disease. *Am. J. Dis. Child.* 97:517–526.

Baatz, J.E., Elledge, B., and Whitsett, J.A. 1990. Surfactant protein SP-B induces ordering at the surface of model membrane bilayers. *Biochemistry* 29:6714–6720.

Bourdos, N., Kollmer, F., Benninghoven, A. et al. 2000. Analysis of lung surfactant model systems with time-of-flight secondary ion mass spectrometry. *Biophys. J.* 79:357–369.

Brown, E.S. 1964. Isolation and assay of dipalmityl lecithin in lung extracts. *Am. J. Physiol.* 207:402–406.

Cai, P., Flach, C.R., and Mendelsohn, R. 2003. An infrared reflection-absorption spectroscopy study of the secondary structure in (KL$_4$)$_4$K, a therapeutic agent for respiratory distress syndrome, in aqueous monolayers with phospholipids. *Biochemistry* 42:9446–9452.

Chang, R., Nir, S., and Poulain, F.R. 1998. Analysis of binding and membrane destabilization of phospholipid membranes by surfactant apoprotein B. *Biochim. Biophys. Acta* 1371:254–264.

Cockshutt, A.M., Absolom, D.R., and Possmayer, F. 1991. The role of palmitic acid in pulmonary surfactant: Enhancement of surface activity and prevention of inhibition by blood proteins. *Biochim. Biophys. Acta* 1085:248–256.

Diemel, R.V., Snel, M.M.E., Waring, A.J. et al. 2002. Multilayer formation upon compression of surfactant monolayers depends on protein concentration as well as lipid composition. An atomic force microscopy study. *J. Biol. Chem.* 277:21179–21188.

Ding, J., Takamoto, D.Y., Von Nahmen, A. et al. 2001. Effects of lung surfactant proteins, SP-B and SP-C, and palmitic acid on monolayer stability. *Biophys. J.* 80:2262–2272.

Ding, J., Doudevski, I., Warriner, H.E. et al. 2003. Nanostructure changes in lung surfactant monolayers induced by interactions between palmitoyloleoylphosphatidylglycerol and surfactant protein B. *Langmuir* 19:1539–1550.

Dluhy, R.A. 1986. Quantitative external reflection infrared spectroscopic analysis of insoluble monolayers spread at the air–water interface. *J. Phys. Chem.* 90:1373–1379.

Dluhy, R.A. and Cornell, D.G. 1985. In situ measurement of the infrared spectra of insoluble monolayers at the air-water interface. *J. Phys. Chem.* 89:3195–3197.

Fleming, B. and Keough, K.M.W. 1988. Surface respreading after collapse of monolayers containing major lipids of pulmonary surfactant. *Chem. Phys. Lipids* 49:81–86.

Furuya, T., Kiyota, T., Lee, S. et al. 2003. Nanotubules formed by highly hydrophobic amphiphilic α-helical peptides and natural phospholipids. *Biophys. J.* 84:1950–1959.

Gericke, A. and Hühnerfuss, H. 1993. In situ investigation of saturated long-chain fatty acids at the air/water interface by external infrared reflection-absorption spectrometry. *J. Phys. Chem.* 97:12899–12908.

Goerke, J. 1998. Pulmonary surfactant: Functions and molecular composition. *Biochim. Biophys. Acta* 1408:79–89.

Gorree, G., Egberts, J., Bakker, G., Beintema, A., and Top, M. 1991. Development of a human lung surfactant, derived from extracted amniotic fluid. *Biochim. Biophys. Acta* 1086:209–216.

Grigoriev, D.O., Kragel, J., Akentiev, A.V. et al. 2003. Relation between rheological properties and structural changes in monolayers of model lung surfactant under compression. *Biophys. Chem.* 104:633–642.

Gustafsson, M., Vandenbussche, G., Curstedt, T., Ruysschaert, J.M., and Johansson, J. 1996. The 21-residue surfactant peptide (LysLeu$_4$)$_4$Lys(KL$_4$) is a transmembrane α-helix with a mixed nonpolar/polar surface. *FEBS Lett.* 384:185–188.

Gustafsson, M., Palmblad, M., Curstedt, T., Johansson, J., and Schurch, S. 2000. Palmitoylation of a pulmonary surfactant protein C analogue affects the surface associated lipid reservoir and film stability. *Biochim. Biophys. Acta* 1466:169–178.

Johansson, J., Curstedt, T., and Jörnvall, H. 1991. Surfactant protein B: Disulfide bridges, structural properties, and kringle similarities. *Biochemistry* 30:6917–6921.

Kirkness, J.P., Eastwood, P.R., Szollosi, I. et al. 2003a. Effect of surface tension of mucosal lining liquid on upper airway mechanics in anesthetized humans. *J. Appl. Physiol.* 95:357–363.

Kirkness, J.P., Madronio, M., Stavrinou, R., Wheatley, J.R., and Amis, T.C. 2003b. Relationship between surface tension of upper airway lining liquid and upper airway collapsibility during sleep in obstructive sleep apnea hypopnea syndrome. *J. Appl. Physiol.* 95:1761–1766.

Kitamura, A., Kiyota, T., Tomohiro, M. et al. 1999. Morphological behavior of acidic and neutral liposomes induced by basic amphiphilic α-helical peptides with systematically varied hydrophobic-hydrophilic balance. *Biophys. J.* 76:1457–1468.

Kiyota, T., Lee, S., and Sugihara, G. 1996. Design and synthesis of amphiphilic α-helical model peptides with systematically varied hydrophobic-hydrophilic balance and their interaction with lipid- and bio-membranes. *Biochemistry* 35:13196–13204.

Krol, S., Ross, M., Sieber, M. et al. 2000. Formation of three-dimensional protein-lipid aggregates in monolayer films induced by surfactant protein B. *Biophys. J.* 79:904–918.

Krüger, P., Baatz, J.E., Dluhy, R.A., and Lösche, M. 2002. Effect of hydrophobic surfactant protein SP-C on binary phospholipid monolayers. Molecular machinery at the air/water interface. *Biophys. Chem.* 99:209–228.

Lee, S., Furuya, T., Kiyota, T. et al. 2001. De novo-designed peptide transforms Golgi-specific lipids into Golgi-like nanotubules. *J. Biol. Chem.* 276:41224–41228.

Lee, S., Sugihara, G., Shibata, O., and Yukitake, H. 2004. JP Patent Appl. P 2004-305006A.

Lipp, M.M., Lee, K.Y., Waring, A., and Zasadzinski, J.A. 1997. Fluorescence, polarized fluorescence, and Brewster angle microscopy of palmitic acid and lung surfactant protein B monolayers. *Biophys. J.* 72:2783–2804.

Lösche, H. and Möhwald, H. 1984. Fluorescence microscopy on monomolecular films at an air/water interface. *Colloids Surf.* 10:217–224.

Ma, J., Koppenol, S., Yu, H., and Zografi, G. 1998. Effects of a cationic and hydrophobic peptide, KL$_4$, on model lung surfactant lipid monolayers. *Biophys. J.* 74:1899–1907.

Mendelsohn, R., Brauner, J.W., and Gericke, A. 1995. External infrared reflection absorption spectrometry of monolayer films at the air–water interface. *Annu. Rev. Phys. Chem.* 46:305–334.

Mitchell, M.L. and Dluhy, R.A. 1988. In situ FT-IR investigation of phospholipid monolayer phase transitions at the air–water interface. *J. Am. Chem. Soc.* 110:712–718.

Nahmen, A.V., Schenk, M., Sieber, M., and Amrein, M. 1997. The structure of a model pulmonary surfactant as revealed by scanning force microscopy. *Biophys. J.* 72:463–469.

Nakahara, H., Nakamura, S., Kawasaki, H., and Shibata, O. 2005a. Properties of two-component Langmuir monolayer of single chain perfluorinated carboxylic acids with dipalmitoylphosphatidylcholine (DPPC). *Colloids Surf. B* 41:285–298.

Nakahara, H., Nakamura, S., Lee, S., Sugihara, G., and Shibata, O. 2005b. Influence of a new amphiphilic peptide with phospholipid monolayers at the air-water interface. *Colloids Surf. A* 270–271:52–60.

Nakahara, H., Lee, S., Sugihara, G., and Shibata, O. 2006a. Mode of interaction of hydrophobic amphiphilic α-helical peptide/dipalmitoylphosphatidylcholine with phosphatidylglycerol or palmitic acid at the air–water interface. *Langmuir* 22:5792–5803.

Nakahara, H., Nakamura, S., Hiranita, T. et al. 2006b. Mode of interaction of amphiphilic α-helical peptide with phosphatidylcholines at the air–water interface. *Langmuir* 22:1182–1192.

Nakahara, H., Lee, S., and Shibata, O. 2009b. Pulmonary surfactant model systems catch specific interaction of an amphiphilic peptide with anionic phospholipid. *Biophys. J.* 96:1415–1429.

Nakahara, H., Lee, S., Sugihara, G., Chang, C.-H., and Shibata, O. 2008. Langmuir monolayer of artificial pulmonary surfactant mixtures with an amphiphilic peptide at the air/water interface: Comparison of new preparations with Surfacten (Surfactant TA). *Langmuir* 24:3370–3379.

Nakahara, H., Dudek, A., Nakamura, Y. et al. 2009a. Hysteresis behavior of amphiphilic model peptide in lung lipid monolayers at the air–water interface by an IRRAS measurement. *Colloids Surf. B* 68:61–67.

Notter, R.H. 2000. *Lung Surfactants: Basic Science and Clinical Applications*, Vol. 1. New York, Basel, Switzerland, Marcel Dekker, Inc., pp. 1–444.

Oosterlaken-Dijksterhuis, M.A., Haagsman, H.P., Van Golde, L.M.G., and Demel, R.A. 1991. Interaction of lipid vesicles with monomolecular layers containing lung surfactant proteins SP-B or SP-C. *Biochemistry* 30:8276–8281.

Piknova, B., Schief, W.R., Vogel, V., Discher, B.M., and Hall, S.B. 2001. Discrepancy between phase behavior of lung surfactant phospholipids and the classical model of surfactant function. *Biophys. J.* 81:2172–2180.

Postle, A.D., Heeley, E.L., and Wilton, D.C. 2001. A comparison of the molecular species compositions of mammalian lung surfactant phospholipids. *Comp. Biochem. Physiol. A* 129:65–73.

Revak, S.D., Merritt, T.A., Cochrane, C.G. et al. 1996. Efficacy of synthetic peptide-containing surfactant in the treatment of respiratory distress syndrome in preterm infant rhesus monkeys. *Pediatr. Res.* 39:715–724.

Serrano, A.G., Cruz, A., Rodriguez-Capote, K., Possmayer, F., and Perez-Gil, J. 2005. Intrinsic structural and functional determinants within the amino acid sequence of mature pulmonary surfactant protein SP-B. *Biochemistry* 44:417–430.

Takamoto, D.Y., Lipp, M.M., Von Nahmen, A. et al. 2001. Interaction of lung surfactant proteins with anionic phospholipids. *Biophys. J.* 81:153–169.

Tanaka, Y., Takei, T., Aiba, T. et al. 1986. Development of synthetic lung surfactants. *J. Lipid Res.* 27:475–485.

Taneva, S. and Keough, K.M. 1994. Pulmonary surfactant proteins SP-B and SP-C in spread monolayers at the air–water interface: I. Monolayers of pulmonary surfactant protein SP-B and phospholipids. *Biophys. J.* 66:1137–1148.

Veldhuizen, R., Nag, K., Orgeig, S., and Possmayer, F. 1998. The role of lipids in pulmonary surfactant. *Biochim. Biophys. Acta* 1408:90–108.

Veldhuizen, E.J.A., Waring, A.J., Walther, F.J. et al. 2000. Dimeric N-terminal segment of human surfactant protein B (dSP-B1–25) has enhanced surface properties compared to monomeric SP-B1–25. *Biophys. J.* 79:377–384.

Wen, X., Lauterbach, J., and Franses, E.I. 2000. Surface densities of adsorbed layers of aqueous sodium myristate inferred from surface tension and infrared reflection absorption spectroscopy. *Langmuir* 16:6987–6994.

Wüstneck, N., Wüstneck, R., Fainerman, V.B., Miller, R., and Pison, U. 2001. Interfacial behaviour and mechanical properties of spread lung surfactant protein/lipid layers. *Colloids Surf. B* 21:191–205.

Wüstneck, R., Wüstneck, N., Moser, B., and Pison, U. 2002. Surface dilatational behavior of pulmonary surfactant components spread on the surface of a pendant drop. 2. Dipalmitoyl phosphatidylcholine and surfactant protein B. *Langmuir* 18:1125–1130.

Yin, C.-L. and Chang, C.-H. 2006. Infrared spectroscopy analysis of mixed DPPC/fibrinogen layer behavior at the air/liquid interface under a continuous compression-expansion condition. *Langmuir* 22:6629–6634.

Yu, S.-H. and Possmayer, F. 2003. Lipid compositional analysis of pulmonary surfactant monolayers and monolayer-associated reservoirs. *J. Lipid Res.* 44:621–629.

Yukitake, K., Nakamura, Y., Kawahara, M. et al. 2008. Development of low cost pulmonary surfactants composed of a mixture of lipids or lipids-peptides using higher aliphatic alcohol or soy lecithin. *Colloids Surf. B* 66:281–286.

3 Supramolecular Assemblies Formed by Self-Assembling Peptides

Christophe Déjugnat and Thomas Zemb

CONTENTS

3.1 INTRODUCTION

Peptide self-assembly is a fast growing research area because self-assembly of proteins and glycolipids control the efficiency of biological processes in intracellular organelles and membranes. Peptides may be considered as structural precursors for more complex protein assemblies and they present self-assembly behaviors at equilibrium, even in solute-buffer simple solutions, that is, at thermodynamic equilibrium in pseudo-binary solutions. They share many similarities with surfactants, copolymers, and nanoparticles. For instance, peptide self-organization could be related to protein folding where different parts of the secondary structure self-assemble to form a 3D arrangement and the tertiary structure of the protein, which is essential to its function. Furthermore, folded proteins can also self-assemble to form oligomers interacting with proteins in a noncovalent way. Two very important cases with widespread medical implication and therefore the attention of molecular biologists are

- Proteins involved in mental degenerative diseases (Alzheimer, Parkinson, etc.); they have the property to self-assemble forming first soluble proto-aggregates that can further inter-act to irreversibly form insoluble fibers, and precipitate and alter the neuronal cells in which they are located.
- Passive self-assembly of virus, as discovered by Caspar and Klug (1962), as well as protein interaction with their membranes: fusion of proteins during viral infection, aggregation of amphipathic helices to form ion channels through biological membranes, etc.

It is therefore of utmost importance, not only as the primary driving force toward protein folding and aggregation but also for the preparation of bio-inspired materials, to understand and to better control the supramolecular assemblies formed by self-assembling peptides. In this chapter, we review pep-tide aggregation from the point of view of different colloidal systems obtained using peptides seen as a succession of more or less flexible, more or less charged, and more or less "hydrophilic" amino acids. All these concepts cannot be attributed in an absolute manner to an amino acid, since they are highly dependent on the "neighborhood" (Hyde et al. 1997). Charge regulation is known for colloids and can be understood as pK_a shifts of several units depending on adjacent amino acids: the analogue for polyelectrolytes is the so-called Manning condensation (Manning 1969). Similarly, "hydropho-bicity/hydrophilicity" balance is understood either as a volume balance in an amphiphilic molecule, or a spontaneous curvature, or distribution coefficients of elements in water/octanol two-phase sys-tems. None of these approaches alone show general predictive power, since they are all dependent on the nature of the added salt or buffer, even at the same ionic strength (Kunz et al. 2004).

In the first part, lipopeptides bearing an aliphatic hydrocarbon chain are described. These can be considered as analogues of conventional surfactants and form a wide variety of aggregates, some of them similar to micelles (Chevalier and Zemb 1990). In the second part, peptides in which amphiphilicity is provided by a special arrangement of the amino acids are considered. In that case, there is no grafting of alkyl chain; the peptides are supposed to be intrinsically amphiphilic. All these different cases will illustrate that, contrary to conventional surfactants, peptides are capable of having a secondary structure that might strongly influence further intermolecular interactions and supramolecular organization. Finally, an outlook and emerging field is the case of binary or multicomponent aggregation: this is the case when two different bricks self-assemble in aggregates depending on the molar ratio between the two bricks. The first well-studied simple case is the case of "catanionic" or acid–base aggregation; but this complex equivalent to "bipolar coacervation of colloids" (Bungenberg de Jong and Kruyt 1930, Bungenberg de Jong 1949) is illustrated in the world of viruses by the case of the viral capsid, when the ratio of protein at vertex (12 per particle) to the number of proteins par face of icosahedra controls the size of the capsid, and hence the resistance toward osmotic pressure of intra-viral content (Zandi et al. 2004). As a conclusion, basic "rules" linking the interaction to the nature of aggregates will be tentatively described.

3.2 ALKYLATED PEPTIDES

Like conventional surfactants, constituted of one "polar" head group linked by a covalent bound to an aliphatic flexible alkyl chain, peptides can also bear such hydrophobic fatty chains to become structurally similar to an amphiphilic molecule. In that case, the peptide sequence represents the polar head group of the surfactant and its self-assembly might be expected to be similar to the observed classical amphiphiles: as shown by Ninham and coworkers, the chemical potential of any molecule confined in an aggregate is related to its effective curvature, averaged over the whole aggregate: thus, the classical sequence is a globular micelle, an elongated micelle, a bicelle, and a continuous network (Israelachvili et al. 1976). The basic parameter describing any amphiphilic structure is the so-called packing parameter (Kunz et al. 2009), which is the ratio of the volume divided by longest chord defining an area, by the area per molecule once embedded in the surfactant film.

Lipopeptides linked to an amphiphilic chain occur naturally with specific functions (antimicrobial, metal sequestration…), but a wide variety of synthetic compounds have also been synthesized, with variations in the number and the position of the alkyl parts, and therefore the class of designed alkylpeptides: monocatenary, bicatenary, bolaforms (i.e., with tow head groups at the end of one chain), geminis (i.e., two surfactants linked by a spacer made from several covalent bounds), and dendritics.

3.2.1 NATURALLY OCCURRING LIPOPEPTIDES

3.2.1.1 Microbial Biosurfactants

Bacterial lipopeptides are cyclic peptides bearing an aliphatic hydrocarbon chain. Several microbial species produce such biosurfactants, among them *Bacillus* species producing the surfactin, iturin, and fengycin series (Figure 3.1 presents *Bacillus* lipopeptides). Their biological role is thought to be important in the formation of biofilms, for cell motion or adhesion on surfaces. They are very strong surfactants and have gained great interest for potential applications as foaming agents, emulsifiers, for oil- or metal recovery, and for plant disease biocontrol (Lang 2002, Mulligan 2005, Lu et al. 2007, Ongena and Jacques 2007, Dexter and Middelberg 2008, Muthusamy et al. 2008, Seydlová and Svobodová 2008). These very attractive compounds remain difficult to synthesize by classical peptide synthesis; production is mainly realized by extraction from cell cultures and remains costly. This is the major limitation for the use of these compounds in industrial processes.

Discovered in 1968 after extraction from *Bacillus subtilis* culture media, the cyclic lipopeptide surfactin is one of the most effective biosurfactant known and has been since largely studied because it appears as a "green" surfactant (Peypoux et al. 1999). Therefore, considerable efforts have been made to decrease production costs and increase production yields. It lowers the surface tension of water from 72 to 27 mN/m, even at very low concentrations (μM range). Surfactin also interacts strongly with phospholipid bilayers in biological membranes which it destabilizes significantly by tilting the phospholipids and with a deep insertion of the peptide moiety in the membrane bilayer (Heerklotz and Seeling 2001, Heerklotz et al. 2004, Brasseur et al. 2007). Studies on phospholipids monolayers have confirmed that surfactin membrane insertion is promoted by its cyclic form and hydrophobic interactions (Eeman et al. 2006).

Moreover, phospholipid vesicles can be "solubilized," that is, converted into globular "mixed" micelles in the presence of surfactin, showing its strong "detergency" power (Kell et al. 2007). Another way to rationalize this observation is to attribute a small packing parameter (of the order of 1/3) to surfactin. Surfactin also strongly adsorbs at interfaces: at the air/water interface or when adsorbed onto a hydrophobic material the peptide adopts a ball-like structure and forms a stable monomolecular film. Neutron reflectivity has been used to show that the film thickness is about 1.5 nm and the area per molecule ranges around 1.5 nm^2, typically twice the value of a standard surfactant. Seen this way, surfactin acts more like an amphipathic nanoparticle rather than a conventional surfactant (Shen et al. 2009).

Depending on the pH, surfactin can be soluble in water when at least one of its two carboxylic groups is partially ionized. In water at pH = 7.5 and at a concentration of 1 mM, spherical micelles have been evidenced with a diameter of 5 nm and an aggregation number of about 20 molecules, which is imposed by sterical packing constraints (similar to those existing in cyclodextrins with grafted chains (Auzély-Velty et al. 1999)) linked to the large volume of the surfactant-head. Surfactin adopts a well-defined core-shell structure when forming micelles: the hydrophilic polar "head groups" being formed by the two negatively charged carboxylate moieties whereas the aliphatic chain and the four leucines are restrained in the micelle core forming the hydrophobic part (Shen et al. 2009).

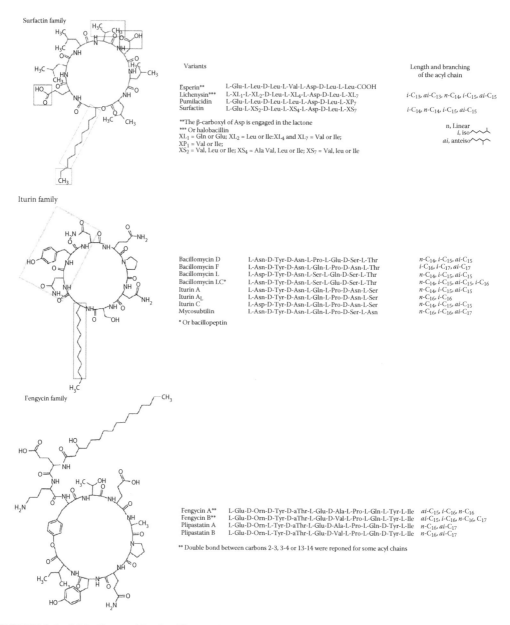

Surfactin family

Variants		Length and branching of the acyl chain
Esperin**	L-Glu-L-Leu-D-Leu-L-Val-L-Asp-D-Leu-L-Leu-COOH	
Lichenysin***	L-XL$_1$-L-XL$_2$-D-Leu-L-XL$_4$-L-Asp-D-Leu-L-XL$_7$	i-C$_{13}$, ai-C$_{13}$, n-C$_{14}$, i-C$_{15}$, ai-C$_{15}$
Pumilacidin	L-Glu-L-Leu-D-Leu-L-Leu-L-Asp-D-Leu-L-XP$_7$	
Surfactin	L-Glu-L-XS$_2$-D-Leu-L-XS$_4$-L-Asp-D-Leu-L-XS$_7$	i-C$_{14}$, n-C$_{14}$, i-C$_{15}$, ai-C$_{15}$

**The β-carboxyl of Asp is engaged in the lactone
*** Or halobacillin
XL$_1$ = Gln or Glu; XL$_2$ = Leu or Ile:XL$_4$ and XL$_7$ = Val or Ile;
XP$_1$ = Val or Ile;
XS$_2$ = Val, Leu or Ile; XS$_4$ = Ala Val, Leu or Ile; XS$_7$ = Val, leu or Ile

n, Linear
i, iso
ai, anteiso

Iturin family

Bacillomycin D	L-Asn-D-Tyr-D-Asn-L-Pro-L-Glu-D-Ser-L-Thr	n-C$_{14}$, i-C$_{15}$, ai-C$_{15}$
Bacillomycin F	L-Asn-D-Tyr-D-Asn-L-Gln-L-Pro-D-Asn-L-Thr	i-C$_{16}$, i-C$_{17}$, ai-C$_{17}$
Bacillomycin L	L-Asp-D-Tyr-D-Asn-L-Ser-L-Gln-D-Ser-L-Thr	n-C$_{14}$, i-C$_{15}$, ai-C$_{15}$
Bacillomycin LC*	L-Asp-D-Tyr-D-Asn-L-Ser-L-Glu-D-Ser-L-Thr	n-C$_{14}$, i-C$_{15}$, ai-C$_{15}$, i-C$_{16}$
Iturin A	L-Asn-D-Tyr-D-Asn-L-Gln-L-Pro-D-Asn-L-Ser	n-C$_{14}$, i-C$_{15}$, ai-C$_{15}$
Iturin A$_L$	L-Asn-D-Tyr-D-Asn-L-Gln-L-Pro-D-Asn-L-Ser	n-C$_{16}$, i-C$_{16}$
Iturin C	L-Asp-D-Tyr-D-Asn-L-Gln-L-Pro-D-Asn-L-Ser	n-C$_{14}$, i-C$_{15}$, ai-C$_{15}$
Mycosubtilin	L-Asn-D-Tyr-D-Asn-L-Gln-L-Pro-D-Ser-L-Asn	n-C$_{16}$, i-C$_{16}$, ai-C$_{17}$

* Or bacillopeptin

Fengycin family

Fengycin A**	L-Glu-D-Orn-D-Tyr-D-aThr-L-Glu-D-Ala-L-Pro-L-Gln-L-Tyr-L-Ile	ai-C$_{15}$, i-C$_{16}$, n-C$_{16}$
Fengycin B**	L-Glu-D-Orn-D-Tyr-D-aThr-L-Glu-D-Val-L-Pro-L-Gln-L-Tyr-L-Ile	ai-C$_{15}$, i-C$_{16}$, n-C$_{16}$, C$_{17}$
Plipastatin A	L-Glu-D-Orn-L-Tyr-D-aThr-L-Glu-D-Ala-L-Pro-L-Gln-D-Tyr-L-Ile	n-C$_{16}$, ai-C$_{17}$
Plipastatin B	L-Glu-D-Orn-L-Tyr-D-aThr-L-Glu-D-Val-L-Pro-L-Gln-D-Tyr-L-Ile	n-C$_{16}$, ai-C$_{17}$

** Double bond between carbons 2-3, 3-4 or 13-14 were reponed for some acyl chains

FIGURE 3.1 Main lipopeptides families produced by *Bacillus* species. (From Ongena, M. and Jacques, P., *TRENDS Microbiol.*, 16, 115, 2007. With permission.)

Depending on the experimental parameters (pH, ionic strength, presence of chelating ions…), globular micelles undergo a transition to larger aggregates such as ellipsoidal elongated micelles (Figure 3.2 represents an elongated surfactin micelle) (Han et al. 2008) with an aggregation number up to about 200 (Ishigami et al. 1995). The transition is not sharp and two kinds of micelles may coexist, a rare situation in the case of flexible surfactants (Figure 3.3 shows a transmission electron microscopy (TEM) micrograph of surfactin aggregates). Moreover, the aggregation number decreases when salt is added: an observation consistent with the known adsorption of chaotropic ions on surfaces due to preferential exclusion from "bulk" water (Knoblich et al. 1995).

The importance of peptide conformation has been evidenced with regard to the aggregation and arrangement of the molecules. The most effective conformation is a β-sheet and it can be stabilized

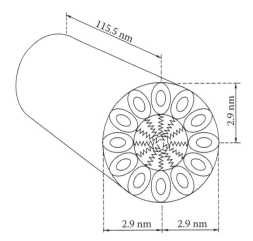

FIGURE 3.2 Schematic representation of a prolate ellipsoid surfactin micelle. (From Ishigami, Y. et al., *Colloids Surf. B Biointerfaces*, 4, 341, 1995. With permission.)

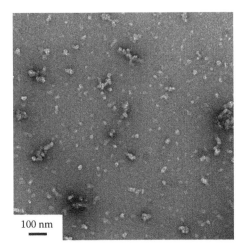

FIGURE 3.3 Transmission electron micrograph of surfactin aggregates formed at 0.3 mM in a pH 7.4 phosphate buffer. (From Han, Y. et al., *J. Phys. Chem. B*, 112, 15195, 2008. With permission.)

depending on external parameters such as complexation with calcium, favoring self-organization (Osman et al. 1998a,b). The stability of β-sheet conformation might be due to a dimerization of surfactin molecules (Ishigami et al. 1995). However, surfactin conformation is strongly dependent on environmental conditions and this molecule has the unique ability of adopting a wide variety of conformations like γ- or β-turns, β-sheet or α-helix (Vass et al. 2001). This is probably the key making possible the coexistence of bilayers and micelles in the same sample of surfactin: shape parameter depends on the conformation of the peptidic moiety. Simulation of surfactin assembly by molecular dynamics at the water/hexane interface has confirmed that the peptidic backbone is very flexible and that surfactin has the tendency to self-associate forming clusters at the interface (Nicolas 2003).

Closely related lipopeptides are lichenysin, iturin A, and fengycin. They also exhibit strong interfacial activity but have been less widely studied than surfactin. For example, iturin A can self-organize in a synergic way with surfactin (Razafindralambo et al. 1997), forming mixed micelles in which complexes between two surfactin molecules and three iturin A molecules have been evidenced. A comparative study of surfactin, iturin A, and fengycin has shown that surfactin is the most efficient in reducing the surface tension of water, whereas iturin A exhibits the best resistance to flocculation and fengycin the highest resistance to coalescence of emulsions (Deleu et al. 1999). Lichenysin is also a very efficient surfactant and a better chelating agent than surfactin for Ca^{2+} and Mg^{2+} ions. In the presence of Ca^{2+}, micellization of lichenysin occurs *via* self-assembly of dimers (Grangemard et al. 2001).

3.2.1.2 Marine Siderophores

Marine siderophores are amphiphilic compounds generated to scavenge and concentrate iron (III) in marine medium. Some of them are peptidic with aliphatic long chains. Among the most interesting siderophores are marinobactins, with a cyclic peptidic head group and one alkyl chain (Figure 3.4 presents chemical structures of marinobactins). In the absence of iron, these amphiphiles self-organize into micelles as probed by light- and neutron scattering experiments. The size of the micelles is 4 nm in the case of marinobactins E. The cmc's are very low, usually between 50 and 75 μM. Upon complexation with Fe^{3+} a phase transition occurs: first the micelles shrink in size when one equivalent of iron is added and then they convert themselves into unilamellar vesicles then multilamellar vesicles upon iron addition (iron in excess). The size of the vesicles is about 200 nm. Based on packing parameter considerations, the authors explained that the first decrease in size upon iron complexation is due to an increase of the polar head group volume compared to the constant hydrophobic aliphatic chain. Further addition of iron above equimolarity implies coordination of the metal by two siderophores. The global shape of the "dimers" then becomes much more cylindrical, favoring the formation of vesicles (Figure 3.5 illustrates the siderophores micelle-to-vesicle transition) (Martinez et al. 2000, Owen et al. 2005, Martinez and Butler 2007).

The effect of competing metals has also been studied: addition of Zn(II), Cd(II), or La(III) on Fe-M micelles induces formation of multilamellar vesicles (100–200 nm) as probed by scattering techniques, cryo-TEM, x-ray diffraction, and absorption spectroscopy. The multi-lamellar vesicles have a low solvent volume fraction and most of the inner space of the vesicle is occupied by lipopeptide multiple bilayers (Owen et al. 2007). The distance between the bilayers is about 5.3 nm. The siderophores interact with Zn or Cd *via* hexadentate coordination (Owen et al. 2008).

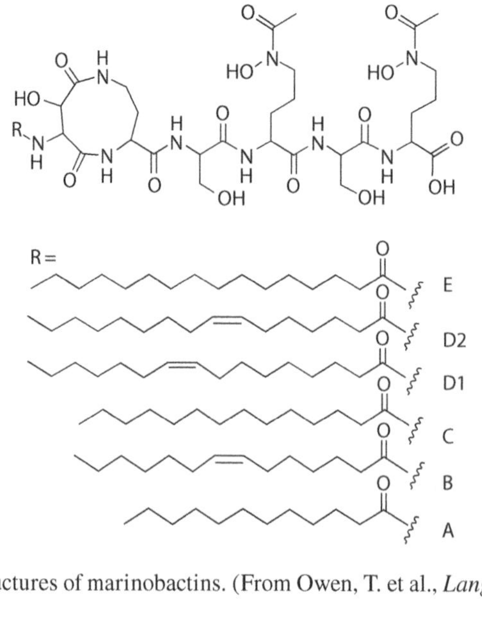

FIGURE 3.4 Chemical structures of marinobactins. (From Owen, T. et al., *Langmuir*, 21, 12109, 2005. With permission.)

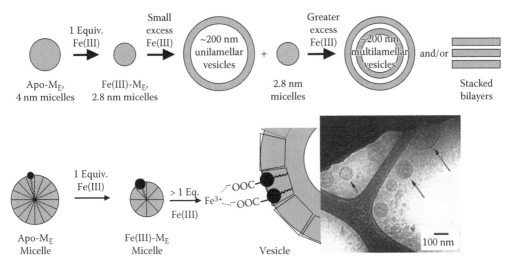

FIGURE 3.5 Micelle-to-vesicle transition of marinobactins aggregates in the presence of Fe^{3+}. (From Owen, T. et al., *Langmuir*, 21, 12109, 2005. With permission.)

3.2.2 Synthetic Lipopeptides

Amphiphilic acylated peptides are usually designed for bioactive applications. The presence of one or more fatty alkyl chains triggers a secondary structure in peptide aggregates. For example, some antimicrobial active peptides are known to remain unfolded in bulk solution but are known to adopt a secondary structure (like α-helix) in contact with a biological membrane (even a simple phospholipid bilayer) to form an amphipathic structure that is thought to be required for membrane binding and lysis. Adding a chain might therefore improve the potential application of such bioactive peptides. Self-assembling alkylpeptides are also used as nanofiber-based scaffolds for the preparation of biomaterials, for tissue engineering or for the preparation of ordered surfaces with a view to control biomineralizations (Cavalli and Kros 2008).

3.2.2.1 Monocatenary (i.e., Single-Chain) Peptides

This kind of amphiphilic peptides is the most widely studied and is closely related to conventional surfactants; here, the polar head group is constituted by the peptidic sequence. Depending on the number of amino acids, the head group can adopt secondary structures that influence the self-organization. Regarding the grafted chain, it is usually a simple fatty alkyl sequence but some examples of PEG-ylated or cholesterylated peptides have also been reported.

3.2.2.1.1 Acylated Short Peptide Sequences

The smallest peptide is made of two amino acids and has no secondary structure. However, when conjugated to a simple alkyl chain, a wide variety of structures can be observed as illustrated by the following few examples.

A simple *N*-acylated dipeptide, Lys-Asp-lauryl has been reported to self-organize in water as evidenced by dynamic light scattering (DLS) and conductivity measurements. Upon increasing concentration, structural transitions starting with small "primary" micelles then bigger "secondary" micelles and finally onion-like vesicles (cvc = 8.5 mM, size = 320 nm) were observed by cryo-TEM (Figure 3.6 shows hierarchical aggregation of Lys-Asp-lauryl in water). The authors explain that salt-bridging between head groups reduced the molecular surface area and allowed a more parallel arrangement of the molecules leading to vesicles, with their effective packing parameter close to one (Jayakumar et al. 2000a).

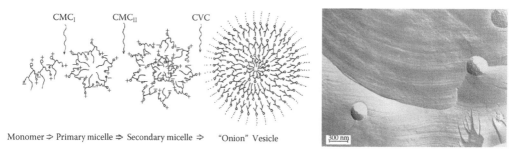

CMC$_I$ CMC$_{II}$ CVC

Monomer ⇒ Primary micelle ⇒ Secondary micelle ⇒   "Onion" Vesicle   300 nm

FIGURE 3.6 Structural transition from micelles to vesicles observed upon increasing concentration of Lys-Asp-Lauryl in water (left) and cryo-TEM micrograph of vesicles. (From Jayakumar, R. et al., *Bioorg. Med. Chem. Lett.*, 10, 1547, 2000a. With permission.)

Another example of *N*-acylated di- and tri-peptides—Gly-Gly and Gly-Gly-Gly—bearing alkyl chains ranging from 10 to 15 carbons has been described (Kogiso et al. 2007). The aggregation was studied in the presence of metal cations (transition metals, Ag$^+$ and La^{3+}). Different kinds of aggregates were observed: nanotubes with Mn^{2+}, Fe^{3+}, and Cu^{2+}, nanofibrils with Ni^{2+} or sheetlike structures using other metals. The typical diameters ranged from 50 to 200 nm and the lengths were of several hundreds of nm. After calcination these structures could even be transformed into metal oxide nanotubes. In such aggregates the packing is ensured first by a polyglycine hydrogen-bond network leading to stacked lipid bilayers membranes. By complexation with the carboxylate end groups of the peptides, metal cations can insert between these bilayers leading to hybrid nanotubes (a scheme of lipopeptide nanotubes is shown in Figure 3.7) (Shimizu 2008). When using Pd^{2+}, the structures appear spherical and the lipopeptide self-assembles into multilayered vesicles. After reduction of palladium, necklace-like chains of hybrid nanospheres are obtained (Zhou et al. 2009).

Vesicular structures were also observed for three *N*-(4-*n*-dodecyloxybenzoyl) peptide surfactants with a polar head group being composed of the sodium salt of Ala-Val, Val-Ala, or Val-Val. These lipopeptides self-assemble in water at very low concentrations (20–40 µM) to form large vesicles in dilute solutions. Increasing the concentration induces a transition from vesicles to rodlike micelles (Khatua and Dey 2007).

Another type of self-assembly can be observed with simple cationic dipeptide amphiphiles. Das et al. have shown that these lipopeptides act as hydrogelators and their efficiency strongly depends

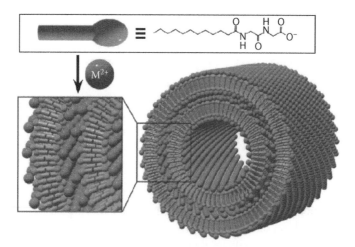

FIGURE 3.7 Schematic representation of a metal-coordinated lipopeptide nanotube. (From Kogiso, M. et al., *Adv. Mater.*, 19, 242, 2007. With permission.)

on the hydrophilic/hydrophobic balance which can be finely tuned by small changes in the polar head group (Mitra et al. 2007).

3.2.2.1.2 Alkylpeptides with Sequences from 4 to 12 Amino Acids

When increasing the peptidic sequence length, secondary conformations can be more easily evidenced. The particular folding of these epitopes can affect the lipopeptides self-assembly process. Cylindrical nanofiber is the most common morphology of the self-organized aggregates encountered: here, the film made by surfactant is no more flexible or even rigid: self-assembly is better understood as from bricks with given contact areas and angles of tilts.

The alternating sequence VEVE was linked to a cetyl chain leading to a peptide that can self-assemble into flat nanobelts with giant dimensions (150 nm width, up to 0.1 mm in length). The peptide sequence, presenting a valine face and a glutamate face, can eliminate all curvature and allow the stacking of peptide amphiphile bilayers. Even with this "short" sequence, the peptide presents a β-sheet arrangement. Upon dilution, twisted nanoribbons are observed and are thought to be the first step in peptide self-organization (Figure 3.8 shows TEM micrographs of twisted nanobelts and nanoribbons). When replacing the VEVE sequence by a VVEE one, flat nanobelts are not present any more and the peptide rather self-assembles into cylindrical fibers (Cui et al. 2009).

Van Hest and Löwick have considered the epitope sequence GANPNAAG (derived from a protein of the Malaria parasite *Plasmodium Falciparum*) coupled to alkyl chains from 0 to 18 carbons. In the case of short chains with less than 14 carbons, no aggregates could be observed and the peptide sequence remained in a random coil form. The C_{14} derivative forms fibrous aggregates in water with an average diameter of 20 nm and few micrometers in length; in that case the peptide adopts a β-sheet conformation which can be converted into a random coil upon heating. Increasing the chain length to 16 then 18 carbons reinforced the β-sheet structure stability and tubular aggregates were observed.

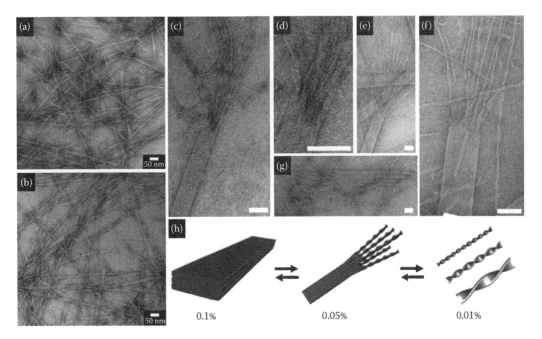

FIGURE 3.8 Twisted nanoribbons at 0.01% wt. solution and intermediate structures of nanobelts transforming into twisted nanoribbons at 0.05% wt. solution. (a,b) At 0.01% wt. narrower nanobelts and twisted nanoribbons are observed. (c-f) Twisted nanoribbons sprouting from one nanobelt end. (d) A closer view of (c). (g) Nanobelts split from both ends into narrower nanobelts. Scale bars of panels c-g: 100 nm. (h) Scheme of morphological transitions with a change in concentration. (From Cui, H. et al., *Nano Lett.*, 9, 945, 2009. With permission.)

Moreover, when these samples were heated at 90°C and cooled again the tubular structures converted into a twisted ribbon configuration. In this example, peptide aggregation is initially controlled by hydrophobic interactions that induce well-defined secondary structures. Amphiphile aggregation and peptide folding become cooperative (Löwik et al. 2005). Introducing polymerizable diacetylene moiety in the alkyl chain did not change the aggregates morphology dramatically: twisted fibers were still formed inducing gel formation at higher concentrations. This dense network of fibers can be aligned using a 20 T magnetic field to form an ordered material. Upon UV irradiation, the ordered gel has been further stabilized by cross-linking (involving the diacetylene groups). This final ordered material can be used as a functional scaffold for biomineralization or cell proliferation (Löwik et al. 2007, van den Heuvel et al. 2008). Using the β-sheet forming peptide KTVIIE as another epitope which is known to mimic amyloid fibril forming proteins, the same group has shown that grafting alkyl chain does not especially promote the β-sheet conformation but increases the stability of fibrillar aggregates against heating or dilution. Peptide fibers have then been stabilized by adding hydrophobic interactions *via* peptide sequences showing β-sheets with intramolecular links *via* hydrogen bonds, with "hydrophobic patches" as a result in the solid aggregate making fibrils (Meijer et al. 2007).

Self-assembling lipopeptides with longer peptide sequences have been designed by the group of Stupp with a view to prepare biocompatible materials for tissue engineering or biomineralization. The amphiphiles are composed by five segments, each of them bringing one specificity: the acyl chain giving a hydrophobic character, few cysteines to enable further stabilizing polymerization by disulfide bonds formation, few glycines acting as a flexible linker, a phosphoserine having complexant properties (especially towards calcium), and finally an epitope region like RGD for promoting cell adhesion (Figure 3.9 represents the lipopeptide and its micelle) (Hartgerink et al. 2001a). These kinds of lipopeptides self-assemble depending on the pH. At neutral pH, the peptide structural charge is −3 and electrostatic repulsions disable aggregation, whereas reducing the charge to 0 in acidic medium promotes the formation of cylindrical elongated micelles arranged in a fibrillar network. These micelles can be further stabilized by a reversible intermolecular cross-linking under oxidizing conditions, the Cys-Cys-Cys-Cys segment acting as a chemical switch. Further complexation with calcium and the phosphoserine promotes hydroxy-apatite mineralization (Hartgerink et al. 2001b). These peptide-amphiphiles appears as very versatile because it is easy to finely tune the contribution of each segment, varying for example the acyl chain length or the peptide sequence. The properties of the resulting nanofiber networks are therefore controllable (Hartgerink et al. 2002). When looking at the intermolecular forces involved in self-assembly and fiber formation, the authors have evidenced several contributions: (1) the screening of electrostatic repulsions by protonation or by counter-ion binding, (2) the van der Waals and hydrophobic interactions between alkyl chains, (3) ionic "bridging" using multivalent metal cations, and (4) the lateral H-bonding between peptide sequences that maintain a β-sheet conformation (Stendahl et al. 2006).

All these interactions can be tuned and the self-assembly is therefore controllable *via* charge effects. For example, incorporating proline residues in the peptide sequence to prevent from β-sheet formation induces the formation of spherical micelles rather than elongated cylindrical ones (Guler and Stupp 2007). Molecular simulations have been performed to study the balance between H-bonding and hydrophobic interactions. The resulting phase diagram shows that hydrophobic interactions favor the formation of spherical micelles whereas increasing H-bonding induces β-sheet structuration resulting in larger aggregates like long cylindrical fibers (Figure 3.10 is the schematic phase diagram obtained by modeling) (Velichko et al. 2008).

A general similar approach has been used to study the pH/salinity phase diagram of the self-assembling peptide amphiphiles. Spherical micelles are favored at low pH and high ionic strength whereas fibers and gel form upon increasing pH and decreasing electrostatic screening (Figure 3.11 presents the pH/salinity phase diagram) (Tsonchev et al. 2008). Using a series of closely related peptide amphiphiles, Hartegrink et al. have evidenced the role of H-bonding and amphiphilic packing on the formation of nanofibers. They have observed that with less than four H-bonding amino acids the lipopeptides self-assemble in spherical micelles leading to nonviscous solutions. With at least

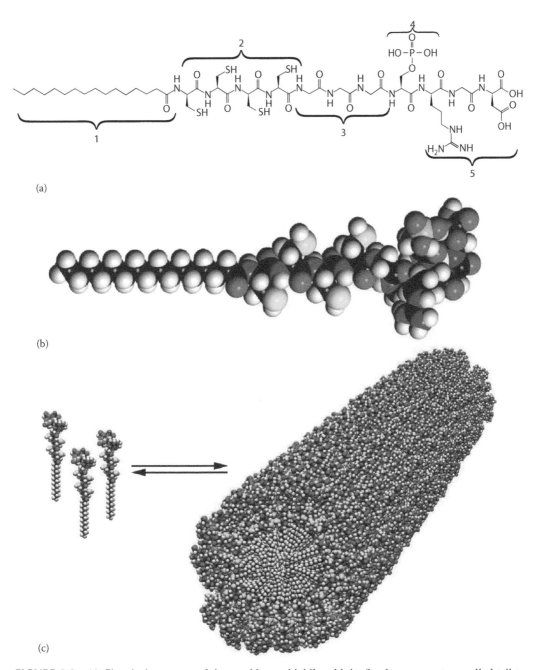

(a)

(b)

(c)

FIGURE 3.9 (a) Chemical structure of the peptide amphiphile with its five key segments: an alkyl tail to provide hydrophobic character, a cysteine series to enable further covalent binding, a glycine series used as flexible spacer, a phosphoserine to make possible calcium complexation, and a RGD sequence to promote cell adhesion. (b) Molecular model of the peptide amphiphile. (c) Schematic view of the peptide self-assembly into cylindrical micelles. (From Hartgerink, J.D. et al., *Science*, 294, 1684, 2001b. With permission.)

four H-bonds the peptide forms β-sheet type interactions leading to an arrangement of fibers and gelification of the solution. Furthermore, the fibers can exist in three different models: straight ribbons, twisted (helical) ribbons, or flat fibers (Paramonov et al. 2006). This sequence follows the universal theory of twisted layers as developed by Boden and colleagues (Aggeli et al. 2001a). Peptide amphiphiles can form even more complex aggregates: when introducing a photo-cleavable lateral

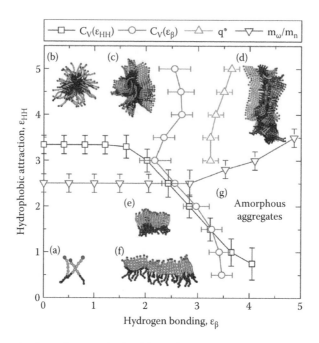

FIGURE 3.10 Schematic phase diagram obtained from simulations and representing (a) the free molecules, (b) spherical micelles, (c) micelles with a β-sheet shell, (d) cylindrical fibers, (e) stacks of parallel β-sheets, (f) single β-sheets, and (g) the amorphous phase. (From Velichko, Y.S. et al., *J. Phys. Chem. B*, 112, 2326, 2008. With permission.)

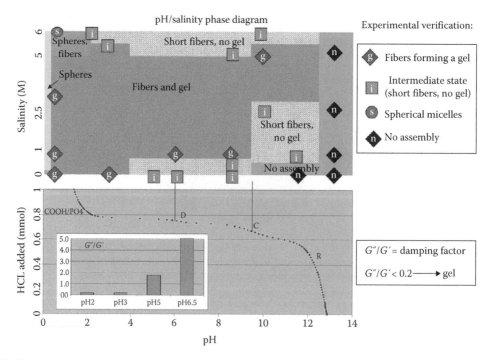

FIGURE 3.11 pH/salinity phase diagram of the self-assembling peptide. In the titration curve of the peptide the inflexion points correspond to the transition lines in the phase diagram. (From Tsonchev, S. et al., *J. Phys. Chem. B*, 112, 441, 2008. With permission.)

bulky group at the junction between the acyl chain and the peptide sequence, the fibers form quadruple helices. The induced steric hindrance favors inter-aggregate interactions and stabilizes the quadruple helix. Under light illumination, the bulky group can be removed and the super-aggregates dissociate into the nonhelical usual single fibers (Muraoka et al. 2008). All these fibrillar materials find a large number of applications, from bioactive substrates for cell attachment, proliferation, and differentiation, to templates for inorganic structured material production and biomineralization.

3.2.2.1.3 Acylated Longer Peptides

Upon increasing the peptide sequence length, protein-like architectures can be observed. Tirrel and Fields have shown that N-acylation of a 23 amino acid residue induces stabilization of α-helix formation, whereas the free peptidic sequence has no defined conformation in water (Yu et al. 1998). Increasing the acyl chain length from C_6 to C_{16} enhances this stabilization and raises the denaturation temperature of the helices. The same kind of stabilization has been achieved for a 39-residue peptide which forms collagen-like triple helices assemblies. Moreover, these stabilized structures self-assemble in larger aggregates that have been supposed to be micelle-like aggregates with aggregation numbers ranging from 15 to 70 (Forns et al. 2000). In the case of a DNA-binding peptide, alkylation promotes self-assembly into tubules driven by interactions between the head groups. Addition of DNA to these aggregates induces a fully helical conformation in the head groups and the tubules convert into flat particles (Bitton et al. 2005).

3.2.2.1.4 PEG-ylated Peptides

When grafting a hydrophilic polyethyleneglycol (PEG) chain to a peptide, the amino acid sequence can become the hydrophobic part of the amphiphile, the PEG group being the polar head moiety. Copolymers composed of MPEG550 or MPEG750 as the hydrophilic segment and GFLGFLEt as the hydrophobic part form stable micelles in water, with a hydrodynamic diameter of 15–17 nm. Upon increasing temperature, this system undergoes phase transition with lower critical solution temperatures (LCST) between 40°C and 60°C (Xu et al. 2007). Using FFKLVFF derived from the hydrophobic KLVFF (Aβ16–20) segment of the amyloid β peptide (Aβ) grafted to PEG hydrophilic parts, Hamley et al. have evidenced the formation of core-shell cylinders (Hamley et al. 2008a,b). The hydrophobic core is constituted by the peptide sequences that stack in a β-sheet conformation, whereas the hydrophilic corona is made of the PEG chains (Figure 3.12 illustrates the core/shell packing of PEG-ylated peptide). Upon increasing the concentration, the cylinders further self-assemble into a series of lyotropic mesophases such as nematic and hexagonal columnar liquid crystals.

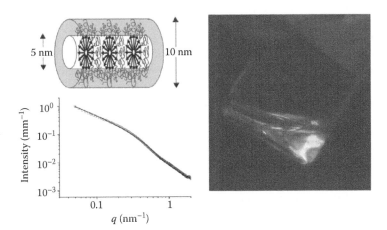

FIGURE 3.12 Core/shell cylinder model describing the SAXS profile of FFKLVFF-PEG (left) and fluid sample imaged between crossed polarizers showing birefringence. (From Hamley, I.W. et al., *Adv. Mater.*, 20, 4394, 2008a. With permission.)

3.2.2.2 Bicatenary (i.e., Double-Chain) Peptides

In this class of peptide amphiphiles, the peptide sequence represents the head group and it is associated with two parallel covalently bound alkyl chains.

A series of bicatenary lipodipeptides were obtained by grafting two alkyl chains on the glutamic side of the Lys-Glu dipeptide. The self-organization properties of these amphiphiles reveal a multistep aggregation: first small spherical micelles are formed. Then, upon increasing concentration, they become elongated and finally transform into spherical liposomes with a mean diameter of 100 nm (Sebyakin et al. 2007).

Peptidic analogues of phospholipids have been prepared using (Val-Glu)$_n$ ($n = 2$–4) oligopeptides as polar head group coupled to a 1,2-dioleoyl-sn-glycero-3-phosphoethanolamine (DOPE) phospholipid tail. The alternating hydrophilic and hydrophobic amino acid residue sequences adopt a β-strand conformation in the β-sheet monolayers assembled at the air/water interface. Grazing incidence x-ray diffraction (GIXD) experiments have shown that the films containing the longest oligopeptides were organized in a new type of 2D structure in which β-sheet ribbons are segregated by lipid tails. These ordered lipopeptide monolayers can be used as templates for the mineralization of calcium carbonate at the air/water interface (Cavalli et al. 2006a,b).

A hybrid lipopeptide, where a PEG core is linked on one side to a nonapeptide and bears at the other extremity two octadecyl chains can self-assemble at very low concentration (about 2 μM). In the formed spherical particles with a size of about 100 nm, the water-exposed peptide sequence adopts a β-sheet conformation. These hybrid nanoparticles can encapsulate hydrophobic drugs and could be employed for targeted drug delivery (Accardo et al. 2006).

The insertion of dialkyl chain tails has also been used for stabilization of larger peptides as collagen-model head groups and these peptide amphiphile self-assemble to form polypro II like triple helical structures. Stabilizing hydrophobic interactions induced by the two alkyl chains confer to the peptide a much higher thermal stability (Yu et al. 1996).

Beyond the formation of phospholipid analogues, there are several ways to derivatize an oligopeptide with fatty chains: the chains can be localized at one extremity of the peptide sequence, or at both sides, and one can even consider a multiple grafting along the peptidic chain. Various self-organization modes are then observed, for example, the formation of helical nanofibers (Matmour et al. 2008, Tsai et al. 2008) or the incorporation into phospholipid membrane to form ion channels (You and Gokel 2008).

3.2.2.3 Gemini-Like Peptides

In contrast to bicatenary structures, gemini peptides are linked to two alkyl chains, one at each extremity of an "hydrophilic" amino acid sequence.

Miravet and Escuder have developed a new series of gemini lipopeptides based on the minimal sequence from natural silks, GAGA. After grafting alkyl chains on both first and fourth amino acids, the peptide amphiphiles self-organize in various solvents to form fibers and gels. The interactions between molecules are ensured by stacking of the peptide cores in β-sheet conformation and by solvophobic interactions between alkyl chains (Figure 3.13 shows antiparallel assembly into β-sheets). Depending on the gemini structures, parallel or antiparallel arrangements are possible (in Figure 3.14 different packing models are proposed) (Escuder and Miravet 2006, Iqbal et al. 2008). This kind of cooperative attractive interactions has been used to induce a secondary structure of a random-coil gemini peptide: by interaction with a liposome membrane, the two alkyl chains became parallel and induced the structuration of the peptide core in a β-hairpin conformation (Figure 3.15 shows the β-hairpin stabilization by interaction with a liposome) (Löwik et al. 2003). By grafting C$_{12}$ to C$_{20}$ alkyl chains on a 25-residue peptide, Privé et al. have shown that the presence of the chains stabilizes a α-helix conformation of the peptide which otherwise remains largely unstructured in water (McGregor et al. 2003). Moreover, the gemini α-helical peptides self-organize in water in micelles with aggregation numbers ranging from 8 (C$_{12}$ chains) to 19 (C$_{20}$ chains). These

Solvophobic,
van der Waals
interactions

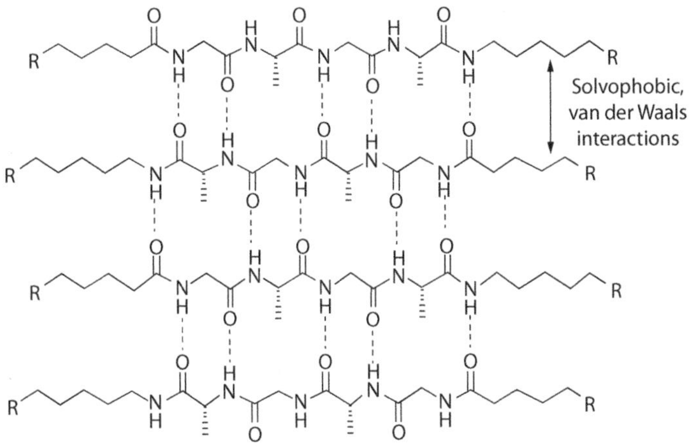

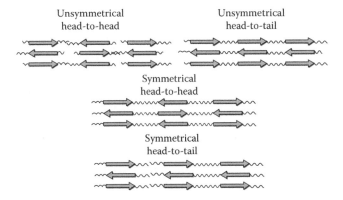

FIGURE 3.13 Antiparallel assembly into β-sheets formed by GAGA-based gemini peptides. (From Iqbal, S. et al., *Eur. J. Org. Chem.*, 4580, 2008. With permission.)

Unsymmetrical
head-to-head
Unsymmetrical
head-to-tail
Symmetrical
head-to-head
Symmetrical
head-to-tail

FIGURE 3.14 Different packing models proposed for the arrangement of antiparallel β-sheets. (From Iqbal, S. et al., *Eur. J. Org. Chem.*, 4580, 2008. With permission.)

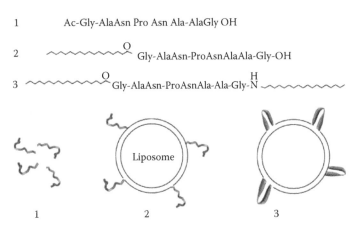

1 Ac-Gly-AlaAsn Pro Asn Ala-AlaGly OH
2 Gly-AlaAsn-ProAsnAlaAla-Gly-OH
3 Gly-AlaAsn-ProAsnAla-Ala-Gly-N
Liposome
1
2
3

FIGURE 3.15 Attachment of terminal alkyl chains stabilizes a β-hairpin conformation. (From Löwik, D.W.P.M. et al., *Org. Biomol. Chem.*, 1, 1827, 2003. With permission.)

lipopeptide detergents have been designed for stabilizing membrane proteins, the α-helix length being comparable to the biological membrane thickness.

3.2.2.4   Bolaform-Like Peptides

These lipopeptides are constituted by a hydrophobic central chain linked to two polar head groups (oligopeptides). Shimizu and coworkers and Matsui and coworkers have studied symmetrical bola-amphiphiles bearing di- or tripeptides at both extremities. Self-organization of these dicarboxylic oligopeptides is mainly driven by intermolecular H-bonding between peptide backbones and also between carboxylic end functions. As a consequence, the aggregation is highly pH-dependent: the dicarboxylic oligoglycine bolaform peptides form rodlike micelles at high pH which transform into microtubules containing vesicles upon decreasing pH (Figure 3.16 illustrates the proton-triggered self-assembly of bolaform peptides) (Kogiso et al. 1998). Crystalline tubules can be formed by the same kind of bolaform peptides in few days, depending on the pH. At high pH, a weak H-bond network is formed and helical ribbons are obtained, whereas at low pH, when strong H-bonds form, crystalline tubules are obtained; they are equivalent to "closed" helical ribbons (Figure 3.17 shows helical ribbons or tubules formed by bolaform peptide) (Matsui and Gologan 2000). The tubules can further be used as templates for metal cation adsorption; after reaction with a reducing agent, metalized wires were obtained (Matsui et al. 2000). Valylvaline bolaamphiphiles self-assemble in various structures depending on the spacer length: short spacers induce formation of crystalline sol-ids whereas longer ones promote self-organization into fiber networks and gels. It clearly shows that hydrophobic interactions between alkyl spacers facilitate the global organization. In fibers, peptides adopt a parallel β-sheet arrangement, favored by bulky isopropyl moieties of valine groups (Kogiso et al. 2000). Complexation with transition metal cations has been used to produce gels by interfiber "crosslinking" (Kogiso et al. 2004). When the spacer is linked to the carboxylic groups of the polar dipeptide heads, it cannot contribute any more to lateral H-bonding. In that case, participation of H-bonds between peptide backbones plays a major role and peptide conformation is affected. In the case of Pro-Val-spacer-Val-Pro bolaamphiphile, fibrillar networks are formed in water under kinetic control with unfolded peptide segments. The supramolecular gel can then self-correct into a thermodynamically more stable state, where the peptide sequences adopt a β-sheet conformation (Figure 3.18 illustrates self-corrected structures) (Rodriguez-Llansola et al. 2009).

3.3   INTRINSICALLY AMPHIPHILIC PEPTIDES

3.3.1   SELF-ASSEMBLY OF NATURAL PEPTIDES

We restrict the scope of this review to natural peptides, and exclude the immense domain of protein aggregation. The distinction is made since protein folding in secondary and tertiary structure can

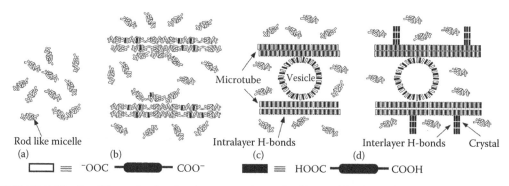

FIGURE 3.16   Possible mechanism for the proton-triggered self-assembly of a vesicle-encapsulated micro-tube and a needle-shaped microcrystal. (From Kogiso, M. et al., *Langmuir*, 14, 4978, 1998. With permission.)

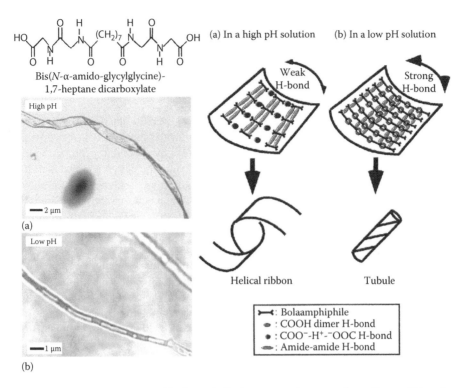

(a) In a high pH solution   (b) In a low pH solution

Bis(*N*-α-amido-glycylglycine)-
1,7-heptane dicarboxylate

Weak H-bond

Strong H-bond

High pH

— 2 μm

(a)

Low pH

— 1 μm

(b)

Helical ribbon   Tubule

⊢•⊣ : Bolaamphiphile
● : COOH dimer H-bond
✦ : COO⁻-H⁺-⁻OOC H-bond
⬭ : Amide-amide H-bond

FIGURE 3.17   Heptane bolaform peptide forms helical structures in water: depending on H-bond strength, helical ribbons or tubules can be observed. (From Matsui, H. et al., *J. Phys. Chem. B*, 104, 3383, 2000. With permission.)

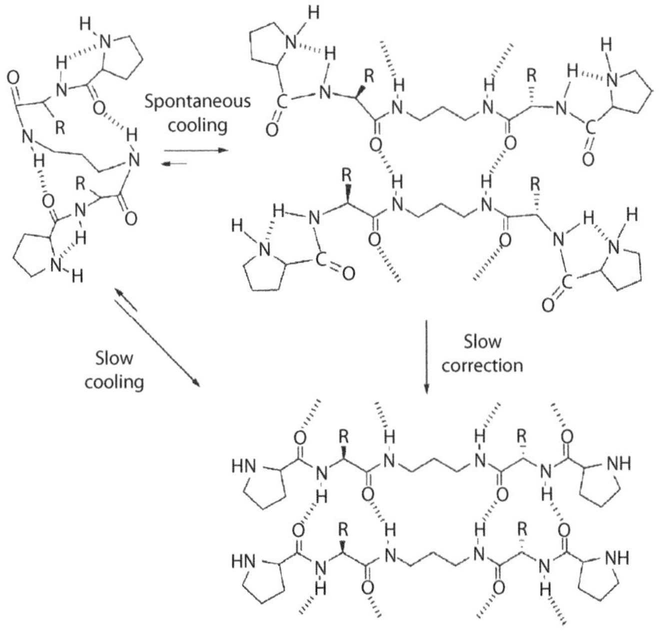

Spontaneous cooling

Slow cooling

Slow correction

FIGURE 3.18   Self-corrections induced by aggregation driven conformational changes. (From Rodriguez-Llansola, F. et al., *Chem. Commun.*, 209, 2009. With permission.)

be followed by protein self-assembly, a process different from coagulation, which can be considered as random binding into large protein-based aggregates.

Upon environmental conditions, conformational changes of a protein or peptide secondary structure can have dramatic consequences: one of the most known—and studied—is the formation of amyloid fibrils due to protein aggregation, during which the β-strand packing induces protein oligomerization and formation of insoluble platelets or fibers.

A peculiar class of peptides with an important biological specific function contains viral fusion peptides: during viral infection, the viral capsid fuses with the "target" cell, prior to any RNA release in the cytoplasm. After anchoring to the cell surface by specific receptors, the membrane fusion is first initiated by the insertion of the fusion peptide into the membrane. Known fusion peptides are made of hydrophobic amino acids and are structured as α-helices that simply solubilize into the lipophilic interior of the membrane, initiating the fusion between the capsid membrane and the host cell membrane. It has been shown that these fusion peptides present self-organizing properties: for example, the simian immunodeficiency virus (SIV)-peptide can form mixed elongated reverse micelles in the presence of phospholipids (El Kirat et al. 2006).

Another class of peptides subject to self-assembly refers to transmembrane ion channels: ion regulation and transport through biological membranes are realized by membrane proteins. Two main classes are known: barrels of locally β–shaped peptides and clusters of amphipathic α-helices. These helices are called amphipathic because they present one hydrophobic face and one hydrophilic face. When inserted perpendicularly into the phospholipid bilayers, they form clusters of few parallel helices: the hydrophobic faces are in contact with the phospholipid bilayer interior, thus isolating the hydrophilic faces from this lipophilic environment. This type of self-assembly results in a hydrophilic channel created through the membrane: this local arrangement of Janus-type cylinders regulates the passage of polar and ionic species.

A third class of self-assembling peptides is the class of the so-called ionophores: these are lipophilic molecules which can specifically complex ions. They insert in the membrane and perturb the ion equilibrium between "inside" and "outside." Therefore, they are used as antibiotics. For example, Gramicidin is a helical peptide forming dimer inserted in the phospholipid membrane and allows passage of small ions within the helix core.

The field of collagen, one of the most important structural protein-based structures, is excluded from this review about peptides since collagen fibers are made from super-triple helices, and this type of self-assembly makes gels and rigid structures only with high molar mass compounds, and not small peptides.

3.3.2 Synthetic Self-Assembling Peptides

3.3.2.1 Amyloidogenic Peptides

Many efforts are currently being made to understand mechanisms implied in amyloid fibrils formation in order to identify efficient ways of inhibiting this process and find curative treatments against neurodegenerative diseases. Short sequences of identified amyloidogenic peptides are especially studied in order to determine the forces and the parameters implied in peptide aggregation.

Gazit et al. have pointed out a significant occurrence of aromatic amino acid residues in peptidic sequences implied in amyloidogenesis. The enhancement of π-stacking could reinforce and stabilize the attractive interactions between peptides and also provide order and directionality in the fiber (Gazit 2002). When studying the aggregation behavior of short peptide sequences issued from amyloid forming human calcitonin, it appears that a large value of global hydrophobicity is not a condition required for the formation of amyloid fibers. For instance, hydrophilic short peptides DKNKF and DFNK segments self-organize into well-defined fibers, whereas FNKF, DFN, or DANKA sequences do not. Specific orientation of aromatic rings and favored aromatic interactions seem to play a key role rather than the hydrophilic/hydrophobic balance in the formation of amyloid fibrils by very short peptides (Reches et al. 2002). The simple Aβ's structural motif dipeptide FF has been found to self-organize into discrete and stiff hollow nanotubes, with a diameter of about 100 nm and about 1 μm in length (Figure 3.19 shows the FF's structure and nanotubes). In this structure, the single amide bound adopts a β-sheet like conformation (Reches and Gazit 2003). These nanotubes, easily obtained with a very simple primary sequence, show thermal and chemical stability (Adler-Abramovich et al. 2006) and have found various applications such as preparation of metallic nanotubes (Reches and Gazit 2003), or as electrodes for biosensor applications (Yemini et al. 2005a,b). Self-assembly of these aromatic dipeptides is very sensitive to chemical modification: for instance, replacing FF by diphenylglycine leads to the formation of spherical nanoparticles instead of nanotubes; in that case, the particles have a diameter of about 50 nm and are stable even at extreme pHs. The same kinds of globular structure were obtained by grafting a cysteine on the nanotube-forming dipeptide FF (Figure 3.20 illustrates nanospheres and nanotubes from aromatic dipeptides) (Reches and Gazit 2004). A model has been proposed for the alternative assembly of nanotubes and nanospheres. Initially driven by π-stacking interactions, a sheet is formed that is stabilized by both aromatic stacking and hydrogen bonding. The extended sheet can then be closed either along one axis (leading to a nanotube) or along its two axes (providing the spherical nano-structures) (in Figure 3.21 a schematic model illustrates the formation of tubes or spheres). Further studies on FF derivatives have shown that a wide variety of nanotubes and spherical globules can be obtained (Mahler et al. 2006, Reches and Gazit 2006a,b) and used for the production of well-organized self-assembled films (Hendler et al. 2007).

Another Aβ fragment has also been shown to form self-organized structures. Aβ(16–20), namely KLVFF, self-assembles in aqueous solutions to form fibrillar structures. Even at very dilute conditions, strong aromatic stacking between phenylalanine residues are implied in the aggregation of

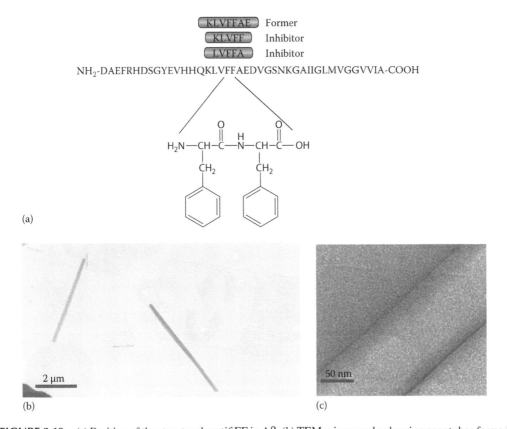

(a)

(b)   (c)

FIGURE 3.19 (a) Position of the structural motif FF in Aβ. (b) TEM micrographs showing nanotubes formed by FF. (From Reches, M. and Gazit, E., *Science*, 300, 625, 2003. With permission.)

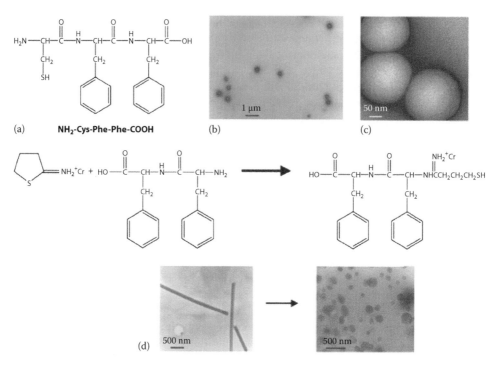

(a)   **NH₂-Cys-Phe-Phe-COOH**   (b)   (c)

FIGURE 3.20 CFF peptide forms nanospheres (a to c) whereas FF sequence induces nanotube formation (d). (From Reches, M. and Gazit, E., *Nano Lett.*, 4, 581, 2004. With permission.)

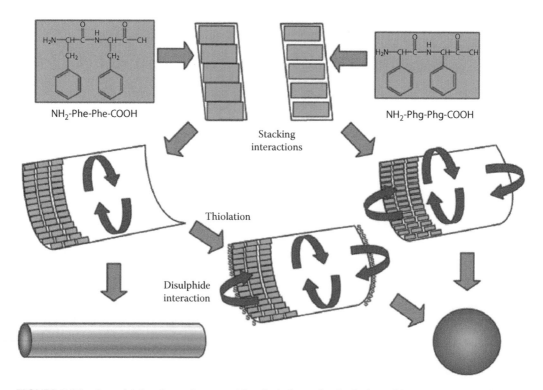

FIGURE 3.21 A model for alternative assembly of tubular and spherical peptide nanostructures, depending on the sheet's closing mode. (From Reches, M. and Gazit, E., *Nano Lett.*, 4, 581, 2004. With permission.)

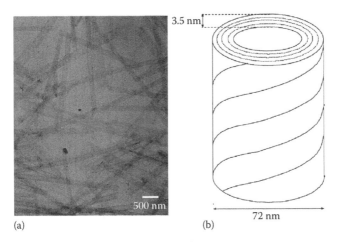

FIGURE 3.22   Nanotubes formed by AAKLVFF in methanol: TEM micrograph (a) and model (b) showing the wrapping of four β-sheets to form the nanotube wall. (From Krysmann, M.J. et al., *Langmuir*, 24, 8158, 2008. With permission.)

β-sheets. Increasing ionic strength screens the electrostatic interactions from lysine and the terminal units: the β-sheets can then self-organize into a fibrillar gel network (Krysmann et al. 2008a). When extending this peptide sequence by two phenylalanines, FFKLVFF forms amyloid-type fibrils even in methanol. Here again π-stacking between aromatic residues seems to drive the aggregation (Krysmann et al. 2007). Extending KLVFF by two alanines, it has been shown that AAKLVFF also can self-organize in methanol. In this case, aromatic interactions still drive the β-sheet self-assembly but, instead of forming fibrils, they wrap helically to form nanotubes. The wall of this nanotube is made of four wrapped β-sheets (Figure 3.22 shows how β-sheets wrap to form nanotubes). At higher concentration the nanotubes further self-align into a nematic phase (Krysmann et al. 2008b). The number of wraps in a nanotube is thermodynamically controlled: increasing the number of wraps around a cylinder causes elastic deformation in curvature. Therefore, the free energy can be minimized by having aggregation between tubes of a defined and finite number of wraps.

Two sequences from fibril-forming prion protein PrP have been studied by Jayakumar et al. PrP(119–126) GAVVGGLG adopts an α-helix conformation in trifluoroethanol (TFE), whereas PrP(121–127) VVGGLGG remains in a random coil. In water, both peptides are initially in a random coil conformation but with time they structure into β-sheets that self-assemble into fibrils (Satheeshkumar et al. 2004). The driving force for the two sequences self-organization seems to be hydrogen bonding rather than charge effects.

Amyloid fiber formation in a peptide containing 16 residues has been studied extensively by Zhang and Rich (1997): in this system, the slow conversion from α-helix to β-sheet could be followed. At a critical ratio, β-sheets are formed: they twist during association in a zipper-like mechanism (Hwang et al. 2004).

3.3.2.2   Surfactant-Like Peptides

A surfactant-like peptide can be obtained when a short sequence of hydrophobic amino acid residues replaces the conventional alkyl fatty chain in surfactants. This lipophilic oligopeptide is linked to the polar part of the peptide sequence made of hydrophilic amino acids: this type of peptide can be considered as macro-surfactants and should share self-assembling properties of surfactants forming 2D films upon lateral "packing." However, peptides are less flexible structures than linear alkyl chains: thus peptide-based films are supposed to be reminiscent of rigid amphiphilic films, such as those made by charged lipid bilayers in the absence of salt or buffer (Demé et al. 2002a,b).

In the 1990s, Jayakumar et al. explored the self-organization properties of short surfactant-like peptides from 2 to 5 amino acids. They have used protected forms (Z, Boc, Bz) of the peptides and have measured in aqueous solutions cmcs ranging from 10 to about $100\,\mu$M (Mandal and Jayakumar 1993, 1994, Mandal et al. 1993, Murugesan et al. 1996, 1999, Ramesh et al. 1998, Jayakumar et al. 2000b, Andrews et al. 2002). They call "cmc" the abbreviation of critical micellar concentration used in the study of simple surfactants, the observed sudden change in physicochemical parameters of the solutions such as fluorescence, UV-Vis absorption, or conductance upon increasing the peptide concentration. The aggregation numbers remain smaller with peptides than in the world of surfactants: 10–20 molecules per aggregate typically. This is coherent with simple packing constraints: the radius of a micelle cannot extend beyond the "length" of a peptide. Since the volume-to-surface ratio of a sphere is imposed, only small numbers of aggregation, typical of micelles with large area per molecule such as micelles made by cyclodextrins can be found. Thermodynamic studies have evidenced that the micellization process is driven by an enthalpic contribution and formation of intermolecular H-bonds stabilize the aggregates. Micellar aggregates present also an anomalous behavior upon heating: above 40°C the micellization is favored and the cmcs are therefore smaller. In fact, the H-bond network undergoes a structural transition during which H-bonds between water and peptides are replaced by H-bonds between amide moieties in the peptidic backbones. This phenomenon has also been observed in Langmuir films of such peptides at the air/water interface: at 40°C, the destruction of H-bonds between peptides and water implies an increase in the film's compactness due to the release of water molecules. Further heating leads to an increase of the molecular area due to newly formed H-bonds between peptides themselves that make them more entangled. This phenomenon can be related to the clouding point of nonionic surfactants, the precipitation of poly(*N*-isopropylacrylamide) above its LCST or the desolvatation temperature of proteins (Jayakumar and Murugesan 2000). The same authors have also performed in parallel a series of studies in organic solvents such as chloroform, methanol, and dimethylformamide (Jayakumar et al. 1993, 1994, 2000c, Murugesan et al. 1997, Ganesh and Jayakumar 2003a,b, Ganesh et al. 2003). Peptide self-assembly was evidenced with higher cmcs (1–3 mM) than observed in aqueous medium. The driving force of self-assembly remains enthalpic in nature, with strong H-bond interactions between peptide molecules. In the reverse micelles they form in chloroform, peptides adopt a stabilizing β-strand conformation. In methanol, very small aggregates are formed (cmc = 0.3 mM, $N_{agg} = 4$) in which the tetrapeptides adopt first a helical turn conformation; upon increasing the water amount in solvated films, a transition to a β-structure is observed and such structural transitions play a role in amyloid fibril formation. Finally, some peptides also form gels in chloroform or dimethylformamide (DMF). In that case the peptides self-organize into reverse micellar fibers in which they adopt a parallel β-sheet conformation.

The most detailed work concerning self-assembling peptides is the one developed by the group of Zhang and coworkers, which has been described in several reviews (Zhang and Altman 1999, Zhang 2002, 2003, Zhang et al. 2002, Santoso et al. 2002a, Zhang and Zhao 2004, Zhao and Zhang 2004, 2007, Gelain et al. 2006). As long as the structure of the aggregates formed is not determined, these authors describe as "cac," the critical aggregation concentration, the lowest concentration at which a thermodynamically stable aggregate forms. The "cac" is a more general term than the "cmc," and should be used each time it has been demonstrated that there is a well-defined concentration threshold in the formation of aggregates, and that aggregates can be reversibly redispersed again upon dilution. In the case of surfactant-like peptides they have considered a series of peptides made of a sequence of hydrophobic amino acid residues (G, A, V, L) as the hydrophobic tail linked to one or few charged amino acids (K, D) representing the hydrophilic head. In water such peptides self-assemble in nanotubes (about 50 nm in diameter) and nanovesicles (about 100 nm) as monitored by DLS and TEM analysis (Figure 3.23 shows aggregates formed by surfactant-like peptides). In the G_4D_2 to $G_{10}D_2$ series the aggregates become more and more polydispersed with sizes ranging from 20 to 200 nm. V_6K and A_6K peptides self-organize in the same type of structures (Santoso et al. 2002b). Including six leucines in the hydrophobic tail leads to a different packing because of the larger size of L compared to G, A, and V and the aqueous solutions of L_6D_2 show a heterogeneous mixture of nanotubes, rodlike micelles, and vesicles (Vauthey et al. 2002). Self-assembling behavior

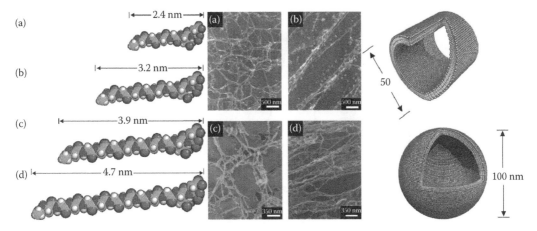

FIGURE 3.23 Left: Molecular models of glycine tail-based surfactant peptides: from G_4D_2 (a) to $G_{10}D_2$ (d). Center: TEM images of self-organized structures formed by the peptides. Right: Molecular modeling of the structures formed by the peptides, nanotubes (a) and nanovesicles (b). (From Santoso, S. et al., *Nano Lett.*, 2, 687, 2002b. With permission.)

of A_6D, V_6D, and V_6D_2 forming nanotubes and vesicles has been completed by fluorescence and DLS experiments to measure the cacs and hydrophobic environments in the aggregates (Yang and Zhang 2006); as expected, aggregation is favored when using more hydrophobic amino acids in the tail or less hydrophilic ones in the head group. These kinds of surfactants have been employed for stabilization of hydrophobic biological molecules like photosystem I (Kiley et al. 2005, Matsumoto et al. 2009) or membrane proteins (Yeh et al. 2005, Zhao et al. 2006) which maintain their conformation and activity in contrast to the use of conventional detergents like *N*-octyl-glucopyranoside. The stabilization of membrane proteins can be linked to the effect of such peptide surfactants on curvature and stability of monoolein bilayers (Yaghmur et al. 2007). Small angle x-ray scattering (SAXS) experiments have shown that the phospholipid bilayers are affected by the peptide/lipid molar ratio, by the peptide structure, and by temperature. Finally, a conic-shaped lipid-like peptide, Ac-GAVILRR-NH$_2$ has been designed and the study of its self-organization in water reveals that this peptide forms nanodonuts (Figure 3.24 is a schematic representation of nanodonut formation by a conical peptide) through fusion or elongation of spherical micelles. This kind of aggregates might be of special interest for membrane protein solubilization (Khoe et al. 2009).

3.3.2.3 Self-Complementary β-Sheet Peptides

The second type of self-assembling peptides developed by Zhang and coworkers refers to the so-called self-complementary peptides. They form β-sheets in water because they present two distinct faces: one is hydrophobic and composed of amino acid residues with lateral alkyl chains, and the other face is hydrophilic. This one is charged and presents alternating complementary ionic residues with regular repeat: +−+−+−+−, ++−−++−−, +++−−−+++−−−, etc. The β-sheets self-assemble in water by association of lipophilic faces due to hydrophobic interactions and by stacking of the hydrophilic faces by charge compensation. This leads usually to the growth of filaments and fibers (with a diameter of 10–20 nm) that make gels. In addition, depending on the designed sequence, this kind of peptide can also adopt stable α-helix conformation in water. For example, a peptide composed of a cluster of negative charges (Asp, Glu) at the N-terminal and a cluster of positively charged residues (Arg, Lys) at the C-terminal can balance the α-helical dipole moment and stabilize the α-helix conformation (Altman et al. 2000). Depending on the environment, a conformational equilibrium can be established between β and α structures, which can be useful as model for protein folding understanding. In order to understand the mechanism of self-complementary peptide self-assembly, Caplan et al. have studied the behavior of KFE12 (FKFEFKFEFKFE).

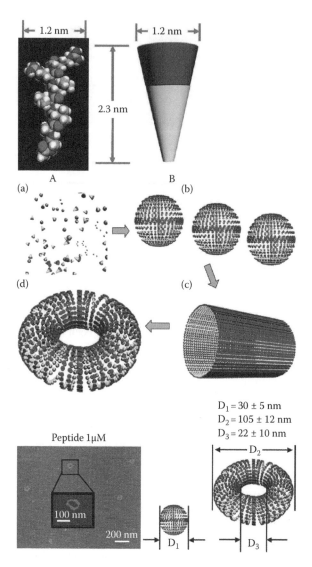

FIGURE 3.24 Top: Molecular model of Ac-GAVILRR-NH$_2$ evidencing the conical shape of the surfactant-like peptide. Middle: Proposed model for the self-assembly process of the nanodonut structure. Bottom: AFM images of the self-organized aggregates. (From Khoe, U. et al., *Langmuir*, 25, 4111, 2009. With permission.)

This peptide can form a gel at neutral pH due to the formation of filaments: all the charges are compensated and thus β-sheets easily stack. The authors show that the self-assembly occurs when intermolecular electrostatic repulsions are screened below van der Waals attractions (DLVO theory) by the addition of salts. This allows the formation of filaments even at low pH, above a critical salt concentration which depends on anion valence (Caplan et al. 2000). A complementary study on (FKFE)$_3$, (IKIE)$_3$, (VKVE)$_3$, (FKFE)$_2$, (FKFE)$_4$, and (FKFQ)$_3$ was conducted to determinate the critical NaCl concentration that is required to observe a transition from a viscous solution to a gel. When increasing the side chains hydrophobicity, the gel is formed at lower NaCl concentrations. When increasing the peptide length, this critical NaCl concentration decreases then increases due to unequal competition between favorable attractions (hydrophobic interactions) and unfavorable repulsions (entropic). It has therefore been possible to design a sequence that remains viscous at neutral pH without NaCl but forms gel in NaCl 0.15 M (physiological conditions) (Caplan et al.

2002). Moreover, it has been demonstrated that the fibrillation of (FKFE)$_3$ involves an intermediate aggregation state: the antiparallel β-sheets first self-assemble into double helical β-sheets before forming fibrils (Figure 3.25 shows intermediate structures formed by KKE8 and imaged by AFM) (Marini et al. 2002). The final fiber network is thermodynamically favored as a final state. In the case of (RADA)$_4$ peptide, the well-defined nanofiber scaffold has been ruptured by sonication into small peptide pieces that spontaneously and quickly reformed the fibrillar network. A sliding diffusion model was proposed to describe the reassembly process involving complementary nanofiber cohesive ends (Yokoi et al. 2005). Chirality is also a parameter that one can play with, especially to produce protease-resistant peptide-based materials. The D-EAK16 of sequence Ac-AEAEAKAKAEAEAKAK-NH$_2$ was synthesized using D-amino acids instead of natural L ones. This peptide, like its enantiomer, forms fibrillar structures in water that can even gelify the solution. In that case, proteases cannot hydrolyze the peptide because unnatural amino acids are not recognized (Luo et al. 2008). All these studies aim at producing biocompatible matrices for cell adhesion and growing, tissue engineering (Genové et al. 2005, Narmoneva et al. 2005, Gelain et al. 2007, Horii et al. 2007), or incorporation and release of bioactive molecules (Nagai et al. 2006, Chau et al. 2008).

Another self-complementary peptide series has been largely described by Aggeli and Boden with a view to understand the mechanism involved in the fibril formation by β-sheet aggregation. They have used models of the transmembrane region of IsK potassium channel protein. The peptides are in a β-sheet conformation and self-assemble in long, semi-flexible polymeric tapes. By varying the polarity and the H-bonding of the solvent, they have determined that in polar solvents the β-sheet structure is mainly stabilized by hydrophobic interactions between peptide side chains and that gel formation requires β-sheet tapes with hydrophilic surfaces. From this study, they have listed few criteria to design gel forming peptides: (1) the peptide should interact through highly cooperative intermolecular H-bonds, (2) attractions between side chains are required (hydrophobic, electrostatic, H-bond), (3) a lateral recognition between adjacent β-strands is involved to constraint the self-assembly in one direction, and (4) the solvent must strongly adhere to the tapes surface to

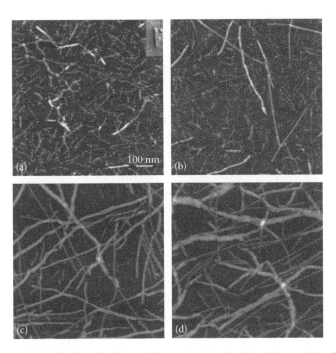

FIGURE 3.25 AFM scans showing the intermediate structures in the self-assembly of FKE8 upon time. (From Marini, D.M. et al., *Nano Lett.*, 2, 295, 2002. With permission.)

control the solubility (Aggeli et al. 1997a). The peptide concentration effect has also been evidenced for the DN1 peptide, Ac-QQRFQWQFEQQ-NH$_2$. It adopts a random coil conformation in dilute aqueous solution, then above 40 mM (the "critical tape concentration") it suddenly structures in β-strands that aggregate in nanotapes (Aggeli et al. 1997b).

A theoretical study of this peptide's aggregation into finite monodisperse fibers has been conducted to understand the self-assembly mechanism. It appears that above the critical aggregation concentration, the peptide in a β-strand conformation starts to form β-sheets further stacking into β-sheet tapes, which are naturally twisted due to chirality. A further face-to-face stacking of these twisted ribbons occurs to produce fibrils in which about 12 tapes are assembled together.

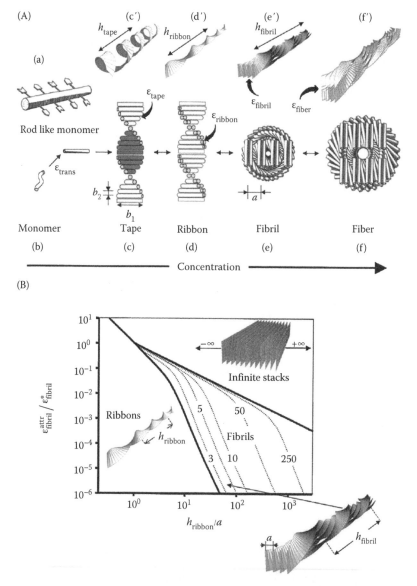

FIGURE 3.26 Model of hierarchical self-assembly of chiral rodlike units with an increase in concentration: from monomers (a) to fibers (f). (b) Phase diagram of a solution of twisted ribbons that form fibrils, representing the relative side-by-side attraction energy between ribbons as a function of the relative helix pitch of isolated ribbons. (From Aggeli, A. et al., *Proc. Natl. Acad. Sci. USA*, 98, 11857, 2001a. With permission.)

This packing implies an untwist and is energy costly, therefore it is limited. Finally, the twisted fibers can wrap around each other to form ropelike fibers (in Figure 3.26 the hierarchical self-assembly of rodlike units is presented) (Nyrkova et al. 2000a,b, Aggeli et al. 2001a). Different peptides have then been studied to clarify in depth the self-assembly mechanism. Using the self-assembling β-sheet peptide K24, the authors have shown that noninteracting fibrils can be stabilized in dilute solutions; above a critical concentration, the fibers interact and a gel is obtained (Aggeli et al. 2001b). The role of side chain interactions also plays a major role in lateral interactions: using a model peptide where few Q have been replaced by F or W, shorter helix pitches are observed because hydrophobic residues enhance the attractive forces between β-tapes in the ribbons (Fishwick et al. 2003).

Another way to control attractive forces between tapes is to play with electrostatics. In a series of peptides with different charges, stable dispersions are obtained when the peptides are fully charged; they exist as monomers because of electrostatic repulsions. A change in pH can reduce this repulsion and pH can be used to switch from an isotropic to a nematic phase (Aggeli et al. 2003a). Moreover, like in the case of pH shifts, ionic strength can be used to control electrostatic repulsions and therefore change the critical tape aggregation concentrations; the addition of salts can broaden the transitions and allow the preparation of nematic gels under physiological concentrations, and they can therefore be used as biocompatible gels (Carrick et al. 2007). Also the hydrophobic/hydrophilic balance can be tuned by the peptide sequence and will affect the aggregation state, the fibrillar length and width (Carrick et al. 2005). These peptide fibrils can self-assemble at the mica/solution interface (Whitehouse et al. 2005) to produce patterned materials, or be used as templates for silica nanotubes formation (Meegan et al. 2004).

Another example of nanotubes obtained *via* β-sheet wrapping is the case of lanreotide. This cyclic octapeptide is used in therapy against acromegaly. When dissolved in water lanreotide can self-organize into well-defined hollow tubes (24.4 nm in diameter, and 1.8 nm in wall thickness) (Valéry et al. 2003). The self-assembly has been shown to be hierarchical: the peptide in a β-hairpin conformation (with aromatic residues on one face, and aliphatic side chains on the other side) forms β-sheets which self-assemble into nanotubes, and the tubes further pack into a hexagonal arrangement (Figure 3.27 illustrates the formation of hollow nanotubes by lanreotide). To get an insight into this mechanism, the phase diagram of lanreotide in water has been established and it shows that noncovalent peptide dimers first form soluble β-sheet filaments. When increasing the peptide content above a critical concentration, 26 β-sheet filaments self-assemble *via* lateral association to produce hollow columns (Valéry et al. 2004). The aromatic residues seem to play a key role in peptide–peptide interactions (Pandit et al. 2008) and a mutational approach has shown that, when these aromatic interactions are modified, lanreotide form less ordered aggregates like amyloid fibers or curved lamellae or even does not self-assemble any more (Valéry et al. 2008). Lanreotide nanotubes are of great interest as templates for biomineralization. They have been used to produce double-walled silica nanotubes (monodisperse in diameter) that pack into highly ordered centimeter-sized fibers (Pouget et al. 2007).

3.3.2.4 Nanotube-Forming Cyclic Peptides

Most synthetic peptides described to self-assemble into nanotubes have been described by the group of Ghadiri (Ghadiri et al. 1993). Cyclic peptides composed of an even number of alternative D- and L-amino acids have been shown to have a flat ring conformation with the amide bonds pointing out of the ring plane. This conformation favors stacking by H-bonding into β-sheet columnar aggregates or nanotubes (Figure 3.28 shows a cyclic peptide and its columnar packing) (Clark et al. 1998, Bong et al. 2001). Depending on the peptide sequence length, the diameter of the inner cavity can be precisely tuned, allowing a steric selectivity toward species circulating inside the tube. For example, this diameter can vary from 7 Å for an octacyclopeptide, allowing the transport of Na^+ or K^+, to 10 Å for a decacyclopeptide—allowing the diffusion of larger species, like glucose—and 13 Å for a dodecacyclopeptide (Kim et al. 1998). When hydrophobic lateral chains are present in

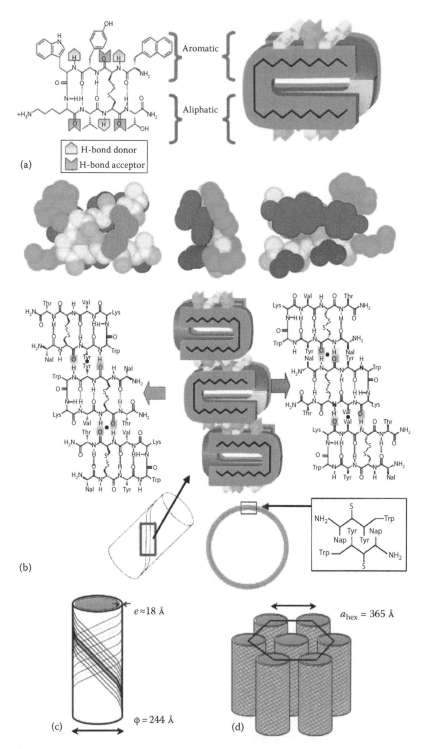

FIGURE 3.27 Schematic representation of lanreotide (a) and its hierarchical self-assembly (b) in hexagonally arranged hollow nanotubes (c). (From Valéry, C. et al., *Proc. Natl. Acad. Sci. USA*, 100, 10258, 2003. With permission.)

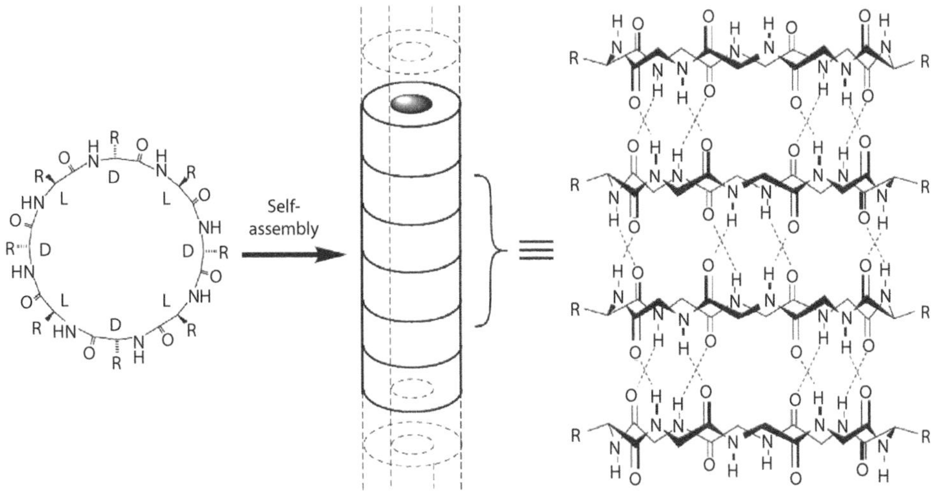

FIGURE 3.28 Schematic representation of a D,L-cyclic octapeptide (left) and its columnar packing into nanotube *via* H-bonding (right). (From Clark, T.D., *J. Am. Chem. Soc.*, 120, 8949, 1998. With permission.)

the sequence, the tube exposes an external hydrophobic surface and can therefore easily insert into biological membranes acting as pore-forming agents or ion channels (Figure 3.29 illustrates peptide nanotubes insertion in lipid bilayer) (Kim et al. 1998). As a consequence they have potential applications for the development of new antibacterial agents (Fernandez-Lopez et al. 2001).

3.4 OUTLOOK: TOWARD MULTICOMPONENT SYSTEMS?

In the case of colloids, this type of aggregation has been considered as bipolar or multipolar coacervation in the classical textbook by Kruyt (Bungenberg de Jong 1949). In the case of micelles, the corresponding domain is considered as the domain of "mixed" micelles. The aggregates can be made by aggregation of two different amphiphilic molecules or by synergetic assembly of an amphiphile and a cosurfactant, that is, a molecule unable to form by itself a stable molecular 2D film. The most common case is random mixing in the film. However, segregation between the two components can occur, giving access to complex shapes ranging form hollow icosahedra to faceted objects (Dubois et al. 2004, Antunes et al. 2007). This type of self-assembly also occurs when the components are peptides: catanionic systems are obtained when combining oppositely charged amphiphiles like association between an anionic peptide and a cationic peptide. Two types of catanionics have to be considered: when the salt obtained by mixing the two peptides is still present, or "true" catanionics, which are three-component systems from the thermodynamic point of view, obtained when excess salt is dialyzed out. The stability of the "ion pair" complex is achieved by electrostatic attractions between the head groups and lateral hydrophobic interactions between the alkyl chains. This is a simple way to obtain multichain amphiphiles without covalent grafting synthesis. For example, catanionic mixtures of amino acid-based surfactants spontaneously self-organize into micelles or vesicles depending on molar ratio and the alkyl chain lengths of the oppositely charged components (Marques et al. 2008). Stupp and coworkers have reported the preparation of peptide amphiphile catanionics by the combination of oppositely charged sequences bearing different epitopes as biological signals (RGD, IKVAV, YIGSR). The surfactant mixture self-organizes into well-defined fibrillar structures (Figure 3.30 shows peptide catanionic micelle and TEM micrograph of nanofibers) that are stable over a wide pH range and a broad concentration range, in contrary to each individual component. In the assembly, peptide sequences are arranged in parallel β-sheet and interact *via* H-bonding (Niece et al. 2003, Behanna et al. 2007).

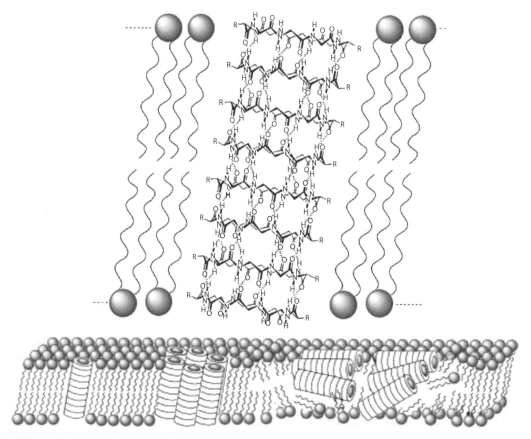

FIGURE 3.29 Peptide nanotube insertion in phospholipid bilayers. (From Kim, H.S. et al., *J. Am. Chem. Soc.*, 120, 4417, 1998. With permission; Fernandez-Lopez, S. et al., *Nature*, 412, 452, 2001. With permission.)

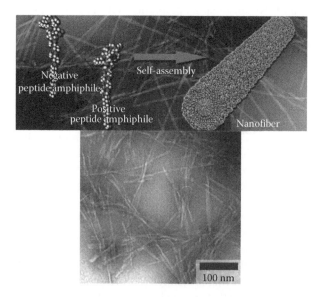

FIGURE 3.30 Catanionic system composed of oppositely charged peptide amphiphiles that self-assemble into cylindrical micelles (a) and TEM micrograph of nanofibers formed by catanionic peptide amphiphiles (b). (From Niece, K.L. et al., *J. Am. Chem. Soc.*, 125, 7146, 2003. With permission.)

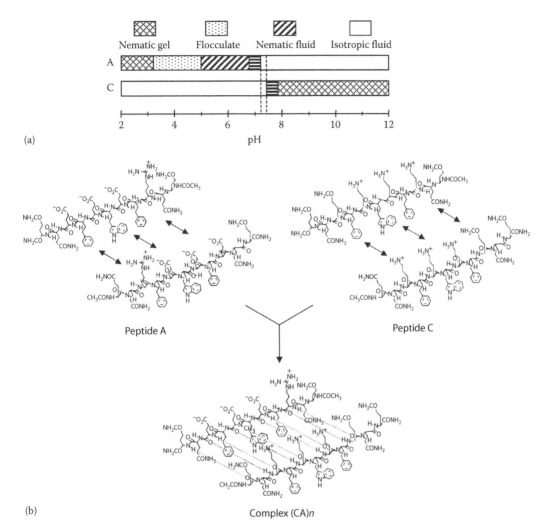

(a)

(b) Complex (CA)*n*

FIGURE 3.31 (a) Phase behavior of oppositely charged peptides as a function of the pH. (b) Molecular structures of the individual peptides and their mixture, showing the electrostatic charge distribution at neutral pH. (From Aggeli, A. et al., *Angew. Chem. Int. Ed.*, 42, 5603, 2003b. With permission.)

Recently, some peptidic catanionics have been obtained by combining the oppositely charged surfactants A_6D and A_6K. Their self-organization is driven by a competition between hydrophobic interactions and electrostatic repulsions and leads to the formation of micelles, elongated micelles, then upon increasing concentration lamellar structures like nanorods (Khoe et al. 2008). The observed cacs are smaller than for individual peptides and it depends on the ratio between the two peptides.

Moreover, by combining oppositely charged peptides, catanionics systems are formed and electrostatic attractions in appropriate pH windows can induce spontaneous formation of fibrillar networks and nematic hydrogels (Figure 3.31 illustrates the catanionic peptide and the phase behavior of its parent oppositely charged peptides). Once formed the gels are very stable over a wide pH range, indicating that the relative pK_a of charged amino acids have been significantly shifted, like in the case of polyelectrolyte complexes (Aggeli et al. 2003b).

To our knowledge, till date, catanionic aggregates of peptides have been studied mainly at equimolarity between the anionic and the cationic component. In surfactant or polyelectrolyte-based catanionic systems, it has been shown that mole ratio is the key parameter: studying mole fractions different from 0.5 of one of the components is a general method of controlling the size of aggregates formed: large sizes of fibrils, tapes, or ribbons are inhibited by electrical charge accumulation at edges. It is likely that studying peptide mixtures at controlled mole ratio will be an emerging domain of experimental investigation in the coming years.

3.5 CONCLUSION: PEPTIDES COMPARED TO OTHER SELF-ASSEMBLING SYSTEMS

Predictive modeling in micellar self-assembling systems started with single-chain surfactants when it was realized that chemical potential is easily accessible experimentally *via* concentration of monomers in equilibrium with aggregates. The free energy of any aggregate could be expressed as a function of average curvature of the aggregate surface (Mitchell and Ninham 1981). The components of the film themselves have a preferred curvature, since the effective area per molecule can be compared to the volume-to-length ratio, in a dimensionless parameter, called the packing parameter. Indeed, packing and hydrophile–lipophile balance are measures of the same property, seen as solubility or as curvature of the aggregate (Kunz et al. 2009). Extension of simple geometrical concepts to temperature-sensitive self-assembly requires introduction of the flexibility (Wennerstroem 1996). Explicit calculations become tractable when applying coarse-graining (Shinoda et al. 2008).

The sequences of aggregate shapes and therefore of lyotropic phases when aggregate self-organize are common to all surfactant systems, and can be transposed to hydrophobic–hydrophilic copolymers: the classical sequence is micelle, elongated, or giant micelle, connected random lamellaes (sponge), and the reverse connected water-in-oil cylinders, the globular reverse micelles. The corresponding lyotropic sequence of phases is hexagonal, cubic, lamellar, and the reverse sequence when apolar solvent is in excess. They are observed for surfactants, lipids, and small copolymers. These sequences cannot easily be observed with peptides. This is because the peptide-building block interacts with neighboring molecules in the aggregate not only *via* long-range interactions, such as electrostatics, steric, van der Waals, and hydration forces, but also *via* localized hydrogen bonding.

The packing between more or less flexible objects is no more dominant: hydrogen bonding becomes then the main constraint. Till date, the most general theory available is in the case of building block self-assembling locally as rigid bilayers, but with intrinsic twisting and rotating angles in the stack. Boden and coworkers have developed a universal theory for homo-aggregation (Aggeli et al. 2001a). This was illustrated in Figure 3.26: the sequences are now ribbons, tubules, and helices. An equivalent theory may exist if the local structure is toward cylindrical shape, but to our knowledge this theory of aggregation of homo-aggregates is not yet available due to lack of clear structural characterization of all bicontinuous networks that can theoretically exist. From the experience of the field of surfactants, with mixed micelles in strong interaction such as catanionics, it is likely that the number of modes of controlled self-aggregation will increase quite dramatically, when off-stoichiometry mixed aggregate hetero-aggregation of a pair of peptides will be systematically investigated. Predicting phenomena such as the ADN-histone self-assembly is still a long way away!

ABBREVIATIONS

Amino Acids

Ala or A   Alanine
Arg or R   Arginine
Asn or N   Asparagine
Asp or D   Aspartic acid
Cys or C   Cysteine
Gln or Q   Glutamine
Glu or E   Glutamic acid

Gly or G   Glycine
Ile or I   Isoleucine
Leu or L   Leucine
Lys or K   Lysine
Phe or F   Phenylalanine
Pro or P   Proline
Ser or S   Serine
Thr or T   Threonine
Trp or W   Tryptophan
Tyr or Y   Tyrosine
Val or V   Valine

OTHER SYMBOLS

Aβ   Amyloid β peptide
AFM   Atomic force microscopy
Boc   Tertiobutyloxycarbonyl
Bz   Benzoyl
Cac   Critical aggregation concentration
Cmc   Critical micellar concentration
Cvc   Critical vesicular concentration
DLS   Dynamic light scattering
DLVO   Theory: theory named after Derjaguin, Landau, Verwey, and Overbeek
DMF   Dimethylformamide
DNA   Deoxyribonucleic acid
DOPE   1,2-Dioleoyl-*sn*-glycero-3-phosphoethanolamine
Et   Ethyl
GIXD   Grazing incidence x-ray diffraction
LCST   Lower critical solution temperature
MPEG   Monomethyl-PEG
PEG   Poly(etheleneglycol)
SAXS   Small angle x-ray scattering
SIV   Simian immunodeficiency virus
TEM   Transmission electron microscopy
TFE   Trifluoroethanol
Z   Carbobenzyloxy

REFERENCES

Accardo, A., Tesauro, D., Mangiapia, G., Pedone, C., and Morelli, G. 2006. Nanostructures by self-assembling peptide amphiphile as potential selective drug carriers. *Pept. Sci.* 88:115–121.

Adler-Abramovich, L., Reches, M., and Gazit, E. 2006. Thermal and chemical stability of diphenylalanine peptide nanotubes: Implications for nanotechnological applications. *Langmuir* 22:1313–1320.

Aggeli, A., Bell, M., Boden, N., Keen, J. N., Knowles, P. F., McLeish, T. C. B., Pitkeathly, M., and Radford, S. E. 1997a. Responsive gels formed by the spontaneous self-assembly of peptides into β-sheet tapes. *Nature* 386:25–262.

Aggeli, A., Bell, M., Boden, N., Keen, J. N., McLeish, T. C. B., Nyrkova, I., Radford, S. E., and Semenov, A. 1997b. Engineering of peptide β-sheet nanotapes. *J. Mater. Chem.* 7:1135–1145.

Aggeli, A., Nyrkova, I. A., Bell, M., Harding, R., Carrick, L., McLeish, T. C. B., Semenov, A. N., and Boden, N. 2001a. Hierarchical self-assembly of chiral rod-like molecules as a model for peptide β-sheet tapes, ribbons, fibrils and fibers. *Proc. Natl. Acad. Sci. USA* 98:11857–11862.

Aggeli, A., Fytas, G., Vlassopoulos, D., McLeish, T. C. B., Mawer, P. J., and Boden, N. 2001b. Structure and dynamics of self-assembling β-sheet peptide tapes by dynamic light scattering. *Biomacromolecules* 2:378–388.

Aggeli, A., Bell, M., Carrick, L. M., Fishwick, C. W. G., Harding, R., Mawer, P. J., Radford, S. E., Strong, A. E., and Boden, N. 2003a. pH as a trigger of peptide β-sheet self-assembly and reversible switching between nematic and isotropic phases. *J. Am. Chem. Soc.* 125:9619–9628.

Aggeli, A., Bell, M., Boden, N., Carrick, L. M., and Strong, A. E. 2003b. Self-assembling peptide polyelectrolyte β-sheet complexes form nematic hydrogels. *Angew. Chem. Int. Ed.* 42:5603–5606.

Altman, M., Lee, P., Rich, A., and Zhang, S. 2000. Conformational behaviour of ionic self-complementary peptides. *Prot. Sci.* 9:1095–1105.

Andrews, M. E., Moses, J. P., Sendhil, S., Rakkappan, C., and Jayakumar, R. 2002. Adiabatic compressibility and intrinsic viscosity studies on peptide aggregates. *Lett. Pept. Sci.* 9:167–172.

Antunes, F. E., Brito, R. O., Marques, E. F., Lindman, B., and Miguel, M. 2007. Mechanisms behind the faceting of catanionic vesicles by polycations: Chain crystallization and segregation. *J. Phys. Chem. B* 111:116–123.

Auzély-Velty, R., Perly, B., Taché, O., Zemb, T., Jéhan, P., Guenot, P., Dalbiez, J.-P., and Djedaïni-Pilard, F. 1999. Cholesteryl-cyclodextrins: Synthesis and insertion into phospholipid membranes. *Carbohydr. Res.* 318:82–90.

Behanna, H. A., Rajangam, K., and Stupp, S. I. 2007. Modulation of fluorescence through coassembly of molecules in organic nanostructures. *J. Am. Chem. Soc.* 129:321–327.

Bitton, R., Schmidt, J., Biesalski, M., Tu, R., Tirrell, M., and Bianco-Peled, H. 2005. Self-assembly of model DNA-binding peptide amphiphiles. *Langmuir* 21:11888–11895.

Bong, D. T., Clark, T. D., Granja, J. R., and Ghadiri, M. R. 2001. Self-assembling organic nanotubes. *Angew. Chem. Int. Ed.* 40:988–1011.

Brasseur, R., Braun, N., El Kirat, K., Deleu, M., Mingeot-Leclercq, M.-P., and Dufrêne, Y. F. 2007. The biologically important surfactin lipopeptide induces nanoripples in supported lipid bilayers. *Langmuir* 23:9769–9772.

Bungenberg de Jong, H. G. 1949. Crystallisation-coacervation-flocculation. In: H. R. Kruyt (Ed.), *Colloid Science*, vol. II (pp. 232–258, Chapter VIII). Amsterdam, the Netherlands: Elsevier.

Bungenberg de Jong, H. G. and Kruyt, H. R. 1930. Koazervation. *Koll. Zeitsch.* 50:39–48.

Caplan, M. R., Moore, P. N., Zhang, S., Kamm, R. D., and Lauffenburger, D. A. 2000. Self-assembly of a β-sheet protein governed by relief of electrostatic repulsion relative to van der Waals attraction. *Biomacromolecules* 1:627–631.

Caplan, M. R., Schwartzfarb, E. M., Zhang, S., Kamm, R. D., and Lauffenburger, D. A. 2002. Control of self-assembling oligopeptide matrix formation through systematic variation of amino acid sequence. *Biomaterials* 23:219–227.

Carrick, L., Tassieri, M., Waigh, T. A., Aggeli, A., Boden, N., Bell, C., Fisher, J., Ingham, E., and Evans, R. M. L. 2005. The internal dynamic modes of charged self-assembled peptide fibrils. *Langmuir* 21:3733–3737.

Carrick, L. M., Aggeli, A., Boden, N., Fisher, J., Ingham, E., and Waigh, T. A. 2007. Effect of ionic strength on the self-assembly, morphology and gelation of pH responsive β-sheet tape-forming peptides. *Tetrahedron* 63:7457–7467.

Caspar, D. L. D. and Klug, A. 1962. Physical principles in the construction of regular viruses. *Cold Spring Harb. Symp. Quant. Biol.* 27:1–24.

Cavalli, S. and Kros, A. 2008. Scope and applications of amphiphilic alkyl- and lipopeptides. *Adv. Mater.* 20:627–631.

Cavalli, S., Handgraaf, J.-W., Tellers, E. E., Popescu, D. C., Overhand, M., Kjaer, K., Vaiser, V., Sommerdijk, N. A. J. M., Papaport, H., and Kros, A. 2006a. Two-dimensional ordered β-sheet lipopeptide monolayers. *J. Am. Chem. Soc.* 128:13959–13966.

Cavalli, S., Popescu, D. C., Tellers, E. E., Vos, M. R. J., Pichon, B. P., Overhand, M., Rapaport, H., Sommerdijk, N. A. J. M., and Kros, A. 2006b. Self-organizing β-sheet lipopeptide monolayers as templates for the mineralization of CaCO$_3$. *Angew. Chem. Int. Ed.* 45:739–744.

Chau, Y., Luo, Y., Cheung, A. C. Y., Nagai, Y., Zhang, S., Kobler, J. B., Zeitels, S. M., and Langer, R. 2008. Incorporation of a matrix metalloproteinase-sensitive substrate into self-assembling peptide—A model for biofunctional scaffolds. *Biomaterials* 29:1713–1719.

Chevalier, Y. and Zemb, T. 1990. The structure of micelles and microemulsions. *Rep. Prog. Phys.* 53:279–371.

Clark, T. D., Buriak, J. M., Kobayashi, K., Isler, M. P., McRee, D. E., and Ghadiri, M. R. 1998. Cylindrical β-sheet peptide assemblies. *J. Am. Chem. Soc.* 120:8949–8962.

Cui, H., Muraoka, T., Cheetham, A. G., and Stupp, S. I. 2009. Self-assembly of giant peptide nanobelts. *Nano Lett.* 9:945–951.

Deleu, M., Razafindralambo, H., Popineau, Y., Jacques, P., Thonart, P., and Paquot, M. 1999. Interfacial and emulsifying properties of lipopeptides from *Bacillus subtilis. Coll. Surf. A: Phys. Eng. Asp.* 152:3–10.

Demé, B., Dubois, M., Gulik-Krzywicki, T., and Zemb, T. 2002a. Giant collective fluctuations of charged membranes at the lamellar-to-vesicle unbinding transition. 1. Characterization of a new lipid morphology by SANS, SAXS, and electron microscopy. *Langmuir* 18:997–1004.

Demé, B., Dubois, M. and Zemb, T. 2002b. Giant collective fluctuations of charged membranes at the lamellar-to-vesicle unbinding transition. 2. Equation of state in the absence of salt. *Langmuir* 18:1005–1013.

Dexter, A. F. and Middelberg, A. P. J. 2008. Peptides as functional surfactants. *Ind. Eng. Chem. Res.* 47:6391–6398.

Dubois, M., Lizunov, V., Meister, A., Gulik-Krzywicki, T., Verbavatz, J. M., Perez, E., Zimmerberg, J., and Zemb, T. 2004. Shape control through molecular segregation in giant surfactant aggregates. *Proc. Natl. Acad. Sci. USA* 101:15082–15087.

Eeman, M., Berquand, A., Dufrêne, Y. F., Paquot, M., Dufour, S., and Deleu, M. 2006. Penetration of surfactin into phospholipid monolayers: Nanoscale interfacial organization. *Langmuir* 22:11337–11345.

El Kirat, K., Dufrêne, Y. F., Lins, L., and Brasseur, R. 2006. The SIV tilted peptide induces cylindrical reverse micelles in supported lipid bilayers. *Biochemistry* 45:9336–9341.

Escuder, B. and Miravet, J. F. 2006. Silk-inspired low-molecular-weight organogelator. *Langmuir* 22:7793–7797.

Fernandez-Lopez, S., Kim, H.-S., Choi, E. C., Delgado, M., Granja, J. R., Khasanov, A., Kraehenbuehl, K., Long, G., Weinberger, D. A., Wilcoxen, K. M., and Ghadiri, M. R. 2001. Antibacterial agents based on the cyclic D,L-α-peptide architecture. *Nature* 412:452–455.

Fishwick, C. W. G., Beevers, A. J., Carrick, L. M., Whitehouse, C. D., Aggeli, A., and Boden, N. 2003. Structures of helical β-tapes and twisted ribbons: The role of side-chain interactions on twist and bend behaviour. *Nano Lett.* 3:1475–1479.

Forns, P., Lauer-Fields, J., Gao, S., and Fields, G. B. 2000. Induction of protein-like molecular architecture by monoalkyl hydrocarbon chains. *Biopolymers* 54:531–546.

Ganesh, S. and Jayakumar, R. 2003a. Circular dichroism and Fourier transform infrared spectroscopic studies on self-assembly of tetrapeptide derivative in solution and solvated film. *J. Pept. Res.* 61:122–128.

Ganesh, S. and Jayakumar, R. 2003b. Structural transitions involved in a novel amyloid-like β-sheet assemblage of tripeptide derivatives. *Biopolymers* 70:336–345.

Ganesh, S., Prakash, S., and Jayakumar, R. 2003. Spectroscopic investigation on gel-forming β-sheet assemblage of peptide derivatives. *Biopolymers* 70:346–354.

Gazit, E. 2002. A possible role for π-stacking in the self-assembly of amyloid fibrils. *FASEB J.* 16:77–83.

Gelain, F., Bottai, D., Vescovi, A., and Zhang, S. 2006. Designer self-assembling peptide nanofiber scaffolds for adult mouse neural stem cell 3-dimensional cultures. *PLoS ONE* 1:e119.

Gelain, F., Horii, A., and Zhang, S. 2007. Designer self-assembling peptide scaffolds for 3-D tissue cell cultures and regenerative medicine. *Macromol. Biosci.* 7:544–551.

Genové, E., Shen, C., Zhang, S., and Semino, C. E. 2005. The effect of functionalized self-assembling peptide scaffolds on human aortic endothelial cell function. *Biomaterials* 26:3341–3351.

Ghadiri, M. R., Granja, J. R., Milligan, R. A., McRee, D. E., and Khazanovich, N. 1993. Self-assembling organic nanotubes based on a cyclic peptide architecture. *Nature* 366:324–327.

Grangemard, I., Wallach, J., Maget-Dana, R., and Peypoux, F. 2001. Lichenysin, a more efficient cation chelator than surfactin. *Appl. Biochem. Biotechnol.* 90:199–210.

Guler, M. O. and Stupp, S. I. 2007. A self-assembled nanofiber catalyst for ester hydrolysis. *J. Am. Chem. Soc.* 129:12082–12083.

Hamley, I. W., Krysmann, M. J., Castelletto, V., and Noirez, L. 2008a. Multiple lyotropic polymorphism of a poly(ethyleneglycol)-peptide conjugate in aqueous solution. *Adv. Mater.* 20:4394–4397.

Hamley, I. W., Krysmann, M. J., Kelarakis, A., Castelletto, V., Noirez, L., Hule, R. A., and Pochan, D. J. 2008b. Nematic and columnar ordering of a PEG-peptide conjugate in aqueous solution. *Chem. Eur. J.* 14:11369–11375.

Han, Y., Huang, X., Cao, M., and Wang, Y. 2008. Micellization of surfactin and its effect on the aggregate conformation of amyloid β (1–40). *J. Phys. Chem. B* 112:15195–15201.

Hartgerink, J. D., Zubarev, E. R., and Stupp, S. I. 2001a. Supramolecular one-dimensional objects. *Curr. Opin. Solid State Mater. Sci.* 5:355–361.

Hartgerink, J. D., Beniash, E., and Stupp, S. I. 2001b. Self-assembly and mineralization of peptide-amphiphile nanofibers. *Science* 294:1684–1688.

Hartgerink, J. D., Beniash, E., and Stupp, S. I. 2002. Peptide-amphiphile nanofibers: A versatile scaffold for the preparation of self-assembling materials. *Proc. Natl. Acad. Sci. USA* 99:5133–5138.

Heerklotz, H. and Seeling, J. 2001. Detergent-like action of the antibiotic peptide surfactin on lipid membranes. *Biophys. J.* 81:1547–1554.

Heerklotz, H., Wieprecht, T., and Seeling, J. 2004. Membrane perturbation by the lipopeptide surfactin and detergents as studied by deuterium NMR. *J. Phys. Chem. B* 108:4909–4915.

Hendler, N., Sidelman, N., Reches, M., Gazit, E., Rosenberg, Y., and Richter, S. 2007. Formation of well-organized self-assembled films from peptide nanotubes. *Adv. Mater.* 19:1485–1488.

Horii, A., Gelain, F., and Zhang, S. 2007. Biological designer self-assembling peptide nanofiber scaffolds significantly enhance osteoblast proliferation, differentiation and 3-D migration. *PLoS ONE* 2:e190.

Hwang, W., Zhang, S., Kamm, R. D., and Karplus, M. 2004. Kinetic control of dimer structure formation in amyloid fibrillogenesis. *Proc. Natl. Acad. Sci. USA* 101:12916–12921.

Hyde, S., Andersson, S., Larsson, K. et al. 1997. *The Language of Shape: The Role of Curvature in Condensed Matter: Physics, Chemistry, and Biology*. Amsterdam, the Netherlands: Elsevier Science B. V.

Iqbal, S., Miravet, J. F., and Escuder, B. 2008. Biomimetic self-assembly of tetrapeptides into fibrillar networks and organogels. *Eur. J. Org. Chem.* 4580–4590.

Ishigami, Y., Osman, M., Nakahara, H., Sano, Y., Ishiguro, R., and Matsumoto, M. 1995. Significance of β-sheet formation for micellization and surface adsorption of surfactin. *Colloids Surf. B Biointerfaces* 4:341–348.

Israelachvili, J. N., Mitchell, D. J., and Ninham, B. W. 1976. Theory of self-assembly of hydrocarbon amphiphiles into micelles and bilayers. *J. Chem. Soc. Faraday Trans. II* 72:1525–1568.

Jayakumar, R. and Murugesan, M. 2000. Anomalous temperature dependence of peptide films at air-water interface. *Bioorg. Med. Chem. Lett.* 10:1055–1057.

Jayakumar, R., Mandal, A. B., and Manoharan, P. T. 1993. Micelle formation of Boc-Val-Val-Ile-OMe tripeptide in chloroform and its conformational analysis. *J. Chem. Soc. Chem. Commun.* 853–855.

Jayakumar, R., Jeevan, R. G., and Mandal, A. B. 1994. Aggregation, hydrogen bonding and thermodynamic studies on Boc-Val-Val-Ile-OMe tripeptide micelles in chloroform. *J. Chem. Soc. Fraday Trans.* 90:2725–2730.

Jayakumar, R., Murugesan, M., and Ahmed, M. R. 2000a. Formation of multilamellar vesicles ('onions') in peptide based surfactant. *Bioorg. Med. Chem. Lett.* 10:1547–1550.

Jayakumar, R., Murugesan, M., Selvi, S., and Scibioh, M. A. 2000b. Aggregational studies on β-turn forming peptide Tyr-Pro-Gly-Asp-Val. *Langmuir* 16:3019–3021.

Jayakumar, R., Murugesan, M., Asokan, C., and Scibioh, M. A. 2000c. Self-assembly of a peptide Boc-(Ile)$_5$-OMe in chloroform and *N, N*-dimtheylformamide. *Langmuir* 16:1489–1496.

Kell, H., Holzwarth, J. F., Boettcher, C., Heenan, R. K., and Vater, J. 2007. Physicochemical studies of the interaction of the lipoheptapeptide surfactin with lipid bilayers of L-α-dimyristoyl phosphatidylcholine. *Biophys. Chem.* 128:114–124.

Khatua, D. and Dey, J. 2007. Fluorescence, circular dichroism, light scattering, and microscopic characterization of vesicles of sodium salts of three *N*-acyl peptides. *J. Phys. Chem. B* 111:124–130.

Khoe, U., Yang, Y., and Zhang, S. 2008. Synergistic effect and hierarchical nanostructure formation in mixing two designer lipid-like peptide surfactants Ac-A$_6$D-OH and Ac-A$_6$K-NH$_2$. *Macromol. Biosci.* 8:1060–1067.

Khoe, U., Yang, Y., and Zhang, S. 2009. Self-assembly of nanodonut structure from a cone-shaped designer lipid-like peptide surfactant. *Langmuir* 25:4111–4114.

Kiley, P., Zhao, X., Vaughn, M., Balso, M. A., Bruce, B. D., and Zhang, S. 2005. Self-assembling peptide detergents stabilize isolated photosystem I on a dry surface for an extended time. *PLoS Biol.* 3:e230.

Kim, H. S., Hartgerink, J. D., and Ghadiri, M. R. 1998. Oriented self-assembly of cyclic peptide nanotubes in lipid membranes. *J. Am. Chem. Soc.* 120:4417–4424.

Knoblich, A., Matsumoto, M., Ishiguro, R., Murata, K., Fujiyoshi, Y., Ishigami, Y., and Osman, M. 1995. Electron cryo-microscopic studies on micellar shape and size of surfactin, an anionic lipopeptide. *Colloids Surf. B: Biointerfaces* 5:43–48.

Kogiso, M., Ohnishi, S., Yase, K., Masuda, M., and Shimizu, T. 1998. Dicarboxylic oligopeptide bola-amphiphiles: Proton-triggered self-assembly of microtubes with loose solid surfaces. *Langmuir* 14:4978–4986.

Kogiso, M., Okada, Y., Hanada, T., Yase, K., and Shimizu, T. 2000. Self-assembled peptide fibers from valylvaline bola-amphiphiles by a parallel β-sheet network. *Biochim. Biophys. Acta* 1475:346–352.

Kogiso, M., Okada, Y., Yase, K., and Shimizu, T. 2004. Metal-complexed nanofiber formation in water from dicarboxylic valylvaline bolaamphiphiles. *J. Colloid Interface Sci.* 273:394–399.

Kogiso, M., Zhou, Y., and Shimizu, T. 2007. Instant preparation of self-assembled metal-complexed lipid nano-tubes that act as templates to produce metal oxide nanotubes. *Adv. Mater.* 19:242–246.

Krysmann, M. J., Castelletto, V., and Hamley, I. W. 2007. Fribrillisation of hydrophobically modified amyloid peptide fragments in an organic solvent. *Soft Matter* 3:1401–1406.

Krysmann, M. J., Castelletto, V., Kelarakis, A., Hamley, I. W., Hule, R. A., and Pochan, D. J. 2008a. Self-assembly and hydrogelation of an amyloid peptide fragment. *Biochemistry* 47:4597–4605.

Krysmann, M. J., Castelletto, V., McKendrick, J. E., Clifton, L. A., and Hamley, I. W. 2008b. Self-assembly of peptide nanotubes in an organic solvent. *Langmuir* 24:8158–8162.

Kunz, W., Lo Nostro, P., and Ninham, B. W. 2004. The present state of affairs with Hofmeister effects. *Curr. Opin. Colloid Interface Sci.* 1–2:1–18.

Kunz, W., Testard, F., and Zemb, T. 2009. Correspondence between curvature, packing parameter, and hydro-philic-lipophilic deviation scales around the phase-inversion temperature. *Langmuir* 25:112–115.

Lang, S. 2002. Biological amphiphiles (microbial biosurfactants). *Curr. Opin. Colloid Interface Sci.* 7:12–20.

Löwik, D. W. P. M., Linhardt, J. G., Adams, P. J. H. M., and van Hest, J. C. M. 2003. Non-covalent stabilization of a β-hairpin peptide into liposomes. *Org. Biomol. Chem.* 1:1827–1829.

Löwik, D. W. P. M., Garcia-Hartjes, J., Meijer, J. T., and van Hest, J. C. M. 2005. Tuning secondary structure and self-assembly of amphiphilic peptides. *Langmuir* 21:524–526.

Löwik, D. W. P. M., Shklyarevskiy, I. O., Ruizendaal, L., Christiansen, P. C. M., Maan, J. C., and van Hest, J. C. M. 2007. A highly ordered material from magnetically aligned peptide amphiphile nanofiber assemblies. *Adv. Mater.* 19:1191–1195.

Lu, J. R., Zhao, X. B., and Yaseen, M. 2007. Biomimetic amphiphiles: Biosurfactants. *Curr. Opin. Colloid Interface Sci.* 12:60–67.

Luo, Z., Zhao, X., and Zhang, S. 2008. Self-organization of a chiral D-EAK16 designer peptide into a 3D nanofiber scaffold. *Macromol. Biosci.* 8:785–791.

Mahler, A., Reches, M., Rechter, M., Cohen, S., and Gazit, E. 2006. Rigid, self-assembled hydrogel composed of modified aromatic dipeptide. *Adv. Mater.* 18:1365–1370.

Mandal, A. B. and Jayakumar, R. 1993. A new micelle-forming peptide. *J. Chem. Soc. Chem. Commun.* 3:237.

Mandal, A. B. and Jayakumar, R. 1994. Aggregation, hydrogen bonding and thermodynamic studies on tetra-peptides micelles. *J. Chem. Soc. Faraday Trans.* 90:161–165.

Mandal, A. B., Dhathathreyan, A., Jayakumar, R., and Ramasami, T. 1993. Characterization of Boc-Lys(Z)-Tyr-NHNH$_2$ dipeptide. *J. Chem. Soc. Faraday Trans.* 89:3075–3079.

Manning, G. S. 1969. Limiting laws and counterion condensation in polyelectrolyte solutions: I. Colligative properties. *J. Chem. Phys.* 51:924–933.

Marini, D. M., Hwang, W., Lauffenburger, D. A., Zhang, S., and Kamm, R. D. 2002. Left-handed helical ribbon intermediates in the self-assembly of a β-sheet peptide. *Nano Lett.* 2:295–299.

Marques, E. F., Brito, R. O., Silva, S. G., Rodríguez-Borges, J. E., do Vale, M. L., Gomes, P., Araújo, M. J., and Söderman, O. 2008. Spontaneous vesicle formation in catanionic mixtures of amino acid-based surfac-tants: Chain length symmetry effects. *Langmuir* 24:11009–11017.

Martinez, J. S. and Butler, A. 2007. Marine amphiphilic siderophores: Marinobactin structure, uptake, and microbial partitioning. *J. Inorg. Biochem.* 101:1692–1698.

Martinez, J. S., Zhang, G. P., Holt, P. D., Jung, H.-T., Carrano, C. J., Haygood, M. G., and Butler, A. 2000. Self-assembling amphiphilic siderophores from marine bacteria. *Science* 287:1245–1247.

Matmour, R., De Cat, I., George, S. J., Adriaens, W., Leclère, P., Bomans, P. H. H., Sommerdijk, N. A. J. M., Gielen, J. C., Christianen, P. C. M., Heldens, J. T., van Hest, J. C. M., Löwik, D. W. P. M., De Feyter, S., Meijer, E. W., and Schenning, A. P. H. J. 2008. Oligo(*p*-phenylenevinylene)-peptide con-jugates: Synthesis and self-assembly in solution and at the solid-liquid interface. *J. Am. Chem. Soc.* 130:14576–14583.

Matsui, H. and Gologan, B. 2000. Crystalline glycylglycine bolaamphiphile tubules and their pH-sensitive structural transformation. *J. Phys. Chem. B* 104:3383–3386.

Matsui, H., Pan, S., Gologan, B., and Jonas, S. H. 2000. Bolaamphiphile nanotube-templated metallized wires. *J. Phys. Chem. B* 104:9576–9579.

Matsumoto, K., Vaughn, M., Bruce, B. D., Koutsopoulos, S., and Zhang, S. 2009. Designer peptide surfactants stabilize functional photosystem-I membrane complex in aqueous solution for extended time. *J. Phys. Chem. B* 113:75–83.

McGregor, C.-L., Chen, L., Pomroy, N. C., Hwang, P., Go, S., Chakrabartty, A., and Privé, G. G. 2003. Lipopeptide detergents designed for the structural study of membrane proteins. *Nat. Biotechnol.* 21:171–176.

Meegan, J. E., Aggeli, A., Boden, N., Brydson, R., Brown, A. P., Carrick, L., Brough, A. R., Hussain, A., and Ansell, R. J. 2004. Designed self-assembled β-sheet peptide fibrils as templates for silica nanotubes. *Adv. Funct. Mater.* 14:31–37.

Meijer, J. T., Roeters, M., Viola, V., Löwik, D. W. P. M., Vriend, G., and van Hest, J. C. M. 2007. Stabilization of fibrils by hydrophobic interaction. *Langmuir* 23:2058–2063.

Mitchell, D. J. and Ninham, B. W. 1981. Micelles, vesicles, and microemulsions. *J. Chem. Soc. Faraday Trans. II* 77:601–629.

Mitra, R. N., Das, D., Roy, S., and Das, P. K. 2007. Structure and properties of low molecular weight amphiphilic peptide hydrogelators. *J. Phys. Chem. B* 111:14107–14113.

Mulligan, C. N. 2005. Environmental applications for biosurfactants. *Environ. Pollut.* 133:183–198.

Muraoka, T., Cui, H., and Stupp, S. I. 2008. Quadruple helix formation of a photoresponsive peptide amphiphile and its light-triggered dissociation into single fibers. *J. Am. Chem. Soc.* 130:2946–2947.

Murugesan, M., Jayakumar, R., and Durai, V. 1996. Self-assembly of a non-ionic peptide surfactant in aqueous medium. *Langmuir* 12:1760–1764.

Murugesan, M., Venugopal, M., and Jayakumar, R. 1997. Aggregation of a tetrapeptide derivative [Boc-Ile-Gly-Met-Thr(Bzl)-OBzl] in chloroform. *J. Chem. Soc. Perkin Trans.* 2:1959–1963.

Murugesan, M., Scibioh, M. A., and Jayakumar, R. 1999. Structural transition of non-ionic peptide aggregates in aqueous medium. *Langmuir* 15:5467–5473.

Muthusamy, K., Gopalakrishnan, S., Ravi, T. K., and Sivachidambaram, P. 2008. Biosurfactants: Properties, commercial production and application. *Curr. Sci.* 94:736–747.

Nagai, Y., Unsworth, L. D., Koustopoulos, S., and Zhang, S. 2006. Slow release of molecules in self-assembling peptide nanofiber scaffold. *J. Control. Release* 115:18–25.

Narmoneva, D. A., Oni, O., Sieminski, A. L., Zhang, S., Gertler, J. P., Kamm, R. D., and Lee, R. T. 2005. Self-assembling short oligopeptides and the promotion of angiogenesis. *Biomaterials* 26:4837–4846.

Nicolas, J. P. 2003. Molecular dynamics simulation of surfactin molecules at the water–hexane interface. *Biophys. J.* 85:1377–1391.

Niece, K. L., Hartgerink, J. D., Donners, J. J. J. M., and Stupp, S. I. 2003. Self-assembly combining two bioactive peptide-amphiphile molecules into nanofibers by electrostatic attraction. *J. Am. Chem. Soc.* 125:7146–7147.

Nyrkova, I. A., Semenov, A. N., Aggeli, A., and Boden, N. 2000a. Fibril stability in solutions of twisted β-sheet peptides: A new kind of micellization in chiral systems. *Eur. Phys. J. B* 17:481–497.

Nyrkova, I. A., Semenov, A. N., Aggeli, A., Boden, N., and McLeish, T. C. B. 2000b. Self-assembly and structure transformations in living polymers forming fibrils. *Eur. Phys. J. B* 17:499–513.

Ongena, M. and Jacques, P. 2007. *Bacillus* lipopeptides: Versatile weapons for plant disease biocontrol. *Trends Microbiol.* 16:115–125.

Osman, M., Høiland, H., and Holmsen, H. 1998a. Micropolarity and microviscosity in the micelles of the heptapeptide biosurfactant "surfactin." *Colloids Surf. B: Biointerfaces* 11:167–175.

Osman, M., Høiland, H., Holmsen, H., and Ishigami, Y. 1998b. Tuning micelles of a bioactive heptapeptide biosurfactant via extrinsically induced conformational transition of surfactin assembly. *J. Pept. Sci.* 4:449–458.

Owen, T., Pynn, R., Martinez, J. S., and Butler, A. 2005. Micelle-to-vesicle transition of an iron-chelating microbial surfactant, marinobactin E. *Langmuir* 21:12109–12114.

Owen, T., Pynn, R., Hammouda, B., and Butler, A. 2007. Metal-dependant self-assembly of a microbial surfactant. *Langmuir* 23:9393–9400.

Owen, T., Webb, S. M., and Butler, A. 2008. XAS study of a metal-induced phase transition by a microbial surfactant. *Langmuir* 24:4999–5002.

Pandit, A., Fay, N., Bordes, L., Valéry, C., Cherif-Cheikh, R., Robert, B., Artzner, F., and Paternostre, M. 2008. Self-assembly of the octapeptide lanreotide and lanreotide-based derivatives: The role of the aromatic residues. *J. Pept. Sci.* 14:66–75.

Paramonov, S. E., Jun, H.-W., and Hartgerink, J. D. 2006. Self-assembly of peptide-amphiphile nanofibers: The roles of hydrogen bonding and amphiphilic packing. *J. Am. Chem. Soc.* 128:7291–7298.

Peypoux, F., Bonmatin, J. M., and Wallach, J. 1999. Recent trends in the biochemistry of surfactin. *Appl. Microbiol. Biotechnol.* 51:553–563.

Pouget, E., Dujardin, E., Cavalier, A., Moreac, A., Valéry, C., Marchi-Artzner, V., Weiss, T., Renault, A., Paternostre, M., and Artzner, F. 2007. Hierarchical architectures by synergy between dynamical template self-assembly and biomineralization. *Nat. Mater.* 6:434–439.

Ramesh, C. V., Yayakumar, R., and Puvanakrishnan, R. 1998. A novel surface-active peptide derivative exhibits *in vitro* inhibition of platelet aggregation. *Peptides* 19:1695–1702.

Razafindralambo, H., Popineau, Y., Deleu, M., Hbid, C., Jacques, P., Thonart, P., and Paquot, M. 1997. Surface-active properties of surfactin/iturin A mixtures produced by *Bacillus subtilis*. *Langmuir* 13:6026–6031.

Reches, M. and Gazit, E. 2003. Casting metal nanowires within discrete self-assembled peptide nanotubes. *Science* 300:625–627.

Reches, M. and Gazit, E. 2004. Formation of closed-cage nanostructures by self-assembly of aromatic dipeptides. *Nano Lett.* 4:581–585.

Reches, M. and Gazit, E. 2006a. Controlled patterning of aligned self-assembled peptide nanotubes. *Nat. Nanotechnol.* 1:195–200.

Reches, M. and Gazit, E. 2006b. Designed aromatic homo-dipeptides: Formation of ordered nanostructures and potential nanotechnological applications. *Phys. Biol.* 3:S10–S19.

Reches, M., Porat, Y., and Gazit, E. 2002. Amyloid fibril formation by pentapeptide and tetrapeptide fragments of human calcitonin. *J. Biol. Chem.* 277:35475–35480.

Rodriguez-Llansola, F., Miravet, J. F., and Escuder, B. 2009. Supramolecular gel formation and self-correction induced by aggregation-driven conformational changes. *Chem. Commun.* 209–211.

Santoso, S. S., Vautey, S., and Zhang, S. 2002a. Structures, function and applications of amphiphilic peptides. *Curr. Opin. Colloid Interface Sci.* 7:262–266.

Santoso, S., Hwang, W., Hartman, H., and Zhang, S. 2002b. Self-assembly of surfactant-like peptides with variable glycine tails to form nanotubes and nanovesicles. *Nano Lett.* 2:687–691.

Satheeshkumar, K. S., Murali, J., and Jayakumar, R. 2004. Assemblages of prion fragments: Novel model systems for understanding amyloid toxicity. *J. Struct. Biol.* 148:176–193.

Sebyakin, Y. L., Budanova, U. A., and Gur'eva, L. Y. 2007. Self-organising aggregates of lipopeptides in an aqueous medium and their complexes with DNA. *Mendeleev Commun.* 17:188–189.

Seydlová, G. and Svobodová, J. 2008. Review of surfactin chemical properties and the potential biomedical applications. *Cent. Eur. J. Med.* 3:123–133.

Shen, H.-H., Thomas, R. K., Chen, C.-Y., Darton, R. C., Baker, S. C., and Penfold, J. 2009. Aggregation of the naturally occurring lipopeptide, surfactin, at interfaces and in solution: An unusual type of surfactant? *Langmuir* 25:4211–4218.

Shimizu, T. 2008. Molecular self-assembly into one-dimensional nanotube architectures and exploitation of their functions. *Bull. Chem. Soc. Jpn.* 12:1554–1566.

Shinoda, W., De Vane, R., and Klein, M. L. 2008. Coarse-grained molecular modeling of non-ionic surfactant self-assembly. *Soft Matter* 4:2454–2462.

Stendahl, J. C., Rao, M. S., Guler, M., and Stupp, S. I. 2006. Intermolecular forces in the self-assembly of peptide amphiphile nanofibers. *Adv. Funct. Mater.* 16:499–508.

Tsai, W.-W., Li, L.-S., Cui, H., Jiang, H., and Stupp, S. I. 2008. Self-assembly of amphiphiles with terthiophene and tripeptide segments into helical nanostructures. *Tetrahedron* 64:8504–8514.

Tsonchev, S., Niece, K. L., Schatz, G. C., Ratner, M. A., and Stupp, S. I. 2008. Phase diagram for assembly of biologically-active peptide amphiphiles. *J. Phys. Chem. B* 112:441–447.

Valéry, C., Paternostre, M., Robert, B., Gulik-Krzywicki, T., Narayanan, T., Dedieu, J.-C., Keller, G., Torres, M.-L., Cherif-Cheikh, R., Calvo, P., and Artzner, F. 2003. Biomimetic organization: Octapeptide self-assembly into nanotubes of viral capsid-like dimension. *Proc. Natl. Acad. Sci. USA* 100:10258–10262.

Valéry, C., Artzner, F., Robert, B., Keller, G., Gabrielle-Madelmont, C., Torres, M.-L., Cherif-Cheikh, R., and Paternostre, M. 2004. Self-association process of a peptide in solution: From β-sheet filaments to large embedded nanotubes. *Biophys. J.* 86:2484–2501.

Valéry, C., Pouget, E., Pandit, A., Verbavatz, J.-M., Bordes, L., Boisdé, I., Cherif-Cheikh, R., Artzner, F., and Paternostre, M. 2008. Molecular origin of the self-assembly of lanreotide into nanotubes: A mutational approach. *Biophys. J.* 94:1782–1795.

van den Heuvel, M., Löwik, D. W. P. M., and van Hest, J. C. M. 2008. Self-assembly and polymerization of diacetylene-containing peptide amphiphiles in aqueous solution. *Biomacromolecules* 9:2727–2734.

Vass, E., Besson, F., Majer, Z., Volpon, L., and Hollosi, M. 2001. Ca^{2+}-induced changes of surfactin conformation: A FTIR and circular dichroism study. *Biochem. Biophys. Res. Commun.* 282:361–367.

Vauthey, S., Santoso, S., Gong, H., Watson, N., and Zhang, S. 2002. Molecular self-assembly of surfactant-like peptides to form nanotubes and nanovesicles. *Proc. Natl. Acad. Sci. USA* 99:5355–5360.

Velichko, Y. S., Stupp, S. I., and Olvera de la Cruz, M. 2008. Molecular simulation study of peptide amphiphile self-assembly. *J. Phys. Chem. B* 112:2326–2334.

Wennerstroem, H. 1996. Thermodynamic theory of surfactant phases. *Curr. Opin. Colloid Interface Sci.* 1:370–375.

Whitehouse, C., Fang, J., Aggeli, A., Bell, M., Brydson, R., Fishwick, C. W. G., Henderson, J. R., Knobler, C. M., Owens, R. W., Thomson, N. H., Smith, D. A., and Boden, N. 2005. Adsorption and self-assembly of peptides on mica substrates. *Angew. Chem. Int. Ed.* 44:1965–1968.

Xu, J. Z., Moon, S. H., Jeong, B., and Sohn, Y. S. 2007. Thermosensitive micelles from PEGylated oligopeptides. *Polymer* 45:3673–3678.

Yaghmur, A., Laggner, P., Zhang, S., and Rappolt, M. 2007. Tuning curvature and stability of monoolein bilayers by designer lipid-like peptide surfactants. *PLoS ONE* 5:e479.

Yang, S. J. and Zhang, S. 2006. Self-assembling behaviour of designer lipid-like peptides. *Supramol. Chem.* 15:389–396.

Yeh, J. I., Du, S., Tortajada, A., Paulo, J., and Zhang, S. 2005. Peptergents: Peptide detergents that improve stability and functionality of a membrane protein, glycerol-3-phosphate dehydrogenase. *Biochemistry* 44:16912–16919.

Yemini, M., Reches, M., Gazit, E., and Rishpon, J. 2005a. Peptide nanotube-modified electrodes for enzyme-biosensor applications. *Anal. Chem.* 77:5155–5159.

Yemini, M., Reches, M., Rishpon, J., and Gazit, E. 2005b. Novel electrochemical biosensing platform using self-assembled peptide nanotubes. *Nano Lett.* 5:183–186.

Yokoi, H., Kinoshita, T., and Zhang, S. 2005. Dynamic reassembly of peptide RADA16 nanofiber scaffold. *Proc. Natl. Acad. Sci. USA* 102:8414–8419.

You, L. and Gokel, G. W. 2008. Fluorescent, synthetic amphiphilic heptapeptide anion transporters: Evidence for self-assembly and membrane localization in liposomes. *Chem. Eur. J.* 14:5861–5870.

Yu, Y.-C., Berndt, P., Tirrell, M., and Fields, G. B. 1996. Self-assembling amphiphiles for construction of protein molecular architecture. *J. Am. Chem. Soc.* 118:12515–12520.

Yu, Y.-C., Tirrell, M., and Fields, G. B. 1998. Minimal lipidation stabilizes protein-like molecular architecture. *J. Am. Chem. Soc.* 120:9979–9987.

Zandi, R., Reguera, D., Bruinsma, R. F., Gelbart, W. M., and Rudnick, J. 2004. Origin of icosahedral symmetry in viruses. *Proc. Natl. Acad. Sci. USA* 101:15556–15560.

Zhang, S. 2002. Emerging biological materials through molecular self-assembly. *Biotechnol. Adv.* 20:321–329.

Zhang, S. 2003. Fabrication of novel biomaterials through molecular self-assembly. *Nat. Biotechnol.* 21:1171–1178.

Zhang, S. and Altman, M. 1999. Peptide self-assembly in functional polymer science and engineering. *React. Funct. Polym.* 41:91–102.

Zhang, S. and Rich, A. 1997. Direct conversion of an oligopeptide from a β-sheet to an α-helix: A model for amyloid formation. *Proc. Natl. Acad. Sci. USA* 94:23–28.

Zhang, S. and Zhao, X. 2004. Design of molecular biological materials using peptide motifs. *J. Mater. Chem.* 14:2082–2086.

Zhang, S., Marini, D. M., Hwang, W., and Santoso, S. 2002. Design of nanostructured biological materials through self-assembly of peptides and proteins. *Curr. Opin. Chem. Biol.* 6:865–871.

Zhao, W. and Zhang, S. 2004. Fabrication of molecular materials using peptide construction motifs. *Trends Biotechnol.* 22:470–476.

Zhao, X. and Zhang, S. 2007. Designer self-assembling peptide materials. *Macromol. Biosci.* 7:13–22.

Zhao, X., Nagai, Y., Reeves, P. J., Kiley, P., Khorana, H. G., and Zhang, S. 2006. Designer short peptide surfactants stabilize G protein-coupled receptor bovine rhodopsin. *Proc. Natl. Acad. Sci. USA* 103:17702–17712.

Zhou, Y., Kogiso, M., and Shimizu, T. 2009. Necklace-like chains of hybrid nanospheres consisting of Pd nanocrystals and peptidic lipids. *J. Am. Chem. Soc.* 131:2456–2457.

4 Physicochemical Characterization of Anisotropic Molecules (Polyelectrolytes and Proteins) in Bulk Solutions

Barbara Jachimska

CONTENTS

4.1 INTRODUCTION

The process of folding in single polymer chains has been attracting interest from biological and physicochemical points of view. In biological science, it is known that long DNA condense into a self-organized folded structure with various kinds of morphologies such as toroid, rod, and spherical. Such a collapse transition, or coil-globule transition, of DNA has been discussed in connection with the tight packing of long DNA chains in phage capsids, and with compact folding in living cells (Berg et al. 2002). However, due to the theoretical limitation of self-consistent field theory for dilute or isolated polymer chain, the intrinsic property of the coil-globule transition still remains unclear. For example, some theoretical studies (Lifshitz et al. 1978) predict the discontinuous nature of the coil-globule transition in a stiff polymer. However, almost all experimental studies have reported a continuous character of the transition of polyelectrolytes including DNA chains. DNA represents semiflexible polyelectrolyte with rigidity that originates from secondary structure stabilized by hydrogen bond and electrostatic interaction that is affected by the Coulomb repulsion, which gives the sum of the persistent length of 50 nm in aqueous environment.

Synthetic polyelectrolyte molecules are flexible and rapidly respond to the change of the charge density resulting in a continuous cascade of coil-globule transitions in the presence of condensation agents (Kiriy et al. 2002, Minko et al. 2002a, Kirwan et al. 2004, Roiter and Minko 2005, Gromer et al. 2008). These synthetic polyelectrolytes are important in many industrial applications. For example, poly(methacrylic acid) (PMA) shows a marked pH-induced conformation transition. This process has been studied by different techniques: viscosimetry titration (Noda et al. 1970), potentiometric titration (Mandel et al. 1967), fluorescent probing (Chu and Thomas 1987), calorimetry (Crescenzi et al. 1972), and Raman spectroscopy (Koenig et al. 1969). The data suggest that at low pH, the macromolecule adopts a hypercoiled form. At a high degree of ionization (higher pH), and in the absence of electrolyte, the PMA chain stretches to a rodlike form. However, the nature of this transition is still controversial. Some authors suggest that the transition is highly cooperative and occurs in one step (Anufrieva et al. 1968), while data from Raman spectroscopy indicate a multiplicity of the structure. Minko et al. (2002) reported the visualization of different conformations (from an elongated wormlike coil to compact globule) of single poly(2-vinylpyridine) (P2VP) using atomic force microscopy (AFM). The conformation can be frozen by metallic nanoparticle assemblies at polyelectrolyte molecules.

The adsorption of polyelectrolytes onto surface is a phenomenon that has attracted the interest of polymer scientists for several decades. Electrostatic interaction between oppositely charged macromolecules is a base of the layer-by-layer (LbL) assembly technique that allows the fabrication of multilayer films from synthetic polyelectrolytes, DNA, proteins, macromolecules, or nanoparticles (Decher et al. 1992). A typical experimental procedure involves the exposure of a solid substrate to dilute solutions of positively or negatively charged species for a period of time optimized for their adsorption. Further film growth is achieved by alternating deposition of polyanions and polycations from their solutions. Inside the film of the oppositely charged molecules, aggregates are formed from multichain complexes (Lvov et al. 1993). The bottom-up strategy for the fabrication of soft nanostructured materials is poised to find practical applications in many areas. Recently, the scope of the multilayer LbL technique has been extended to such materials as polysaccharides, proteins, polynucleotides, enzymes, polypeptides, and various other biopolymers, with the aim of creating functional biocompatible surfaces (Thierry et al. 2003). The adsorption of polyelectrolyte onto charged patterned surface is important for the development of biosensor, the understanding of protein complexation, or the design of patterned polyelectrolyte multilayer. The complex formation between proteins (histones) and DNA is another example of the interaction between a polyelectrolyte (DNA) and a charged patterned surface. LbL adsorption of polyelectrolytes onto charged surface has been extensively studied by Decher and coworkers (1992, 1994, 1997). These experimental studies concern multichain adsorption, and the results are reported in terms of the amount of polyelectrolyte adsorbed. Since the bottom-up approach utilized in the self-assembly of these nanostructures and multilayers begins with the adsorption of polyelectrolyte molecules onto a substrate, which will affect the conformation of adjacent structures and layers, adsorption processes could be better controlled if one could manipulate the conformation of individual adsorbed molecules.

4.2 CHARACTERIZATION OF POLYMER IN BULK SOLUTION

Polyelectrolytes are macromolecules with many ionizable groups that, in aqueous solution and under appropriate conditions, dissociate into ionized charged groups on the polyion backbone and counterions in the bulk solution (Hara 1993). Interaction between the polyions and the counterions, which definitely characterize the polyelectrolyte character in the system, may be described by means of various coefficients. One of the significant parameters is the electrical conductivity, which takes into account the movement of any charged entity present in aqueous system. The equivalent conductivity of polyelectrolyte solution is given by

$$\Lambda = \frac{\kappa - \kappa_0}{c} = f_c(\lambda_p + \lambda_c^0) \tag{4.1}$$

where
κ and κ_0 are the specific conductivity of the solution and the solvent
λ_c^0 is the equivalent conductivity of the counterion in an infinitely diluted solution in the absence of the polyions

λ_p is the equivalent conductivity of the polyion

f_c is the interaction parameter, which includes the electrostatic interactions among the polyions, the counterions, and the degree of ionization

The various definitions of f_c result from the difference in the theoretical models. On the basis of counterion condensation theory the Manning theory was established. In this theory, a polyion chain is defined of contour length l of a linear conformation (infinitely thin) of N monomers, each of them bearing a charge $|z_p|e$ with counterions of charge $|z_c|e$ partially condensed onto the polyion and partially noncondensed and distributed in the neighborhood of the polyion according to the linearized Poisson–Boltzmann equation (Manning 1969). In the limit of very high dilution (any polyion–polyion interaction is neglected), the conductivity behavior of the whole solution is characterized by the charge density parameter q defined as

$$q = \frac{e^2}{\varepsilon k T l_c} \tag{4.2}$$

where

e is the elementary charge

ε is the dielectric constant

kT is the thermal energy

$l_c = l/N$ is the average distance between charged groups on the polyion chain

Manning has calculated f_c in salt-free solutions for charge distance l_c less than the Bjerrum length l_B, and consequently

$$f_c = 0.866 \left(\frac{l_B}{l_c} \right)^{-1} \tag{4.3}$$

Manning assumes that the nonbounded counterions are influenced by the Debye–Huckel potential of the polyion. The fraction of the condensed counterions is $1 - q^{-1} \neq 1 - f_c$. Hence, f_c cannot be understood as the fraction of free counterions. In the condensation theory of Manning, it has been suggested that the condensed counterion are immobilized, but recent model assigned that the counterions are mobile. The conductance–anisotropy effect clearly indicates that the atmospherically bound (condensed) ions retain a high degree of translational mobility parallel to the polyion chain (Vink 1995). Most of the existing theories are based on the assumption that the effective charge on the polyion chain does not change with the polymer concentration. Bordi et al. (1999) develop the Manning model, taking into account the different chain conformations that dominate in the low concentration limit from those that become prominent with increasing concentration (semidilute and concentrated regimes). They found scaling prediction in good agreement with experimental data of electrical conductivity on dilute and semidilute poly(L-lysine) solution (Bordi et al. 2002). In the recent work of Muthukumar (2004) and Beer et al. (1997), it was shown that the degree of ionization of the polymer decreases continuously with $1/\varepsilon T$, depending sensitively on the local dielectric heterogeneity. The degree of counterion condensation increases with an increase in salt concentration, monomer concentration, or chain flexibility. These predictions were consistent with all the trends observed earlier in simulations (Stoll and Chodanowski 2002), and were distinctly different from the Manning argument for rodlike chains.

The comparison of experimental results with theoretical predictions reveals only a partial agreement. However, the deviations between experiment and theory often result from the fact that the experimental conditions are not consistent with theoretical assumptions. The most frequent discrepancies are that the concentrations are far from being high diluted, water quality does not corresponding to "salt-free" conditions, and the samples are polydispersed.

Electrophoresis measurements seem to be the most frequently exploited electrokinetic technique. The electrophoretic mobility of a molecule is determined by the balance between the electrical and viscosity forces acting on a charged colloid particle placed in an applied electric field. Knowing the mobility, one can calculate the mobile charge on the polymer molecule using the basic electrostatic relationship (Adamczyk 2006b)

$$\mathbf{F} = q\,\mathbf{E} \tag{4.4}$$

where

$\mathbf{F}$ is the force acting on the molecule
q is the effective charge
$\mathbf{E}$ is the electric field

Considering that particle velocity $\mathbf{U}$ due to the force $\mathbf{F}$ is defined as $\mathbf{U} = \mathbf{MF}$, one can convert Equation 4.4 to the form

$$\mathbf{U} = q\,\mathbf{ME} \tag{4.5}$$

were $\mathbf{M}$ is mobility matrix of molecule.

For uniform electric fields, Equation 4.5 can be rearranged to the form

$$q = \frac{\langle U \rangle}{\mu_e E} = \frac{kT}{D}\mu_e = 6\pi\eta R_H \mu_e \tag{4.6}$$

where

μ_e is the polymer mobility
$\mu_e = \langle U \rangle / E$ and $\langle U \rangle$ is the averaged migration velocity of polymer molecules in the uniform electric field E
R_H is the hydrodynamic radius
η is the dynamic viscosity of fluid

Equation 4.6 can be directly used for calculations of the average number of elementary charges per molecule $N_c(e)$ considering that $e = 1.602 \times 10^{-19}\,C$, thus

$$N_c = \frac{6\pi\eta}{1.602 \times 10^{-19}} R_H \mu_e \tag{4.7}$$

Accordingly, the effective ionization degree α_i^* of the molecule is given by

$$\alpha_i^* = \frac{N_c}{N_m} = \frac{6\pi\eta}{1.602 \times 10^{-19}} \frac{R_H \mu_e}{N_m} \tag{4.8}$$

where $N_m(e)$ is the nominal number of charges per molecule.

As showed in Yoon and Kim (1989), Equation 4.6 remains valid for spheroids and cylinders if $2\kappa d_e < 1$, where κ^{-1} is the electric double layer thickness, given by the expression

$$\kappa^{-1} = \left(\frac{\varepsilon\, kT}{2e^2 I^*} \right)^{1/2} \tag{4.9}$$

where I^* is the overall ionic strength of the polyelectrolyte solution, which can be expressed as (Rushing and Hester 2004)

$$I^* = \frac{1}{2} \sum z_i n_i^2 + \frac{1}{2} N_c Z_c^2 n_p \qquad (4.10)$$

where
z_i is the valency of the ions of the background electrolyte
n_i is their concentration in the bulk
N_c is the number of free charges per molecule
Z_c is the valency of the counter ions
n_p is the concentration of the polyelectrolyte in the solution

Another important parameter characterizing polymer solution dynamics, especially the rate of deposition on interfaces, is the diffusion coefficient D, which is accessible experimentally from the dynamic light scattering (DLS) measurements. However, a proper interpretation of these measurements is complicated by the fact that the D (evaluated from the autocorrelation function) is dependent in a strongly nonlinear manner on polymer concentration and ionic strength of the solution (Koene and Mandel 1983). This is because of the highly elongated shape, the pair interactions between chains become important for bulk polymer concentration above 100 ppm, especially for low ionic strength of the solution (Koene and Mandel 1983). Usually, the repulsion between the chains leads to significant increase in the apparent diffusion coefficient with polymer concentration (Koene and Mandel 1983, Sedlak and Amis 1992).

The experimental data, in the dilute polymer solution limit, can well be interpreted in terms of simple hydrodynamic considerations derived for creeping flow conditions. This seems to be justified because flows associated with polyelectrolyte motion in solutions are characterized by very low Reynolds number $Re = LV/\nu$ (where L is the characteristic length scale of the motion, V is the characteristic velocity, and ν is the kinematic viscosity of the solution). It can be easily deduced that for the nanometer length scale involved and aqueous solution (ν of the order of $0.01\ cm^2 s^{-1}$ and less), the Reynolds number remains much smaller than unity even for the molecule translation velocities of the order of $m\,s^{-1}$. As a consequence of low Re number, also the relaxation time of establishing the stationary conditions of the flow distribution around a moving polyelectrolyte molecules remains exceedingly small. The relaxation time τ_h can be estimated from the simple formula

$$\tau_h = \frac{L^2}{\nu} \qquad (4.11)$$

For comparison, the mass diffusion relaxation time τ_d can be estimated from the formula

$$\tau_d = \frac{L^2}{D} \qquad (4.12)$$

Thus, the ratio of these two times equals (Equations 4.11 and 4.12) D/ν that is a very small quantity because, as discussed above, D is of the order of $10^{-7} cm^2 s^{-1}$. In other words, because of much faster momentum diffusion in comparison with mass diffusion, the flow distribution is adjusting instantaneously to the polyelectrolyte molecule motion. As a consequence, its resistance coefficient could well be evaluated by using the geometrical shape of the molecule.

For spheroids, as well as for similar bodies having symmetry like, e.g., cylinders, bent spheroids, disks, rings, etc., the translation and rotation diffusion matrices assume the form (Brenner 1974, van de Ven 1989)

$$\mathbf{D} = \frac{kT}{\eta}\mathbf{M}^{-1} = \frac{kT}{\eta}\begin{vmatrix} 1/K_{\parallel} & & 0 \\ & 1/K_{\perp} & \\ 0 & & 1/K_{\perp} \end{vmatrix} \tag{4.13}$$

$$\mathbf{D_r} = \frac{kT}{\eta}\mathbf{M_r}^{-1} = \frac{kT}{\eta}\begin{vmatrix} 1/K_{r\parallel} & & 0 \\ & 1/K_{r\perp} & \\ 0 & & 1/K_{r\perp} \end{vmatrix}$$

where

$\mathbf{M}$ and $\mathbf{M_r}$ are the translation and rotation mobility matrices

$K_{\parallel}$, $K_{\perp}$ are the hydrodynamic resistance coefficients for the translation motion along the symmetry axes

$K_{r\parallel}$, $K_{r\perp}$ are the coefficients for the rotary motion

The average translation diffusion coefficient (scalar), which can be determined experimentally by, e.g., the DLS, is given by (Brenner 1974, Koene and Mandel 1983)

$$\langle D \rangle = \frac{kT}{3\eta}\left(\frac{1}{K_{\parallel}} + \frac{2}{K_{\perp}} \right) = \frac{kT}{6\pi\eta R_H} \tag{4.14}$$

where the quantity

$$R_H = \frac{1}{2\pi\left(\dfrac{1}{K_{\parallel}} + \dfrac{2}{K_{\perp}} \right)} \tag{4.15}$$

is usually defined as the hydrodynamic radius of the body for translational motion.

The coefficients $K_{\parallel}, K_{\perp}$, $K_{r\parallel}, K_{r\perp}$, and R_H are known in analytical form for prolate and oblate spheroids. They are explicitly given by (Harding 1995)

$$K_{\parallel} = \frac{16\pi a}{\lambda^2(2\beta + \alpha_{\parallel})} \qquad\qquad K_{\perp} = \frac{16\pi a}{2\lambda^2\beta + \alpha_{\perp}}$$

$$K_{r\parallel} = \frac{2}{3\alpha_{\perp}}v_s \qquad\qquad K_{r\perp} = \frac{2(\lambda^2 + 1)}{3(\lambda^2\alpha_{\parallel} + \alpha_{\perp})}v_s \tag{4.16}$$

$$R_H = \frac{a}{\lambda^2\beta}$$

where λ is the aspect ratio parameter, which equals $a/b > 1$ for prolate spheroids and $b/a < 1$ for oblate spheroids (a is the longer semiaxis and b the shorter semiaxis of spheroids whose volume v_s equals $\frac{4}{3}\pi ab^2$ for prolate and $\frac{4}{3}\pi a^2 b$ for oblate spheroids).

$$\alpha_{\parallel}(\lambda) = \frac{2}{\lambda^2 - 1}(\lambda^2\beta - 1) \qquad \alpha_{\perp}(\lambda) = \frac{\lambda^2}{\lambda^2 - 1}(1 - \beta) \tag{4.17}$$

$$\beta = \frac{\cos^{-1}\lambda}{\lambda(1 - \lambda^2)^{1/2}} \qquad \beta = \frac{\cosh^{-1}\lambda}{\lambda(\lambda^2 - 1)^{1/2}} = \ln\left[\lambda + \left(\lambda^2 - 1\right)^{1/2}\right] \tag{4.18}$$

oblate spheroids   prolate spheroids

Interesting limiting forms can be derived from the above equations for prolate spheroids when $\lambda \gg 1$ (slender bodies limit). In this case, the expression for the hydrodynamic radius (Equation 4.15) becomes

$$R_H = \frac{L_e}{2\left(c_1 \ln \lambda - c_2\right)} \tag{4.19}$$

where
  $L_e = 2a$ is the characteristic length of the spheroids
  $c_1 = 1$, $c_2 = 0$ are coefficients

It was shown that Equation 4.19 applies for other slender bodies (van de Ven 1989), with the coefficients $c_1 = 1$, $c_2 = 0.11$ for cylinders (rods), $c_1 = 11/12$, $c_2 = 0.31$ for bend spheroids forming a semicircle (semi-torus), and $c_1 = 11/12$, $c_2 = 1.20$ for spheroids bent to the form of a torus (ring).

It can be also noticed from Equation 4.19 that the hydrodynamic radius increases for $\lambda \gg 1$ proportionally to $L_e/\ln L_e$, which is a function rather insensitive to the particle linear dimension (contour length of the polyelectrolyte molecule).

In Table 4.1, the expressions for calculating the hydrodynamic radius of molecules of various shapes are collected. R_H calculated for poly(allylamine hydrochloride) (PAH) and poly(sodium 4-styrenesulfonate) (PSS) polyelectrolytes are collected in Tables 4.2 and 4.3. One can notice that R_H (resistance coefficient) for a particle having the shape of a rod is the smallest. Bending the object leads to the increase in R_H and in consequence to the increase in the hydrodynamic resistance coefficient. This allows one to interpret quantitatively the experimental data obtained for various ionic strength of the polymer solution. In the case of low ionic strength, the polymer chain is expected to be rather extended [as the numerical simulations suggest (Fuoss 1951, Dobrynin and Rubinstein 2005)], resembling a straight rod. On the other hand, for increased ionic strength the polyelectrolyte molecule will bent, thus the experimental R_H value corresponds closely to the semicircle case. It is interesting to note that in accordance with numerical simulations, the experimental R_H value never remains smaller than the theoretical value predicted for circle.

Intrinsic viscosity is extensively used for the analysis or characterization of synthetic polymer, biological macromolecules, nanoparticles, and colloids (Yamanaka et al. 1990, Rushing and Hester 2004, Adamczyk et al. 2006, Jachimska et al. 2010). Indeed, viscosity provides information about fundamental properties of the solute and its interaction with the solvent that can be used for theoretical or computation simulations. It can be precisely related to the conformation of the flexible (linear and nonlinear) chains, wormlike macromolecules, and rigid particles of different shape (Adamczyk et al. 2004b, Jachimska and Adamczyk 2007).

However, because of the variety of parameters influencing polyelectrolyte viscosity, these results are often misinterpreted using, for example, the scaling theories of de Gennes et al. (1976) applicable only for high molecular weight polyelectrolytes in solutions with no added salt. A better description

TABLE 4.1

Analytical and Approximate Expressions for R_H for Particles of Various Shapes

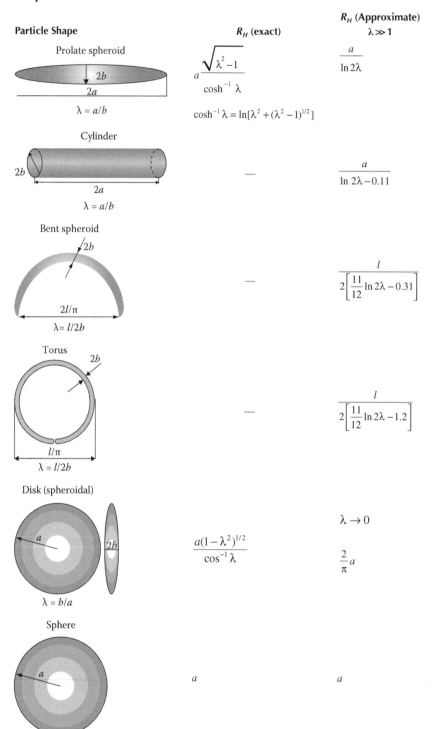

Particle Shape	R_H (exact)	R_H (Approximate) $\lambda \gg 1$
Prolate spheroid $\lambda = a/b$	$a\dfrac{\sqrt{\lambda^2-1}}{\cosh^{-1}\lambda}$ $\cosh^{-1}\lambda = \ln[\lambda^2 + (\lambda^2-1)^{1/2}]$	$\dfrac{a}{\ln 2\lambda}$
Cylinder $\lambda = a/b$	—	$\dfrac{a}{\ln 2\lambda - 0.11}$
Bent spheroid $\lambda = l/2b$	—	$\dfrac{l}{2\left[\dfrac{11}{12}\ln 2\lambda - 0.31\right]}$
Torus $\lambda = l/2b$	—	$\dfrac{l}{2\left[\dfrac{11}{12}\ln 2\lambda - 1.2\right]}$
Disk (spheroidal) $\lambda = b/a$	$\dfrac{a(1-\lambda^2)^{1/2}}{\cos^{-1}\lambda}$	$\lambda \to 0$ $\dfrac{2}{\pi}a$
Sphere	a	a

TABLE 4.2
Values of the Hydrodynamic Radius, R_H,
Predicted for Various PAH ($M_w = 15,000$) and PSS
($M_w = 15,800$) Molecule Shapes

Assumed Molecule Shape	PSS 15,800 R_H (nm)	PAH 15,000 R_H (nm)
Prolate spheroids	2.55	4.82
Cylinder	2.64	4.95
Bent spheroid	3.13	5.73
Torus (ring)	4.83	7.63
Equivalent sphere	1.54	1.75

of viscosity of polyelectrolytes in real solutions can be attained using the electrostatic wormlike chain theory developed by Odjik (1977) and Skolnick and Fixman (1977), known as the Odjik–Skolnick–Fixman (OSF) theory. This approach was based on the concept of persistence length (L_p), whose electrostatic contribution was calculated by solving the linearized Poisson–Boltzmann equation for a uniform line charge. It was predicted that because of electrostatic repulsion, this electrostatic component of L_p increases proportionally to the square of Debye screening length κ^{-1}, i.e., inversely proportional to the ionic strength I of polyelectrolyte solution. The L_p expression has been corrected by Davis and Russel (1987) who calculated the correction valid for arbitrary κd_e (where d_e is the polymer chain diameter).

By exploiting the concept of the L_p, Yamakawa and Fujii (1974) formulated a viscosity theory for the wormlike polymers, which predicts that the intrinsic viscosity increases as $L_p^{3/2}$, i.e.,

TABLE 4.3
Values of the Hydrodynamic Radius R_H
Predicted for Various PAH ($M_w = 70,000$) and PSS
($M_w = 70,000$) Molecule Shapes

Assumed Molecule Shape	PAH 70,000 R_H (nm)	PSS 70,000 R_H (nm)
Prolate spheroids	16.09	9.44
$2b$ $2a$ $\lambda = a/b$		
Cylinder	16.40	9.66
$2b$ $2a$ $\lambda = a/b$		
Bent spheroid	18.63	11.07
$2b$ $2l/\pi$ $\lambda = l/2b$		
Torus (ring)	22.66	14.13
a $\lambda = l/2b$		
Equivalent sphere	2.9	2.86
a		

proportionally to the cube of the Debye screening length. The OSF theory and Yamakawa–Fujii model was modified by Rushing and Hester (2004), who proposed a semiempirical model, which correlates the intrinsic viscosity of polyelectrolytes with its molecular weight and the screening length in the limit of high electrolyte concentration. They found, by analyzing extensively existing experimental data for various polyelectrolytes, that the slope of the curve of the intrinsic viscosity increase with the screening length.

The wormlike model was modified by Davis and Russel (1987) who considered the excluded volume effect, which introduced additional stiffness of the polymer chain. This allowed them to renormalize the L_p occurring in the Yamakawa–Fujii model, which properly reflects the effect of the molecular weight and the ionic strength on the intrinsic viscosity for potassium poly(sodium 4-styrenesulfonate) solutions.

Dobrynin et al. (1995) proposed a necklace model of polyelectrolyte chain, in which its conformations were resulting from intrachain electrostatic interactions and counterion condensation on the polyelectrolyte backbone (Jeon and Dobrynin 2007). The necklace structure was obtained as balance of the correlation-induced attraction of the condensed counterions, charged monomers, and electrostatic repulsion between uncompensated charges. According to this model, the transitions of polymer conformations (chain-necklace-globule) depend on the value of the Bjerrum length, determining the strength of the electrostatic interactions and the value of the Lennard–Jones interaction parameter, which controls the solvent quality for the polymer backbone.

The intrinsic viscosity data for the wormlike regime can be quantitatively interpreted in terms of the hydrodynamic model proposed originally by Kirkwood and Auer (1951) who assumed an extended, rodlike shape of polyelectrolytes. In the more general case of ellipsoidal or spheroidal shape of macromolecules, one can use the well-established results of Brenner (1974) and Harding (1995) who derived analytical solutions for the intrinsic viscosity as a function of the axis ratio parameters. However, these theoretical results can only be applied if the shape of the molecule is known as a function of various physicochemical parameters.

Harding (1995) described five modeling strategies concerning the behavior of macromolecular conformation in solution such as the following: ellipsoids of revolution, general triaxial ellipsoid, bead models, general structures, and flexibility analysis.

4.3 CONFORMATIONAL TRANSITION IN POLYELECTROLYTE MOLECULE IN SOLUTION

Two kinds of polyelectrolytes were selected for presentation of conformational transition in polyelectrolyte molecules. Methodology of these measurements was presented only in few publications (Adamczyk et al. 2004, 2006). PAH and PSS sodium having low molecular weight (15,000) and higher molecular weight (70,000), respectively, were chosen (Adamczyk et al. 2008, Jachimska et al. 2010). These polymers represented weak and strong polyelectrolytes. Moreover, these polymers are very often used in the formation of multilayer film (Decher et al. 1992).

The specific density ρ_p of PAH and PSS in the crystalline state determined by the density matching method was 1.15 and 1.18 g cm^{-3}, respectively (Adamczyk et al. 2004). Thus, the molecular volume $v_m = M/\rho_p A_v$ (where M is the molecular mass of the polymer and A_v is the Avogadro number) of PAH and PSS equals to 22.0 and 22.3 nm^3, respectively. For calculation of the extended length of polyelectrolytes L_e, a parameter of primary importance, the effective chain cross section should be estimated. The calculations performed by Donath et al. (1997) using the molecular modeling package Insight II suggested that the equivalent diameter of the PAH chain in the isotactic configuration equals to 1.0 nm, which gives the cross-section area of 0.78 nm^2. On the other hand, in the syndiotactic configuration the effective cross-section area of the PAH chain (assumed to resemble an ellipse with the axis length equal to 0.51 and to 0.87 nm) was 0.35 nm^2. Hence, the average value of both configurations is 0.56 nm^2. Analogously, the cross-section shape of the PSS chain, most probable being of syndiotactic configuration, was found to be 0.96 nm^2. Using dynamic molecular simulation the effective diameter of the PAH and PSS was calculated as being 0.90 and 1.17 nm, respectively (Jasiński et al. 2007). This gives the effective chain cross-section area equal to 0.54 nm^2 for PAH and to 1.07 nm^2 for PSS. As can be noticed, these values agree quite well with the calculations of Donath et al. (1997). Using these cross-section areas one can calculate the extended chain length as

TABLE 4.4
Physicochemical Characteristics of PAH
($M_w = 15,000$) and PSS ($M_w = 15,800$),
$T = 293\,K$

Property	Polyelectrolyte	
	PAH	PSS
Density (g/cm³)	1.15	1.18
Molecular weight	15,000	15,800
Molecular weight monomer	93.5	206.2
Number of monomers/charges	160	77
Volume per molecule (nm³)	22	22.3
Volume per monomer (nm³)	0.135	0.29
Bare cross section (nm²)	0.54	1.07
Extended length L (nm)	40.7	16.0
Equivalent chain diameter d_e (nm)	0.90	1.17
Hydrated chain diameter	1.20	1.39
$\quad d_h$ (nm) $= d_e + 2 \times 0.145$ (nm)		
Aspect ratio $\lambda_e = L/d_e$	45.2	14.5
$\lambda_h = L/d_h$	34	11.5
R_g (nm)	11.9	4.6
R_S (nm)	1.75	1.54

being 40.7 and 16.0 nm for PAH and PSS, respectively. Hence, the length-to-width ratio, λ, of the polymers was 45.2 and 14.5 for PAH and PSS, respectively. This axis ratio parameter is of primary importance for predicting the hydrodynamic behavior of these molecules. Another length scale of primary importance for polymer solutions is the radius of gyration of a chain that can be calculated in the limit of a rigid rod shape from the simple formula $R_g = \sqrt{L_e^2/12}$ (Mandel 1993). Using this equation one obtains $R_g = 11.9$ and 4.6 nm, for PAH and PSS, respectively. It also can be of interest to compare these length scales with the equivalent sphere radius $R_s = \sqrt{3v_m/4\pi}$, i.e., the radius of the sphere having the same volume as the polymer molecule. It was found equal to 1.75 and to 1.54 nm for PAH and PSS, respectively. For the sake of convenience, all the relevant parameters determined for both polymers have been collected in Table 4.4 (PAH and PSS low molecular weight) and Table 4.5 (PAH and PSS high molecular weight).

The hydrodynamic radius determined by the DLS for PAH molecules was obtained by extrapolation of the dependence of the diffusion coefficient on the suspension concentration (measured between 500 and 5000 ppm) to zero concentration. This procedure is more accurate than direct measurements of the diffusion coefficient of PAH in the limit of low concentrations. R_H values varied between 4.8 nm at pH = 6.0 and 6.0 nm at pH = 9.5 for ionic strength 1×10^{-2} M. With increased ionic strength, the measured value of R_H increased as well, becoming 6.15 nm for pH = 6.0 and 6.5 nm for pH = 9.5 at $I = 0.15$ M. These experimental data are consistent with a hydrodynamic model postulating a flexible rod behavior of the chain in electrolyte solutions. Hence, at low ionic strength, when the molecule is fully charged, it assumes an extended shape, and its hydrodynamic resistance coefficient can be approximated by the straight rod with aspect ratio, λ, close to 36 (or to spheroid). On the other hand, for high electrolyte concentration, the charge of the molecule is compensated by the counterions, and it assumes the shape resembling a circular ring (torus).

The value R_H calculated from Equation 4.19 for PAH are shown in Table 4.5. It can be seen that R_H (resistance coefficient) for a particle having the shape of a rod is the smallest. Bending the object leads to an increase in R_H and, consequently, to an increase in the hydrodynamic resistance coefficient. In case of PAH, a molecule that is a weak polyelectrolyte, it is also possible to observe

TABLE 4.5
Values of the Hydrodynamic Radius R_H Predicted for Bare and Hydrated PAH ($M_w = 15,000$) Molecule of Various Shapes

Assumed Molecule Shape	Bare Molecule $d = 0.90$ nm, $\lambda = 45.2$ R_H (nm)	Hydrated Molecule $d = 1.20$ nm, $\lambda = 34$ R_H (nm)
Prolate spheroids	4.52	4.82
Cylinder	4.63	4.95
Bent spheroid	5.33	5.73
Torus (ring)	6.97	7.63

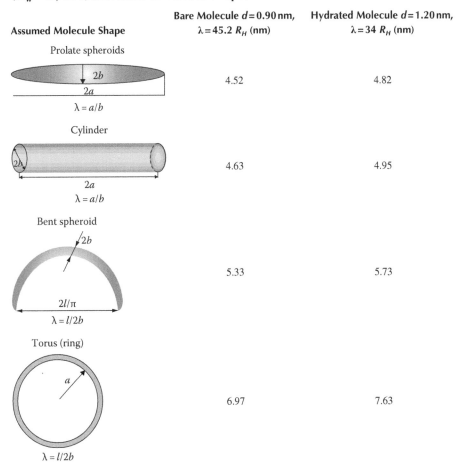

changes of conformation depending on the pH of the solution, which reflects the degree of ionization of PAH molecules.

Other bulk characteristics of polyelectrolyte solutions, which have basic significance for interpretation of their adsorption and deposition phenomena, is the effective charge, determined as a function of pH and ionic strength. These data can be directly derived from the electrophoretic mobility measurements. The experimental results are shown in Figure 4.1 in the form of the dependence of μ_e on pH of the PAH solution for two ionic strength 1×10^{-2} M and 0.15 M. Electrophoretic mobility, μ_e, decreases monotonically from 3.8×10^{-8} m^2 (V s)$^{-1}$ at pH = 6.5 to 0.0 m^2 (V s)$^{-1}$ at pH = 10.8 for $I = 1 \times 10^{-2}$ M. A zero value of the electrophoretic mobility is usually referred to as the izoelectric point (i.e.p.). Moreover, they can be exploited for estimating the number of uncompensated (free) charges on the polyelectrolyte chains as a function of the ionic strength. Knowing the electrophoretic mobility and the hydrodynamic radius, one can calculate the effective $N_c(e)$ (uncompensated) charge of PAH molecules q using the Lorenz–Stokes equation (Equation 4.6).

Substituting experimental data, i.e., $\mu_e = 3.8 \times 10^{-8}$ m^2 (V s)$^{-1}$, $R_H = 4.8$ nm ($I = 10^{-2}$ M), one obtains $N_c = 27.0$ (e) as the average number of free charges (of positive sign) per PAH molecule. This gives the effective ionization degree $\alpha_i^* = 0.17$ (considering that the nominal number of charges equals

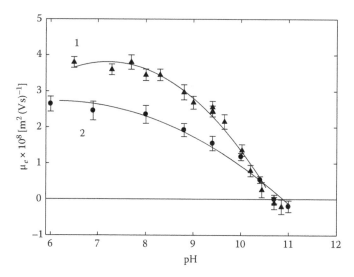

FIGURE 4.1 The dependence of the microelectrophoretic mobility $\mu_e \times 10^8 \, m^2 \, (V \, s)^{-1}$ on pH determined experimentally for PAH solution: 1. $I = 1 \times 10^{-2} \, M$, 2. $I = 0.15 \, M$.

to 160 (e)). For $I = 0.15 \, M$, taking $\mu_e = 2.50 \times 10^{-8} \, m^2 \, (V \, s)^{-1}$ and $R_H = 6.15 \, nm$, one can calculate $N_c = 20.8$ (e) as the maximum number of mobile charges and $\alpha_i^* = 0.14$.

The nominal ionization degree derived from Equation 4.8 is compared in Figure 4.2. The effective ionization degree is very strongly dependent on pH and remains much smaller than unity, which indicates that the nominal charge of PAH molecules is to a considerable extent compensated. This effect can be explained by specific adsorption of counterions (Na$^+$ in our case), which is often referred to as the ion condensation phenomenon (Manning 1969). According to Manning–Ossawa theory (Manning 1969, Ossawa 1970), counterion condensation occurs if the Manning parameter $q_0 = l_B/l_c$ is greater than unity. Here, $l_B = e^2/(4\pi\varepsilon kT)$ the Bjerrum's length (defined as the length scale

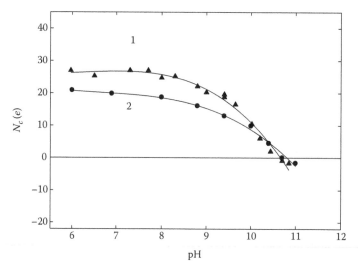

FIGURE 4.2 Number of uncompensated changes $N_c(e)$ of PAH on pH calculated from the experimental results derived from microelectrophoretic measurements using Equation 4.8: 1. $I = 1 \times 10^{-2} \, M$, 2. $I = 0.15 \, M$.

at which the Coulomb interaction between two elementary charges e in a dielectric medium with the dielectric constant ε is equal to the thermal energy kT), for our experimental conditions, equals to 0.72 nm; and l_c (the average distance between charged groups at the fully extended polymer chain) equals to 0.22 nm for PAH molecule.

The effective ionization degree of PAH using Fourier transform infrared spectroscopy was reported by Choi and Rubner (2005). They estimated that the pK_a value (the pH at which 50% of the polymer functional groups are ionized) of PAH was 8.8. However, the values reported in their work varied from 0.9 at pH = 6.0 to 0 at pH 11.3.

The same procedure was used for PSS 15,000 Mw. The hydrodynamic radius determined by DLS for PSS sample was 3.1 nm (for pH = 6.5 and ionic strength 5×10^{-3} M). This value was obtained by extrapolation of the dependence of the diffusion coefficient on the suspension concentration (measured between 100 and 5000 ppm) to zero concentration. With the increase of the ionic strength, the measured value of R_H increased, becoming 3.4 nm for $I = 10^{-2}$ M and 4.0 nm for $I = 0.15$.

Electrophoretic mobility, μ_e, measurements of PSS molecules are shown in Figure 4.3. As can be seen, μ_e was fairly independent of pH (changed within 6.5–10) and assumed as an averaged value of -3.8×10^{-8} m^2 (V s)$^{-1}$ for $I = 10^{-2}$ M and -2.7×10^{-8} m^2 (V s)$^{-1}$ for $I = 0.15$ M. It can be calculated from Equation 4.9 that for $T = 293$ K and $I = 10^{-3}$ M, $\kappa^{-1} = 9.6$ nm, which gives $\kappa\, d_e = 0.11$. For $I = 10^{-2}$ M, $\kappa^{-1} = 3.4$ nm, thus $\kappa\, d_e = 0.32$. Because these values are smaller than unity, one can expect that Equations 4.7 and 4.8 are fully applicable. Substituting experimental data, i.e., $\mu_e = -3.8 \times 10^{-8}$ m^2 (V s)$^{-1}$, $R_H = 3.1$ nm ($I = 10^{-2}$ M), one obtains $N_c = 15.2$ (e) as the average number of free charges (of negative sign) per PSS molecule. This gives the effective ionization degree $\alpha_i^* = 0.20$ (considering that the nominal number of charges equals 77 (e)) for $I = 0.15$ M, taking $\mu_e = -1.50 \times 10^{-8}$ m^2 (V s)$^{-1}$ and $R_H = 4.0$ nm, which gives $N_c = 14.5$ (e) as the maximum number of mobile charges and $\alpha_i^* = 0.19$. The effective ionization degree is practically independent of ionic strength and pH and remains much smaller than unity, which indicates that the nominal charge of PSS molecules is to a considerable extent compensated (Figure 4.4). This same effect was observed for PAH. The average distance between nominal charges $l_c = 0.21$ nm for PSS sample.

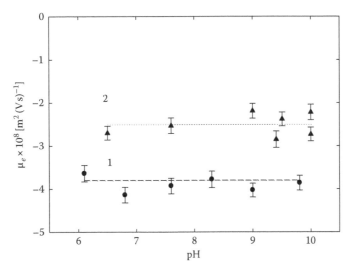

FIGURE 4.3 The dependence of the microelectrophoretic mobility $\mu_e \times 10^8$ m^2 (V s)$^{-1}$ on pH determined experimentally for PSS solution: 1. $I = 1 \times 10^{-2}$ M, 2. $I = 0.15$ M.

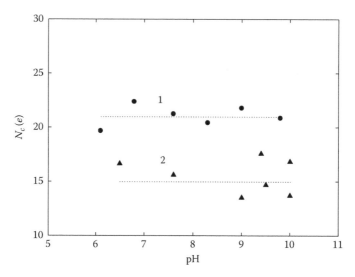

FIGURE 4.4 Number of uncompensated changes $N_c(e)$ of PSS on pH calculated from the experimental results derived from microelectrophoretic measurements using Equation 4.8: 1. $I = 1 \times 10^{-2}$ M, 2. $I = 0.15$ M.

The effective charge of strongly ionized flexible polyelectrolyte in water measured by electrophoretic method was found to be much lower than the chemical charge determined by titration curves. The effective charge of polyvinyl pyridinium and PSS salts in water at room temperature was 0.25 and 0.30, respectively, of the charge of fully ionized polyelectrolyte (Beer et al. 1997).

Dynamic viscosity of polyelectrolyte solutions was measured using capillary viscometer in vertical orientation with an automatic detection of suspension level via electric conductivity (Adamczyk et al. 2004). It was calibrated using pure liquids of known viscosity such as water, butyl, and amyl alcohols, ethylene glycol, etc. The accuracy of viscosity determination for the range 1–10 cP [g (cm·s)$^{-1}$] was estimated to be 0.5% with the main source of error stemming from the suspension density measurements. It is interesting to note that for this viscosity range, a typical time of the measurement t was about 50 s for the capillary having the radius $R = 0.05$ cm and the suspension volume V_{sus} about 12 cm^3. Thus, the maximum suspension velocity in the capillary $v_{max} = 2V_{sus}/\pi R^2 t$ was about 60 cm s^{-1}. Accordingly, the mean shear rate in the capillary $\langle G \rangle = 8V_{sus}/3\pi R^3 t$ was 1.6×10^3 s^{-1}. One can estimate that the Peclet number $Pe = \langle G \rangle a^3/D$ (where a is the averaged particle radius and D is the diffusion coefficient in the bulk) characterizes the ratio of flow to Brownian diffusion effects. Very small values of Pe indicate that the particle diffusion in suspensions dominated over the shearing flow effects. According to theoretical and experimental evidences (Kruif et al. 1985), for a low Pe range suspension rheology exhibits a fully Newtonian behavior with the dynamic viscosity independent of the shear rate. Indeed, experiments performed for other capillary diameter (leading to different shear rates) produced very similar viscosity data. Therefore, the results discussed hereafter are of an universal, shear independent character. For molecules of nonspherical shape, the increase in the intrinsic viscosity of suspensions is expected because of the increase in fluid velocity gradients due to particle rotation. This effect was described quantitatively by Brenner (1974) for spheroidal particles, having both a prolate and an oblate shape.

In the case of spheroidal particles immersed in simple shear flows, Brenner (1974) has shown that the intrinsic viscosity is described by the general expression

$$[\eta] = 5Q_1(\lambda) - Q_2(\lambda) + 2Q_3(\lambda) = F_v(\lambda) \tag{4.20}$$

where

$$Q_1 = \frac{1}{5\alpha_{\parallel}'}$$

$$Q_3 = \frac{1}{5\alpha_{\parallel}'} \left[\frac{\lambda(\alpha_{\parallel} + \alpha_{\perp})}{\lambda^2 \alpha_{\parallel} + \alpha_{\perp}} \left(\frac{\alpha_{\parallel}''}{\alpha_{\perp}'} \right) - 1 \right] \tag{4.21}$$

$$Q_2 = \frac{2}{15\alpha_{\parallel}' \left(1 - \frac{\alpha_{\parallel}''}{\alpha_{\perp}''} \right)}$$

$$\alpha_{\perp}'' = \frac{\lambda^2}{(\lambda^2 - 1)^2} \left[\left(\frac{2\lambda^2 + 1}{\beta} \right) - 3 \right] \qquad \alpha_{\parallel} = \frac{\lambda^2}{4(\lambda^2 - 1)^2} (3\beta + 2\lambda^2 - 5)$$

$$\alpha_{\perp}' = \frac{\lambda}{(\lambda^2 - 1)^2} (\lambda^2 - 3\lambda^2 \beta + 2) \qquad \alpha_{\parallel}'' = \frac{\lambda^2}{4(\lambda^2 - 1)^2} \left[2\lambda^2 + 1 - \left(4\lambda^2 - 1 \right) \beta \right]$$

For prolate spheroids, in the limit of $a \gg b$, Equations 4.20 and 4.21 reduce to the simpler form

$$[\eta] = \frac{\lambda^2}{15} \left[\frac{3}{\ln 2\lambda - 0.5} + \frac{1}{\ln 2\lambda - 1.5} \right] + \frac{8}{5} \tag{4.22}$$

It is interesting to observe that the intrinsic viscosity of prolate spheroids in the limit $\lambda \gg 1$ can be more sensitive to the length of the body than the hydrodynamic radius R_H, because it increases proportionally to $L_e^2 / \ln L_e$.

The intrinsic viscosity $[\eta]$ PAH 15,000 was determined from the slope of the dependence of the relative viscosity $[\eta_{rel}] = \eta_s / \eta$ on the polyelectrolyte volume fraction Φ_V. The results obtained for the PAH volume fraction range $\Phi_V < 0.05$ and the ionic strength of $I = 1 \times 10^{-3} M$, $5 \times 10^{-3} M$, $1 \times 10^{-2} M$, and $0.15 M$ at pH 6.5 are presented in Figure 4.5. As can be seen, the dependence of $[\eta_{rel}]$ on Φ_V for PAH for all ionic strength can well be fitted by a linear regression. The slope of these lines $d[\eta_{rel}]/d\Phi_V$ was found equal to 105 for $I = 5 \times 10^{-3} M$ and to 35 for $0.15 M$, respectively. The results are much higher than what the Einstein formula $\eta_s / \eta = 2.5 \Phi_V$ predicts for spherically shaped particles, equal to 2.5. Usually, the deviation from the Einstein formula occurring for the low-volume fraction of suspensions is interpreted in terms of the primary electroviscous effect stemming from shear-induced deformations of ionic double layers surrounding suspended particles. This leads to increased dissipation of mechanical energy resulting in increased viscosity of suspensions that exceeds values predicted by the Einstein formula. There are two other molecular contributions to the intrinsic viscosity: one from the shape, second from size or volume as summarized by the formula (Brenner 1974, Adamczyk et al. 2006, 2008)

$$[\eta_{rel}] = 1 + \eta_V \Phi_V \tag{4.23}$$

where
η_V is Simha factor
Φ_V is the volume fraction

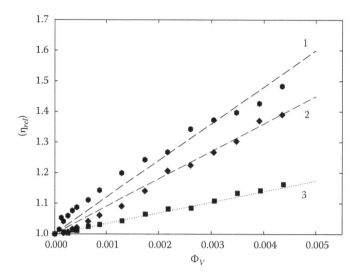

FIGURE 4.5 The dependence of the relative viscosity [η_{rel}] of PAH volume fraction Φ_V, $T=293$ K, pH$=6.5$, 1. $I=10^{-3}$ M, 2. $I=10^{-2}$ M, 3. $I=0.15$ M.

It seems that a proper interpretation of these data can be sought in the highly nonspherical shape of PAH molecules in solutions. Again, quantitative data on the intrinsic viscosity of nonspherical molecules of spheroidal shape can be extracted from the work of Brenner (1974). In the case of PAH suspensions, by substituting value of $\lambda=45.2$, one obtains from Equation 4.22 intrinsic viscosity value [η_V] $=115$, which is the maximum value for any physicochemical conditions. As can be seen, [η_V] values for low ionic strength are larger than the theoretical value predicted for fully extended, bare molecule. This deviation can be in a most simple way explained in terms of the hydration degree of the molecule, which increases its apparent size in the solution and consequently its volume fraction (Adamczyk et al. 2006). The significance of this effect can be quantitatively accounted by considering that the hydrated volume of the PAH molecule $v_m^* = v_m \left(d_h/d_c\right)^2$ is 1.8 times larger than the bare volume. Accordingly, the effective intrinsic viscosity of PAH solutions is to be reduced by the factor of 0.55. These corrected intrinsic viscosity varies between 57.7 and 19.2 for I changing within 1×10^{-3} and 0.15 M.

These experimental data were successfully interpreted in terms of the Yamakawa (Yamakawa and Fujii 1974) model predictions with the correction for the volume excluded effect. Using these corrected experimental data of intrinsic viscosity, one can recalculate new dimensionless aspect ratio parameter λ_c (longer to shorter axis ratio of the equivalent spheroid having the same intrinsic viscosity as the experimental value). λ_c varies between 25.5 (for $I=5 \times 10^{-3}$ M) to 12.5 (for $I=0.15$ M) at pH$=6.5$, and 13.2 (for $I=5 \times 10^{-3}$ M) to 10.1 (for $I=0.15$ M) at pH$=9.5$. Assuming the hydrated chain diameter $d_h=1.20$ nm, one can calculate that this corresponds to the equivalent length of the molecule $L_e^* = (4v^* \lambda_c^2 / \pi)^{1/3}$ varying between 32.0 to 20 nm (for I changed between 5×10^{-3} and 0.15 M) at pH$=6.5$, and 20.7 to 17.3 nm at pH$=9.5$.

The same methodology was used for PSS molecule. The intrinsic viscosity was determined for $I=5 \times 10^{-3}$ M, [η_V] $=24.8$; for $I=10^{-2}$ M, [η_V] $=21$; and for $I=0.15$ M, [η_V] $=8$ (Figure 4.6). The intrinsic viscosity [η_V] values for low ionic strength are larger than the theoretical value predicted for fully extended, bare molecule, equal to 23.8. This deviation can be, in a most simple way, explained in terms of the hydration degree of the molecule. The hydrated volume of the PSS molecule $v_m^* = v_m \left(d_h/d_c\right)^2$ is 1.6 times larger than the bare volume. The effective intrinsic viscosity of PSS solutions is to be reduced by the factor of 0.63. The corrected intrinsic viscosity values varied between 15.4 and 5 (for I changing within 5×10^{-3}–1×10^{-2} M). It is interesting to observe that the corrected values of the intrinsic viscosity for PSS changes with the ionic strength proportionally to

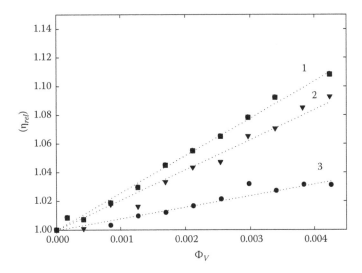

FIGURE 4.6 The dependence of the relative viscosity $[\eta_{rel}]$ of PSS volume fraction Φ_V, $T = 293$ K, pH $= 6.5$, 1. $I = 5 \times 10^{-3}$ M, 2. $I = 10^{-2}$ M, 3. $I = 0.15$ M.

$I^{-1/3} \sim \kappa^{-2/3}$. This agrees quite well with the experimental results of Davis and Russel (1987) obtained for potassium poly(styrenesulfonate) having the molecular weight of 50 kD.

For PSS molecule λ_c varies between 11.8 (for $I = 2 \times 10^{-3}$ M) and 4.5 (for $I = 0.15$ M). Assuming the hydrated chain diameter $d_h = 1.39$ nm, one can calculate that this corresponds to the equivalent length of the molecule $L_e^* = (4v^* \lambda_c^2 / \pi)^{1/3}$ varying between 16.3 to 8.5 nm (for I varying between 2×10^{-3} and 0.15 M).

The same procedure as for PAH and PSS with molecular weight 15,000 was used for high molecular weight 70,000 to confirm that conformation transition occurs (Table 4.6). The ionization degree of the PAH (Mw 70,000) chain can be calculated from the electrophoretic mobility measurements as a function of the pH and ionic strength. The data concerning electrophoretic mobility are shown in Figure 4.7. Electrophoretic mobility, μ_e, decreases monotonically from 5.8×10^{-8} m^2 (V s)$^{-1}$ at pH $= 6.5$ to 0.0 m^2 (V s)$^{-1}$ at pH $= 10.8$ for $I = 1 \times 10^{-3}$ M and from 3.0×10^{-8} m^2 (V s)$^{-1}$ at pH $= 6.5$ to 0.0 m^2 (V s)$^{-1}$ at pH $= 10.8$ for $I = 0.15$ M. The ionization degree of the PAH chain can be calculated from the electrophoretic mobility, and the data are presented in Figure 4.8. These results suggest a significant reduction in the number of ionic groups at the PAH molecule surface. Substituting experimental data, i.e., $\mu_e = -5.8 \times 10^{-8}$ m^2 (V s)$^{-1}$, $R_H = 13.4$ nm (pH $= 6.5$ and $I = 1 \times 10^{-3}$ M) result in the average number of free charges (of positive sign) per PAH molecule $N_c = 188$ (e) and gives the effective ionization degree $\alpha_i^* = 0.25$ (considering that the nominal number of charges equals to 750 (e)). For $I = 0.15$ M at pH $= 6.5$, taking $\mu_e = -3.0 \times 10^{-8}$ m^2 (V s)$^{-1}$ and $R_H = 18.7$ nm, the maximum number of free charges was calculated to be $N_c = 53$ (e) and consequently $\alpha_i^* = 0.07$.

The results of viscosity measurements of PAH 70,000 solutions performed for volume fraction range $\Phi_V < 0.0045$ and three different ionic strength of $I = 1 \times 10^{-3}$ M, $I = 5 \times 10^{-3}$ M, and 0.15 M are presented in Figure 4.9. The slope of these lines $d[\eta_{rel}]/d\Phi_V$ was found to be equal to 1200 for $I = 1 \times 10^{-3}$ M, 420 for $I = 5 \times 10^{-3}$ M, and 90 for $I = 0.15$ M. It can be calculated from Equation 4.22 for $\lambda = 167$ (the value pertinent to the extended PAH molecule) to be equal to $[\eta_V] = 1483$. The experimental values $[\eta_V] = 1200$ obtained for $I = 1 \times 10^{-3}$ M are slightly lower than calculated for prolate spheroids. The results obtained for PAH 70,000 are in accordance with results obtained for PAH 15,000.

The extended length of the PSS polyelectrolyte chain having the molecular weight of 70,000 attains 91 nm, and the length-to-diameter ratio of the chain $\lambda = 78$. The ionization degree of the PSS chain can be calculated from the electrophoretic mobility measurements as a function of the

TABLE 4.6

Physicochemical Characteristics of PAH
($M_w = 70,000$) and PSS ($M_w = 70,000$). $T = 293$ K

Property	Polyelectrolyte	
	PAH	PSS
Density (g/cm³)	1.15	1.18
Molecular weight	70,000	70,000
Molecular weight monomer	93.5	206.2
Number of monomers/charges	750	340
Volume per molecule (nm³)	101	98.5
Volume per monomer (nm³)	0.135	0.290
Bare cross section (nm²)	0.54	1.07
Extended length L (nm)	187	91
Equivalent chain diameter d_e (nm)	0.83 ± 0.07	1.17 ± 0.04
Hydrated chain diameter	1.12	1.46
d_h (nm) $= d_e + 2 \times 0.145$ (nm)		
Aspect ratio $\lambda_e = L/d_e$	225	78
$\lambda_h = L/d_h$	167	62
Charge distance L_{ex}/N (nm)	0.249	0.267
R_g	54	26
R_S	2.9	2.86

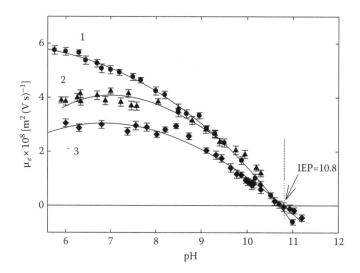

FIGURE 4.7 The dependence of the microelectrophoretic mobility $\mu_e \times 10^8$ m² (V s)⁻¹ on pH determined experimentally for PAH solution: 1. $I = 1 \times 10^{-3}$ M, 2. $I = 1 \times 10^{-2}$ M, 3. $I = 0.15$ M.

pH and ionic strength. The data are shown at the Figure 4.10, the electrophoretic mobility of PSS increased from $\mu_e = -4.5 \times 10^{-8}$ m² (V s)⁻¹ for $I = 1 \times 10^{-3}$ M to $\mu_e = -3.0 \times 10^{-8}$ m² (V s)⁻¹ for $I = 0.15$. These results suggest a significant reduction in the number of ionic groups at the PSS molecule surface. As can be seen, for the pH range studied, i.e., 5–10 the electrophoretic mobility of PSS 15,000 (Figure 4.5) and 70,000 (Figure 4.10) remains fairly constant, which suggests a pH-independent ionization degree of surface groups of this molecule.

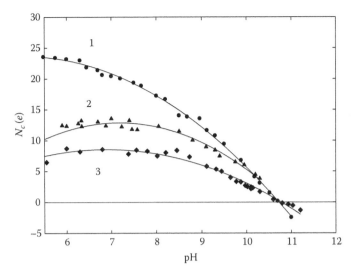

FIGURE 4.8 Number of uncompensated changes $N_c(e)$ of PAH on pH calculated from the experimental results derived from microelectrophoretic measurements using Equation 4.8: 1. $I = 1 \times 10^{-3}$ M, 2. $I = 1 \times 10^{-2}$ M, 3. $I = 0.15$ M.

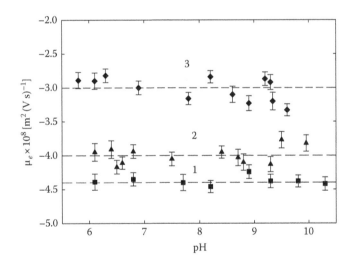

FIGURE 4.9 The dependence of the microelectrophoretic mobility on pH determined for PSS solution: 1. $I = 5 \times 10^{-3}$ M, 2. $I = 1 \times 10^{-2}$ M, 3. $I = 0.15$ M. (From Weroński, P., Kinetics and topology of irreversible adsorption of anisotropic particles at homogeneous interfaces. PhD thesis, Institute of Catalysis and Surface Chemistry, Polish Academy of Science, Cracow, Poland, 2001. With permission.)

The effective $N_c(e)$ (uncompensated) charge of PSS molecules q was calculated using the Lorenz–Stokes equation (Equation 4.6). Substituting experimental data, i.e., $\mu_e = -4.4 \times 10^{-8}$ m^2 (V s)$^{-1}$, $R_H = 13.5$ nm ($I = 5 \times 10^{-3}$ M), one obtains $N_c = 75$ (e) as the average number of free charges (of negative sign) per PSS molecule. This gives the effective ionization degree $\alpha_i^* = 0.21$ (considering that the nominal number of charges equals to 340). For $I = 0.15$ M, taking $\mu_e = -3.0 \times 10^{-8}$ m^2 (V s)$^{-1}$ and $R_H = 13.8$ nm the maximum number of mobile charges was calculated to be $N_c = 51$ (e), and consequently $\alpha_i^* = 0.15$. The nominal ionization degree derived from Equation 4.8 is compared in Figure 4.11. It is interesting to observe that the effective charge of PSS molecule for both molecular weight have the same dependence on pH and ionic strength.

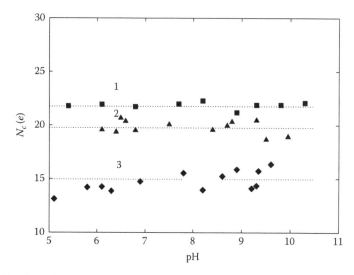

FIGURE 4.10 Number of uncompensated changes of PSS on pH calculated for the experimental results derived from microelectrophoretic measurements using Equation 4.8: 1. 1. $I = 5 \times 10^{-3}$ M, 2. $J = 1 \times 10^{-2}$ M, 3. $I = 0.15$ M.

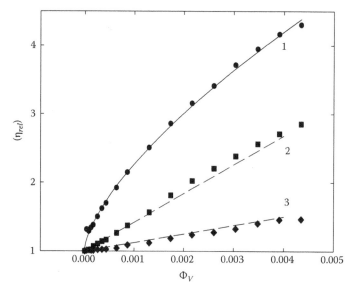

FIGURE 4.11 The dependence of the relative viscosity of PAH volume fraction, $T = 293$ K, pH $= 6.5$: 1. $I = 1 \times 10^{-3}$ M, 2. $J = 5 \times 10^{-3}$ M, 3. $I = 0.15$ M.

The results of viscosity measurements of PSS 70,000 solutions performed for volume fraction range $\Phi_V < 0.004$ and three different ionic strength of $I = 1 \times 10^{-3}$ M, $I = 5 \times 10^{-3}$ M, and 0.15 M are presented in Figure 4.12. As can be seen, the dependence of $[\eta_{rel}]$ on Φ_V can well be fitted by a linear regression. The slope of these lines $d[\eta_{rel}]/d\Phi_V$ was found to be equal to 80 for $I = 1 \times 10^{-3}$ M, 60 for $I = 5 \times 10^{-3}$ M, and 26 for $I = 0.15$ M. This is again much higher than what the Einstein formula predicts for spherically shaped particles. Again, these data can be interpreted in terms of the Brenner model (1974). It can be calculated from Equation 4.22 where for $\lambda = 78$ (the value pertinent to the extended PSS molecule) $[\eta_V] = 383$, for $\lambda = 50$ (the molecule bend to the form of a semicircle) $[\eta_V] = 177$, and for $\lambda = 25$ (the molecule bend to the form of a circle) $[\eta_V] = 55$. As can be

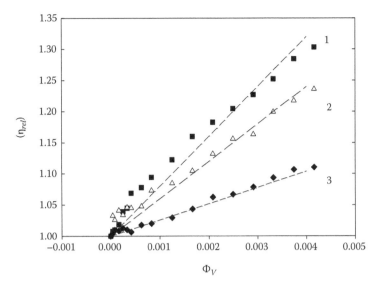

FIGURE 4.12 The dependence of the relative viscosity [η_{rel}] of PSS volume fraction Φ_V, $T = 293$ K, pH $= 6.5$, 1. $I = 10^{-3}$ M, 2. $I = 5 \times 10^{-3}$ M, 3. $I = 0.15$ M.

seen, the intrinsic viscosity agrees well with the experimental value obtained for the ionic strength of 5×10^{-3} M. However, a more precise comparison between experimental and theoretical results would require formulating of the hydrodynamic theories concerning the intrinsic viscosity for particles having more complicated shapes than spheroids, for example, cylinders bent to the form of semicircles or rings (toruses).

These experimental data suggested that the most probable shape of PAH and PSS molecules for low ionic strength is a flexible rod bent to the form of a semicircle or a torus and for high ionic strength (lower ionization degree) bent to the form of a helicoidal structure. Therefore, it can be concluded that dynamic viscosity and DLS measurements can be exploited as a convenient tool for determining the effective length of polyelectrolytes in electrolyte solutions.

4.3.1 Deposition on Surface to Design Media of Desired Architecture

AFM has been used successfully for studying the morphology of polyelectrolytes adsorbed on the surface. Images of necklace globules of poly(2-vinylpyridine) and poly(methacryloxyethyl dimethylbenzylammonium chloride) adsorbed at mica surface were reported by Kiriy et al. (2002) and by Minko et al. (2002). These images clearly show that with increasing ionic strength of the solution an abrupt conformational transition from elongated chains to compact globules occurs. Gromer et al. (2008) used liquid cell to study dynamic properties of PSS. They compared the structure of PSS chains of three different charge densities: 32%, 67%, and 92%. Adsorption and conformation of the chains were studied directly in solution, so that their real conformation was observed. The adsorption was studied on two types of surfaces: mica and cationic phospholipid bilayers. The PSS molecule formed globular structure independent of the substrate at low charged density. At high charged density (67% and 92%), the observed difference in conformation of PSS molecules depends on the kind of surface at which the adsorption occurs. This experiment's measurement visualizes prediction obtained from viscosity measurements.

Polyelectrolytes very often are used to form self-assembly polyelectrolyte films (self-assembled monolayer, SAM). It is important to know the structure of SAM in order to understand and control how compounds are incorporated into the film. However, it is especially important to know how they are attached on the surface during the SAM buildup. Much effort has been made to elucidate the internal structure and the growing rate of these films. The following most important techniques were employed: scanning angle reflectometry (Ladam et al. 2000), optical waveguide lightmode spectroscopy (OWLS) (Picard et al. 2001), Neutron (Losche et al. 1998), and x-ray diffraction (Kayushina et al. 1996), quartz microbalance (QCM) (Lvov et al. 1995), and attenuated total reflection-Fourier transform infrared (ATR-FTIR) (Schwinte et al. 2001). Nevertheless, there are only a few techniques to study surface structure of SAM, for example, scanning electron microscopy (SEM) and AFM. However, they have certain inconvenience of requiring pretreatment of samples that does not allow observing the surface structure in solution, where the process of deposition occurs. Liquid-cell AFM could be an appropriate technique to study the surface structure of SAM. AFM has already been used to investigate different issues of SAM: fabrication of microporous films for PAA/PAH by a pH-induced phase separation (Mendelsohn et al. 2000), salt effect on polyelectrolyte multilayer film morphology (McAloney et al. 2001), kinetics of film formation of PAA/PSS on a positively charged SAM (Tsukruk et al. 1997), thickness measurements (Lobo et al. 1999), and comparison of surface structure of polyelectrolyte films for a linear (PSS/PAH) and exponential (PGA/PLL) growth regime (Lavalle et al. 2002). However, all the reported images have been obtained in air after a natural drying process or in water solution, but after a transfer from the film preparation devise to the liquid-cell AFM (Onda et al. 1997). Therefore, the observed surfaces are only an approximation of wet surfaces, where the real deposition takes place. Menchaca et al. (2003) for the first time used AFM liquid cell where the SAM of the polyelectrolyte PSS/PAH films are built up and are imaged *in situ*. They observed a granulate structure on the surface of SAM dependent on buffer pH (presented at Figure 4.13). Granulate structure has been already reported with multilayer films on solid macroscopic substrates (Schlenoff et al. 1998, Kim et al. 1999, Lvov et al. 1999, Leporatti et al. 2000), on melamin formaldehyde, and on biological cell (Castelnovo and Joanny 2000). However, all these studies have also been done in air where drying process occurs. There are evidences that granulate structure is related to conformation of polyelectrolytes in solution that are employed during film construction. Grain size, roughness, and film thickness grow as the salt concentration increases in a PDDA/PSS system (McAloney et al. 2001). These results were interpreted as a conformation transition of polyelectrolytes from extended rod to globular coil in polymer solution as a function of salt concentration (McAloney et al. 2001). There are also evidences that polyelectrolyte complexes are formed inside the multilayer (Castelnovo and Joanny 2000). Neutron reflectometry has shown that two consecutive layers in the SAM are strongly interacted (Losche et al. 1998). The precise form of these polyelectrolyte complexes inside SAM has not been elucidated. Therefore, granulate structure observed by AFM may be related to the complex size that should be formed if positive and negative polyelectrolytes were mixed.

4.4 STABILITY OF BIOLOGICAL POLYELECTROLYTES (GLOBULAR PROTEIN) IN SOLUTION AND STRUCTURAL REARRANGEMENT ON THE SURFACE

Adsorption of proteins at solid/liquid interfaces is a process of major importance for biomedical technologies such as biosensors, biochips, and biomaterials for medical implants. A proper description and control of protein deposition phenomena requires a thorough knowledge of their structural and transport properties in the bulk, especially their shape, conformation, hydration degree, charge, diffusion coefficients and aggregation phenomena in relation to protein bulk concentration, electrolyte composition (ionic strength), pH, and temperature. Despite a rapid progress in the field of protein structural characterization (Tanaka et al. 1997, Stradner et al. 2004, 2006) there exist few experimental methods that can be efficiently used for determining these parameters in dilute aqueous solutions of proteins.

One class of the most rapid, noninvasive, and direct methods is represented by the hydrodynamic methods, involving the DLS, dynamic viscosity, sedimentation velocity, fluorescence depolarization, and circular dichroism (CD) measurements (Harding et al. 1995).

Solution properties are among the main sources of information about the conformation of biological macromolecules. For quasi-rigid macromolecules, the essential structure features are the dimensions and shape of the macromolecular particle. However, many macromolecules of biological

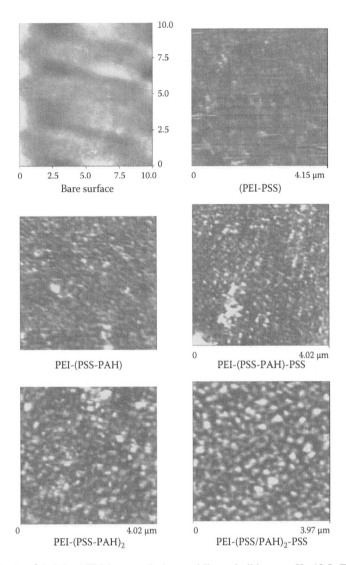

FIGURE 4.13 $4 \times 4\,\mu m^2$ height AFM images during multilayer buildup at pH = 10.5. First image corresponds to flat glass surface. Polyelectrolyte films are shown from (PEI/PPS) layer to PEI-(PSS/PAH)$_2$-PSS layer. Images are presented at the same z scale of 20 nm to better appreciate the evolution. (Reprinted from Menchaca, J.L. et al., *Colloids Surf. A*, 222, 185, 2003. With permission.)

relevance in solution are flexible entities and do not have a defined shape. For them, the essential structure feature is the flexibility itself, which may have an essential role in their function, and which also determined their solution properties.

Norde and Anusiem (1992) introduced two classes of proteins: "hard" and "soft." The "hard" proteins have a strong internal coherence, and structural rearrangements do not significantly contribute to the adsorption process. These proteins adsorb on hydrophobic surfaces, while on hydrophilic surfaces they adsorb only if they are electrostatically attracted. Proteins of much lower structural stability, the "soft" proteins, adsorb even under the seemingly unfavorable conditions of a hydrophilic, electrostatically repelling surface. These proteins with a large driving force for adsorption result from structural rearrangements (Norde 2008).

The most extensively studied and applied proteins are lysozyme (LSZ), bovim serum albumin (BSA), and human serum albumin (HSA). The structure and properties of these proteins are well known, including amino acid sequences and x-ray data for special structures. LSZ have the two-domain structure. In the first domain, the polypeptide chain is folded with β-sheets while the second domain is represented by the α-helix structure (RCSB Protein Data Bank). Four S-S bridges in LSZ provide high pH and thermal stability of this protein. Polar and nonpolar patches are inhomogeneously distributed along the surface of the molecule. The conformation of LSZ is highly stable in the pH region from 4 to 11. The high value of i.e.p. for pH 10–11 was caused mainly by the high value of pK_a of Arg. 12.5 and Lys. 10.8, that these groups in the native protein are immersed in hydrophobic environment and are not accessible for water molecules up to very high pH values.

Spatial structure of albumins resembles the heart, and consists of six subdomains (three domains) repeating the α-helix pattern. The BSA molecule is characterized by an asymmetric charge distribution: −9 (e), −7.8 (e), and −1.3 (e) charges, respectively, for I, II, and III domains theoretically calculated for pH 7. Domains I and II result together in a net charge of −16.8 (e) when domain III has only −1.3 (e) net charge at pH 7. A similar asymmetric charge distribution was reported for HSA. The structure of albuminHSA is analogous to that of BSA because the amino acid sequences of these two proteins match by 80%. Albumins have some conformational flexibility depending on the pH. Foster (1977) reported that BSA has several isomeric forms at different pH media, and they correspond to different α-helix contents. The highest α-helix was observed between pH = 4.3 and 8 (normal form), with a larger decrease for the pH lower than pH = 4.3 and smaller decrease for the pH larger than pH = 8. Vermonden et al. (2001) also reported that BSA in the normal form has the most compact state. Serum albumins are the main components of blood, having the highest concentration in the circulation system, and provide colloid osmotic blood pressure. This protein is one of the most extensively studied and applied proteins in biochemistry. Physicochemical properties of selected proteins are given in Tables 4.7 and 4.8.

For characterization properties of proteins in bulk solution the dependence of the diffusion coefficient on the concentration of proteins was determined. The specific density of BSA and HSA proteins is $1.35 \, g \cdot cm^{-3}$, which corresponds to the range of volume fraction of proteins Φ_v 3.7×10^{-4} -5.92×10^{-3} and the range of weight fraction $5 \times 10^{-4} - 4 \times 10^{-3}$. The results obtained for BSA and for HSA are shown in Figure 4.14. As can be seen, the diffusion coefficient of BSA was practically independent of its bulk concentration, assuming an average value of $6 \times 10^{-7} cm^2 \, s^{-1}$ for ionic strength range $10^{-2} - 0.15 M$ (pH = 6.3). Similar values of the diffusion coefficient of BSA, which prove to be independent of pH, were reported by Wang and Yu (1988) for BSA at the ionic strength varying between 10^{-2} and $5 M$.

TABLE 4.7
Physicochemical Properties of Proteins

Property	Protein		
	BSA	HSA	LSZ
Molecular weight (Da)	67,000	69,000	14,000
Specific density (g cm⁻³)	1.35	1.35	1.35
Specific volume (nm³)	82.5	85.0	17.2
Equivalent sphere radius (nm)	2.70	2.72	1.6
Hydrodynamic radius R_H (nm)	3.3–4.3	3.3–4.1	2.0
Structure type	α-structure	α-structure	α+β-structure
Shape	Prism (like heart)	Prism (like heart)	Ellipsoid
Geometrical dimensions, spheroid (nm)	9.5×5×5	9.5×5×5	4×3×3
Geometrical volume for spheroid (nm³)	124	124	18.8
Porosity	0.33	0.33	0.25

TABLE 4.8
Amino Acid Composition and pK Values for BSA and LSZ

Amino Acid	Charge	pK	BSA	LSZ
Arg	+	12.5	23	11
His	+	6.0	17	1
Lys	+	10.8	59	10
Asp, Glu	−	4.1	100	9
Tyr	−	10.9	19	3
Cys	−	8.3	35	8
i.e.p.			5.51	9.54

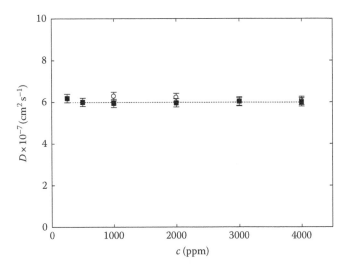

FIGURE 4.14 The dependence of the diffusion coefficient of BSA and HSA solutions on the bulk concentration c [ppm] determined experimentally for pH = 6.3, T = 298 K, I = 1 × 10⁻² M, ● BSA, ■ HSA. The line shows the linear fit of experimental points, i.e., D = 6 × 10⁻⁷ cm² s⁻¹.

From the diffusion coefficient measurements, one can determine the Stokes hydrodynamic radius of proteins, R_H. These values calculated from Equation 4.14 using measured diffusion coefficients of BSA are plotted in Figure 4.15 as a function of pH (for I = 0.15 M). As can be seen, R_H calculated by exploiting the entire scattering peak and its maximum value (volume average) are different, although they seem to be pH independent. Accordingly, the average R_H value for pH 4–9 calculated from the entire peak was 4.3 nm, and from the maximum was 3.4 nm (the ratio of both being 1.26). Moreover, Park et al. (1992) determined similar values of the averaged R_H = 3.9 nm for pH range 5–10 in their measurements. Longsworth (1954) determined the approximate value R_H = 3.7 nm for BSA molecule.

The difference in R_H calculated from the average peak and from the maximum suggests quite unequivocally that BSA and HSA suspensions contain a significant fraction of protein dimmers (Figure 4.15). This is in accordance with previous experimental evidences (Rezwan et al. 2004, Desroches et al. 2007).

A more quantitative analysis of protein aggregation can be performed by considering their shape and dimensions. The molecular volume of BSA can be calculated from the formula $v_m = M/\rho_p A_v$, where M is the molecular mass of the protein, ρ_p is the protein density, and A_v is the Avogadro number.

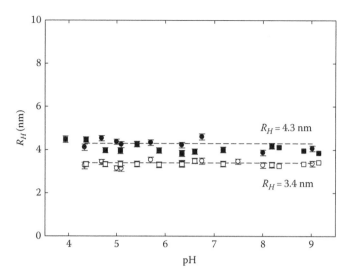

FIGURE 4.15 The hydrodynamic diameter R_H of BSA and HSA determined by the DLS method, $c = 1000$ ppm, $I = 0.15$ M. The dashed lines denote the limiting analytical solutions calculated from Equation 2.14 for the dimmer (upper line $R_H = 4.3$ nm, ● the R_H value calculated from the averaged scattering intensity) and monomer (lower line $R_H = 3.4$ nm, ○ the R_H value calculated from the maximum of the scattering peak).

v_m calculated from this equation for BSA and HSA equals to 82.5 and to 85.0 nm³, respectively. The equivalent sphere radius $R_s = \sqrt[3]{3v_m/4\pi}$, i.e., was 2.70 and 2.72 nm for BSA and HSA, respectively.

This simple estimation, compared with the above experimental values of R_H suggests that the shape of both proteins deviates significantly from a spherical shape. This is in accordance with structural data reported in the protein data bank (Rezwan et al. 2004, Desroches et al. 2007, RCSB Protein Data Bank). The physical form of BSA monomer assumes a rather complex cardioidal shape, which, according to Rezwan et al. (2004) and others (Kaufman et al. 2007) can be approximated by a prolate spheroid having the dimensions 9×5.5×5.5 nm. Moreover, there are three differently charged domains in the protein structure, depending on the pH of the solution. Using these dimensions it is possible to calculate the molecular volume of proteins (assuming a prolate spheroid shape), which can be calculated from the formula $v_s = 4\pi a^2 b/3 = 142$ nm³. As can be noticed, this value exceeds 1.73 and 1.67 times previously calculated values of 82.5 and 85.0 nm³ stemming from molecular mass. This discrepancy could suggest that both molecules form porous structures, having void ratio $1 - v_m/v_s = 0.42$ for BSA and 0.43 for HSA. Since these values are significantly larger than the usually accepted data for globular proteins (being close to 0.25 (Haynes and Norde 1994)), this could suggest that the effective dimension of the shorter axis of these proteins is slightly smaller. This hypothesis seems to be further supported by the fact that in most of the protein deposition experiments (Fitzpatrick et al. 1992, Liebmann-Vinson et al. 1996, Su et al. 1999, Terashima and Tsuji 2002) where the thickness of the monolayer was determined, it varied between 3.5 and 5 nm.

In accordance with these calculations, the shape of these proteins was approximated in the literature (Golander and Kiss 1988) by a prolate spheroid having the dimensions 14.1×3.8×3.8 nm. Assuming this, one can calculate that $v_s = 106$ nm³, which exceeds 1.28 and 1.25 times previously calculated values of 82.5 and 85.0 nm³ stemming from molecular mass, and gives for the void ratio the value of 0.29. This agrees better with the suggested limit for globular proteins. However, the postulated length of BSA is definitely too large and its thickness too small in comparison with dimensions acquired from the data bank (Table 4.7). Therefore, the "compromise" dimensions of these proteins approximated by a prolate spheroid to be 9.5×5×5 nm. In this case, $v_s = 124$ nm³, which exceeds 1.51 and 1.50 times the molecular volume giving for the void ratio the acceptable value of 0.33. It is advantageous to approximate the true protein shape by a prolate spheroid shape because its hydrodynamic radius and intrinsic viscosity can be calculated analytically from the hydrodynamic Brenner's theory (1974). Using his expression, which is an extension of the Stokes–Einstein relationship, Equation 4.14, the analytical expression for R_H for prolate spheroids can be derived:

$$R_H = \frac{a_s(\lambda^2 - 1)^{1/2}}{\cosh^{-1}\lambda} \quad \text{prolate spheroids, } \lambda = a/b > 1 \tag{4.24}$$

$$R_H = \frac{a_s(1 - \lambda^2)^{1/2}}{\cos^{-1}\lambda} \quad \text{oblate spheroids, } \lambda = b/a < 1$$

where

   a, b are the longer and the shorter semiaxes of the spheroid
   λ is the aspect ratio parameter having a major significance for predicting the hydrodynamic behavior of spheroidal particles

It can be calculated from Equation 4.24 that for the prolate spheroid of the dimensions $9.5 \times 5 \times 5$ nm, aspect ratio $\lambda = 1.9$ and $R_H = 3.22$ nm. This theoretical value agrees within experimental error bounds of about 5% with $R_H = 3.4$ nm, determined experimentally from the maximum of the scattering intensity peak.

It is possible to calculate the theoretical value of R_H for the dimer of BSA because its structural data are available from the data bank (Table 4.8). As can be noticed, the dimer is composed of two BSA molecules attached along their hydrophobic parts (Lassen and Malmsten 1996, Rezwan et al. 2004). The shape of the dimer is rather irregular having approximate dimensions of $10 \times 9.5 \times 5$ nm. This can be replaced by an oblate spheroid (of the same cross section) having the dimensions of $11 \times 11 \times 5$ nm. It can be calculated from Equation 4.24 that for such a prolate spheroid, aspect ratio $\lambda = 2.3$ and $R_H = 4.4$ nm. This value is close to the experimental result of 4.3 nm determined using the volume-averaged scattering intensity.

The protein molecule is surrounded by a more or less strongly adsorbed layer of solvent. When the macromolecule moves under the influence of an external effect, or as a consequence of Brownian motion, the layer moves rigidly attached to the macromolecules. Thus, the effective hydrodynamic particle includes the hydrodynamic water, and the hydrodynamic size of the macromolecules should be larger than the one that corresponds to the anhydrous macromolecules. Garcia de la Torre (2001) obtained hydration parameters (thickness of hydration layer t_h and hydration ratio δ as a gram water per gram protein) for 19 proteins, using atomic-level bead-modeling. They estimated the following parameters: $t_h = 0.3$ Å and $\delta = 0.17$ for BSA, and $t_h = 0.9$ Å and $\delta = 0.33$ for LSZ.

As comes out from Figure 4.15, the pH of BSA solutions did not exert practically any effect on the hydrodynamic radius, what suggests that the aggregation degree remains practically independent of this parameter. On the other hand, the temperature exerted pronounced influence on the hydrodynamic radius of BSA as can be seen in Figure 4.16. For temperatures higher than 333 K (60°C), the hydrodynamic diameter of BSA increased significantly, suggesting the appearance of large aggregates due to the denaturation process. The temperature at which denaturation occurs is often referred to the protein melting point, denoted by T_M. Previous literature data indicate that T_M for BSA was 63°C (Norde and Anusiem 1992). The LSZ have more stable structure than BSA (Figure 4.16).

Other bulk characteristics of protein solutions, which have basic significance for interpretation of their adsorption and deposition phenomena, are the effective charge, determined as a function of pH and ionic strength. The experimental values of the electrophoretic mobility, μ_e, and the net number of uncompensated charges per molecule, $N_c(e)$, are collected in Figures 4.17 through 4.20. For pH < 5, both proteins acquire positive charges increasing monotonically with the decrease of pH. The maximum number of positive charges for pH = 3 calculated from Equation 4.7 was rather

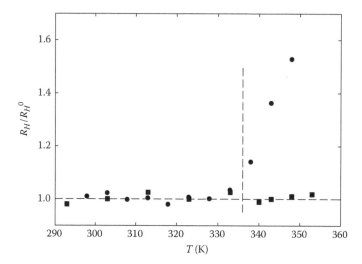

FIGURE 4.16   The relative hydrodynamic diameter R_H/R_H^0 (where R_H^0 is hydrodynamic diameter in $T = 298\,\mathrm{K}$) ● of BSA, ■ of LSZ determined by the DLS method as a function of the absolute temperature T, $I = 1 \times 10^{-2}\,\mathrm{M}$, pH = 6.5. The dashed line denotes the initial value and the solid line denotes the nonlinear regression.

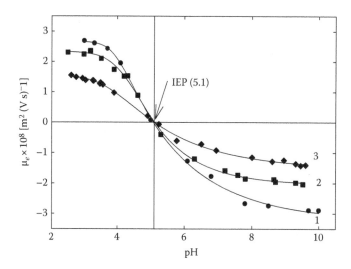

FIGURE 4.17   Electrophoretic mobility, μ_e, of BSA as a function of pH, $T = 298\,\mathrm{K}$. The points denote experimental values determined for 1. $I = 1 \times 10^{-3}\,\mathrm{M}$, 2. $I = 1 \times 10^{-2}\,\mathrm{M}$, 3. $I = 0.15\,\mathrm{M}$.

low and equal to 7.9 (*e*) for both proteins (at the ionic strength of $5 \times 10^{-3}\,\mathrm{M}$). On the other hand, for pH > 5, the proteins acquire negative charge, for BSA elementary charges equal to −5 (*e*), and −7.9 (*e*) for HSA for pH = 7 and $I = 5 \times 10^{-3}\,\mathrm{M}$. The uncompensated charge determined from the microelectrophoretic measurements is much smaller than the one predicted theoretically from the dissociation constant. Rezwan et al. (2004) predicted effective charges $N_c = -19.1$ (*e*) for pH = 7 (the ionic strength was unspecified). Hence, the effective charge is only 26 (*e*), which constitutes 41% of the limiting value predicted from the dissociation equilibrium. This phenomenon, observed experimentally for polyelectrolytes, is due to the adsorption of counterions, often referred to as ion condensation phenomenon. Obviously, in the case of adsorption of counterions increases with the ionic strength of the supporting electrolyte as it was observed. For $I = 0.15\,\mathrm{M}$ and pH = 7, the value of free charges for BSA molecule was only −3 (*e*), which amounts to 16% of the overall charge predicted from dissociation

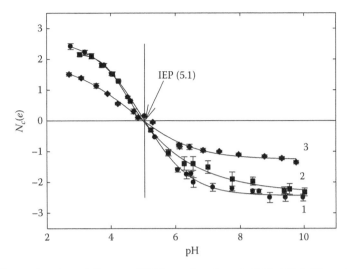

FIGURE 4.18 Electrophoretic mobility, μ_e, of HSA as a function of pH, $T = 298\,K$. The points denote experimental values determined for 1. $I = 1 \times 10^{-3}\,M$, 2. $I = 1 \times 10^{-2}\,M$, 3. $I = 0.15\,M$.

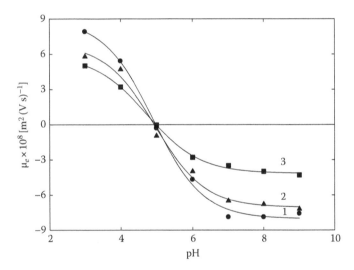

FIGURE 4.19 Number of uncompensated charges $N_c(e)$ of BSA as a function of pH, 1. $I = 1 \times 10^{-3}\,M$, 2. $I = 5 \times 10^{-3}\,M$, 3. $I = 0.15\,M$ ($N_C = [(6\pi\eta R_H 10^8)/1.602]\mu_e$, $T = 298\,K$, $c = 1000\,ppm$, $\eta = 8.9 \times 10^{-3}\,g$ (cm s)$^{-1}$, $R_H = 3.4 \times 10^{-7}\,cm$).

equilibrium. Since these values for uncompensated charges are rather small, they probably cannot prevent protein association into dimers as predicted by the hydrodynamic radius measurements.

It is often more convenient to analyze the electrokinetic behavior of proteins in terms of electrophoretic mobility, which is commonly used for interpretation of colloid, polyelectrolyte, and protein suspension stability (Adamczyk 2006).

The dependence of electrophoretic mobility on pH is shown in Figures 4.17 through 4.18. As can be seen in Figure 4.17, electrophoretic mobility of BSA decreased from $2.8 \times 10^{-8}\,m^2$ (V s)$^{-1}$ at pH 3 to $-2.8 \times 10^{-8}\,m^2$ (V s)$^{-1}$ at pH 10 for $I = 1 \times 10^{-3}\,M$. For higher ionic strength, $I = 0.15$, obtained electrophoretic mobility is equal to $1.5 \times 10^{-8}\,m^2$ (V s)$^{-1}$ at pH 3 and $-1.5 \times 10^{-8}\,m^2$ (V s)$^{-1}$ at pH 10. However, independently of the ionic strength, the zero value of electrophoretic mobility was attained for pH = 5.1 for both albumins, which is usually referred to as the i.e.p. The i.e.p. for BSA

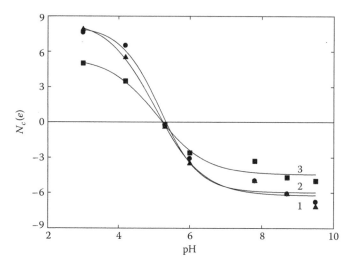

FIGURE 4.20 Number of uncompensated charges $N_c(e)$ of HSA as a function of pH, 1. $I = 1 \times 10^{-3}$ M, 2. $I = 5 \times 10^{-3}$ M, 3. $I = 0.15$ M ($N_C = [(6\pi\eta R_H 10^8)/1.602]\mu_e$, $T = 298$ K, $c = 1000$ ppm, $\eta = 8.9 \times 10^{-3}$ g (cm s)$^{-1}$, $R_H = 3.4 \times 10^{-7}$ cm).

and HSA agrees well with the data reported in the literature: 4.7–5.1 for BSA (Rezwan et al. 2004, 2005) and 4.7–4.9 for HSA (Abramson and Gorin 1942, Dawson 1986). The small deviations can be explained by the use of various buffers for preparing protein solutions used in these experiments. For example, the i.e.p. of BSA is shifted from pH = 5.1 in the Mes buffer to pH = 4.7 when the Mes-Tris buffer is used (Jachimska et al. 2008).

The experimental data of i.e.p. can be compared with theoretical predictions taking into account the amino acid composition of these proteins and their pK values of the side chains (Burton et al. 1988, Berg et al. 2002, Rezwan et al. 2004, 2005). These parameters are presented in Table 4.8. Summing up contributions stemming from various amino acids as a function of pH, it becomes possible to calculate the theoretical charge and i.e.p. of the proteins. This simple approach assumes that all ionizable amino acid groups are accessible for water molecules. In our calculations, the pK values were taken from literature, the amino acid sequence for BSA from Hirayama et al. (1990), and the protein data from RCSB Protein Data Bank. From these theoretical calculations, the i.e.p. of BSA was predicted to be 5.5. Our experimental result, 5.1, is slightly smaller than the theoretical estimation. This can be explained by the fact that a small part of charged amino acids is not accessible for water.

For a quantitative analysis of LSZ adsorption, information about the diffusion coefficient, the shape, and the effective (uncompensated) charges of the molecules is needed. From DLS measurements, it has been determined that the averaged diffusion coefficient of LSZ was 12.2×10^{-7} M cm^2 s^{-1} (for the concentration range 100–5000 ppm) at the ionic strength $I = 0.15$ M (Figure 4.21). Using the experimental value of the diffusion coefficient of LSZ, hydrodynamic radius, R_H, was calculated to be equal to 2.0 nm (Figure 4.22). It can be calculated from Equation 4.24 that for the prolate spheroid with dimensions $4 \times 3 \times 3$ nm^3, $\lambda = 1.3$ obtained $R_H = 2.09$ nm. The molecular volume of LSZ $v_m = 17.22$ nm^3 and $v_s = 18.80$ nm^3 (assuming prolate shape). Accordingly, the nominal aspect ratio (length to diameter) $\lambda = 1.3$, which means that the LSZ molecule is less elongated than BSA molecule with $\lambda = 1.9$. The electrophoretic mobility, μ_e, for LSZ is shown in Figure 4.23 in the form of the dependence of μ_e on pH. As can be seen, μ_e decreases monotonically from 2.8×10^{-8} m^2 (V s)$^{-1}$ at pH 3.0 to 0.0 m^2 (V s)$^{-1}$ at pH = 10.0 for $I = 1 \times 10^{-3}$ M. From theoretical calculations i.e.p. of LSZ was predicted to be 9.54, which is slightly smaller than the experimental result measured at pH 10.0. Identical value of i.e.p. was reported by Pan et al. (2007). The LSZ molecule is positive charged in the range from pH = 3.0 to 10.0.

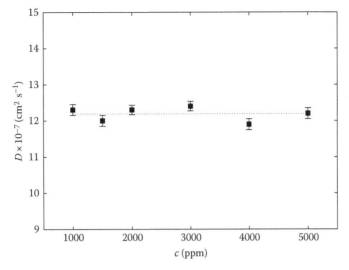

FIGURE 4.21 The dependence of the diffusion coefficient of LSZ as a function of the bulk concentration determined experimentally for pH = 6.5, $T = 298$ K: ■ $I = 0.15$ M. The line shows the linear fit of experimental points, i.e., $D = 12.2 \times 10^{-7}$ cm² s⁻¹.

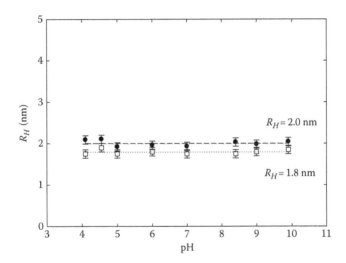

FIGURE 4.22 The hydrodynamic diameter R_H of LSZ determined by the DLS method, $c = 1500$ ppm, $I = 1 \times 10^{-2}$ M. The dashed lines denote the limiting analytical solutions calculated from Equation 4.14, ● the R_H value calculated from the averaged scattering intensity (upper line $R_H = 2.0$ nm), o the R_H value calculated from the maximum of the scattering peak (lower line $R_H = 1.8$ nm).

As suggested by previous measurements for particles (Adamczyk et al. 2004, Jachimska and Adamczyk 2007) and polyelectrolytes (Adamczyk et al. 2006), additional information on the shape of particles forming suspensions and their aggregation degree can be extracted from dynamic viscosity measurements carried out for low range of volume fractions. The slope of the relative viscosity of a suspension η_s/η versus its volume fraction Φ_V, called the intrinsic viscosity, can be quantitatively related to the shape of the molecules or aggregates. The results obtained for BSA are presented in Figure 4.24, where the dependence of the relative viscosity of the BSA solution protein $[\eta_{rel}]$ on $\Phi_V < 0.008$ and on $I = 1 \times 10^{-2}$ M is shown. As can be seen, the results obtained for pH range 4–9 could be well reflected by the linear dependence with the slope (intrinsic viscosity) equal to 4.5. It is

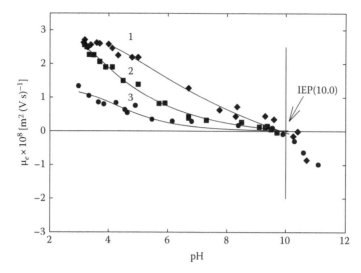

FIGURE 4.23 Electrophoretic mobility, μ_e, of LSZ as a function of pH, $T = 298\,K$. The points denote experimental values determined for 1. $I = 1 \times 10^{-3}\,M$, 2. $I = 1 \times 10^{-2}\,M$, 3. $I = 0.15\,M$.

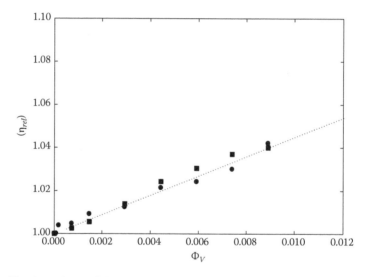

FIGURE 4.24 The dependence of the relative viscosity of BSA solutions [η_{rel}]on the volume fraction Φ_V determined for $T = 298\,K$, $I = 1 \times 10^{-2}\,M$, and various pH (• pH = 6.5, ■ pH = 9.0).The solid line denotes linear fits with slop equal to 4.5.

interesting to compare this result with the theoretical predictions stemming from the hydrodynamic theories of dilute suspension viscosity. As it is well known for rigid, spherical particles, the Einstein model predicts that the intrinsic viscosity [η_V] is equal to 2.5. This theory has been generalized by Brenner (1974), who considered the case of spheroidal particles immersed in linear velocity fields. In the case of prolate or oblate spheroids in simple shear flow (pertinent to our measurements), the intrinsic viscosity of their suspensions can be evaluated from the analytical dependence. Assuming as before that BSA has the prolate spheroid shape of the dimensions $9.5 \times 5 \times 5\,nm$ ($\lambda = 1.9$), one can predict from Equation 4.20 that the theoretical value of the intrinsic viscosity, corrected for the effective volume, equals to 4.4. For the compact dimer, having the shape of an oblate spheroid of the dimensions $11 \times 11 \times 5\,nm$, the theoretical value of the intrinsic viscosity equals to 4.5. As can be noticed, the average value of monomer and compact dimer is very close to what was experimentally measured for the entire pH range studied.

For higher rank aggregates, the intrinsic viscosity values predicted from Equation 4.20 are much higher, for example, 9.7 (for a linear aggregate formed of three BSA molecules) and 13.8 (for a four-molecule aggregate). Hence, the dynamic viscosity measurements confirmed association of BSA into a compact dimer structure, and excluded the possibility of forming the higher-order structures.

A very interesting effect was demonstrated by Komatsu et al. (2005) for the viscosity of aqueous BSA solution which was measured in the neutral and acidic region. With decreasing pH, the viscosity values of BSA increased abruptly for two different pH values that correspond to the structure transitions N (normal)–F (fast) at pH = 4.3 and F-E (expanded) at pH = 3.0. These measurements confirm that the N-BSA form is predominated between pH 4.3 to basic conditions. This form has the most compact state and the highest content of α-helix (Vermonden et al. 2001).

It is well known that variations in pH or salt concentrations result in changed protein behaviors upon aggregation and subsequent gelation. Weijers et al. (2008) presented the effect of ovalbumin net charge on the aggregate morphology. Using cryo-TEM and SEC-MALLS, they found that strong electrostatic repulsion does not entirely prevent the aggregation process. Cryo-TEM images showed that with increasing net charge the degree of branching and flexibility of the aggregates decreased. Not only the net charge but also the distribution of this charge on an unfolded protein molecule may play a crucial role in determining the fibril-forming properties.

The relative viscosity of the LSZ protein solutions η_s/η on $\Phi_V < 0.009$ is demonstrated in Figure 4.25. Results obtained for ionic strength $I = 0.15\,\text{M}$ (pH = 6.5 and pH 9.5) and $I = 1 \times 10^{-2}\,\text{M}$ (pH 6.5) could be well reflected by the linear dependence with the slope equal to 4.0. The correction for hydration effect leads to the increase of the apparent volume of LSZ, and the slope becomes equal to 3.0 ($[\eta]_c = 0.75[\eta]$). The intrinsic viscosity $[\eta_V]$ values are similar to the theoretical value predicted for the bare molecule ($4 \times 3 \times 3\,\text{nm}^3$) obtained from Equation 3.1. Using the method previously applied in the case of polyelectrolytes, one can calculate the effective length, L_e, of the molecule from the formula $L_e^* = \left(4v^*\lambda_c^2/\pi\right)^{1/3}$, where λ_c was calculated using the experimental values of the intrinsic viscosity, $v_m^* = v_m\left(d_h/d_c\right)^2$. In electrolyte solutions, the LSZ molecules are strongly hydrated as most proteins, which means that their effective diameter and consequently the aspect ratio is slightly larger. Assuming that on an average one monolayer of water dipoles having

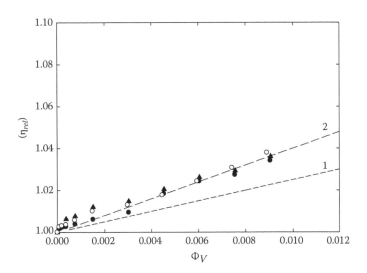

FIGURE 4.25 The dependence of the relative viscosity of LSZ solutions [η_{rel}] on the volume fraction Φ_V determined for $T = 298\,\text{K}$, $I = 0.15\,\text{M}$, and various pH (• pH = 6.5, ▲ pH = 9.5), and ionic strength $I = 1 \times 10^{-2}\,\text{M}$ (○ pH = 6.5). The solid line 2 denotes linear fits with the slop equal to 4.0 and line 1 with slope 2.5.

the length of 0.145 nm is adsorbed uniformly over the LSZ molecule, one can calculate that the hydrated diameter of $d_h = 3.29$ nm and the effective aspect ratio $\lambda^* = 1.3$. In this way, it was found that $L_e = 5.6$ nm, which is larger than the nominal length of LSZ in the crystalline state, 4.0 nm.

The literature reported that at relatively low concentration of LSZ, the system consists mostly of monomers or dimmers, while at high concentrations, large dynamic clusters dominate. These results are obtained by Kim et al. (2002), who adsorbed LSZ molecules at mica surface. The authors observed the cluster formations by the LSZ molecules with approximately five molecules per cluster. Analogous aggregation process for LSZ molecule was observed by Stradner et al. (2006).

All protein surfaces are composed of the mixture of hydrophilic and hydrophobic residues. The polar and apolar residues are more or less evenly distributed over the protein surface. This complex structure will be reflected in adsorption. The principle factors involved in protein adsorption at solid/liquid interfaces are the following: (a) structural rearrangements in the protein molecule, (b) dehydration of protein surface (part of), (c) redistribution of charged groups in the interfacial layer, and (d) polarity of protein surface. The adsorption mechanism is a multistep process, where the proteins can be found in many different conformational states (Vermonden et al. 2001, Norde 2008). This change of conformation upon adsorption is thought to be dependent on the initial protein concentration. For example, the BSA and HSA monomer structure is prolate spheroids (heart shapes), having the dimensions $9.5 \times 5 \times 5$ nm. Hence, two different adsorption modes are possible: side-on and unoriented. The adsorption of BSA at the solution/solid interface has been studied by a variety of experimental methods including the measurement of the difference in solution concentration before and after the adsorption using elipsometry, OWLS, total internal reflection (TIR), radiolabeling technique, quartz crystal microbalance (QCM), x-ray photoelectron spectroscopy (XPS), and AFM (Höök et al. 2002, Kingshott and Höcker 2002). However, a broader understanding of protein adsorption still remains a challenge to improving existing materials. For example, from the literature we found that BSA molecules are adsorbed side-on (Fitzpatrick et al. 1992), end-on (Gallinet and Gauthier-Manuel 1992), or form multilayer films (Terashima and Tsuji 2002). Interesting results for LSZ were described by Su et al. (1998). They report that the structural conformation of LSZ molecules adsorbed at the hydrophilic silica surface is determined by the lateral electrostatic repulsion within the layer, the magnitude of which is determined by the surface concentration in the layer and the net charge on the LSZ molecule.

Individual methods cannot give a complete description of the adsorption process; therefore, complementary methods should be used to provide different types of information, including the amount of adsorbed protein, the rate of adsorption, the conformation, and the orientation or aggregation of the adsorbed protein molecules.

Quantitative analyses of particle adsorption furnish interesting information on specific interactions under dynamic conditions. Furthermore, by measuring particle adsorption in model systems (monodisperse colloid suspension) gives important information concerning the mechanisms and kinetics of molecular adsorption, which is difficult to be obtained from direct experimental studies. The statistical–mechanical theories test the irreversible (colloidal) and reversible (molecular) systems. Adsorption of protein and colloid particles is a process that occurs via a more complicated path than molecular adsorption mainly due to irreversibility effects. This means that the adsorption kinetics, the maximum monolayer density, and its structure depend on the particle transport mechanism, either diffusion, convection, or migration rather than on particle concentration. The irreversible adsorption of particles is often analyzed in terms of the random sequential adsorption (RSA) model (Adamczyk 2000). This model assumes that particles adsorb randomly to the substrate. If the particle is incident at an empty region of the surface, it adsorbs irreversibly. If another particle is incident at a position with any overlap with previously adsorbed particles, it fails to adsorb. As the process continues, the surface coverage increased until it is saturated. This saturation threshold is referred to as the jamming limit Θ_∞. RSA model is applied for simulating adsorption of particles of various shapes. This phenomenon was demonstrated in Weroński (2001), the applicability of the RSA approach for predicting adsorption of spheroidal particles at solid/liquid interface. These theoretical results are relevant for modeling irreversible protein adsorption whose shape deviated from a spherical one (Adamczyk and Weroński 1997, Adamczyk et al. 2002). Using the RSA model, it is possible to determine adsorption kinetics, the structure of monolayer, and the jamming coverage Θ_∞ being the parameter of primary practical interest; see Table 4.9. The influence of the particle shape on the mass of particle monolayer is shown in Figure 4.26 (Weroński 2001) as the dependence

TABLE 4.9
Jamming Coverage for Adsorption of Hard Spheroids Particles Driven from RSA Model

Axis Ratio $\lambda = a/b$	Prolate Spheroids (Unoriented) $\Theta_\infty = \pi b^2 N_\infty$	Prolate Spheroids (side-On) $\Theta_\infty = \pi ab N_\infty$	Oblate Spheroids (Unoriented) $\Theta_\infty = \pi ab N_\infty$	Oblate Spheroids (Side-On) $\Theta_\infty = \pi a^2 N_\infty$
2	0.95	0.477	1.01	0.505
3	1.38	0.461	1.38	0.461
4	1.78	0.445	1.71	0.428
5	2.17	0.434	2.02	0.404
10	3.86	0.386	3.19	0.319

Source: Weroński, P., Kinetics and topology of irreversible adsorption of anisotropic particles at homogeneous interfaces. PhD thesis, Institute of Catalysis and Surface Chemistry, Polish Academy of Science, Cracow, Poland, 2001.

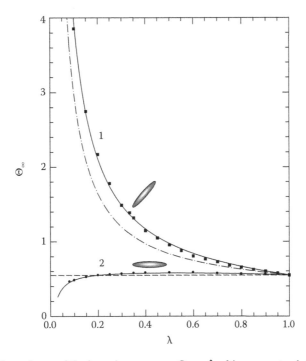

FIGURE 4.26 The dependence of the jamming coverage Θ_∞ on $\lambda = b/a$ parameter determined for hard prolate spheroids. The points denote exact numerical calculations performed for 1. Unoriented adsorption and 2. Side-on adsorption, and the solid line shows the analytical results. The dashed-dotted line shows the averaged results from the side-on and edge-on orientations. (From Adamczyk, Z. et al., *Colloids Surf. A*, 208, 29, 2002. With permission from Elsevier.)

of Θ_∞ on the $\lambda = a/b$ (longer to shorter axis ratio) parameter. The calculations were carried out for both the side-on adsorption of spheroids, when all particles lie flat at the interface, and for the unoriented adsorption. In the limit of $a/b \sim 1$ (spheres) the jamming coverage in both cases attains the value of 0.547, which is much smaller than the hexagonal packing of the spheres in two dimensions, $\pi / 2\sqrt{3} = 0.91$. The jamming coverage for the unoriented adsorption becomes much larger than for the side-on adsorption, which can be attributed to the possibility of a close to the perpendicular orientation of the particles that minimalize the area required for their adsorption. In the case of the side-on adsorption, the jamming coverage increases with a/b. This result concerned the case of hard particle adsorption when specific interactions among the particles were negligible. This condition is fulfilled for concentrated electrolyte solution when the electrostatic interactions are effectively eliminated, e.g., for protein adsorption under physiological conditions $I > 0.15\,M$. However, for dilute electrolyte or small particles, the repulsive electrostatic interaction may exert a significant influence on particle adsorption, especially on the jamming coverage. The range of electrostatic interactions is usually characterized by the dimensionless parameter κa. The inverse of the κa parameter has a physical interpretation as the dimensionless Debye screening length; for $(\kappa a)^{-1} < 1$ the range of electrostatic interaction becomes smaller than particle dimension. The RSA simulations of interacting spheroidal particles for side-on adsorption are shown in Figure 4.27 (Weroński 2001) and for unoriented adsorption in Figure 4.28 (Weroński 2001). The reduced maximum coverage is defined as $\bar{\Theta}_{max} = \Theta_{max}/\Theta_\infty$. The maximum coverage is considerably reduced for $\kappa a > 5$ values as a result of the lateral interactions among adsorbed particles. The electrostatic interaction among adsorbed particles influences not only the maximum coverage but also the structure of the adsorbed particle monolayer characterized by the pair correlation function g presented in Figure 4.29. The correlation function is accessible experimentally, e.g., by direct imaging of adsorbed particles using optical microscopy (particles, ranging from 40 nm to several hundreds of nanometers in diameter) (Adamczyk et al. 2005) or AFM (smaller entities like gold nanoparticles or dendrymers) (Kooij et al. 2002,

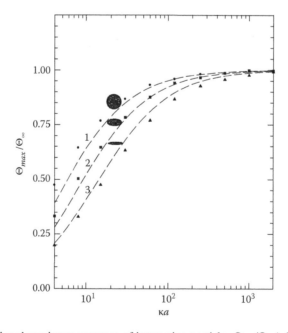

FIGURE 4.27 The reduced maximum coverage of interacting particles $\Theta_{max}/\Theta_\infty$ (where Θ_∞ is the jamming coverage of hard particles) versus κa. The points denote the numerical RSA simulations performed for side-on adsorption: 1. $\lambda = 1$ (spheres), 2. $\lambda = 2$, 3. $\lambda = 5$. The dashed lines denote the analytical results. (From Adamczyk, Z. et al., *Colloids Surf. A*, 208, 29, 2002. With permission from Elsevier.)

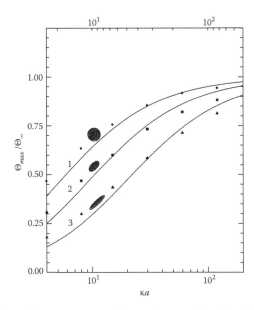

FIGURE 4.28 The reduced maximum coverage of interacting particles $\Theta_{max}/\Theta_\infty$ (where Θ_∞ is the jamming coverage of hard particles) versus κa. The points denote the numerical RSA simulations performed for unoriented adsorption: 1. $\lambda=1$ (spheres), 2. $\lambda=2$, 3. $\lambda=5$. The dashed lines denote the analytical results. (Reprinted from Weroński, P., Kinetics and topology of irreversible adsorption of anisotropic particles at homogeneous interfaces. PhD thesis, Institute of Catalysis and Surface Chemistry, Polish Academy of Science, Cracow, Poland, 2001. With permission.)

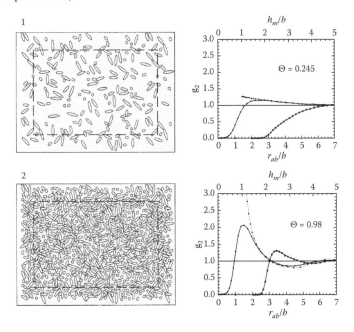

FIGURE 4.29 Configurations of interacting prolate spheroids adsorbing under the unoriented regime with corresponding pair correlation function g_2 generated numerically for $\lambda=5$, $H^*=0.1$, part 1, $\Theta=0.245$, part 2, $\Theta=0.98$ (maximum coverage). The continuous lines denote the surface-to-surface and the center-to-center pair correlation functions, and the dotted line represents the results obtained for hard spheroids. (Reprinted from Weroński, P., Kinetics and topology of irreversible adsorption of anisotropic particles at homogeneous interfaces. PhD thesis, Institute of Catalysis and Surface Chemistry, Polish Academy of Science, Cracow, Poland, 2001. With permission.)

Pericet-Camara et al. 2004, Cahill et al. 2008, Lundgren et al. 2008). From the analysis of the pair correlation function in limit for low coverage, it is possible to estimate directly the range and magnitude of the lateral interaction between adsorbed and adsorbing particles.

4.5 OUTLOOK

For a good understanding of the behavior of the anisotropic particles like polyelectrolytes or proteins, reliable information is necessary, which comes not only from the characterization of the bulk solution, but also from the adsorption of these molecules on the surface. Moreover, Norde (2008), who is an expert in the field of proteins, proposed further development of the research in the direction of "novel approach which should zoom in on discrete events that occur at a sub-molecular level, taking into account nanoscopic heterogeneity of the surface and special details of the molecules."

ACKNOWLEDGMENT

This work was supported by the Grant MNiSzW N 204 028536

LIST OF SYMBOLS

A_v	Avogadro number
a	Longer semiaxis of spheroids
b	Shorter semiaxis of spheroids
c	Concentration
c_1, c_2	Coefficient
$\mathbf{D}$	Translation diffusion coefficient matrix
$\mathbf{D_r}$	Rotation diffusion coefficient matrix
D	Diffusion coefficient
d	Diameter
d_e	Polymer chain diameter
d_h	Hydrated chain diameter
$\mathbf{E}$	Electric field
e	Elementary charge
$\mathbf{F}$	Force vector
f_c	Interaction parameter
G	Shear rate
g_2	Pair correlation functions
I	Ionic strength
I^*	Overall ionic strength of the polyelectrolyte solution
$K_\perp, K_\parallel$	Hydrodynamic resistance coefficients for the translation motion
$K_{r\parallel}, K_\perp$	Hydrodynamic resistance coefficients for the rotary motion
k	Boltzmann constant
L	Characteristic length scale of the motion
L_e	Characteristic length
L_p	Persistence length
l	Contour length
l_B	Bjerrum length
l_c	Distance between charged groups
$\mathbf{M}$	Translation mobility matrix
$\mathbf{M_r}$	Rotation mobility matrix
M	Molecular mass
N	Number of monomers

N_c	Number of elementary charges per molecule
N_m	Nominal number of charges per molecule
n_i	Concentration of ion in the bulk
n_p	Concentration of the polyelectrolyte in the bulk
Pe	Peclet number
p	Primary electroviscous function
Q_1, Q_2, Q_3	Coefficients
q	Electric charge
q_0	Manning parameter
R	Capillary radius
Re	Reynolds number
R_g	Radius of gyration
R_H	Hydrodynamic radius
R_s	Equivalent sphere radius
T	Absolute temperature
t	Time
$\mathbf{U}$	Migration velocity
V	Characteristic velocity
V_{sus}	Suspension volume
ν	Kinematic viscosity
v_m	Molecular volume
v^*	Hydrated volume of molecules
v_{max}	Maximum suspension velocity
z_i	Valency of the ions
z_c	Valency of the counterions

GREEK

α	Parameter describing ellipsoid resistance coefficient
α_i^*	Effective ionization degree
β	Parameter describing ellipsoid resistance coefficient
ε	Dielectric constant
κ	Specific conductivity of the solution
κ_0	Specific conductivity of the solvent
κ^{-1}	Electric double layer thickness
Θ_{max}	Maximum surface coverage
Θ_∞	Jamming coverage of particles
λ	Aspect ratio
λ_c	Dimensionless aspect ratio
λ_c^0	Equivalent conductivity of the counterion
λ_p	Equivalent conductivity of the polyion
μ_e	Electrophoretic mobility
ρ	Density
ρ_p	Specific density
η	Dynamic viscosity of fluid
$[\eta]$	Intrinsic viscosity
$[\eta_{rel}]$	Relative viscosity
η_s	Solution viscosity
τ_d	Mass diffusion relaxation time
τ_h	Relaxation time
Φ_V	Volume fraction

REFERENCES

Abramson, H. A. and Gorin, M. L. S., 1942. *Electrophoresis of Proteins*, Hafner Publishing Company, New York.

Adamczyk, Z., 2000. Kinetics of diffusion-controlled adsorption of colloid particles and proteins. *J. Colloid Interface Sci.*, 229, 477–489.

Adamczyk, Z., 2006a. Particle adsorption and deposition: Role of electrostatic interactions. *Adv. Colloid Interface Sci.*, 100–102, 267–347.

Adamczyk, Z., 2006b. *Particles at Interfaces: Interactions, Deposition, Structure*, Elsevier, London, U.K.

Adamczyk, Z. and Weroński, P., 1997. Unoriented adsorption of interacting spheroidal particles. *J. Colloid Interface Sci.*, 189, 348–360.

Adamczyk, Z., Weroński, P., and Musiał, E., 2002. Particle adsorption under irreversible conditions: Kinetics and jamming coverage. *Colloids Surf. A*, 208, 29–40.

Adamczyk, Z., Jachimska, B., and Kolasińska, M., 2004a. Structure of colloid silica determined by viscosity measurements. *J. Colloid Interface Sci.*, 273, 668–674.

Adamczyk, Z., Zembala, M., Warszyński, P., and Jachimska, B., 2004b. Characterization of polyelectrolyte multilayers by the streaming potential method. *Langmuir*, 20, 10517–10525.

Adamczyk, Z., Jaszczółt, K., Michna, A., Siwek, B., Szyk-Warszyńska, L., and Zembala, M., 2005. Irreversible adsorption of particles on heterogeneous surfaces. *Adv. Colloid Surf. Sci.*, 118, 25–42.

Adamczyk, Z., Bratek, A., Jachimska, B., Jasiński, T., and Warszyński, P., 2006. Structure of poly(acrylic acid) in electrolyte solutions determined from simulations and viscosity measurements. *J. Phys. Chem. B*, 110, 22426–22435.

Adamczyk Z., Jachimska, B., Jasiński, T., Warszyński, P., and Wasilewska, M., 2009. Structure of poly(sodium 4-styrenesulfonate) (PSS) in electrolyte solutions: Theoretical modeling and measurements. *Colloids Surf. A*, 343, 96–103.

Anufrieva, E. V., Birshtein, T. M., Nekrasova, T. N., Ptitsyn, C. B., and Sheveleva, T. V., 1968. *J. Polym Sci. Part C Polym. Symp.*, 16, 3514.

Beer, M., Schmidt, M., and Muthukumar M., 1997. The electrostatic expansion of linear polyelectrolytes: Effects of gegenions, co-ions and hydrophobicity. *Macromolecules*, 30, 8375–8385.

Berg, J. M., Tymoczko, J. L., and Stryer, L., 2002. *Biochemistry*, 5th ed., Freeman & Co, New York.

Bordi, F., Cametti, C., Motta, A., and Paradossi, G., 1999. Electrical conductivity of dilute and semidilute aqueous polyelecrolyte solutions. A scaling theory approach. *J. Phys. Chem. B*, 103, 5092–5099.

Bordi, F., Colby, R. H., Cametti, C., de Lorenzo, L., and Gili, T., 2002. Electrical conductivity of polyelectrolyte solutions in the semidilute and concentrated regime: The role of counterion condensation. *J. Phys. Chem. B*, 106, 6887–6893.

Brenner, H., 1974. Rheology of a dilute suspension of axisymmetric Brownian particles. *J. Multiphase Flow*, 1, 195–341.

Burton, W. G., Nugent, K. D., Slattery, T. K., Summers, B. R., and Snyder, L. R. 1988, Separation of proteins by reversed-phase HPLC, optimizing the column. *J. Chromatogr. A*, 443, 363–379.

Cahill, B. P., Papastavrou, G., Koper, G. J. M., and Borkovec M., 2008. Adsorption of poly(amido amine) (PAMAM) dendrymers on silica: Importance of electrostatic three-body attraction. *Langmuir*, 24, 465–473.

Castelnovo, M. and Joanny, J.-F., 2000. Formation of polyelectrolyte multilayers. *Langmuir*, 16, 7524–7532.

Choi, J. and Rubner, M., 2005. Influence of the degree of ionization on weak polyelectrolyte multilayer assembly. *Macromolecules*, 38, 116–124.

Chu, D. Y. and Thomas, J. K., 1987. Photophysical and photochemical studies on a polymeric intramolecular micellar system PA-18 K2. *Macromolecules*, 20, 2133–2138.

Crescenzi, V., Quadrifogio, F., and Delben, F., 1972. Calorimetric investigation of poly(methacrylic acid) and poly(acrylic acid) in aqueous solution. *J. Polym. Sci. Part A*, 2(10), 347–368.

Dawson, R. M., 1986. *Data for Biochemical Research*, Clarendon Press, Oxford, NY.

Davis, R. M. and Russel, W. B., 1987. Intrinsic viscosity and Huggins coefficients for potassium poly(styrenesulfonate) solutions. *Macromolecules*, 20, 518–525.

Decher, G., 1997. Fuzzy nanoassemblies: Toward layered polymeric multicomposites. *Science*, 277, 1232–1237.

Decher, G. and Schmitt, J., 1992. Fine-tuning of the film thickness of ultrathin multilayer films composed of consecutively alternating layers of anionic and cationic polyelectrolytes. *Prog. Polym. Sci.*, 89, 160–164.

Decher, G., Honig, J. D., and Schmitt, J., 1992. Buildup of ultrathin multilayer films by a self-assembly process: III consecutively alternating adsorption of anionic and cationic polyelectrolytes on charged surfaces. *Thin Solid Films*, 210, 831–835.

Decher, G., Lvov, Y., and Schmitt, J., 1994. Proof of multilayer structural organization in self-assembled polycation-polyanion molecular films. *Thin Solid Films*, 244, 772–777.

de Gennes, P. G., Pincus, P., Velasco, R. M., and Brochard, F., 1976. Remarks on polyelectrolyte conformation. *J. Phys.*, 37, 1461–1476.

Desroches, M., Chaudhary, N., and Omanovic, S., 2007. PM-IRRAS investigation of the interaction of serum albumin and fibrinogen with a biomedical grade stainless steel 316 LVM surface. *Biomacromolecules*, 8, 2836–2844.

Dobrynin, A. V. and Rubinstein, M., 2005. Theory of polyelectrolytes in solutions and at surfaces. *Prog. Polym Sci.*, 30, 1049–1118.

Dobrynin, A. V., Colby, R. H., and Rubinstein, M., 1995. Scaling theory of polyelectrolyte solutions. *Macromolecules*, 28, 1859–1871.

Donath, E., Walther, D., Shilov, V. N., Knippel, E., Budde, A., Lowack, K., Helm, C. A., and Moehwald, H., 1997. Nonlinear hairy layer theory of electrophoretic fingerprinting applied to consecutive layer by layer polyelectrolyte adsorption onto charged polystyrene latex particles. *Langmuir*, 13, 5294–5305.

Fitzpatrick, H., Luckham, P. F., Eriksen, S., and Hammond, K., 1992. Bovine serum albumin adsorption to mica surfaces. *Colloids Surf.*, 65, 43–49.

Foster, J. F., 1977, *Albumin Structure, Function and Uses*, Rosenoer, V. M., Oratz, M., and Rothschild, M. A. (Eds.), Pergamon, Oxford, NY, pp. 53–84

Fuoss, R. M., 1951. Polyelectrolytes. *Discuss. Faraday. Soc.*, 11, 125.

Gallinet, J. P. and Gauthier-Manuel, B., 1992. Adsorption–desorption of serum albumin on bare mica surfaces. *Colloids Surf.*, 68, 189–193.

Golander, C. G. and Kiss, E., 1988. Protein adsorption on functionalized and ESCA characterized polymer films studied by ellipsometry. *J. Colloid Interface Sci.*, 121, 240–253.

Gromer, A., Rawiso, M., and Maaloum, M., 2008. Visualization of hydrophobic polyelectrolytes using atomic Force microscopy in solution. *Langmuir*, 24(16), 8950–8953.

Hara, M., 1993. *Polyelectrolytes*, Marcel Dekker, New York.

Harding, S. E., 1995. On the hydrodynamic analysis of macromolecular conformation. *Biophys. Chem.*, 55, 69–93.

Haynes, C. A. and Norde, W., 1994. Globular proteins at solid/liquid interfaces. *Colloids Surf. B*, 2, 517–566.

Hirayama, K., Akashi, S., Furuya, M., and Fukuhara, K., 1990. Rapid confirmation and revision of the primary structure of bovine serum albumin by ESIMS and frit-FAB LC/MS. *Biochem. Biophys. Res. Commun.*, 173(2), 639–646.

Höök, F., Vörös, J., Rodahl, M., Kurrat, R., Böni, P., Ramsden, J. J., Textor, M., Spencer, N. D., Tengvall, P., Gold, J., and Kasemo, B., 2002. A comparative study of protein adsorption on titanium oxide surfaces using in situ ellipsometry, optical waveguide ligthmode spectroscopy and quartz crystal microbalance/dissipation. *Colloids Surf. B*, 24, 155–170.

Jachimska, B. and Adamczyk, Z., 2007. Characterization of rheological properties of colloidal zirconia. *J. Eur. Ceram. Soc.*, 27(5), 2209–2015.

Jachimska, B., Wasilewska, M., and Adamczyk, Z., 2008. Characterization of globular protein solutions by dynamic light scattering, electrophoretic mobility and viscosity measurements. *Langmuir*, 24(13), 6866–6872.

Jachimska, B., Jasiński, T., Warszyński, P., and Adamczyk, Z., 2010. Structure of polyallylamine hydrochloride (PAH) in electrolyte solutions: Theoretical modeling and measurements. *Colloids Surf. A*, 355, 7–15.

Jasiński, T., Cuisinier, F., and Warszyński, P., 2007. Molecular dynamic simulations of polyelectrolytes: Effect of force field parameters on the conformations and effective charge. In *Surfactants and Dispersed Systems in Theory and Practice*, Wilk, K. A. (Ed.), Wrocław, Poland, p. 65.

Jeon, J. and Dobrynin, A. V., 2007. Necklace globule and counterion condensation. *Macromolecules*, 40, 7695–7706.

Kaufman, E. D., Beleyea, J., Johnson, M. C., Nicholson, Z. M., Ricks, J. L., Shah, P. K., Bayless, M. et al., 2007. Probing protein adsorption onto mercaptoundecanoic acid stabilized gold nanoparticles and surfaces by quartz crystal microbalance and ξ-potential measurements. *Langmuir*, 23, 6053–6062.

Kayushina, R., Lvov, Y., Stepina, N., Belyaev, V., and Khurgin, Y., 1996. Construction and x-ray reflectivity study of self-assembled lysozyme/polyion multilayers. *Thin Solid Films*, 284, 246–248.

Kim, D., Han S. W., Kim, C. H., Hong, J. D., and Kim, K., 1999. *Thin Solid Films*, 350, 153.

Kim, D. T., Blanch, H. W., and Radke, C. J., 2002. Direct imaging of lysozyme adsorption onto mica by atomic force microscopy. *Langmuir*, 18, 5841–5850.

Kingshott, P. and Höcker, H., 2002. Methods of assessing protein adsorption. In *Encyclopedia of Surface and Colloid Science*, Hubbard, A. T. (Ed.), Marcel Dekker, New York, pp. 3342–3365.

Kiriy, A., Gorodyska, G., Minko, S., Jaeger, W., Stepanek, P., and Stamm, M., 2002. Cascade of coil-globule conformational transitions of single flexible polyelectrolyte molecules in poor solvent. *J. Am. Chem. Soc.* 124, 13454–13462.

Kirkwood, J. and Auer, P. L., 1951. The visco-elastic properties of solution of rod-like molecules. *J. Chem. Phys.*, 19, 281–285.

Kirwan, L. J., Papastavrou, G., and Borkovec, M., 2004. Imaging the coil-to-globule conformational transition of a weak polyelectrolyte by tuning the polyelectrolyte charge density. *Nano Lett.*, 4, 149–152.

Koene, R. S. and Mandel, M., 1983. Scaling relations for aqueous polyelectrolyte-salt solutions. 1. Quasi-elastic light scattering as a function of polyelectrolyte concentration and molar mass. *Macromolecules*, 16, 220–227.

Koenig, J. L., Angoud, A. C., Semen, J., and Lando, J. B., 1969. Laser-excited Raman studies of the conformational transition of syndiotactic polymethacrylic acid in water. *J. Am. Chem. Soc.*, 91, 7250–7254.

Komatsu, U., Matsuki, H., Kansshina, S., and Ogli, K., 2005. Effect of an inhalation anesthetic on the viscosity of aqueous bovine serum albumin solutions. *Int. Congr. Ser.*, 1283, 322–323, www.ics-elsevier.com.

Kooij, E. S., Brouwer, E. A. M., Wormeester, H., and Poelsema, B., 2002. Ionic strength mediated self-organization of gold nanocrystals: An AFM study. *Langmuir*, 18, 7677–7680.

Kruif, C. G., van Iersel, E. M. F., and Vrij, A., 1985. Hard sphere colloidal dispersions: Viscosity as a function of shear rate and volume fraction. *J. Chem. Phys.*, 83, 4717–4723.

Ladam, G., Schaad, P., Voegel, J.-C., Schaaf, P., Decher, G., and Cuisinier, F. J. G., 2000. In situ determination of the structural properties of initially deposited polyelectrolyte multilayers. *Langmuir*, 16, 1249–1255.

Lassen, B. and Malmsten, M., 1996. Structure of protein layers during competitive adsorption. *J. Colloid Interface Sci.*, 180, 339–349.

Lavalle, Ph., Gergely, C., Cuisinier, F. J. G., Decher, G., Schaaf, P., Voegel, J.-C., and Picart, C., 2002. Comparison of the structure of polyelectrolyte multilayer films exhibiting a linear and an exponential growth regime: An in situ atomic force microscopy study. *Macromolecules*, 35, 4458–4465.

Leporatti, S., Voigt, A., Mitlohner, R., Sukhorukov, G., Donath, E., and Möhwald H., 2000. Scanning force microscopy investigation of polyelectrolyte nano- and microcapsule wall texture. *Langmuir*, 16, 4059–4063.

Liebmann-Vinson, A., Lander, L. M., Foster, M. D., Brittain, W. J., Vogler, E. A., Majkrzak, C. F., and Satija, S., 1996. A neutron reflectometry study of human serum albumin adsorption in situ. *Langmuir*, 12, 2256–2262.

Lifshitz, M., Grosberg, A., and Khokhlov, A. R., 1978. Some problems of the statistical physics of polymer chains with volume interaction. *Rev. Mod. Phys.*, 50, 683–713.

Lobo, R. F. M., Pereira-da-Silva, M. A., Raposo, M., Faria, R. M., and Oliveira, O. N., 1999. In situ thickness measurements of ultra-thin multilayer polymer films by atomic force microscopy. *Nanotechnology*, 10, 389–393.

Longsworth, L. F., 1954. Temperature dependence of diffusion in aqueous solutions. *J. Phys. Chem.*, 58, 770–773.

Losche, M., Schmitt, J., Decher, G., Bouwman, W. G., and Kjear, K., 1998. Detailed structure of molecularly thin polyelectrolyte multilayer films on solid substrates as revealed by neutron reflectometry. *Macromolecules*, 31, 8893–8906.

Lundgren, A. O., Björefors, F., Olofsson, L. G. M., and Elwing, H., 2008. Self-arrangement among charge-stabilized gold nanoparticles on a dithiothreitol reactivated octanedithiol monolayer. *Nano Lett.*, 8(11), 3989–3992.

Lvov, Y., Decher, G., and Möhwald, H., 1993. Assembly, structural characterization and thermal behaviour of layer-by-layer deposited ultrathin films of poly(vinyl sulfate) and poly(allylamine). *Langmuir*, 9, 481–486.

Lvov, Y., Ariga, K., Ichinose, I., and Kunitake, T., 1995. Assembly of multicomponent protein films by means of electrostatic layer by layer adsorption. *J. Am. Chem Soc.*, 117, 6117–6123.

Lvov, Y., Onda, M., Ariga, K., and Kunitake, T., 1998. Ultrathin films of charged polysaccharides assembled alternately with linear polyions. *J. Biomater. Sci., Polym.*, 9(4), 345–355.

Lvov, Y., Ariga, K., Onda, M., Ichionose, I., and Kunitake, T., 1999. A careful examination of the adsorption step in the alternate layer-by layer assembly of linear polyanion and polycation. *Colloids Surf. A*, 146, 337–346.

Mandel, M., 1993. *Polyelectrolytes*, Hara, M. (Ed.), Marcel Dekker, New York.

Mandel, M., Leyte, J. C., and Stadhonder, M. G., 1967. The conformational transition of poly(methacrylic acid) in solution. *J. Phys. Chem.*, 71, 603.

Manning, G. S., 1969. Limiting laws and counterion condensation in polyelectrolyte solutions 1. Colligative properties. *J. Phys. Chem.*, 51, 924–933.

McAloney, R. A., Sinyour, M., Dudnik, V., and Goh, M. C., 2001. Atomic force microscopy studies of salt effects on polyelectrolyte multilayer film morphology. *Langmuir*, 17, 6655–6663.

Menchaca, J. L., Jachimska, B., Perez, E., and Cuisinier, F. J. G., 2003. In situ surface structure study of polyelectrolyte multilayers by liquid-cell AFM. *Colloids Surf. A*, 222, 185–194.

Mendelsohn, J. D., Barrett, C. J., Chan, V. V., Pal, A. J., Mayes, A. M., and Rubner M. F., 2000. Fabrication of microporous thin films from polyelectrolyte multilayers. *Langmuir*, 16, 5017.

Minko, S., Kiriy, A., Gorodyska, G., and Stamm, M., 2002a. Single flexible hydrophobic polyelectrolyte molecules adsorber on solid substrate: Transition between a stretched chain, necklace-like conformation and a globule. *J. Am. Chem. Soc.* 124, 3218–3219.

Minko, S., Kiriy, A., Gorodyska, G., and Stamm, M., 2002b. Mineralization of single flexible polyelectrolyte molecules. *J. Am. Chem. Soc.*, 124, 10192–10197.

Muthukumar, M., 2004. Theory of counter-ion condensation on flexible polyelectrolytes: Adsorption mechanism. *J. Chem. Phys.* 120, 9343–9350.

Noda, L., Tsuge, T., and Nagasawa, M., 1970. The intrinsic viscosity of polyelectrolytes. *J. Phys. Chem.*, 74, 710–719.

Norde, W., 2008. My voyage of discovery to proteins in flatland... and beyond. *Colloids Surf. B*, 61, 1–9.

Norde, W. and Anusiem, A. C. I., 1992. Adsorption, desorption and re-adsorption of proteins on solid surfaces. *Colloids Surf.*, 66, 73–80.

Odjik, T., 1977. Polyelectrolytes near the rod limit. *J. Polym. Sci. Polym. Phys. Ed.*, 15, 477.

Onda, M., Lvov, Y., Ariga, K., and Kunitake, T., 1997. Sequential reaction and product separation on molecular films of glucoamylase and glucose oxidase assembled on an ultrafilter. *J. Appl. Phys.*, 36, 608.

Ossawa, F., 1970. *Polyelectrolytes*, Marcel Dekker, New York.

Pan, X., Yu, S., Yao, P., and Shao, Z., 2007. Self-assembling of β-casein and lysozyme. *J. Colloid Interface Sci.*, 316, 405–412.

Park, J. M., Muhoberac, B. B., Dubin, P. L., and Xia, J., 1992. Effects of protein charge heterogenity in protein-polyelectrolyte complexation. *Macromolecules*, 25, 290–295.

Pericet-Camara, R., Papastavrou, G., and Borkovec, M., 2004. Atomic force microscopy study of the adsorption and electrostatic self-organization of poly(amidoamine) dendrymers on mica. *Langmuir*, 20, 3264–3270.

Picart, C., Ladam, G., Senger, B., Voegel, J.-C., Schaaf, P., Cuisinier, F. J. G., and Gergely, C., 2002. Determination of structural parameters characterizing thin films by optical methods: A comparison between scanning angle reflectometry and optical waveguide lightmode spectroscopy. *J. Chem. Phys.*, 115, 1086–1094.

RCSB Protein data Bank, http://www.pdb.mde-berlin.de

Rezwan, K., Meier, L. P., Rezwan, M., Voros, J., Textor, M., and Gauckler, L. J., 2004. Bovine serum albumin adsorption onto colloidal Al_2O_3 particles: A new model based on zeta potential and UV-Vis measurements. *Langmuir*, 20, 10055–10061.

Rezwan, K., Meier, L. P., and Gauckler, L. J., 2005. A prediction method for the isoelectric point of binary protein mixtures of bovine serum albumin and lysozyme adsorbed on colloidal titania and alumina particles. *Langmuir*, 21, 3493–3497.

Roiter, Y. and Minko, S., 2005. AFM single molecule experiments at the solid-liquid interface: In situ conformation of adsorbed flexible polyelectrolyte chains. *J. Am. Chem. Soc.*, 127, 15688–15689.

Rushing, T. S. and Hester, R. D., 2004. Semi-empirical model for polyelectrolyte intrinsic viscosity as a function of solution ionic strength and polymer molecular weight. *Polymer*, 45, 6587–6594.

Schlenoff, J. B., Ly, H., and Li, M., 1998. Charge and mass balance in polyelectrolyte multilayers. *J. Am Chem. Soc.*, 120, 7626–7634.

Schwinte, P., Voegel, J.-C., Picart, C., Haikel, Y., Schaaf, P., and Szalontai, B., 2001. Stabilizing effects of various polyelectrolyte multilayer films on the structure of adsorbed, embedded fibrinogen molecules an ATR-FTIR study. *J. Phys. Chem. B*, 105, 11906–11916.

Sedlak, M. and Amis E. J., 1992. Dynamics of moderately concentrated salt-free polyelectrolyte solutions:molecular weight dependence. Phase diagram of salt-free polyelectrolyte solutions as studied by light scattering. *J. Chem. Phys.*, 96, 817–834.

Skolnick, J. and Fixman, M., 1977. Electrostatic persistence length of a wormlike polyelectrolyte. *Macromolecules*, 10, 944–948.

Stoll, S. and Chodanowski, P., 2002. Polyelectrolyte adsorption on an oppositely charged spherical particle. Chain rigidity effects. *Macromolecules*, 35, 9556–9562.

Stradner, A., Sedgwick, H., Cardinaux, F., Poon, W. C., Egelhaaf, S. U., and Schurtenberger P., 2004. Equilibrium cluster formation in concentrated protein solutions and colloids. *Nature*, 432, 492–495.

Stradner, A., Cardinaux, F., and Schurtenberger, P., 2006, A small-angle scattering study on equilibrium clusters in lysozyme solutions. *J. Phys. Chem. B*, 110, 21222–21231.

Su, T. J., Lu, J. R., Thomas, R. K., Cui, T. A., and Penfold, J., 1998. The adsorption of lysozyme at the silica–water interface: A neutron reflection study. *J. Colloid Interface Sci.*, 203, 419–429.

Su, T. J., Lu, J. R., Thomas, R. K., and Cui, Z. F., 1999. Effect of pH on the adsorption of bovine serum albumin at the silica/water interface studied by neutron reflection. *J. Phys. Chem. B*, 103, 3727–3736.

Tanaka, N., Nishizawa, H., and Kunugi, S., 1997. In conducting research utilizing recombination DNA technology. *Biochim. Biophys. Acta*, 1366, 13–20.

Terashima, H. and Tsuji, T., 2002. Adsorption of bovine serum albumin onto mica surfaces studied by a direct weighing technique. *Colloids Surf. B*, 27, 115–122.

Thierry, B., Winnik, F. M., Merhi, Y., and Tabrizian, M., 2003. Nanocoatings onto arteries layer-by-layer deposition toward the *in vivo* repair of damaged blood vessels. *J. Am. Chem. Soc.*, 125, 7494–7495.

de la Torre, G. J., 2001. Hydration from hydrodynamics. General considerations and applications of bead modeling to globular proteins. *Biophys. Chem.* 93, 159–170.

Tsukruk, V. V., Bliznyuk, V. N., Visser, D., Campbell, A. L., Bunning, T. J., and Adams, W. W., 1997. Electrostatic deposition of polyionic monolayers on charged surfaces. *Macromolecules*, 30, 6615–6625.

Wang, L. and Yu, H., 1988. Chain conformation of linear polyelectrolyte in salt solutions: Sodium poly(styrene sulfonate) in potassium chloride and sodium chloride solutions. *Macromolecules*, 21, 3498–3501.

Weijers, M., Broersen, K., Barneveld, P. A, Cohen Stuart, M. A., Hamer, R. J., de Jongh, H. H. J., and Visschers, R. W., 2008. Net charge affects morphology and visual properties of ovalbumin aggregates. *Biomacromolecules*, 9(11), 3165–3172.

Weroński, P., 2001. Kinetics and topology of irreversible adsorption of anisotropic particles at homogeneous interfaces. PhD thesis, Institute of Catalysis and Surface Chemistry, Polish Academy of Science, Cracow, Poland.

van de Ven, T. G. M., 1989. *Colloidal Hydrodynamics*, Academic Press, London, U.K.

Vermonden, T., Giacomelli, C. E., and Norde, W., 2001. Reversibility of structural rearrangements in bovine serum albumin during homomolecular exchange from AgI particles. *Langmuir*, 17, 3734–3740.

Vink, H., 1995. Renewed studies of the conductance–anisotropy effect in polyelectrolyte solutions. *J. Colloid Surf. Sci.*, 173, 211–214.

Yamakawa, A. and Fujii, M., 1974. Intrinsic viscosity of wormlike chains. Determination of the shift factor. *Macromolecules*, 7, 128–135.

Yamanaka, J., Matsuoka, H., Kitano, H., Hasegawa, M., and Ise, N., 1990. Revisit to the intrinsic viscosity-molecular weight relationship of ionic polymers. 2. Viscosity behaviour of salt-free aqueous solutions of sodium poly(styrenesulfonates). *J. Am. Chem. Soc.*, 112, 587–592.

Yoon, B. J. and Kim, S., 1989. Electrophoresis of spheroidal particles. *J. Colloid Interface Sci.* 128, 275–288.

5 Lysine-Based Surfactants

Ramon Pons, María del Carmen Morán,
María Rosa Infante, Aurora Pinazo, and Lourdes Pérez

CONTENTS

5.1 INTRODUCTION

There is today a strong trend to replace conventional surfactants with more environmentally benign compounds. Manufacturers and consumers demand for novel environmentally friendly surfactants from renewable resources produced by clean and sustainable technologies (bio-based surfactants). The challenge is to find molecules that meet the mildness, biodegradability, as well as performance and cost–benefit requirements. The use of renewable raw materials such as amino acids and vegetable oil derivatives to prepare novel "natural" surfactants is an exciting and attractive research activity to conciliate the sustainable issues with the industrial development.

Amino acid-based surfactants constitute an important class of natural surface-active bio-molecules of great interest to organic and physical chemists as well as to biologists, with an unpredictable number of basic and industrial applications (Xia and Nnanna 2001). Structurally, lipoaminoacids are a very heterogeneous group of compounds but with a common advantage—they are relatively easy to design and synthesize. Often these molecules combine charged, or non-charged residues [i.e., glutamic acid (Glu), lysine (Lys), arginine (Arg), serine (Ser), leucine (Leu), phenylalanine (Phe), and alanine (Ala)] as the hydrophilic head group with a hydrophobic tail of a different structure, length, and number (i.e., fatty acids, fatty alcohols, fatty amines) as synthons for the amphiphilic structure (Takehara 1989, Boyat et al. 2000, Gerova et al. 2008, Vijay et al. 2008). This fact explains the diversity of amino acid/peptide-based surfactants and the variety of their physicochemical and biological properties (Das et al. 2006, Roy and Dey 2006, Varka et al. 2006, Capone et al. 2008, Ohta et al. 2008).

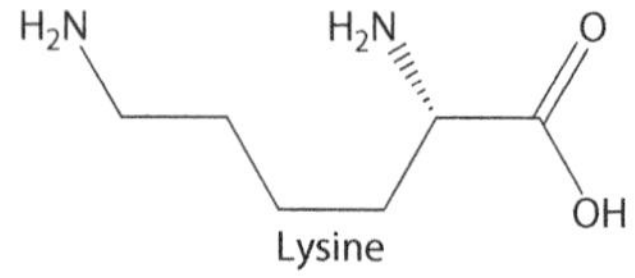

For more than 25 years, our group has synthesized and studied new monodisperse, amino acid-based surfactants of diverse structure [monocatenary (Molinero et al. 1988, Vinardell et al. 1990, Infante and Moses 1994, Infante et al. 1997, Piera et al. 1998), gemini (Pinazo et al. 1993, 1994, 1998, Pérez, et al. 1996, 2002a, David et al. 2002), and glycerolipid (Morán et al. 2001, 2002, 2004a, Pérez et al. 2002b, 2004)] with multifunctional properties. These compounds were obtained from the condensation of vegetable oil derivatives (fatty acids, amines and alcohols, and mono- and diacyl glycerides) with different amino acids for applications in the food and cosmetics sector. Our research in this field has contributed to basic science as well as to technology developments, and it has been an interdisciplinary research, drawing on the expertise of researchers from chemical synthesis, biocatalysis, physical chemistry of colloids and surfaces, ecotoxicity, toxicology, microbiology, and numerous industrial collaborators. Furthermore, we have developed green chemistry (Clapés et al. 1999, Morán et al. 2004b)—synthesis using nonorganic solvents (i.e., performed in aqueous media), biotechnological methods in the enzymatic catalysis for amide and ester bond formation (Clapés and Infante 2002), and analytical methods for the purification of the surfactants on semipreparative and preparative scales (Piera et al. 1999, Torres et al. 2001). In this chapter, the authors will give special attention to the lysine-based surfactants.

FIGURE 5.1 Chemical structure of lysine.

Lysine (Lys) is an essential amino acid and belongs to the aspartate biosynthetic pathway. Lysine can be produced by bacterial fermentation (Kinoshita et al. 1957, Parekh et al. 2000). Since the middle of the last century, a lot of lysine surfactant molecules of different structures have been described, although only a few have been commercialized (Xia et al. 2001). The fatty chain can be introduced into the lysine acid structure (Figure 5.1) through acyl, ester, alkyl, or amide linkages. The synthetic surfactants can yield anionic, cationic, amphoteric, and nonionic derivatives. Given the chemical duality (amino and carboxy groups) of the amino acid building block, the ionic nature of the amphiphile lysine derivatives depends on pH and on the specific structure modification of the surfactant.

Amphiphiles with different structures can be designed by hydrophobic modulation, although most of them are N^α-acyl lysine (Figure 5.2, **1**) or N^ε-acyl lysine (Figure 5.2, **2**) salt or ester derivatives. Thus, monocatenary—one-lysine-residue-bearing one hydrophobic tail (Figure 5.3, **1**), dicatenary—one-lysine-residue-bearing two hydrophobic chains (Figure 5.3, **2**), gemini—two lysine polar heads and two hydrophobic tails per molecule (Figure 5.3, **3**), and glycerolipid derivatives—one lysine polar head and one or two hydrophobic moieties linked together through a glycerol skeleton (Figure 5.3, **4**), characterized by the presence of weak amide and/or ester bonds anywhere in the molecule have been described (Infante et al. 1984, Pérez et al. 1996, Pegiadou et al. 2000, Morán et al. 2004b, Brito et al. 2006, Tan and Xiao 2008).

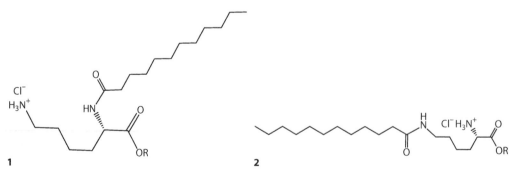

FIGURE 5.2   Chemical structure of N^α-acyl lysine (**1**) and N^ε-acyl lysine (**2**) derivatives, where R is Me, Et, H, Na⁺.

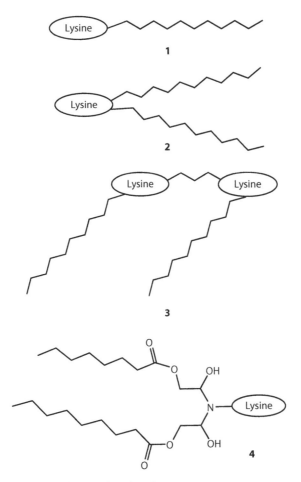

FIGURE 5.3 Schematic structure of lysine-based surfactants.

In this chapter, we will discuss some synthetic aspects of the preparation of these surfactants as well as the aggregation and uses of the main families present in the literature. Oligo- and polypeptide surfactants are out of the scope of this review.

5.2 SYNTHETIC ASPECTS

The structure of the lysine amino acid allows for a versatile use as a building block in surfactant synthesis (Figure 5.1). The presence of one carboxylate (anionic) and two amino (cationic) moieties in the molecule of lysine makes the synthesis of surfactants with different ionic characters in the polar head groups (anionic, cationic, nonionic, and amphoteric derivatives) possible by introducing hydrophobic groups (fatty acid/fatty amine or fatty alcohol) into the molecule and also possible capping groups (Figure 5.3).

5.2.1 MONOCATENARY LYSINE DERIVATIVES

The structure is characterized by a fatty acyl radical linked to the α- (Figure 5.2, **1**) or ε-amino (Figure 5.2, **2**) group of lysine through an amide linkage that has a strong hydrogen-bonding ability (Takehara 1989). The synthesis of these compounds has been carried out following the methodologies for the formation of amide functions in the liquid phase, including peptide chemistry (Bodarszky and Bodanszky 1984). Ideal chemical conditions for the preparation of *N*-acyl amino acids would allow acyl bond formation to be carried out rapidly and quantitatively under mild conditions, avoiding side reactions while maintaining all the adjacent chiral centers. In practice, however, diverse methodologies have been devised to overcome the problems related to the reactivity and purity processes. Recently, our group synthesized cationic N^α-acyl lysine methyl ester and N^ε-acyl lysine methyl ester to study the influence of the cationic charge nature on the biological properties. Compared with the N^α-acyl lysine derivatives, the N^ε-acyl lysine compounds show moderate antimicrobial power and better hemolytic activity (Pérez et al. 2009).

Alkyl-ester and alkyl-amide lysine derivatives were also described as good antimicrobial agents. In this case, the alkyl chain is linked to the carboxylic group of the lysine through an amide or ester bond, and the polar head contains two positive charges (Nakamiya et al. 1976). Using classical reactions in peptide chemistry, Gryc et al. also prepared 2-aminoethyl esters of N^α-palmitoyl-L-lysine derivatives with one, two, or three lysine residues in the polar head (Gryc et al. 1979). Amphoteric lysine derivatives of the type $N^\alpha,N^\alpha,N^\alpha$-trimethyl-$N^\varepsilon$-acyl lysine were prepared from N^ε-acyl lysine and methyl iodide. The introduction of the methyl groups considerably increases the solubility of the compounds (Yokota et al. 1985).

5.2.2 DICATENARY LYSINE DERIVATIVES

Symmetrical (two equal fatty acid chains) (Figure 5.4, **1a**) and asymmetrical (two different fatty acid chains) (Figure 5.4, **2b**) nonionic double-chain surfactants of the type N^α,N^ε-diacyl lysine polyoxyethylene glycol amide compounds, with a structural resemblance to natural lecithin phospholipids, have been reported by our group (Seguer et al. 1994a,b, 1996, Macian et al. 1996) to determine the effect of several structural parameters (hydrophobic chain length, polyoxyethylene (POE) chain length, and number of POE chains) on the physicochemical properties and biological performance of these natural mimics.

The structure of these nonionic amphiphilic compounds is based on N^α,N^ε-diacyl lysine and contains no active hydrogen in the oxyethylene (OE) hydrophilic head [they possess a methoxy-capped oxyethylene (EO$_n$) chain], with the result that they are more chemically inert. As with lecithins, these compounds can have two hydrophobic tails of identical length and one hydrophilic head. However, in the latter compounds the polar head is of the nonionic type (one or two chains of monomethyl ether oxyethylene glycol with different EO units), whereas in lecithins it is of the zwitterionic type. The central pivot in the structure of lecithins, i.e., the glycerol, is mimicked by the natural trifunctional amino acid lysine. The fatty acids and methyl ether oxyethylene glycol amine residues are introduced into the α and ε amino or α carboxylic functions of the lysine through amide bonds in place of the ester bonds in the lecithins. These amide bonds offer greater resistance to hydrolysis than the ester of lecithins. They have been prepared by the condensation of fatty acid

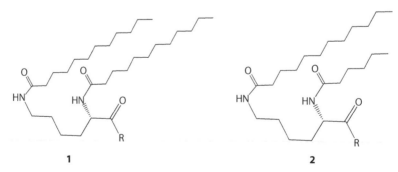

1   **2**

FIGURE 5.4 Chemical structure of symmetrical N^α,N^ε-dilauroyl lysine (**1**) and asymmetrical N^α-butyl, N^ε-lauroyl lysine (**2**) derivatives, where R is (a) –NH–(CH$_2$–CH$_2$–O)$_n$Me, –N–[(CH$_2$–CH$_2$–O)$_n$Me]$_2$, OH or OMe for nonionic derivatives, and (b) –ONa, –OLys, –OK, –OLi, OTris for anionic derivatives.

to the conveniently protected lysine in organic media. The formation of *N*-acyl lysine derivatives takes place by the reaction of the corresponding fatty acyl chloride with lysine, but this procedure requires prior protection of the second free amino group of the amino acid. The usual protecting groups for the *N*-amino function of Lys and homologues are *tert*-butoxycarbonyl (Boc) and benzyloxycarbonyl (Z), which can be easily removed by acidolysis and hydrogenolysis, respectively (Bodarszky and Bodanszky 1984).

Similarly, symmetrical anionic double-chain surfactants of the type $N^{\alpha},N^{\varepsilon}$-diacyl lysine have been chemically synthesized by our group (Figure 5.4, **1b**), and later by others (Gomes et al. 2008). However, the rapid advance of biotechnology has led to produce, via mediation of enzymatic or chemo-enzymatic catalysts, an interesting number of structures. The main advantages associated with the use of biocatalysts are mild reaction conditions and high enzymatic specificity, which often eliminates the need for regioselective protection of multifunctional starting materials. This point has been clearly demonstrated by a number of recent reports dealing with the application of enzymes to synthesis and/or modification of amino acids, sugar fatty acid esters, phospholipids, and alkyl glycosides (Valivety et al. 1997, Wada et al. 2002, Soo et al. 2004, Zhang et al. 2005, Villeneuve 2007).

Suzuki et al. have synthesized by enzymatic methodologies different lysine derivatives that can form stable hydrogels at very low concentrations. They have prepared cationic derivatives of the type $N^{\alpha},N^{\varepsilon}$-diacyl by the introduction of positively charged pyridinium and imidazolinium groups (Suzuki et al. 2005a). Asymmetric N^{α}-hexanoyl-N^{ε}-lauroyl lysine as carboxylic acid derivatives and alkali metal salts have also been prepared by these authors to study their ability to form supramolecular gels (Suzuki et al. 2008). Double-tailed biocompatible nonionic glycolipids based on lysine or glycine were also synthesized in order to obtain surfactants for biomedical applications. In the case of the lysine, the α-amino and α-carboxy functionalities are condensed with fatty acids and amines, respectively, by conventional peptide-coupling methods, and the ε-amino group is condensed to lactono- or maltonolactase in boiling methanol using slightly alkaline conditions (Pucci et al. 1993).

5.2.3 Bisglycidol Lysine-Based Surfactants

A novel class of lysine-based cationic amphiphilic derivatives of the type $N^{\varepsilon},N^{\varepsilon'}$-bis(*n*-acyloxypropyl)-L-lysine methyl ester salts combining several hydroxyl functions and aliphatic chains of 12 or 14 carbon atoms has been prepared in our lab. The main structural characteristics of these new cationic amphiphilic molecules are a lysine methyl ester residue as a polar group, which gives the cationic character to the molecule, linked to a bisglycidol chain and a residue of bis(2,3-dihydroxypropyl) that can carry one (Figure 5.5a) or two (Figure 5.5b) aliphatic chains as a part of the hydrophobic moiety. The lysine is bonded to the polyol skeleton through an *N*-alkyl amine linkage. They are obtained by the reaction of lysine to glycidol and subsequent chemical acylation of the hydroxyl functions with fatty acyl chlorides as acylating agents (Pinazo et al. 2009, Pérez et al. 2010).

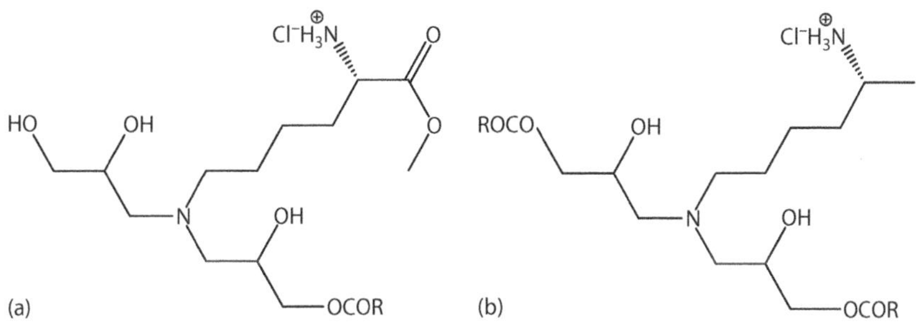

FIGURE 5.5 Chemical structure of the $N^{\varepsilon}, N^{\varepsilon'}$-bis(*n*-acyloxypropyl)-L-lysine methyl ester salts with one (a) and two (b) alkyl chains.

5.2.4 GEMINI LYSINE-BASED SURFACTANTS

Gemini surfactants, characterized by two hydrophobic chains with polar heads that are linked by a hydrophobic bridge, have very interesting properties compared to their monomeric equivalents. They are very effective in reducing interfacial tension between oil- and water-based liquids and have very low critical concentrations (Menger and Littau 1991). Long, alkyl-chain, cationic gemini surfactants from lysine with different structures have been proposed for gene transfection (Camilleri et al. 2004). Asymmetrical gemini surfactants from lysine, which have two quaternary ammonium groups linked to the fluorinated or hydrocarbon alkyl chains through amide bonds and the lysine as spacer chain, have been successfully synthesized (Figure 5.6). The compounds exhibited good anti-microbial activity against a broad spectrum of bacteria (Tan and Xiao 2008). The synthesis pathway consists on three steps: N^α,N^ϵ-bisbromoacetyl-L-lysine ethyl ester was obtained through the reaction of L-lysine ethyl ester with bromoacetic bromide. Fluorinated or hydrocarbon fatty acid (3-dimethyl amino-propyl) amides were prepared by the reaction of 1,3-(dimethylamino)-1-propylamine with the corresponding fatty acid chloride. Finally, lysine gemini compounds were prepared with the two compound analogues described above in the isopropanol refluxing.

Recently, our group has synthesized a new class of gemini cationic surfactants derived from lysine to study the effect of several structural parameters (hydrophobic chain length, number and type of the cationic charge, and spacer chain nature) on their physicochemical properties and cellular toxicity. These compounds can be considered dimers of the long-chain N^α- or N^ϵ-acyl lysine derivatives. They consist of two symmetrical long-chain N^α/N^ϵ-acyl lysine residues linked by amide bonds through an α,ω-alkylendiamine spacer chain of varying length and chemical nature (Figure 5.7). They were obtained with a purity of 99% by chemical condensation of the previously protected single N-acyl-L-lysine to the corresponding spacer in the presence of an activating agent. A final deprotection reaction was carried out to obtain the cationic gemini compounds (Colomer et al. 2009).

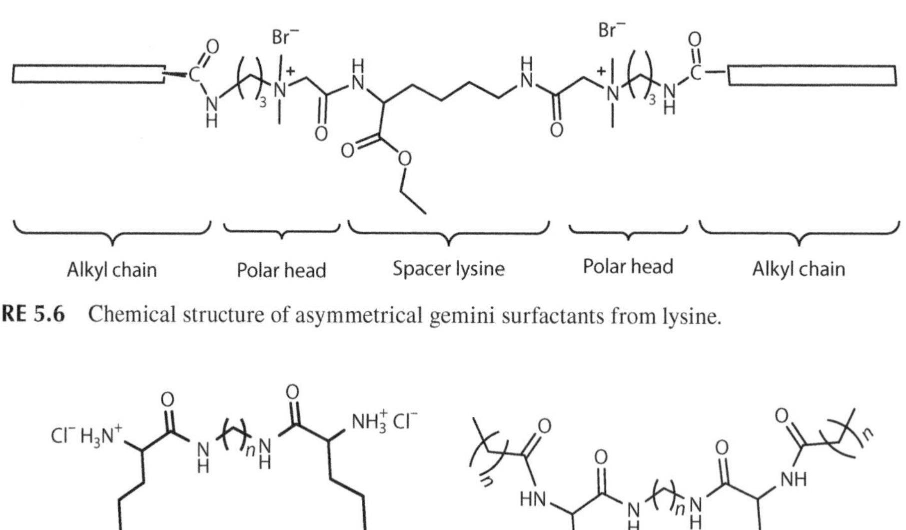

FIGURE 5.6 Chemical structure of asymmetrical gemini surfactants from lysine.

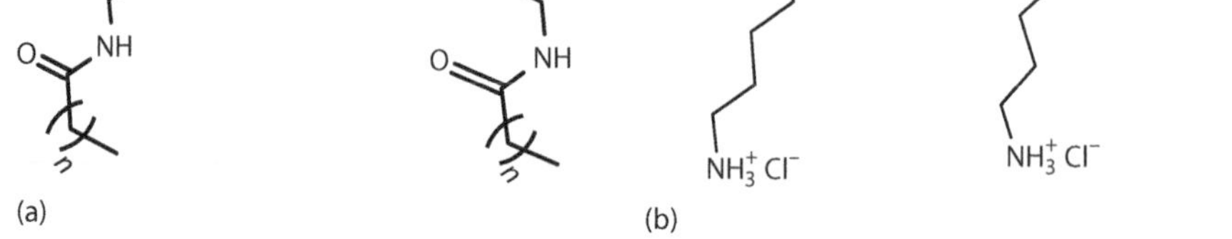

FIGURE 5.7 Chemical structure of cationic gemini surfactants of the type N^α- (b) or N^ϵ-acyl lysine derivatives (a).

5.3 PHYSICOCHEMICAL PROPERTIES

5.3.1 Micellization

Our group has reported the physicochemical properties of N^α-acyl lysine salts, both as cationic surfactants and amphoteric derivatives (Infante et al. 1997) (Figure 5.2, **1**). The N^α-lauroyl lysine methyl ester derivative has good water solubility and a CMC (critical micellar concentration) around 4 mmol L^{-1}, which is reasonable for a cationic surfactant with a C12 hydrophobic chain. Amphoteric derivative (Takehara 1989) water solubility depends strongly on the net charge, which will be modulated by pH, and therefore the aggregation properties will also depend on pH. Quaternization of the free amino group improves solubility because of the presence of at least one cationic charge, whatever the pH conditions be. The quaternized lauroyl derivative showed in unbuffered media a CMC value of 3 mmol L^{-1} (Infante et al. 1997).

N^ε-acyl lysine esters (Figure 5.2, **2**) have been studied recently (Pérez et al. 2009, Pinazo et al. 2009). The CMC values of these cationic surfactants are slightly smaller than those of the corresponding quaternary ammonium conventional surfactants—5.5 mmol L^{-1} for the lauroyl lysine compared to the 16 mmol L^{-1} of dodecyl trimethylammonium bromide. From diffusion NMR studies, it was found that the N^ε-lauroyl-L-lysine methyl ester derivative formed spherical micelles while the myristoyl tends to form elongated micelles, and the palmitoyl derivative has low solubility at 25°C but forms also spherical micelles at 50°C. Concerning the effect of the position of the linkage between the lysine group and the hydrophobic chain, the differences are small or insignificant on the CMC—for instance, the N^α-lauroyl derivative presented a CMC around 6–8 mmol L^{-1} (Infante et al. 1985), which is very close to that of the N^ε-lauroyl derivative.

The micellization N^α,N^ε-diacyl lysine compounds with symmetric diacyl chains have been studied recently, both as nonionic and anionic compounds (Figure 5.4, **1a** and **1b**) (Seguer et al. 1994a,b, Brito et al. 2008, Pinazo et al. 2008). Seguer et al. (1994b, 1996) studied the micellization properties of the same basic structure N^α,N^ε-diacyl lysine oxyethylene (OE) derivatives bonded to the acid group via an amide bond (Figure 5.4, **1**). These ethylene oxide chains increase the water solubility of the products and allow for CMC determination in aqueous (unbuffered) solutions. Two series of compounds were studied—a single OE chain (Figure 5.4, **1**, R=–NH–(CH$_2$–CH$_2$–O)$_n$ Me) and two OE chains (Figure 5.4, **1**, R=–N–[(CH$_2$–CH$_2$–O)$_n$Me]$_2$) bonded to the same amino group. The CMC and surface tension reduction for the two ethylene oxide derivatives was similar to that of single-chain alcohol ethoxylates, with the same hydrocarbon chain length and a similar number of ethylene oxide units in the head group, and much bigger than the corresponding lecithin with the same chain length. The CMC of the single ethylene oxide chain per head group derivatives was smaller than those with two chains in the head group. Dihexanolyl derivatives with a single ethylene oxide chain in the head group presented CMC values of 30–50 and 57 mmol L^{-1} for the double ethylene oxide derivative. The dioctanoyl derivative single ethylene oxide chain CMC was 1.6 mmol L^{-1}, while that of double ethylene oxide chain derivatives was around 6 mmol L^{-1} and didecanoyl derivative double ethylene oxide chain was around 1 mmol L^{-1}. The areas per head group are relatively large (between 0.80 and 1.02 nm^2).

Several short-chain anionic homologues were described before (Takehara 1989, George et al. 1998); however, only the CMC of the dioctanoyl derivative in the form of sodium salt was reported to be 5 mmol L^{-1}, completely agreeing with recent results (Pinazo et al. 2008). The nonionic compounds (with the same structure as that shown in Figure 5.4, **1a** with R=OH) present very low solubilities in water; at basic pH, they acquire charge and present non-negligible solubilities. When acting as anionic surfactants by solubilizing in 60 mmol L^{-1} NaOH, the CMC has been determined for dihexanoyl (30 mmol L^{-1}), dioctanoyl (2 mmol L^{-1}), didecanoyl (0.04 mmol L^{-1}), and didodecanoyl (0.005 mmol L^{-1}) (Soza et al. 2009). It was found that the short-chain N^α,N^ε-dioctyl lysine derivatives formed small spherical micelles in an aqueous solution (Pinazo et al. 2008). The area per molecule, as obtained from surface tension measurements, gives a high value for the dioctyl lysine derivatives (Brito 2008, Marques et al. 2008, Pinazo et al. 2008), agreeing with the small micellar sizes. No significant influence of the counterion size was obtained for lithium, sodium, potassium, lysine, and trishydroxyl methyl amino methane. Longer derivatives like $N^{\alpha},N^{\varepsilon}$-dilauroyl lysine salt, however, form lamellar aggregates (Brito et al. 2006, Brito 2008, Marques et al. 2008) and correspondingly show small area per molecule from surface area measurements (Soza et al. 2009).

Diacyl lysine sodium salt surfactants' CMCs have also been reported in aqueous media at 25°C for didecanoyl (3.7 mmol L^{-1}) (Brito 2008), and dioctanoyl (11 mmol L^{-1}) (Marques et al. 2008) originating from a slightly different synthetic route (Gomes et al. 2008). These values roughly agree with those found for the acid derivatives in excess NaOH. However, the didodecanoyl derivative (CMC measured in 10 mmol L^{-1} NaOH at 43°C) was found to have an extremely low CMC—0.00055 mmol L^{-1} (Brito 2008, Marques et al. 2008). The extremely small value for the didodecanoyl compound compared with the results of Soza et al. could be due to taking an apparent value below the real one, which could be that of the second break on the figure. The first break, with strong slope decrease of surface tension, could be due to the fact that at these low concentrations the surfactant is significantly partitioned between surface and bulk. A second break in surface tension around 1–2 μm concentration could correspond to the true CMC. Additional warning should be made concerning the possible chemical unstability of these products in harsh conditions like the strongly basic conditions needed to solubilize these products.

Gemini lysine-based surfactants' CMC with C12 hydrophobic chains has been found to be 0.8 mmol L^{-1}; the presence of an additional cationic charge in the head group only slightly increases this value to 1.0 mmol L^{-1} (Pérez et al. 2009). As expected, there is a reduction of around one order of magnitude of the CMC for the gemini structure as compared to the corresponding single-chain derivatives. Concerning the bisglycidol derivatives, single-chain derivatives have CMC values similar to conventional cationic surfactants with the same chain length, while the diacyl derivatives have a further CMC value reduction of an order of magnitude (Pinazo et al. 2009).

Data of CMC as a function of carbon number allows for the calculation of transfer energy from the aqueous to the micellar environment. The slope of log (CMC) as a function of carbon number for lecithins is almost double than that of the two-OE-chains family, while that of the single-chain family is intermediate. This implies that the energy of transfer of a methylene group ($-CH_2$) from a hydrocarbon to an aqueous environment is smaller for those lysine derivatives than for usual surfactants (the energy transfer per methylene can be evaluated as 3.0 kJ mol^{-1} for the lecithins, 1.6 kJ mol^{-1} for the double-EO-chain surfactants, and 1.8 kJ mol^{-1} for the diacyl lysine sodium salt surfactants). This last value is close to that evaluated for ionic single-chain surfactants. It is reported in the literature that amphoteric and nonionic surfactants present transfer energies of the same order (around 3.0 kJ mol^{-1}). Therefore, the transfer energy in the double-EO-chain surfactants is relatively small compared to what would be expected. It was also shown that the homologues containing two short alkyl chains and two methyl ether oxyethylene glycol chains were interesting nonhemolytic molecules that could be applied as water-soluble surfactants in the biological field. By contrast, the ones with higher alkyl chain lengths (even with two OE chains) showed no water solubility, yielding compounds, in principle, with little technological interest in aqueous media but some potential in nonaqueous environments. The asymmetrical homologues resulted in higher surface active properties and a greater capacity for micellization when compared with their symmetrical analogues, although with a slight increase in toxicity (Infante et al. 1999). The interaction of oxyethylene diacyl lysine with short-chain phospholipids (dilauroyl phosphatidyl choline (DLPC)) was studied by surface tension methods, including Langmuir isotherm, equilibrium surface tension, and pulsating surface tension (Pinazo et al. 1997). Although the lysine-based surfactants were less surface active than DLPC, a synergistic effect produced higher surface pressure values.

5.3.2 Thermotropic Phase Behavior

Pinazo et al. studied the effect of alkyl chain length on the dry symmetric diacyl lysine nonionic derivatives in the acid form (Figure 5.4, **1a** R = OH); cubic to lamellar transition was observed as the hydrophobic chain length increased (Pinazo et al. 2008). The shorter (N^α,N^ε-dihexanoyl lysine) product did not follow this trend and crystallized in a reverse hexagonal phase; the reason for this difference was discussed in terms of the difficulty of bringing together two short chains. The thermotropic phase behavior of nonionic methyl esters was also studied as a function of hydrophobic chain length by Brito et al. (2008). The same anomaly in the crystallization of the shorter chain analogues was also found. The gel to liquid crystal and melting transitions of these compounds were determined. The formation of reverse hexagonal structures was also found at higher temperatures for the decanoyl compound. Concerning the diacyl sodium salts (Figure 5.4, **1b**), in the dry state they show a similar trend with chain length as the nonionic (both free acid and ester derivatives). The influence of the counterion volume was also studied for the dioctanoyl derivative; big counterions induce the formation of lamellar phases, while small counterions induce the formation of the *pn3m* cubic phase (Pinazo et al. 2008). The comparison of the methyl ester thermotropic behavior with that of the sodium salts showed significant differences that were attributed to different contributions of van der Waals and electrostatic forces in the crystallization behavior of both series (Brito et al. 2008).

5.3.3 Gelation

Suzuki et al. have extensively studied the ability to form hydrogels and organogels with lysine-derived surfactants. By changing their charge and hydrophilic–hydrophobic balance, the gelation properties can be enhanced. The molecules studied correspond to a wide range of structures: N^ε-lauroyl lysine hydrocarbon terminated with a positive group and with the carboxylic group, esterified as hydrogelators (Suzuki et al. 2002, 2005a); gemini surfactants from N^ε-lauroyl lysine with free carboxylic acid groups, with partial salts, or with these groups, esterified with fatty alcohols as organogelators (Suzuki et al. 2003, 2005b); asymmetric N^α,N^ε-diacyl lysine as mixtures of free carboxylic derivatives and salts, as both hydrogelators and organogelators (Suzuki et al. 2004, 2008); and N^α-glucoheptonamide N^ε-lauroyl lysine, as hydrogelators (Suzuki et al. 2007). On the other hand, α-lysine ω amino bolaamphiphiles were shown to aggregate forming monolayer nanotubes (Furhop et al. 1993). Lysine–glutamine diacyl surfactants (Sebyakin and Budanova 2006) in aqueous dispersions formed tubular structures with short hydrophobic chains, while longer chain length induced the formation of vesicles. Takehara also mentioned the organogel properties of dihexanoyl lysine (Takehara 1989). On the other hand, amino acid surfactants are chiral compounds, and because of chirality they can form gels containing helical fibers that are stabilized by amide hydrogen bonds. Gemini surfactants with two L-lysine derivatives linked by different alkylene chain lengths through the amide bond are good organogelators that gel most organic solvents such as alcohols, cyclic ethers, aromatic solvents, and acetonitrile (Suzuki et al. 2003).

5.4 BIOLOGICAL PROPERTIES

In general, the amino acid-based surfactants can be classified as biocompatible compounds—they are ready biodegradable molecules and have low toxic effects in the aquatic environment. Depending on their chemical structure, they can act as emulsifiers, detergents, wetting agents, or foaming or dispersing compounds. Therefore, they can be used in a wide spectrum of industrial and biomedical applications, and the marketplace projection of these surfactants is encouraging. Salts of long-chain N-acyl-amino acid are currently used as detergents, foaming agents, and shampoos because they are nonirritating to the human skin and highly biodegradable.

Lysine-based surfactants show moderate to good anitimicrobial properties if a cationic charge is present in the head group. If we consider the nonionic compounds, their antimicrobial activity is very low and, at the same time, their biocompatibility is maximum. Compared with the N^α-acyl lysine derivatives, the N^ε-acyl lysine compounds show moderate antimicrobial power and better hemolytic activity (Pinazo et al. 2007, Pérez et al. 2009). Alkyl ester and alkyl amide lysine derivatives were also described as good antimicrobial agents (Nakamiya et al. 1976, Gryc et al. 1979). Asymmetrical gemini surfactants from lysine (Figure 5.6) exhibited good antimicrobial activity against a broad spectrum of bacteria (Tan and Xiao 2008).

Biocompatible cationic surfactants from the amino acid lysine (hydrochloride salts of N^ε-lauroyl-lysine methyl ester, N^ε-miristoyl-lysine methyl ester, and N^ε-palmitoyl-lysine methyl ester) show moderate antimicrobial activity against the gram-positive bacteria. The hemolytic activity of these compounds is considerably lower than that reported for other cationic N^α-acyl-amino acid analogues (Pérez et al. 2005). Taking into account the high biodegradation level and the low hemolytic activity, these compounds could be considered safe surfactants in relation to the cell of the human body. These properties make them suitable candidates for biological and medical applications (Pérez et al. 2007).

Diacyl lysine anionic surfactants, N^α-octanoyl N^ε-octanoyl lysine with different counterions (Lysine, Na, K, Li) (Figure 5.4, **1**, R=−O Na⁺, −OLysine⁺, −OK⁺, −OLi⁺) showed less cytotoxicity than SDS. Moreover, they were less eye-irritating than SDS, and none showed phototoxic effects. These surfactants are a promising alternative to commercial anionic surfactants given their low ocular and dermal irritancy. These properties offer great potential for topical preparations (Sanchez et al. 2006).

The efficient delivery of DNA to cells *in vivo* has been a major goal for some years, and there is a continuing need to develop novel low-toxicity surfactant molecules to facilitate the effective transfer of polynucleotides into cells. We studied the interaction of the single-chain arginine-based surfactant—the salt of the lauryl amide of arginine (ALA) with DNA (Brito et al. 2006). The ability of this surfactant alone to compact DNA is compared by fluorescence microscopy studies to classical cationic surfactants. Furthermore, toxicity studies revealed that the incorporation of ALA in catanionic vesicles systems transformed them into cell-viable systems, therefore extending their use to drug- and gene-delivery systems. Nontoxic catanionic vesicles can be also obtained with diacyl lysine derivatives compounds (Brito et al. 2006). Cationic diether lipids based in arginine, tryptophan, histidine, and lysine have also been proposed as gene transfer agents (Heyes et al. 2002).

Gemini surfactants show greatly enhanced surfactant properties relative to the corresponding monovalent compounds, and this makes them of special interest for biomedical applications. In the last 10 years, different gemini cationic surfactants with amino acids in the polar head have been proposed as new synthetic vectors for gene transfection (Rosin et al. 2001, Buijnsters et al. 2002, Kirby et al. 2003, Sen and Chaudhuri 2005).

5.5 CONCLUSION

As it has been described in this chapter, lysine is a versatile building block to form biocompatible surfactants. This family comprises several hydrophobic structures (monocatenary, dicatenary, and gemini) together with a flexible range of hydrophilic head group character (anionic, nonionic, cationic, and amphoteric). Because of their possible green-chemistry origin, biodegradability, biocompatibility, some antimicrobial properties, and self-aggregation properties, this family of surfactants can have multiple application fields.

ABBREVIATIONS

ALA   Lauryl amide of arginine
CMC   Critical micellar concentration
DNA   Desoxy ribonucleic acid

Et   Ethyl
Lys   Lysine amino acid
Me   Methyl
OE   Oxyethylene chain
Pn3m   Cubic bicontinuous phase
SDS   Sodium dodecyl sulfate
Tris   Trishydroxyl methyl amino methane

ACKNOWLEDGMENTS

The authors acknowledge the financial support from the Ministerio de Ciencia e Innovación, projects CTQ2006-01582 and CTQ2007-60409/BQU, and Generalitat de Catalunya, project SGR-00066-2005.

REFERENCES

Bodarszky, M., Bodanszky, A. 1984. *The Practice of Peptide Synthesis*, p. 21. Berlin, Germany: Springer.

Boyat, C., Rolland-Fulcrand, V., Roumestant, M.L., Viallefont, P., Martinez, J. 2000. Chemo-enzymatic synthesis of new non ionic surfactants from unprotected carbohydrates, *Prep. Biochem. Biotechnol.*, 30:281–294.

Brito, R.O. 2008. PhD thesis, Universidade de Porto, Porto, Portugal.

Brito, R.O., Marques, E.F., Gomes, P., Falcão, S., Söderman, O. 2006. Self-assembly in a catanionic mixture with an aminoacid-derived surfactant: From mixed micelles to spontaneous vesicles, *J. Phys. Chem. B*, 110:18158–18165.

Brito, R.O., Marques, E.F., Gomes, P., Araújo, M.J., Pons, R. 2008. Structure/property relationships for the thermotropic behavior of lysine-based amphiphiles: From hexagonal to smectic phases, *J. Phys. Chem. B*, 112:14877–14887.

Buijnsters, P.J.J.A., Rodriguez, C.L.G., Willighagen, E.L., Sommerdijk, N.A.J.M., Kremer, A., Camilleri, P., Feiters, M.C., Nolte, R.J.M., Zwanenburg, B.L. 2002. Cationic gemini surfactants based on tartaric acid: Synthesis, aggregation, monolayer behaviour, and interaction with DNA, *European Journal of Organic Chemistry* 8:1397–1406.

Camilleri, P., Feiters, C.M., Kirbi, A.J., Ronsin, B. 2004. Patent, International Publication Number WO 03/082809.

Capone, S., Walde, P., Seebach, D., Ishikawa, T., Caputo, R. 2008. pH-sensitive vesicles containing a lipidic beta-amino acid with two hydrophobic chains, *Chem. Biodivers.*, 5:16–30.

Clapés, P., Infante, M.R. 2002. Amino acid-based surfactants: Enzymatic synthesis, properties and potential applications, *Biocatal. Biotransform.*, 20:215–233.

Clapés, P., Morán, C., Infante, M.R. 1999. Enzymatic synthesis of arginine- based cationic surfactants, *Biotechnol. Bioeng.*, 63:333–343.

Colomer, A., Pérez, L., Pinazo, A., Vinardell, M.P., Mitjans, M., Infante, M.R., Ribosa, I., García, M.T. 2009. Sintesis y propiedades de nuevos tensioactivos catiónicos derivados de lisina, in *Proceedings of the 39 Jornadas del CED*, Barcelona, Spain.

Das, D., Dasgupta, A., Roy, S., Mitra, R.N., Debnath, S., Das, P.K. 2006. Water gelation of an amino acid-based amphiphile, *Chem. A Eur. J.*, 12:5068–5074.

David, S., Pérez, L., Infante, M.R. 2002. Sequestration of bacterial lipopolysaccharide by bis(Args) gemini compounds, *Bioorg. Med. Chem. Lett.*, 12:357–360.

Furhop, J.H., Spiroski, D., Boettcher, C. 1993. Molecular monolayer rods and tubules made of alpha-(L-lysine),omega-(amino) bolaamphiphiles, *J. Am. Chem. Soc.*, 115:1600–1601.

George, A., Modi, J., Jain, N., Bahadur, P. 1998. A comparative study on the surface activity and micellar behaviour of some *N*-acylamino acid based surfactants, *Ind. J. Chem.*, 37A:985–992.

Gerova, M., Rodrigues, F., Lamere, J.F., Dobrev, A., Fery-Forgues, S. 2008. Self-assembly properties of some chiral *N*-palmitoyl amino acid surfactants in aqueous solution, *J. Colloid Interface Sci.*, 319:526–533.

Gomes, P., Araujo, M.J., Marques, E.F. 2008. Straightforward method for the preparation of lysine-based double-chained anionic surfactants, *Synth. Commun.*, 38:2025–2036.

Gryc, W., Dabrowska, M., Tomicka, B., Perkowska, D., Kupryszewski, G. 1979. Antibacterial peptide derivatives. Part V. Hydrochlorides of 2-aminoethyl esters of N^α-palmitoyl-L-lysine and its peptides, *Pol. J. Chem.*, 53:1085–1093.

Heyes, J.A., Niculescu-Duvaz, D., Cooper, R.G., Sprinter, C. J. 2002. Synthesis of novel cationic lipids: Effect
of structural modification on the efficiency of gene transfer, *J. Med. Chem.*, 45:99–114.

Infante, M.R., Moses, V. 1994. Synthesis and surface activity properties of hydrophobic/hydrophilic peptides,
Int. J. Pept. Protein Res., 43:173–179.

Infante, M.R., García Domínguez, J.J., Erra, P., Juliá, M.R., Prats, M. 1984. Surface active molecules:
Preparation and properties of long chain N^αacyl-L-α-amino-ω-guanidine alkylic acid derivatives, *Int. J. Cosmet. Sci.*, 6:275–282.

Infante, M.R., Molinero, J., Erra, P., Juliá, M.R., García Domínguez, J.J. 1985. A comparative study on surface
active and antimicrobial properties of some $N\alpha$-lauroil-L-α,ωdibasic amino acid derivatives, *Fette Seifen Anstrichmittel*, 8:309–313.

Infante, M.R., Pinazo, A., Seguer, J. 1997. Non-conventional surfactants from amino acids and glycolipids:
Structure, preparation and properties, *Colloids Surf. A*, 123–124:49–70.

Infante, M.R., Seguer, J., Pinazo, A., Vinardell, M.P. 1999. Synthesis and properties of asymmetrical nonionic
double chain surfactants from lysine, *J. Dispers. Sci. Technol.*, 20:621–642.

Kinoshita, S., Udaka, S., Shimono, M. 1957. Studies on the amino acid fermentation. Part I. Production of
L-glutamic acid by various microorganisms, *J. Gen. Microbiol.*, 3:193–205.

Kirby, A.J., Camilleri, P., Engberts, J.B.F.N., Feiters, M.C., Nolte, R.J.M., Soderman, O., Bergsma, M., Bell,
P.C., Fielden, M.L., Rodriguez, C.L.G., Guedat, P., Kremer, A., McGregor, C., Perrin, C., Ronsin, G.,
van Eijk, M.C.P. 2003. Gemini surfactants: New synthetic vectors for gene transfection, *Angewandte Chemie-International Edition*, 42:1448–1457.

Macian, M., Seguer, J., Infante, M.R., Selve, C., Vinardell, P. 1996. Preliminary studies of the toxic effects of
non-ionic surfactants derived from lysine, *Toxicology*, 106:1–9.

Marques, E.F., Brito, R.O., Silva, S.G., Rodríguez-Borges, J.E., do Vale, M.L., Gome, S.P., Araújo, M.J.,
Söderman, O. 2008. Spontaneous vesicle formation in catanionic mixtures of amino acid-based surfactants: Chain length symmetry effects, *Langmuir*, 24:11009–11017.

Menger, F.M., Littau, C.A. 1991. Gemini surfactants—Synthesis and properties, *J. Am. Chem. Soc.*,
113:1451–1452.

Molinero, J., Julia, M.R., Erra, P., Robert, M., Infante, M.R. 1988. Synthesis and properties of $N\alpha$-lauroyl-L-
arginine dipeptides from collagen-lauroyl-L-arginine dipeptides from collagen, *J. Am. Oil Chem. Soc.*,
65:975–978.

Morán, C., Infante, M.R., Clapés, P. 2001. Synthesis of glycero amino acid-based surfactants. Part 1. Enzymatic
preparation of rac-1-O-($N\alpha$-acetyl-L-aminoacyl)glycerol derivatives, *J. Chem. Soc. Perkin Trans.*,
1:2063–2070.

. Morán, C., Infante, M.R., Clapés, P. 2002. Synthesis of glycero amino acid-based surfactants. Part 2. Lipase-
catalysed synthesis of 1-O-lauroyl-rac-glycero-3-O-($N\alpha$-acetyl-L-amino acid) and 1,2-di-O-lauroyl-rac-
glycero-3-O-(N^α-acetyl-L-amino acid) derivatives, *J. Chem. Soc. Perkin Trans.*, 1:1124–1134.

Morán, C., Pinazo, A., Pérez, L., Clapés, P., Angelet, M., García, M.T., Vinardell, M.P., Infante, M.R. 2004a.
"Green" amino acid-based surfactants, *Green Chem.*, 6:233–240.

Morán, C., Pinazo, A., Pérez, L., Clapés, P., Pons, R., Infante, M.R. 2004b. Enzymatic synthesis and physicochemical characterization of glycero arginine-based surfactants, *Comp. Ren. Chim.*, 7:169–176.

Nakamiya, T., Mizuno, H., Meguro, T., Ryono, H., Takinami, K. 1976. Antibacterial activity of lauryl ester of
DL-lysine, *Ferment. Technol.*, 54:369–373.

Ohta, A., Toda, K., Morimoto, Y., Asakawa, T., Miyagishi, S. 2008. Effect of the side chain of N-acyl amino,
acid surfactants on micelle formation: An isothermal titration calorimetry study, *Colloids Surf. A*,
317:316–322.

Parekh, S., Vinci, V.A, Strobel, R.J. 2000. Improvement of microbial strains and fermentation processes, *Appl.
Microbiol. Biotechnol.*, 54:287–301.

Pegiadou, S., Perez, L., Infante, M.R. 2000. Synthesis, characterization and surface properties of 1-N-L-
tryptophan-glycerol-ether surfactants, *J. Surfact. Deterg.*, 3:517–525.

Pérez, L., Torres, J.L., Manresa, A., Solans, C., Infante, M.R. 1996. Synthesis, aggregation, and biological
properties of a new class of gemini cationic amphiphilic compounds from arginine, bis(Args), *Langmuir*,
12:5296–5301.

Pérez, L., García, M.T., Ribosa, I., Vinardell, M.P., Manresa, A., Infante, M.R. 2002a. Biological properties of
arginine-based gemini cationic surfactants, *Environ. Toxicol. Chem.*, 21:1279–1285.

Pérez, L., Pinazo, A., Vinardell, P., Clapés, P., Angelet, M., Infante, M.R. 2002b. Synthesis and biological properties of dicationic arginine-diglycerides, *New J. Chem.*, 26:1221–1227.

Pérez, L., Infante, M.R., Pons, R., Morán, C., Vinardell, P., Mitjans, M., Pinazo, A. 2004. A synthetic alternative to natural lecithins with antimicrobial properties, *Colloids Surf. B Biointerfaces*, 35:235–242.

Pérez, N., Pérez, L., Infante, M.R., García, M.T. 2005. Biological properties of arginine-based glycerolipidic cationic surfactants, *Green Chem.*, 7:540–546.

Pérez, L., Pinazo, A., Garcia, M.T., Lozano, M., Angelet, M., Infante, M.R. 2007. Cationic surfactants from lysine, in *Book of Proceedings 2nd Iberic Meeting of Colloids and Interfaces*, Coimbra, Portugal, A. Valente and J. Seixas de Melo (eds.), pp. 341–349. Coimbra, Portugal: SPQ.

Pérez, L., Pinazo, A., García, M.T., Lozano, M., Manresa, A., Angelet, M., Vinardell, M.P., Mitjans, M., Pons, R., Infante, M.R. 2009. Cationic surfactants from lysine: Synthesis, micellization and biological evaluation, *Eur. J. Med. Chem.*, 44:1884–1892.

Pérez, L., Pinazo, A., Infante, M.R., Angelet, M. 2010. Compuestos del tipo n^ε-$n^{\varepsilon'}$-aciloxipropil lisina metil éster y n^ε-$n^{\varepsilon'}$-bis(n-aciloxipropil) lisina metil éster, Patent numbers: ES2265245, WO5070158ES, W02006056636, ES2265245(B1), EP1813600(A1), JP2008520631 (T), WO2006056636 (A1).

Piera, E., Comelles, F., Erra, P., Infante, M.R. 1998. New alquil amide type cationic surfactants from arginine, *J. Chem. Soc. Perkin Trans.*, 2:335–342.

Piera, E., Domínguez, C., Clapés, P., Erra, P., Infante, M.R. 1999. Qualitative and quantitative analysis of new alkyl amide arginine surfactants by high-performance liquid chromatography and capillary electrophoresis, *J. Chromatogr. A*, 852:499–506.

Pinazo, A., Diz, M., Solans C., Pés, M.A., Erra, P., Infante, M.R. 1993. Synthesis and properties of cationic surfactants containing a disulfide bond, *J. Am. Oil Chem. Soc.*, 70:37–42.

Pinazo, A., Infante, M.R., Chang, C.-H., Franses, E.I. 1994. Surface tension properties of aqueous solutions of disulfur betaine derivatives, *Colloids Surf. A Physicochem. Eng. Asp.*, 87:117–123.

Pinazo, A., Seguer, J., Infante, M.R., Park, S.Y., Franses, E.I. 1997. Surface properties of aqueous systems of new nonionic double-chain surfactants and their mixtures with dilauroylphosphatidylcholine, *Colloids Surf. A Physicochem. Eng. Asp.*, 126:49–58.

Pinazo, A., Wen, X., Pérez, L., Infante, M.R., Franses, E.I. 1998. Aggregation behavior in water of monomeric and gemini cationic surfactants derived from arginine, *Langmuir*, 15:3134–3142.

Pinazo, A., Pons, R., Angelet, M., Lozano, M., Infante, M.R., Perez L. 2007. Propiedades de agregación de tensioactivos monocatenarios derivados del aminoácido lisina, in *Book of Proceedings 2nd Iberic Meeting of Colloids and Interfaces*, Coimbra, Portugal, A. Valente and J. Seixas de Melo (eds.), pp. 321–329. Coimbra, Portugal: SPQ.

Pinazo, A., Pérez, L., Lozano, M., Angelet, M., Infante, M.R., Vinardell, M.P., Pons, R. 2008. Aggregation properties of diacyl lysine surfactant compounds: Hydrophobic chain length and counterion effect, *J. Phys. Chem. B*, 112:8578–8585.

Pinazo, A., Angelet, M., Pons, R., Lozano, M., Infante, M.R., Perez, L. 2009. Lysine-bisglycidol conjugates as novel lysine cationic surfactants, *Langmuir*, 25:7803–7814.

Pucci, B., Guedj, C., Pavia, A.A. 1993. Biocompatible non-ionic surfactants for biomedical utilization: Double tailed hydrocarbon and/or fluorocarbon synthetic glycolipids, *Biorg. Med. Chem. Lett.*, 3:1003–1006.

Ronsin, G., Perrin, C., Guedat, P., Kremer, A., Camilleri, P., Kirby, A.J. 2001. Novel spermine-based cationic gemini surfactants for gene delivery, *Chemical Communications*, 21:2234–2235.

Roy, S., Dey, J. 2006. Self-organization properties and microstructures of sodium N-(11-acrylamidoundecanoyl)-L-valinate and -L-threoninate in water, *Bull. Chem. Soc. Jpn.*, 79:59–66.

Sanchez, L., Mitjans, M., Infante, M.R., Vinardell, M.P. 2006. Potential irritation of lysine derivative surfactants by hemolysis and HaCaT cell viability, *Toxicol. Lett.*, 161:53–60.

Sebyakin, Y.L., Budanova, U.A. 2006. pH-sensitive cationic lipopeptides for the design of drug-delivery systems, *Russ. J. Bioorg. Chem.*, 32:407–412.

Seguer, J., Infante, M.R., Allouch, M., Vinardell, P., Mansuy, L., Selve, C. 1994a. Synthesis and evaluation of nonionic amphiphilic compounds from amino-acids—Molecular mimics of lecithins, *New J. Chem.*, 18:765–774.

Seguer, J., Selve, C., Infante, M.R. 1994b. New nonionic surfactants from lysine and their performance, *J. Dispers. Sci. Technol.*, 15:591–610.

Seguer, J., Selve, C., Allouch, M., Infante, M.R. 1996. Nonionic amphiphilic compounds from lysine as molecular mimics of lecithins, *J. Am. Oil Chem. Soc.*, 73:79–86.

Sen, J., Chaudhuri, A. 2005. Gene transfer efficacies of novel cationic amphiphiles with alanine, beta-alanine, and serine headgroups: A structure-activity investigation, *Bioconjugate Chemistry*, 16:903–912.

Soo, E.L., Salleh, A.B., Basri, M., Rahman, R.N.Z.A., Kamaruddin, K. 2004. Response surface methodological study on lipase-catalyzed synthesis of amino acid surfactants, *Process Biochem.*, 39:1511–1518.

Soza, K., Pérez, L., Pons, R. 2010.

Suzuki, M., Yumoto, M., Kimura, M., Shirai, H., Hanabusa, K. 2002. Novel family of low molecular weight hydrogelators based on L-lysine derivatives, *Chem. Commun.*, 884–885.

Suzuki, M., Nigawara, T., Yumoto, M., Kimura, M., Shirai, H., Hanabusa, K. 2003. L-Lysine based gemini organogelators: Their organogelation properties and thermally stable organogels, *Org. Biomol. Chem.*, 1:4124–4131.

Suzuki, M., Yumoto, M., Kimura, M., Shirai, H., Hanabusa, K. 2004. New low-molecular-mass gelators based on L-lysine: Amphiphilic gelators and water-soluble organogelators, *Helv. Chim. Acta*, 87:1–10.

Suzuki, M., Yumoto, M., Shirai, H., Hanabusa, K. 2005a. L-Lysine-based supramolecular hydrogels containing various inorganic ions, *Org. Biomol. Chem.*, 3:3073–3078.

Suzuki, M., Nanbu, M., Yumoto, M., Shirai, H., Hanabusa, K. 2005b. Novel dumbbell-form low-molecular-weight gelators based on L-lysine: Their hydrogelation and organogelation properties, *New J. Chem.*, 29:1439–1444.

Suzuki, M., Owa, S., Shirai, H., Hanabusa, K. 2007. Supramolecular hydrogel formed by glucoheptonamide of L-lysine: Simple preparation and excellent hydrogelation ability, *Tetrahedron*, 63:7302–7308.

Suzuki, M., Yumoto, M., Shirai, H., Hanabusa, K. 2008. Supramolecular gels formed by amphiphilic low-molecular-weight gelators of N^α,N^ε-diacyl-L-lysine derivatives, *Chem. Eur. J.*, 14:2133–2144.

Takehara, M. 1989. Properties and applications of amino-acid based surfactants, *Colloids Surf.*, 38:149–167.

Tan, H., Xiao, H. 2008. Synthesis and antimicrobial characterization of novel L-lysine gemini surfactants pended with reactive groups, *Tetrahedron Lett.*, 49:1759–1761.

Torres, J.L., Piera, E., Infante, M.R., Clapés, P. 2001. Purification of non-toxic, biodegradable arginine-based gemini surfactants, bis(Args), by ion exchange chromatography, *Prep. Biochem. Biotechnol.*, 31:259–274.

Valivety, R., Jauregi, P., Gill, I., Vulfson, E. 1997. Chemo-enzymatic synthesis of amino acid-based surfactants, *J. Am. Oil Chem. Soc.*, 74:879–886.

Varka, E.M.A., Heli, M.G., Coutouli-Argyropoulou, E., Pegiadou, S.A. 2006. Synthesis and characterization of nonconventional surfactants of aromatic amino acid–glycerol ethers: Effect of the amino acid moiety on the orientation and surface properties of these soap-type amphiphiles, *Chem. A Eur. J.*, 12:8305–8311.

Vijay, R., Angayarkanny, S., Bhasker, G. 2008. Amphiphilic dodecyl ester derivatives from aromatic amino acids: Significance of chemical architecture in interfacial adsorption characteristics, *Colloids Surf. A*, 317:643–649.

Villeneuve, P. 2007. Lipases in lipophilization reactions, *Biotechnol. Adv.*, 25:515–536.

Vinardell, M.P., Molinero, J., Parra, J.L., Infante, M.R. 1990. Comparative ocular test of lipopeptidic surfactants, *Int. J. Cosmet. Sci.*, 12:13–20.

Wada, E., Handa, M., Imamura, K., Sakiyama, T., Adachi, S., Matsuno, R., Nakanishi, K. 2002. Enzymatic synthesis of N-acyl-L-amino acids in a glycerol-water system using acylase I from pig kidney, *J. Am. Oil Chem. Soc.*, 79:41–46.

Xia J. and Nnanna I.A. 2001. Synthesis, physicochemical properties, and applications, in *Protein-Based Surfactants*, I.A. Nnanna and J. Xia (eds.), pp. 1–14. New York: Marcel Dekker.

Xia, J., Nnanna, I.A., Sakamoto K. 2001. Aminoacid surfactants: Chemistry, synthesis and properties, in *Protein-Based Surfactants*, I.A. Nnanna and J. Xia (eds.), p. 105. New York: Marcel Dekker.

Yokota, H., Sagawa, K., Eguchi, C., Takehara, M. 1985. New amphoteric surfactants derived from lysine. 1. Preparation and properties of N-epsilon-acyllysine derivatives, *J. Am. Oil Chem. Soc.*, 62:1716–1719.

Zhang, X.M., Adachi, S., Watanabe, Y., Matsuno, R. 2005. Lipase-catalyzed synthesis of O-lauroyl L-serinamide and O-lauroyl L-threoninamide, *Food Res. Int.*, 38:297–300.

6 Synthesis of Proteins in Cell-Free Systems

Mingrong Wang and Mingyue He

CONTENTS

6.1 INTRODUCTION

Proteomics studies require high throughput tools for the large-scale expression of proteins. Currently, the most preferred *in vivo* expression system is *Escherichia coli*, which still labors under the limitation of producing insoluble aggregates for most eukaryotic proteins despite recent improvements (Braun and LaBaer, 2003; Langlais et al., 2007). Protein production in cell-free systems is thus becoming a widely used alternative since it provides an open and flexible system for the rapid synthesis of folded proteins (Goshima et al., 2008). The direct use of PCR fragments as the template to express proteins obviates the need for DNA cloning, giving greater usefulness for the parallel expression of large numbers of different proteins. The open nature of cell-free systems also allows external molecules and components to be added to create a favorable environment for the synthesis of required proteins, allowing production of "difficult-to-express" proteins (Jackson et al., 2004; Spirin, 2004). Recently, cell-free systems have been exploited to develop novel protein technologies such as ribosome display and cell-free protein arrays. They are powerful tools for *in vitro* selection and evolution of proteins (Mattheakis et al., 1994; Hanes and Plückthun, 1997; He and Taussig, 1997, 2001, 2007; Roberts and Szostak, 1997; Tawfik and Griffiths, 1998; Ramachandran et al., 2004; He et al., 2008a,b). This chapter reviews recent developments in cell-free systems and their applications in biotechnology and proteomics.

6.2 CELL-FREE SYSTEMS

Cell-free protein synthesis makes use of cellular lysates containing all the necessary biological and chemical components required to direct protein synthesis from added DNA or mRNA template(s) (Figure 6.1). In general, extracts from any cells can be made or reconstituted for synthesis of proteins. To date, the successful cell-free lysates prepared for protein synthesis include *E. coli*; yeast; rabbit reticulocyte; wheat germ; a hyperthermophile; *Drosophila* embryo; hybridoma; *Xenopus* oocyte or egg; and insect, mammalian, and human cells (Endoh et al., 2006; Mikami et al., 2006). Some of them (*E. coli*, rabbit reticulocyte, wheat germ, and insect cells) are commercially available. Cell-free systems have been made in either coupled or uncoupled format. While coupled systems generally use DNA as the template, uncoupled systems only translate mRNA molecule(s). Lysates from different cell origins may contain specific cellular factors to promote protein expression, folding, or posttranslational modifications, and a wide range of co- or posttranslational modification activities has already been demonstrated in eukaryotic lysates (Jackson et al., 2004). The availability of a variety of cell-free lysates makes it possible to produce proteins under a choice of different cellular environments (prokaryotic or eukaryotic) or conditions.

Remarkably, a cell-free system can be reconstituted *in vitro* using purified components/elements required for transcription and translation. A system called PURE (protein synthesis using recombinant elements) was first developed by mixing individually purified 31 recombinant proteins, together with 46 tRNAs, essential substrates, and the corresponding enzymes (Shimizu et al., 2001). This system provides a clearly defined environment/element for monitoring protein synthesis. Recently, different cell-free systems have been compared, which provide basic insights into their biological differences and thus can be used as a guide for their applications (Hillebrecht and Chong, 2008; Hino et al., 2008). The choice of a cell-free system should be based on the protein origin and the downstream application as it may affect the production of a particular protein. For example, the parallel expression of five different coding sequences of both bacterial and eukaryotic origin in either *E. coli* S30, wheat germ extract, or rabbit reticulocyte lysate systems has shown that while full-length products were predominantly produced in the two eukaryotic systems, *E. coli* S30 produced many incomplete nascent polypeptides, which was possibly due to the pausing of the *E. coli* ribosome at some genetic codons (Ramachandiran et al., 2000).

6.3 FEATURES OF CELL-FREE SYNTHESIZED PROTEINS

Proteins produced in cell-free systems have demonstrated to fold co-translationally on ribosomes in a manner similar to protein folding *in vivo*, i.e., a growing peptide starts to fold as it emerges from the large ribosomal subunit or immediately at the end of translation prior to release from the ribosome (Netzer and Hartl 1997; Kolb et al., 2000; Ellis et al., 2008). The co-translational protein-folding mechanism has been supported by structural studies (Agrawal and Frank 1999). Molecular chaperones are also involved in the folding process (Ying et al., 2006; Merz et al., 2008). Ribosomes themselves or ribosomal RNA, from both prokaryotic and eukaryotic source, can contribute to the folding process (Das et al., 2008; Choi et al., 2009).

The nascent proteins in a cell-free system can be modified during translation. Using engineered tRNAs carrying nonnatural or chemically modified amino acids, novel protein molecules can be synthesized with chemical diversity at the defined sites (Rothschild and Gite, 1999). This potential has been explored to label newly synthesized proteins with isotopic, fluorescent

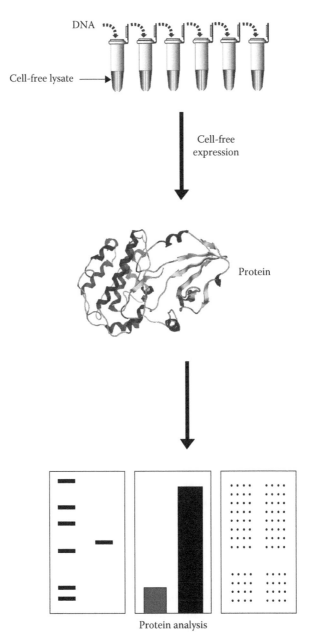

FIGURE 6.1   A diagram showing the process of cell-free generation and analysis of proteins.

or biotin, or photo-reactive groups for sensitive detection and functional analysis of proteins (Jackson et al., 2004). With a modified initiator tRNA, the N-terminal methionine can be replaced by a conjugated amino acid, with labeling efficiencies of 67% in an *E. coli* cell-free system (Mamaev et al., 2004). C-terminal labeling proceeds through an incorporation of coupled puromycin derivatives with labeling efficiencies of up to 90% (Nemoto et al., 1999; Yamaguchi et al., 2001). Modifications at internal positions are achieved by using stop codon suppression methodology, which introduces a stop codon at the defined position of the target gene and supplies the cell-free lysate with a complementary suppressor tRNA carrying a modified amino acid (Goerke and Swartz, 2009).

A variety of co- and posttranslational protein modifications can also be carried out in eukaryotic cell-free systems (Sakurai et al., 2007; Denman, 2008). These include signal peptide cleavage, glycosylation, acetylation, phosphorylation, isoprenylation, myristoylation, proteolytic processing, and the action of degradation pathways. In addition, cell-free systems can correctly assemble protein subunits into active complexes, providing a simple tool for generating protein complexes/virus particles (Jackson et al., 2004).

6.4 FACTORS AFFECTING PROTEIN EXPRESSION

Efforts have been made to increase protein yield in cell-free systems by identifying key factors affecting *in vitro* transcription and translation (Sawasaki et al., 2002; Spirin, 2004). These improvements include the composition of the system itself, e.g., extracts of genetically engineered bacterial strains, various energy resources or amino acid concentrations, or the use of defined components. Second, various expression conditions have been employed, such as dialysis, continuous-flow, continuous exchange, hollow fiber, and bilayer systems (Sawasaki et al., 2002; Goshima et al., 2008; He et al., 2008b). The DNA construct itself also affects protein expression. In addition to codon optimization, the fusion of a well-expressed N-terminal tag sequence to the target gene can significantly increase protein expression (Son et al., 2006). Interestingly, the human Cκ domain fused at the C-terminus of the target gene has been able to enhance expression of proteins that are either poorly or not expressed in the *E. coli* S30 system (Palmer et al., 2006).

Protein yield in the mg/mL range is now achievable in *E. coli* S30 and wheat germ systems (Kigawa et al., 1999; Madin et al., 2000). Large-scale production of functional membrane proteins and cysteine-rich polypeptides has also been achieved in cell-free systems (Klammt et al., 2006; Ezure et al., 2007). The high level production of proteins in combination with direct labeling of the nascent proteins greatly facilitates the downstream applications such as protein detection, purification, characterization, and structural studies (Kigawa et al., 2002).

6.5 TECHNOLOGIES BASED ON CELL-FREE EXPRESSION

A number of cell-free protein technologies, such as ribosome display and protein *in situ* arrays, have been developed for a rapid selection or screening of target proteins from large DNA libraries.

6.5.1 CELL-FREE DISPLAY METHODS

Display technologies link genetic information (genotype) to their encoded proteins (phenotype) for selection. In contrast to *in vivo* display systems such as phage display and yeast surface display, cell-free methods avoid cell transformation, thus allowing for display of larger libraries with possible continuous expanding of new diversity during the selection process. This feature makes the cell-free methods ideal for molecular evolution *in vitro*. They have been widely used for protein *in vitro* selection and evolution in which protein variants are created and selected on the basis of their activity.

6.5.1.1 Ribosome Display

Ribosome display selects required proteins from very large PCR libraries. The gene–protein linkage is achieved by generating protein–ribosome–mRNA (PRM) complexes through cell-free expression of DNA constructs in the absence of the stop codon (Figure 6.2a). The formation of PRM complexes allows affinity isolation, by immobilized ligand, of the nascent protein together with its translated mRNA, which can then be converted and amplified as DNA by RT-PCR (Hanes and Plückthun, 1997; He and Taussig, 1997). The process can be reiterated to enrich the targets initially presented as rare species from a large population. Both eukaryotic and prokaryotic ribosome display systems with their own DNA recovery protocols have been published (He and Khan, 2005), indicating the

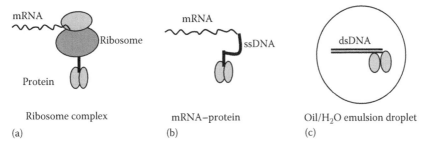

FIGURE 6.2 Protein-gene linkages in cell-free protein display technologies. (a) Ribosome display. (b) mRNA display; ssDNA: single-stranded DNA. (c) *In vitro* compartmentalization; dsDNA: double-stranded DNA.

difference in the stability of ribosome complexes produced by prokaryotic and eukaryotic cell-free systems. Recently, intracellular PRM complexes have been produced inside living cells using *E. coli* SecM translation arrest mechanism (Contreras-Martínez and DeLisa, 2007).

Ribosome display has been widely applied in particular for *in vitro* antibody selection, evolution, and humanization, leading to the generation of unique antibody fragments with either high affinity ($K_d = 10^{-12}$ M) or having catalytic activities or recognizing conformational epitopes (He and Khan, 2005). Other proteins such as peptides, ligand-binding domains/motifs, transcription factors, proteases, receptors, enzymes and as well as vaccine candidates have also been successfully displayed on the surface of ribosome (Weichhart et al., 2003; He and Khan, 2005; Ihara et al., 2006; Quinn et al., 2008). Remarkably, full-length membrane proteins can be functionally produced as ribosome complexes for screening molecular interactions (Schimmele and Plückthun, 2005). In addition, ribosome display is capable of isolating protease-resistant scaffolds or antibody mimics (Matsuura and Plückthun, 2003; Binz et al., 2004). Recently a 5′-untranslated sequence (5′-UTR) was identified for promoting translation efficiency (Mie et al., 2008).

6.5.1.2 mRNA Display

Similar to ribosome display, mRNA display selects proteins linked to their encoding mRNA, but is covalently joined without the presence of the translating ribosome (Roberts and Szostak, 1997) (Figure 6.2b). In this technology, hybrid RNA-DNA molecules are first generated *in vitro* by chemically attaching 3′ end of the mRNA to a short ssDNA oligonucleotide, which also carries a puromycin moiety. When the hybrid RNA–DNA molecules are used as template, the ribosome stalls upon reaching the RNA–DNA junction and the attached puromycin moiety enters the ribosome peptidyl transferase site, forming a covalent linkage with the growing nascent polypeptide (Roberts and Szostak, 1997). The resultant protein-mRNA fusion molecules are then used as the selection entities. mRNA display has been used to select functional proteins and peptides including cyclic peptides from either natural or synthetic libraries (Austin et al., 2008; Litovchick and Szostak, 2008). The application to cDNA libraries has identified Bcl-X_L binders with affinities ranging from 2 nM to 20 μM (Hammond et al., 2001). Surprisingly, both ATP-binding domains and novel enzymes (ligases) have been generated from completely artificial random-sequence libraries (Keefe and Szostak, 2001; Seelig and Szostak, 2007). In recent papers, mRNA display has been demonstrated for the isolation of antigen-binding proteins based on the scaffold fibronectin type III and the recovery of high-affinity molecules that bound TNF-α (Xu et al., 2002) or recognized specifically endogenous phosphorylated IκBα (Olson et al., 2008). mRNA display selected "adnectins" against VEGF receptor R2 are currently in clinical trials (www.adnexustx.com).

6.5.1.3 *In Vitro* Compartmentalization

In vitro compartmentalization generates DNA-protein linkage for *in vitro* selection by creating artificial compartments/localizations, which contain individual DNA molecules in such a way that each newly synthesized protein binds to the DNA itself either via a tag sequence or is captured by an immobilized antibody localized at the same surface (Tawfik and Griffiths, 1998; Sepp et al., 2002) (Figure 6.2c). The "compartments" are usually made either as encapsulated aqueous microdroplets through oil/water emulsions (Tawfik and Griffiths, 1998) or on the surface of microbeads (Sepp et al., 2002). *In vitro* compartmentalization is particularly useful for engineer enzymes such as restriction endonucleases, methylase, and polymerase.

6.5.2　Cell-Free Protein Array Technologies

Protein microarrays are high throughput tools for miniaturized, highly parallel analysis of proteins. Conventional cell-based methods for generating protein arrays are usually time-consuming, requiring separate expression and purification of individual proteins to be arrayed, and with limitations of maintaining functional proteins on the array surfaces during storage. These problems can be avoided by *in situ* synthesis of protein through cell-free expression on the array surfaces (He and Taussig, 2001; He et al., 2008a,b). This allows protein arrays to be generated directly from arrayed DNA molecules as and when required.

6.5.2.1　Protein *In Situ* Array

Protein *in situ* array (PISA) is designed by cell-free expression of proteins on a surface pre-coated by a protein-capturing reagent, so that the newly synthesized protein is captured *in situ* onto the array surface immediately after its release from ribosome (Figure 6.3a) (He and Taussig, 2001). After washing off other proteins/reagents, the captured proteins form protein arrays for immediate analysis. Through multiplexing, *PISA* rapidly generates a protein array in a single step from a set of PCR fragments. Miniaturization by directly spotting droplets containing 350 pL cell-free lysate onto pre-coated glass slides offers the potential for generating 13,000 spots per slide (Angenendt et al., 2006). *PISA* first demonstrated the advantages of "on-chip" synthesis of proteins, notably the ability to convert DNA into functional protein in an array format, without separate *E. coli* cloning, expression, and purification. It also obviates the problem of maintaining protein stability on the array surface in long-term storage, since the proteins are only expressed as and when the protein array is needed.

6.5.2.2　Nucleic Acid Programmable Protein Array

Nucleic acid programmable protein array (NAPPA) was the first method to demonstrate the direct conversion of DNA arrays into protein arrays through on-chip synthesis of proteins. Here, DNA was first arrayed onto a glass slide together with a protein-capturing antibody. The surface of the DNA array was then covered with a cell-free transcription/translation system to express proteins, which became trapped by the antibody co-localized in each spot (Figure 6.3b) (Ramachandran et al., 2004). NAPPA has been applied to produce high-density protein arrays to identify immune

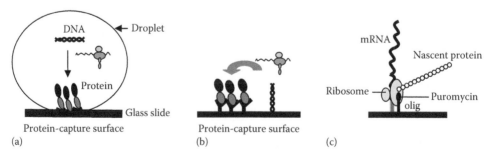

FIGURE 6.3 Arraying proteins through *in situ* protein synthesis on the surfaces. (a) PISA. (b) NAPPA. (c) Puromycin capture; ssDNA: single-stranded DNA.

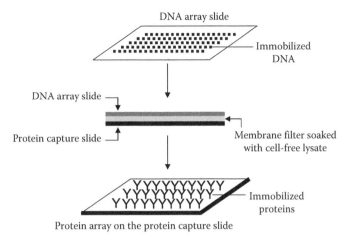

DNA array slide

Immobilized DNA

DNA array slide

Protein capture slide

Membrane filter soaked with cell-free lysate

Immobilized proteins

Protein array on the protein capture slide

FIGURE 6.4 Principle of the DAPA technology.

response signatures of breast cancer autoantibodies in patient sera (Anderson et al., 2008), as well as isolating target proteins with agonist activities from a cDNA library (Rolfs et al., 2008). As the proteins immobilized by this procedure are on the same DNA array surface, NAPPA in fact generates arrayed proteins co-localized with their DNA.

6.5.2.3 DNA Array to Protein Array

DNA array to protein array (DAPA), a recent novel system, has been reported to convert a reusable DNA array into multiple copies of a protein array (He et al., 2008a,b). Unlike NAPPA, the DAPA technology generates pure protein arrays on a separate surface (Figure 6.4). The principle of DAPA is to carry out cell-free protein synthesis within a membrane filter sandwiched between two glass slides, on one of which is an array of immobilized PCR fragments while the other is coated with a reagent to capture translated proteins. Individual proteins synthesized in parallel within the membrane diffuse through the membrane filter and become immobilized on the opposite surface through interaction with the protein-capturing reagent, forming a protein array with the same layout as the DNA array. This method can produce at least 20 copies of a protein array by repeated use of a single DNA array template (He et al., 2008a), making it useful for repeated printing of protein arrays, particularly for laboratories without access to microarray spotters. DAPA has been demonstrated for a number of proteins including antibody fragments, transcription factors, and a GTP-binding protein (He et al., 2008a).

6.5.2.4 *In Situ* Puromycin Capture

In situ puromycin capture is a method using an adaptation of mRNA display (Roberts and Szostak, 1997) to fabricate protein arrays through *in situ* puromycin capture of nascent polypeptides (Tao and Zhu, 2006) (Figure 6.3c). This method requires extra manipulation steps to make mRNA–ssDNA hybrid templates before printing, and protein yields are limited by the amount of the template spotted.

6.6 APPLICATIONS

6.6.1 High-Throughput Expression

Large-scale expression of proteins holds particular promise in proteomics and protein engineering for identifying proteins from diversified libraries. Cell-free systems can be easily adapted for such high-throughput (HT) platforms for the parallel production of many proteins (Spirin, 2004). For example, DNA molecules can be transcribed and translated in microplate wells and the synthesized protein assayed immediately. Cell-free systems can express protein population in a single reaction, making it ideal for developing deconvolution strategies for the comprehensive screening and identification of cloned proteins from libraries. This concept has been demonstrated by a technology known as "*in vitro* expression cloning" (IVEC) in which a large collection of clones was first divided into pools of 50–100 cloned plasmids in multiple wells followed by cell-free expression and activity analysis (King et al., 1997). Pools giving a positive signal were then further subdivided and re-screened and the process repeated until single clones were obtained. The success of IVEC requires a sensitive assay method since each represented protein in the population only occupies a small proportion. In combination with an *in vitro* PCR cloning method, cell-free expression has been able to rapidly screen functional mutants from diversified PCR libraries without the need for *E. coli* cloning (Burks et al., 1997; Rungpragayphan et al., 2003).

Cell-free protein synthesis has been used to discover novel genes from *Arabidopsis* by screening whole genome libraries (Sawasaki et al., 2002). It has also been demonstrated that the expression conditions identified in a cell-free system can be directly transferred to *in vivo* protein production (Lamla et al., 2006).

For medical applications, cell-free protein expression allows for both screening of translation-terminating mutations (Roest et al., 1993) and identification of effective vaccine candidates (Kanter et al., 2007). Toxic reagents or translation inhibitors can also be rapidly detected by using cell-free systems (Wei et al., 2005; Brandi et al., 2007).

6.6.2 STUDYING MOLECULAR INTERACTIONS

Molecular interactions, such as protein–protein, protein–DNA, protein–RNA, protein–ligand, DNA–RNA, and even RNA–RNA, can be rapidly studied by using cell-free expression. For identifying protein interactions, one labeled entity (protein, nucleotide, or ligand) is incubated with the interacting protein partner synthesized in a cell-free system, followed by isolation of the interaction complex, using either immunoprecipitation (Derbigny et al., 2000) or directly detected by electrophoretic mobility shift assay in which protein complexes are retarded in comparison to unbound molecules (Lee and Chang, 1995). Nucleic acid interactions are usually performed by hybridization using a designed nucleic acid at a target sequence and analyzing the effects on the transcription or translation of the target (Keum et al., 2009).

6.6.3 PRODUCTION OF "DIFFICULT-TO-EXPRESS" PROTEINS

Without the membrane barriers, cell-free protein synthesis provides a flexible system for producing "difficult" proteins such as toxic or membrane proteins. It has been shown that many proteins that fail to express by *in vivo* systems have been functionally produced in cell-free systems (Jackson et al., 2004). Recent improvements have been able to produce functional membrane proteins in large amounts in an *E. coli* cell-free lysate (Klammt et al., 2006). Furthermore, a complete heterotetrameric and multiple disulfide-bridged IgG has been successfully assembled in a coupled cell-free lysate under defined folding and oxidation conditions (Frey et al., 2008). An MS2 phage coat protein and a C-terminally truncated hepatitis B core protein, after being produced as virus-like particles (VLP) in an *E. coli* cell-free system, have shown the comparable characteristics to those produced *in vivo* (Bundy et al., 2008). The ability to produce infectious encephalomyocarditis virus (EMCV) in HeLa, 293-F and Krebs-2 S10 cell-free extracts (Svitkin et al., 2005; Svitkin and Sonenberg, 2007) provides a useful *in vitro* tool for both understanding the mechanism of viral replication and screening of antiviral drugs. Recently, the entire ORF of hepatitis C virus (HCV) RNA has been translated and correctly processed in the Krebs-2 system when supplemented with canine microsomal membranes (Svitkin et al., 2005; Svitkin and Sonenberg, 2007).

6.6.4 BRIDGING CHEMISTRY AND BIOLOGY

The possibility of easily incorporating unnatural or chemically diversified amino acids into nascent proteins during cell-free translation raises new opportunities for creating nonnatural polymers and chemically diversified drug-like peptides (Josephson et al., 2005; Goto et al., 2008). This idea has been exploited to produce both proteins with an array of chemical groups and novel nonbiological polypeptides of pharmaceutical interest. The PURE system has been used to produce cyclic peptides containing very similar characteristics to their chemically synthesized products (Goto et al., 2008).

6.7 FUTURE OUTLOOK

Cell-free systems will be continuously optimized and improved, particularly in the expression level and correct protein folding. The development of automation and miniaturization platforms would significantly increase the capability for high throughput expression of proteins. Protein technologies derived from cell-free systems will be applied increasingly for isolating engineered molecules with important biotechnological and medical applications, as well as for basic research. They will provide versatile tools for proteomics studies.

ABBREVIATIONS

K_d	Equilibrium dissociation constant
M	Molar concentration (e.g., 10^{-3} M = 1 mM)
PCR	Polymerase chain reaction
RT-PCR	Coupled reverse transcription and polymerase chain reaction
tRNA	Transfer RNA

ACKNOWLEDGMENTS

The Babraham Institute is an institute of the Biotechnology and Biological Sciences Research Council (BBSRC), United Kingdom. The authors would like to thank the State High-Tech Research and Development Plan (863) for its support.

REFERENCES

Agrawal, R.K. and Frank, J. 1999. Structural studies of translational apparatus. *Curr. Opin. Struct.* **9**: 215–221.

Anderson, K.S., Ramachandran, N., Wong, J., Raphael, J.V., Hainsworth, E. et al. 2008. Application of protein microarrays for multiplexed detection of antibodies to tumor antigens in breast cancer. *J. Proteome Res.* **7**: 1490–1499.

Angenendt, P., Kreutzberger, J., Glokler, J., and Hoheisel, J.D. 2006. Generation of high density protein microarrays by cell-free *in situ* expression of unpurified PCR products. *Mol. Cell Proteomics* **5**: 1658–1666.

Austin, R.J., Ja, W.W., and Roberts, R.W. 2008. Evolution of class-specific peptides targeting a hot spot of the Galphas subunit. *J. Mol. Biol.* **377**: 1406–1418.

Binz, H.K., Amstutz, P., Kohl A. et al. 2004. High-affinity binders selected from designed ankyrin repeat protein libraries. *Nat. Biotechnol.* **22**: 575–582.

Brandi, L., Fabbretti, A., Milon, P., Carotti, M., Pon, C.L., and Gualerzi, C.O. 2007. Review: Methods for identifying compounds that specifically target translation. *Methods Enzymol.* **431**: 229–267.

Braun, P. and LaBaer, J. 2003. High throughput protein production for functional proteomics. *Trends Biotechnol.* **21**: 383–388.

Bundy, B.C., Franciszkowicz, M.J., and Swartz, J.R. 2008. *Escherichia coli*-based cell-free synthesis of virus-like particles. *Biotechnol. Bioeng.* **100**: 28–37.

Burks, E.A., Chen, G., Georgiou, G., and Iverson, B.L. 1997. *In vitro* scanning saturation mutagenesis of an antibody binding pocket. *Proc. Natl. Acad. Sci. USA* **94**: 412–417.

Choi, S., Ryu, K., and Seong B.L. 2009. RNA-mediated chaperone type for de novo protein folding. *RNA Biol.* **6**: 1–4.

Contreras-Martínez, L.M. and DeLisa, M.P. 2007. Intracellular ribosome display via SecM translation arrest as a selection for antibodies with enhanced cytosolic stability. *J. Mol. Biol.* **372**: 513–524.

Das, D., Das, A., Samanta, D., Ghosh, J., Dasgupta, S. et al. 2008. Role of the ribosome in protein folding. *Biotechnol. J.* **3**: 999–1009.

Derbigny, W.A., Kim, S.K., Caughman, G.B., and O'Callaghan, D.J. 2000. The EICP22 protein of equine herpesvirus I physically interacts with the immediate early protein and with itself to form dimers of higher order complexes. *J. Virol.* **74**: 1425–1435.

Denman, R.B. 2008. Protein methyltransferase activities in commercial *in vitro* translation systems. *J. Biochem.* **144**: 223–233.

Ellis, J.P., Bakke, C.K., Kirchdoerfer, R.N., Jungbauer, L.M., and Cavagnero, S. 2008. Chain dynamics of nascent polypeptides emerging from the ribosome. *ACS Chem. Biol.* **3**: 555–566.

Endoh, T., Kanai, T., Sato, Y.T., Liu, D.V., Yoshikawa, K., Atomi, H., and Imanaka, T. 2006. Cell-free protein synthesis at high temperatures using the lysate of a hyperthermophile. *J. Biotechnol.* **126**: 186–195.

Ezure, T., Suzuki, T., Shikata, M., Ito, M., Ando, E., Nishimura, O., and Tsunasawa, S. 2007. Expression of proteins containing disulfide bonds in an insect cell-free system and confirmation of their arrangements by MALDI-TOF MS. *Proteomics* **7**: 4424–4434.

Frey, S., Haslbeck, M., Hainzi, O., and Buchner, J. 2008. Synthesis and characterization of a functional intact IgG in a prokaryotic cell-free expression system. *Biol. Chem.* **389**: 37–45.

Goerke, A.R. and Swartz, J.R. 2009. High-level cell-free synthesis yields of proteins containing site-specific non-natural amino acids. *Biotechnol. Bioeng.* **102**: 400–416.

Goshima, N., Kawamura, Y., Fukumoto, A. et al. 2008. Human protein factory for converting the transcriptome into an *in vitro*-expressed proteome. *Nat. Methods* **5**: 1011–1017.

Goto, Y., Ohta, A., Sako, Y., Yamagishi, Y., Murakami, H., and Suga, H. 2008. Reprogramming the translation initiation for the synthesis of physiologically stable cyclic peptides. *ACS Chem. Biol.* **3**: 120–129.

Hammond, P.W., Alpin, J., Rise, C.E., Wright, M., and Kreider, B.L. 2001. *In vitro* selection and characterisation of Bcl-X(L)-binding proteins from a mix of tissue-specific mRNA display libraries. *J. Biol. Chem.* **276**: 20899–20906.

Hanes, J. and Plückthun, A. 1997. *In vitro* selection and evolution of functional proteins by using ribosome display. *Proc. Natl. Acad. Sci. USA* **94**: 4937–4942.

He, M. and Khan, F. 2005. Review: Ribosome display: Next-generation of display technologies for production of antibodies *in vitro. Expert Rev. Proteomics* **2**: 421–430.

He, M. and Taussig, M.J. 1997. Antibody-ribosome-mRNA (ARM) complexes as efficient selection particles for *in vitro* display and evolution of antibody combining sites. *Nucleic Acids Res.* **25**: 5132–5134.

He, M. and Taussig, M.J. 2001. Single step generation of protein arrays from DNA by cell-free expression and *in situ* immobilisation (PISA method). *Nucleic Acids Res.* **29**: e73.

He, M. and Taussig, M.J. 2007. Eukaryotic ribosome display with *in situ* DNA recovery. *Nat. Methods* **4**: 281–288.

He, M., Stoevesandt, O., Palmer, E.A., Khan, F., Ericsson, O., and Taussig, M.J. 2008a. Printing protein arrays from DNA arrays. *Nat. Methods* **5**: 175–177.

He, M., Stoevesandt, O., and Taussig, M.J. 2008b. *In situ* synthesis of protein arrays. *Curr. Opin. Biotechnol.* **19**: 4–9.

Hillebrecht, J.R. and Chong, S. 2008. A comparative study of protein synthesis in *in vitro* systems: From the prokaryotic reconstituted to the eukaryotic extract-based. *BMC Biotechnol.* **8**: 58–66.

Hino, M., Kataoka, M., Kajimoto, K., Yamamoto, T., Kido, J., Shinohara, Y., and Baba Y. 2008. Efficiency of cell-free protein synthesis based on a crude cell extract from *Escherichia coli*, wheat germ, and rabbit reticulocytes. *J. Biotechnol.* **133**: 183–189.

Ihara, H., Mie, M., Funabashi, H., Takahashi, F., Sawasaki, T., Endo, Y., and Kobatake, E. 2006. *In vitro* selection of zinc finger DNA-binding proteins through ribosome display. *Biochem. Biophys. Res. Commun.* **345**: 1149–1154.

Jackson, A.M., Boutell, J., Cooley, N., and He, M. 2004. Review: Cell-free protein synthesis for proteomics. *Brief. Funct. Genomic Proteomic* **2**: 308–319.

Josephson, K., Hartman, M.C., and Szostak, J.W. 2005. Ribosomal synthesis of unnatural peptides. *J. Am. Chem. Soc.* **127**: 11727–11735.

Kanter, G., Yang, J., Voloshin, A., Levy, S., Swartz, J.R., and Levy, R. 2007. Cell-free production of scFv fusion proteins: An efficient approach for personalized lymphoma vaccines. *Blood* **109**: 3393–3399.

Keefe, A.D. and Szostak, J.W. 2001. Functional proteins from a random-sequence library. *Nature* **410**: 715–718.

Keum, J.W., Ahn, J.H., Kang, T.J., and Kim, D.M. 2009. Combinatorial, selective and reversible control of gene expression using oligodeoxynucleotides in a cell-free protein synthesis system. *Biotechnol. Bioeng.* **102**: 577–582.

Kigawa, T., Yabuki, T., Yoshida, Y., Tsutsui, M., Ito, Y., Shibata, T., and Yokoyama, S. 1999. Cell-free production and stable-isotope labelling of milligram quantities of proteins. *FEBS Lett.* **442**: 15–19.

Kigawa, T., Yamaguchi-nunokawa, E., Kodama, K., Matsuda, T., Yabuki, T. et al. 2002. Selenomethionine incorporation into a protein by cell-free synthesis. *J. Struct. Funct. Genomics* **2**: 29–35.

King, R.W., Lustig, K.D., Stukenberg, P.T., McGarry, T.J., and Kirschner, M.W. 1997. Expression cloning in the test tube. *Science* **277**: 973–974.

Klammt, C., Schwarz, D., Lohr, F., Schneider, B., Dotsch, V., and Bernhard, F. 2006. Cell-free expression as an emerging technique for the large scale production of integral membrane protein. *FEBS J.* **273**: 4141–4153.

Kolb, V.A., Makeyev, E.V., and Spirin, A.S. 2000. Co-translational folding of an eukaryotic multidomain protein in a prokaryotic translation system. *J. Biol. Chem.* **275**: 16597–16601.

Langlais, C., Guilleaume, B., Wermke, N., Scheuermann, T., Ebert, L., LaBaer, J., and Korn, B. 2007. A systematic approach for testing expression of human full-length proteins in cell-free expression systems. *BMC Biotechnol.* **7**: 64–75.

Lamla, T., Hoerer, S., and Bauer, M.M. 2006. Screening for soluble expression constructs using cell-free protein synthesis. *Int. J. Biol. Macromol.* **39**: 111–121.

Lee, H.J. and Chang, C. 1995. Identification of human TR2 orphan receptor response element in the transcriptional initiation site of the simian virus 40 major late promoter. *J. Biol. Chem.* **270**: 5434–5440.

Litovchick, A. and Szostak, J.W. 2008. Selection of cyclic peptide aptamers to HCV IRES RNA using mRNA display. *Proc. Natl. Acad. Sci. USA* **105**: 15293–15298.

Madin, K., Sawasaki, T., Ogasawara, T., and Endo, Y. 2000. A highly efficient and robust cell-free protein synthesis system prepared from wheat embryos: Plants apparently contain a suicide system directed at ribosomes. *Proc. Natl. Acad. Sci USA* **97**: 559–564.

Mamaev, S., Olejnik, J., Olejnik, E.K., and Rothschild, K.J. 2004. Cell-free N-terminal protein labeling using initiator suppressor tRNA. *Anal. Biochem.* **326**: 25–32.

Matsuura, T. and Plückthun, A. 2003. Selection based on the folding properties of proteins with ribosome display. *FEBS Lett.* **539**: 24–28.

Mattheakis, L.C, Bhatt, R.R., and Dower W.J. 1994. An *in vitro* polysome display system for identifying ligands from very large peptide libraries. *Proc. Natl. Acad. Sci. USA* **91**: 9022–9026.

Merz, F., Boehringer, D., Schaffitzel, C., Preissler, S., Hoffmann, A. et al. 2008. Molecular mechanism and structure of Trigger Factor bound to the translating ribosome. *EMBO J.* **27**: 1622–1632.

Mie, M., Shimizu, S., Takahashi, F., and Kobatake, E. 2008. Selection of mRNA 5′-untranslated region sequence with high translation efficiency through ribosome display. *Biochem. Biophys. Res. Commun.* **373**: 48–52.

Mikami, S., Kobayashi, T., Yokoyama, S., and Imataka, H. 2006. A hybridoma-based *in vitro* translation system that efficiently synthesizes glycoproteins. *J. Biotechnol.* **127**: 65–78.

Nemoto, N., Miyamoto-Sato, E., and Yanagawa, H. 1999. Fluorescence labeling of the C-terminus of proteins with a puromycin analogue in cell-free translation systems. *FEBS Lett.* **462**: 43–46.

Netzer, W.J. and Hartl, F.U. 1997. Recombination of protein domains facilitated by cotranslational folding in eukaryotes. *Nature* **388**: 343–349.

Olson, C.A., Liao, H.I., Sun, R., and Roberts, R.W. 2008. mRNA display selection of a high-affinity, modification-specific phospho-ikappabalpha-binding fibronectin. *ACS Chem. Biol.* **3**: 480–485.

Palmer, E., Liu, H., Khan, F., Taussig, M.J., and He, M. 2006. Enhanced cell-free protein expression by fusion with immunoglobulin Ckappa domain. *Protein Sci.* **15**: 2842–2846.

Quinn, D.J., Cunningham, S., Walker, B., and Scott, C.J. 2008. Activity-based selection of a proteolytic species using ribosome display. *Biochem. Biophys. Res. Commun.* **370**: 77–81.

Ramachandiran, V., Kramer, G., and Hardesty, B. 2000. Expression of different coding sequences in cell-free bacterial and eukaryotic systems indicates translational pausing on *Escherichia coli* ribosomes. *FEBS Lett.* **482**: 185–188.

Ramachandran, N., Hainsworth, E., Bhullar, B., Eisenstein, S., Rosen, B., Lau, A.Y., Walter, J.C., and LaBaer, J. 2004. Self-assembling protein microarrays. *Science* **305**: 86–90.

Roberts, R.W. and Szostak, J.W. 1997. RNA-peptide fusions for the *in vitro* selection of peptides and proteins. *Proc. Natl. Acad. Sci. USA* **94**: 12297–12302.

Roest, P.A., Roberts, R.G., Sugino, S., van Ommen, G.J., and den Dunnen, J.T. 1993. Protein truncation test (PTT) for rapid detection of translation-terminating mutations. *Hum. Mol. Genet.* **2**: 1719–1721.

Rolfs, A., Montor, W.R., Yoon, S.S., Hu, Y., Bhullar, B. et al. 2008. Production and sequence validation of a complete full length ORF collection for the pathogenic bacterium *Vibrio cholerae*. *Proc. Natl. Acad. Sci. USA* **105**: 4364–4369.

Rothschild, K. and Gite, S. 1999. tRNA-mediated protein engineering. *Curr. Opin. Biotechnol.* **10**: 64–70.

Rungpragayphan, S., Nakano, H., and Yamane, T. 2003. PCR-linked *in vitro* expression: A novel system for high-throughput construction and screening of protein libraries. *FEBS Lett.* **540**: 147–150.

Sakurai, N., Moriya, K., Suzuki, T., Sofuku, K., Mochiki, H., Nishimura, O., and Utsumi, T. 2007. Detection of co- and posttranslational protein N-myristoylation by metabolic labeling in an insect cell-free protein synthesis system. *Anal. Biochem.* **362**: 236–244.

Sawasaki, T., Ogasawara, T., Morishita, R., and Endo, Y. 2002. A cell-free protein synthesis system for high-throughput proteomics. *Proc. Natl. Acad. Sci USA* **99**: 14652–14657.

Schimmele, B. and Plückthun, A. 2005. Identification of a functional epitope of the Nogo receptor by a combinatorial approach using ribosome display. *J. Mol. Biol.* **352**: 229–241.

Seelig, B. and Szostak, J.W. 2007. Selection and evolution of enzymes from a partially randomized non-catalytic scaffold. *Nature* **448**: 828–831.

Sepp, A., Tawfik, D.S., and Griffiths, A.D. 2002. Microbead display by *in vitro* compartmentalisation: Selection for binding using flow cytometry. *FEBS Lett.* **532**: 455–458.

Shimizu, Y., Inoue, A., Tomari, Y., Suzuki, T., Yokogawa, T., Nishikawa, K., and Ueda, T. 2001. Cell-free translation reconstituted with purified components. *Nat. Biotechnol.* **19**: 751–755.

Son, J.M., Ahn, J.H., Hwang, M.Y., Park, C.G., Choi, C.Y., and Kim, D.M. 2006. Enhancing the efficiency of cell-free protein synthesis through the polymerase-chain-reaction-based addition of a translation enhancer sequence and the in situ removal of the extra amino acid residues *Anal. Biochem.* **351**: 187–192.

Spirin, A. 2004. High-throughput cell-free systems for synthesis of functionally active proteins. *Trends Biotechnol.* **22**: 538–545.

Svitkin, Y.V. and Sonenberg, N. 2007. A highly efficient and robust *in vitro* translation system for expression of picornavirus and hepatitis C virus RNA genomes. *Methods Enzymol.* **429**: 53–82.

Svitkin, Y.V., Pause, A., Lopez-Lastra, M., Perreault, S., and Sonenberg, N. 2005. Complete translation of the hepatitis C virus genome *in vitro*: Membranes play a critical role in the maturation of all virus proteins except for NS3. *J. Virol.* **79**: 6868–6881.

Tao, S.-C. and Zhu, H. 2006. Protein chip fabrication by capture of nascent polypeptides. *Nat. Biotechnol.* **24**: 1253–1254.

Tawfik, D.S. and Griffiths, A.D. 1998. Man-made cell-like compartments for molecular evolution. *Nat. Biotechnol.* **16**: 652–656.

Wei, Q., Fredrickson, C.K., Jin, S., and Fan, Z.H. 2005. Toxin detection by a miniaturised *in vitro* protein expression array. *Anal. Chem.* **77**: 5494–5500.

Weichhart, T., Horky, M., Sollner, J. et al. 2003. Functional selection of vaccine candidate peptides from *Staphylococcus aureus* whole-genome expression libraries *in vitro*. *Infect. Immun.* **71**: 4633–4641.

Xu, L., Aha, P., Gu, K. et al. 2002. Direct evolution of high affinity antibody mimics using mRNA display. *Chem. Biol.* **9**: 933–942.

Yamaguchi, J., Nemoto, N., Sasaki, T. et al. 2001. Rapid functional analysis of protein-protein interactions by fluorescent C-terminal labeling and single-molecule imaging. *FEBS Lett.* **502**: 79–83.

Ying, B.W., Taguchi, H., and Ueda, T. 2006. Co-translational binding of GroEL to nascent polypeptides is followed by posttranslational encapsulation by GroES to mediate protein folding. *J. Biol. Chem.* **281**: 21813–21819.

7 Nonfood Applications of Milk Proteins: A Review

Dorota Kalicka, Dorota Najgebauer-Lejko, and Tadeusz Grega

CONTENTS

7.1 INTRODUCTION

Milk proteins are represented by two major families of proteins: caseins (CN) and whey proteins. In the past, milk casein was used mainly for technical applications such as wood glue, paper coating, leather finishing, synthetic fibers, and rigid plastics. There are also historical records about the use of whey protein fraction for cosmetic preparations and as pharmaceutical agents. Nowadays, milk proteins find their use as valuable food ingredients. However, they have also many nonfood applications due to their specific structural and functional properties, which additionally can be easily modified. Novel applications of these components include biodegradable coatings and biomaterials, for encapsulation of active substances, natural cosmetic formulations, as well as health-care products and pharmaceutical preparations (Marshall 2004, Southward 1992). Many of these new applications are still under research.

Besides its technical application, CN production and recovery of protein from whey yield high profits for dairy companies. The possibility of using dairy wastes, by-products, and milk, which do not meet the requirements of food applications is also of great importance.

In that light, this chapter reviews the past and present applications of proteins derived from milk as well as their great potential for future uses. The composition and characteristics of caseins and whey proteins as well as methods of their isolation are described.

7.2 THE COMPOSITION AND CHARACTERISTIC OF MILK PROTEINS

Proteins constitute about 95% of nitrogen compounds present in milk. Cow's milk contains approximately 3.2% of proteins, but this number varies in some range due to many factors, the genetic ones being the most important. Milk proteins can be divided into two groups: CNs (about 2.6% of cow's milk), which precipitate at pH 4.6 at 20°C and whey or serum proteins (0.6%), which remain soluble under these conditions (Fox and Kelly 2004, Walstra et al. 1999). The average composition of bovine milk proteins is given in Table 7.1.

Milk also contains other proteins in minor quantities (0.8 g/kg milk): membrane proteins, enzymes (lactoperoxidase, lysozyme), and iron-binding proteins (lactoferrin and transferrin) (Walstra et al. 1999).

TABLE 7.1
The Composition of Milk Proteins

	Milk (g/L)	Total Protein (w/w, %)
Total proteins	30–35	100
CNs	24–28	78.5
α_{s1}-CN	12–15	31
α_{s2}-CN	3–4	8
β-CN	9–11	30.4
κ-CN	2–4	10
Whey proteins	5–7	19
β-Lactoglobulin	2–4	9.8
α-Lactoalbumin	0.6–1.7	3.7
Immunoglobulins	0.5–1.8	2.4
Bovine serum albumin	0.2–0.4	1.2

Sources: Walstra, P. et al., *Dairy Technology. Principles of Milk Properties and Processes*, Marcel Dekker Inc., New York, 1999; Whitney, R.McL. In: Wong, N.P. et al. (Eds.), *Fundamentals of Dairy Chemistry*, Aspen Publishers, Inc., Gaithersburg, MA, 1999.

7.2.1 Caseins

Casein (CN), the main protein of bovine milk, is not a homogeneous fraction as it consists of four different proteins: α_{s1}-CN, α_{s2}-CN, β-CN, and κ-CN. Their average proportions in one whole CN micelle are as follows: 38%:10%:36%:12% (Fox and Kelly 2004). All of these CN families differ in chemical composition and properties (Table 7.2). Moreover, they exist in several genetic variants, which differ in the degree of phosphorylation, glycosylation, and the deletion or substitutions of single amino acids. Detailed characteristics of all CNs as well as whey proteins are given in the report by Farrel et al. (2004).

In fresh milk (pH 6.6–6.8) CNs exist in the form of micelles (spherical particles with a diameter of 50–300 nm) in colloidal dispersion. There are several different models of the CN micelle structure but all of them are consistent in the fact that the micelle is built up of a hydrophobic core covered by a soluble layer of κ-CN molecules (Dalgleish 1998). They have negative net charge and are quite stable. During acidification, the isoelectric potential diminishes. At 4.6 pH value, known as isoelectric point, the negative and positive charges on the CN are equal and the CN precipitates from the solution.

As CN contains a relatively high number of proline residues, it is characterized by changes in secondary and tertiary structures (β-sheets and α-helices). This feature affects the high thermal stability of this protein. CNs are also strongly hydrophobic and as a result poorly soluble in water. High surface activity confers good foaming and emulsifying properties to the CNs. CN's flexible, open structure, as well as amphipathic character enhances the film-forming and coating abilities. The high amount of polar groups in CN produces good adhesion to different substances such as wood, glass, or paper (Audic et al. 2003, Fox and Kelly 2004).

7.2.2 Whey Proteins

Whey is a coproduct of cheese production and CN manufacture in dairy industry. After the CN curd separates from the milk, the remaining yellow/green color watery liquid is called whey

---

TABLE 7.2
Selected Properties of Bovine Milk Caseins

| | CN | | | |
Property	α_{s1}-Casein (α_{s1}-CN)	α_{s2}-Casein (α_{s2}-CN)	β-Casein (β-CN)	κ-Casein (κ-CN)
Character	Phosphoprotein	Phosphoprotein	Phosphoprotein	"Glycoprotein"
Major genetic variants	**B, C**	**A**	A^1, A^2, **B**	**A**, B
Molecular weight[a]	23,615	25,226	23,983	19,037
Amino acids residues[a]	199	207	209	169
Phosphoserine (residues/mol)[a]	8	11	5	1
–S–S– linkages/mol[a]	0	1	0	–
Hydrophobicity (kJ/residue)[a]	4.9	4.7	5.6	5.1
Net charge/residue[a]	−0.10	−0.07	−0.06	−0.02
Isoelectric point[a]	4.44–4.76	–	4.83–5.07	5.45–5.77
Association tendency[a]	Strong	Strong	Poor at $t < 5$°C Strong at 37°C	Strong

Sources: Farrell, H.M. Jr. et al., *J. Dairy Sci.*, 87, 1641, 2004; Walstra, P. et al., *Dairy Technology. Principles of Milk Properties and Processes*, Marcel Dekker Inc., New York, 1999.

[a] Properties are given for one selected (in bold) genetic variant.

(Zadow 1994). This by-product retains 55% of milk nutrients. Among them are lactose, soluble proteins, lipids, mineral salts, and vitamins. Additionally, whey contains variable amounts of lactic acid and nonsoluble nitrogen (Kosikowski 1979). Two main whey varieties are produced. Sweet whey, with pH at least 5.6 originates from rennet-coagulated cheese production. Acid whey, with a pH no higher than 5.1 comes from the manufacture of acid-coagulated cheese (Jelen 2003). Acid wheys typically have higher ash and lower protein contents than sweet wheys. Because of its low concentration of milk constituents (only 6%–7% dry matter), whey has commonly been considered a waste product. Therefore, cheese whey represents an important environmental problem because of the high volumes produced and its high organic matter content, exhibiting a BOD_5 30,000–50,000 ppm and COD 60,000–80,000 ppm (González Siso 1996). Until recently, whey has been discarded or used as animal feed. Recovering the solid components of whey is attractive for two main reasons: reduction of organic pollution created by whey waste when they are discarded and, mostly, for optimal utilization of the nutritional and functional properties offered by whey proteins (Bonnaillie and Tomasula 2008).

Whey proteins due to their high digestibility, nutritional value, and physicochemical properties are currently perceived as very valuable ingredients of traditional and functional food (USDEC 2003), and especially novel food, popular in well-developed countries (Ha and Zemel 2003). Their abilities also evoke the interest of scientists from outside the nutritional field.

The major whey proteins are α-lactalbumin (α-LA), β-lactoglobulin (β-LG), bovine serum albumin (BSA), and heavy- and light-chain immunoglobulins (Igs). Other important proteins found in whey, but present in minor quantities, are lactoferrin (LF) and lactoperoxidase. Whey may also include the proteose-peptone components—glycomacropeptides (GMPs) formed by the enzymatic degradation of the CNs during cheese production (Bonnaillie and Tomasula 2008).

7.2.2.1 β-Lactoglobulin

β-LG, the main protein in whey, is present in the milk of most ruminants, and is known to be absent in human milk. It accounts for around 50% of total whey protein (Lucena et al. 2006). β-LG is a small, soluble, and globular protein. At a pH lower than 3 it exists as a monomer with a molecular mass of about 18 kDa. At pH of between 3 and 7, β-LG exists in solution as a dimer with an effective molecular mass of about 36 kDa. Native β-LG has five cysteine residues existing as two disulfide bonds, with one free thiol group which is very reactive. β-LG is very hydrophobic and it is not soluble in pure water. Its solubility closely depends on pH and ionic strength, but it does not precipitate on acidification of milk (Walstra et al. 1999). The native conformation is sensitive to heat and pH and the free thiol group is of major significance on denaturation. The dimmer dissociates at high temperature (Mehra and O'Kennedy 2008). β-LG exists in a large number of genetically determined variants but it is generally present as either the A or the B variant. Although isolated over 60 years ago, the biological function of this protein is still not fully understood but it does bind calcium and zinc and a variety of hydrophobic small molecules (Walzem et al. 2002). It has been speculated that the function of β-LG may be related to vitamin A transport, because this protein is similar in sequence to retinol-binding proteins. β-LG is stable against stomach acids and proteolytic enzymes, is a rich source of the essential acid cysteine, and may be responsible for carrying the vitamin A precursor retinol from the cow to its calf (Tunick 2008). β-LG may also function as a fatty acid or lipid-binding protein (Perez and Calvo 1995).

7.2.2.2 α-Lactalbumin

α-LA is the second most important whey protein comprising approximately 20% of the whey proteins. It has a lower molecular weight of 14.2 kDa and a globular structure in aqueous solution. This protein is a calcium-binding protein and has a high affinity to other metal ions, including Zn^{2+}, Mn^{2+}, Cd^{2+}, Cu^{2+}, and Al^{3+} (Walzem et al. 2002). However, binding of Ca^{2+} is of utmost importance for maintaining the structure of protein. In the presence of saturating amounts of calcium, α-LA is characterized by being quite thermostable but in the absence of calcium is very unstable

(Chatterton et al. 2006). It has four disulfide bonds consisting of 123 amino acids in a single peptide chain (Konrad and Kleinschmidt 2008). α-LA is a particularly good source of essential amino acids Trp and Cys as these amino acids are precursors of serotonin and glutathione. α-LA appears in all milk types in which the presence of lactose can be demonstrated. Its biological function is coenzyme action in the synthesis of lactose (Madureira et al. 2007). Bovine α-LA is structurally and compositionally similar to the major whey protein in human milk, and is thus used in infant formula (Lucena et al. 2006).

7.2.2.3 Bovine Serum Albumin, Immunoglobulins

The other proteins found in whey are blood proteins. BSA and Igs pass in to the whey by "leakage" from blood serum. BSA is a large molecule with a molecular mass of about 66 kDa, with many −S−S− linkages and much α-helix. BSA is an elongated molecule, about 3×12 nm in size (Walstra et al. 1999). The function of this protein in milk is unknown, but it does bind fatty acids and other small molecules (Kinsella and Whitehead 1989, Walzem et al. 2002). Igs, also called antibodies are the largest proteins in whey, with molecular weight of 150 kDa. The Igs in whey are divided into classes having different physiochemical structure and biological activities (i.e., IgG1, IgG2, IgGA, and IgM). Igs are made up of 90% protein and 10% carbohydrates. The basic unit of all Igs is similar to IgG, which is composed of four polypeptide chains: two light (23 kDa) and two heavy (52 kDa) (El-Loly 2007). This group of whey proteins provides passive immunity for infants and may stimulate immune function in adults (Mehra et al. 2006).

7.2.2.4 Lactoferrin

LF is a globular multifunctional protein found in milk as well as other exocrine secretions such as tears and saliva, and neutrophil granules in mammals. It is an 80 kDa nonheme, iron-binding glycoprotein of the transferring family (Wakabayashi et al. 2006). There are 689 amino acids in the molecule. LF sequesters and solubilizes iron, thus controlling the amount of iron available for gut metabolism. Lactoferrin is the major nonspecific disease-resistance factor in the mammary gland. In general, the properties of lactoferrin include antibacterial and antiviral properties, prevention of the growth of pathogenic organisms in the gut, stimulation of the immune system, regulation of iron metabolism, and control of cell or tissue damage and wound healing (Bockman and Guidon 1996, Walzem et al. 2002).

7.2.2.5 Lactoperoxidase

Lactoperoxidase (LPO) makes up approximately 0.5% of the whey protein. It is a glycoprotein consisting of a single peptide chain of 608 amino acids, with a molecular weight of approximately 78 kDa. It has a much higher isoelectric point (pH 9.6) than most of the other whey proteins (Kussendrager and van Hooijdonk 2000). The enzyme contains a heme structure, with one iron molecule per mole of LPO. The conformation of the protein is stabilized by a strongly chelated calcium ion. LPO is also present and active in many more secretory fluids in various part of the body. LPO has no antibacterial effect on its own but has the ability to oxidize the thiocyanate ion (SCN⁻) in the presence of hydrogen peroxide (H_2O_2). The resulting chemical compound has an antibacterial effect (Boots and Floris 2006).

7.2.2.6 Glycomacropeptides

GMP, also called caseinomacropeptide (CMP), is a low-molecular-weight product (8.6 kDa) formed by chymosin cleavage of κ-CN between Phe_{105} and Met_{106} during the manufacture of cheese. Hydrophilic GMP is released into whey whereas hydrophobic fragment, named *para*-κ-CN becomes a part of the cheese curd. Therefore, this protein, present in sweet whey and absent in acid whey, is formed when CNs are precipitated by lowering the pH to 4.6 (Manso and López-Fandiño 2004). This peptide is a 64-amino acid peptide that contains varying amounts (0–5 U) of *N*-acetylneuraminic acid, commonly known as sialic acid. CMP is rich in branched-chain amino acids (Ile and Val) and low in Met. The uniqueness of GMP is the low levels of aromatic amino acids (Phe, Trp, and Tyr). This makes GMP suitable for diets in the treatment of patients with hepatic diseases and phenylketonuria (Thomä-Worringer et al. 2006). GMP or peptides derived from whey have important biological functions, including stimulation of cholecystokinin (a hormone regulating energy and food intake) release from intestinal cells, inhibition of platelet aggregation, and support of beneficial intestinal bacteria (i.e., bifidobacteria) (Abd El-Salam et al. 1996).

7.3 MANUFACTURE OF MILK PROTEINS

7.3.1 METHODS OF CASEIN EXTRACTION FROM MILK

From the technological point of view, CN and CNates (caseinates) are specific dairy preserves. They can be defined as washed and dried highly proteinous products made by means of milk coagulation. CN can be extracted from skim milk by enzyme treatment (so-called rennet CN) or by acid precipitation (acid CN).

In the case of rennet treatment, coagulation takes place under the unchanged pH. After heating to 32°C–35°C, a specific enzyme, chymosin, is added to the milk, which cleaves the peptide chain of κ-CN between residues 105 (Phe) and 106 (Met). As a result, the water-soluble portion of κ-CN called glycomacropeptide (GMP) or caseinomacropeptide (CMC) is liberated to the whey and the remaining hydrophobic part called *para*-κ-CN loses its stabilizing ability. As a consequence, a 3D gel network is formed in the presence of calcium ions. After the formation, which takes 15–20 min, the clot is cut and heated to 60°C, which promotes curd shrinkage and whey syneresis. Then the curd is separated from the whey and washed several times with water to remove the remaining lactose and minerals. After mechanical dewatering (pressing or centrifugation), the curd is dried, cooled, milled, sieved, and the obtained CN granules packed into the bags. The quality and storage stability of CN preparations depends on the degree of separation of fat, lactose, and mineral salts (Alfa Laval 1987, Early 1992).

Acid CN is made by milk acidification to isoelectric point using hydrochloric, sulfuric, or lactic acids. Precipitation can be also achieved by means of lactose fermentation using lactic acid cultures of *Lactococcus lactis, Streptococcus thermophilus, Lactobacillus delbrueckii* ssp. *bulgaricus.* Whereas the properly diluted acid is added directly to the warm milk during mixing and after heating the resulting curd in the form of fine, separate particles are separated from the whey and processed in a manner similar to that described for rennet CN, but the biological process requires more time. The milk after pasteurization is cooled to 20°C–27°C and inoculated with the proper strains of lactic acid bacteria. After coagulation to 4.6–4.7 pH, which takes 10–16 h, the obtained coagulum is heated to 50°C–60°C, separated from the whey, washed, and dried. The properties of the resulting CN depend highly on applied pH and temperature (Early 1992).

CNates can be produced from fresh acid curd or from dried granular CN by reaction with one of the alkalis (e.g., NaOH, KOH, Ca(OH)$_2$, Mg(OH)$_2$) under aqueous conditions. The CNate solution is then dried to the moisture level of 3%–8%. CNates in comparison to acid and rennet CN are highly soluble in water (Varnam and Sutherland 1994).

The chemical composition and properties significantly depend on the type of CN preparation. In the case of acidification, the significant amount of calcium salts dissociate to the whey. Thus resulting CN contains less mineral components and is characterized with lower viscosity than the rennet one (Nielsen and Ullum 1989). Table 7.3 presents the chemical composition and functional properties of different CN preparations.

7.3.2 RECOVERY OF PROTEINS FROM WHEY

Until the 1970s, whey protein was available only in the heat denatured form, a water-insoluble, gritty, yellowish-brown powder that found limited use (Wingerd 1971). Advances in processing

TABLE 7.3
Chemical Composition and Selected Properties of Different Casein Preparations

	Acid CN (%)	Rennet CN (%)	Sodium CNate (%)
Max. moisture	10	10	3.8
Min. protein	90	84	91.4
Max. fat in TS	2.25	2.0	1.0
Max. lactose	1.0	1.0	0.1
Milk ash	Max. 2.5	Min. 7.5	3.7
Functional properties	Good solubility at pH 5.5, good water absorption	Insoluble below pH 9.0, suspensions of >15% gel at 25°C	Gels at concentrations above 17%

Sources: Nielsen, E.W. and Ullum, J.A. *Dairy Technology 2*, Danish Turkney Dairies Ltd., Arhus, Denmark, 1989; Early, R. (ed.), *The Technology of Dairy Products*, VCH Publishers Inc., New York, 1992; Varnam, A.H. and Sutherland, J.P., *Milk and Milk Products. Technology, Chemistry and Microbiology*, Chapman & Hall, London, U.K., 1994.

technologies have led to the industrial production of different products with varying protein contents from liquid cheese whey. Various methods for the isolation of native whey proteins have been developed. Many of these processes are based on the varying particle sizes of the milk constituents.

Membrane filtration is a molecular sieving technique that employs a 150 μm-thick semipermeable surface supported by a more porous layer of similar material on a reinforcing base. The permeate (soluble compounds of low molecular weight) flows through while passage of the retentate is blocked. The dividing line between permeate and retentate is known as the molecular weight cutoff (MWCO) (Tunick 2008). Membrane filtration allowed for the separation and fractionation of whey proteins while retaining their solubility. Whey processors employ six types of membrane filtration, sometimes in combination: microfiltration (MF), ultrafiltration (UF), diafiltration (DF), electrodialysis (ED), nanofiltration (NF), and reverse osmosis (RO). All are followed by spray drying to obtain a dry (<5% moisture) product, and combinations of these processes are utilized to create whey protein powders with different protein contents.

RO is the tightest possible membrane process (the MWCO is 150 Da) in liquid–liquid separation. Water is in principle the only material passing through the membrane; essentially all dissolved and suspended material is rejected. Two-thirds of the water in whey can be removed by RO, leaving a concentrate.

NF, sometimes called ultra-osmosis, removes particles smaller than 300–1000 Da, and is also suitable for desalting and demineralizing whey. Separation of whey by UF is normally performed at temperatures below 55°C, with inlet pressure around 300 kPa and membrane pore size of 250 nm (Wagner 2001). UF of whey leads to a selective concentration of the protein, which when dried is called whey protein concentrate (WPC). WPC may contain between 20% and 89% protein. WPC with 35% protein is a common product. The lactose and mineral content in the WPC can be further reduced using a subsequent DF in which deionized water is continually added to the retentate while lactose and minerals are simultaneously removed in the filtrate. This combined UF–DF yields a high-value retentate containing approximately 85% protein (Zydney 1998). MF is similar to UF, but with a MWCO of 200 kDa. MF has been used to reduce the total number of lactic acid bacteria and other microorganisms in the permeate. The process also resulted in defatting of the whey and can be considered as a gentle sterilization method (Rektor and Vatai 2004).

ED is another method for demineralizing whey. In this electrochemical process, direct current is passed through whey inside chambers with ion-permeable walls (Sienkiewicz and Riedel 1990). It results in a product called whey protein isolate (WPI) with higher concentration of whey proteins (>90%) and virtually all the lactose removed. An ion-exchange tower, which separates components by ionic charge instead of molecular size, is often used in conjunction with membrane filtration. WPC and WPI can provide a vast array of functional, textural, and nutritional properties but it is only a portion of the potential of whey proteins.

With increasing knowledge of the physicochemical properties of the individual whey proteins, development of techniques for the cost-effective fractionations of these proteins became feasible (Smithers 2008). Techniques for purified fractions may combine both a physical process such as filtration and an enzymatic pretreatment such as hydrolysis and/or ion exchange chromatography (Gerberding and Byers 1998).

7.4 TECHNICAL APPLICATIONS OF MILK PROTEINS

7.4.1 BIODEGRADABLE MATERIALS

Plastics are used worldwide in packaging applications due to their stability, excellent mechanical properties, processability, transparency, and low cost. However, the amount of postconsumer plastic waste steadily rises creating a significant disposal problem. One of the best long-term solutions in waste-management problem is to develop and use packagings that are biodegradable. Biodegradable packagings are defined as materials that are degradable under the action of microorganisms to the CO_2, H_2O, and small particles (humus) (Zhang et al. 2002).

7.4.1.1 Edible Films and Coatings from Milk Proteins

Edible films and coatings are good examples of totally biodegradable packaging materials. They are traditionally used as sausage coatings, wax coatings for fruits, chocolate coatings, for confectionary, etc. Edible films can be defined as thin layers of edible materials placed on the food surface or between food components, which prevents the transfer of moisture, oxygen, carbon dioxide, aromas/flavors, and lipids. Their main role is to extend shelf life and to improve the quality and safety of food products. They offer an attractive alternative for popular, synthetic packagings which are not biodegradable and cause pollution problem as they are accumulated year by year in the landfills (Kester and Fennema 1986, Khwaldia et al. 2004, Schou et al. 2005).

Edible films and coatings are made from both plant and animal agricultural productions (polysaccharides, proteins, lipids) such as starch, soy protein, wheat gluten, maize zein, collagen (gelatin), and milk proteins (CN/CNates, whey proteins, or their mixtures). They are usually prepared from the mixtures of biopolymers in the solvent (water or ethanol) with the addition of plasticizer. They can also contain other additional components such as antioxidants, antibacterial agents, aromas, and coloring agents (Arvanitoyannis 1999, Fishman et al. 2000, Guilbert 2000, Kester and Fennema 1986, Khwaldia et al. 2004, Metzger 1997, Yoshida et al. 2002, Wang et al. 2003). Edible films are usually characterized with good barrier properties against oxygen and aromas and under the conditions of low or medium humidity are also resistant to fats and oils. On the other hand, they do not possess sufficient barrier abilities against moisture (Hernández-Muñoz et al. 2004).

The properties of edible films depend on the nature of main components, their concentration (film thickness), additives used (plasticizers, crosslinking or texture agents), and preparation conditions (pH, temperature). In the case of milk proteins, the barrier and mechanical properties differ significantly depending on the kind of protein used. CNate films are weak, have low viscosity, and are easily water soluble in contrast to whey protein films. Addition of plasticizers makes the films more flexible but also weaker and more permeable to water vapor (Khwaldia et al. 2004, Longares et al. 2004, Schou et al. 2005). According to Fabra et al. (2008), who studied tensile and water vapor permeability (WVP) properties of sodium CNate–based films, glycerol was more effective as plasticizer than sorbitol in CNate matrices.

Longares et al. (2005) studied the physical properties of edible films made from sodium CNate, whey protein isolate (WPI), or the mixtures of both components, all plasticized with glycerol. They found that films containing WPI were characterized with higher maximum load and elastic modulus (EM) during the mechanical test as well as significantly lower solubility in water. The authors suggested that it can result from stronger bond energy of disulfide bonds induced by heat treatment in WPI films compared to much weaker forces present in the CNate films (electrostatic and hydrophobic interactions and hydrogen bonding). It is consistent with the findings of Floris et al. (2008), who stated that solubility of whey protein coatings and films can be controlled by controlling either the accessibility of thiol groups or the accessibility of disulfide bonds within the whey protein aggregates.

Chick and Ustunol (1998) compared the mechanical and barrier properties of edible films obtained from rennet and lactic acid CN using sorbitol or glycerol in different proportions as plasticizers. The investigations revealed that lactic acid CN plasticized with sorbitol had the most effective mechanical and barrier properties against moisture and oxygen.

The preparation of milk protein–based films can be achieved by simple solvent casting method. Yoshida and authors (2002) produced films from 6.5% aqueous whey protein concentrate (WPC) solution with 3% glycerol as plasticizer, heated at 90°C for 30 min prior to air drying at room temperature. Such films have hydrophilic nature but are quite effective as gas (O_2, CO_2) and aroma barriers (McHugh and Krochta 1994, Yoshida et al. 2002). The process of solvent evaporation can be accelerated using microwave drying, which additionally improves elongation (EL) and tensile strength (TS) values without affecting water vapor transmission rate (Kaya and Kaya 2000). The protein-based films can be also produced by compression molding. The advantage of the latter process over solvent casting is that it requires less processing time (2–3 min) and space (Sothornvit et al. 2003). Frinault et al. (1997) developed also an alternative procedure for continuous CN film formation adapted from the wet-spinning process of the CN fiber manufacture.

The mechanical and barrier properties and other features of milk protein films can be modified using many methods. The use of high-pressure CO_2 to precipitate CN from milk limits the dissociation of calcium phosphates linkages resulting in more water-resistant films (Tomasula and Parris 1998). Moisture barrier properties can also be improved by incorporating hydrophobic substances like lipids, waxes, long-chain saturated fatty acids to the film matrix or on their surface (Chick and Hernandez 2002, Sohail et al. 2006). Wax application to the CN films significantly decreased WVP but on the other hand decreased the TS and Young's Modulus of the films (Sohail et al. 2006). Oleic acid, pure or mixed with beeswax, has a plasticizing effect on the sodium CNate films, increasing their elasticity, flexibility, and stretchability, and reduces WVP (Fabra et al. 2008). However, lipids also impart opacity and loss of gloss to the sodium CNate films (Fabra et al. 2009). Another way of WVP modification is cross-linking with γ-irradiation, which induces a substantial increase of high molecular weight protein components in film forming solutions thus influencing the reduction of the WVP of protein films (Cieśla et al. 2006, Ouattara et al. 2002). Better mechanical properties can be achieved using cross-linking agents like Ca^{2+} or transglutaminase (Fang et al. 2002, Mahmoud and Savello 1992, Sohail et al. 2006).

Many additional components can be added to milk protein–based films to modify their mechanical characteristic, water solubility, barrier properties, or to give them some additional value. The possible modifications found in the literature are summarized in Table 7.4.

CN and whey protein films can be used as edible sausage casings on the surface of fruits, vegetables, cheeses, candies, bakery products, or as packaging materials for nonhygroscopic, powdery or fat-containing products, tablets, or aromatic dry products. (Ahvenainen et al. 1997, Metzger 1997).

Wrapping bread samples in the CNate films reduced hardness during 6 h storage at ambient temperatures relative to the unwrapped controls (Schou et al. 2005). Milk protein–based films can be successfully used as fruit and vegetable coatings. The investigations of Le Tien et al. (2001) revealed that calcium CNate and whey protein coatings efficiently inhibited browning of apple and potato slices due to their antioxidant activity, which was stronger for whey proteins than calcium CNate. WPI coatings protected the peanuts from oxygen permeation and oxidation (Han et al. 2008). Certel et al. (2004) reported that CNate- or milk-protein-concentrate-based edible coatings reduced water loss and had a beneficial effect on the sensory quality, soluble solids, titratable acidity, and pH of Bing cherries.

TABLE 7.4
Selected Modifications of Milk Protein–Based Films

Film Composition	Modification	References
Sodium CNate with glycerol or sorbitol	Improved tensile properties	Fabra et al. (2008)
Sodium CNate with oleic acid or mixture of oleic acid with beeswax	Increased elasticity, flexibility, stretchability, and reduced WVP	Fabra et al. (2008)
WPI with glycerol or soya oil	Oil addition—increased percent EL, glass transition (T_g) temperature, and decreased moisture content (MC), TS, and EM, glycerol—increased EL, MC, decreased TS, EM, T_g, and film opacity	Shaw et al. (2002)
Whey protein with sorbitol, beeswax, and potassium sorbate	Modification of WVP, water solubility, stickiness, and appearance	Ozdemir and Floros (2008)
WPI (whey protein isolate): glycerol: 10 mM Ca^{2+}	Moderate gel network, improved tensile properties, UV-light barrier properties, moderate WVP	Fang et al. (2002)
γ-Irradiated calcium CNate: WPI and WPC (whey protein concentrate) or WPI: SPI (soya protein isolate)	Increased molecular mass and modified conformation of proteins, improved puncture strength, decreased WVP, evaluated microbial resistance	Lacroix et al. (2002)
γ-Irradiated calcium CNate:WPI:glycerol	Creation of a crosslinked β-structure, improvement of barrier properties, and TS	Cieśla et al. (2006)
Calcium CNate: WPI or WPC cross-linked with γ-irradiation	Increased molecular weight of protein components, reduction of the WVP	Ouattara et al. (2002)
WPI:pullulan	Good appearance, greatest values of oxygen permeability (OP), WVP, MC, film solubility (FS), and transmittance	Gounga et al. (2007)
WPI:mesquite gum	Improved flexibility (significantly lower TS and higher EL at break)	Osés et al. (2009)
Whey protein laminated with zein protein with glycerol or olive oil as plasticizer	2–3-fold higher ultimate TS, higher barrier properties	Ghanbarzadeh and Oromiehi (2008)
WPI:acetylated monoglyceride laminate and emulsion	Decreased WVP, enhanced fracture properties—increased maximum strain at break	Anker et al. (2002)
WPI with glycerol and spelt bran	Decreased WVP, increased EM and complex modulus (E^*), changes in color and rheological properties	Mastromatteo et al. (2008)
CN:gelatin blend films	Greater EL values	Chambi and Grosso (2006)
CN:gelatin (75:25) cross-linked with transglutaminase	Decreased level of WVP	Chambi and Grosso (2006)
WPI:nano-clay (Cloisite 30B)	Bacteriostatic effect against *Listeria monocytogenes*	Sothornvit et al. (2009)
WPI:lysozyme	Retardation of *L. monocytogenes growth*, higher EM values, lower percentage of EL and OP values	Min et al. (2005a)
WPI with 1.0%–4.0% of oregano, rosemary, and garlic essential oils	Antimicrobial effect against *E. coli*, *Staphylococcus aureus*, *Salmonella enteritidis*, *L. monocytogenes*, and *Lactobacillus plantarum* (in the order oregano > garlic > rosemary)	Seydim and Sarikus (2006)
Na-CNate plasticized with sorbitol with K sorbate, Na lactate, nisin	Antimicrobial action against *L. monocytogenes*, moderate MC, WVP, Tg, Young modulus, max. TS, EL	Kristo et al. (2007)

Recently, there is a growing interest in the use of edible films and coatings not only to reduce the use of nondegradable synthetic packagings but also to create new-quality products (Khwaldia et al. 2004). Especially, the so-called active packagings, incorporating functional ingredients, more often antimicrobial or antioxidant substances, have received attention. Addition of antimicrobial substances into food packaging materials helps to control harmful microorganisms in food during storage and distribution and their action is much more effective than their direct application onto the food surface. Active, antimicrobial substances incorporated into edible films comprise organic acids and salts, nisin, plant extracts or essential oils, lysozyme, lactoferrin, lactoperoxidase system, etc. (Kristo et al. 2007, Min et al. 2005b,c, Uysal et al. 2007). For example milk protein–based edible films supplemented with herb essential oils or lysozyme retard the growth of pathogenic bacteria and increase the shelf life during storage of beef muscle or fish (Min et al. 2005a, Oussalah et al. 2004). Some other examples with respect to milk protein–based films are given in Table 7.4. Edible milk protein–based films and coatings have a great potential for application on the food products susceptible to microbiological spoilage such as meat products, seafood, fresh fruits and vegetables, cheeses, and bakery products. They exhibit many advantages like perfect biodegradability and availability. They are derived from renewable resources, easily formed from water solutions, and can be made from milk surpluses or dairy by-products.

7.4.1.2 Casein Plastics

The CN-based plastic under the trade name of galalith (gala = milk, lithos = stone) was developed at the turn of the nineteenth century in Germany. The most popular dry process of its manufacturing comprises grinding of rennet precipitated protein, which is then moistened with water containing appropriate colorants if needed, kneaded, and formed into rods by extrusion or pressed into sheets. The final hardening of protein material was achieved by immersing the material in formaldehyde solution. As galalith has a low thermoplasticity, only products with simple shapes can be made from it (buttons sliced from rods or punched from sheets, shallow bowls, etc.). It has also high water absorption (7%–14%) and becomes soft under the action of alkalis. On the other hand, CN plastics are characterized with good thermal stability (nonflammable, withstand temperatures up to 150°C) and susceptibility to dyeing and polishing. The main commercial utilization of galalith comprise buttons, fountain pens, knitting needles, knife handles, and fashion accessories like combs, fans, boxes, necklaces, etc. This CN plastic is often processed to imitate tortoiseshell and horn products. The plastic material made from CN was known in Europe and United States also under the trade names of erinoid, lactoid, neolite, aladdinite, karolith, and kyloid. CN plastic was widely used in the past; however, it is still produced for buttons in some countries like New Zealand (Brother 1940, Feldman and Akovali 2005, Morgan 1999, Paris et al. 2005, Vaccari 2002).

Recently, some experiments have been conducted to incorporate CN into plastics in the form of composites or by means of grafting. Modified urea-p-aminophenol-formaldehyde resins containing CN were synthesized by Raval et al. (2005). Raval et al. (2006) prepared modified melamine-formaldehyde resin with different ratios of CN achieving some improvement in chemical resistancy and flexibility of the composite materials. The blend of waterborne polyurethane and CN cross-linked with ethanedial provided a new protein plastic with good water resistance (Wang et al. 2004). New plastics were also obtained by grafting CN with acrylates or acrylonitrile (Mohan et al. 1984, Somanathan et al. 1987, 1989). It is worth maintaining that incorporating natural polymer such as CN in the synthetic plastic is also important in the aspect of its biodegradability.

Another way to improve the characteristic of CN plastic is to make composites with other natural materials. The investigations of Fossen (2000) revealed that incorporating lignocellulosic fibers (wood pulp and flax bast fibers) improved mechanical properties of the CNate plastic. Injection-molded biodegradable and bioactive thermoplastics based on CN and soybean protein reinforced with Al_2O_3 and Ca_3PO_4 were designed with properties that made them suitable for applications in the biomedical field (Vaz et al. 2003).

7.4.2 COSMETICS

Milk and dairy products have been used in cosmetic skin care applications for hundreds of years. Because of their natural origin and functional properties; i.e., solubility, viscosity, cohesion and adhesion, emulsifying properties, binding of water, making strong foam and gel-forming, conditioning effect on hairs, whey proteins are used in the formulation of nonaggressive creams and shampoos as substitutes of synthetic surfactants (Audic et al. 2003). These properties are very attractive to the cosmetic industry. Also, according to the consumer demands, natural ingredients are very popular in modern cosmetic products. Whey shampoo could be a valuable source of proteins and minerals, and shows the positive influence of whey proteins on foaming ability and consistency (Kalicka et al. 2008). Especially, acid whey is suitable for shampoo formulation. Higher amount of lactic acid caused pH to be more favorable for skin, and acid whey shampoo has also higher viscosity in comparison to that with rennet whey (Grega et al. 2004).

Many attempts have been made to provide compositions for maintaining or improving the condition of skin and hair. Specific properties of single whey proteins like β-LG and α-LA are used in skin care products as hydrating and antiwrinkle agents (Cotte 1991). LF can prevent formation of free radicals through its iron-chelating property. Cosmetics containing LF include lotions, creams, and face washes, and are expected to contribute to hygiene, moistening, and antioxidation in the skin. Antibacterial properties of LF are used in the formulation of antiacne cosmetics. LF is sometimes combined with other mucosal defense proteins such as LPO and lysozyme, and used in oral care products include toothpaste, mouth gels and washes, and chewing gum. These formulations are expected to contribute to hygiene, moistening in the mouth (Wakabayashi et al. 2006). GMP has been reported to inhibit the adhesion of cariogenic bacteria such as *Streptococcus mutans*, *S. sanguis*, and *S. sobrinus* to the oral cavity (Neeser 1991) and modulate composition of the dental plaque microbiota (Manso and López-Fandiño 2004). This could help to control acid formation in the dental plaque, in turn reducing hydroxyapatite dissolution from tooth enamel and promoting remineralization (Thomä-Worringer et al. 2006). This activity would make a GMP suitable component in oral hygiene products such as mouthwash and toothpaste. Whey could be the main ingredient of many beauty products like soaps, body lotions, shower gel, bath foam, intimate wash, and detergents (Herbert 1995, Schmitt 2007, Staples 1982).

Studies have reported on the physiological effects of fermented milk products on the skin. One report showed that fermented whey has the potential to enhance the production of filaggrin-related natural moisturizing factor, because of its effect on the induction of epidermal differentiation, and is expected to be a useful skin moisturizing agent (Baba et al. 2006). Another studies show that whey extract of bovine milk stimulates wound repair activity *in vitro* and promotes healing of rat skin incisional wounds (Rayner et al. 2000).

7.4.3 PHARMACEUTICALS

There are historical references to the use of whey for medicinal purposes (Smithers 2008). Hippocrates is thought to have praised the health benefits of whey, and in Europe in the Middle Ages whey was considered not only as a medicine, but also even as an aphrodisiac and skin balm (Madureira et al. 2007). In the seventeenth and eighteenth centuries, whey became a fashionable drink, served in so-called whey houses, especially for "stomach disease." Whey baths were popular in nineteenth century spas served to patients suffering from a variety of ailments (Holsinger et al. 1974).

Whey proteins are an excellent source of the essential amino acids and branched chain amino acids (BCAAs) (Leu, Ile, Val). Whey protein is also a rich and balanced source of the sulfur amino acids (Met, Cys) (Ha and Zemel 2003, Walzem et al. 2002). Specific biological properties of whey proteins make them potential ingredients of pharmaceutical products. Current evidence for the potential of whey proteins to have health benefits beyond basic nutrition arises from a number of sources. These include the ability of whey proteins, whey protein hydrolysates (WPHs), and their associated peptides to beneficially impact (I) *in vitro* biomarkers associated with a particular disease state or condition, (II) human cell culture studies, (III) *in vivo* studies with small animals, and (IV) human trials/studies (Morris and FitzGerald 2008).

The biological components of whey, including LF, β-LG, α-LA, GMP, and Igs, demonstrate a range of immune-enhancing properties. Whey also has the ability to act as an antioxidant, antihypertensive, antitumor, hypolipidemic, antiviral, antibacterial, and chelating agent. The primary mechanism by which whey is thought to exert its effects is by intracellular conversion of the amino acid cysteine to glutathione, a potent intracellular antioxidant. A number of clinical trials have successfully been performed using whey in the treatment of cancer, HIV, hepatitis B, cardiovascular disease, osteoporosis, and as an antimicrobial agent (Marshall 2004).

7.4.3.1   Anticancer Effect

WPC have been researched extensively in the prevention and treatment of cancer. Glutathione stimulation is thought to be the primary immune-modulating mechanism. The amino acid precursors to glutathione available in whey might: (I) increase glutathione concentration in relevant tissues, (II) stimulate immunity, and (III) detoxify potential carcinogens (Bounous 2000). Other authors conclude the iron-binding capacity of whey may also contribute to anticancer potential, as iron may act as a mutagenic agent causing oxidative damage to tissues (Weinberg 1996). Evidence is growing that specific whey proteins and peptides have the antitumor and anticarcinogenic effect against certain tumors (colon, breast, skin, prostate). Most evidence to date has been based on *in vitro* cell culture, *in vivo* animal studies, and some epidemiological studies. For example *in vitro* study (Kent et al. 2003) demonstrated that an isolate of whey protein, when compared to a CN-based protein, increased glutathione synthesis and protected human prostate cells against oxidant-induced cell death. Bovine serum albumin (10%–15% of total whey protein) has demonstrated inhibition of growth in human breast cancer cells *in vitro* (Laursen et al. 1990). Whey proteins were shown to be more effective than other proteins (CN, meat, soy) in reducing the incidence and burden of colon tumors in a rat animal model of the disease (Hakkak et al. 2001).

7.4.3.2   Cardiovascular Health (Hypocholesterolomic Effect, Opioid Agonist)

Hypertension, or high blood pressure is a controllable risk factor in the development of a range of cardiovascular disease states. Bioactive whey peptides may protect against hypertension through angiotensin-converting enzyme (ACE) inhibition and opioid-like activity. Several milk-derived peptides have been shown to have ACE inhibitory activity (from CN are called casokinins and from whey—lactokinins). ACE inhibitory peptides have been identified in the tryptic hydrolysates of bovine α_{s2}-CN (Korhonen and Pihlanto-Leppälä 2006), in bovine, ovine and caprine GMP (Manso and López-Fandiño 2004, Miguel et al. 2007) and in β-LG, α-LA derived peptides (Chatterton et al. 2006). Some of these peptides demonstrated also opioid properties with pharmacological activity at micro-molecular concentrations (Morris and FitzGerald 2008). Thrombosis is another major risk factor in cardiovascular disease. Milk peptides are believed to inhibit platelet aggregation and reduce blood cholesterol levels. Studies demonstrated that a GMP and LF-derived peptides may be involved in platelet binding (Gerdes et al. 2001). Studies on rats reported the hypocholesterolomic effect of whey peptides, particularly, lactostatin and β-lactotensin (Morris and FitzGerald 2008). Peptides derived from food are considered to be safer than the drugs currently used. Whey peptides show great promise and further clinical trials are expected to further substantiate their efficacy.

7.4.3.3   Gastrointestinal Health

Some whey proteins survive passage through the stomach and small intestine and arrive as intact proteins in the large intestine, where they exert their biological effects. For example, GMP has been shown to support the growth of bifidobacteria, and LF supports growth of both bifidobacteria and lactobacilli (Causey and Thomson 2003). Whey proteins demonstrate antibacterial, antiviral, and antifungal properties and may have therapeutic value against gastrointestinal infections in humans. α-LA and GMP inhibit growth of the *Escherichia coli* and *Salmonella* strains (Bruck et al. 2003). LF was reported to reduce the colonization density of *Helicobacter pylori* (Marshall 2004). GMP can inhibit the adhesion of *H. pylori* and rotavirus to the cell membrane by binding to pathogen-receptor sites (Manso and López-Fandiňo 2004). Antiviral activity has been described for several whey components including LF, LPO, Igs, and minor whey proteins (USDEC 2003, Walzem et al. 2002). LF has significant antiviral activity against human immunodeficiency virus (HIV), herpes viruses, and hepatitis C virus (Bastian and Harper 2003, Harper 2000, Wakabayashi et al. 2006). Recent clinical trials have demonstrated that intake of cysteine-rich whey protein formulas benefits patients with HIV/AIDS. Whey protein formulas increased plasma glutathione levels in patients with HIV infections, increased weight gain, reduced the occurrence of gastrointestinal side effects, and improved tolerance to highly active antiretroviral therapy in HIV patients (Cribb 2005).

7.4.3.4 Immunomodulatory Properties

The ability to avoid many forms of illnesses and diseases depends largely upon strong immunity. The biological components of whey, including LF, β-LG, α-LA, GMP, and Igs, demonstrate a range of immune-enhancing properties (Lucas 1999). Although the exact mechanisms are not yet fully understood, whey proteins appear to modulate immune function by boosting glutathione (GSH) production in various tissues and preserving the muscle glutamine reservoir. GSH is the centerpiece of the body's antioxidant defense system that regulates many aspects of immune function. Muscle glutamine is the essential fuel of the immune system (Cribb 2004). Whey may also enhance antioxidant capacity by contributing cysteine-rich proteins that are pivotal in the synthesis of glutathione, a major intracellular antioxidant (Ha and Zemel 2003). GSH is the main intracellular defense against oxidative stresses (Madureira et al. 2007).

7.4.3.5 Wound Care and Repair (Growth Factor Activity)

Growth factors are proteins that promote growth and healing of connective and support tissues (Bockman and Guidon 1996). Whey growth factor extract has been shown to have potent mammalian cell growth activity, notably for fibroblast cell lines and remarkable wound healing effects. A hamster study demonstrated an efficacious remedy of this extract in preventing and treating chemotherapy-induced oral mucositis. This protection is thought to occur via induction of tumor growth factor-beta (TGF-(β)), which reduces basal epithelial cell proliferation (Clarke et al. 2002).

7.4.3.6 Memory and Stress

The whey protein α-LA has a high tryptophan content compared to other whey proteins. Tryptophan is the precursor of serotonin and under stressful conditions their levels in brain are exhausted to below needs. Decline in serotonin activity is associated with depression and anxiety. Drugs like Prozac increases serotonin level in brain (Brink 2002). The studies showed that the α-LA increases the plasma ratio of tryptophan to the other large neutral amino acids, and in vulnerable subjects raises brain serotonin activity, reduces cortisol concentration, and improves mood under stress (Markus et al. 2000). The sialic acid content of CMP is also interesting in terms of bioactivity. Large amounts of this carbohydrate are found in the brain and in the central nervous system in the form of gangliosides and glycoproteins, contributing to the functioning of cell membranes and membrane receptors and to normal brain development. An *in vivo* experiment with laboratory animals has shown that exogenous administration of sialic acid increases the production of ganglioside sialic acid in the brain, improving learning ability (Wang et al. 2001).

7.4.3.7 Oral Cavity Infections

The LPO-containing products have been clinically proven to inhibit harmful microorganisms associated with gingivitis and oral irritation, to promote the healing of bleeding gums and reduce inflammation, and to combat both the causes and effects of halitosis (bad breath) (Tenovuo 2002).

LPO is finding application in oral health-care products directed to prevention and treatment of xerostomia (dry mouth) (Gil-Montoya et al. 2008).

7.4.3.8 Prevention and Treatment of Osteoporosis

Milk has been proposed as a nutritional food that aids in the prevention of osteoporosis due to its bioavailable calcium content. Researchers have begun to examine the different components of milk to determine if a particular isolate is responsible for the bone-protective effects. Initially, *in vitro* and animal studies have shown that milk basic protein (MBP), a component of whey, has the ability to stimulate proliferation and differentiation of osteoblastic cells as well as suppress bone resorption (Toba et al. 2001) and increase radial bone density (Yamamura et al. 2002).

7.4.4 ADHESIVES AND COATINGS

Adhesives are defined as materials, in the form of liquid, paste, powder, or dry films, used for sticking or adhering one surface to another. The commercial adhesives comprise pastes, glues, pyroxylin cements, rubber cements, latex cements, special cements of chlorinated rubber, synthetic resins, and natural mucilages. Natural adhesives include adhesives of vegetable and animal origin or natural gums and are usually used to join paper, cardboard, foil, and lightwood pieces. They are inexpensive, easy to apply, have a long shelf life, develop tack quickly, but are characterized with low strength of joints (Vaccari 2002).

CN glue, a particularly good adhesive for wood, was used in the eighteenth century in the construction of chalets in Switzerland and for joining the wooden frame parts in the construction of aircraft during World War I (Gerritsen 2001).

CN glues cannot be used in open air due to their poor behavior in moist conditions and susceptibility to mould and fungal attack (Serrano and Källander 2005). Water-resistant CN glue sets to a gel via a slow, chemical reaction by converting sodium CNate into calcium CNate. The CN-lime product readily dispersed in cold water is often used as a wood glue (Gooch 1997). CN-latex adhesive is used to bond metal to wood for panel construction and to join laminated plastics and linoleum to wood and metal (Vaccari 2002).

CN films can be used as washable wallpapers. The procedure of preparation of such films, described by Metzger in U.S. Patent (1997), comprises spraying CN powder with the solution of acetic acid, glycerol, and water. The swollen mixture is then plasticized in an extruder, discharged through a gap, pre-stretched, solidified with a hardening solution (glycerol, calcium acetate, calcium hydroxide, aluminum potassium bisulfate, ammonium hydroxide in deionized water), sprayed, dried on the stainless steel strip to a water content of 25%, removed from the strip, and post stretched. Wallpapers laminated with the resultant film can be cleaned with water or aqueous solutions of a stain-removing salt. The same procedure can be employed for lamination of papers to obtain glossy labels.

Paper can also be coated with whey proteins. Han and Krochta (1999) reported that WPI coating increased water vapor resistance as well as homogeneity and smoothness of the paper surface by filling the porous structure of the paper and enhanced printability.

CN can also be used as a plasticizer for concrete (Smith 2005).

7.4.5 TEXTILE INDUSTRY

The first successfully commercialized CN fiber was invented in 1935 by Antonio Ferretti under the name of "*lanital*" (lana = wool; ital = Italy). It was manufactured by Snia Viscosa (Milan) in the early 1940s followed by *merinova* (CN fiber blended with rabbit hair) produced in the late 1960s. CN fiber is made by dissolving CN crystals in alkaline solution, spun into continuous filaments by forcing it through spinnererets and coagulating them into an acid bath. These are then spun alone or blended with other fibers to form mixed yarns. *Lanital* was intended to replace the wool at a lower cost (especially during World War II). It resembles wool, and has practically the same chemical composition, the same odor during burning, and can be dyed by the same method. It is silky and smooth and, unlike wool, has a perfect round cross section when viewed under the microscope. *Lanital* exhibits some disadvantages: it damages readily under the action of alkalies and is susceptible to mildewing. On the other hand, it is resistant to moths. It was most often blended with other natural (wool, cotton, silk, rabbit hair, mohair) or synthetic (rayon) fibers for woven and knitted fibers in a variety or weaves, textures, and prints or felts for hats. In the United States, CN fiber was developed in 1937 under the trade name of *aralac*. Other names for CN fiber are *R-53*, *caslen*, and *fibrolane* (Fletcher 1942, Génin 1941, Kiplinger 2003, Vaccari 2002).

In recent years, some attention has been paid on incorporating CN by means of graft copolymerization to synthetic fibers to enhance their properties. The investigations of Jia et al. (2007) revealed that CN-grafted acrylic fiber exhibited improved surface properties like hygroscopicity, antistatic characteristic, and spinnability.

7.4.6 PAINTS AND PIGMENTS

It is reported that ancient Egyptians used the CN as a fixing agent for pigments in wall paintings. It has been used as a binder for paints since the Renaissance (Gerritsen 2001, Morgan 1999). However, CN paints (consisting of CN solution with pigments and extenders) are still used by contemporary artists. They can be treated as any other water-based media or as an underpainting for oils and pastels. As CN paints are inflexible, they must be used on rigid panels, boards, or cardboards (Sutherland 2003, Vaccari 2002).

CN is also used as a vehicle for water-based flexographic inks for paper, which are characterized by good press stability, printability, and economy of water but also with low gloss and slow drying which limits its use (Gooch 1997).

7.4.7 ENCAPSULATION AND CONTROLLED DELIVERY APPLICATIONS

Encapsulation is a rapidly expanding technology with a lot of potential in different areas including pharmaceutical and food industries, for the protection of enzymes or health ingredients, the taste masking of encapsulated drugs, controlled release, the encapsulation of flavors for food and drinks, and in home and personal care products. It is a process by which small particles of core materials are packaged within a wall material to form microcapsules. Efforts have been made to develop systems for controlled release of pharmaceuticals and other active compounds. In recent years, there has been interest in milk proteins as microencapsulating agents and drug carriers for controlled core-release applications.

Microcapsules are defined as particles with sizes in the range of 50 nm to 2 mm consisting of a polymer matrix and an "encapsulated" or bound active component (Arshady 1999). Compared to many synthetic polymers that are employed for effecting sustained delivery of drugs by the oral route, natural polymers have better biocompatibility and nontoxicity (Latha et al. 1995). The concept of using whey proteins as microencapsulating agents has been established (Rosenberg 1997). A series of studies has indicated that whey proteins exhibit excellent physicochemical properties and are suitable for microencapsulation of volatile and nonvolatile core materials. Results indicated also that whey-derived proteins could be successfully used as highly functional wall materials in water-soluble as well as water-insoluble microcapsules designed for food and nonfood applications (Heelan and Corrigan 1998, Keogh and O'Kennedy 1999, Lee and Rosenberg 2000, Satpathy and Rosenberg 2003).

CN can also be employed for the preparation of functional microspheres (empty microcapsules). Hydrophilic polystyrene microspheres with CN molecules on the surface were obtained by means of suspension polymerization. The goal was to introduce functional (hydroxyl, carboxyl, and amino) groups of CN onto the surface of polystyrene microspheres. Such functional microspheres have a great potential for applications in life sciences and medicine, e.g., enzyme immobilization, targeted drug and hormone delivery, immunoassay, bioseparation, advanced cosmetics, etc. (Ai and Wei 2008, Knepp et al. 1993, Latha and Jayakrishnan 1994, Latha et al. 2000, Willmott et al. 1992).

Whey protein concentrate (WPC) and whey protein isolate (WPI) have been shown to exhibit excellent gelling and emulsification properties. Wall matrices consisting of WPI have been reported to provide effective protection against oxidation of encapsulated lipids in storage conditions that are known to promote lipid oxidation (Bae and Lee 2008, Jafari et al. 2008, Weinbreck et al. 2004). Whey proteins can be used as hydrogels and nanoparticle systems for encapsulation and controlled delivery of bioactive compounds. β-LG is the main whey protein component and its principal gelling agent. A hydrogel can be defined as a 3D network that exhibits the ability to swell in water and retains a significant fraction of water within its structure. WPC hydrogels exhibit pH-sensitive swelling behavior with minimum swelling ratio near the isoelectric point (pI) of whey proteins (~5.1). The advantages of using whey protein-based gels as potential devices for controlled release of bioactives is that they are entirely biodegradable and there is no need for any chemical cross-linking agents in their preparation. These are two of the major requirements for wide use of hydrogels not only in the pharmaceutical area but also in many food and bioprocessing applications.

Whey proteins can also be formed into nanoparticle–matrix systems of a dense polymeric network under 100 μm in size in which an active molecule may be dispersed throughout the matrix (Gunasekaran et al. 2007). Nowadays, when nanotechnology is one of the most rapidly growing disciplines, the investigations are conducted to develop nanoparticles or nanofibers for both food and nonfood applications. The nanoparticles offer the feasibility to entrap drugs or bioactive compounds within but not chemically bound to them (Gunasekaran 2008). Emulsion and desolvation methods have been used for nanoparticle formation of proteins such as BSA and human serum albumin HAS (Arnedo et al. 2002). But β-LG is smaller and less hydrophobic than BSA; therefore it is a better candidate for preparing nanoparticles (Ko and Gunasekaran 2006).

Among several methods used to produce nanofibers, electrospinning is one of the most effective and simple methods. CN, due to high number of intra- and intermolecular hydrogen bonding cannot be in its native form easily subjected to electrospinning (Kriegel et al. 2008). Xie and Hsieh (2003) made some efforts to process CN into fibrous form by mixing it with poly(ethylene oxide) or poly(vinyl alcohol) in the aqueous solution of triethanolamine. Fibers with sizes ranging from 0.1 to 1 μm were fabricated by the blend composition.

7.4.8 Other Industrial Uses of Milk Proteins

Suzuki and Maruyama (2005) took advantage of the CN-foaming ability and its function as an excellent collector, and developed a new method of emulsified oil separation from oily wastewater using CN. The authors reported that adding CN before the foam separation process dramatically improved removal of oil. Oily wastewater, generated by various industries such as petroleum refining, steel manufacturing, and vehicle repair, is characterized with low biodegradability and causes serious pollution problems.

Other technical and agricultural uses of milk proteins found in the literature include leather finishing, horticultural spreaders, photoresistive materials, absorbing and recovering chromate in manufacturing wastes, reinforcing and stabilizing agent in rubber tires, dish-washing liquids (Audic et al. 2003, de Kruif 2003, Southward 1992).

7.5 CONCLUSIONS AND PERSPECTIVES

This chapter deals with past, present, and potential nonfood uses of proteins derived from milk. On the one hand, many of these applications are historical, but on the other hand there are also many new possibilities taking advantage of their bioavailability, biocompatibility, biodegradability, and other excellent functional properties.

In recent times, attempts have been made to replace packagings of synthetic and/or environmentally harmful materials by biodegradable films, preferably based on natural and renewable products. Of the polymers available, CNs and/or whey proteins seem to be an excellent choice material. CN readily forms films because of its open structure and flexibility, and forms intermolecular interactions through hydrogen, electrostatic, and hydrophobic bonds. The materials of improved properties can be made because milk proteins are susceptible to many modifications e.g., cross-linking. Furthermore, considerable interest exists in finding new uses for milk components due to their industrial surplus as well as in the management of dairy waste streams. Especially, a number of investigations on the development of edible films suggest that there is a great potential for milk proteins. The future trend in that field is to design so-called active packagings, which besides being biodegradable, provide prolonged shelf life or some new functional properties to the foods. They also have great potential in the field of nanotechnology for producing nanoparticles, or nanofibers with a wide range of applications not only in food products but also in pharmacology, biotechnology, for cosmetic preparations, or other applications. Milk protein microspheres might be used for encapsulation of flavors, enzymes, for controlled drug delivery or biotechnology-selective sensors.

ABBREVIATIONS

ACE	Angiotensin converting enzyme
BCAAs	Branched chain amino acids
BSA	Bovine serum albumin
CMC	CNomacropeptide
CN	Casein
Cnate	Caseinate
Cys	Cysteine
DF	Diafiltration
ED	Electrodialysis
EL	Percent elongation
EM	Elastic modulus
FS	Film solubility
GMPs	Glycomacropeptides
GSH	Glutathione
HAS	Human serum albumin
HIV	Human immunodeficiency virus
Igs	Immunoglobulins
Ile	Isoleucine
Leu	Leucine
LF	Lactoferrin
LPO	Lactoperoxidase
MBP	Milk basic protein
MC	Moisture content
Met	Methionine
MF	Microfiltration
MWCO	Molecular weight cutoff
NF	Nanofiltration
OP	Oxygen permeability
Phe	Phenylalanine
RO	Reverse osmosis
T_g	Glass transition temperature

Trp   Trypthophan
Tyr   Tyrosine
TS   Tensile strength
UF   Ultrafiltration
Val   Valine
WPC   Whey protein concentrate
WPHs   Whey protein hydrolysates
WPI   Whey protein isolate
WVP   Water vapor permeability
α-LA   α-Lactalbumin
α_{s1}-CN α_{s1}-Casein
α_{s2}-CN α_{s2}-Casein
β-CN   β-Casein
β-LG   β-Lactoglobulin
κ-CN   κ-Casein

REFERENCES

Abd El-Salam M.H., El-Shibiny S., Buchheim W. 1996. Characteristics and potential uses of the CN macropeptide: A review. *International Dairy Journal*, 6:327–341.

Ahvenainen R., Myllarinen P., Poutanen K. 1997. Prospects of using edible and biodegradable protective films for foods. *European Food & Drink Review*, 73(75):77–80.

Ai Y., Wei D. 2008. Preparation of hydrophilic polystyrene microspheres with CN molecules on the surface. *Journal of Macromolecular Science, Part A: Pure and Applied Chemistry*, 45:456–461.

Alfa Laval. 1987. *Dairy Handbook, Food Engineering AB*. Lund, Sweden: Alfa LavalCo.

Anker M., Berntsen J., Hermansson A.M., Stading M. 2002. Improved water vapor barrier of whey protein films by addition of an acetylated monoglyceride. *Innovative Food Science & Emerging Technologies*, 3(1):81–92.

Arnedo A., Espuelas S., Irache J.M. 2002. Albumin nanoparticles as carriers for a phosphodiester oligonucleotide. *International Journal of Pharmacy*, 244:59–72.

Arshady R. 1999. *Microspheres, Microcapsules & Liposomes, Vol. 1: Preparation & Chemical Applications*. London, U.K.: Citus Books, ISBN 0953218716.

Arvanitoyannis I.S. 1999. Totally and partially biodegradable polymer blends based on natural and synthetic macromolecules: Preparation, physical properties, and potential as food packaging materials. *Journal of Macromolecular Science-Reviews in Macromolecular Chemistry & Physics*, C39(2):205–271.

Audic J.L., Chaufer B., Daufin G. 2003. Non-food applications of milk components and dairy co-products: A review. *Le Lait*, 83:417–438.

Baba H., Masuyama A., Takano T. 2006. Short Communication: Effects of *Lactobacillus helveticus*-fermented milk on the differentiation of cultured normal human epidermal keratinocytes. *Journal of Dairy Science*, 89:2072–2075.

Bae E.K., Lee S.J. 2008. Microencapsulation of avocado oil by spray drying using whey protein and maltodextrin. *Journal of Microencapsulation*, 1–12, iFirst.

Bastian E.D., Harper W.J. 2003. Emerging health benefits of whey. *The Dairy Council Digest*, 74:31–36. http://www.nationaldairycouncil.org/

Bockman R., Guidon P. 1996. Methods of enhancing wound healing and tissue repair. U.S. Patent, 5,556,645.

Bonnaillie L.M., Tomasula P.M. 2008. Whey protein fraction. In: C.I. Onwulata, and P.J. Huth (Eds.), *Whey Processing, Functionality and Health Benefits*, pp. 15–38. Ames, IA: Wiley-Blackwell, IFT Press.

Boots J.W., Floris R. 2006. Lactoperoxidase: From catalytic mechanism to practical applications. *International Dairy Journal*, 16:1272–1276.

Bounous G. 2000. Whey protein concentrate (WPC) and glutathione modulation in cancer treatment. *Anticancer Research*, 20:4785–4792.

Brink W. 2002. The new faces of whey. *LE Magazine*. http://www.lef.org/magazine/mag2002/jan2002_report_whey_01.html

Brother G.H. 1940. *Casein plastic. Industrial and Engineering Chemistry*, 32(1):31–33.

Bruck W.M., Graverholt G., Gibson G.R. 2003. A two-stage continuous culture system to study the effect of supplemental α-lactalbumin and glycomacropeptide on mixed cultures of human gut bacteria challenged with enteropathogenic *Escherichia coli* and *Salmonella* serotype *typhimurium*. *Journal of Applied Microbiology*, 95(1):44–53.

Causey J., Thomson K. 2003. The whey to intestinal health. *Today's Dietitian*. http://www.wheyoflife.org/news/WheyFeature.pdf

Certel M., Uslu M.K., Ozdemir F. 2004. Effects of sodium caseinate- and milk protein concentrate-based edible coatings on the postharvest quality of Bing cherries. *Journal of the Science of Food and Agriculture*, 84(10):1229–1234.

Chambi H., Grosso C. 2006. Edible films produced with gelatin and casein cross-linked with transglutaminase. *Food Research International*, 39(4):458–466.

Chatterton D.E.W., Smithers G., Roupas P., Brodkorb A. 2006. Bioactivity of β-lactoglobulin and α-lactalbumin—Technological implications for processing: A review. *International Dairy Journal*, 16:1229–1240.

Chick J., Hernandez R.J. 2002. Physical, thermal and barrier characterization of casein-wax-based edible films. *Journal of Food Science*, 67(3):1073–1079.

Chick J., Ustunol Z. 1998. Mechanical and barrier properties of lactic acid and rennet precipitated casein-based edible films. *Journal of Food Science*, 63(6):1024–1027.

Cieśla K., Salmieri S., Lacroix M. 2006. Modification of the properties of milk protein films by gamma radiation and polysaccharide addition. *Journal of the Science of Food and Agriculture*, 86(6):908–914.

Clarke J., Butler R., Howarth G. et al. 2002. Exposure of oral mucosa to bioactive milk factors reduces severity of chemotherapy-induced mucositis in the hamster. *Oral Oncology*, 38:478–485.

Cotte J. 1991. La lait, une matiere d'avenir pour la cosmetique. *Le Lait*, 71:213–224.

Cribb P. 2004. Whey proteins and immunity. In: *Applications Monograph*. U.S. Dairy Export Council, Arlington, VA, ww.usdec.org

Cribb P. 2005. Whey proteins and body composition. In: *Applications Monograph*. U.S. Dairy Export Council, Arlington, VA, ww.usdec.org

Dalgleish D.G. 1998. Casein micelles as colloids: Surface structure and stabilities. *Journal of Dairy Science*, 81:3013–3018.

de Kruif C.G. 2003. Non-food applications of caseins. In: R.J. Aalberg, R.J. Hamer, P. Jasperese, H.H.J. deJongh, C.G. de Kruif, P. Walstra, F.A. de Wolf (Eds.), *Progress in Biotechnology: Industrial Proteins in Perspective*, vol. 23. Amsterdam, the Netherlands: Elsevier Science Ltd.

Early R. (Ed.). 1992. *The Technology of Dairy Products*. New York: VCH Publishers Inc.

El-Loly M.M. 2007. Bovine milk immunoglobulins in relation to human health. *International Journal of Dairy Science*, 2(3):183–195.

Fabra M.J., Talens P., Chiralt A. 2008. Tensile properties and water vapor permeability of sodium CNate films containing oleic acid–beeswax mixtures. *Journal of Food Engineering*, 85(3):393–400.

Fabra M.J., Talens P., Chiralt A. 2009. Microstructure and optical properties of sodium CNate films containing oleic acid-beeswax mixtures. *Food Hydrocolloids*, 23(3):676–683.

Fang Y., Tung M.A., Yada S., Dalgleish D.G. 2002. Tensile and barrier properties of edible films made from whey proteins. *Journal of Food Science*, 67(1):188–193.

Farrell H.M. Jr., Jimenez-Flores R., Bleck G.T. et al. 2004. Nomenclature of the proteins of cows' milk—Sixth revision. *Journal of Dairy Science*, 87:1641–1647.

Feldman D., Akovali G. 2005. The use of polymers in construction: Past and future trends. In: G. Akovali (Ed.), *Handbook of Polymers in Construction*. Shrewsbury, GBR: Smithers Rapra.

Fishman M.L., Coffin D.R., Konstance R.P., Onwulata C.J. 2000. Extrusion of pectin/starch blends plasticized with glycerol. *Carbohydrate Polymers*, 41(4):317–325.

Fletcher H.M. 1942. *Synthetic Fibers and Textiles. Bulletin 300*. Kansas State College of Agriculture and Applied Science, Manhattan, KS.

Floris R., Bodnár I., Weinbreck F., Alting A.C. 2008. Dynamic rearrangement if disulphide bridges influences solubility of whey protein coatings. *International Dairy Journal*, 18(5):566–573.

Fossen M. 2000. Lignocellulosic fibre reinforced caseinate plastic. *Applied Composite Materials*, 7(5–6):433–437.

Fox P.F., Kelly A.L. 2004. The caseins. In: R.Y. Yada (Ed.), *Proteins in Food Processing*. Cambridge, U.K.: Woodhead Publishing Limited.

Frinault A., Gallant D.J., Bouchet B., Dumont J.P. 1997. Preparation of casein films by a modified wet spinning process. *Journal of Food Science*, 62(4):744–747.

Génin G. 1941. L'industrie laitière à l'étranger. *Le Lait*, 22(211–213):38–42.

Gerberding S.J., Byers C.H. 1998. Preparative ion-exchange chromatography of proteins from dairy whey. *Journal of Chromatography A*, 808:141–151.

Gerdes S.K., Harper W.J., Miller G. 2001. Bioactive components of whey and cardiovascular health. In: *Applications Monograph*. U.S. Dairy Export Council, Arlington, VA, www.usdec.org

Gerritsen V.B. 2001. Of buttons, digestion and glue. *Protein Spotlight*, 16.

Ghanbarzadeh B., Oromiehi A.R. 2008. Biodegradable biocomposite films based on whey protein and zein: Barrier, mechanical properties and AFM analysis. *International Journal of Biological Macromolecules*, 43(2):209–215.

Gil-Montoya J.A., Guardia-López I., González-Moles M.A. 2008. Evaluation of the clinical efficacy of a mouthwash and oral gel containing the antimicrobial proteins lactoperoxidase, lysozyme and lactoferrin in elderly patients with dry mouth—A pilot study. *Gerodontology*, 25:3–9.

González Siso M.I. 1996. The biotechnological utilization of cheese whey: A review. *Bioresource Technology*, 57:1–11.

Gooch J.W. 1997. *Analysis and Deformulation of Polymeric Materials: Paints, Plastics, Adhesives, and Inks*. Hingham, MA: Kluwer Academic Publishers.

Gounga M.E., Xu S.Y., Wang Z. 2007. Whey protein isolate-based edible films as affected by protein concentration, glycerol ratio and pullulan addition in film formation. *Journal of Food Engineering*, 83(4):521–530.

Grega T., Bonczar G., Domagała J., Najgebauer D. 2004. Use of whey for shampoo production. *Polish Veterinary Medicine*, 60(7):766–769.

Guilbert S. 2000. Potential of the protein based biomaterials for the food industry. In: *Proceedings of the Food Biopack Conference*, Copenhagen, Denmark, August, 27–29, pp. 13–18.

Gunasekaran S. 2008. Whey protein hydrogels and nanoparticles for encapsulation and controlled delivery of bioactive compounds. In: C.I. Onwulata, and P.J. Huth (Eds.), *Whey Processing, Functionality and Health Benefits*, pp. 227–284. Ames, IA: Wiley-Blackwell, IFT Press.

Gunasekaran S., Ko S., Xiao L. 2007. Use of whey proteins for encapsulation and controlled delivery applications. *Journal of Food Engineering*, 83:31–40.

Ha E., Zemel M.B. 2003. Functional properties of whey, whey components and essential amino acids: Mechanisms underlying health benefits for active people. *Journal of Nutritional Biochemistry*, 14:251–258.

Hakkak R., Korourian S., Martin S., Ronis J.J., Johnston J.M. 2001. Dietary whey protein protects against azoxymethane-induced colon tumours in male rats. *Cancer Epidemiology Biomarkers and Prevention*, 10:555–558.

Han J.H., Krochta J.M. 1999. Wetting properties and water vapor permeability of whey protein coated paper. *Transactions of the ASAE*, 42(5):1375–1382.

Han J.H., Hwang H.M., Min S., Krochta J.M. 2008. Coating of peanuts with edible whey protein film containing alpha-tocopherol and ascorbyl palmitate. *Journal of Food Science*, 73(8):E349–E355.

Harper W.J. 2000. *Biological Properties of Whey Components. A Review*. Chicago, IL: The American Dairy Products Institute.

Heelan B.A., Corrigan O.I. 1998. Preparation and evaluation of microspheres prepared from whey protein isolate. *Journal of Microencapsulation*, 15:93–105.

Herbert G. 1995. Production process for whey based detergents, soaps and cosmetics. Canada Patent: CA 2,101,622.

Hernández-Muñoz P., López-Rubio A., Del-Valle V., Almenar G., Gavara R. 2004. Mechanical and water barrier properties of glutenin films influenced by storage time. *Journal of Agricultural and Food Chemistry*, 52:79–83.

Holsinger V.H., Posati L.P., DeVilbiss E.D. 1974. Whey beverages: A review. *Journal of Dairy Science*, 57:849–859.

Jafari S.M., Assadpoor E., Bhandari B., He Y. 2008. Nano-particle encapsulation of fish oil by spray drying. *Food Research International*, 41:172–183.

Jelen P. 2003. Whey processing: Utilization and products. In: H. Roginski, J.W. Fuquay, and P.F. Fox (Eds.), *Encyclopedia of Dairy Sciences*, pp. 2739–2745. New York: Academic Press.

Jia Z., Du S., Tian G. 2007. Surface modification of acrylic fiber by grafting of casein. *Journal of Macromolecular Science, Part A: Pure and Applied Chemistry*, 44:299–304.

Kalicka D., Sikora E., Grega T., Ogonowski J. 2008. Application of cheese whey as a raw material in formulation of shampoos. *ECIS 2008, Conference Materials*, P68:306.

Kaya S., Kaya A. 2000. Microwave drying effects on properties of whey protein isolate edible films. *Journal of Food Engineering*, 43(2):91–96.

Kent K.D., Harper W.J., Bomser J.A. 2003. Effect of whey protein isolate on intracellular glutathione and oxidant-induced cell death in human prostate epithelial cells. *Toxicology In Vitro*, 17:27–33.

Keogh M.K., O'Kennedy B.T. 1999. Milk fat microencapsulation using whey proteins. *International Dairy Journal*, 9:657–663.

Kester J.J., Fennema O.R. 1986. Edible films and coatings: A review. *Food Technology*, 40:47–58.

Khwaldia K., Perez C., Banon S., Desobry S., Hardy J. 2004. Milk proteins for edible films and coatings. *Critical Reviews in Food Science and Nutrition*, 44:239–251.

Kinsella J.E., Whitehead D.M. 1989. Proteins in whey: Chemical, physical and functional properties. *Advances in Food and Nutrition Research*, 33:343–438.

Kiplinger J. 2003. Meet the azlons from A to Z: Regenerated & rejuvenated. www.fabrics.net

Knepp W.A., Jayakrishnan A., Quigg J.M., Sitren H.S., Bagnall J.J., Goldberg E.P. 1993. Synthesis, properties and intratumoral evaluation of mitoxantrone-loaded casein microspheres in Lewis lung carcinoma. *Journal of Pharmacy and Pharmacology*, 45:887–891.

Ko S., Gunasekaran S. 2006. Preparation of sub-100-nm beta-lactoglobulin (BLG) nanoparticles. *Journal of Microencapsulation*, 23:887–898.

Konrad G., Kleinschmidt T. 2008. A new method for isolation of native α-lactalbumin from sweet whey. *International Dairy Journal*, 18:47–54.

Korhonen H., Pihlanto-Leppälä A. 2006. Bioactive peptides: Production and functionality. *International Dairy Journal*, 16(9):945–960.

Kosikowski F.V. 1979. Whey utilization and whey products. *Journal of Dairy Science*, 62:1149–1160.

Kriegel C., Arrechi A., Kit K., McClements D.J., Weiss J. 2008. Fabrication, functionalization, and application of electrospun biopolymer nanofibres. *Critical Reviews in Food Science and Nutrition*, 48:775–797.

Kristo E., Koutsoumanis K.P., Biliaderis C.G. 2007. Thermal, mechanical and water vapor barrier properties of sodium caseinate films containing antimicrobials and their inhibitory action on *Listeria monocytogenes*. In: *Proceedings of the Fifth International Congress on Food Technology*, Thessaloniki, Greece, March 9–11, vol. 2, pp. 276–284.

Kussendrager K.D., van Hooijdonk A.C.M. 2000. Lactoperoxidase: Physico-chemical properties, occurrence, mechanism of action and applications. *British Journal of Nutrition*, 84(Suppl. 1):19–25.

Lacroix M., Le T.C., Ouattara B., Yu H., Letendre M., Sabato S.F., Mateescu M.A., Patterson G. 2002. Use of γ-irradiation to produce films from whey, casein and soya proteins: Structure and functionals characteristics. *Radiation Physics and Chemistry*, 63(3–6):827–832.

Latha M.S., Jayakrishnan A. 1994. Glutaraldehyde cross-linked bovine casein microspheres as a matrix for the controlled release of theophylline: *In vitro* studies. *Journal of Pharmacy and Pharmacology*, 46:8–13.

Latha M.S., Rathinam K., Mohanan P.V., Jayakrishnan A. 1995. Bioavailability of theophylline from glutaraldehyde cross-linked casein microspheres in rabbits following oral administration. *Journal of Controlled Release*, 34(1):1–7.

Latha M.S., Lal A.V., Kumary T.V., Sreekumar R., Jayakrishnan A. 2000. Progesterone release from glutaraldehyde cross-linked casein microspheres: *In vitro* studies and *in vivo* response in rabbits. *Contraception*, 61:329–334.

Laursen I., Briand P., Lykkesfeldt A.E. 1990. Serum albumin as a modulator on growth of the human breast cancer cell line MCF-7. *Anticancer Research*, 10:343–351.

Le Tien C., Mateescu M.A., Lacroix M. 2001. Milk protein coatings prevent oxidative browning of apples and potatoes. *Journal of Food Chemistry*, 66(4):512–516.

Lee S.J., Rosenberg M. 2000. Preparation and some properties of water-insoluble, whey protein-based microcapsules. *Journal of Microcapsulation*, 17(1):29–44.

Longares A., Monahan F.J., O'Riordan E.D., O'Sullivan M. 2004. Physical properties and sensory evaluation of WPI films of varying thickness. *LWT*, 37 (5):545–550.

Longares A., Monahan F.J., O'Riordan E.D., O'Sullivan M. 2005. Physical properties of edible films made from mixtures of sodium caseinate and WPI. *International Dairy Journal*, 15:1255–1260.

Lucas D.O. 1999. Breakthrough technology products concentrated whey protein with bioactive immunoglobulins. *Clinical Nutrition Insights*, 6(21):1–4.

Lucena M.E., Alvarez S., Menéndez C., Riera F.A., Alvarez R. 2006. Beta-lactoglobulin removal from whey protein concentrates. Production of milk derivatives as a base for infant formulas. *Separation and Purification Technology*, 52:310–316.

Madureira A.R., Pereira C.I., Gomes A.M.P., Pintado M.E., Malcata F.X. 2007. Bovine whey proteins—Overview on their main biological properties. *Food Research International*, 40:1197–1211.

Mahmoud R., Savello P. 1992. Mechanical properties of and water vapor transferability through whey protein films. *Journal of Dairy Science*, 75(4):942–946.

Manso M.A., López-Fandiño R. 2004. κ-Casein macropeptides from cheese whey: Physicochemical, biological, nutritional, and technological features for possible uses. *Food Reviews International*, 20(4):329–355.

Markus C.R., Olivier B., Panhuysen G.E.M. et al. 2000. The bovine protein α-lactalbumin increases the plasma ratio of tryptophan to the other large neutral amino acids, and in vulnerable subjects raises brain serotonin activity, reduces cortisol concentration, and improves mood under stress. *American Journal of Clinical Nutrition*, 71(6):1536–1544.

Marshall K. 2004. Therapeutic applications of whey protein. *Alternative Medicine Review*, 9:136–156.

Mastromatteo M., Chillo S., Buonocore G.G. et al. 2008. Influence of spelt bran on the physical properties of WPI composite films. *Journal of Food Engineering*, 92(4):467–473.

McHugh T.H., Krochta J.M. 1994. Milk protein based edible films and coatings. *Food Technology*, 48(1):97–103.

Mehra R., Marnila P., Korhonen H. 2006. Milk immunoglobulins for health promotion: A review. *International Dairy Journal*, 16:1262–1271.

Mehra R., O'Kennedy B.T. 2008. Separation of β-lactoglobulin from whey: Its physico-chemical properties and potential uses. In: C.I. Onwulata, and P.J. Huth (Eds.), *Whey Processing, Functionality and Health Benefits*, pp. 39–62. Ames, IA: Wiley-Blackwell, IFT Press.

Metzger W. 1997. Method of producing CN film. U.S. Patent, 5,681,517.

Miguel M., Manso M.A., López-Fandiño R., Alonso J.A., Salaices M. 2007. Vascular effects and antihypertensive properties of κ-CN macropeptide. *International Dairy Journal*, 17:1473–1477.

Min S., Harris L.J., Han J.H., Krochta J.M. 2005a. *Listeria monocytogenes* inhibition by whey protein films and coatings incorporating lysozyme. *Journal of Food Protection*, 68(11):2317–2325.

Min S., Harris L.J., Krochta J.M. 2005b. *Listeria monocytogenes* inhibition by whey protein films and coatings incorporating the lactoperoxidase system. *Journal of Food Science*, 70(7):317–324.

Min S., Harris L.J., Krochta J.M. 2005c. Antimicrobial effects of lactoferrin, lysozyme, and the lactoperoxidase and edible whey protein films incorporating the lactoperoxidase system against *Salmonella enterica* and *Escherichia coli* O157:H7. *Journal of Food Science*, 70(7):332–338.

Mohan D., Radhakrishnan G., Rajadurai S., Nagabhushanam T., Thomas Joseph K. 1984. Studies on graft copolymerization of acrylate monomers onto casein. *Journal of Applied Polymer Science*, 29(1):329–339.

Morgan J. 1999. The centenary of casein. *Gloucestershire Society for Industrial Archeology Journal*, 44–50.

Morris P.E., FitzGerald R.J. 2008. Whey proteins and peptides in human health. In: C.I. Onwulata, and P.J. Huth (Eds.), *Whey Processing, Functionality and Health Benefits*, pp. 285–343. Ames, IA: Wiley-Blackwell, IFT Press.

Neeser J.R. 1991. Dental anti-plaque and anti-caries agent. U.S. Patent, 4,992,420.

Nielsen E.W., Ullum J.A. 1989. *Dairy Technology 2*. Arhus, Denmark: Danish Turkney Dairies Ltd.

Osés J., Fabregat-Vázquez M., Pedroza-Islas R., Tomás S.A., Cruz-Orea A., Maté J.I. 2009. Development and characterization of composite edible films based on whey protein isolate and mesquite gum. *Journal of Food Engineering*, 92(1):56–62.

Ouattara B., Canh L.T., Vachon C., Mateescu M.A., Lacroix M. 2002. Use of γ-irradiation cross-linking to improve the water vapor permeability and the chemical stability of milk protein films. *Radiation Physics and Chemistry*, 63(3–6):821–825.

Oussalah M., Caillet S., Salmiéri S., Saucier L., Lacroix M. 2004. Antimicrobial and antioxidant effects of milk protein-based film containing essential oils for the preservation of whole beef muscle. *Journal of the Agricultural and Food Chemistry*, 52(18):5598–5605.

Ozdemir M., Floros J.D. 2008. Optimization of edible whey protein films containing preservatives for water vapor permeability, water solubility and sensory characteristics. *Journal of Food Engineering*, 86(2):215–224.

Paris C., Lecomte S., Coupry C. 2005. ATR-FTIR spectroscopy as a way to identify natural protein-based materials, tortoiseshell and horn, from their protein-based imitation, galalith. *Spectrochimica Acta Part A: Molecular and Biomolecular Spectroscopy*, 62(1–3):532–538.

Perez M.D., Calvo M. 1995. Interaction of beta-lactoglobulin with retinol and fatty acids and its role as a possible biological function for this protein: A review. *Journal of Dairy Science*, 78:978–988.

Raval D.K., Narola B.N., Patel A.J. 2005. Synthesis, characterization, and composite properties of casein incorporated p-aminophenol-urea-formaldehyde copolymers. *International Journal of Polymeric Materials*, 54:731–741.

Raval D.K., Patel A.J., Narola B. 2006. A study on composites from casein modified melamine-formaldehyde resin. *Polymer-Plastics Technology and Engineering*, 45:293–299.

Rayner T.E., Cowin A.J., Robertson J.G., Coter R.D., Harries R.C. 2000. Mitogenic whey extract stimulates wound repair activity *in vitro* and promotes healing of rat incisional wounds. *American Journal of Physiology—Regulatory, Integrative and Comparative Physiology*, 278:R1651–R1660.

Rektor A., Vatai G. 2004. Membrane filtration of Mozzarella whey. *Desalination*, 162:279–286.

Rosenberg M. 1997. Milk derived whey protein-based microencapsulating agents and a method of use. U.S. Patent, 5,601,760.

Satpathy G., Rosenberg G. 2003. Encapsulation of chlorothiazide in whey proteins: Effects of wall-to-core ratio and cross-linking conditions on microcapsule properties and drug release. *Journal of Microencapsulation*, 20(2):227–245.

Schmitt C.J.E. 2007. Cosmetic use of whey protein micelles. European Patent Application, EP 1,844,758 A1.

Schou M., Longares A., Montesinos-Herrero C., Monahan F.J., O'Riordan D., O'Sullivan M. 2005. Properties of edible sodium caseinate films and their application as food wrapping. *LWT*, 38:605–610.

Serrano E., Källander B. 2005. Building and construction—Timber. In: R. Adams (Ed.), *Adhesive Bonding*. Cambridge, MA: Woodhead Publishing Ltd.

Seydim A.C., Sarikus G. 2006. Antimicrobial activity of whey protein based edible films incorporated with oregano, rosemary and garlic essential oils. *Food Research International*, 39(5):639–644.

Shaw N.B., Monahan F.J., O'Riordan E.D., O'Sullivan M. 2002. Effect of soya oil and glycerol on physical properties of composite WPI films. *Journal of Food Engineering*, 51(4):299–304.

Sienkiewicz T., Riedel C.L. 1990. *Whey and Whey Utilization*. Gelsenkirchen-Buer, Germany: Verlag Th. Mann.

Smith R. 2005. *Biodegradable Polymers for Industrial Applications*. Cambridge, GBR: Woodhead Publishing.

Smithers G.W. 2008. Whey and whey proteins-from 'gutter-to-gold.' *International Dairy Journal*, 18:695–704.

Sohail S.S., Wang B., Biswas M.A.S., Oh J.H. 2006. Physical, morphological, and barrier properties of edible casein films with wax applications. *Journal of Food Science*, 71(4):C255–C259.

Somanathan N., Arumugam V., Sanjeevi R., Narasimhan V. 1987. Mechanical properties of grafted casein films. *Journal of Applied Polymer Science*, 34(6):2299–2311.

Somanathan N., Arumugan V., Naresh M.D., Sanjeevi R. 1989. Effect of extension rate on the stress–strain characteristics of grafted casein film. *Journal of Applied Polymer Science*, 37(5):1311–1317.

Sothornvit R., Olsen C.W., McHugh T.H., Krochta J.M. 2003. Formation conditions, water-vapor permeability, and solubility of compression-molded whey protein films. *Journal of Food Science*, 68(6):1985–1989.

Sothornvit R., Rhim J.W., Seok-In Hong S.I. 2009. Effect of nano-clay type on the physical and antimicrobial properties of whey protein isolate/clay composite films. *Journal of Food Engineering*, 91(3):468–473.

Southward C.R. 1998. Casein products. *Chemical Processes in New Zealand*, www.nzic.org.nz

Staples L.C. 1982. Whey protein containing cosmetic formulations, process of preparing the same and a method for setting hair by using a non-hydrolyzed whey product. European Patent Application: EP 0,046,326 A2.

Sutherland J. 2003. Casein: An overview. *American Artist*, 67(733):10.

Suzuki Y., Maruyama T. 2005. Removal of emulsified oil from water by coagulation and foam separation. *Separation Science and Technology*, 40:3407–3418.

Tenovuo J. 2002. Clinical applications of antimicrobial host proteins lactoperoxidase, lysozyme and lactoferrin in xerostomia: Efficacy and safety. *Oral Diseases*, 8:23–29.

Thomä-Worringer C., Sørensen J., López-Fandino R. 2006. Health effect and technological features of casein-macropeptide: A review. *International Dairy Journal*, 16:1324–1333.

Toba Y., Takada Y., Matsuoka Y. et al. 2001. Milk basic protein promotes bone formation and suppresses bone resorption in healthy adult men. *Bioscience, Biotechnology & Biochemistry*, 65:1353–1357.

Tomasula P.M., Parris N.Y. 1998. Properties of films made of CO_2 precipitated casein. *Journal of Agricultural and Food Chemistry*, 46:4470–4474.

Tunick M.H. 2008. Whey protein production and utilization: A brief history. In: C.I. Onwulata, and P.J. Huth (Eds.), *Whey Processing, Functionality and Health Benefits*, pp. 1–13. Ames, IA: Wiley-Blackwell, IFT Press.

USDEC. 2003. Reference Manual for U.S. Whey and Lactose Products. U.S. Dairy Export Council, Arlington, VA. http://www.usdec.org/Library/ProductInfo.cfm?Category=Manuals&navItemNumber=82612

Uysal İ., Korel F., Yemenicioğlu A., Yener F.Y.G. 2007. Application of protein-based edible films to various foods. In: *Proceedings of the Fifth International Congress on Food Technology*, Thessaloniki, Greece, March 9–11, vol. 3, pp. 563–569.

Vaccari J.A. 2002. *Materials Handbook*, 2nd edn. New York: McGraw-Hill.

Varnam A.H., Sutherland J.P. 1994. *Milk and Milk Products. Technology, Chemistry and Microbiology*. London, U.K.: Chapman & Hall.

Vaz C.M., Fossen M., van Tuil R.F., de Graaf L.A., Reis R.L., Cunha A.M. 2003. Casein and soybean protein-based thermoplastics and composites as alternative biodegradable polymers for biomedical applications. *Journal of Biomedical Materials Research*, 65A:60–70.

Wagner J. 2001. *Membrane Filtration Handbook: Practical Hints and Tips*, 2nd edn. Minnetonka, MN: Osmonics, Inc.

Wakabayashi H., Yamauchi K., Mitsunori T. 2006. Lactoferrin research, technology and applications. *International Dairy Journal*, 16:1241–1251.

Walstra P., Geurts T.J., Noomen A., Jellema A., van Boekel M.A.J.S. 1999. *Dairy Technology. Principles of Milk Properties and Processes*. New York: Marcel Dekker Inc.

Walzem R.L., Dillard C.J., German J.B. 2002. Whey components: Millennia of evolution create functionalities for mammalian nutrition: What we know and what we may be overlooking. *Critical Reviews in Food Science and Nutrition*, 42(4):353–375.

Wang B., Brand-Miller J., McVeagh P., Petocz P. 2001. Concentration and distribution of sialic acid in human milk and infant formulas. *American Journal of Clinical Nutrition*, 74:510–515.

Wang N., Zhang L., Lu Y., Du Y. 2004. Properties of crosslinked casein/waterborne polyurethane composites. *Journal of Applied Polymer Science*, 91(1):332–338.

Wang Y., Rakotonirainy A.M., Padua G.W. 2003. Thermal behaviour of zein-based biodegradable films. *Starch/Stärke*, 55 (1):25–29.

Weinberg E.D. 1996. The role of iron in cancer. *European Journal of Cancer Prevention*, 5:19–36.

Weinbreck F., Minor M., De Kruif C.G. 2004. Microencapsulation of oils using whey protein/gum arabic coacervates. *Journal of Microencapsulation*, 21(6):667–679.

Whitney R.McL. 1999. Proteins of milk. In: N.P. Wong, R. Jenness, M. Keeney, E.H. Marth (Eds.), *Fundamentals of Dairy Chemistry*. Gaithersburg, MA: Aspen Publishers, Inc.

Willmott N., Magee G.A., Cummimgs J., Halbert G.W., Smyth J.F. 1992. Doxorubicin-loaded casein microspheres: Protean nature of drug incorporation. *Journal of Pharmacy and Pharmacology*, 44:472–475.

Wingerd W.H. 1971. Lactalbumin as a food ingredient. *Journal of Dairy Science*, 54:1234–1236.

Xie J., Hsieh Y.L. 2003. Ultra-high surface fibrous membranes from electrospinning of natural proteins: Casein and lipase enzyme. *Journal of Material Science*, 38(10):2125–2133.

Yamamura J., Aoe S., Toba Y. et al. 2002. Milk basic protein (MBP) increases radial bone mineral density in healthy adult women. *Bioscience, Biotechnology & Biochemistry*, 66:702–704.

Yoshida C.M.P., Antunes A.C.B., Antunes A.J. 2002. Moisture adsorption by milk whey protein films. *International Journal of Food Sciences and Technology*, 37:329–332.

Zadow J.G. 1994. Utilization of milk components: Whey. In: R.K. Robinson (Ed.), *Modern Dairy Technology, Advances in Milk Processing*, vol. 1, 2nd edn., pp. 313–373. London, UK: Chapman & Hall.

Zhang P., Huang F., Wang B. 2002. Characterization of biodegradable aliphatic/aromatic copolyesters and their starch blends. *Polymer-Plastics Technology and Engineering*, 41(2):273–283.

Zydney A.L. 1998. Protein separations using membrane filtration: New opportunities for whey fractionation. *International Dairy Journal*, 8:243–250.

8 The Role of Colloidal Delivery Systems in DNA Genetic Therapeutics

Jenny Ho, Michael K. Danquah, Yi Huang, and Shan Liu

CONTENTS

8.1 GENE THERAPY WITH DNA: THE NEXT GENERATION OF BLOCKBUSTER THERAPEUTICS

Gene therapy represents a new paradigm of therapy for diseases, where the disease is treated at the molecular level by restoring defective biological functions or reconstituting homeostatic mechanisms within cells. The ability to engineer the *in vivo* production of therapeutic gene-based products compartmentally within the cells would help in combating the unmet medical needs for certain genetic and end-stage diseases such as cancer and acquired immunodeficiency syndrome (AIDS) (Liu et al. 2006). The clinical trials of gene therapy were initiated nearly two decades ago, and underwent several cycles of ups and downs (Plautz et al. 1993, Blaese et al. 1995, MacGregor et al. 1998, Hacein-Bey-Abina et al. 2003, Raper et al. 2003, Edelstein et al. 2004). However, significant hurdles continue to remain, and emphasis is now on improving the safety of viral vectors and the efficiency of nonviral systems (Rolland 2005). The limitations of viral vectors, especially those pertaining to its safety concerns have prompted the development

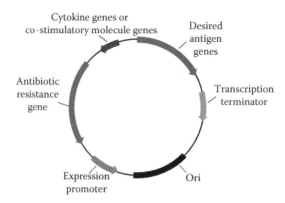

FIGURE 8.1 Schematic representation of construction for a plasmid DNA.

of synthetic vectors based on nonviral systems. Deoxyribonucleic acid (DNA) has the potential to lead a new generation of reverse engineered biopharmaceuticals as it has inherent advantages over other biomolecules such as protein and ribonucleic acid (RNA) in terms of simplicity in production and its high thermal stability. Plasmid-based approaches to gene therapy often termed "nonviral" involve the recombination of gene sequence, encoding a therapeutic protein into closed circular piece of DNA, and administered directly to patients to induce gene expression (Figure 8.1). The ability of genetic DNA vaccines to generate both B-cell and T-cell responses has been identified as a promising technique to prevent cancer, and intracellular bacterial and viral infections. Effective gene therapy requires that the DNA successfully gets access to the target cell, is taken up for internalization into the cell, is trafficked through the cell after escaping the degradative pathway to the nucleus, and is subsequently transcribed and translated to produce the desired gene product (Ledley 1996). Due to the size and charge of naked DNA and the enzymatic and membrane barriers imposed by the cell, the entry of DNA molecules into cells and subsequent expression is a very wasteful process (Liu 2003). This is proved by the fact that the progress of naked plasmid DNA in clinical trials from phase I to phase III was highly unsuccessful (Figure 8.2) (Edelstein 2009). The rapidly rising demand for therapeutic grade DNA molecules requires associated improvements in encapsulation and delivery technologies. This includes the formulation of DNA molecules into synthetic delivery systems for enhanced cellular transformation efficiencies.

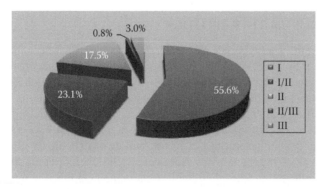

FIGURE 8.2 The progress of naked plasmid DNA as vector in clinical trial stages. (From Edelstein, M. 2009. *Gene Therapy Clinical Trials Worldwide.* Wiley Inter Science, September 2008 [updated by December, 2009]. Available from http://www.wiley.co.uk/genmed/clinical/)

8.2 COLLOIDAL DELIVERY SYSTEMS FOR DNA

A colloidal system refers to a type of mixture where internal phase is dispersed evenly throughout a continuous phase. The internal phase may consist of a particular kind of solid, liquid, or gas composite or a combination in the medium. If all composite in a colloidal system are nearly the same size, the system is called monodisperse; and the opposite is called heterodisperse (Burgess 2006). Colloids are defined as the materials with length scales below a micron, and often in the sub-100 nm regime. The huge surface area of colloidal materials offers the flexibility to tailor their surface properties and particle behavior to achieve new delivery modes for drugs and improve their therapeutic profiles (Bawarski et al. 2008). The development of colloidal delivery systems is multidisciplinary, which is a combination of pharmaceutics, bioconjugate chemistry, and molecular biology. The commonly used colloidal systems, especially where human physiology is concerned, are colloidal suspension and colloidal emulsion (Besseling et al. 2009). A colloidal suspension is one in which of the size of the solid particles in a liquid lies in the colloidal range, and a colloidal emulsion refers to one in which the droplets are dispersed in a liquid. In emulsions, the droplets often exceed the usual limits of colloids in size. An emulsion is denoted by the symbol O/W if the continuous phase is an aqueous solution and by W/O if the continuous phase is an organic liquid. More complicated emulsions such as aqueous droplets that have finer organic liquid droplets dispersed within them, while the aqueous droplets themselves are dispersed in organic phase (O/W/O) are also possible (Pal 1996, Tamilvanan 2004). When the internal phase is smaller, the surface of the colloid is large and the charge is greater (Gupta and Moulik 2008). This gives colloids their unique nature. The most important characteristic of colloidal delivery system is the tremendous charge on their surface that repels each other to overcome the tendency to aggregate and to remain dispersed (Burgess 2006).

Research works on colloidal delivery systems in genetic therapeutics are based on the molecular level focusing on the interdisciplinary development of pharmaceutical DNA delivery approaches. Colloidal delivery systems modify many physicochemical properties, aiming to protect the DNA from degradation, minimize DNA loss, prevent harmful side effects, enhance DNA targeting, increase drug bioavailability, and stimulate the immune systems (Ledley 1996, O'Hagan et al. 2004, Patil et al. 2005). Various colloidal delivery systems have been studied for decades and these have been providing promising approaches in improving the delivery of problematic DNA candidates. In this chapter, the maturation pathway and recent advances for the major colloidal delivery carriers are reviewed. With all the ongoing efforts, improved colloidal delivery techniques have become one of the promising delivery methods, which open up new markets and offer great potential benefits to DNA genetic therapeutics. A solution to current problems of DNA molecules delivery can be achieved provided the colloidal delivery systems are carefully designed and developed with reference to the target and route of administration. To improve the overall efficiency of nonviral gene delivery systems to resemble that of viruses is achievable.

8.2.1 POLYMERIC NANOCARRIERS

Polymeric particles, including natural and synthetic ones, are gaining interest as the potential controlled delivery carriers for introducing DNA into cells for several reasons including the following:

1. The ability to protect pDNA payload from extracellular degradation
2. The ability to accommodate large size plasmids and other immunostimulatory agents
3. The ability to offer a phagocytosis-based passive targeting to APCs
4. The ability to be conjugated with appropriate functionalities to enhance target and uptake (Elfinger et al. 2008, Nguyen et al. 2009)

Cationic polymers such as poly-L-lysine (PLL) (Zauner et al. 1998), polyethylenimines (PEI) (Boussif et al. 1995, Lungwitz et al. 2005), polyamidoamine (PAMAM) dendrimers (KukowskaLatallo et al. 1996, Dufes et al. 2005), and chitosan (MacLaughlin et al. 1998, Borchard 2001) can be applied as carriers for converting gene vectors into complex polyplexes to form defined sizes similar to the virus-like forms. When applied to cells, the positively charged polyplexes mediates the transfection via a multistage process that includes cationic binding to the negatively charged cell membrane, to facilitate entrance into the cytoplasm (Petersen et al. 2002, Lungwitz et al. 2005). As a result of this, formulated or encapsulated therapeutics have been shown to exhibit greater therapeutic efficacy compared to unformulated biomolecules (Pouton and Seymour 2001). With the assistance of cationic polymers, it is noticeable that transfection activity parallels the membrane toxicity (Wagner et al. 1991, Wolfert and Seymour 1996). The general association between cytotoxicity and transfection suggests that a degree of membrane damage must be caused for the cells to gain access to the cytoplasm (Godbey et al. 2001, Panyam and Labhasetwar 2003). Successful transfection relies on achieving the correct balance between gaining adequate accesses by the DNA molecules into the cytoplasm and causing excessive and lethal damage to the cells. Cell culture experiments (Figure 8.3) showed that PEI-induced high transfection efficiency, however, entailed higher cell death rate (Figure 8.4). The high density of amino groups confers significant buffering capacity to the PEI, especially in the endosome when pH decreases from 7 to 5. The "proton sponge effect" explains the high transfection efficiencies obtained with PEI, as it appears as an endosomolytic reagent. However, the large number of positive charges leads to high toxicity profile, and is directly affected by many factors such as molecular weight, degree of branching, ionic strength of the solution, zeta potential, and particle size (Elfinger et al. 2008). Coupling the poly(ethylene glycol) (PEG) to mask the surface charge of the PEI/pDNA polyplexes is a popular approach to lower the cellular toxicity by reducing the nonspecific interactions of polyplexes in the bloodstream (Tang et al. 2003). Additionally, poly(β-amino-ester)s have been developed by incorporating moieties that can be hydrolyzed allowing this type of cationic polymer to readily degrade into nontoxic metabolites for long term and repeated application in *in vivo* use. This biodegradable cationic polymer provides a mechanism for polymer-DNA dissociation following cellular uptake, which is critical for efficient transfection and gene expression.

Polymeric gene carriers can accommodate large size DNA, can be conjugated with appropriate functionalities, and can be administered repeatedly. Functionalization of polymeric carriers with specific ligands is a very promising and highly selective strategy to improve specificity for the

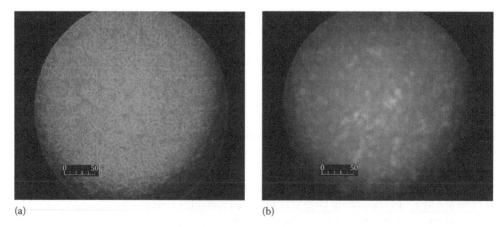

(a)   (b)

FIGURE 8.3   (a) Bright-field image of immortalized human endothelial cells culture (EaHy 926) incubated with 100 μL of PEI/plasmid DNA (pEGFP-N1) polyplexes solution with nitrogen to phosphorus ratio (*N/P*) of 15; (b) Fluorescence image of the same cells culture after 4 days, the green dots represented successfully transfected cells expressing green fluorescent protein (GFP).

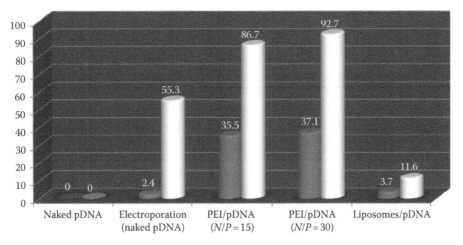

FIGURE 8.4 Comparison of transfection efficiency (gray) and cell death rate (white) for immortalized human endothelial cells culture (EaHy 926) after 4 days incubated with plasmid DNA (pDNA) encoded GFP by using different delivery systems.

target sites of interest (Elfinger et al. 2008). The ligands could be recognized by receptors on the surface of the cells of interest. On the other hand, block copolymers, especially poly(lactide-*co*-glycolide) (PLGA), attract DNA delivery applications. This is because their chemical composition, total molecular weight, and block length ratios can easily be changed to allow control of the size and morphology of the polymeric carriers (Nguyen et al. 2009). Polymeric particles with encapsulated DNA are vesicular systems in which the DNA is confined to a cavity surrounded by polymers, whereas particles adsorbed with DNA are matrix systems in which the DNA is physically and uniformly dispersed on the surface (O'Hagan et al. 2006). Encapsulation of pDNA protects it from nuclease degradation, and the controlled gene delivery system can be readily designed to exhibit varying degradation times and release kinetics of pDNA for prolonged gene expression over a required duration. PLGA undergoes ester hydrolysis in the physiological environment forming biocompatible monomers. Although early studies with PLGA were promising, it is not an optimal gene delivery material. Harsh formulation conditions, poor loading efficiency, acidic microenvironment during degradation, slow release, and subsequently low transfection efficiency are the main aspects (Nguyen et al. 2009). Thus, significant efforts have been made by changing polymer physical chemistry, formulation methods, and addition of cationic transfection agents (Jilek et al. 2005). Modest efficiency and significant toxicity are the key issues in restricted broader *in vivo* therapeutic applications of polymeric carriers. For clinical use, small and monodisperse in size, absence of toxic free polymers, increased storage stability, biodegradability, and bioresponsiveness are among the characteristics to be pursued for improving extra- and intracellular genetic therapeutic improvements.

8.2.2 LIPOSOMES

Liposomes are spherical vesicles composed of natural or synthetic amphiphilic phospholipids, which self-associate into bilayers (McNeil and Perrie 2006). Majority of the liposomes are <400 nm, and are classified into multilamellar vesicles, large unilamellar vesicles, and small lamellar vesicles, based on the size and number of layers. The potency of liposomes depends on composition, electric charge, and methods of formulation, and can be categorized as conventional, pH-sensitive, cationic, immuno-, and long-circulating liposomes (Smith et al. 1997, Bawarski et al. 2008). There is no limit to the length of DNA that can be incorporated during liposomal composition, in contrast to viral vectors. However, conventional liposomes prepared with a neutral zwitterionic lipid or an anionic lipid showed low entrapment efficiency for pDNA with high molecular weight (Mannino et al. 1979, Lindner et al. 2006). This property was improved by the inclusion of cationic lipids to promote the pDNA condensation by forming lipoplexes and to encourage cellular uptake (Felgner 1996, May et al. 2000, de Lima et al. 2001). Cationic lipids as gene transfer agents *in vitro* were pioneered in 1987 by Felgner and colleagues in which the polar characteristic of the liposomal core enables polar DNA molecules to be encapsulated (Felgner et al. 1987). There are *in vitro* and *in vivo* evidence that cationic liposomes can enhance both humoral and cell-mediated immune responses induced in mice by plasmid DNA immunogens (Hofland et al. 1996, 1997, Perrie et al. 2001a). The formation of complex between DNA and cationic liposomes is based on the electrostatic interaction with the hydrophobic bilayers of membrane lipids oriented outward into the solvent (Ulrich 2002). This complex is, therefore, very hydrophobic, poorly soluble in water, and is capable of interacting effectively with cell taken up into the membrane vesicles following fusion with endosomes (Nakanishi and Noguchi 2001). The lipoplex formation is dependent on the physical conditions such as pH level, electrical charges, and ionic strength of medium as well as the structural characteristics of the liposomes. Widely used cationic lipid such as DOTMA (1,2-dioleyloxypropyl-3-trimethyl ammonium bromide) or DOTAP (1,2-bis(oleoyloxy)-3-trimethylammonio propane) have been used successfully *in vivo* (Brigham et al. 1989, Canonico et al. 1994). The use of ester bonds linking the cationic head group to the lipid anchor would reduce toxicity by facilitating degradation in eukaryotic cells (Gould-Fogerite et al. 1998). Unfortunately, lipoplexes progress slowly in clinical trials because its polycationic nature pronounces the tendency to bind to proteins in serum, thus resulting in the release of the associated pDNA after loss of cationic surface charge (McNeil and Perrie 2006). The transfection efficiencies of liposomal systems are significantly lower than the viral vectors. Some *in vivo* studies have revealed that the gene transduction responses obtained by liposomal were transient and short-lived (Wheeler et al. 1996, Liu et al. 1997). The number of gene therapy clinical trials using liposomes as vector has dropped drastically since 2001 (Figure 8.5) (Edelstein 2009). An important aspect that needs to be considered when developing lipoplexes is maintain their stability over a long period of time, which translates into long shelf life. Liposomes in general are relatively unstable as aggregates form within the suspension and reduce the transfection efficiency considerably. However, this property can be altered by the incorporation of various molecules such as cholesterol, acetylphosphate, and stearylamine into the phospholipid bilayer (Allison and Gregoriadis 1974, Elfinger et al. 2008). Continuing efforts such as incorporating agents to enhance plasmid endosomal escape, peptides to promote nuclear localization, steric coating to stabilize the vectors in sera, and targeting moieties to improve cellular specificity for liposome systems have been summarized elsewhere (McNeil and Perrie 2006).

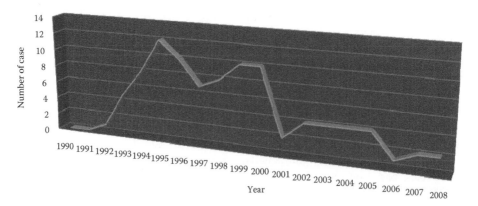

FIGURE 8.5 The number of clinical trials using liposomes as transfection vector in gene therapy from year 1990 till 2008. (From Edelstein, M. 2009. *Gene Therapy Clinical Trials Worldwide.* Wiley Inter Science, [updated by December, 2009]. Available from http://www.wiley.co.uk/genmed/clinical/)

8.2.3 Hydrogels

Hydrogels are water-saturated turgid crosslinked or interwoven networks, which mimic the 3-D environments of various cells in native cartilaginous tissues. They represent an improvement as a class of biodegradable polymeric materials with excellent biocompatibility, and allow protracted localized gene expression (Vinogradov 2006). Hydrogels are soft materials and classified as neutral, anionic, or cationic. They have the unique property of swelling in an aqueous environment to function as an intelligent delivery system. The swelling of hydrogel is controlled by both (1) the hydrogel structure (chemical structure of polymer, degree of crosslinking, charge density) and (2) the environment parameters (pH, ionic strength, and temperature) (Li et al. 2003, Oishi et al. 2007, Kabanov and Vinogradov 2009). Encapsulation of pDNA into the flexible polymer networks of microgel or nanogel (effective diameter <100 nm) can protect the pDNA from enzymatic degradation by extracellular and intracellular nucleases. Loading of biological agents is usually achieved spontaneously through electrostatic, van der Waals, or hydrophobic interactions between the agent and the polymer matrix (Lemieux et al. 2000b). Nanogels are very promising carriers for their high loading capacity and high stability. The macroporous networks of swollen nanogels with low number of crosslinks can accommodate larger biomolecules and form polymeric envelopes (Kabanov and Vinogradov 2009). The nanocarrier surface is often modified with biospecific targeting groups to enhance the delivery of gene into specific cells; and also with inert hydrophilic polymers, such as PEG, to extend the circulation time in bloodstream. A human transferrin-modified PEG-*cl*-PEI nanogel loaded with pDNA was shown to increase the transfection efficiency of human breast carcinoma cells in the presence of serum (Vinogradov 2006). Indeed, the delivery of pDNA from a wide range of hydrogel materials has been investigated, from natural biopolymers such as gelatin (Fukunaka et al. 2002, Kushibiki et al. 2003, 2006), hyaluronan (Kim et al. 2003, Segura et al. 2005, Wieland et al. 2007), pullulan (Gupta and Gupta 2004), silk-elastinlike protein (SELP) (Megeed et al. 2004) to poly(ethylene glycol)-based synthetic polymers (Kasper et al. 2005, Vinogradov 2006, Oishi and Nagasaki 2007). PEGylated nanogels that bear a lactose group at the PEG end showed significant endosomolytic abilities, achieving the pronounced transfection efficiency of the PEG-*b*-PLL/pDNA polyplex without any significant toxicity (Oishi and Nagasaki 2007). Precise control of pDNA release from natural polymer hydrogels can be difficult to achieve, because it depends on the enzymatic degradation and ion exchange that may vary within subjects and sites. Alternatively, synthetic polymer hydrogels offer a higher degree of control over pDNA dosing through tailoring the properties of gel such as mesh size and degradation kinetics (Kasper et al. 2005).

8.2.4 Polymeric Micelles

Micelles are nanosized, spherical colloidal particles with a hydrophobic interior (core) and a hydrophilic exterior (shell), formed by self-assembly of block or graft copolymers in aqueous. Modifiable properties of micelles are of great interest for gene and drug delivery applications. For example, their ability to attenuate toxicities, enhance delivery to desired biological sites and improve the therapeutic efficacy of active pharmaceutical ingredients (Kataoka et al. 2001). Their individual particle size is normally <100 nm in diameter, and is designed to be thermodynamically stable and biocompatible so that it may circulate for prolonged periods in the blood. The micellar delivery systems including graft, diblock, or multiblock copolymers possess numerous advantages over liposomes, and attracted attentions as promising gene carriers (Choi et al. 2000, Nishiyama and Kataoka 2006). Specifically, the graft copolymers have a comb-like structure with hydrophilic segments attached on the side of the cationic segments to improve the solution properties of polyion complex (Kakizawa and Kataoka 2002). The block copolymer is a linear copolymer with one segment covalently joined to the head of another segment, and provides a core-shell structure, where pharmaceutical complexes are surrounded by a protective and stabilizing hydrophilic corona (Kakizawa and Kataoka 2002). Until now, several micelles gene carriers have been intensively studied in preclinical and clinical trials. A series of block copolymers containing a poly(ethylene glycol) block were synthesized using a combination of click chemistry and ring-opening metathesis polymerization (Wigglesworth et al. 2008). A triblock copolymer, consisting of lactosylated poly(ethylene glycol), a pH-responsive polyamine segment, and a DNA-condensing polyamine segment was associated with plasmid DNA to form three-layered polyplex micelles as a targetable and endosome disruptive nonviral gene vector (Oishi et al. 2006). Incorporation of PEG segment with a degradable cationic polyphosphoramidate (PPA) segment to form PEG-*b*-PPA/DNA micelles upon condensation with pDNA has exhibited uniform and reduced particle size ranging from 80 to 100 nm. After intrabiliary infusion in rats, these PEG-*b*-PPA/DNA micelles mediated more uniform transgene expression throughout the liver and showed four times higher of the overall transgene expression than PPA/DNA complexes without any significant level of liver toxicity (Jiang et al. 2007). In addition, a polycation-based micelles gene delivery system which introduces a lysine (Lys) unit in poly(ethylene glycol)-*b*-cationic poly(*N*-substituted asparagine) with a flanking *N*-(2-aminoethyl)-2-aminoethyl group (PEG-*b*-Asp(DET)) resulting in PEG-*b*-P[Lys/Asp(DET)] has displayed improved stability, cellular uptake, and high transfection efficacy. The lysine unit was functioning as DNA anchoring moiety and the Asp(DET) unit acted as a buffering moiety inducing endosomal escape with minimal cytotoxicity (Miyata et al. 2007).

8.2.5 Inorganic Nanoparticles

Inorganic nanoparticles represent a new alternative to previous categories of gene carriers; they show low toxicity and promising properties for controlled delivery. The advantage of using inorganic nanoparticles for gene delivery results from their versatile properties, such as wide availability, rich functionality, and good biocompatibility. Their small size (1 nm to a few 100 nm) enables the penetration into small capillaries and enhances the intracellular uptake that subsequently allows an efficient pDNA accumulation at the targeted sites in the body. Inorganic nanoparticles allow sustained release at the targeted site over a period of days or even weeks after administration. These particles can be engineered to accumulate preferentially at tumor sites using targeting ligands, and provide a possible solution for cancer gene therapy. However, the relatively small size of these systems limits their use, as only small quantities of DNA can be involved. Here we review two different types of inorganic nanoparticles; these are iron oxide and gold nanoparticles.

Magnetofection is gene delivery to cells driven and site-specifically guided by the magnetic force acting on the vectors, which are associated with magnetic nanoparticles (MNP). Generally, polycation coated superparamagnetic iron oxide nanoparticles (SPIONs) bound to pDNA and guided along a magnetic gradient field show several fold enhancement in transfection efficiencies (Schillinger et al. 2005, Chorny et al. 2007). They also have the potential to accumulate in tumor cells and tumor-associated macrophages for detection at magnetic resonance (MR) imaging (Moore et al. 2000). Examples of polymer used in modifying the surfaces of MNP are PLL (Xiang et al. 2003), PEI (Huth et al. 2004), pullulan (Gupta and Gupta 2005), and PEI-PEG-chitosan copolymer (Kievit et al. 2009). These coatings can suppress the toxic effect and enhanced cellular uptake.

Gold nanoparticles (AuNPs) provide attractive candidate for pDNA delivery because of their good biocompatibility, ease of bioconjugation, and high electron density (Li et al. 2009). AuNPs are attractive scaffolds for the creation of transfection agents (Sandhu et al. 2002, Rosi et al. 2006), and they are bioinert (Connor et al. 2005). Functional diversity of monolayer protected cluster systems on the surfaces of AuNPs allows the tuning of charge and hydrophobicity to maximize transfection efficiency while minimizing toxicity (Ghosh et al. 2008). At the optimized branched PEI (2 kDa) to AuNPs ratio, this transfection vector is more potent than the unmodified polycation itself and its 25 kDa counterpart at the same concentration (Thomas and Klibanov 2003). Other recent studies demonstrated the ability of AuNPs to deliver plasmid DNA into breast cancer cells (MCF-7) (Wang et al. 2007) and human embryonic kidney cells (HEK 293) (Li et al. 2008), with modified coating biomaterials. Figure 8.6 illustrates the colloidal delivery approaches for nonviral DNA therapy.

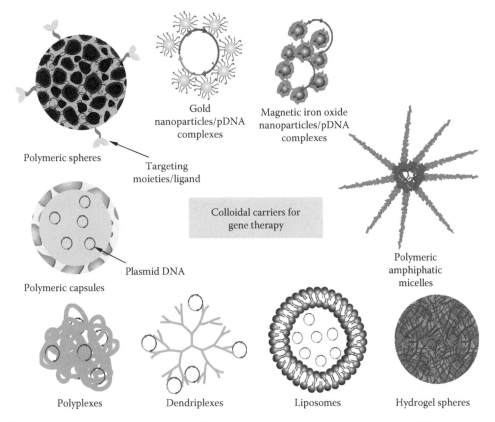

FIGURE 8.6 Schematic representation of different colloidal carriers for nonviral plasmid DNA gene therapy.

8.3 ROUTES OF ADMINISTRATION

Gene delivery is a compelling challenge to scientists as the transfection efficiency of nonviral plasmid-based therapeutics is relatively low when compared to other vectors such as viruses, viral-like particles, and cellular vectors. Different delivery systems and approaches for gene therapy have been conducted since two decades ago. However, despite these efforts, only 3.2% have progressed into phase III for gene therapy systems that entered into clinical trials by using naked pDNA. The failure rate is as high as 72.5% from phase I to phase II (Edelstein 2009). Some of the key challenges include degradation of plasmid DNA by serum enzymes such as endonucleases and in the endosome–lysosome compartment (Thierry et al. 2003, Patil et al. 2005); aggregation that are subsequently cleared from the blood by liver and spleen for colloidal delivery systems after their interaction with various compounds in the body fluid, such as serum proteins (Opanasopit et al. 2002); relatively few cells take up the plasmid molecules (1%–3%) (Kumar et al. 2004) due to their high molecular weight and polyanionic nature that leads to reduced expression of the encoded protein (Luo 2005). In practice, intended route of administration and location of the target cells impose significant biological barriers for biodistribution. The systemic circulation of gene medicine is complex and sensitive toward the colloidal properties and surface chemistry of the different delivery formulations.

8.3.1 INTRAMUSCULAR INJECTION

A plethora of studies have reported that muscle cells can take up and express genes after intramuscular injection of DNA formulations (Mumper et al. 1996, Dupuis et al. 2000, Richard et al. 2005, Ratanamart and Shaw 2006, Chang et al. 2007). Therefore, plasmid-based muscle-targeted gene therapy offers safe, cost-effective, and high patient compliance method to express therapeutic proteins. Four major factors that affect the levels of gene expression during intramuscular injection preclinical studies are the following: formulation, injection technique, species, and pretreatment of the muscle with myotoxic agents to induce muscle damage (Mumper and Rolland 1998). Lipoplexes and polyplexes were designed to protect plasmid from nuclease degradation, disperse and retain plasmid molecules intact in muscle, and facilitate its uptake by muscle cells. However, results have shown that the complexes are often localized at the injection site and induce low transfection efficiency (~1%) after intramuscular injection (Mumper and Rolland 1998, Pouton and Seymour 2001). Further studies showed that formulations of plasmids with hydrophilic and noncondensing polymers can promote dispersion and transfection within solid tissues. This supports the fact that uniformity and widespread of colloidal particles is critical for intramuscular injection (Mumper et al. 1996, 1998). Furthermore, an increase in gene expression using plasmid complexes with nonionic amphiphilic block copolymers together with electroporation has been demonstrated in mice (Lemieux et al. 2000a, Pitard et al. 2002, Riera et al. 2004).

Nearly all of the measures taken so far are focused on construction of new vectors, usage of enhancing reagents, and development of novel formulations to improve the delivery to and expression from muscle cells. However, little is known about how myocytes take up DNA complexes, or how the cellular transcriptional and translational mechanisms control the transgene expression. Also, how the therapeutic protein products are transferred from muscle fiber cytoplasm into the systemic circulation is yet to be known (Lu et al. 2003). As reviewed, gradual advancement to address the muscle-specific gene expression is needed to broaden the horizon of gene therapy clinical opportunities via intramuscular injection.

8.3.2 Mucosal Administration

The mucous membranes covering all body passages such as the respiratory and alimentary tracts are endowed with powerful protection network against foreign matters through the immunocytes that form mucosa-associated lymphoid tissues (MALT) (Holmgren and Czerkinsky 2005). Most infectious pathogens such as human immunodeficiency virus (HIV) and influenza virus enter the body through mucosal tissues; therefore, induction of humoral, mucosal, and cellular immune response via noninvasive DNA therapeutic conduit is gaining popularity. Mucosal immunization was reported to induce stronger antigen-specific responses in comparison to parenteral immunization (Goya et al. 2008). Mucoadhesive properties of colloidal delivery systems have been intensively investigated for their ability to prolong retention time at the mucosal surfaces for enhanced gene expression (Alpar et al. 2005, Goya et al. 2008). The most common mucosal administration routes, including oral, nasal, and pulmonary, will be described in Sections 8.3.2.1 through 8.3.2.3.

8.3.2.1 Oral Route

Oral gene therapy is an attractive mode for its significant potentials to treat local and systemic diseases (Romano et al. 2000, Prieto et al. 2002), and to provide both mucosal and systemic immunity (Jones et al. 1996, Dubensky et al. 2000, Chew et al. 2003, Li et al. 2009). Oral delivery is highly acceptable among patients due to its easy and noninvasive administration. In addition, large surface areas exist for mucoadhesive materials adsorption; gut epithelium is present for DNA uptake and expression; and large number of stem cells in the intestinal crypts for long-lasting gene expression has endowed the gastrointestinal tract an interesting site for gene therapy (Bhavsar and Amiji 2007). Nevertheless, oral delivery of nonviral gene medicine has proven to be extremely challenging, owing to the poor stability of the biomolecules in the acidic and enzyme-rich environment of the gastrointestinal tract. The formidable barriers such as proteolytic enzymes and endogenous nucleases in the gut lumen, gut flora, mucus layer, and cell epithelial lining in gastrointestinal organs caused low transfection efficiency after oral delivery. These physiological and anatomical hurdles prevent tissue-specific localization, cellular uptake, and could degrade the translated protein (Bhavsar and Amiji 2007).

Several strategies including liposome and polymeric nanoparticle delivery systems have been adapted to efficiently bypass the barriers in gastrointestinal tract, and protect the delicate gene vector so selectively targeting a specific cell type and express the therapeutic genes in a regulated fashion. Liposomes have been widely investigated as mucosal vaccine vehicles (Perrie et al. 2001b, Jain and Vyas 2006). Unfortunately, conventional liposomes are fragile in the gastric environment; therefore, options such as change in liposomal composition (Kersten and Crommelin 2003), coating with polysaccharide to increase liposomes stability (Venkatesan and Vyas 2000), and polymerization of liposomal surface (Chen et al. 1996) were explored. These improved and cross linked liposomes membrane are strong and stable enough to prevent leakages in the gastrointestinal tract, even if the phospholipids undergo hydrolysis at low pH and/or enzymatic degradation by phospholipases (Goya et al. 2008). Encapsulation of nonviral gene medicine in biodegradable polymers such as PLGA and chitosan has been successfully used for oral gene therapy (Roy et al. 1999, Chen et al. 2004, Bowman and Leong 2006) and DNA vaccination (Jones et al. 1996, He et al. 2005). Polymeric carriers not only provide adjuvant and sustained release effects, but also protect the payload from hostile mucosal environment and increase their uptake by the microfold cells and antigen presenting cells (Bhavsar and Amiji 2007). Despite the advantages offered by polymeric delivery technologies, toxicity, irritancy, allergenicity, and biodegradability are the primary issues to be considered when designing carrier systems.

The efficiency of oral formulations has been established in animal studies. However, attempt to scale up the results to humans is generally problematic as substantially high doses are required and the clinical experience has been mixed (Mirchamsy et al. 1996, Lambert et al. 2001, Katz et al. 2003). The future success of oral formulations for DNA gene therapy will depend on their ability to bypass the barriers in gastrointestinal tract, to enhance payload uptake and absorption by gut-associated lymphoid tissue (GALT) and to reduce the doses needed in inducing significant therapeutic effects.

8.3.2.2 Nasal Route

In the last decade, intranasal delivery has been recognized as a very promising route because it offers a faster and more effective therapeutic effect. The nasal route is an important arm of local mucosal and systemic immune system, since it is often the first point of contact for inhaled antigens. Both humoral and cellular immune responses can be induced via nasal route (Davis 2001). In addition, the loss of drug by first-pass metabolism in the liver can be avoided because the venous blood from the nose passes directly into the systemic circulation. The mucosal surface in the nasal cavity is highly vascularized and interspersed with immunologically active tissues, identified as nasopharyngeal lymphoid tissue (NALT), which are responsible for the induction and mediation of immune responses against the potential pathogenic organisms (Boyaka et al. 2003). The nasal mucosa is also lined with extensive pseudostratified columnar epithelium, which includes ciliated cells and mucous-secreting goblet cells that enhance the uptake and absorption of colloidal carrier systems. After the intranasal immunization, the antigen is sampled and passed on to underlying lymphoid cells in the submucosa, where the antigen is processed and the presentation takes place. This resulted in the activation of T-cells that help B-cells to develop into immunoglobulin A (IgA) plasma cells (Partidos 2000). Induction of antigen-specific immunoglobulin G (IgG) antibodies in serum, saliva, and respiratory tract secretions has also been reported in human trials (Bergquist et al. 1997, Kozlowski et al. 2002).

The major limitation of nasal delivery is inadequate absorption of colloidal carriers. Factors such as physiochemical properties of gene medicine, nasal mucociliary clearance, and nasal absorption enhancers greatly affect absorption in the nasal mucosa. Molecular weight of biomolecules, perfusion rate, perfusate volume, concentration, and solution pH also contribute to the rate and extent of absorption (Turker et al. 2004). As a result, several enhancers have been utilized to improve the nasal absorption. These include enhancers that can alter the physiochemical properties of biomolecules (such as solubility, partition coefficient, and ionic interaction) and enhancers that can affect the mucosa surface (Davis and Illum 2003). Examples of such enhancers include mucoadhesive polymers (Vila et al. 2004, Alpar et al. 2005, Di Colo et al. 2008), liposomes (Vadolas et al. 1995, de Jonge et al. 2004), and gels (Turker et al. 2004). Nasal delivery of DNA using nanoparticles blending of poloxamer and PLGA showed the ability to elicit strong humoral response (Csaba et al. 2006). Nasal administration using chitosan formulated pDNA vaccine has also showed promising *in vivo* cytotoxic T lymphocytes (CTL) responses against respiratory syncytial virus (RSV) infection (Kumar et al. 2002, Iqbal et al. 2003). Deposition of particles in the nasal cavity region is most common for larger particles, and it takes place via inertial impaction. Nevertheless, both theoretical and experimental studies have indicated that particles of 10–20 µm in diameter favor deposition in the upper airways and minimize deposition in the lungs and gastrointestinal tract during nasal inhalation (Clark and Egan 1994, Kippax and Fracassi 2003, Newman et al. 2004). Moreover, nasal delivery generally requires much lower doses than oral delivery owing to lower enzymatic activity in the nasal cavity.

8.3.2.3 Pulmonary Route

Immediate accessibility by inhalation to provide gene therapy in lung attribute to the extensive dendritic cell network lining the airway epithelium is believed to have advantages over alternative mucosal routes. The immediate pharmacological effect after the translocation of antigen from the lung to the lymph node and the clearance of colloidal carriers from the lung by alveolar phagocytosing cells proved to be prominent in generating immune responses (Lu and Hickey 2007, Helson et al. 2008). In addition, the bronchus-associated lymphoid tissue (BALT) and larynx-associated lymphoid tissue of the lower respiratory tract are also considered to play a crucial role in the protection against infection via the pulmonary system (Alpar et al. 2005). Several preliminary clinical gene therapies using variety of formulations include lipoplexes and polyplexes to treat variety of diseases have been completed and more studies are in progress (Gorman et al. 1997, Goula et al. 1998, Li et al. 2000, Stern et al. 2003, Bivas-Benita et al. 2004, Tahara et al. 2007, Hyde et al. 2008).

Regardless of its eminent advantages for gene therapy, the lung epithelium has a number of features that discourage the uptake of formulations, for instance the tight junctions between cells, mucus secreted by goblet cells, and apical surface coated with cilia. All these features act as mechanical and diffusional barriers that expelled the particles from the lung and swallowed into the esophagus (Pouton and Seymour 2001). Most recently, the incorporation of positively charged moieties such as PEI to increase the loading of plasmid DNA as well as to formulate highly porous PLGA microspheres may offer interesting strategy for the pulmonary delivery of DNA (Alpar et al. 2005, Ho et al. 2008). The porous structure confers the porogenous formulation with less density and aerodynamic diameter, but increases its propensity for deep lung deposition (Edwards et al. 1997). Highly defined chitosan oligomers that formed polyplexes with DNA are more easily dissociated, and mediated higher and comparable gene expression than high–molecular-weight chitosan and 25 kDa PEI *in vivo* after intratracheal application in mice (Koping-Hoggard et al. 2004).

8.4 CONCLUDING REMARKS

Over the last two decades, gene therapy has brought human medical prospect into a new phase, whereby genetic defects in cells can be regulated and also a range of diseases can be prevented. DNA-based molecules are being employed to prevent, treat, and cure diseases by changing the expression of genes that are responsible for the pathology. Since its inception, plasmid DNA-mediated gene therapy has seen significant growth and brings fruitful clinical trials. However, the major underlying challenge is the development of a carrier system that can facilitate the safe and efficient delivery of plasmid DNA to the target site, followed by cellular uptake, internalization and processing, and the production of the therapeutic level of gene products for a desired period. Colloidal delivery systems are being recognized as a viable carrier to provide solution in gene therapy from the perspective of formulation and clinical problem, by improving the therapeutic outcomes. Colloidal carriers offer the opportunity to design surface properties to enable them traverse biological barriers such as skin, mucous membranes, and leaky vasculature. The smart design of the colloidal carrier may protect the DNA-based molecules from deleterious degradation, and may provide sustained release of payload in a therapeutically advantageous fashion. Successful gene delivery systems should be carefully tailored based on realistic application. Currently, the efficiency of genetic therapeutic that utilizes nonviral systems remains less encouraging, and will rely on major improvements in the design of effective carriers. A continuous effort focused on improving safety, feasibility, and efficacy of colloidal carriers for DNA gene therapy will enable the successful translation of these bench works to clinically meaningful therapeutic products, which will be suitable for widespread implementation in helping reduce morbidity and mortality.

ABBREVIATIONS

AIDS	Acquired immunodeficiency syndrome
Asp	Asparagine
AuNP	Gold nanoparticle
BALT	Bronchus-associated lymphoid tissue
DET	*N*-(2-aminoethyl)-2-aminoethyl
DNA	Deoxyribonucleic acid
DOTAP	1,2-bis(oleoyloxy)-3-trimethylammonio propane
DOTMA	1,2-dioleyloxypropyl-3-trimethyl ammonium bromide
GALT	Gut-associated lymphoid tissue
GFP	Green fluorescent protein
HIV	Human immunodeficiency virus
IgA	Immunoglobulin A
IgG	Immunoglobulin G
Lys	Lysine
MALT	Mucosa-associated lymphoid tissue
MNP	Magnetic nanoparticle
NALT	Nasopharyngeal lymphoid tissue
pDNA	Plasmid DNA
PAMAM	Polyamidoamine
PEG	Polyethylene glycol
PEI	Polyethylenimine
PLGA	Poly(lactide-*co*-glycolide)
PLL	Poly-L-lysine
PPA	Polyphosphoramidate
RNA	Ribonucleic acid
RSV	Respiratory syncytial virus
SPION	Superparamagnetic iron oxide nanoparticle

REFERENCES

Allison, A. C. and G. Gregoriadis. 1974. Liposomes as immunological adjuvants. *Nature* 252:252–258.

Alpar, H. O., S. Somavarapu, K. N. Atuah, and V. Bramwell. 2005. Biodegradable mucoadhesive particulates for nasal and pulmonary antigen and DNA delivery. *Advanced Drug Delivery Reviews* 57(3):411–430.

Bawarski, W. E., E. Chidlowsky, D. J. Bharali, and S. A. Mousa. 2008. Emerging nanopharmaceuticals. *Nanomedicine—Nanotechnology Biology and Medicine* 4(4):273–282.

Bergquist, C., E. L. Johansson, T. Lagergard, J. Holmgren, and A. Rudin. 1997. Intranasal vaccination of humans with recombinant cholera toxin B subunit induces systemic and local antibody responses in the upper respiratory tract and the vagina. *Infection and Immunity* 65(7):2676–2684.

Besseling, R., L. Isa, E. R. Weeks, and W. C. K. Poon. 2009. Quantitative imaging of colloidal flows. *Advances in Colloid and Interface Science* 146(1–2):1–17.

Bhavsar, M. D. and M. M. Amiji. 2007. Polymeric nano- and microparticle technologies for oral gene delivery. *Expert Opinion on Drug Delivery* 4(3):197–213.

Bivas-Benita, M., K. E. van Meijgaarden, K. L. M. C. Franken, H. E. Junginger, G. Borchard, T. H. M. Ottenhoff, and A. Geluk. 2004. Pulmonary delivery of chitosan-DNA nanoparticles enhances the immunogenicity of a DNA vaccine encoding HLA-A*0201-restricted T-cell epitopes of *Mycobacterium tuberculosis*. *Vaccine* 22(13–14):1609–1615.

Blaese, R. M., K. W. Culver, A. D. Miller, C. S. Carter, T. Fleisher, M. Clerici, G. Shearer et al. 1995. T-Lymphocyte-directed gene-therapy for ADA-SCID—Initial trial results after 4 years. *Science* 270(5235):475–480.

Borchard, G. 2001. Chitosans for gene delivery. *Advanced Drug Delivery Reviews* 52(2):145–150.

Boussif, O., F. Lezoualch, M. A. Zanta, M. D. Mergny, D. Scherman, B. Demeneix, and J. P. Behr. 1995. A versatile vector for gene and oligonucleotide transfer into cells in culture and *in-vivo*: Polyethylenimine. *Proceedings of the National Academy of Sciences of the United States of America* 96(16):7297–7301.

Bowman, K. and K. W. Leong. 2006. Chitosan nanoparticles for oral drug and gene delivery. *International Journal of Nanomedicine* 1(2):117–128.

Boyaka, P. N., A. Tafaro, R. Fischer, K. Fujihashi, E. Jirillo, and J. R. McGhee. 2003. Therapeutic manipulation of the immune system: Enhancement of innate and adaptive mucosal immunity. *Current Pharmaceutical Design* 9(24):1965–1972.

Brigham, K. L., B. Meyrick, B. Christman, M. Magnuson, G. King, and L. C. Berry. 1989. *In vivo* transfection of murine lungs with a functioning prokaryotic gene using a liposome vehicle. *American Journal of the Medical Sciences* 298(4):278–281.

Burgess, D. J. 2006. Colloids and colloid drug delivery system. In *Encyclopedia of Pharmaceutical Technology*, eds. J. Swarbrick and J. C. Boylan, Informa Healthcare USA, Inc., New York.

Canonico, A. E., J. T. Conary, B. O. Meyrick, and K. L. Brigham. 1994. Aerosol and intravenous transfection of human alpha-1-antitrypsin gene to lungs of rabbits. *American Journal of Respiratory Cell and Molecular Biology* 10(1):24–29.

Chang, C. W., D. H. Choi, W. J. Kim, J. W. Yockman, L. V. Christensen, Y. H. Kim, and S. W. Kim. 2007. Non-ionic amphiphilic biodegradable PEG-PLGA-PEG copolymer enhances gene delivery efficiency in rat skeletal muscle. *Journal of Controlled Release* 118(2):245–253.

Chen, H. M., V. Torchilin, and R. Langer. 1996. Lectin-bearing polymerized liposomes as potential oral vaccine carriers. *Pharmaceutical Research* 13(9):1378–1383.

Chen, J., W. L. Yang, G. Li, J. Qian, J. L. Xue, S. K. Fu, and D. R. Lu. 2004. Transfection of mEpo gene to intestinal epithelium *in vivo* mediated by oral delivery of chitosan-DNA nanoparticles. *World Journal of Gastroenterology* 10(1):112–116.

Chew, J. L., C. B. Wolfowicz, H. Q. Mao, K. W. Leong, and K. Y. Chua. 2003. Chitosan nanoparticles containing plasmid DNA encoding house dust mite allergen, Der p 1 for oral vaccination in mice. *Vaccine* 21(21–22):2720–2729.

Choi, J. S., D. K. Joo, C. H. Kim, K. Kim, and J. S. Park. 2000. Synthesis of a barbell-like triblock copolymer, poly(L-lysine) dendrimer-block-poly(ethylene glycol)-block-poly(L-lysine) dendrimer, and its self-assembly with plasmid DNA. *Journal of the American Chemical Society* 122(3):474–480.

Chorny, M., B. Polyak, I. S. Alferiev, K. Walsh, G. Friedman, and R. J. Levy. 2007. Magnetically driven plasmid DNA delivery with biodegradable polymeric nanoparticles. *FASEB Journal* 21(10):2510–2519.

Clark, A. R. and M. Egan. 1994. Modeling the deposition of inhaled powdered drug aerosols. *Journal of Aerosol Science* 25(1):175–186.

Connor, E. E., J. Mwamuka, A. Gole, C. J. Murphy, and M. D. Wyatt. 2005. Gold nanoparticles are taken up by human cells but do not cause acute cytotoxicity. *Small* 1:325–327.

Csaba, N., A. Sanchez, and M. J. Alonso. 2006. PLGA: Poloxamer and PLGA: Poloxamine blend nanostructures as carriers for nasal gene delivery. *Journal of Controlled Release* 113(2):164–172.

Davis, S. S. 2001. Nasal vaccines. *Advanced Drug Delivery Reviews* 51(1–3):21–42.

Davis, S. S. and L. Illum. 2003. Absorption enhancers for nasal drug delivery. *Clinical Pharmacokinetics* 42(13):1107–1128.

de Jonge, M. I., H. J. Hamstra, W. Jiskoot, P. Roholl, N. A. Williams, J. Dankert, L. van Alphen, and P. van der Ley. 2004. Intranasal immunisation of mice with liposomes containing recombinant meningococcal OpaB and OpaJ proteins. *Vaccine* 22(29–30):4021–4028.

de Lima, M. C. P., S. Simoes, P. Pires, H. Faneca, and N. Duzgunes. 2001. Cationic lipid-DNA complexes in gene delivery: From biophysics to biological applications. *Advanced Drug Delivery Reviews* 47(2–3):277–294.

Di Colo, G., Y. Zambito, and C. Zaino. 2008. Polymeric enhancers of mucosal epithelia permeability: Synthesis, transepithelial penetration-enhancing properties, mechanism of action, safety issues. *Journal of Pharmaceutical Sciences* 97(5):1652–1680.

Dubensky, T. W., M. A. Liu, and J. B. Ulmer. 2000. Delivery systems for gene-based vaccines. *Molecular Medicine* 6(9):723–732.

Dufes, C., I. F. Uchegbu, and A. G. Schatzlein. 2005. Dendrimers in gene delivery. *Advanced Drug Delivery Reviews* 57(15):2177–2202.

Dupuis, M., K. Denis-Mize, C. Woo, C. Goldbeck, M. J. Selby, M. C. Chen, G. R. Otten, J. B. Ulmer, J. J. Donnelly, G. Ott, and D. M. McDonald. 2000. Distribution of DNA vaccines determines their immunogenicity after intramuscular injection in mice. *Journal of Immunology* 165(5):2850–2858.

Edelstein, M. 2009. *Gene Therapy Clinical Trials Worldwide*. Wiley Inter Science, September 2008 [cited 30 March 2009]. Available from http://www.wiley.co.uk/genmed/clinical/

Edelstein, M. L., M. R. Abedi, J. Wixon, and R. M. Edelstein. 2004. Gene therapy clinical trials worldwide 1989–2004: An overview. *Journal of Gene Medicine* 6(6):597–602.

Edwards, D. A., J. Hanes, G. Caponetti, J. Hrkach, A. BenJebria, M. L. Eskew, J. Mintzes, D. Deaver, N. Lotan, and R. Langer. 1997. Large porous particles for pulmonary drug delivery. *Science* 276(5320):1868–1871.

Elfinger, M., S. Uezguen, and C. Rudolph. 2008. Nanocarriers for gene delivery: Polymer structure, targeting ligands and controlled-release devices. *Current Nanoscience* 4(4):322–353.

Felgner, P. L. 1996. Improvements in cationic liposomes for *in vivo* gene transfer. *Human Gene Therapy* 7(15):1791–1793.

Felgner, P. L., T. R. Gadek, M. Holm, R. Roman, H. W. Chan, M. Wenz, J. P. Northrop, G. M. Ringold, and M. Danielsen. 1987. Lipofection: A highly efficient, lipid-mediated DNA-transfection procedure. *Proceedings of the National Academy of Sciences of the United States of America* 84(21):7413–7417.

Fukunaka, Y., K. Iwanaga, K. Morimoto, M. Kakemi, and Y. Tabata. 2002. Controlled release of plasmid DNA from cationized gelatin hydrogels based on hydrogel degradation. *Journal of Controlled Release* 80(1–3):333–343.

Ghosh, P. S., C. K. Kim, G. Han, N. S. Forbes, and V. M. Rotello. 2008. Efficient gene delivery vectors by tuning the surface charge density of amino acid-functionalized gold nanoparticles. *ACS Nano* 2(11):2213–2218.

Godbey, W. T., K. K. Wu, and A. G. Mikos. 2001. Poly(ethylenimine)-mediated gene delivery affects endothelial cell function and viability. *Biomaterials* 22(5):471–480.

Gorman, C. M., M. Aikawa, B. Fox, E. Fox, C. Lapuz, B. Michaud, H. Nguyen, E. Roche, T. Sawa, and J. P. WienerKronish. 1997. Efficient *in vivo* delivery of DNA to pulmonary cells using the novel lipid EDMPC. *Gene Therapy* 4(9):983–992.

Goula, D., C. Benoist, S. Mantero, G. Merlo, G. Levi, and B. A. Demeneix. 1998. Polyethylenimine-based intravenous delivery of transgenes to mouse lung. *Gene Therapy* 5(9):1291–1295.

Gould-Fogerite, S., M. T. Kheiri, F. Zhang, Z. Wang, A. J. Scolpino, E. Feketeova, M. Canki, and R. J. Mannino. 1998. Targeting immune response induction with cochleate and liposome-based vaccines. *Advanced Drug Delivery Reviews* 32:273–287.

Goya, A. K., K. Khatri, N. Mishra, and S. P. Vyas. 2008. New patents on mucosal delivery of vaccines. *Expert Opinion on Therapeutic Patents* 18(11):1271–1288.

Gupta, M. and A. K. Gupta. 2004. Hydrogel pullulan nanoparticles encapsulating pBUDLacZ plasmid as an efficient gene delivery carrier. *Journal of Controlled Release* 99(1):157–166.

Gupta, A. K. and M. Gupta. 2005. Cytotoxicity suppression and cellular uptake enhancement of surface modified magnetic nanoparticles. *Biomaterials* 26(13):1565–1573.

Gupta, S. and S. P. Moulik. 2008. Biocompatible microemulsions and their prospective uses in drug delivery. *Journal of Pharmaceutical Sciences* 97(1):22–45.

Hacein-Bey-Abina, S., C. von Kalle, M. Schmidt, F. Le Deist, N. Wulffraat, E. McIntyre, I. Radford, J. L. Villeval, C. C. Fraser, M. Cavazzana-Calvo, and A. Fischer. 2003. A serious adverse event after successful gene therapy for x-linked severe combined immunodeficiency. *New England Journal of Medicine* 348(3):255–256.

He, X. W., F. Wang, L. Jiang, J. Li, S. K. Liu, Z. Y. Xiao, X. Q. Jin et al. 2005. Induction of mucosal and systemic immune response by single-dose oral immunization with biodegradable microparticles containing DNA encoding HBsAg. *Journal of General Virology* 86:601–610.

Helson, R., W. Olszewska, M. Singh, J. Z. Megede, J. A. Melero, D. O. Hagan, and P. J. M. Openshaw. 2008. Polylactide-*co*-glycolide (PLG) microparticles modify the immune response to DNA vaccination. *Vaccine* 26(6):753–761.

Ho, J., H. T. Wang, and G. M. Forde. 2008. Process considerations related to the microencapsulation of plasmid DNA via ultrasonic atomization. *Biotechnology and Bioengineering* 101(1):172–181.

Hofland, H. E. J., L. Shephard, and S. M. Sullivan. 1996. Formation of stable cationic lipid/DNA complexes for gene transfer. *Proceedings of the National Academy of Sciences of the United States of America* 93(14):7305–7309.

Hofland, H. E. J., D. Nagy, J. J. Liu, K. Spratt, Y. L. Lee, O. Danos, and S. M. Sullivan. 1997. *In vivo* gene transfer by intravenous administration of stable cationic lipid DNA complex. *Pharmaceutical Research* 14(6):742–749.

Holmgren, J. and C. Czerkinsky. 2005. Mucosal immunity and vaccines. *Nature Medicine* 11(4):S45–S53.

Huth, S., J. Lausier, S. W. Gersting, C. Rudolph, C. Plank, U. Welsch, and J. Rosenecker. 2004. Insights into the mechanism of magnetofection using PEI-based magnetofectins for gene transfer. *Journal of Gene Medicine* 6(8):923–936.

Hyde, S. C., I. A. Pringle, S. Abdullah, A. E. Lawton, L. A. Davies, A. Varathalingam, G. Nunez-Alonso et al. 2008. CpG-free plasmids confer reduced inflammation and sustained pulmonary gene expression. *Nature Biotechnology* 26(5):549–551.

Iqbal, M., W. Lin, I. Jabbal-Gill, S. S. Davis, M. W. Steward, and L. Illum. 2003. Nasal delivery of chitosan-DNA plasmid expressing epitopes of respiratory syncytial virus (RSV) induces protective CTL responses in BALB/c mice. *Vaccine* 21(13–14):1478–1485.

Jain, S. and S. P. Vyas. 2006. Mannosylated niosomes as adjuvant-carrier system for oral mucosal immunization. *Journal of Liposome Research* 16(4):331–345.

Jiang, X., H. Dai, C. Y. Ke, X. Mo, M. S. Torbenson, Z. P. Li, and H. Q. Mao. 2007. PEG-b-PPA/DNA micelles improve transgene expression in rat liver through intrabiliary infusion. *Journal of Controlled Release* 122(3):297–304.

Jilek, S., H. P. Merkle, and E. Walter. 2005. DNA-loaded biodegradable microparticles as vaccine delivery systems and their interaction with dendritic cells. *Advanced Drug Delivery Reviews* 57(3):377–390.

Jones, D. H., S. Corris, S. McDonald, J. C. S. Clegg, and G. H. Farrar. 1996. Poly(DL-lactide-*co*-glycolide)-encapsulated plasmid DNA elicits systemic and mucosal antibody responses to encoded protein after oral administration. Paper read at *International Meeting on Nucleic Acid Vaccines for the Prevention of Infectious Diseases*, February 05–07, Bethesda, MD.

Kabanov, A. V. and S. V. Vinogradov. 2009. Nanogels as pharmaceutical carriers: Finite networks of infinite capabilities. *Angewandte Chemie—International Edition* 48(30):5418–5429.

Kakizawa, Y. and K. Kataoka. 2002. Block copolymer micelles for delivery of gene and related compounds. *Advanced Drug Delivery Reviews* 54:203–222.

Kasper, F. K., S. K. Seidlits, A. Tang, R. S. Crowther, D. H. Carney, M. A. Barry, and A. G. Mikos. 2005. *In vitro* release of plasmid DNA from oligo(poly(ethylene glycol) fumarate) hydrogels. *Journal of Controlled Release* 104(3):521–539.

Kataoka, K., A. Harada, and Y. Nagasaki. 2001. Block copolymer micelles for drug delivery: Design, characterization and biological significance. *Advanced Drug Delivery Reviews* 47(1):113–131.

Katz, D. E., A. J. DeLorimier, M. K. Wolf, E. R. Hall, F. J. Cassels, J. E. van Hamont, R. Newcomer, M. A. Davachi, D. N. Taylor, and C. E. McQueen. 2003. Oral immunization of adult volunteers with microencapsulated enterotoxigenic *Escherichia coli* (ETEC) CS6 antigen. *Vaccine* 21(5–6):341–346.

Kersten, G. F. A. and D. J. A. Crommelin. 2003. Liposomes and ISCOMs. *Vaccine* 21(9–10):915–920.

Kievit, F. M., O. Veiseh, N. Bhattarai, C. Fang, J. W. Gunn, D. Lee, R. G. Ellenbogen, J. M. Olson, and M. Q. Zhang. 2009. PEI-PEG-chitosan-copolymer-coated iron oxide nanoparticles for safe gene delivery: Synthesis, complexation, and transfection. *Advanced Functional Materials* 19(14):2244–2251.

Kim, A., D. M. Checkla, P. Dehazya, and W. L. Chen. 2003. Characterization of DNA-hyaluronan matrix for sustained gene transfer. *Journal of Controlled Release* 90(1):81–95.

Kippax, P. and J. Fracassi. 2003. Particle size characterisation in nasal sprays and aerosols. In *LabPlus International* February/March.

Koping-Hoggard, M., K. M. Varum, M. Issa, S. Danielsen, B. E. Christensen, B. T. Stokke, and P. Artursson. 2004. Improved chitosan-mediated gene delivery based on easily dissociated chitosan polyplexes of highly defined chitosan oligomers. *Gene Therapy* 11(19):1441–1452.

Kozlowski, P. A., S. B. Williams, R. M. Lynch, T. P. Flanigan, R. R. Patterson, S. Cu-Uvin, and M. R. Neutra. 2002. Differential induction of mucosal and systemic antibody responses in women after nasal, rectal, or vaginal immunization: Influence of the menstrual cycle. *Journal of Immunology* 169(1):566–574.

KukowskaLatallo, J. F., A. U. Bielinska, J. Johnson, R. Spindler, D. A. Tomalia, and J. R. Baker. 1996. Efficient transfer of genetic material into mammalian cells using Starburst polyamidoamine dendrimers. *Proceedings of the National Academy of Sciences of the United States of America* 93(10):4897–4902.

Kumar, M., A. K. Behera, R. F. Lockey, J. Zhang, G. Bhullar, C. P. De La Cruz, L. C. Chen, K. W. Leong, S. K. Huang, and S. S. Mohapatra. 2002. Intranasal gene transfer by chitosan-DNA nanospheres protects BALB/c mice against acute respiratory syncytial virus infection. *Human Gene Therapy* 13(12):1415–1425.

Kumar, M. N. V. R., U. Bakowsky, and C. M. Lehr. 2004. Nanoparticles as non-viral transfection agents. In *Nanobiotechnology: Concepts, Applications and Perspectives*, eds. C. M. Niemeyer and C. A. Mirkin, Wiley-VCH, Weinheim, Germany.

Kushibiki, T., R. Tomoshige, Y. Fukunaka, M. Kakemi, and Y. Tabata. 2003. *In vivo* release and gene expression of plasmid DNA by hydrogels of gelatin with different cationization extents. *Journal of Controlled Release* 90(2):207–216.

Kushibiki, T., R. Tomoshige, K. Iwanaga, M. Kakemi, and Y. Tabata. 2006. Controlled release of plasmid DNA from hydrogels prepared from gelatin cationized by different amine compounds. *Journal of Controlled Release* 112(2):249–256.

Lambert, J. S., M. Keefer, M. J. Mulligan, D. Schwartz, J. Mestecky, K. Weinhold, C. Smith et al. 2001. A Phase I safety and immunogenicity trial of UBI (R) microparticulate monovalent HIV-1 MN oral peptide immunogen with parenteral boost in HIV-1 seronegative human subjects. *Vaccine* 19(23–24):3033–3042.

Ledley, F. D. 1996. Pharmaceutical approach to somatic gene therapy. *Pharmaceutical Research* 13(11):1595–1614.

Lemieux, P., N. Guerin, G. Paradis, R. Proulx, L. Chistyakova, A. Kabanov, and V. Alakhov. 2000a. A combination of poloxamers increases gene expression of plasmid DNA in skeletal muscle. *Gene Therapy* 7(11):986–991.

Lemieux, P., S. V. Vinogradov, C. L. Gebhart, N. Guerin, G. Paradis, H. K. Nguyen, B. Ochietti et al. 2000b. Block and graft copolymers and Nanogel (TM) copolymer networks for DNA delivery into cell. *Journal of Drug Targeting* 8(2):91–105.

Li, S., Y. D. Tan, E. Viroonchatapan, B. R. Pitt, and L. Huang. 2000. Targeted gene delivery to pulmonary endothelium by anti-PECAM antibody. *American Journal of Physiology—Lung Cellular and Molecular Physiology* 278(3):L504–L511.

Li, Z. H., W. Ning, J. M. Wang, A. Choi, P. Y. Lee, P. Tyagi, and L. Huang. 2003. Controlled gene delivery system based on thermosensitive biodegradable hydrogel. *Pharmaceutical Research* 20(6):884–888.

Li, P. C., D. Li, L. X. Zhang, G. P. Li, and E. K. Wang. 2008. Cationic lipid bilayer coated gold nanoparticles-mediated transfection of mammalian cells. *Biomaterials* 29(26):3617–3624.

Li, D., P. C. Li, G. P. Li, J. Wang, and E. K. Wang. 2009a. The effect of nocodazole on the transfection efficiency of lipid-bilayer coated gold nanoparticles. *Biomaterials* 30(7):1382–1388.

Li, G. P., Z. G. Liu, B. Liao, and N. S. Zhong. 2009b. Induction of Th1-type immune response by chitosan nanoparticles containing plasmid DNA encoding house dust mite allergen Der p 2 for oral vaccination in mice. *Cellular & Molecular Immunology* 6(1):45–50.

Lindner, L. H., R. Brock, D. Arndt-Jovin, and H. Eibl. 2006. Structural variation of cationic lipids: Minimum requirement for improved oligonucleotide delivery into cells. *Journal of Controlled Release* 110(2):444–456.

Liu, M. A. 2003. DNA vaccines: A review. *Journal of Internal Medicine* 253(4):402–410.

Liu, F., H. Qi, L. Huang, and D. Liu. 1997. Factors controlling the efficiency of cationic lipid-mediated transfection *in vivo* via intravenous administration. *Gene Therapy* 4(6):517–523.

Liu, M. A., B. Wahren, and G. B. K. Hedestam. 2006. DNA vaccines: Recent developments and future possibilities. *Human Gene Therapy* 17(11):1056–1061.

Lu, D. M. and A. J. Hickey. 2007. Pulmonary vaccine delivery. *Expert Review of Vaccines* 6(2):213–226.

Lu, Q. L., G. Bou-Gharios, and T. A. Partridge. 2003. Non-viral gene delivery in skeletal muscle: A protein factory. *Gene Therapy* 10(2):131–142.

Lungwitz, U., M. Breunig, T. Blunk, and A. Gopferich. 2005. Polyethylenimine-based non-viral gene delivery systems. *European Journal of Pharmaceutics and Biopharmaceutics* 60(2):247–266.

Luo, D. 2005. Nanotechnology and DNA delivery. *MRS Bulletin* 30(9):654–658.

MacGregor, R. R., J. D. Boyer, K. E. Ugen, K. E. Lacy, S. J. Gluckman, M. L. Bagarazzi, M. A. Chattergoon et al. 1998. First human trial of a DNA-based vaccine for treatment of human immunodeficiency virus type 1 infection: Safety and host response. *Journal of Infectious Diseases* 178(1):92–100.

MacLaughlin, F. C., R. J. Mumper, J. J. Wang, J. M. Tagliaferri, I. Gill, M. Hinchcliffe, and A. P. Rolland. 1998. Chitosan and depolymerized chitosan oligomers as condensing carriers for *in vivo* plasmid delivery. *Journal of Controlled Release* 56(1–3):259–272.

Mannino, R. J., E. S. Allebach, and W. A. Strohl. 1979. Encapsulation of high molecular-weight DNA in large unilamellar phospholipid-vesicles: Dependence on the size of the DNA. *FEBS Letters* 101(2):229–232.

May, S., D. Harries, and A. Ben-Shaul. 2000. The phase behavior of cationic lipid-DNA complexes. *Biophysical Journal* 78(4):1681–1697.

McNeil, S. E. and Y. Perrie. 2006. Gene delivery using cationic liposomes. *Expert Opinion on Therapeutic Patents* 16(10):1371–1382.

Megeed, Z., M. Haider, D. Q. Li, B. W. O'Malley, J. Cappello, and H. Ghandehari. 2004. *In vitro* and *in vivo* evaluation of recombinant silk-elastinlike hydrogels for cancer gene therapy. *Journal of Controlled Release* 94(2–3):433–445.

Mirchamsy, H., H. Manhouri, M. Hamedi, P. Ahourai, G. Fateh, and Z. Hamzeloo. 1996. Stimulating role of toxoids-laden liposomes in oral immunization against diphtheria and tetanus infections. *Biologicals* 24(4):343–350.

Miyata, K., S. Fukushima, N. Nishiyama, Y. Yamasaki, and K. Kataoka. 2007. PEG-based block catiomers possessing DNA anchoring and endosomal escaping functions to form polyplex micelles with improved stability and high transfection efficacy. *Journal of Controlled Release* 122(3):252–260.

Moore, A., E. Marecos, A. Bogdanov, and R. Weissleder. 2000. Tumoral distribution of long-circulating dextran-coated iron oxide nanoparticles in a rodent model. *Radiology* 214:568–574.

Mumper, R. J. and A. P. Rolland. 1998. Plasmid delivery to muscle: Recent advances in polymer delivery systems. *Advanced Drug Delivery Reviews* 30(1–3):151–172.

Mumper, R. J., J. G. Duguid, K. Anwer, M. K. Barron, H. Nitta, and A. P. Rolland. 1996. Polyvinyl derivatives as novel interactive polymers for controlled gene delivery to muscle. *Pharmaceutical Research* 13(5):701–709.

Mumper, R. J., J. J. Wang, S. L. Klakamp, H. Nitta, K. Anwer, F. Tagliaferri, and A. P. Rolland. 1998. Protective interactive noncondensing (PINC) polymers for enhanced plasmid distribution and expression in rat skeletal muscle. *Journal of Controlled Release* 52(1–2):191–203.

Nakanishi, M. and A. Noguchi. 2001. Confocal and probe microscopy to study gene transfection mediated by cationic liposomes with a cationic cholesterol derivative. *Advanced Drug Delivery Reviews* 52:197–207.

Newman, S. P., G. R. Pitcairn, and R. N. Dalby. 2004. Drug delivery to the nasal cavity: *In vitro* and *in vivo* assessment. *Critical Reviews in Therapeutic Drug Carrier Systems* 21(1):21–66.

Nguyen, D. N., J. J. Green, J. M. Chan, R. Langer, and D. G. Anderson. 2009. Polymeric materials for gene delivery and DNA vaccination. *Advanced Materials* 21(8):847–867.

Nishiyama, N. and K. Kataoka. 2006. Current state, achievements, and future prospects of polymeric micelles as nanocarriers for drug and gene delivery. *Pharmacology & Therapeutics* 112(3):630–648.

O'Hagan, D. T., M. Singh, and J. B. Ulmer. 2004. Microparticles for the delivery of DNA vaccines. *Immunological Reviews* 199(1):191–200.

O'Hagan, D. T., M. Singh, and J. B. Ulmer. 2006. Microparticle-based technologies for vaccines. *Methods* 40(1):10–19.

Oishi, M. and Y. Nagasaki. 2007. Synthesis, characterization, and biomedical applications of core-shell-type stimuli-responsive nanogels: Nanogel composed of poly[2-(*N*,*N*-diethylamino)ethyl methacrylate] core and PEG tethered chains. *Reactive & Functional Polymers* 67:1311–1329.

Oishi, M., K. Kataoka, and Y. Nagasaki. 2006. pH-Responsive three-layered PEGylated polyplex micelle based on a lactosylated ABC triblock copolymer as a targetable and endosome-disruptive nonviral gene vector. *Bioconjugate Chemistry* 17(3):677–688.

Oishi, M., H. Hayashi, K. Itaka, K. Kataoka, and Y. Nagasaki. 2007. pH-responsive PEGylated nanogels as targetable and low invasive endosomolytic agents to induce the enhanced transfection efficiency of nonviral gene vectors. *Colloid and Polymer Science* 285(9):1055–1060.

Opanasopit, P., M. Nishikawa, and M. Hashida. 2002. Factors affecting drug and gene delivery: Effects of interaction with blood components. *Critical Reviews in Therapeutic Drug Carrier Systems* 19(3):191–233.

Pal, R. 1996. Multiple O/W/O emulsion rheology. *Langmuir* 12(9):2220–2225.

Panyam, J. and V. Labhasetwar. 2003. Biodegradable nanoparticles for drug and gene delivery to cells and tissue. *Advanced Drug Delivery Reviews* 55(3):329–347.

Partidos, C. D. 2000. Intranasal vaccines: Forthcoming challenges. *Pharmaceutical Science & Technology Today* 3(8):273–281.

Patil, S. D., D. G. Rhodes, and D. J. Burgess. 2005. DNA-based therapeutics and DNA delivery systems: A comprehensive review. *AAPS Journal* 7(1):E61–E77.

Perrie, Y., P. M. Frederik, and G. Gregoriadis. 2001a. Liposome-mediated DNA vaccination: The effect of vesicle composition. *Vaccine* 19:3301–3310.

Perrie, Y., M. Obrenovic, D. McCarthy, and G. Gregoriadis. 2001b. Liposome (Lipodine (TM))-mediated DNA vaccination by the oral route. Paper read at *5th International Conference on Liposome Advances*, December 17–21, London, U.K.

Petersen, H., K. Kunath, A. L. Martin, S. Stolnik, C. J. Roberts, M. C. Davies, and T. Kissel. 2002. Star-shaped poly(ethylene glycol)-block-polyethylenimine copolymers enhance DNA condensation of low molecular weight polyethylenimines. *Biomacromolecules* 3(5):926–936.

Pitard, B., H. Pollard, O. Agbulut, O. Lambert, J. T. Vilquin, Y. Cherel, J. Abadie, J. L. Samuel, J. L. Rigaud, S. Menoret, I. Anegon, and D. Escande. 2002. A nonionic amphiphile agent promotes gene delivery *in vivo* to skeletal and cardiac muscles. *Human Gene Therapy* 13(14):1767–1775.

Plautz, G. E., Z. Y. Yang, B. Y. Wu, X. Gao, L. Huang, and G. J. Nabel. 1993. Immunotherapy of malignancy by *in vivo* gene-transfer into tumors. *Proceedings of the National Academy of Sciences of the United States of America* 90(10):4645–4649.

Pouton, C. W. and L. W. Seymour. 2001. Key issues in non-viral gene delivery. *Advanced Drug Delivery Reviews* 46(1–3):187–203.

Prieto, J., M. Herraiz, B. Sangro, C. Qian, G. Mazzolini, I. Melero, and J. Ruiz. 2002. The promise of gene therapy in gastrointestinal and liver diseases. Paper read at *European Institute of Healthcare Gastroenterology Symposium*, April, Barcelona, Spain.

Raper, S. E., N. Chirmule, F. S. Lee, N. A. Wivel, A. Bagg, G. P. Gao, J. M. Wilson, and M. L. Batshaw. 2003. Fatal systemic inflammatory response syndrome in a ornithine transcarbamylase deficient patient following adenoviral gene transfer. *Molecular Genetics and Metabolism* 80(1–2):148–158.

Ratanamart, J. and J. A. M. Shaw. 2006. Plasmid-mediated muscle-targeted gene therapy for circulating therapeutic protein replacement: A tale of the tortoise and the hare? *Current Gene Therapy* 6(1):93–110.

Richard, P., H. Pollard, C. Lanctin, M. Bello-Roufai, L. Desigaux, D. Escande, and B. Pitard. 2005. Inducible production of erythropoietin using intramuscular injection of block copolymer/DNA formulation. *Journal of Gene Medicine* 7(1):80–86.

Riera, M., M. Chillon, J. M. Aran, J. M. Cruzado, J. Torras, J. M. Grinyo, and C. Fillat. 2004. Intramuscular SP1017-formulated DNA electrotransfer enhances transgene expression and distributes hHGF to different rat tissues. *Journal of Gene Medicine* 6(1):111–118.

Rolland, A. 2005. Gene medicines: The end of the beginning? *Advanced Drug Delivery Reviews* 57(5):669–673.

Romano, G., P. Mitcheli, C. Pacilio, and A. Giordano. 2000. Latest developments in gene transfer technology: Achievements, perspectives, and controversies over therapeutic applications. *Stem Cells* 18(1):19–39.

Rosi, N. L., D. A. Giljohann, C. S. Thaxton, A. K. R. Lytton-Jean, M. S. Han, and C. A. Mirkin. 2006. Oligonucleotide-modified gold nanoparticles for intracellular gene regulation. *Science* 312:1027–1030.

Roy, K., H. Q. Mao, S. K. Huang, and K. W. Leong. 1999. Oral gene delivery with chitosan-DNA nanoparticles generates immunologic protection in a murine model of peanut allergy. *Nature Medicine* 5(4):387–391.

Sandhu, K. K., C. M. McIntosh, J. M. Simard, S. W. Smith, and V. M. Rotello. 2002. Gold nanoparticle-mediated transfection of mammalian cells. *Bioconjugate Chemistry* 13(1):3–6.

Schillinger, U., T. Brill, C. Rudolph, S. Huth, S. Gersting, F. Krotz, J. Hirschberger, C. Bergemann, and C. Plank. 2005. Advances in magnetofection: Magnetically guided nucleic acid delivery. *Journal of Magnetism and Magnetic Materials* 293:501–508.

Segura, T., P. H. Chung, and L. D. Shea. 2005. DNA delivery from hyaluronic acid-collagen hydrogels via a substrate-mediated approach. *Biomaterials* 26(13):1575–1584.

Smith, J., Y. L. Zhang, and R. Niven. 1997. Toward development of a non-viral gene therapeutic. *Advanced Drug Delivery Reviews* 26(2–3):135–150.

Stern, M., K. Ulrich, D. M. Geddes, and E. W. F. W. Alton. 2003. Poly (D, L-lactide-*co*-glycolide)/DNA microspheres to facilitate prolonged transgene expression in airway epithelium *in vitro*, *ex vivo* and *in vivo*. *Gene Therapy* 10(16):1282–1288.

Tahara, K., H. Yamamoto, H. Takeuchi, and Y. Kawashima. 2007. Development of gene delivery system using PLGA nanospheres. *Yakugaku Zasshi—Journal of the Pharmaceutical Society of Japan* 127:1541–1548.

Tamilvanan, S. 2004. Oil-in-water lipid emulsions: Implications for parenteral and ocular delivering systems. *Progress in Lipid Research* 43(6):489–533.

Tang, G. P., J. M. Zeng, S. J. Gao, Y. X. Ma, L. Shi, Y. Li, H. P. Too, and S. Wang. 2003. Polyethylene glycol modified polyethylenimine for improved CNS gene transfer: Effects of PEGylation extent. *Biomaterials* 24(13):2351–2362.

Thierry, A. R., E. Vives, J. P. Richard, P. Prevot, C. Martinand-Mari, I. Robbins, and B. Lebleu. 2003. Cellular uptake and intracellular fate of antisense oligonucleotides. *Current Opinion in Molecular Therapeutics* 5(2):133–138.

Thomas, M. and A. M. Klibanov. 2003. Conjugation to gold nanoparticles enhances polyethylenimine's transfer of plasmid DNA into mammalian cells. *Proceedings of The National Academy of Sciences of the United States of America* 100:9138–9143.

Turker, S., E. Onur, and Y. Ozer. 2004. Nasal route and drug delivery systems. *Pharmacy World & Science* 26(3):137–142.

Ulrich, A. S. 2002. Biophysical aspects of using liposomes as delivery vehicles. *Bioscience Reports* 22(2):129–150.

Vadolas, J., J. K. Davies, P. J. Wright, and R. A. Strugnell. 1995. Intranasal immunization with liposomes induces strong mucosal immune-responses in mice. *European Journal of Immunology* 25(4):969–975.

Venkatesan, N. and S. P. Vyas. 2000. Polysaccharide coated liposomes for oral immunization: Development and characterization. *International Journal of Pharmaceutics* 203(1–2):169–177.

Vila, A., A. Sanchez, C. Evora, I. Soriano, J. L. V. Jato, and M. J. Alonso. 2004. PEG-PLA nanoparticles as carriers for nasal vaccine delivery. *Journal of Aerosol Medicine—Deposition Clearance and Effects in the Lung* 17(2):174–185.

Vinogradov, S. V. 2006. Colloidal microgels in drug delivery applications. *Current Pharmaceutical Design* 12(36):4703–4712.

Wagner, E., M. Cotten, R. Foisner, and M. L. Birnstiel. 1991. Transferrin polycation DNA complexes: The effect of polycations on the structure of the complex and DNA delivery to cells. *Proceedings of the National Academy of Sciences of the United States of America* 88(10):4255–4259.

Wang, H., Y. Chen, X. Y. Li, and Y. Liu. 2007. Synthesis of oligo(ethylenediamino)-beta-cyclodextrin modified gold nanoparticle as a DNA concentrator. *Molecular Pharmaceutics* 4(2):189–198.

Wheeler, C. J., P. L. Felgner, Y. J. Tsai, J. Marshall, L. Sukhu, S. G. Doh, J. Hartikka et al. 1996. A novel cationic lipid greatly enhances plasmid DNA delivery and expression in mouse lung. *Proceedings of the National Academy of Sciences of the United States of America* 93(21):11454–11459.

Wieland, J. A., T. L. Houchin-Ray, and L. D. Shea. 2007. Non-viral vector delivery from PEG-hyaluronic acid hydrogels. *Journal of Controlled Release* 120(3):233–241.

Wigglesworth, T. J., F. Teixeira, F. Axthelm, S. Eisler, N. S. Csaba, H. P. Merkle, W. Meier, and F. Diederich. 2008. Dendronised block copolymers as potential vectors for gene transfection. *Organic & Biomolecular Chemistry* 6(11):1905–1911.

Wolfert, M. A. and L. W. Seymour. 1996. Atomic force microscopic analysis of the influence of the molecular weight of poly(L)lysine on the size of polyelectrolyte complexes formed with DNA. *Gene Therapy* 3(3):269–273.

Xiang, J. J., J. Q. Tang, S. G. Zhu, X. M. Nie, H. B. Lu, S. R. Shen, X. L. Li, K. Tang, M. Zhou, and G. Y. Li. 2003. IONP-PLL: A novel non-viral vector for efficient gene delivery. *The Journal of Gene Medicine* 5(9):803–817.

Zauner, W., M. Ogris, and E. Wagner. 1998. Polylysine-based transfection systems utilizing receptor-mediated delivery. *Advanced Drug Delivery Reviews* 30(1–3):97–113.

9 Affinity Magnetic Beads for Chemical Biology and Medicine

Yosuke Iizumi, Yasuaki Kabe, Mamoru Hatakeyama,
Satoshi Sakamoto, and Hiroshi Handa

CONTENTS

9.1 INTRODUCTION

Biological events in cells are caused by specific interactions among a wide variety of biomolecules, including proteins, nucleotides, chemicals, and so on. Chemical biology is focused on elucidating the mechanism and network of biological reactions by using chemical compounds. Several chemical compounds show specific inhibition or activation against target proteins, mediated by a specific binding to the functional domain of proteins. Therefore, the identification of target proteins for chemical compounds provides us with valuable information about biological and biochemical reactions, and also a new insight for developing therapeutic agents by an analysis of structure–activity relationship. Despite the importance of identifying target molecules, some difficulties are encountered in a conventional affinity purification system, due to their high nonspecific binding and weakness of physical properties of the carrier. To resolve such problems, we have developed a novel affinity nanoparticle for high-performance purification. In Section 9.2, we introduce the construction and the properties of our affinity particles.

9.2 DEVELOPMENT OF NOVEL NANOCARRIERS FOR AFFINITY PURIFICATION

First, we developed a novel affinity matrix composed of SG beads, which are produced by seed polymerization methods. An SG particle consists of a polystyrene (poly-St)/polyglycidylmethacrylate (poly-GMA) core and a surface completely covered with poly-GMA (Figure 9.1). SG beads have a number of advantageous properties as an affinity carrier: submicron size, monodispersion in water, extremely low adsorption of nonspecific protein binding, and reactive epoxide groups on

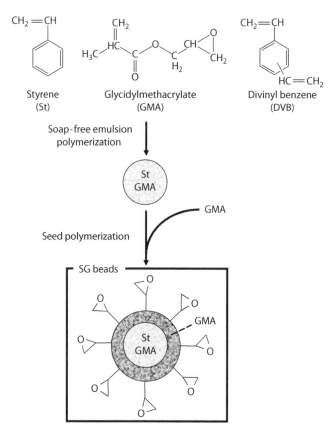

FIGURE 9.1 Preparation of SG beads.

the surface of carriers for coupling with various types of ligands. These properties dramatically overcome the limitations of agarose beads, which have been used conventionally in affinity purification (Shimizu et al., 2000). SG beads have been used successfully for the purification of the specific interactors for various biomolecules directly from crude cell extracts (Inomata et al., 1992, 1995; Hayashida et al., 1995; Imai et al., 1996; Tomohiro et al., 2002; Yamamoto et al., 2007; Kang et al., 2009). Although SG beads contain excellent properties for affinity purification, they are not suitable for using high-throughput screening (HTS) of drug receptors, because the beads are hard to be separated quickly from the dispersion solution. We further studied the preparation of magnetic nanocarriers together with the advantageous characteristics of SG beads.

Magnetite nanoparticles of approximately 10 nm are generally used as the magnetic core of carriers. Their superparamagnetic property offers high dispersibility without magnetic aggregation. On the other hand, it is difficult to produce magnetically collectable carriers of nanometer size because of their small magnetization value. Therefore, we designed magnetic nanometer-sized beads composed of a core with several dozen magnetite particles (MPs) showing enough magnetic response. Our initial effort was focused on preparing uniformly sized MPs greater than 10 nm, which are known to have good magnetic response property. The MPs were separated from the solution with a magnet and washed several times with pure water. MPs of approximately 40 nm were obtained in single crystalline form, and a vibrating sample magnetometer (VSM) (Riken Denshi custom model) analysis revealed that they exhibited a high magnetization saturation value ($M_s = 88.6$ emu/g) (Figure 9.2A).

Next, we attempted to encapsulate the MPs with polymer by several methods. When the mini-emulsion polymerization method was used, self-aggregation of large-sized magnetites induced

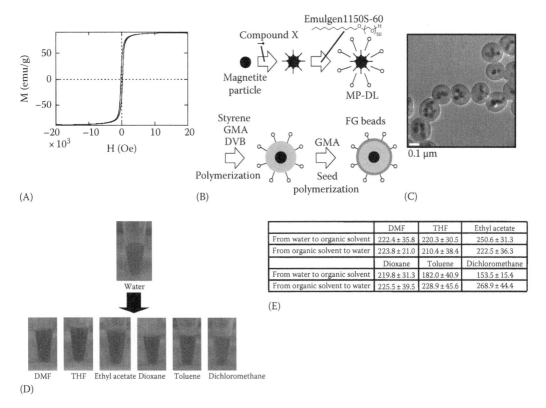

FIGURE 9.2 (A) H-M curve of synthesized MPs. (B) Synthetic scheme of FG beads. (C) FE-TEM image of isolated FG beads. (D) Photo image of FGNE beads dispersed in DMF; THF; ethyl acetate; 1,4-dioxane; toluene; and dichloromethane. These dispersions contain 0.4 mg of FGNE beads. (E) DLS analyses of FGNE beads in organic solvents and water-resuspended beads from each organic solvent.

large clusters due to their strong magnetic dipole interaction, which prevented the formation of stable mini-emulsions. Therefore, we selected the encapsulation method of admicellar polymerization. The MPs were covered with a double layer of surfactant: the primary layer was adsorbed onto the MPs' surface and changed the surface property from hydrophilic to hydrophobic; the secondary layer covered the first layer and facilitated dispersion in a water-based medium. After examining several surfactant candidates, an ionic surfactant was selected as the hydrophobic primary layer. This ionic surfactant was soluble in an alkaline solution and was uniformly adsorbed onto the MPs. A sodium salt surfactant (0.1 M) was mixed with MPs (180 mg) dispersed in distilled water. After the addition of 0.1 M of HCl, MPs were flocculated by overlaying with the surfactant.

The second surfactant layer facilitates the dispersion of MPs coated with the first surfactant layer. Since ionic surfactants like sodium dodecyl sulfate and sodium dodecyl benzene sulfonate lose the dispersion of the MPs, Emulgen 1150S-60 (Kao Corp.), a nonionic surfactant with polyoxyethylene, was selected, because it could minimize ionic effects and increase steric repulsion. Emulgen 1150S-60 in an aqueous solution was added to the MPs coated with the primary surfactant layer. Resultant MPs with the double surfactant layer (MP-DL) were obtained after an ultrasonic treatment. The size distribution of MP-DL was 91.3 ± 18.3 nm, determined by a dynamic laser scattering (DLS) analysis (FPAR-1000, Otsuka Electronics Co., Ltd.). The MPs' crystal size, surfactant thicknesses, and hydrodynamic size indicate that each micelle is composed of several MPs. Their high dispersibility was sustained for at least several dozen minutes.

Admicellar polymerization was carried out according to the following procedure. Divinyl benzene (DVB) was used as a bridged reagent because it renders a rigid polymer structure, suppresses swelling in organic solvents, and inhibits the decapsulation of magnetic cores. Monomer mixtures of styrene, GMA, and DVB were added into a continuous aqueous-phase MP-DL at room temperature. After the mixture was stirred at 200 rpm for 20 min at 70°C, a water-soluble radical initiator, 2,2′-Azobis (2-methylpropionamidine) dihydrochloride, was added to the suspension. After 16 h of stirring at 70°C, the synthesized magnetic beads were collected and washed with distilled water. After re-dispersion into distilled water, the bead surface was additionally covered with poly-GMA by seed polymerization (Figure 9.2B). Figure 9.2C shows an field-emission transmission electron microscope (FE-TEM) (TECNAI F20; FEI Company, Hillsboro, Oregon) image of the magnetic beads (FG beads), and several MPs encapsulated by the homogenous polymer shell. The magnetic beads show a discrete size distribution (184.5 ± 9.0 nm). Monodispersibility was confirmed by a DLS measurement of their hydrodynamic size in distilled water (203.6 ± 28.7 nm), and the analysis of thermogravimetry (TG) showed that the MP content was approximately 30%.

For affinity purification of drug targets, linkers are immobilized on carrier beads to prevent steric hindrance for binding with target proteins. After amination of epoxy groups on the surface of FG beads, ethyleneglycoldiglycidyl ether (EGDE) was immobilized as the linker (FGNE beads) (Hatakeyama et al., 2007). High dispersibility was confirmed in a variety of organic solvents: N,N-dimethylformamide (DMF); tetrahydrofuran (THF); ethyl acetate; 1,4-dioxane; toluene; and dichloromethane. In some cases, shrinkage and swelling of beads was observed (Figure 9.2D and E), but the bead core/shell structure and dispersibility were conserved after returning beads to aqueous conditions (data not shown), confirming the excellent amphiphilic property of FG beads in various kinds of organic solvents (Nishio et al., 2008).

9.3 IDENTIFICATION OF DRUG RECEPTORS

To examine the availability of SG beads for affinity purification of drug receptors, we first compared the properties of SG beads and the conventional agarose resin system by using an immunosuppressive agent, FK506 (Figure 9.3A), as a model ligand. The amino derivative of FK506 was immobilized on SG beads with an epoxy group of these beads. The FK506-fixed SG beads were incubated with Jurkat cell cytoplasmic extracts, and bound proteins to these beads were analyzed by sodium dodecyl sulfate-polyacrylamide gel electrophoresis (SDS-PAGE). Although the FK506-fixed agarose beads showed low specificity, a 12 kDa protein was purified specifically by FK506-fixed SG beads. The protein purified by the SG beads was identified as the FK506 binding protein (FKBP12) by an immunoblot analysis. FKBP12 was also observed in the fraction purified by agarose beads;

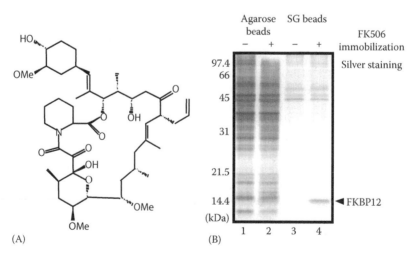

FIGURE 9.3   (A) Structure of FK506. (B) Purification of FKBP12 using agarose beads or SG beads.

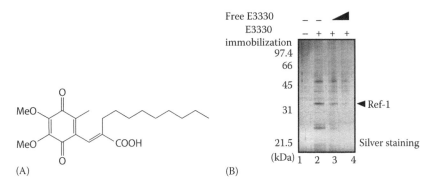

FIGURE 9.4 (A) Structure of E3330. (B) Purification of E3330 binding proteins.

however, the recovery rate of FKBP12 from crude lysate was lower (Figure 9.3B). These results demonstrate that SG beads allow higher binding efficiency of ligand-binding proteins without non-specific binding. The FKBP12 yield by using SG beads was around 8–10 times higher than the yield by using conventional agarose beads (Shimizu et al., 2000).

Next, we performed the identification of the target protein for quinone derivative E3330 (Figure 9.4A), a compound originally developed as an anti-inflammatory agent, which specifically suppresses the nuclear factor kappa B (NF-κB)-mediated transcriptional activation in Jurkat cells (Hiramoto et al., 1998). To identify the molecular target of E3330, an amino derivative of E3330 (NH_2-E3330) was immobilized on SG beads. Jurkat cell nuclear extracts were incubated with NH_2-E3330-immobilized SG beads, and bound proteins were analyzed by SDS-PAGE. Three protein bands of 55, 38, and 27 kDa were specifically bound to the E3330-fixed beads (Figure 9.4B). Amino acid sequencing by a quadrupole time-of-flight mass spectrometry (Q-TOF MS) analysis identified the 38 kDa protein as a redox-sensitive factor, Ref-1. We confirmed the E3330 binding with Ref-1 by a drug far-Western analysis and a surface plasmon resonance (SPR) analysis. By an electrophoretic mobility shift assay (EMSA), Ref-1 was shown to enhance the DNA binding activity of NF-κB through regulating the NF-κB redox state by direct interaction with the p50 subunit of NF-κB. E3330 inhibited the Ref-1-mediated NF-κB activation. Thus, we demonstrated that SG beads can be used to identify the unknown drug target proteins (Shimizu et al., 2000).

Moreover, we attempted to identify novel receptors of methotrexate (MTX). MTX is the anti-cancer and anti-rheumatoid drug that is believed to block nucleotide synthesis and cellular proliferation by inhibiting the dihydrofolate reductase (DHFR) activity. While high-dose MTX is used for the treatment of some malignancies, low-dose MTX, which shows no substantial anticarcinogenic effects, is successfully used for the treatment of rheumatoid arthritis. This raises the possibility that there is another MTX target besides DHFR, but little is known about an alternative mechanism.

To understand the molecular actions of MTX, we attempted to identify the novel MTX-binding proteins. We prepared two types of MTX-fixed beads. First, an MTX amino derivative bearing an amino group instead of the α-carboxyl group was fixed on COOH beads by amidation (Figure 9.5A). Second, MTX was fixed on OH beads by esterification (Figure 9.5B). These beads were mixed with cytoplasmic extracts of the monocytic leukemia cell line THP-1, and MTX-binding proteins were purified (Figure 9.5A and B). We found two different bands specifically binding to the MTX amino derivative–fixed beads and the MTX-fixed beads, and these were identified as DHFR and dCK, respectively. An *in vitro* enzymatic assay revealed that MTX enhanced the activity of ara-C phosphorylation by dCK at high concentration of the substrate. Furthermore, MTX enhanced the ara-C-induced cytotoxicity by facilitating ara-C incorporation into the cellular DNA, independent of its effects on DHFR. These results indicate that MTX facilitates the ara-C cytotoxicity by activating the dCK activity (Figure 9.5C). This mechanism can explain the synergistic effect of MTX

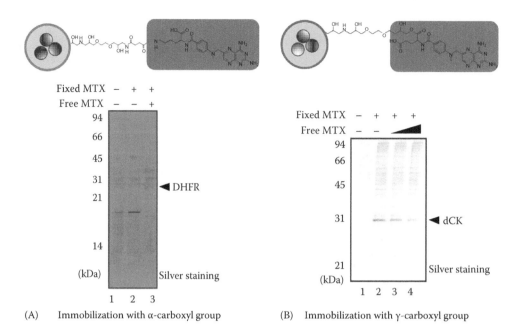

Fixed MTX   −   +   +
Free MTX   −   −   +
94
66
45
31 ◄ DHFR
21

14

(kDa)   Silver staining
1   2   3
(A)   Immobilization with α-carboxyl group

Fixed MTX   −   +   +   +
Free MTX   −   −
94
66
45

31 ◄ dCK

21
(kDa)   Silver staining
1   2   3   4
(B)   Immobilization with γ-carboxyl group

FIGURE 9.5   (A) Purification of DHFR using SG beads conjugated to the α-carboxyl group of MTX. (B) dCK purification using SG beads conjugated to the γ-carboxyl group of MTX.

and ara-C in the treatment of leukemia (Uga et al., 2006). We also identified the molecular targets of drugs and chemicals such as FR-225659, TAS-103, atrazine, and capsaicin (Hatori et al., 2004; Hase et al., 2008; Yoshida et al., 2008; Kuramori et al., 2009).

9.4   IDENTIFICATION OF THE TARGETS OF HEME-RELATED PORPHYRINS

Porphyrins consist of a tetrapyrrole ring structure. Heme (Fe-protoporphyrin IX) is the most common biomolecule of the porphyrin derivative, which is a prosthetic molecule for several heme proteins and plays an essential role in various biological processes, such as oxygen transport, respiration, and signal transduction. For the biosynthesis of heme, a heme biosynthetic precursor porphyrin (protoporphyrin IX) must be transported into mitochondria from cytosol. However, the mechanism of translocation of heme precursors in the mitochondrial inner membrane is unclear.

First, we found that palladium *meso*-tetra (4-carbocyphenyl) porphyrin (PdTCPP) (Figure 9.6A), a phosphorescent porphyrin derivative, accumulated in the mitochondria of several cell lines. To identify the translocators for porphyrins, we performed the purification of the PdTCPP-binding proteins. The succinated PdTCPP was fixed to the amino-modified beads, and these beads were mixed with the mitochondrial extracts of HeLa cells. Approximately 30 kDa protein was purified, and the protein was identified as a 2-oxoglutarate carrier (OGC) (Figure 9.6B). OGC is the mitochondrial transporter of 2-oxoglutarate. Further studies revealed that the mitochondrial accumulation of porphyrin derivatives was inhibited by the addition of an OGC inhibitor (phenylsuccinate) and 2-oxoglutarate. These results suggest that OGC transports porphyrin derivatives into the mitochondria (Figure 9.6D) (Kabe et al., 2006).

Next, we purified heme and protoporphyrin IX (PPIX)-binding proteins. The succinated heme or PPIX was coupled to the amino-modified beads; these beads were mixed with the mitochondrial extracts of rat liver. Approximately 30 kDa protein was purified, and the protein was identified as an adenine nucleotide translocator (ANT) (Figure 9.6C). ANT is an inner-membrane transporter that facilitates the exchange of ATP and ADP. The mitochondrial uptake of PPIX was inhibited by the addition of ADP and ATP, that is, the substrates of ANT. Furthermore, the disruption of ANT

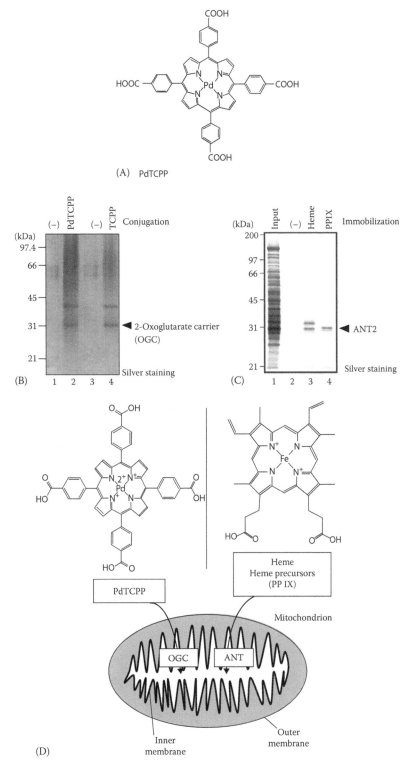

FIGURE 9.6 (A) Structure of PdTCPP. (B) Purification of PdTCPP-binding proteins. (C) Identification of mitochondrial heme-binding proteins. (D) Model of the translocations of PdTCPP and heme precursors to mitochondria.

genes in yeast resulted in a reduction of heme biosynthesis by blocking the translocation of heme precursors into the matrix. These results indicate that ANT transports the heme precursor PPIX into mitochondria (Figure 9.6D) (Azuma et al., 2008).

9.5 INVESTIGATION OF PROTEIN–PROTEIN INTERACTIONS

Enteropathogenic *Escherichia coli* (EPEC) causes severe diarrhea and is a major cause of infantile diarrhea worldwide. A related species of EPEC, enterohemorrhagic *E. coli* (EHEC), causes hemorrhagic colitis and hemolytic uremic syndrome. These pathogens attach themselves to the surface of the intestinal epithelium and cause the destruction of epithelial brush border microvilli and rearrangement of the host cell cytoskeleton. This results in lesions that are termed attaching and effacing (A/E) lesions. In addition, EPEC inhibits macrophage phagocytosis, and thereby escapes the mucosal immune system. These pathogenic processes are mediated by the locus of enterocyte effacement (LEE) pathogenicity island. LEE encodes transcriptional regulators, components of a type-III secretion system (TTSS), effectors that are injected by TTSS into infected cells, chaperones for these effectors, and outer-membrane protein intimin.

Enteropathogenic *E. coli* secreted protein B (EspB), which is one of the effectors secreted by EPEC, is essential for EPEC infection in humans and animal models. To elucidate the physiological roles of EspB plays inside cells, we attempted to identify the EspB-binding proteins. A recombinant histidine-tagged EspB protein (His-EspB) was immobilized on the tosyl-activated beads. These beads were mixed with HeLa cell cytoplasmic extracts, and the EspB-binding proteins were purified (Figure 9.7A). The protein bound specifically to the EspB-fixed beads and was identified as myosin-1c by tandem mass spectrometry.

Myosin-1c is a member of myosin superfamily proteins, which move along actin filaments with their motor domains and participate in various cellular processes. We examined the interaction between EspB and myosin superfamily proteins. EspB bound to the actin-interacting regions of myosin-1a, -1c, -2, -5, -6, and -10, and inhibited the interaction between these myosins and actin. Since myosin-1a, -1c, -2, -5, -6, and -10 were required for phagocytosis and the microvillus structure formation of enterocytes, we confirmed whether EspB participates in the EPEC-mediated phagocytosis and microvilli destruction. EPEC *espB*-Δmid, that is, the strain expressing an EspB mutant that lacks the myosin-binding region, could not prevent phagocytosis

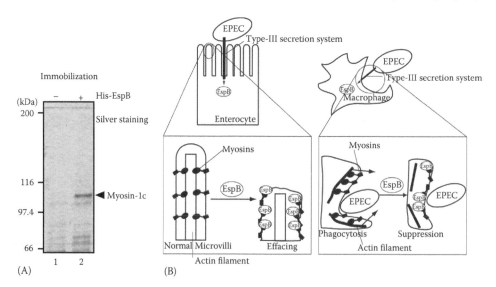

FIGURE 9.7 (A) Purification of EspB-binding proteins. (B) Molecular mechanisms of EPEC-mediated antiphagocytosis and microvilli effacing.

of macrophage and destroy microvilli of enterocytes. Our results suggest that EspB facilitates microvillus effacing and antiphagocytosis by inhibiting a myosin function (Figure 9.7B) (Iizumi et al., 2007). We also identified the molecular target of adeno-associated virus type 2 Rep68 protein (Han et al., 2004).

9.6 CONCLUSION

We have developed SG beads and FG beads, which can be used for discovering targets for proteins, chemical compounds (therapeutic agents, heme, and toxins), and nucleotides. A rapid and convenient identification system of chemical targets contributes not only to advances in the scientific understanding of unknown biological processes but also to the development of drug discovery and medicine. Currently, we are developing an automated high-throughput screening system of drug target identification by using affinity magnetic nanobeads (Hanyu et al., 2009). In future, it should be possible to develop a comprehensive identification system of chemical target proteins.

ABBREVIATIONS

%	Percentage
2,2′-Azobis	2-Methylpropionamidine
A/E lesions	Attaching and effacing lesions
ANT	Adenine nucleotide translocator
dCK	Deoxycytidine kinase
DHFR	Dihydrofolate reductase
DLS	Dynamic laser scattering
DMF	N,N-Dimethylformamide
DNA	Deoxyribonucleic acid
DVB	Divinyl benzene
emu	Electromagnetic unit
EGDE	Ethyleneglycoldiglycidyl ether
EHEC	Enterohemorrhagic *E. coli*
EMSA	Electrophoretic mobility shift assay
EPEC	Enteropathogenic *E. coli*
EspB	Enteropathogenic *E. coli* secreted protein B
FE-TEM	Field-emission transmission electron microscope
FG beads	High-performance affinity magnetic beads
FGNE beads	EGDE-conjugated FG beads
FKBP12	FK506 binding protein 12 kDa
h	Hour
H-M curve	Magnetization curve
His-EspB	Recombinant histidine-tagged EspB protein
HTS	High-throughput screening
kDa	Kilodalton
LEE	Locus of enterocyte effacement
mg	Milligram
min	Minute
M	Mole/liter
M_s	Saturation magnetization
MP	Magnetite particle
MP-DL	MPs with double surfactant layer
MTX	Methotrexate
nm	Nanometer

NF-κB	Nuclear factor kappa B
NH$_2$-E3330	Amino derivative of E3330
OGC	2-Oxoglutarate carrier
poly-GMA	Polyglycidylmethacrylate
poly-St	Polystyrene
PdTCPP	Palladium *meso*-tetra (4-carbocyphenyl) porphyrin
PPIX	Protoporphyrin IX
Q-TOF MS	Quadrupole time-of-flight mass spectrometry
rpm	Revolutions per minute
Ref-1	Redox factor 1
RNA	Ribonucleic acid
SG beads	High-performance affinity latex beads
SDS-PAGE	Sodium dodecyl sulfate-polyacrylamide gel electrophoresis
SPR	Surface plasmon resonance
TG	Thermogravimetry
THF	Tetrahydrofuran
TTSS	Type-III secretion system
VSM	Vibrating sample magnetometer

REFERENCES

Azuma, M., Y. Kabe, C. Kuramori, M. Kondo, Y. Yamaguchi, and H. Handa. Adenine nucleotide translocator transports haem precursors into mitochondria. *PLoS ONE* 3(8) (2008): e3070.

Han, S. I., M. A. Kawano, K. Ishizu, H. Watanabe, M. Hasegawa, S. N. Kanesashi, Y. S. Kim, A. Nakanishi, K. Kataoka, and H. Handa. Rep68 protein of adeno-associated virus Type 2 interacts with 14-3-3 proteins depending on phosphorylation at serine 535. *Virology* 320(1) (2004): 144–155.

Hanyu, N., K. Nishio, M. Hatakeyama, H. Yasuno, T. Tanaka, M. Tada, T. Nakagawa, A. Sandhu, M. Abe, and H. Handa. High-throughput bioscreening system utilizing high-performance affinity magnetic carriers exhibiting minimal non-specific protein binding. *J. Magn. Magn. Mater.* 321 (2009): 1625–1627.

Hase, Y., M. Tatsuno, T. Nishi, K. Kataoka, Y. Kabe, Y. Yamaguchi, N. Ozawa, M. Natori, H. Handa, and H. Watanabe. Atrazine binds to F1F0-ATP synthase and inhibits mitochondrial function in sperm. *Biochem. Biophys. Res. Commun.* 366(1) (2008): 66–72.

Hatakeyama, M., K. Nishio, M. Nakamura, S. Sakamoto, Y. Kabe, T. Wada, and H. Handa. Polymer particles as the carrier for affinity purification. *Koubunshi Ronbunshu* 1 (2007): 9–20.

Hatori, H., T. Zenkoh, M. Kobayashi, Y. Ohtsu, N. Shigematsu, H. Setoi, M. Hino, and H. Handa. FR225659-binding proteins: Identification as serine/threonine protein phosphatase PP1 and PP2A using high-performance affinity beads. *J. Antibiot. (Tokyo)* 57(7) (2004): 456–461.

Hayashida, N., Y. Sumi, T. Wada, H. Handa, and K. Shinozaki. Construction of a cDNA library for a specific region of a chromosome using a novel cDNA selection method utilizing latex particles. *Gene* 165(2) (1995): 155–161.

Hiramoto, M., N. Shimizu, K. Sugimoto, J. Tang, Y. Kawakami, M. Ito, S. Aizawa, H. Tanaka, I. Makino, and H. Handa. Nuclear targeted suppression of NF-kappa B activity by the novel quinone derivative E3330. *J. Immunol.* 160(2) (1998): 810–819.

Iizumi, Y., H. Sagara, Y. Kabe, M. Azuma, K. Kume, M. Ogawa, T. Nagai, P. G. Gillespie, C. Sasakawa, and H. Handa. The enteropathogenic *E. coli* effector EspB facilitates microvillus effacing and antiphagocytosis by inhibiting myosin function. *Cell Host Microbe* 2(6) (2007): 383–392.

Imai, T., Y. Sumi, M. Hatakeyama, K. Fujimoto, H. Kawaguchi, N. Hayashida, K. Shiozaki, K. Terada, H. Yajima, and H. Handa. Selective isolation of DNA or RNA using single-stranded DNA affinity latex particles. *J. Colloid Interface Sci.* 177(1) (1996): 245–249.

Inomata, Y., H. Kawaguchi, M. Hiramoto, T. Wada, and H. Handa. Direct purification of multiple ATF/E4TF3 polypeptides from HeLa cell crude nuclear extracts using DNA affinity latex particles. *Anal. Biochem.* 206(1) (1992): 109–114.

Inomata, Y., Y. Kasuya, K. Fujimoto, H. Handa, and H. Kawaguchi. Purification of membrane receptors with peptide-carrying affinity latex particles. *Colloids Surf. B Biointerfaces* 4 (1995): 231–241.

Kabe, Y., M. Ohmori, K. Shinouchi, Y. Tsuboi, S. Hirao, M. Azuma, H. Watanabe, I. Okura, and H. Handa. Porphyrin accumulation in mitochondria is mediated by 2-oxoglutarate carrier. *J. Biol. Chem.* 281 (2006): 31729–31735.

Kang, J., M. Gemberling, M. Nakamura, F. G. Whitby, H. Handa, W. G. Fairbrother, and D. Tantin. A general mechanism for transcription regulation by Oct1 and Oct4 in response to genotoxic and oxidative stress. *Genes Dev.* 23(2) (2009): 208–222.

Kuramori, C., M. Azuma, K. Kume, Y. Kaneko, A. Inoue, Y. Yamaguchi, Y. Kabe, T. Hosoya, M. Kizaki, M. Suematsu, and H. Handa. Capsaicin binds to prohibitin 2 and displaces it from the mitochondria to the nucleus. *Biochem. Biophys. Res. Commun.* 379(2) (2009): 519–525.

Nishio, K., Y. Masaike, M. Ikeda, H. Narimatsu, N. Gokon, S. Tsubouchi, M. Hatakeyama, S. Sakamoto, N. Hanyu, A. Sandhu, H. Kawaguchi, M. Abe, and H. Handa. Development of novel magnetic nano-carriers for high-performance affinity purification. *Colloids Surf. B Biointerfaces* 64(2) (2008): 162–169.

Shimizu, N., K. Sugimoto, J. Tang, T. Nishi, I. Sato, M. Hiramoto, S. Aizawa, M. Hatakeyama, R. Ohba, H. Hatori, T. Yoshikawa, F. Suzuki, A. Oomori, H. Tanaka, H. Kawaguchi, H. Watanabe, and H. Handa. High-performance affinity beads for identifying drug receptors. *Nat. Biotechnol.* 18(8) (2000): 877–881.

Tomohiro, T., J. Sawada Ji, C. Sawa, H. Nakura, S. Yoshida, M. Kodaka, M. Hatakeyama, H. Kawaguchi, H. Handa, and H. Okuno. Total analysis and purification of cellular proteins binding to cisplatin-damaged DNA using submicron beads. *Bioconjug. Chem.* 13(2) (2002): 163–166.

Uga, H., C. Kuramori, A. Ohta, Y. Tsuboi, H. Tanaka, M. Hatakeyama, Y. Yamaguchi, T. Takahashi, M. Kizaki, and H. Handa. A new mechanism of methotrexate action revealed by target screening with affinity beads. *Mol. Pharmacol.* 70(5) (2006): 1832–1839.

Yamamoto, N., M. Suzuki, M. A. Kawano, T. Inoue, R. U. Takahashi, H. Tsukamoto, T. Enomoto, Y. Yamaguchi, T. Wada, and H. Handa. Adeno-associated virus site-specific integration is regulated by TRP-185. *J. Virol.* 81(4) (2007): 1990–2001.

Yoshida, M., Y. Kabe, T. Wada, A. Asai, and H. Handa. A new mechanism of 6-((2-(dimethylamino)ethyl) amino)-3-hydroxy-7h-indeno(2,1-C)quinolin-7-one dihydrochloride (TAS-103) action discovered by target screening with drug-immobilized affinity beads. *Mol. Pharmacol.* 73(3) (2008): 987–994.

10 Extracellular Biological Synthesis of Gold Nanoparticles

K. Badri Narayanan and N. Sakthivel

CONTENTS

10.1 INTRODUCTION

Nanotechnology is the term used to describe the creation and exploitation of materials with structural features in between those of atoms and bulk materials with at least one dimension in the nanometer range ($1\,nm = 10^{-9}\,m$). Nano comes from nanos, a Greek word meaning "dwarf"; it refers to things of one-billionth 10^{-9} of a meter in size. The word "nanotechnology" was first used by Norio Taniguchi in 1974 in a paper entitled "On the basic Concept of Nano-Technology" presented at the international conference on production engineering, held in Tokyo, Japan. Man has been learning from nature since nomadic times and certain efficient substitutes in manufacturing and design have been found. In the early sixteenth century Roman times, the Lycurgus cup was found to be coated with colloidal gold and silver, and "A labor of the Months" in Norwich England was coated with gold nanoparticles. Richard Feynman and Eric Drexler popularized this concept of nanotechnology as a new and developing technology in which man manipulates objects whose dimensions are approximately between 1 and 100 nm. In the last decade of the twentieth century, research in nanotechnology has burgeoned and has been employed in other disciplines like physics, chemistry, biology, material science, and medicine.

Nanotechnology deals with atoms, molecules, and other objects with a size of 0.1–1000 nm. Nanoparticles may be synthesized by the top-down and bottom-up approaches. The top-down approach involves building up materials from the molecular levels to form macrosized shapes and structures. This method is widely used in photolithography and electron beam lithography. In the bottom-up approach, the atomic or molecular units are used to assemble molecular structures, ranging from atomic dimensions up to supramolecular structures in nanometer range. This approach is used in templating, chemical, electrochemical, sonochemical, thermal, and photochemical reduction techniques. This bottom-up approach is a commonly used method for chemical and biological synthesis of nanoparticles. In order to build materials using a bottom-up approach, the first requirement is to have clusters of the material consisting of a few ($3–10^7$) molecules. One such system of clusters of particles is the colloidal system. Colloids can be defined as a mixture with properties between those of a solution and fine suspension. Colloids can be composed of particles of a wide range of sizes from a few micrometers ($10^{-6}\,m$) to a few nanometers ($10^{-9}\,m$). But for the bottom-up approach of building materials, only the colloidal systems with particle dimensions less than 50 nm are generally used.

Nanoparticles in a colloidal system do have negligible molecular collision with each other, an effect called "Brownian motion." This Brownian movement is a rather random motion, causing the nanoparticles to collide with each other to form larger particles by aggregation and settles down. Nanoparticles in colloidal suspension remain stable resisting aggregation by two means: (1) Electrostatic stabilization: when nanoparticles are formed, an electrical double layer is created from ions adsorbed on the surface of the particle and associated counterions that surround the particle. Thus, if the electric potential associated with double layer is high, the coulombic repulsion between the particles will prevent agglomeration. (2) Steric hindrance: when large molecules get adsorbed on the surface of the particle, two mechanisms restrict these particles from agglomeration. First, the adsorbed molecules are restricted in motion, which causes a decrease in the configurational entropic contribution to the free energy (ΔG). Second, the local increase in concentration of large molecules (e.g., polymers) between approaching particles results in an osmotic repulsion, which separates these particles.

Nanoparticles have distinctive (chemical, electronic, electrical, mechanical, magnetic, thermal, dielectric, optical, and biological) properties as opposed to bulk materials. Some of the advantages of gold nanoparticles include its surface plasma resonance, enhanced Rayleigh scattering, and surface-enhanced Raman scattering (SERS). Therefore, gold nanoparticles are the building blocks of the next generation of optoelectronics, electronics, and various chemical and biochemical sensors. Since the size and shape determines the optoelectronic, physicochemical, and electronic properties of nanoparticles, numerous methodologies have been formulated. Several physical and chemical methods are commonly used to produce monodispersity and selectivity of shape nanoparticles. In case of gold nanoparticles, even though they are considered as biocompatible, various chemical methods may still lead to the coating of toxic chemicals on the surface of the particles that may have adverse effects in medical applications. Considering the problems associated with chemical synthesis, there is an increasing need to develop clean, nontoxic, and environmentally benign nanoparticles synthesis that do not use toxic chemicals in the synthesis protocols to avoid adverse effects in medical applications. An efficient alternative for nanoparticles synthesis and assembly has turned toward the use of biological systems such as plants, algae, and microbes. This chapter describes the recent advancements made on the synthesis of gold nanoparticles of different chemical compositions, sizes, shapes, and controlled monodispersity through reliable, eco-friendly, green chemistry approaches.

10.2 PROPERTIES OF GOLD NANOPARTICLES

Decreasing the dimension of nanoparticles has pronounced effect on the physical properties that significantly differ from the bulk material. These physical properties are caused because of large surface atoms, large surface energy, spatial confinement, and reduced imperfections. Some of these properties of nanoparticles such as surface plasmon resonance (SPR), quantum size effect (QSE), and SERS have been studied in detail.

10.2.1 SURFACE PLASMON RESONANCE

When metal nanoparticles absorb light of a resonant wavelength, it causes the electron cloud on the surface to vibrate (plasmon) and dissipates the energy. This is called surface plasmon resonance. Simply, it is called electron cloud vibration. Plasmons are the oscillations of electron cloud. For nanogold, surface area is more compared to the bulk material. So the surface plasma resonance potential will also be higher and these particles experience the SPR in visible portion of the spectrum. This means that certain portion of the visible wavelengths will be absorbed, while another portion will be reflected. The portion reflected will impart color to the nanoparticles. Gold nanoparticles of size ~400–500 nm absorb light in blue-green portion and reflect red light (~700 nm) giving deep red color to the gold nanoparticles. As the size of the particle increases, the wavelength of SPR-related absorption shifts to longer, redder wavelengths. Therefore, plasmon-associated absorption wavelengths will shift from blue to red, and reflected light will shift from red to blue. Therefore, particle absorption will change from red to blue on aggregation. The frequency of the surface plasmons depends on the size, shape, as well as on the dielectric constant of the metal nanoparticles and its surrounding medium (Kerker 1969, Bohren and Huffman 1983, Kreibig and Vollmer 1995).

10.2.2 QUANTUM SIZE EFFECT

The optical property of nanomaterials is also due to QSE, which arises primarily because of confinement of electrons within particles of dimension smaller than the bulk electron delocalization length. This effect is seen in metal nanoparticles when the particle is less than 2 nm in diameter (Perenboom et al. 1981, Halperin 1986). The average electronic energy level spacing of successive quantum levels, also known as Kubo gap,

$$\text{Kubo gap } (\delta) = \frac{4E_f}{3n}, \tag{10.1}$$

where
E_f is the Fermi energy of the bulk material
n is the total number of valence electrons in the nanocrystal

can be tuned to make a system either metallic or nonmetallic. Thus, the properties such as electric conductivity and magnetic susceptibility exhibit QSEs due to the presence of the Kubo gap in individual nanoparticles (Aiyer et al. 1994). Therefore, the properties of traditional materials change at nanolevel due to the quantum effect, and the behavior of surfaces start to dominate the behavior of bulk materials.

10.2.3 ENHANCED SURFACE RAMAN SCATTERING

SERS is a surface phenomena that results in the enhancement of Raman scattering by molecules adsorbed on rough metal surfaces, which allows to detect single molecules (Nie and Emory 1997). SERS signal is enhanced by 10^6–10^7 several orders of magnitude, which enhances the otherwise weak Raman scattering signal by the electromagnetic field near the particle's surface. The mechanism for the enhancement of SERS is the chemical interaction of the surface-adsorbed molecules with the metal. This enhancement is highly observed for aggregated colloidal gold nanoparticles (Hildebrandt and Stockburger 1984, Moshovists 1985, Campion 1998). Chemisorbed molecules experience an additional enhancement of about 10^2 compared to physisorbed particles (Schatz 1984, Zeman and Schatz 1987, Otto 1991, Otto et al. 1992, Campion 1998), which may result from a resonant scattering process caused by an absorption-induced metal-to-molecule or molecule-to-metal charge transfer electronic transition. This property can make gold nanoparticles as nanoamplifiers for increasing the sensitivity of weak Raman signal.

10.3 EXTRACELLULAR GOLD SYNTHESIS OF GOLD NANOPARTICLES

10.3.1 BIOSYNTHESIS OF GOLD NANOPARTICLES BY PLANTS

In recent years, new green chemistry approaches for the synthesis of metal nanoparticles have received much attention. Biological systems such as plants, algae, and microbes have been considered as alternative systems than physical and chemical methods because of their simplicity and environmentally benign nature. The use of plants in the synthesis of nanoparticles is more advantageous over microbes, which needs elaborate process of maintaining microbial cultures. The capping and stabilizing agents in plants are mostly derived from the alkaloids and terpenoids (Rao et al. 2002). Even plant polymers like gum arabic and gelatin can be used for the stabilization of gold nanoparticles. These biocompatible particles have tremendous applications in diagnostics and therapeutics (Kattumuri et al. 2007, Lu et al. 2008). Identification and isolation of these biomolecules responsible for the observed shape control and stabilization may lead to the development of economically viable methods for synthesizing metal nanoparticles with various shapes and sizes as an alternative to chemical and physical methods, which could be biocompatible to be used in medical applications.

10.3.1.1 Geranium (*Pelargonium graveolens*)

For the first time, geranium plant parts such as leaf, stem, and root have been used for the extracellular production of gold nanoparticles. Shankar et al. (2003) reported the biological reduction of gold ions to gold nanoparticles on exposure to geranium leaf broth and using ultraviolet-visible (UV-Vis) spectroscopy identified the SPR band at 551 nm. Transmission electron microscopy (TEM) analysis showed that the particles are decahedral and icosahedral in shape, ranging in size from 20 to 40 nm, which exhibits multiply twinned particles (MTPs). During reduction process of gold ions, the alcohol groups of terpenoids are oxidized to carbonyl groups, and amide (II) band of proteins acts as stabilizing molecule in the formation of nanoparticles. Using geranium stem extract, Shankar et al. (2004c) showed the formation of spherical nanoparticles in the size range of 8.3–23.8 nm, average of about 14 nm. X-ray diffraction (XRD) pattern showed broad diffraction peaks indicating crystalline and nanoscale dimensions of particles. The Bragg reflection also indexed on the basis of face-centered cubic (fcc) structure of gold. The gold nanoparticles are reduced and stabilized by biomolecules such as alkaloids or terpenoids present in the geranium stem extract. Similarly, geranium root extract was used for the synthesis of polydispersed spherical gold nanoparticles in size from 11.4 to 34 nm of average size of 20.6 nm (Table 10.1). Powder XRD studies showed the Bragg's reflection of fcc structure of crystalline gold nanoparticles (Shankar et al. 2004c).

10.3.1.2 Neem (*Azadirachta indica*)

The leaf extracts of neem was reported to produce gold nanoparticles when added with $HAuCl_4$ and incubated for 2 h, the SPR band centered at 550 nm (Shankar et al. 2004a) with color change to pink-ruby indicating the formation of gold nanoparticles. Flavones and terpenoids are involved in the reduction of gold ions. These nanoparticles appear to have propensity to form thin planar structures rather than just spherical particles. These planar particles are predominantly triangular with a very small percentage of hexagonal-shaped particles with size of 50–100 nm (Table 10.1).

10.3.1.3 Lemongrass (*Cymbopogon flexuosus*)

Shankar et al. (2004b) produced gold nanotriangles and spherical nanotriangles of size 0.05–1.8 μm from lemongrass extract (Table 10.1). Atomic force microscopy (AFM) imaging of gold nanoparticles showed a thickness of 14 nm and an edge length of 440 nm. XRD and selected area electron diffraction (SAED) analyses confirmed the fcc structure of crystalline gold nanoparticles. The gold nanoparticles are capped and stabilized by binding of aldehydes/ketones. These gold nanotriangles have absorption spectrum in NIR region, which could be used in hyperthermia of tumors.

10.3.1.4 Indian Gooseberry (*Emblica officinalis*)

The fruit extracts of Indian gooseberry was used to reduce gold ions to gold nanoparticles extracellularly, which was highly stable (Ankamwar et al. 2005a). TEM analysis of gold nanoparticles indicated the particle size in the range from 15 to 25 nm (Table 10.1).

10.3.1.5 Tamarind (*Tamarindus indica*)

Using tamarind leaf extract, gold nanotriangles and hexagons were synthesized (Ankamwar et al. 2005b). The edge length of nanotriangles was 100–500 nm with thickness in the range of 20–40 nm (Table 10.1). They exhibited large absorption near infrared region. SAED pattern shows highly crystalline fcc structure of gold nanoparticles. Fourier transform infrared (FTIR) spectroscopy analysis shows a characteristic carbonyl stretch vibration possibly from the acid groups

TABLE 10.1
Extracellular Biological Synthesis of Gold Nanoparticles

Biological Material	Shape	Size (nm)	References
Plants			
Pelargonium graveolens (leaf)	Decahedral, icosahedral	20–40	Shankar et al. (2003)
P. graveolens (stem)	Spherical	8.3–23.8	Shankar et al. (2004c)
P. graveolens (root) (geranium)	Spherical, triangle	11.4–34	Shankar et al. (2004c)
Azadirachta indica (neem)	Triangle, hexagonal	50–100	Shankar et al. (2004a)
Cymbopogon flexuosus (lemongrass)	Triangle, spherical	0.05–1.8	Shankar et al. (2004b)
Emblica officinalis (Indian gooseberry)	—	15–25	Ankamwar et al. (2005a)
Tamarindus indica (tamarind)	Flat triangle, hexagonal	20–40	Ankamwar (2005b)
Aloe barbadensis (*Aloe vera*)	Spherical, triangle	15.2 ± 4.2	Chandran et al. (2006)
Cinnamomum camphora (camphor tree)	Flat, platelike triangle	55–80	Huang et al. (2007)
Coriandrum sativum (coriander)	Spherical, triangle, truncated triangle	20.65 ± 7.09	Badri Narayanan and Sakthivel (2008)
Algae			
Sargassum wightii Greville (marine algae)	Thin planar	8–12	Singaravelu et al. (2007)
Fungi			
Fusarium oxysporum	Spherical, triangle	20–40	Mukherjee et al. (2002)
Colletotrichum sp.	Spherical	20–40	Shankar et al. (2003)
Trichothecium sp.	Triangle, hexagonal	5–200	Ahmad et al. (2005)
Bacteria			
Pseudomonas aeruginosa	—	15–30	Husseiny et al. (2007)
Rhodopseudomonas capsulata	Spherical	10–20 (pH 7)	He et al. (2007)
	Triangular nanoplates, spherical	50–400 (pH 4)	
Bacillus megatherium D01	Spherical	1.9 ± 0.8	Wen et al. (2009)
Actinomycete			
Thermomonospora sp.	Spherical	8	Ahmad et al. (2003)

present in the tamarind leaf extract, tartaric acid. These nanotriangles have potential applications in vapor sensing.

10.3.1.6 *Aloe vera* (*Aloe barbadensis*)

Chandran et al. (2006) demonstrated the formation of gold nanoparticles from *Aloe vera* using UV-Vis-NIR spectroscopy showing a relatively increased intensity of transverse band in comparison with longitudinal band. The electron diffraction pattern was also indexed on the basis of fcc crystalline structure of gold. FTIR spectrum showed the presence of carbonyl groups in *Aloe vera*, which played an important role in stabilizing and capping the nanoparticles. TEM analysis showed the average size of spherical and nanotriangles as 15.2 ± 4.2 nm (Table 10.1).

10.3.1.7 Camphor (*Cinnamomum camphora*)

Similarly, the sun-dried leaves of *Cinnamomum camphora* were used for the first time in the synthesis of gold nanotriangles with flat and platelike morphology, which showed absorbance in the near-infrared (NIR) region. TEM analysis showed the size of nanoparticles in the range 55–80 nm. AFM study showed the thickness of nanotriangles as 7 nm, and FTIR spectrum revealed the presence of water-soluble heterocyclic compounds like alkaloids, flavones, and anthracenes act as capping agents of gold nanoparticles (Huang et al. 2007). XRD pattern shows the crystalline nature of nanoparticles with fcc structure. The difference of shape control and size control is mainly attributed to the comparative presence of protective biomolecules and reductive biomolecules (Table 10.1).

10.3.1.8 Coriander (*Coriandrum sativum*)

For the first time, gold nanoparticles were synthesized using leaf extract of coriander (Badri Narayanan and Sakthivel 2008). UV-Vis spectroscopic analysis showed SPR band at 536 nm with color change to pinkish-ruby color indicating the formation of nanoparticles. TEM analysis showed the formation of triangular, truncated triangular, and spherical nanoparticles of size 20.65 ± 7.09 nm. FTIR spectroscopic analysis showed the presence of amide (I) and (II) bands of protein that capped and stabilized the nanoparticles, and diffraction studies like XRD and SAED confirmed the fcc structure of crystalline gold nanoparticles.

10.3.2 BIOSYNTHESIS OF GOLD NANOPARTICLES BY ALGAE

Sargassum wightii Greville, a marine algae, dried powder was exposed to gold ions, which reduced 95% of $AuCl_4^-$ ions within 12 h at stirring conditions. TEM micrograph of the gold nanoparticles showed the formation of monodispersed thin planar structures nanoparticles ranging from 8 to 12 nm in size (Singaravelu et al. 2007).

10.3.3 BIOSYNTHESIS OF GOLD NANOPARTICLES BY MICROBES

Microbial resistance to most toxic heavy metals is not only because of chemical detoxification, but also from energy-dependent ion efflux from the cell by membrane proteins that function either as ATPase, or as chemiosmotic cation or proton anti-transporters. Therefore, microbial systems can detoxify the metal ions by either reduction and/or precipitation of these soluble toxic inorganic metal ions to insoluble nontoxic metal nanoclusters. The redox potential being higher for Au^{3+} to Au^0, the bioreduction of gold ions is considerably favorable for microbes. Thereby, it can synthesize biogenic zero-valent gold nanoparticles (Au^0). Gold ions are less toxic metal ions compared to other metal ions. It can be detoxified easily either by extracellular biomineralization or intracellular bioaccumulation. However, extracellular production of zero-valent gold nanoparticles (Au^0) has more commercial applications than intracellular accumulation (Table 10.1).

10.3.3.1 Bacteria-Mediated Biosynthesis

Rhodopseudomonas capsulata, a prokaryotic bacterium was found to reduce Au^{3+} to Au^0 at room temperature (He et al. 2007). UV-Vis spectra showed two SPR bands centered at 540 nm (band I) and 900 nm (band II). TEM analysis showed that particles are mainly spherical in the size range of 10–20 nm at pH 7.0. But with change in pH of the solution, various shapes and sizes are formed. At pH 4.0, triangular nanoparticles also appeared with spherical nanoparticles. These triangular nanoplates are in the size of 50–400 nm and spherical with a size of 10–50 nm. Thus, synthesis of gold nanoparticles of different sizes and shapes is of great importance for their applications in optical devices, electronics, and catalysis (Rao and Cheetham 2001, Moreno-Manas and Pleixats 2003). Husseiny et al. (2007) demonstrated that *Pseudomonas aeruginosa* ATCC 90271, *P. aeruginosa* (1), and *P. aeruginosa* (2) synthesized gold nanoparticles extracellularly with particle size distribution in the order of 40 ± 10, 25 ± 15, and 15 ± 5 nm, respectively. As the particle size increases, the color was found to be shifted from pink to blue color due to the SPR of gold nanoparticles (Table 10.1). Similarly, the dried powder of *Bacillus megatherium* D01 was used to reduce the gold salts into monodispersed gold nanoparticles with dodecanethiol as capping ligand to stabilize the particle at 26°C. TEM analysis showed the effect of thiol on the shape, size, and dispersity of gold nanoparticles. The presence of thiol during biosynthesis indicated the formation of small, spherical gold nanoparticles with 1.9 ± 0.8 nm in size (Wen et al. 2009).

10.3.3.2 Fungi-Mediated Biosynthesis

The synthesis of metal nanoparticles especially gold nanoparticles by eukaryotic microorganisms like fungi is considered as natural nanofactories, which has advantage in processing and handling of the biomass (Mukherjee et al. 2001, 2002, Sastry et al. 2003). Shankar et al. (2003) found an endophytic fungus, *Colletotrichum* sp. isolated from the leaves of geranium plant (*Pelargonium graveolens*), which rapidly reduced gold ions to zero-valent gold nanoparticles. It was spherical in shape and polydispersed. These particles exhibited a mixture of flat disk and rodlike morphology. They also found that the polypeptide capping the gold nanoparticles are glutathiones, which can bind to gold nanoparticles either through free amine group or cysteine residues in the protein and stabilized gold nanoparticles (Gole et al. 2001). TEM analysis showed the size of the particles in the range of 8–40 nm. Another fungus, *Fusarium oxysporum,* also synthesized spherical and triangular gold nanoparticles extracellularly in the size range of 20–40 nm (Mukherjee et al. 2002). FTIR spectrum showed the presence of amide (I) and (II) bands from carbonyl and amine stretch vibrations in proteins, respectively. Electrophoresis revealed the protein of molecular mass between 66 and 10 kDa involved in nanoparticles stabilization. When fungus *Trichothecium* sp. was cultured in static condition, it reduced Au^{3+} to form gold nanoparticles (Ahmad et al. 2005). TEM images showed the shapes of triangles and hexagonals with smaller highly polydispersed spheres and rod-like structures. The average size of dimension was found to be 5–200 nm (Table 10.1). The Scherrer ring pattern was characteristic of fcc gold. The release of some loosely bound specific enzymes or proteins by the fungal *Trichothecium* sp. mat in the solution was found to be responsible for the synthesis of nanoparticles of different morphology.

10.3.3.3 Actinomycete-Mediated Biosynthesis

Actinomycetes are microorganisms that have the characteristics of eukaryotic fungi and prokaryotic bacteria. They are known as ray fungi. They are reported mostly for the production of antibiotics in general. A novel extremophilic actinomycete, *Thermomonospora* sp., was found to synthesize extracellular monodispersed spherical gold nanoparticles at an average size of 8 nm (Ahmad et al. 2003). FTIR spectral analysis showed the presence of amide (I) and (II) bands of protein as capping and stabilizing agents on the surface of nanoparticles. These particles are stable for more than 6 months. Electrophoresis analysis showed the protein of molecular weight from 80 to 10 kDa was involved in the stabilization of nanoparticles (Table 10.1).

10.4 CHARACTERIZATION OF METAL NANOPARTICLES

Several microscopic and spectroscopic methods such as XRD, various electron microscopy (EM) including scanning electron microscopy (SEM) and high-resolution transmission microscopy (HRTEM), scanning probe microscopy (SPM), optical spectroscopy, and electron spectroscopy have been used extensively to characterize the nanoparticles.

10.4.1 Structural Characterization

10.4.1.1 X-Ray Diffraction

XRD is the most widely used technique for characterizing the various nanomaterials. Powder XRD is used for studying particles of polycrystalline solids especially metal nanoparticles. It is used to solve the structure of crystals including their lattice constants and geometry, identification of unknown materials (Segmuller and Murakami 1988, Cullity and Stock 2001), orientation of polycrystals, defects, and stresses. In XRD, a collimated beam of short wavelength, hard x-rays in the range of 0.7–2 Å, of photon energy between 1 and 120 keV is used. Common targets used in x-ray tubes include Cu and Mo, which emit 8 and 14 keV x-rays with corresponding wavelengths of 1.54 and 0.8 Å, respectively. This incident beam is diffracted by the crystalline phases in the specimen according to

$$\text{Bragg's law: } \lambda = 2d \sin \theta, \tag{10.2}$$

where

d is the interplane distance in the crystalline phase

λ is the x-ray wavelength

The measurement of diffraction pattern allows to deduce the distribution of atoms in a material. The peaks are directly related to the atomic distances. The positions and intensities of the peaks are used for identifying the structure (or phase) of the material. This diffraction pattern is used to identify the specimen's crystalline phases and to measure its structural properties. The shift in the peak positions is used to calculate the change in d-spacing, which is the result of the change of lattice constants under a strain. In a homogenous strain, the crystallite size, D, can be estimated from the peak width with Scherrer's formula (Borchert et al. 2005):

$$D = \frac{K\lambda}{\beta_{cor} \cos \theta}, \quad \text{with } \beta_{cor} = \left(\beta_{sample}^2 - \beta_{ref}^2 \right)^{1/2}, \tag{10.3}$$

where

D is the average crystal size

K is the Scherrer coefficient (0.89)

λ is the x-ray wavelength ($\lambda = 1.5406$ Å)

θ is Bragg's angle (2θ)

β_{cor} the corrected of the full width at half-maximum (FWHM) in radians

β_{sample} and β_{ref} are the FWHM of the reference and sample peaks, respectively

XRD can provide the collective information of the particle sizes and usually requires a sizable amount of powder (Segmuller and Murakami 1985). The main disadvantage of XRD is that it needs large material for acquiring small diffraction intensities and the information is an average over a large amount of material.

10.4.1.2 Small-Angle X-Ray Scattering

The small-angle x-ray scattering (SAXS) is a small-angle scattering (SAS) technique used to determine the structure of microscale or nanoscale particle systems in terms of parameters such as averaged particle sizes, shapes, distribution, and surface-to-volume ratio. SAS of inhomogeneity samples of 10 nm or larger by elastic x-ray scattering of wavelength 0.1–0.2 nm is recorded at very low angles of typically $2\theta < 5°$. Strong diffraction peaks result from constructive interference of x-rays scattered from ordered arrays of atoms and molecules. A lot of information can be obtained from the angular distribution of scattered intensity at low angles. These variations can be from differences in density, composition, or both, and do not need to be periodic (Guinier and Fournet 1955, Glatter and Kratky 1982). The amount and angular distribution of scattered intensity provides information, such as the size of very small particles or their surface area per unit volume, regardless of whether the sample or particles are crystalline or amorphous (Templeton et al. 2000, Nakamura et al. 2003). Applications of SAXS include colloids of all types of metal, macromolecules like proteins, and in pharmaceuticals.

10.4.1.3 Scanning Electron Microscopy

The SEM is another common tool used to study nanoparticles and other nanostructures. This technique is most widely used to gather information about the surface topography, composition, and electrical conductivity of nanoparticles by scanning it with high-energy beam electrons ranging from few hundred eV to 50 keV in a raster scan pattern (Lawes 1987).

The theoretical limit to an instrument's resolving power is determined by the wavelengths of the electron beam used and the numerical aperture of the system. The resolving power, R, of an instrument is defined as

$$R = \frac{\lambda}{2NA},$$

(10.4)

where

λ is the wavelength of electrons used

NA is the numerical aperture

As the electrons strike and penetrate the surface, a number of interactions occur that result in the emission of electrons and photons from the sample. Three types of SEM detections such as secondary electron images, backscattered electron images, and elemental x-ray maps have been used. When a high-energy primary electron interacts with an atom, it undergoes either inelastic scattering with atomic electrons or elastic scattering with the atomic nucleus.

In an inelastic collision with an electron, the primary electron transfers part of its energy to the other electron. When the energy transferred is large enough, the other electron will emit from the sample. If the emitted electron has energy of less than 50 eV, it is referred to as a secondary electron. These can produce very high-resolution images of a sample surface, about 1–5 nm with a magnification of about ~10 to over 300,000. Backscattered electrons (BSE) are high-energy electrons that are elastically scattered and essentially possess the same energy as the incident or primary electrons. The probability of backscattering increases with the atomic number (Z) of the sample material. BSE can provide information about the distribution of different elements in the sample. SEM not only provides the image of the morphology and microstructures of bulk and nanostructured materials and device, but can also provide detailed information of chemical composition and distribution (Newbury et al. 1986).

10.4.1.4 Transmission Electron Microscopy

In TEM, electrons have a typical wavelength of 0.0025 nm at accelerated voltage of 100–200 keV. Since the electron wavelength is smaller, the greater the accelerating voltage, higher is the resolution. The greatest advantages that TEM offers are the high magnification ranging from 50 to 10^6 and its ability to provide both image and diffraction information for a single sample. The high magnification or resolution of all TEM is a result of the small effective electron wavelengths, λ, which is given by the de Broglie relationship:

$$\lambda = \frac{h}{\sqrt{2mqV}},$$

(10.5)

where

m and q are the electron mass and charge

h is Planck's constant

V is the potential difference through which electrons are accelerated

SAED offers a unique capability to determine the crystal structure of individual nanoparticles, nanomaterials, and the crystal structures of different parts of a sample. In SAED, the condenser lens is defocused to produce parallel illumination at the specimen and a selected-area aperture is used to limit the diffracting volume. SAED patterns are often used to determine the Bravais lattices and lattice parameters of crystalline materials by the same procedure used in XRD (Cullity and Stock 2001). The rings in the diffraction picture were measured and then the lattice spaces were calculated by

$$\lambda L = d_{hkl}R, \qquad (10.6)$$

where λL is the microscope camera constant (wavelength, λ, in Angstroms and camera length, L, in centimeters). R corresponds to ring radius in centimeters. The lattice spaces d_{hkl} obtained were compared with the crystallographic data for silver. TEM is also useful in determining the melting points of nanocrystals, in which an electron beam is used to heat up the nanocrystals and the melting points are determined by the disappearance of electron diffraction (Goldstein et al. 1992).

10.4.1.5 Scanning Probe Microscopy

SPM is unique among imaging techniques; it forms three-dimensional (3D) images of surface using a physical probe, which is one atom in diameter, to scan the specimen. An image is obtained mechanically moving the probe in a raster scan, line by line, and recording probe–surface interaction as a function of position. SPM is a general term for a family of microscopes depending on the probing forces used. Scanning tunneling microscopy (STM) and AFM are the members of SPM, which is used to characterize metal nanoparticles and other nanostructures.

STM technique is used for viewing the nanoparticles' surfaces at the atomic level. It can probe the density of states of a material using tunneling current. STM was first developed by Binnig and his coworkers in 1981 (Binnig et al. 1982). It has the lateral resolution of 0.1 nm and depth resolution of 0.01 nm (Bai 1999). When the conducting probe is brought very near to a metallic or semiconducting surface, a bias between the two can allow electrons to tunnel through the vacuum between them. For low voltages, this tunneling current is a function of the local density of states (LDOS) at the Fermi level, E_f, of the sample. Variations in current as the probe passes over the surface are translated into an image. This probe can also be used to rotate the individual bonds within single molecules, i.e., to change the topography of the sample. The limitation of STM is that only electrically conductive sample surface can be used (Howland and Kirk 1992).

AFM is a technique used for 3D imaging, measuring, and manipulating matters at nanoscale. It is mainly to determine the topography of flat samples. It can scan any surface of insulators, semiconductors, conductors, magnetic, transparent, and opaque materials unlike STM. It can image only to a maximum height on the order of micrometers and maximum scanning area of around 150 by 150 μm.

10.4.2 CHEMICAL CHARACTERIZATION

The chemical characterization of nanoparticles helps to determine the surface and interior atoms and compounds as well as their spatial distributions. These include optical spectroscopy, electron spectroscopy, and ion spectrometry.

10.4.2.1 Optical Spectroscopy

Various optical spectroscopies like absorption (UV-Vis) and emission (fluorescence) and vibrational (infrared and Raman) spectroscopies have been widely used for the characterization of nanoparticles. The former determines the electronic structures of atoms, ions, molecules, or crystals through exciting electrons from the ground to excited states (absorption) and relaxing from the excited to ground states (emission). The vibrational techniques may be summarized as involving the interactions of photons with species in a sample that results in energy transfer to or from the sample via vibrational excitation or de-excitation. The vibrational frequencies provide the information of chemical bonds in the detecting samples, which are called "fingerprints" of the material for its identification.

10.4.2.1.1 UV-Vis Spectroscopy

UV-Vis absorption spectroscopy deals with the study of electronic transitions between orbitals or bands of atoms, ions, or molecules in gaseous, liquid, and solid state (Jorgensen 1962). Gold nanoparticles spectra exhibit an SPR peak at ~520 nm that broadens and decreases in intensity as the particle size decreases. These resonances occur only in the case of nanoparticles and not in the case of bulk metallic particles. Hence, UV-Vis can be utilized to study the unique optical properties of nanoparticles.

10.4.2.1.2   Fluorescence Spectroscopy

When light of some wavelength is directed onto a specimen, it prompts the transition of electron from the ground to excited state, which then undergoes a non-radiative internal relaxation and the excited electron moves to a more stable excited level (Colvard 1992). After a characteristic lifetime in the excited state, the electron returns to the ground state by emitting the characteristic wavelength in the form of light. This emitted energy can be used to provide qualitative and sometime quantitative information about chemical composition, structure, impurities, kinetic process, and energy transfer.

10.4.2.1.3   Infrared Spectroscopy

Infrared spectroscopy exploits the fact that molecules have specific frequencies at which they rotate or vibrate corresponding to discrete energy levels (vibrational modes). The far-infrared, approximately 400–10 cm^{-1}, has low energy and may be used for rotational spectroscopy. The mid-infrared, approximately 4000–400 cm^{-1}, may be used to study the fundamental vibrations and associated rotational–vibrational structure. In order for a vibrational mode in a molecule to be IR active, it must be associated with changes in the permanent dipole. These absorption frequencies represent excitations of vibrations of the chemical bonds and, thus, are specific to the type of bond and the group of atoms involved in the vibration (Griffiths and De Haseth 1986).

10.4.2.1.4   Raman Spectroscopy

Raman spectroscopy is a vibrational and rotational spectroscopic technique for studying chemical bonding (Orhring 1992). When the incident photon interacts with the chemical bond, the chemical bond is excited to a higher energy state. Most of the energy would be reradiated at the same frequency as that of the incident exciting light, which is known as the Rayleigh scattering. A small portion of the energy is transferred and results in exciting the vibrational modes; this is called Stokes Raman scattering. The subsequent reradiation has a frequency lower than that of the incident exciting light, and if the molecule was already in an elevated vibrational energy state, the Raman scattering is called anti-Stokes Raman scattering. However, Stokes scattering spectra are mostly used since they are less temperature sensitive. The Raman effect is extremely weak and, thus, intense monochromatic continuous gas lasers are used as the exciting light. Since vibrational information is specific for the chemical bonds in molecules, it therefore provides a fingerprint by which the molecule can be identified.

10.4.2.2   Electron Spectroscopy

10.4.2.2.1   Energy Dispersive X-Ray Spectroscopy

The energy dispersive x-ray spectroscopy (EDX) is an analytical technique used for the elemental analysis of nanoparticles and nanomaterials. When high-energy beam of charged particles like electrons or x-rays is incident on the sample, an electron from an inner shell is ejected and leaves a hole, i.e., electron vacancy in the inner shell. An electron from an outer, higher energy shell then fills the hole, and the difference in energy between the higher energy shell and the lower energy shell may be released in the form of an x-ray. The number and energy of the x-rays emitted from a specimen can be measured by an energy dispersive spectrometer. As the energy of the x-rays is characteristic of the difference in energy between the two shells and of the atomic structure of the element from which they were emitted, this allows the elemental composition of the specimen to be measured. EDX can detect elements with $Z > 11$. EDX is mostly found on scanning electron microscopes (SEM-EDX) and electron microprobes for elemental analysis.

10.4.2.2.2　Auger Electron Spectroscopy

The Auger electron spectroscopy (AES) is also a common analytical technique for characterization of nanoparticles surface in material science. When an atom is probed by a beam of electrons with energies in the range of 2–50 keV, a core state electron can be removed leaving behind a hole. As this is an unstable state, the core hole can be filled by an outer shell electron, whereby the electron moving to the lower energy level loses an amount of energy equal to the difference in orbital energies. Since orbital energies are unique to an atom of a specific element, analysis of the ejected electrons can yield information about the chemical composition of a surface.

10.4.2.2.3　X-Ray Photoelectron Spectroscopy

The x-ray photoelectron spectroscopy (XPS) is a quantitative technique to determine the chemical composition, empirical formula, chemical state, and electronic state of metal nanoparticles present within the top 10 nm of the surface of nanomaterials (Fadley 1978). It is a surface chemical analysis where the relatively low-energy x-rays are used to eject the electrons from an atom via the photoelectric effect. It can detect elements with $Z > 3$. The energy of the ejected electron, E_E, is determined by both the energy of the incident photon, $h\nu$, and the bound electron state:

$$E_B : E_E = h\nu - E_B I. \tag{10.7}$$

Since the values of the binding energy are element-specific, atomic identification is possible through measurement of photoelectron energies. It can provide information on the nature of chemical bonding and valence states of gold nanoparticles.

10.5　APPLICATIONS OF GOLD NANOPARTICLES

Gold nanoparticles are the most commonly used nanomaterials in biosensing, biolabeling, delivering, heating, and catalysis due to their unique properties of optical absorption, fluorescence, Raman scattering, atomic and magnetic force, and electrical conductivity.

10.5.1　Biolabeling with Gold Nanoparticles

Various biological materials can be attached to gold nanoparticles in various ways. Biological molecules like antibodies, oligonucleotides, peptides, or polyethylene glycol have functional groups (like thiols), which can bind to the gold surface. These biological molecules can replace some of the stabilizing molecules, which surround the colloidal gold nanoparticles and makes linkage with gold nanoparticles (Zanchet et al. 2001, Levy et al. 2006). In bioconjugation, the biological molecules can be attached to the surface of stabilizing molecules around the gold nanoparticles. These bioconjugation occurs mostly with $(-NH_2)$ amine groups on biological molecules with $(-COOH)$ carboxyl groups at the free ends of stabilizer molecules by using EDC (1-ethyl-3-(3-dimethylaminopropyl)-carbodiimide-HCl) (Hermanson 1996, Sperling et al. 2006).

The rapid detection of *Staphylococcus aureus*, the most important human pathogens causing nosocomial and community-acquired infections in patient specimen, is essential. Here, Protein A, which is a component of cell wall of *S. aureus*, is utilized as a target for the detection of *S. aureus*. The preparation of antiprotein A IgG antibodies using labels of gold nanoparticles and the development of immunochromatographic assay based on gold nanoparticles is used to detect *S. aureus* in clinical specimens (Huang 2006).

10.5.1.1　Single-Particle Tracking

Gold nanoparticles are conjugated with ligands or antibodies specific for membrane-bound molecules or structure. These conjugated gold nanoparticles are added to living cells. Thus, the labeled individual membrane-bound molecules diffusion within the cell membrane can be traced via observation of gold nanoparticles. Mostly, membrane-bound receptor molecules are investigated by single-particle tracking. By time-resolved imaging of the receptors that are labeled with gold nanoparticle, their movements within the cell membrane can be observed (Kusumi et al. 2005). Movement of gold nanoparticles >40 nm can be traced directly with phase contrast or differential interference contrast microscopy (Felsenfeld et al. 1996). Gold nanoparticles >20–30 nm can be traced by dark field microscopy (Cang et al. 2006). Movement of receptors with even smaller gold nanoparticles (~5 nm) has been visualized with photothermal imaging (Cognet et al. 2003, Lasne et al. 2006). There is no limitation in observation time by photobleaching as no fluorescence detection method is used (Wang et al. 1994).

10.5.2 Delivery with Gold Nanoparticles

10.5.2.1 Gene Gun

Gold nanoparticles can be used for delivering molecules into cells. These biomolecules will detach the nanoparticles and function after they enter into the cell. Gene gun techniques are mainly used to introduce DNA into the cells (Chen et al. 2000). DNA is adsorbed onto the surface of gold nanoparticles. These microparticles or nanoparticles are fired at the cells using particle gun by a technique called biolistics (Yang et al. 1990). Mostly, it is used for the introduction of plasmid DNA into plant cells. However, it can also be used for delivering DNA into animal cells, which do not possess cell walls (Kuriyama et al. 2000).

10.5.3 Gold Nanoparticles as Thermal Source

When gold nanoparticles are illuminated with light, they absorb light and the free electrons get excited. This excitation at SPR frequency causes a collective oscillation of free electrons. When electrons interact with themselves and the crystal lattice of gold particles, the electrons relax and the thermal energy is transformed to lattice. This heat energy can be used in various ways for manipulating the surrounding tissues (Pissuwan et al. 2006).

10.5.3.1 Hyperthermia

This is an anticancer therapy where colloidal gold nanoparticles are conjugated with ligands that are specific to receptors that are overexpressed on cancer cells compared to healthy cells. The cancer cells are enriched with the gold nanoparticles when illuminated with light, the light energy is converted to thermal energy. This temperature inside the cells rises above normal 37°C (T_0) to few degrees (T) (i.e., $T > T_0$) leading to cell death of cancer cells (Huang et al. 2006, Huff et al. 2007). Tissues can absorb light in visible region and even infrared (IR) light can only penetrate relatively thin tissue. For this reason, gold nanoparticles are needed, which absorb light in IR rather than in the visible range such as gold rods or hollow structures (Huang et al. 2008). So hyperthermia by gold nanoparticles by photoinduced heating will work best for tissue near skin surface. For tissues deep inside the body, heating with magnetic particles is used.

10.5.4 Gold Nanoparticles as Biosensors

10.5.4.1 Surface-Enhanced Raman Scattering

SERS can be used for the detection of analytes. The surface of gold nanoparticles is modified with ligands that can specifically bind the analyte. Upon binding to the gold nanoparticles, the Raman signal to the analyte is enhanced drastically and allows its detection (Nie and Emory 1997). Recent developments include the detection of DNA (Cao et al. 2002) or proteins (Ni et al. 1999) and two-photon excitation (Kneipp et al. 2006) by gold nanoparticles modified with Raman-active reporter molecules.

10.5.4.2 Gold Stains

Gold nanoparticles can be used in ELISA-like assays (enzyme-linked immunosorbent assays) instead of fluorophores or absorbing dyes for qualitative or quantitative detection of analytes. Here, the analyte is immobilized on a surface either by simple adsorption or specific binding by a capture antibody. The analyte-specific antibodies are conjugated to the surface of gold nanoparticles. The binding of gold nanoparticles to the surface shows the presence of analyte, and the concentration of analyte molecules can be quantified by the optical absorption of the gold spot (Wang et al. 2005). Sensitivity can be increased by involving secondary antibodies and silver enhancement. Danscher (1981) first reported that silver enhancement can be made using silver acetate and hydroquinone at a pH 3.5. In the presence of gold nanoparticles, which act as catalysts, the silver ions are reduced to metallic silver by the hydroquinone. The silver atoms formed are deposited in layers on the gold surface, resulting in significantly larger particles and more intense macroscopic signal. This is called immunogold silver staining (IGSS) similar to that obtained with alkaline phosphatase using colorimetric detection and several times more sensitive than ^{125}I-labeled antibodies. These gold particles can also be used as a general stain for proteins on blots.

10.5.5 GOLD NANOPARTICLES IN CATALYSIS

Bulk gold is chemically inert and cannot be used in catalysis (Hammer and Norskov 1995). However, gold nanoparticles can be an excellent catalyst. The excellent catalytic property of gold nanoparticles is a combination of size effect and the unusual properties of individual gold atom. The unusual properties of gold atom are attributable to the so-called relativistic effect that stabilizes the $6s^2$ electron pairs, and thus catalytic properties of the novel group elements is determined by the high energy and reactivity of the 5d electrons (Pyykko 1988). The excellent catalytic activity of supported gold nanoparticles for partial and complete oxidation of hydrocarbons, oxidation of carbon monoxide, nitric oxide, and unsaturated hydrocarbons has been reported (Haruta 1997). Thiol-stabilized gold nanoparticles have also been exploited in asymmetric hydroxylation reactions (Li et al. 1999), carboxylic ester cleavage (Pasquato et al. 2000), and particle-bound ring opening metathesis polymerization (Bartz et al. 1998). Therefore, the main factors for attaining high catalytic activity are smaller particle size and support.

10.5.5.1 Oxidation of Cyclohexane

The extracellularly synthesized gold nanoparticles using fungus *Fusarium oxysporum* (Mukherjee et al. 2002) have been supported on amorphous (fumed) silica, and also intracellularly synthesized gold nanoparticles by fungus, *Verticillium* sp. (Mukherjee et al. 2001), have been explored for oxidation reaction. The aerial oxidation of cyclohexane to adipic acid can be achieved at 120°C and 4.3 MPa without using any solvent. The conversion of cyclohexane increases with increasing amount of gold nanoparticles (Au–SiO_2) as catalyst. The aerial oxidation of cyclohexane in the absence of catalyst at 120°C and 2.1–4.3 MPa results in the formation of only cyclohexanol and cyclohexane. The use of gold nanoparticles as catalyst under similar condition along with cyclohexanol and cyclohexanone forms acid products such as adipic acid (AA), glutaric acid (GA), and succinic acid (SA). The oxidation of cyclohexane to adipic acid involves 1,2-cyclohexanedione as an intermediate, which results due to the oxidation of cyclohexanone. The oxidation of cyclohexanone to 1,2-cyclohexanedione may involve relatively high activation energy pathway and may not be achievable in the absence of catalyst.

10.6 SUMMARY

Among various metal nanoparticles, gold nanoparticles receive much greater attention due to their viable applications in biosensing, diagnostics, *in vivo* imaging, gene delivery, and catalysis. The extracellular synthesis of gold nanoparticles by biological systems such as plants, algae, and microorganisms as sources of reducing agents and stabilizing molecules that catalyze specific reactions is a simple and environment-friendly approach. The secretion of extracellular enzymes or molecules by microorganisms offer the advantage of production of pure nanoparticles devoid of other cellular components or proteins associated with the microbes. Identification of genes encoding these reducing agents and capping molecules makes the possibility of cloning these genes and overexpression of these molecules in certain host systems. Thereby, the control of size and shape of the biogenic nanoparticles can be achieved. This green chemistry approach and the process have many advantages such as economic viability, scaling up, and eco-friendly nature.

ACKNOWLEDGMENT

We thank the Department of Biotechnology (DBT), Ministry of Science and Technology, Government of India.

ABBREVIATIONS AND SYMBOLS

AES	Auger electron spectroscopy
AFM	Atomic force microscopy
ATPase	Adenine triphosphatase
Au	Gold
BSE	Backscattered electrons
DNA	Deoxyribonucleic acid
EDS	Energy dispersive x-ray spectroscopy
ELISA	Enzyme-linked immunosorbent assay
EM	Electron microscopy
fcc	Face-centered cubic
FTIR	Fourier transform infrared
FWHM	Full width at half maximum
^{121}I	121 Iodine
HRTEM	High-resolution transmission electron microscopy
IgG	Immunoglobulin G
IGSS	Immunogold silver staining
MTP	Multiply twinned particles
NIR	Near-infrared
QSE	Quantum size effect
SAED	Selected area electron diffraction
SAS	Small angle scattering
SEM	Scanning electron microscopy
SERS	Surface-enhanced Raman scattering
Sp	Species
SPM	Scanning probe microscopy
SPR	Surface plasmon resonance
STM	Scanning tunneling microscopy
TEM	Transmission electron microscopy
UV-Vis	Ultraviolet-visible
XRD	X-ray diffraction
XPS	X-ray photoelectron spectroscopy
E_f	Fermi energy
ΔG	Gibbs free energy
h	Planck's constant
q	Electron charge

Z	Atomic number
Å	Angstrom
°C	Celsius temperature
kDa	Kilodalton
keV	Kiloelectron volt
m	Meter
m	Electron mass
MPa	Mega pascal
nm	Nanometer
V	Volt
δ	Delta
λ	Lambda
θ	Theta
μm	Micrometer
ν	Nu

BASIC SI UNITS

Physical Quantity	Name of SI Unit	Symbol
Length	Meter	m
Mass	Kilogram	kg
Time	Second	s
Electric current	Ampere	A
Thermodynamic temperature	Kelvin	K
Pressure	Pascal	Pa
Electric charge	Coulomb	C
Electric potential difference	Volt	V
Frequency	Hertz	Hz

POWERS OF UNITS—PREFIXES

Multiple	Prefix	Symbol
10^3	Kilo	k
10^{-3}	Milli	M
10^{-6}	Micro	μ
10^{-9}	Nano	n
10^{-10}	Angstrom	Å

Values of some physical constants in SI units

Planck constant (*h*): 6.63×10^{-34} J s

GLOSSARY

Absorbance: It is the amount of light absorbed by a substance suspended in gas, liquid, or solid matrix. It is used to quantitatively measure the concentration of substance.

American Type Culture Collection (ATCC): Independent nonprofit organization for preservation and distribution of reference cultures.

Angstrom (Å): It is the unit of length to measure atoms and molecules, 10^{-10} m or 0.1 nm.

Antibody: It is a group of immunoglobulins that encounters foreign substances or toxins in the body.

Bacteria: Greek *bakterion* means stick. They are unicellular prokaryotic microscopic organisms with different shapes such as round, rodlike, spiral, and filamentous.

Biosensor: It is a device used to detect the presence of certain chemicals or molecules such as DNA, antigen, chemicals, and toxins.

Biosynthesis: The production of a chemical or compounds by a living organism.

Bottom-up: Building of larger objects with smaller molecules or atoms.

Brownian motion: This is named after scientist Robert Brown. It is the random inherent motion of particles in a fluid due to thermal agitation.

Catalysis: A process of chemical reaction with increased rate. The entity that catalyzes the reaction is called catalyst.

Dalton: A unit of mass very nearly equal to that of a hydrogen atom (1.000 on atomic mass scale).

Electron microscopy: It is a technique to magnify and visualize very small entities such as viruses and large molecules using electron beams instead of light rays.

Enzymes: These proteins function as organic catalyst in biological systems and are very specific for chemical reactions.

Eukaryote: Living organism with one or more cells with significant nucleus and cytoplasm. All viruses and bacteria do not belong to this class.

Flavonols: These are phytochemicals mainly present in citrus plants. It is antioxidant in nature.

Fluorophores: Molecules known as fluorophores absorb specific wavelength of light and emit light of longer wavelength than the wavelength of absorbed light.

Fungus: A major group of saprophytic and parasitic eukaryotic organisms such as yeasts, mushrooms, molds, mildews, ergots, rusts, and toadstools that lack chlorophyll.

Gene therapy: Insertion of genes into selected cells in the body with a purpose of therapeutic applications.

In vivo: Latin for "within the living" cell or organism.

Kilodalton (kDa): A unit of mass equal to 1000 Da.

Nanocrystals: It is the crystal line structure possessing dimensions (e.g., overall width) measured in nanometers.

Pathogen: Pathogen refers to virus, bacterium, parasitic protozoans, or other microbes that causes infection by invading the host (plants, animals, etc.).

Terpenoids: Phytochemicals containing cyclic hydrocarbon molecules derived from plants especially conifers, oranges, and certain blue-green algae.

REFERENCES

Ahmad, A., Senapati, S., Khan, M. I., Kumar, R., and Sastry, M. 2003. Extracellular biosynthesis of monodisperse gold nanoparticles by a novel extremophilic actinomycete, *Thermomonospora* sp. *Langmuir* 19:3550–3553.

Ahmad, A., Senapati, S., Khan, M. I., Kumar, R., and Sastry, M. 2005. Extra-/intracellular biosynthesis of gold nanoparticles by an alkalotolerant fungus, *Trichothecium* sp. *J Biomed Nanotechnol* 1:47–53.

Aiyer, H. N., Vijayakrishnan, V., Subbanna, G. N., and Rao, C. N. R. 1994. Investigations of Pd clusters by the combined use of HREM, STM, high-energy spectroscopies and tunneling conductance measurements. *Surf Sci* 313:392–398.

Ankamwar, B., Chinmay, D., Absar, A., and Murali, S. 2005a. Biosynthesis of gold and silver nanoparticles using *Emblica officinalis* fruit extract, their phase transfer and transmetallation in an organic solution. *J Nanosci Nanotechnol* 5:1665–1671.

Ankamwar, B., Chaudhary, M., Sastry, M. 2005b. Gold nanotriangles biologically synthesized using tamarind leaf extract and potential application in vapor sensing. *Synth React Inorg Metal-Org Nanometal Chem* 35:19–26.

Badri Narayanan, K. and Sakthivel, N. 2008. Coriander leaf mediated biosynthesis of gold nanoparticles. *Mater Lett* 62:4588–4590.

Bai, C. 1999. *Scanning Tunneling Microscopy and Its Applications*. New York: Springer Verlag.

Bartz, M., Kuther, J., Seshadri, R., and Tremel, W. 1998. Colloid-bound catalysts for ring-opening metathesis polymerization: A combination of homogenous and heterogeneous properties. *Angew Chem Int Ed Engl* 37:2466–2468.

Binnig, G., Rohrer, H., Gerber, C., and Weibel, E. 1982. Surface studies by scanning tunneling microscopy. *Phys Rev Lett* 49:57–61.

Bohren, C. F. and Huffman, D. R. 1983. *Absorption and Scattering of Light by Small Particles*. New York: John Wiley & Sons.

Borchert, H., Schevchenko, E. V., Robert, A. et al. 2005. Determination of nanocrystal sizes: A comparison of TEM, SAXS, and XRD studies of highly monodisperse $CoPt_3$ particles. *Langmuir* 21:1931–1936.

Campion A. 1998. Surface-enhanced Raman scattering. *Chem Soc Rev* 27:241–250.

Cang, H., Wong, C. M., Xu, C. S., Rizvi, H. H., and Yang, H. 2006. Confocal three dimensional tracking of a single nanoparticle with concurrent spectroscopic readouts. *Appl Phys Lett* 88:223901.

Cao, Y. C., Jin, R., and Mirkin, C. A. 2002. Nanoparticles with Raman spectroscopic fingerprints for DNA and RNA detection. *Science* 297:1536–1540.

Chandran, S. P., Chaudhary, M., Pasricha, R., Ahmad, A., and Sastry, M. 2006. Synthesis of gold nanotriangles and silver nanoparticles using *Aloe vera* plant extract. *Biotechnol Prog* 22:577–583.

Chen, D. R., Wendt, C. H., and Pui, D. Y. H. 2000. A novel approach for introducing bio-materials into cells. *J Nanopart Res* 2:133–139.

Cognet, L., Tardin, C., Boyer, D., Choquet, D., Tamarat, P., and Lounis, B. 2003. Single metallic nanoparticle imaging for protein detection in cells. *Proc Natl Acad Sci USA* 100:11350–11355.

Colvard, C. 1992. *Encyclopedia of Materials Characterization*, C. R. Brundle, C. A. Evans Jr., and S. Wilson (eds.), p. 373. Stoneham, MA: Butterworth-Heinemann.

Cullity, B. D. and Stock, S. R. 2001. *Elements of X-Ray Diffraction*. Upper Saddle River, NJ: Prentice Hall.

Danscher, G. 1981. Histochemical demonstration of heavy metals, a revised version of the sulphide silver method suitable for both light and electron microscopy. *Histochemistry* 71:1–16.

Fadley, C. S. 1978. *Electron Spectroscopy: Theory, Techniques and Applications*, C. R. Brundle and A. D. Baker (eds.), pp. 1–156. New York: Academic Press.

Felsenfeld, D. P., Choquet, D., and Sheetz, M. P. 1996. Ligand binding regulates the directed movement of β1 integrins on fibroblasts. *Nature* 383:438–440.

Glatter, O. and Kratky, O. 1982. *Small Angle X-Ray Scattering*. New York: Academic Press.

Goldstein, A. N., Echer, C. M., and Alivisatos, A. P. 1992. Melting in semiconductor nanocrystals. *Science* 256:1425–1427.

Gole, A., Dash, C. V., Ramachandran, V. et al. 2001. Pepsin-gold colloid conjugates: Preparation characterization and enzymatic activity. *Langmuir* 17:1674–1679.

Griffiths, P. R. and De Haseth, J. A. 1986. *Fourier Transform Infrared Spectroscopy*. New York: John Wiley & Sons.

Guinier, A. and Fournet, G. 1955. *Small Angle Scattering of X-Rays*. New York: John Wiley & Sons.

Halperin, W. P. 1986. Quantum size effects in metal particles. *Rev Mod Phys* 58:533–606.

Hammer, B. and Norskov, J. 1995. Why gold is the noblest of all the metals. *Nature* 376:238–240.

Haruta, M. 1997. Size and support dependency in the catalysis by gold. *Catal Today* 36:153–166.

He, S., Guo, Z., Zhang, Y., Zhang, S., Wang, J., and Gu, N. 2007. Biosynthesis of gold nanoparticles using the bacteria *Rhodopseudomonas capsulata*. *Mater Lett* 61:3984–3987.

Hermanson, G. T. 1996. *Bioconjugate Techniques*. San Diego, NY: Academic Press.

Hildebrandt, P. and Stockburger, M. 1984. Surface-enhanced resonance Raman spectroscopy of rhodamine 6G adsorbed on colloidal silver. *J Phys Chem* 88:5935–5944.

Howland, R. S. and Kirk, M. D. 1992. *Encyclopedia of Materials Characterization*, C. R. Brundle, C. A. Evans Jr., and S. Wilson (eds.), p. 85. Stoneham, MA: Butterworth-Heinemann.

Huang, S. H. 2006. Gold nanoparticle-based immunochromatographic test for identification of *Staphylococcus aureus* from clinical specimens. *Clin Chim Acta* 373:139–143.

Huang, X. H., Jain, P. K., El-Sayed, I. H., and El-Sayed, M. A. 2006. Determination of the minimum temperature required for selective photothermal destruction of cancer cells with the use of immunotargeted gold nanoparticles. *Photochem Photobiol* 82:412–417.

Huang, J., Li, Q., Sun, D. et al. 2007. Biosynthesis of silver and gold nanoparticles by novel sundried *Cinnamomum camphora* leaf. *Nanotechnol* 18:105104–105114.

Huang, X., Jain, P., El-Sayed, I., and El-Sayed, M. 2008. Plasmonic photothermal (PPTT) using gold nanoparticles. *Lasers Med Sci* 23:217–228.

Huff, T. B., Tong, L., Zhao, Y., Hansen, M. N., Cheng, J. X., and Wei, A. 2007. Hyperthermic effects of gold nanorods on tumor cells. *Nanomedicine* 2:125–132.

Husseiny, M. I., Abd El-Aziz, M., Badr, Y., and Mahmoud, M. A. 2007. Biosynthesis of gold nanoparticles using *Pseudomonas aeruginosa*. *Spectrochim Acta A* 67:1003–1006.

Jorgensen, C. K. 1962. *Absorption Spectra and Chemical Bonding in Complexes*. New York: Pergamon.

Kattumuri, V., Katti, K., Bhaskaran, S. et al. 2007. Gum arabic as a phytochemical construct for the stabilization of gold nanoparticles *in vivo* pharmacokinetics and x-ray-contrast-imaging studies. *Small* 3:333–341.

Kerker, M. 1969. *The Scattering of Light and Other Electromagnetic Radiation*. New York: Academic Press.

Kneipp, J., Kneipp, H., and Kneipp, K. 2006. Two-photon vibrational spectroscopy for biosciences based on surface-enhanced hyper-Raman scattering. *Proc Natl Acad Sci USA* 103:17149–17153.

Kreibig, U. and Vollmer, M. 1995. *Optical Properties of Metal Clusters*. Berlin, Germany: Springer.

Kuriyama, S., Mitoro, A., Tsujinoue, H. et al. 2000. Particle-mediated gene transfer into murine livers using a newly developed gene gun. *Gene Ther* 7:1132–1136.

Kusumi, A., Nakada, C., Ritchie, K., Murase, K., Suzuki, K., Murakoshi, H., Kasai, R. S., Kondo, J., and Fujiwara, T. 2005. Paradigm shift of the plasma membrane concept from the two-dimensional continuum fluid to the partitioned fluid: High-speed single-molecule tracking of membrane molecules. *Biophys Biomol Struct* 34:351–378.

Lasne, D., Blab, G. A., Berciaud, S. et al. 2006. Single nanoparticle photothermal tracking (SNaPT) of 5-nm gold beads in live cells. *Biophys J* 91:4598–4604.

Lawes, G. 1987. *Scanning Electron Microscopy and X-Ray Microanalysis*. Chichester, U.K.: John Wiley & Sons.

Levy, R., Wang, Z. X., Duchesne, L., Doty, R. C., Cooper, A. I., Brust, M., and Fernig, D. G. 2006. A generic approach to monofunctionalized protein-like gold nanoparticles based on immobilized metal ion affinity chromatography. *ChemBioChem* 7:592–594.

Li, H., Luk, Y. Y., and Mrksich, M. 1999. Catalytic asymmetric dihydroxylation by gold colloids functionalized with self-assembled monolayers. *Langmuir* 15:4957–4959.

Lu, L., Ai, K., Ozaki, Y. et al. 2008. Environmentally friendly synthesis of highly monodisperse biocompatible gold nanoparticles with urchin-like shape. *Langmuir* 24:1058–1063.

Moreno-Manas, M. and Pleixats, R. 2003. Formation of carbon–carbon bonds under catalysis by transition-metal nanoparticles. *Acc Chem Res* 36:638–643.

Moshovists M. 1985. Surface enhanced spectroscopy. *Rev Mod Phys* 57:783–826.

Mukherjee, P., Ahmad, A., Mandal, D. et al. 2001. Bioreduction of $AuCl_4^-$ ions by the fungus, *Verticillium* sp. and surface trapping of the gold nanoparticles formed. *Angew Chem Int Ed* 40:3585–3588.

Mukherjee, P., Senapati, S., Mandal, D. et al. 2002. Extracellular synthesis of gold nanoparticles by the fungus *Fusarium oxysporum*. *ChemBioChem* 3:461–463.

Nakamura, K., Kawabata, T., and Mori, Y. 2003. Size distribution analysis of colloidal gold by small-angle x-ray scattering and light absorbance. *Powder Technol* 131:120–128.

Newbury, D. E., Joy, D. C., Echlin, P., Fiori, C. E., and Goldstein, J. I. 1986. *Advanced Scanning Electron Microscopy and X-Ray Microanalysis*. New York: Plenum press.

Ni, J., Lipert, R. J., Dawson, G. B., and Porter, M. D. 1999. Immunoassay readout method using extrinsic Raman labels adsorbed on immunogold colloids. *Anal Chem* 71:4903–4908.

Nie, S. and Emory, S. R. 1997. Probing single molecules and single nanoparticles by surface-enhanced Raman scattering. *Science* 275:1102–1106.

Orhring, M. 1992. *The Materials Science of Thin Films*. San Diego, NY: Academic Press.

Otto A. 1991. Surface enhanced Raman scattering of adsorbates. *J Raman Spectr* 22:743–752.

Otto, A., Mrozek, I., Grabhorn, H., and Akemann, W. 1992. Surface-enhanced Raman scattering, *J Phys Condens Matter* 4:1143–1212.

Pasquato, L., Rancan, F., Scrimin, P., Mincin, F., and Frigeri, C. 2000. *N*-Methylimidazole-functionalized gold nanoparticles as catalysts for cleavage of a carboxylic acid ester. *Chem Commun* 22:2253–2254.

Perenboom, J. A. A., Wyder, P., and Meier, P. 1981. Electronic properties of small metallic particles. *Phys Rep* 78:173–292.

Pissuwan, D., Valenzuela, S. M., and Cortie, M. B. 2006. Therapeutic possibilities of plasmonically heated gold nanoparticles. *Trends Biotechnol* 24:62–67.

Pyykko, P. 1988. Relativistic effects in structural chemistry. *Chem Rev* 88:563–594.

Rao, C. N. R. and Cheetham, A. K. 2001. Science and technology of nanomaterials: Current status and future prospects. *J Mater Chem* 11:2887–2893.

Rao, B. R. R., Kaul, P. N., Syamasundar, K. V. et al. 2002. Water soluble fractions of rose-scented geranium (*Pelargonium* species) essential oil. *Bioresour Technol* 84:243–246.

Sastry, M., Ahmad, A., Khan, M. I., and Kumar, R. 2003. Biosynthesis of metal nanoparticles using fungi and actinomycete. *Curr Sci* 85:162–170.

Schatz, G. C. 1984. Theoretical studies of surface enhanced Raman scattering. *Acc Chem Res* 17:370–376.

Segmuller, A. and Murakami, M. 1985. *Thin Films from Free Atoms and Particles*, K. J. Klabunde (ed.), p. 325. Orlando, FL: Academic Press.

Segmuller, A. and Murakami, M. 1988. *Analytical Techniques for Thin Films*, K. N. Tu and R. Rosenberg (eds.), p. 143. San Diego, NY: Academic Press.

Shankar, S. S., Ahmad, A., Pasricha, R., and Sastry, M. 2003. Bioreduction of chloroaurate ions by geranium leaves and its endophytic fungus yields gold nanoparticles of different shapes. *J Mat Chem* 13:1822–1826.

Shankar, S. S., Rai, A., Ahmad, A., and Sastry, M. 2004a. Rapid synthesis of Au, Ag, and bimetallic Au core-Ag shell nanoparticles using neem (*Azadirachta indica*) leaf broth. *J Colloid Interface Sci* 275:496–502.

Shankar, S. S., Rai, A., Ankamwar, B., Singh, A., Ahmad, A., and Sastry, M. 2004b. Biological synthesis of triangular gold nanoprisms. *Nat Mater* 3:482–488.

Shankar, S. S., Rai, A., Ahmad, A., and Sastry, M. 2004c. Biosynthesis of silver and gold nanoparticles from extracts of different parts of the geranium plant. *Appl Nanosci* 1:69–77.

Singaravelu, G., Arockiamary, J. S., Ganesh Kumar, V., and Govindaraju, K. 2007. A novel extracellular synthesis of monodisperse gold nanoparticles using marine alga, *Sargassum wightii* Greville. *Colloids Surf B Biointerfaces* 57:97–101.

Sperling, R. A., Pellegrino, T., Li, J. K., Chang, W. H., and Parak, W. J. 2006. Electrophoretic separation of nanoparticles with a discrete number of functional groups. *Adv Funct Mater* 16:943–948.

Templeton, A. C., Wuelfing, W. P., and Murray, R. W. 2000. Monolayer-protected cluster molecules. *Acc Chem Res* 33:27–36.

Wang, Y. L., Silverman, J. D., and Cao, L. G. 1994. Single particle tracking of surface receptor movement during cell division. *J Cell Biol* 127:963–971.

Wang, Z., Lee, J., Cossins, A. R., and Brust, M. 2005. Microarray-based detection of protein binding and functionality by gold nanoparticle probes. *Anal Chem* 77:5770–5774.

Wen, L., Lin, Z., Gu, P. et al. 2009. Extracellular biosynthesis of monodispersed gold nanoparticles by a SAM capping route. *J Nanopart Res* 11:279–288.

Yang, N., Burkholder, J., Roberts, B., Martinell, B., and McCabe, D. 1990. *In vivo* and *in vitro* gene transfer to mammalian somatic cells by particle bombardment. *Proc Natl Acad Sci USA* 87:9568–9572.

Zanchet, D., Micheel, C. M., Parak, W. J., Gerion, D., and Alivisatos, A. P. 2001. Electrophoretic isolation of discrete Au nanocrystal/DNA conjugates. *Nano Lett* 1:32–35.

Zeman, E. J. and Schatz, G. C. 1987. An accurate electromagnetic theory study of surface enhancement factors for silver, gold, copper, lithium, sodium, aluminum, gallium, indium, zinc, and cadmium. *J Phys Chem* 91:634–643.

11 Gold Nanoparticles Conjugated to Peptides

*Guillermina Ferro-Flores, Blanca E. Ocampo-García,
Flor de María Ramírez, Claudia E. Gutierrez-Wing,
Consuelo Arteaga de Murphy, and Clara L. Santos-Cuevas*

CONTENTS

11.1 PEPTIDES FOR *IN VIVO* DIAGNOSIS AND THERAPY

11.1.1 PRINCIPLE OF *IN VIVO* PEPTIDE RECEPTOR CANCER TARGETING

Peptides are molecules consisting of several amino acids linked together with peptide bonds. Their size can vary from molecules with only two amino acids to as much as 50. Peptides do not only exist in a natural form but can also be designed synthetically as novel molecules.

Regulatory peptides represent a group of different families of molecules known to act on multiple targets in the human body. They control and modulate the function of almost all key organs and metabolic processes and include neuropeptides present in the brain, gut peptide hormones, as well as peptides present in vascular (vasoactive peptides) and endocrine systems (Reubi, 2003).

Regulatory peptide receptors are proteins overexpressed in numerous human cancer cells. These receptors have been used as molecular targets for radiolabeled peptides to localize cancer tumors (Reubi and Maecke, 2008). The useful clinical results achieved during the last decade with somatostatin receptor–expressing neuroendocrine tumor imaging have been extended to the study of other peptides to target alternative cancer-associated peptide receptors such as gastrin-releasing peptide (GRP), cholecystokinin, peptide ligands for integrin receptors, or neurotensin (NT). The improvement of peptide analogues allows specific clinical imaging and therapy of different tumor types, including breast, prostate, lung, intestine, pancreas, and brain tumors (Ferro-Flores et al., 2006a; de Visser et al., 2008; Reubi and Maecke, 2008). Therefore, specific cancer targeting through selective peptides for diagnostic and therapeutic purposes is considered to be a promising strategy in oncology.

Molecular imaging comprises noninvasive monitoring of functional and spatiotemporal processes at molecular and cellular levels in humans and other living systems. Imaging techniques such as magnetic resonance imaging (MRI), single-photon emission computed tomography (SPECT), positron emission tomography (PET), and optical imaging (OI: bioluminescence and fluorescence) have been used to monitor such processes. *In vivo* peptides for cancer imaging using the different medical diagnostic techniques such as those mentioned above rely mostly on the use of radiopeptides and peptides conjugated to near-infrared (NIR) fluorochromes, metallic nanoparticles, or quantum dots (nanocrystals). Conjugating gold nanoparticles (AuNPs) to peptides is particularly interesting because of its feasibility to produce biocompatible and stable multifunctional systems with target-specific molecular recognition.

11.1.2 COMMON PEPTIDES USED AS DIAGNOSTIC AGENTS AND DRUGS

The term molecular imaging implies the *in vivo* characterization and measurement of biological processes at a cellular and molecular level. In contrast to "classical" diagnostic imaging, molecular imaging endeavors to probe the molecular abnormalities that are the basis of disease rather than to image the end effects of these molecular alterations (Weissleder and Mahmood, 2001).

Regulatory peptide analogues represent a class of molecules developed for specific cancer targeting. In spite of having a relatively short history of about two decades, they are now receiving increasing interest, as they often advantageously compare with immunotargeting using antibodies (Reubi and Maecke, 2008).

11.1.2.1 Somatostatin Analogues

Somatostatin is a cyclic peptide comprised of 14 amino acids and plays an important role in the secretion of hormones, such as growth hormone, insulin, and glucagon. It is now recognized that there are five somatostatin receptor subtypes. Octreotide (OC) was developed as a somatostatin analogue for hypersecretion suppression to control the symptoms of neuroendocrine diseases. OC contains eight amino acids retaining an internal disulfide crosslink to constrain the geometry of the four essential amino acids, and is stable against enzymatic degradation *in vivo*. Pituitary adenomas and several neuroendocrine tumors overexpress somatostatin receptors such as carcinoid, pancreas, pheochromocytomas, medullary thyroid carcinoma, paragangliomas, gastrinoma, glucagonoma, neuroblastoma, meningioma, insulinoma, and small cell lung cancer. In somatostatin-based cancer imaging, a stable somatostatin analogue is linked to a bifunctional chelating agent (BFCA) that can bind radioactive metals such as ^{111}In, ^{99m}Tc, or ^{68}Ga. For example, ^{99m}Tc-Tyr^3-OC (^{99m}Tc-TOC) is currently used as a stable complex to detect neuroendocrine tumors by molecular imaging in nuclear medicine (Figure 11.1A) (Plachcinska et al., 2004; Gonzalez-Vazquez et al., 2006; Czepczynski et al., 2007).

11.1.2.2 Bombesin/Gastrin-Releasing Peptide Analogues

The small peptide bombesin (BN, 14 amino acids) was isolated from frog skin and it belongs to a large group of neuropeptides with many biological functions. The human equivalent is the gastrin-releasing peptide (GRP, 27 amino acids) and its receptors (GRP-r) are overexpressed in the tumor cell membrane. GRP and bombesin differ by only one of the 10 carboxy-terminal residues and this explains the similar biological activity of the two peptides. The strong, specific BN-GRP-r binding

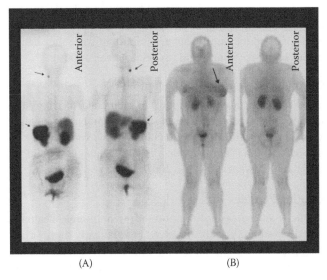

(A)   (B)

FIGURE 11.1 Anterior and posterior whole-body gamma images obtained 2 h after administration of (A) ^{99m}Tc-HYNIC-Tyr^3-octreotide (patient with metastatic gastrinoma) and (B) ^{99m}Tc-HYNIC-Bombesin (patient with breast cancer).

is the basis for labeling BN with radionuclides (Smith et al., 2003; Alves et al., 2006; Ferro-Flores et al., 2006b; Kunstler et al., 2007). BN receptor subtype 2 (GRP receptor) is overexpressed in various human tumors including breast, prostate, small cell lung, and pancreatic cancers (Figure 11.1B) (de Visser et al., 2007; Santos-Cuevas et al., 2008; van de Wiele et al., 2008).

11.1.2.3 Peptide Ligands for Cholecystokinin (CCK) and Gastrin Receptors

Cholesytokinin (CCK) and gastrin act as neurotransmitters in the central nervous system, as regulators of various functions in the gastrointestinal tract, and as stimulatory growth factors in several tumors, such as colon and gastric cancers. They are structurally related in that they share the same C-terminal five amino acids (-Gly-Trp-Met-Asp-Phe-NH2), which is the active site for binding to cholecystokinin-2 (CCK-2) receptor. CCK-2 receptor protein has been identified in cell membranes of medullary thyroid carcinomas (92%), whereas it is absent in differentiated thyroid cancers (Reubi, 2007; Reubi and Maecke, 2008).

11.1.2.4 GLP-1 Receptors in Insulinomas

Glucagon-like peptide-1 (GLP-1) is an intestinal hormone that stimulates insulin secretion through receptors expressed on islet cells. GLP-1 has been found to have receptors that are massively overexpressed in virtually all insulinomas (Reubi and Waser, 2003; Wild et al., 2008).

11.1.2.5 Peptide Ligands for Integrin αvβ3: Markers of Tumor Angiogenesis

"Angiogenesis represents the formation of new capillaries by cellular outgrowth from existing microvessels" (Bhattacharya and Mukherjee, 2008). Integrins are cell-adhesion receptors able to convert extracellular ligand binding into activation of intracellular processes (outside-in signaling) as well as employing intracellular processes to activate cellular responses (inside-out signaling). The αvβ3 integrin is involved in tumor-induced angiogenesis and tumor metastasis. The high binding specificity to αvβ3 integrins of peptides containing Arg-Gly-Asp (RGD) residues has been used to radiolabel RGD peptides, which are used as tumor-specific imaging agents.

A considerable number of synthetic peptides containing RGD have been developed. It has been found that constraining the RDG mobility in a cyclized pentapeptide increased the potency *in vitro*. These data suggest that the binding site in the αvβ3 receptor is limited. The peptide ligand for αvβ3 receptors mostly used in the labeling with technetium-99m is the c-RGDfK (cyclic-Arg-Gly-Asp-D-Phe-Lys) in both forms, as a monomer or as a dimer. In this peptide, the lysine side chain provides primary amine functionality for coupling BFCAs. It is important to mention that RGD peptides do not belong to the regulatory peptide family, but are important since they can target neoangiogenic vessels through integrin receptors (Kenny et al., 2008; Liu et al., 2009).

11.1.2.6 Peptide Ligands for Neurotensin Receptors

NT is a 14-amino acid linear peptide that is found in high concentration in the ileum and hypothalamus, and induces various physiologic effects such as hypotension, analgesia, gut contraction, and an increase of vascular permeability. Receptors of NT are expressed in pancreatic and prostate cancer (Reubi and Maecke, 2008). Neuropeptide-Y (NPY) analogues are also useful for detecting sarcomas (Korner et al., 2008; Zwanziger et al., 2008).

11.1.2.7 Tat Penetrating Peptides

Targeted entry into cells is an increasingly important research area. Diagnoses and treatment of disease by novel methods would be greatly enhanced by efficiently transporting materials to living cell nuclei. Penetrating peptides are attractive drug delivery tools. The HIV Tat-derived peptide is a small basic peptide called "trojan horse" because it has been successfully used to deliver a large variety of cargoes into cells such as nanoparticles, proteins, peptides, and nucleic acids. The "transduction domain" or region conveying cell-penetrating properties appears to be confined to a small stretch of basic amino acids with the sequence RKKRRQRRR and known as Tat(49–57). (Dietz and

Bähr, 2004; Deshayes et al., 2005; Cornelissen et al., 2008; Costantini et al., 2008; Kersemans et al., 2008). Tat peptides do not belong to the regulatory peptide family.

11.1.3 RADIOLABELED PEPTIDES FOR TUMOR THERAPY

An important benefit of receptor-specific peptides is their use for targeted radiotherapy. A peptide labeled with an imaging radionuclide such as ^{99m}Tc could be used for identifying malignant tumors and metastatic sites and for the treatment-planning dosimetry (energy delivered per unit of targeted mass, J/kg = Gray = Gy). The same molecule labeled with a particle-emitting radionuclide could be used for therapy (e.g., yttrium-90, ^{90}Y and lutetium-177, ^{177}Lu).

In general, a target-specific radiopharmaceutical, designed through the "bifunctional approach," is composed of three parts: the radiometal ion, a BFCA, and a bioactive targeting fragment (e.g., peptides). The BFCA located between the radionuclide and the targeting molecule strongly coordinates the metal ion and is covalently bound to the receptor-specific molecule either directly or through a linker molecule. The bioactive fragment serves as the delivering system, which carries the radionuclide to the receptor site at the target cell. Examples of BFCA are HYNIC (6-hydrazinopyridine-3-carboxylic acid) and DOTA (1,4,7,10-tetraazacyclododecane-N,N',N'',N'''-tetraacetic acid) (Ferro-Flores et al., 2006a).

The most notable success in peptide development has been the use of radiolabeled somatostatin analogues that can be linked, via a chelator, to ^{90}Y or ^{177}Lu for the diagnosis and treatment of somatostatin receptor–positive tumors. Some of the reported analogues are DOTA-Tyr^3-octreotide (DOTA-TOC), DOTA-lanreotide (DOTA-LAN), and DOTA-Tyr^3-octreotate (DOTA-TATE) (de Jong et al., 2004). Patients with metastatic neuroendocrine tumors have received 1.7–27 GBq during 1–5 cycles of ^{90}Y-DOTA-TOC, leading to renal radiation doses of 5.9–26.9 Gy/cycle and a total of 18.3–38.7 Gy. Other patients with metastatic neuroendocrine tumors have received 3–7 cycles (3.7–7.4 MBq) of ^{177}Lu-DOTA-TATE up to a final cumulative activity of 22.2–29.6 GBq, leading to renal radiation doses of 1.8–7.8 Gy/cycle and a total of 7.3–26.7 Gy. The absorbed dose by tumors has reached up to 600 Gy. The results obtained are very encouraging in terms of tumor regression. If kidney protective agents are used, the side effects of this therapy are few and mild, and the therapy response duration for both radiopharmaceuticals is more than 2 years. These data compare favorably with those for the limited number of alternative treatment approaches (Kwekkeboom et al., 2005).

11.2 GOLD NANOPARTICLES (AUNPS) IN MEDICAL APPLICATIONS

11.2.1 BIOLOGICAL AND PHYSIOLOGICAL PROPERTIES OF AuNPs

Gold has a long history of use. The therapeutic use of red colloidal gold can be traced back to the Chinese in 2500 BC (Higby, 1982). Gold has been used for rejuvenation, nervous conditions, and the treatment of tuberculosis, epilepsy, syphilis, and rheumatic diseases (Bhattacharya and Mukherjee, 2008). In many cases, gold was used as Au(I) and Au(III) complexes causing nephrotoxicity, mouth ulcers, skin reactions, blood disorders, and liver toxicity. Interesting is the antitumor activity of bis(diphos)gold(I) complexes but with cardiovascular toxicity as a major side effect (Berners-Price and Sadler, 1987).

AuNPs exploit their unique chemical and physical properties for transporting pharmaceuticals. First, the Au(0) gold core is essentially inert and nontoxic, since AuNPs are taken up by human cells but do not cause acute cytotoxicity (Connor et al., 2005). A second advantage is their ease of synthesis. Further versatility is imparted by their ready functionalization with biological molecules to make them interact to a specific biological target. Moreover, their photophysical properties could trigger drug release at a remote site (Bhattacharya et al., 2007). It is believed nowadays that reduced Au(0) under biological conditions may be the anticancer active species because of its anti-angiogenic property (Mukherjee et al., 2005; Bhattacharya and Mukherjee, 2008).

Two approaches have been developed for targeting, namely, "passive" and "active" (Brigger et al., 2002). Passive targeting depends on homing of the vectors in unhealthy tissues due to extravasation through leaky blood vessels (gaps ~600–800 nm).

An important aspect of carrier systems in the 5–20 nm scale is their ability to take advantage of the enhanced permeation and retention (EPR) effect (Maeda et al., 2000). On the other hand, active targeting presents ligands on the carrier surface for specific recognition by cell surface receptors. The ligands could be small molecules, peptides, or proteins. Combinations of both types of targeting will render an ideal carrier for *in vivo* delivery (Ghosh et al., 2008).

From the point of view of molecular recognition, peptides have excellent properties that allow them to participate in ligand-receptor molecular interactions. AuNPs coated with peptides increase their stability and biocompatibility allowing them to be directed to the desired target (Olmedo et al., 2008).

11.2.2 BIOCOMPATIBILITY AND STABILITY OF AuNPs IN BIOLOGICAL MEDIA

As mentioned before, AuNPs have been examined for uptake and acute toxicity in human cells (Connor et al., 2005). The nanoparticles (average diameter = 18 nm), which possessed various surface modifiers, were not toxic to cells during continuous exposure. Citrate-capped nanoparticles were further examined for their cellular uptake by ultraviolet visible (UV-Vis) spectrometry and transmission electron microscopy. Results indicate that although some precursors of nanoparticles may be toxic, the nanoparticles themselves are not necessarily detrimental to cellular function. Goodman et al. (2004) also demonstrated that anionic AuNPs (2 nm) are usually nontoxic. Therefore, AuNPs provide nontoxic carriers for drug delivery applications.

A drug delivery system should provide positive attributes to a "free" drug by improving solubility, *in vivo* stability, and biodistribution. In AuNPs, the gold core imparts stability to the assembly, while the monolayer allows tuning surface properties. If the monolayer is a regulatory peptide analogue, most of the biological properties will correspond to that conferred by the peptide itself, including the biological stability. Thus, a macromolecular AuNP capped by peptide produces stable and uniform multifunctional systems in biological media (Wang et al., 2005; Surujpaul et al., 2008; Cheung et al., 2009). Attaching multiple units of a receptor-specific peptide to one AuNP (1000 or more peptide molecules to one 20 nm AuNP) provide useful tools in improving imaging or therapy of tumors overexpressing peptide receptors. Their combination of low inherent toxicity, ample surface area, and biological stability provides them with unique attributes that should enable new cancer therapy strategies.

As described earlier, under *in vivo* conditions, AuNP peptides are delivered to tumors by both passive and active-targeting mechanisms. The passive mode comprises two factors: (1) angiogenic tumors produce vascular endothelial growth factors (VEGF) that hyperpermeabilize the tumor-associated neovasculatures (from 600 up to 800 nm) and cause leakage of circulating macromolecules and small particles and (2) tumors lack an effective lymphatic drainage system, which leads to macromolecule accumulation. For active tumor targeting, the AuNP peptide targets specific receptors overexpressed in cancer cells.

11.2.3 POTENTIAL APPLICATIONS OF AuNPs IN CANCER AND ARTHRITIS DUE TO THEIR ANTI-ANGIOGENIC PROPERTIES

Angiogenesis, the formation of new blood vessels from existing ones, plays an important role in the growth and spread of cancer. New blood vessels "feed" the cancer cells with oxygen and nutrients, allowing these cells to grow, invade nearby tissue, spread to other parts of the body, and form solid tumors (Bussolino et al., 1997).

Recently, Mukherjee et al. (2005) demonstrated that AuNPs inhibit the vascular endothelial growth factor 165-isoform (VEGF165). VEGF165 is an endothelial cell (EC) mitogen and a prime mediator of angiogenesis that plays an important role in pathological neovascularization including rheumatoid arthritis (RA), chronic inflammation, and neoplastic disorders. This is the first example of an inorganic compound that is anti-angiogenic in nature and opens up a new area of research using inorganic nanoparticles as anti-angiogenic agent (Bhattacharya and Mukherjee, 2008).

Specifically, AuNP binds to the heparin-binding domain in VEGF165 (Mukherjee et al., 2005). Direct binding of VEGF165 to AuNPs was confirmed by x-ray photoelectron spectroscopy (XPS) analysis. The presence of a single Au 4f7/2 peak at 83.9 eV clearly demonstrated the presence of only one form of Au in solution: Au(0).

It is now recognized that angiogenesis plays a key role in the formation and maintenance of pannus in RA (Koch, 2003). The new blood capillaries promote synovial inflammation by transporting the inflammatory cells to the site of synovitis in RA. It is found that VEGF levels in serum, synovial fluid, and in inflamed synovium of RA patients are elevated (Lee et al., 2001). Therefore, AuNP therapy could be a potential treatment option for RA.

11.2.4 PHOTOTHERMAL PROPERTIES OF AuNPs IN THERAPY

From a therapeutic point of view, there are two properties of gold that are most relevant: resistance to oxidation and plasmon resonance with light (Pissuwan et al., 2006). The plasmon resonance for ordinary gold nanospheres is at 520 nm, in the middle of the visible spectrum, but this can be red-shifted to the NIR region from 800 to 1200 nm. This is useful because body tissue is moderately transparent to NIR light (Sharma et al., 2006), thereby providing an opportunity for therapeutic effects in deep tissues.

Local application of heat is a known concept in therapeutic medicine that has been extensively explored for cancer treatment and other conditions. Excitation sources, such as infrared lamps, ultrasound or lasers, can be used in the process, but there is always the problem of limiting the heat generated to just the region of the target tissue (Pissuwan et al., 2006). As mentioned earlier, this problem can be solved, in part, by using AuNPs designed to absorb in the NIR spectrum so that the resulting localized heating causes irreversible thermal cellular destruction (Hirsch et al., 2003; West and Halas, 2003).

The first use of gold particles in hyperthermal therapy was reported by Hirsch and West (Hirsch et al., 2003; West and Halas, 2003). In 2004, gold on silica (core–shell or nanoshells) conjugated to HER2 antibody were reported to target breast carcinoma cells (Loo et al., 2004). NIR irradiation led to a rise in the temperature of the target regions from 40°C to 50°C, which selectively destroyed the carcinomas. Pitsillides et al. (2003) selectively destroyed CD8C lymphocytes in a mixed CD8C and CD8C-lymphocyte culture using ordinary colloidal AuNPs. The particles were conjugated to a CD8C-specific antibody and used together with a 532 nm laser, resulting in highly selective apoptosis of the CD8C lymphocyte population.

11.3 ROUTES FOR AUNP SYNTHESIS AS REQUIRED FOR PREPARING NANOCONJUGATES

Many routes for synthesizing AuNPs have been described, but most start from Au(III) salts, which are then reduced to Au(0). Almost all methods use $HAuCl_4$ as the starting reagent dissolved in H_2O. However, methods that involve other solvents are also available.

11.3.1 CITRATE REDUCTION OF Au(III) TO Au(0)

The sodium citrate reduction technique, first described by Turkevich et al. (1951), is one of the most widely used and oldest methods. It consists of an aqueous $HAuCl_4$ reduction with trisodium citrate as the reducing and stabilizing agent. The resulting NPs acquire a citrate layer over their surface, which confers stability. Citrate can be easily displaced by several species that form stronger interactions with Au. Transmission electron microscopy (TEM) has shown that these NPs are approximately 20 nm and show little dispersion. It has also been reported that size can be controlled by changing the $HAuCl_4$:citrate ratio (Frens, 1973).

11.3.2 Surfactant Method for Direct AuNP-Peptide Synthesis

In this technique, the surfactant forms micelles, which stabilize the AuNPs. Reduction reagents stronger than citrate are usually applied, such as $NaBH_4$. Many surfactants have been used for this purpose, most of which are quaternary ammonium salts, such as cetyltrimethylammonium bromide (CTAB), didodecyldimethylammonium bromide (DDAB), and tetradodecylammonium bromide (TTAB). Anionic surfactant syntheses have also been reported (Zhang et al., 2006).

The resulting AuNPs usually have smaller mean sizes than their citrate counterparts, although they have broader size dispersion, unless a seeding procedure is used. In this procedure, smaller AuNPs (usually less than 6 nm in diameter) are synthesized first, and they are known as seeds. Later, these seeds are grown under another $HAuCl_4$-reduction stage (Busbee et al., 2003). AuNPs can be stabilized by thiols and other sulfur compounds. In this case, the AuNPs are obtained with a strong reducing agent, as in the previous method, but the strong interaction between sulfur and Au allows thiol compounds to form a self-assembled monolayer (SAM) adsorbed over the NP surface, thereby conferring the AuNPs increased stability (Brust et al., 1995).

11.3.3 Microwave and Ultrasound in AuNPs Preparation

Microwaves have been applied to increase the solvent temperature via microwave irradiation. In this case, a slightly narrower size range is achieved.

Doolittle and Dutta (2006) reported a H_2O-sodium bis(2-ethylhexyl) sulfosuccinate (AOT)-heptane reverse micelle system for the synthesis of Au particles by hydrazine reduction of $HAuCl_4$ in the presence and absence of microwave radiation. The duration of the microwave radiation was limited to a 2 min duration at a power of 300 W, thereby ensuring that the reverse micelle phase was maintained during the synthesis. For all hydrazine concentrations studied (0.5–2 M), the presence of microwave radiation led to an increase in the Au particle size by 33%–58%. In a nonmicrowaved reverse micelle system, Au growth continued with time (~15 min), whereas, upon microwaving, growth was complete in 2 min.

Shen et al. (2006) prepared hydrophobic AuNPs protected with octadecylamine, tetra-decylamine, or decylamine by using n-heptane/ethanol reduction in reverse micelle through microwave dielectric heating. The various alkylamine-stabilized AuNPs obtained through this method were characterized and analyzed by UV-Vis spectroscopy, X-ray diffraction (XRD), TEM, thermogravimetric analysis (TGA), and elemental analysis. When the molar ratio of alkylamine to $HAuCl_4$ was 40:1, the average diameters and standard deviations for octadecylamine-, tetradecylamine-, and decylamine-capped AuNPs were 4.53 ± 0.79, 5.22 ± 1.66, and 4.09 ± 1.22 nm, respectively.

The ultrasound technique is used to generate H and hydroxyl radicals. Similar to traditional methods, ultrasound methods start from Au salts, but H radicals are used as the reducing agents and a range of sizes and shapes are obtained (Okitsu et al., 2005; Park et al., 2006). The rate of sonochemical reduction of Au(III) to produce Au nanoparticles in aqueous solutions containing 1-propanol has been found to be strongly dependent upon the ultrasound frequency. The size and distribution of the Au nanoparticles produced can also be correlated with the rate of Au(III) reduction, which in turn is influenced by the applied frequency (Okitsu et al., 2005).

11.3.4 Laser Ablation and X-ray Synthesis to Stabilize AuNPs with Biocompatible Molecules

Colloidal AuNPs with a broad size distribution have been prepared by laser ablation of a gold metal plate in an aqueous solution of sodium dodecyl sulfate (SDS), and were fragmented under irradiation of a 532 nm laser at different SDS concentrations and laser fluences. AuNPs with a desired average size (1.7–5.5 nm in diameter) were prepared by properly tuning the surfactant concentration and laser fluence (Mafune et al., 2002). Thus, laser irradiation can be used through direct ablation of a metallic Au surface or to control the standard reduction of HAuCl$_4$ (Bjerneld et al., 2003). In the former case, no reduction is performed, the process is completely physical and laser simply destroys Au bulk material leaving Au nanosized pieces (Kogan et al., 2007). The resulting particles show more irregular shapes, as shown by TEM micrographs.

Laser-ablation synthesis has been used to stabilize AuNPs with known biocompatible molecules, remarkably, cyclodextrins, 3-mercaptopropionic acid, and *N*-propylamine (Kabashin, and Meunier, 2006). In this study, authors remarked that traditional chemical techniques require the presence of toxic by-products, therefore, laser ablation and x-rays are proposed as alternatives for the synthesizing of biocompatible AuNPs.

11.3.5 ISOASCORBIC ACID REDUCTION OF AUNPS

A recent study describes a simple synthesis using isoascorbic acid as a reducing reagent (Andreescu et al., 2006). The study describes a convenient, rapid, and reproducible method for the synthesis of stable dispersions of uniform AuNPs by mixing HAuCl$_4$ and isoascorbic acid aqueous solutions at room temperature. It was found that the size of the resulting nanoparticles is affected by the concentration and the pH of the gold solution, while the stability of the electrostatically stabilized final colloid is strongly dependent on reductant excess in the system, the ionic strength, and precipitation temperature. Since the preparation process does not require adding a dispersing agent, the surface of the resulting AuNPs can be easily functionalized to make them suitable for applications in medicine and biology.

It is important to mention that AuNPs can be synthesized with one of the methods described above and then functionalized with regulatory peptides.

11.4 PEPTIDE DESIGN FOR CONJUGATION TO GOLD NANOPARTICLES

The basic design principle is to create a ligand peptide that would readily attach to the surface of the gold particle and form a well-packed layer with a hydrophilic terminus. Extremely stable water/0.9% NaCl-soluble AuNP-peptide conjugates have to be obtained.

11.4.1 GOLD NANOPARTICLES AGGREGATION BY PEPTIDE INDUCTION

Gold colloids in an aqueous media have to be electrostatically stabilized. When HAuCl$_4$ reduction is carried out using trisodium citrate, AuNPs are surrounded by an electrical double layer due to adsorbed citrate and chloride anions and the cations attracted to them. An increase in ionic strength of the medium compresses the double layer and shortens repulsion range leading to an irreversible particle aggregation (Stoeva et al., 2007). Formation of large aggregates caused a change in color of the AuNP suspension from red (dispersed AuNPs) to blue (aggregated networks) and finally colorless. The citrate-capped AuNPs can be functionalized by other ligands for improving its stability. The monolayer-protected gold clusters (MPCs) can contain modified thiol ligands such as straight-chain alkanethiolates of different length, glutathione (Schaaff et al., 1998), tiopronin (Templeton et al., 1999; Kohlmann et al., 2001), thiolated poly(ethylene glycol) (PEG) (Wuelfing et al., 1998; Kanaras et al., 2002; Liu et al., 2007) *p*-mercaptophenol (Brust et al., 1995), or peptides (Levy et al., 2004; Porta et al., 2007).

In the case of AuNP stabilized by peptides, an electrolyte-induced aggregation depends on the concentration, sequence, length, hydrophobicity, and peptide charge; the AuNP size and solution pH. During aggregate formation, the UV spectra show that a new peak appearing at longer wavelength intensifies and redshifts from the original peak position (Basu et al., 2007).

11.4.2   Steric Effect and Influence of the First and Last Amino Acid in a Peptide Chain

In 2004, Levy et al. reported an interesting study on the design of peptide-capping ligands for AuNPs: "The design strategy of the peptide took into account the need to have a strong affinity for gold, ability to self-assemble into a dense layer that excludes water, and a hydrophilic terminus, which would ensure solubility and stability in water." In the designed pentapeptide CALNN (cysteine-alanine-leucine-aspargine-aspargine), the thiol group in the side chain of the N-terminal cysteine (C) is able to make a covalent bond to the gold surface. This interaction may be additive to that of the N-terminal primary amine, since amino groups are also known to have a strong interaction with gold surfaces (Zhang et al., 2000). It has been shown by Bellino et al. (2004) that the presence of a positively charged ammonium group in the vicinity of the thiol significantly accelerates the adsorption kinetics of thiols onto citrate-stabilized AuNPs. Alanine (A) and leucine (L) in positions 2 and 3 possess hydrophobic side chains and were chosen to promote peptide self-assembly. The leucine side chain is larger than that of alanine to take into account nanoparticle curvature. Asparagine (N) in positions 4 and 5 is an uncharged, but hydrophilic amino acid due to the amide group on the side chain, and the C-terminal asparagine in position 5 can bear a negative charge due to the terminal carboxylic group (Levy et al., 2004).

During the design of AuNP-CLPFFD-NH$_2$ conjugate, Olmedo et al. (2008) demonstrated that the peptide sequence, steric effect, and charge and disposition of hydrophilic and hydrophobic residues are crucial parameters when considering the design of AuNP peptide conjugates for biomedical applications.

11.4.3   Influence of Peptide Length, Hydrophobicity, and Charge

Levy et al. (2004) also demonstrated that for a two amino acid peptide (CA), the aggregation parameter increases rapidly with NaCl concentration. As the length of the peptide increases from CA to CAL, CALN, and finally to CALNN, the NaCl-induced aggregation is displaced to increasingly higher concentrations of NaCl, suggestive of a direct correlation between peptide length and stability of the peptide-capped nanoparticles. When KALNN (N-terminal lysine, K) is used, aggregation is also observed, clearly indicating that the thiol group plays a major role in stabilization. However, Porta et al. (2007) obtained red solutions of AuNP-GK without observable aggregation. The lyophilized AuNP-GK colloidal solution was redissolved in water without any UV-Vis spectra change.

Another critical factor is the negative charge position in the peptide. The AuNp-peptide negative charge should be located in an external position contributing to repulsion between AuNPs and, therefore, increasing the colloidal stability (Olmedo et al., 2008).

The biogenic glutathione (Gly-Cys-Glu) peptide (GSH) is principally located at intracellular level (concentration inside cells: 10 mM) while in blood plasma its concentration is low (2 μM). If AuNPs are functionalized with a specific thiol-drug, a thiol exchange can occur inside cells producing *in vivo* AuNP-GSH conjugates and "free" drug. This is known as "drug delivery system using AuNP" (Ghosh et al., 2008). Chompoosor et al. (2007) found that the stability of AuNPs toward GSH or thiols is governed by the AuNPs' surface charge. They observed that the rate of drug release was very slow with anionic particles and faster with cationic analogues.

The first AuNP-regulatory peptide conjugate with a cyclized peptide and containing aromatic groups was reported by Surujpaul et al. (2008). The Tyr^3-octreotide (TOC)(H-D-Phe-Cys-Tyr-Trp-Lys(Boc)-Thr-Cys-Thr(ol)[disulfide bridge:2–7]) peptide contains two hydrophilic and neutral side chains in 6th and last positions (2 Thr); hydrophobic, neutral, and aromatic side chains in 4th and 1st position (Phe and Trp); and hydrophilic, neutral, and aromatic side chain in third position (Tyr). Cysteine side chains are forming the disulfide bridge and lysine side chain was Boc-protected since it is the key biological active site. Spectroscopy analyses suggested that AuNPs were functionalized with TOC through interactions with the N-terminal amine of the phenylalanine and possibly with the indole group of the tryptophan residue promoting the self-assembly of the peptide. As result of this interaction, an emission band was observed in the NIR region (692 nm).

11.5 PREPARATION OF AUNP-PEPTIDE CONJUGATES

11.5.1 SPONTANEOUS REACTION OF A PEPTIDE-THIOL (CYS) WITH THE AUNP SURFACE (DIRECT METHOD)

In this strategy, the conjugation of a peptide with biological activity to one AuNP is by means of spontaneous reaction of a thiol (Cys) or a N-terminal primary amine with AuNP surface. Thiols are the most important type of stabilizing molecules for AuNPs of any size. It is an accepted assumption that the use of thiols leads to the formation of strong Au–S bonds (Kogan et al., 2007). In general, peptides are mixed with AuNPs in a molar ratio of 5000:1 (peptide:particle) and a volume ratio of 10:1. For example, 100 μL of 50 μM peptide in water is added to 1 mL of 20 nm diameter citrate-capped AuNPs (1 nM). The mixture is stirred at room temperature from 1 to 16 h to allow complete exchange of citrate with thiol on the particle surface. The AuNP-peptide conjugate can be purified by centrifugation (11,500g for 30 min), dialysis, or size exclusion chromatography (Figure 11.2). The number of peptide molecules bound to one AuNP depends on particle size. For example, Levy et al. (2004) reported 1.67 peptides/nm², as they used a 12.3 nm AuNP size (surface area = 475 nm²), and the total number of peptides bound to one AuNP was 791.

Applying the direct method, hybrid AuNP-peptide conjugates can also been prepared. It means that two different peptides can be bound to one AuNP. In our group, we have obtained radiolabeled AuNPs conjugated to two different peptides, one to be used as BFCA to link the radionuclide and the other one as regulatory peptide analogue: ^{99m}Tc-EDDA/HYNIC-Gly-Gly-Cys-AuNP-Lys³-bombesin (Figure 11.3). This system could be useful for imaging the sentinel lymph node in breast cancer patients by means of nuclear medicine techniques (data not published).

11.5.2 AUNP-BIFUNCTIONAL MOLECULE ACTIVATION FOR PEPTIDE CONJUGATION

In this strategy, AuNPs are capped with a linker, which is then activated and functionalized with the biologically active peptide. The linker is a bifunctional molecule containing a thiol, which

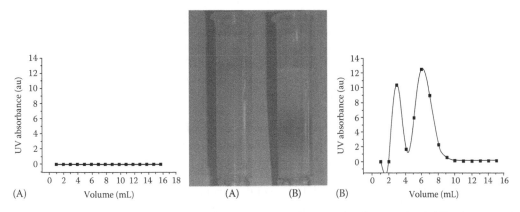

FIGURE 11.2 Purification by size-exclusion chromatography (PD-10 columns) (A) AuNPs-citrate (dark color at the column top): most of AuNPs remain at the column top (B). A first peak (3.0–3.5 mL) corresponds to the void volume of the column and contains the AuNP-peptide conjugate (dark color represents the red solution); a second peak (6 mL) corresponds to the free peptide (determined by UV-Vis analyses).

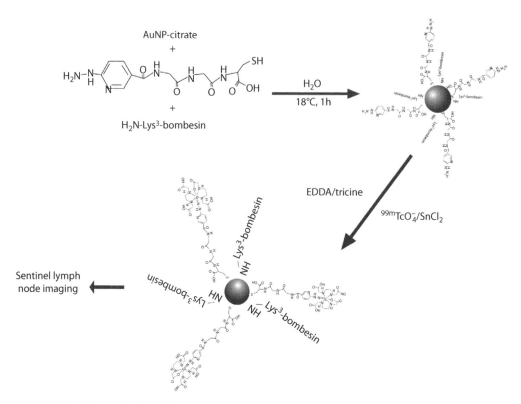

FIGURE 11.3 Schematic view of the spontaneous or direct method used to prepare AuNP-peptide conjugates.

allows binding to the Au surface, and a functional group (e.g., carboxyl group) that is bound to the peptide. One example of this strategy is the functionalization of AuNPs with tiopronin and the conjugation of these functionalized AuNP-tiopronin with a Tat protein–derived peptide sequence (48–57, GRKKRRQRRR) (De la Fuente and Berry, 2005). In the general procedure, hydrogen tetrachloroaureate(III) trihydrate and N-(2-mercaptopropionyl)-glycine (tiopronin) are co-dissolved in methanol/acetic acid, giving a ruby red solution. $NaBH_4$ in H_2O is added and the solvent is removed under vacuum. AuNP-tiopronin is completely insoluble in methanol but quite soluble in water. (N-(3-Dimethylaminopropyl)-N'-ethylcarbodiimide hydrochloride (EDC) and N-hydroxysuccinimide (NHS) are added to AuNP-tiopronin solution, and after 30 min the Tat peptide is added. Finally, the AuNP-tiopronin-Tat(48–57) conjugate is purified by dialysis (De la Fuente and Berry, 2005). This technique has been used to prepare AuNPs conjugated to RGD (De la Fuente et al., 2006) and can be extended to the preparation of AuNP functionalized with any regulatory peptide such as bombesin, OC, and GLP-1 (Figure 11.4).

The free carboxyl groups in AuNP-CALNN conjugates have also been carbonyl activated and bound to biologically active biomolecules (Wang et al., 2005; Cheung et al., 2009).

Recently, Nitin et al. (2007) reported two complementary thiol-modified oligonucleotide-based approaches conjugated to AuNPs, which can be functionalized with a variety of biomolecules: small molecules, peptides, and antibodies. The approach provides flexibility to control the orientation of conjugated biomolecules, and to control the size of the functionalized particles.

The copper(I)-catalyzed cycloaddition of azides and terminal alkynes ("click" chemistry) has also been proposed to functionalize AuNPs with peptides. von Maltzahn et al. (2008) found that "click" chemistry allows cyclic LyP-1 targeting peptides to be specifically linked to azido-nanoparticles and to direct their binding to p32-expressing tumor cells *in vitro*.

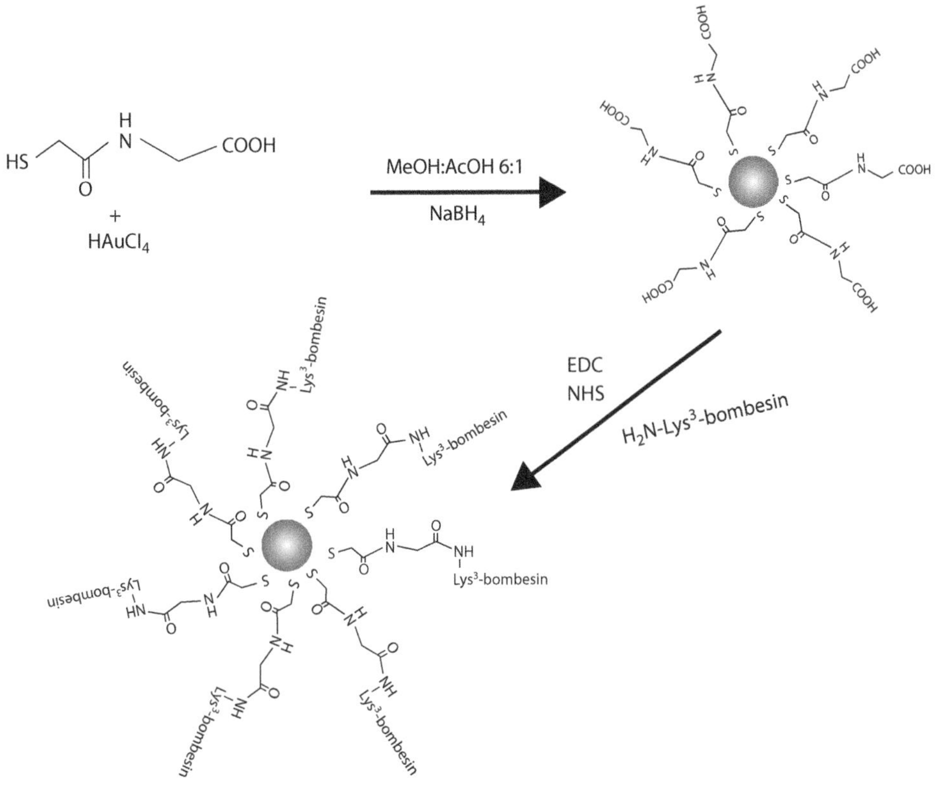

FIGURE 11.4 Schematic view of the AuNP-bifunctional molecule activation for peptide conjugation (indirect method).

11.6 PHYSICAL AND CHEMICAL CHARACTERIZATION OF AUNPS CONJUGATED TO PEPTIDES

11.6.1 Techniques Concerning AuNP-Peptide Conjugation: UV-Vis, XPS, IR(FT-IR), Luminescence

11.6.1.1 UV-Vis Spectroscopy

UV-Vis or UV/Vis spectroscopy involves photon spectroscopy in the UV-Vis region. It uses light in the visible and adjacent near ultraviolet (UV) and NIR ranges. In this region of the electromagnetic spectrum, molecules undergo electronic transitions. This technique is complementary to fluorescence spectroscopy, in that fluorescence deals with transitions from the excited state to the ground state, while absorption measures transitions from the ground state to the excited state. A preliminary evaluation of the AuNP-peptide conjugation can be performed using UV-Vis spectroscopy. This technique provides information concerning AuNP capping.

AuNPs display optical properties that could be exploited in optoelectronic devices. The source of the optical absorption is the surface plasmon resonance (SPR). Small shifts in the position of the SPR occur as a result of changes in the dielectric properties of the medium in which the AuNPs are found or the presence of materials adsorbed on the surface of the Au particle. Mie's scattering theory is frequently used to explain shifts in SPR (Creighton and Eadon, 1991). The theory predicts that shifts in this resonance can also occur when particles deviate from spherical geometry. In this scenario, the transverse and longitudinal dipole polarizability no longer produces equivalent

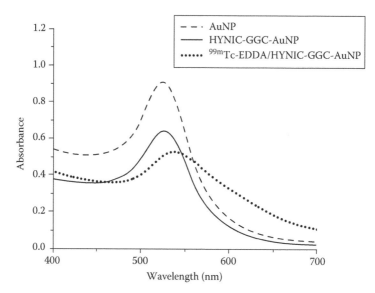

FIGURE 11.5 UV-Vis spectra of AuNP-citrate, AuNP-peptide conjugate, and radiolabeled AuNP peptide.

resonances. Consequently, two plasmon resonances appear, a broadened and red-shifted longitudinal resonance and a transverse resonance whose absorbance remains centered around 540 nm (Link and El-Sayed, 1999). Aggregation causes coupling of the plasma modes of AuNPs, which results in a red-shift and broadening of the longitudinal plasmon resonance in the optical spectrum. Levy et al. (2004) established a practical parameter to quantify AuNP-aggregation by UV-Vis spectroscopy: "aggregation parameter (AP)."

The AP is defined as follows: $AP = (A - A_0)/A_0$, where A is the integrated absorbance between 600 and 700 nm of AuNP peptide and A_0 is the integrated absorbance between 600 and 700 nm of AuNPs. In Figure 11.5, it is observed that the area under the curve between 600 and 700 nm is almost the same within AuNPs (maximum 524 nm) and AuNP peptides (maximum 526 nm), suggesting that conjugation of peptides to AuNP produced negligible aggregation. However, the AP increases after radiometal addition (maximum 537 nm). At the same time, a red-shift in the maximum absorbance is representative of the conjugation process. The small shifts in the SRP position occur as result of the presence of peptide adsorbed on the AuNPs' surface. It is important to mention that the SRP is absent for AuNPs with core diameter less than 2 nm, as well as for bulk gold (De la Fuente and Berry, 2005).

11.6.1.2 XPS Spectroscopy

XPS is a quantitative spectroscopic technique that measures elemental composition, empirical formula, chemical state, and electronic state of the elements that exist within a material. XPS spectra are obtained by irradiating a material with a beam of aluminum or magnesium x-rays while simultaneously measuring the kinetic energy and number of electrons that escape from the top 1 to 10 nm of the material being analyzed. XPS requires ultra-high vacuum (UHV) conditions.

XPS can be used to characterize and determine the presence of Au–S bonds. The oxidation state of AuNP shows binding energies (BE) of the doublet for Au $4f_{7/2}$ (83.8 eV) and Au $4f_{5/2}$ (87.5 eV) characteristic of Au(0) and no band is found for Au(I) at 84.9 eV. AuNP-S-peptide conjugates should present a clear edge at 165 eV, which is indicative of the presence of sulfur bounded atoms. High-resolution data can also be recorded in the S 2p, S 2s, and Au 4f spectral regions. The S 2p signal consists of a broad band with maximum at 162.2 eV that corresponds to the chemisorptions of sulfur grafted onto gold. The S 2s photoelectron BE from bulk peptide with the free thiol and AuNP peptide thiolate are found at 228.2 and 227.3 eV, respectively. The accuracy of these BE values is estimated to be ±0.2 eV (Olmedo et al., 2008). After a study of several AuNP-peptide conjugates, it has been fixed a BE of the Au $4f_{7/2}$ to 84.0 eV as typical value expected for thiolate monolayers to self-assemble onto AuNP surface (Fabris et al., 2006).

11.6.1.3 IR(FT-IR) Spectroscopy

Vibrational spectroscopy methods, such as infrared (IR) spectroscopy, are well suited for the identification of functional groups by means of their band positions (vibration frequencies) in the IR spectrum. The positions can be affected by the surrounding. This technique is particularly interesting when the conjugation of the citrate-capped nanoparticles to peptides is investigated. From the IR spectra, it is possible to determine if peptides displace the citrate from the AuNP surface and if nanoparticles are functionalized with peptides through covalent character bonding (e.g., with the N-terminal amine) or/and electrostatic bindings (Porta et al., 2007; Surujpaul et al., 2008).

11.6.1.4 Luminescence

Luminescence means "cold light" while the phenomenon of light from hot materials called incandescence means "hot light." Luminescence is defined as emission of light from an electronically excited state usually produced by excitation with light (photoluminescence), electric current or electric field (electroluminescence), or by a chemical or biochemical reaction (chemi- or bioluminescence), to name a few. In the case of organic molecules bearing aromatic residue, light absorption (excitation) leads to a singlet excited state, and the return to the ground state is allowed, so that the emission of light is fast and the lifetime of the excited state is from 1 to 100 ns (fluorescence). If the excited molecule relaxes to a triplet state, the transitions to the ground state are forbidden by selection rules so that emission of light is slow and the lifetime is long from 1 ms to 1 s (phosphorescence). The phosphorescence emission has to involve a change of spin otherwise could not occur (Bünzli, 2009).

The fluorescence of AuNP rises from the surface plasmon resonance and can be enhanced, quenched, or photobleached by the AuNP size, the nature of bounded cap to AuNP, as well as the surrounding of the capped AuNP. Ideally, the efficiency of the energy transfer to the AuNP from its organic cap only depends on the extent of the overlap of the cap band emission and the surface plasmon resonance band of the nanoparticle, which would be translated in a fluorescence resonance energy transfer (FRET). However, contribution from the medium, for instance, a vibrational process can partially or totally quench the energy transfer (Surujpaul et al., 2008), and a full FRET is difficult to be completed or observed by a conventional luminescent technique. Fluorescence and phosphorescence emissions are very important in the study of AuNP conjugated to peptides since in general peptides contain aromatic residues which transfer their energy to the AuNP following the pathway: fluorescence level (peptide) to phosphorescence level (peptide) to surface plasmon resonance level (AuNP) and the absorbed light reemitted, usually in or close to the NIR range if the NIR fluorescence (NIRF; Lee et al., 2008) emission is not quenched or masked by the luminescence background, in particular, in tissues. NIR light (700–1000 nm) can penetrate several centimeters into tissue and its imaging offers a potentially noninvasive means of characterizing diseased tissues (Ke et al., 2003).

It has been widely demonstrated that AuNPs conjugated to peptides have unlimited future in biomedical applications, in particular for enhancing the bioimaging properties of the AuNPs. When FRET is not completed, a good choice of the experimental conditions allows the evaluation of the extent and nature of the interaction AuNP peptide, by locating the shift of the maximum peak of the fluorescence or phosphorescence (commonly) emission of the peptide before and after conjugation. A red-shift of the peak implies a substantial covalent character in the AuNP-peptide bonding while blue-shift suggests a major contribution of ionic character to such bonding (Surujpaul et al., 2008). Time-resolved luminescence is a good tool for overcoming the limitations of a conventional luminescent technique for bioimaging (Bünzli, 2009).

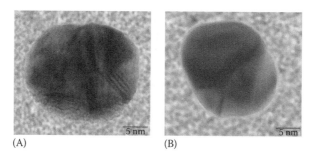

FIGURE 11.6 TEM image of (A) AuNP-citrate and (B) AuNP-peptide conjugate (particle size ~20 nm).

11.6.2 Size, Shape, and Aggregation of Nanoconjugates: TEM and DLS

One of the main characterization techniques used for the analysis of nanoparticles is TEM, where electrons are accelerated to 100 keV or higher (up to 1 MeV) and projected onto a thin specimen. The electron beam penetrates the sample creating a high magnification image, ranging from 50 up to 10^6 (Cao, 2004). TEM and dynamic light scattering (DLS) allow determination of the size, shape, and aggregation state of the conjugates. The core size distribution histogram of the AuNPs and their nanoconjugates can also be obtained. The increase of the particle hydrodynamic diameter by effect of the peptide conjugation can be observed by TEM as a low electronic density around the AuNP due to the poor electron interaction with the peptide molecules, which have a low electron density in contrast with the strong scattering of the electron beam as it interacts with the metallic nanoparticles (Figure 11.6).

11.6.3 Peptide Molecules per AuNP: Amino Acid Analyses Combined with HPLC

UV-Vis spectrometry and analysis of amino acids by high-performance liquid chromatography (HPLC) (which determines the amino acid composition of a peptide or protein) allow the determination of the number of peptide molecules per AuNP. The AuNPs in the solution (before conjugation) can be characterized by the number of particles per milliliter and gold concentration, from which the AuNP molecular weight and concentration can be calculated. Free peptide concentration can be evaluated considering its molar absorption coefficient (Olmedo et al., 2008). After conjugation the AuNP-peptide solution can be purified by centrifugation and the amino acid composition (HPLC) and UV-Vis absorbance of the supernatant (free peptide) determined, obtaining with these sufficient data to calculate the number of peptide molecules per AuNP. Another easy alternative is the peptide titration using increasing AuNP concentration (Levy et al., 2004).

11.7 BIOLOGICAL RECOGNITION OF NANOCONJUGATES TO SPECIFIC RECEPTORS

11.7.1 Receptor-Binding Studies in Cancer Cells

Receptor binding studies should be performed in order to determine the receptor affinity of the AuNP-peptide conjugates for the specific receptor. These binding assays are carried out on cell membranes, tissue membranes, or intact cells in tissue culture. The two primary types of receptor binding experiments are saturation and competition studies. The general methodology can be carried out as in the case of radiolabeled peptides (Ferro-Flores et al., 2006a).

AuNP-peptide conjugate fluorescence emission can also be studied by incubation of the nanoconjugate with human cancer cell lines and then examined by fluorescence microscopy (Surujpaul et al., 2008). The nanoconjugate cell internalization can be studied by TEM.

11.7.2 Uptake of Nanoconjugates in Tumors Induced in Athymic Mice

The preclinical evaluation of AuNP conjugated to peptides includes *in vivo* studies of both biodistribution and toxicity in animals. Biodistribution studies use normal animals to identify organ distribution patterns and excretion. Animals with induced malignant tumors (usually athymic mice) are used to determine specific binding to receptors, which are overexpressed in cancer cells. Blocking studies, co-injecting the AuNP peptide, and an excess of free peptide, are useful to determine the degree of receptor-specific-mediated uptake or lack of binding to other receptors (Ferro-Flores et al., 2006a). Animals with induced malignant tumors are essential to evaluate AuNP-peptide diagnostic or therapeutic efficacy.

11.8 THERAPEUTIC AND DIAGNOSTIC POSSIBILITIES OF AUNP-PEPTIDE CONJUGATES

Although most evidence suggests that colloid AuNP ("naked" AuNP) are chemically inert *in vivo*, they could be ingested by the immune system unless unrecognized as "strange" substances. If AuNP are capped with regulatory or natural peptides, the AuNP-peptide conjugate may be "invisible" to the immune system. At the same time, the AuNP peptide with specific recognition by cancer cells (active targeting) could be a strategy to concentrate nanoparticles in the tissue to be selected for photothermal therapeutic ablation.

Local heat, delivered by AuNP selectively attached to a target, was used as molecular surgery to remove toxic and clogging aggregates of the Aβ protein involved in Alzheimer's disease (Kogan et al., 2006). Researchers remotely redissolved Aβ deposits by using the local heat dissipated by AuNPs conjugated to the CLPFFD-NH$_2$ peptide. The resulting conjugate was selectively attached to the aggregates of Aβ, and the system was irradiated with low gigahertz electromagnetic fields.

Latent membrane protein (LMP)-2 epitope is expressed in the tumor tissue of nasopharyngeal carcinoma strongly associated with Epstein–Barr virus (EBV). The activation of cytotoxic T-lymphocytes with the restricted SSCSSCPLSK peptide epitope in LMP-2 can induce an antiviral response. Cheung et al. (2009) prepared stable AuNP-peptide (*N*-cysteinated LMP-2 epitope) conjugates in the presence of a CALNN-capping peptide. In this way, they obtained a highly immunogenic multiple antigen peptide system. Dendritic cells treated with the conjugate could effect CD8+ T-cell activation leading to epitope-specific cytotoxic T-lymphocytes killing responses *in vitro*.

The Kahalalide F peptide is an anticancer drug that causes alteration in cell lysosomes. Two Cys-containing analogues of this peptide were synthesized and conjugated to 20 and 40 nm AuNPs. The AuNP-Cys-Kahalalide F system could act as a concentrator for the delivery of drugs, thereby increasing bioactivity. The nanoconjugates were located subcellularly at lysosome-like bodies, which may be related to the action mechanism of Kahalalide F. The results suggested that the selective delivery and activity of Kahalalide F analogues can be improved by conjugating the peptides to AuNPs (Hosta et al., 2009).

Surujpaul et al. (2008) prepared a stable multifunctional system of AuNPs conjugated to TOC peptide which was characterized by TEM, UV-Vis, infrared, and fluorescence spectroscopy. AuNP and AuNP-TOC fluorescence emission spectra were obtained both in solution and in murine AR42J-tumor tissues. The fluorescence analyses in tissue revealed a recognition of the AuNP-TOC conjugate for the neuroendocrine tumor because of the lower energy position of the fluorescence resonance (692 nm) with respect to that of the AuNP in the same tumor tissue (684 nm). The emission band observed in the NIR region (692 nm) opens the possibility for AuNP-TOC use in bioimaging.

Radiolabeled AuNPs conjugated to regulatory peptides could improve molecular images routinely obtained in nuclear medicine because of the use of multiple selective molecules (peptides) available in one AuNP for specific targeting of tumors.

In cancer therapy, one of the most important things to keep in mind is that "the magic bullet does not exist," and in order to increase therapeutic response, the application of combined modalities with different therapeutic agents is necessary. Radiolabeled AuNPs conjugated to regulatory peptides, for example, ^{177}Lu-DOTA-Gly-Gly-Cys-AuNP-Lys^3-bombesin (size = 10 nm) could represent a unique multifunctional target-specific pharmaceutical that administered as a single drug would be capable of acting as a combined therapy system: targeted radiotherapy, photothermal therapy, angiogenesis inhibition, and apoptosis induction in prostate and breast cancers.

ABBREVIATIONS

AuNP	Gold nanoparticle
αvβ3 integrin	Vitronectin receptor = glycoprotein containing two distinct chains: α (alpha) and β (beta) subunits
BFCA	Bifunctional chelating agent
BN	Bombesin peptide
Boc	*tert*-butyloxycarbonyl
CALNN	C = Cysteine (Cys), A = alanine (Ala), L = leucine (Leu), N = aspargine (Asp)
CLPFFD	C = Cysteine (Cys), L = leucine (Leu), P = proline (Pro), F = phenylalanine (Phe), D = aspartic acid (Asp)
CCK	Cholecystokinin peptide
CD8C	Cluster of differentiation 8C (non-glycosylated phosphoprotein)
CTAB	Cetyltrimethylammonium bromide
DDAB	Didodecyldimethylammonium bromide
DOTA	1,4,7,10-tetraazacyclododecane- N', N'', N'''-tetraacetic acid
EDC	N-(3-dimethylaminopropyl)-N'-ethylcarbodiimide hydrochloride
EDDA	Ethylendiamine-N,N'-diacetic acid
^{68}Ga	Gallium-68 radionuclide
GBq	Gigabecquerel (radioactivity unit)
GLP-1	Glucagon-like peptide-1
GK	G = Glycine (Gly), K = lysine (Lys)
GRP	Gastrin-releasing peptide
GRP-r	Gastrin-releasing peptide receptor
GSH	Glutathione peptide
Gy	Gray = J/Kg
HIV	Human immunodeficiency virus
HPLC	High-performance liquid chromatography
HYNIC	6-hydrazinopyridine-3-carboxylic acid
^{111}In	Indium-111 radionuclide
LAN	Lanreotide peptide
LMP	Latent membrane protein
^{177}Lu	Lutetium-177 radionuclide
MBq	Megabecquerel (radioactivity unit)
MRI	Magnetic resonance imaging
NHS	N-hydroxysuccinimide
NIR	Near infrared
NPY	Neuropeptide Y
NT	Neurotensin
OC	Octreotide peptide
OI	Optical imaging
PET	Positron emission tomography
RGD	R = Arginine (Arg), G = glycine (Gly), D = aspartic acid (Asp)
RKKRRQRRR	R = Arginine (Arg), K = lysine (Lys), Q = glutamine (Gln)
SPECT	Single-photon emission computed tomography

SPR	Surface plasmon resonance
SSCSSCPLSK	S = Serine (Ser), C = cysteine (Cys), P = proline (Pro), L = leucine (Leu), K = lysine (Lys)
Tat	Trans-activating transcriptional activator peptide
TATE	[Tyr^3]octreotate peptide
TEM	Transmission electron microscopy
Tiopronin	N-(2-mercaptopropionyl)-glycine
^{99m}Tc	Technetium-99 metastable radionuclide
TOC	[Tyr^3]octreotide peptide
TTAB	Tetradodecylammonium bromide
VEGF	Vascular endothelial growth factor
XPS	X-ray photoelectron spectroscopy
^{90}Y	Yttrium-90 radionuclide

REFERENCES

Alves, S., Correia, J.D., Santos, I., Veerendra, B., Sieckman, G.L., Hoffman, T.J., Rold, T.L. et al., 2006. Pyrazolyl conjugates of bombesin: A new tridentate ligand framework for the stabilization of fac-[M(CO)3]+ moiety. *Nucl. Med. Biol.*, 33, 625–634.

Andreescu, D., Sau, T.K., Goia, D.V., 2006. Stabilizer-free nanosize gold sols. *J. Colloid Interface Sci.*, 298, 742–751.

Basu, S., Ghosh, S.K., Kundu, S., Panigrahi, S., Praharaj, S., Pande, S., Jana, S., Pal, T., 2007. Biomolecule induced nanoparticle aggregation: Effect of particle size on interparticle coupling. *J. Colloid Interface Sci.*, 313, 724–734.

Bellino, M.G., Calvo, E.J., Gordillo, G., 2004. Adsorption kinetics of charged thiols on gold nanoparticles. *Phys. Chem. Chem. Phys.*, 6, 424–428.

Berners-Price, S.J., Sadler, P.J., 1987. Interaction of the antitumor Au(I) complex [Au(Ph2P(CH2)2PPh2)2]Cl with human blood plasma, red cells, and lipoproteins: 31P and 1H NMR studies. *J. Inorg. Biochem.*, 31, 267–281.

Bhattacharya, R., Mukherjee, P., 2008. Biological properties of "naked" metal nanoparticles. *Adv. Drug Deliv. Rev.*, 60, 1289–1306.

Bhattacharya, R., Patra, C.R., Earl, A., Wang, S., Katarya, A., Lu, L., Kizhakkedathu, J.N. et al., 2007. Attaching folic acid on gold nanoparticles using noncovalent interaction via different polyethylene glycol backbones and targeting of cancer cells. *Nanomedical*, 3, 224–238.

Bjerneld, E.J., Svedberg, F., Kall, M., 2003. Laser induced growth and deposition of noble metal NPs for surface-enhanced Raman scattering. *Nano Lett.*, 3, 593–596.

Brigger, I., Dubernet, C., Couvreur, P., 2002. Nanoparticles in cancer therapy and diagnosis. *Adv. Drug Deliv. Rev.*, 54, 631–651.

Brust, M., Fink, J., Bethell, D., Schiffrin, D.J., Kiely, C., 1995. Synthesis and reactions of functionalised gold nanoparticles. *J. Chem. Soc. Chem. Commun.*, 16, 1655–1656.

Bünzli, J.-C.G., 2009. Highlight review: Lanthanide luminescent bioprobes (LLBs). *Chem. Lett.*, 38, 104–109.

Busbee, B.D., Obare, S.O., Murphy, C.J., 2003. An improved synthesis of high-aspect-ratio gold nanorods. *Adv. Mater.*, 15, 414–416.

Bussolino, F., Mantovani, A., Persico, G., 1997. Molecular mechanisms of blood vessel formation. *Trends. Biochem. Sci.*, 22, 251–256.

Cao, G., 2004. *Nanostructures and Nanomaterials: Synthesis, Properties and Applications.* Imperial College Press, London, U.K.

Cheung, W.H., Chan, V.S.F., Pang, H.W., Wong, M.K., Guo, Z.H., Tam, P.K.H., Che, C.M., Lin, C.L., Yu, W.Y., 2009. Conjugation of latent membrane protein (LMP)-2 epitope to gold nanoparticles as highly immunogenic multiple antigenic peptides for induction of Epstein–Barr virus-specific cytotoxic T-lymphocyte responses *in vitro*. *Bioconjug. Chem.*, 20, 24–31.

Chompoosor, A., Han, G., Rotello, V.R., 2007. Charge dependence of ligand release and monolayer stability of gold nanoparticles by biogenic thiols. *Bioconjug. Chem.*, 19, 1342–1345.

Connor, E.E., Mwamuka, J., Gole, A., Murphy, C.J., Wyatt, M.D., 2005. Gold nanoparticles are taken up by human cells but do not cause acute cytotoxicity. *Small*, 1, 325–327.

Cornelissen, B., McLarty, K., Kersemans, V., Scollard, D., Reilly, R., 2008. Properties of [^{111}In]-labeled HIV-1 tat peptide radioimmunoconjugates in tumor-bearing mice following intravenous or intratumoral injection. *Nucl. Med. Biol.*, 35, 101–110.

Costantini, D.L., Hu, M., Reilly, R., 2008. Peptide motifs for insertion of radiolabeled biomolecules into cells and routing to the nucleus for cancer imaging or radiotherapeutic applications. *Cancer Biother. Radiopharm.*, 23, 3–23.

Creighton, J.A., Eadon, D.G., 1991. Ultraviolet visible absorption spectra of the colloidal metallic elements. *Faraday Trans. J. Chem. Soc.*, 87, 3881–3883.

Czepczynski R., Parisella M.G., Kosowicz J., Mikolajczak R., Ziemnicka K., Gryczynska M., Sowinski J., Signore, A., 2007. Somatostain receptor scintigraphy using ^{99m}Tc-EDDA/HYNIC-TOC in patients with medullary thyroid carcinoma. *Eur. J. Nucl. Med. Mol. Imaging* 34, 1635–1645.

de Jong, M., Kwekkeboom, D.J., Valkema, R., Krenning, E.P., 2004. Tumor therapy with radiolabelled peptides: Current status and future directions. *Dig. Liver Dis.*, 36, S48–S54.

De la Fuente, J.M., Berry, C.C., 2005. Tat Peptide as an efficient molecule to translocate gold nanoparticles into the cell nucleus. *Bioconjug. Chem.*, 16, 1176–1180.

De la Fuente, J.M., Berry, C.C., Riehle, M.O., Curtis, S.G., 2006. NPs targeting at cells. *Langmuir*, 22, 3286–3293.

Deshayes, S., Morris, M.C., Divita, G., Heitz, F., 2005. Interactions of primary amphipathic cell penetrating peptides with model membranes: Consequences on the mechanisms of intracellular delivery of therapeutics. *Curr. Pharm. Des.*, 11, 3629–3638.

de Visser, M., van Weerden, W.M., de Ridder, C.-M., 2007. Androgen-dependent expression of the gastrin-releasing peptide receptor in human prostate tumor xenografts. *J. Nucl. Med.*, 48, 88–93.

de Visser, M., Verwijnen, S.M., de Jong, M., 2008. Improvement strategies for peptide receptor scintigraphy and radionuclide therapy. *Cancer Biother. Radiopharm.*, 23, 137–157.

Dietz, G.P., Bähr, M., 2004. Delivery of bioactive molecules into the cell: The Trojan horse approach. *Mol. Cell. Neurosci.*, 27, 85–131.

Doolittle, J.W., Dutta, P.K., 2006. Influence of microwave radiation on the growth of AuNPs and microporous zincophosphates in a reverse micellar system. *Langmuir*, 22, 4825–4831.

Fabris, L., Antonello, S., Armelao, L., 2006. Gold nanoclusters protected by conformationally constrained peptides. *J. Am. Chem. Soc.*, 128, 326–336.

Ferro-Flores, G., Arteaga de Murphy, C., Melendez-Alafort, L., 2006a. Third generation radiopharmaceuticals for imaging and targeted therapy. *Curr. Pharm. Anal.*, 2, 339–352.

Ferro-Flores, G., Arteaga de Murphy, C., Rodriguez-Cortes, J., Pedraza-Lopez, M., Ramirez-Iglesias, MT., 2006b. Preparation and evaluation of ^{99m}Tc-EDDA/HYNIC-[Lys^3]-Bombesin for imaging of GRP receptor-positive tumours. *Nucl. Med. Commun.*, 27, 371–376.

Frens, G., 1973. Controlled nucleation for the regulation of the particle size in monodisperse gold suspensions. *Nat. Phys. Sci.*, 241, 20–22.

Ghosh, P., Han, G., De, M., Kyu-Kim, C., Rotello, V., 2008. Gold nanoparticles in delivery applications. *Adv. Drug Deliv. Rev.*, 60, 1307–1315.

Gonzalez-Vazquez A., Ferro-Flores G., Arteaga de Murphy C., Gutierrez-Garcia, Z., 2006. Biokinetics and dosimetry in patients of ^{99m}Tc-EDDA/HYNIC-Tyr^3-octreotide prepared from lyophilized kits. *Appl. Radiat. Isot.*, 64, 792–797.

Goodman, C.M., McCusker, C.D., Yilmaz, T., Rotello, V.M., 2004. Toxicity of gold nanoparticles functionalized with cationic and anionic side chains. *Bioconjug. Chem.*, 15, 897–900.

Higby, G.J., 1982. Gold in medicine: A review of its use in the West before 1900. *Gold Bull.*, 15, 130–140.

Hirsch, L.R., Stafford, R.J., Bankson, J.A., Sershen, S.R., Rivera, B., Price, R.E., Hazle, J.D., Hals, N.J., 2003. Nanoshell-mediated near-infrared thermal therapy of tumors under magnetic resonance guidance. *Proc. Natl. Acad. Sci. USA*, 100, 13549–13554

Hosta, L., Pla-Roca, M., Arbiol, J., Lopez-Iglesias C., Samitier, J., Cruz, L.J., Kogan, M.J., Albericio, F., 2009. Conjugation of Kahalalide F with gold nanoparticles to enhance *in vitro* antitumoral activity. *Bioconjug. Chem.*, 20, 138–146.

Kabashin, A.V., Meunier, M., 2006. Laser ablation based synthesis of functionalized colloidal nanomaterials in biocompatible solutions. *J. Photochem. Photobiol. A*, 182, 330–334.

Kanaras, A.G., Kamounah, F.S., Schaumburg, K., Kiely, C.J., Brust, M., 2002. Thioalkylated tetraethylene glycol: A new ligand for water soluble monolayer protected gold clusters. *Chem. Commun.*, 20, 2294–2295.

Ke, S., Wen, X., Gurfinkel, M., Charnsangavej, C., Wallace, S., Sevick-Muraca, E.-M., Li, C., 2003. Near-infrared optical imaging of epidermal growth factor receptor in breast cancer xenografts. *Cancer Res.*, 63, 7870–7875.

Kenny, L.M., Coombes, R.C., Oulie, I., Contractor, K.B., Miller, M., Spinks, T.J., McParland, B. et al., 2008. Phase I trial of the positron-emitting Arg-Gly-Asp (RGD) peptide radioligand ^{18}F-AH111585 in breast cancer patients. *J. Nucl. Med.*, 49, 879–886.

Kersemans, V., Kersemans, K., Cornelissen B., 2008. Cell penetrating peptides for *in vivo* molecular imaging applications. *Curr. Pharm. Des.*, 14, 2415–2427.

Koch, A.E., 2003. Angiogenesis as a target in rheumatoid arthritis. *Ann. Rheum. Dis.*, 62(Suppl 2), ii60–ii67.

Kogan, M.J., Bastus, N.G., Amigo, R., Grillo-Bosch, D., Araya, E., 2006. Nanoparticle-mediated local and remote manipulation of protein aggregation. *Nano Lett.*, 6, 110–115.

Kogan, M.J., Olmedo, I., Hosta, L., Guerrero, A.R., Cruz, L.J., Albericio, F., 2007. Peptides and metallic nanoparticles for biomedical applications. *Nonomedicine*, 2, 287–306.

Kohlmann, O., Steinmetz, W.E., Mao, X.-A., Wuelfing, W.P., Templeton, A.C., Murray, R.W., Johnson C.S., 2001. NMR diffusion, relaxation, and spectroscopic studies of water soluble, monolayer-protected gold nanoclusters. *J. Phys. Chem. B*, 105, 8801–8809.

Korner, M., Waser, B., Reubi, J.C., 2008. High expression of neuropeptide Y1 receptors in Ewing sarcoma tumors. *Clin. Canc. Res.*, 14, 5043–5049.

Kunstler, J.U., Veerendra, B., Figueroa, S.D., Sieckman, G.L., Rold, T.L., Hoffman, T.J., Smith, C.J., Pietzsch, H.J., 2007. Organometallic (99m)Tc(III) '4 + 1' bombesin(7–14) conjugates: Synthesis, radiolabeling, and *in vitro/in vivo* studies. *Bioconjug. Chem.*, 18, 1651–1661.

Kwekkeboom, D.J., Mueller-Brand, J., Paganelli, G., Anthony, L.B., Pauwels, S., Kvols, L.K., O'Dorisio, T.M. et al., 2005. Overview of results of peptide receptor radionuclide therapy with 3 radiolabeled somatostatin analogs. *J. Nucl. Med.*, 46, 62S–66S.

Lee, S., Cha, E.-J., Park, K., Lee, S.-Y., Hong, J.K., Sun, I.-C., Sang, S.-Y. et al., 2008. A near-infrared-fluorescence-quenched gold-nanoparticle imaging probe for *in vivo* drug screening and protease activity determination. *Angew. Chem. Int. Ed.*, 47, 2804–2807.

Lee, S.S., Joo, Y.S., Kim, W.U., Min, D.J., Min, J.K., Park, S.H., Cho, C.S., Kim, H.Y., 2001. Vascular endothelial growth factor levels in the serum and synovial fluid of patients with rheumatoid arthritis. *Clin. Exp. Rheumatol.*, 19, 321–324.

Levy, R., Thanh, N.T.K., Doty, R.C., Hussain, I., Nichols, R.J., Schiffrin, D.J., Brust M., Ferning, D.G., 2004. Rational and combinatorial design of peptide capping ligands for gold nanoparticles. *J. Am. Chem. Soc.*, 126, 10076–10084.

Link, S., El-Sayed, M., 1999. Spectral properties and relaxation dynamics of surface plasmon electronic oscillations in gold and silver nanodots and nanorods. *J. Phys. Chem. B*, 103, 8410–8426.

Liu, Y., Shipton, M.K., Ryan, J., Kaufman, E.D., Franzen, S., Feldheim, D.L., 2007. Synthesis, stability, and cellular internalization of gold nanoparticles containing mixed peptide-poly(ethylene glycol) monolayers. *Anal. Chem.*, 79, 2221–2229.

Liu, Z., Yan, Y., Chin, F.T., Wang, F., Chen, X., 2009. Dual integrin and gastrin-releasing peptide receptor targeted tumor imaging using ^{18}F-labeled PEGylated RGD-bombesin heterodimer 18F-FB-PEG3-Glu-RGD-BBN. *J. Med. Chem.*, 52, 425–432.

Loo, C., Lin, A., Hirsh, L., Lee, M.H., Barton, J., Kalas, N., West, J., Dresek, R., 2004. Nanoshell-enabled photonics-based imaging and therapy of cancer. *Technol. Cancer Res. Treat.*, 3, 33–40.

Maeda, A., Wu, J., Sawa, T., Matsumura Y., Hori, K., 2000. Tumor vascular permeability and the EPR effect in macromolecular therapeutics: A review. *J. Control. Release*, 65, 271–284.

Mafune, F., Kohno, J.Y., Takeda, Y., Kondow, T., 2002. Full physical preparation of size-selected gold nanoparticles in solution: Laser ablation and laser-induced size control. *J. Phys. Chem. B*, 106, 7575–7577.

Mukherjee, P., Bhattacharya, R., Wang, P., Wang, L., Basu, S., Nagy, J.A., Atala, A., Mukhopadhyay, D., Soker, S., 2005. Antiangiogenic properties of gold nanoparticles. *Clin. Cancer Res.*, 11, 3530–3534.

Nitin, N., Javier, D.J., Richards-Kortum, R., 2007. Oligonucleotide-coated metallic nanoparticles as a flexible platform for molecular imaging agents. *Bioconjug. Chem.*, 18, 2090–2096.

Okitsu, K., Ashokkumar, M., Grieser, F., 2005. Sonochemical synthesis of AuNPs: Effects of ultrasound frequency. *J. Phys. Chem. B*, 109, 20673–20675.

Olmedo, I., Araya, E., Sanz, F., Medina, E., Arbiol, J., Toledo, P., Alvarez-Lueje, A., Giralt, E., Kogan, M.J., 2008. How changes in the sequence of the peptide CLPFFD-NH$_2$ can modify the conjugation and stability of gold nanoparticles and their affinity for beta-amyloid fibrils. *Bioconjug. Chem.*, 19, 1154–1163.

Park, J.E., Atobe, M., Fuchigami, T. 2006. Synthesis of multiple shapes of AuNPs with controlled sizes in aqueous solution using ultrasound. *Ultrason. Sonochem.*, 13, 237–241.

Pitsillides, C.M., Joe, E.K., Wei, X., Anderson R.R., Lin, C.P., 2003. Selective cell targeting with light absorbing microparticles and nanoparticles. *Biophys. J.*, 84, 4023–4032.

Pissuwan, D., Valenzuela, S.M., Cortie, M.B., 2006. Therapeutic possibilities of plasmonically heated gold nanoparticles. *Trends Biotechnol.*, 24, 62–67.

Plachcinska, A., Mikolajczak, R., Maecke, H.R., Michalski, A., Rzeszutek, K., Kozak, J., Kusmierek, J., 2004. ^{99m}Tc-EDDA/HYNIC-TOC scintigraphy in the differential diagnosis of solitary pulmonary nodules. *Eur. J. Nucl. Med. Mol. Imaging*, 31, 1005–1010.

Porta, F., Speranza, G., Krpetic, Z., Santo, V.D., Francescato P., Scari, G., 2007. Gold nanoparticles capped by peptides. *Mater. Sci. Eng. B*, 140, 187–194.

Reubi, J.C., 2003. Peptide receptors as molecular targets for cancer diagnosis and therapy. *Endocrine Rev.*, 24, 389–427.

Reubi J.C., 2007. Targeting CCK receptors in human cancers. *Curr. Top. Med. Chem.*, 7, 1239–1242.

Reubi, J.C., Maecke, H.R., 2008. Peptide-based probes for cancer imaging. *J. Nucl. Med.*, 49, 1735–1738.

Reubi, J.C., Waser, B., 2003. Concomitant expression of several peptide receptors in neuroendocrine tumors as molecular basis for *in vivo* multireceptor tumor targeting. *Eur. J. Nucl. Med.*, 30, 781–793.

Santos-Cuevas, C.L., Ferro-Flores, G., Arteaga de Murphy, C., Pichardo-Romero, P., 2008. Targeted imaging of GRP receptors with ^{99m}Tc-EDDA/HYNIC-[Lys^3]-bombesin: Biokinetics and dosimetry in women. *Nucl. Med. Commun.*, 29, 741–747.

Schaaff, T.G., Knight, G., Shafigullin, M.N., Borkman, R.F., Whetten, R.L., 1998. Isolation and selected properties of a 10.4 kDa gold: Glutathione cluster compound. *J. Phys. Chem. B*, 102, 10643–10646.

Sharma, P., Brown, S., Walte, G., Santra, S., Moudgil, B., 2006. Nanoparticles for bioimaging. *Adv. Colloid Interface Sci.*, 123, 471–485.

Shen, M., Du, Y.K., Hua, N.P., Yang, P., 2006. Microwave irradiation synthesis and self assembly of alkylamine-stabilized AuNPs. *Powder Technol.*, 162, 64–72.

Smith, C.J., Sieckman, G.L., Owen, N.K., Hayes, D.L., Mazuru, D.G., Volkert, W.A., Hoffman, T.J., 2003. Radiochemical investigations of [^{188}Re(H$_2$O)(CO)$_3$-diaminopropionic acid-SSS-bombesin(7–14)NH$_2$]: Synthesis, radiolabeling and *in vitro/in vivo* GRP receptor targeting studies. *Anticancer Res.*, 23, 63–70.

Stoeva, S.I., Smetana, A.B., Sorensen, C.M., Klabunde, K.J., 2007. Gram-scale synthesis of aqueous gold colloids stabilized by various ligands. *J. Colloid Interface Sci.*, 309, 94–98.

Surujpaul, P.P., Gutierrez-Wing, C., Ocampo-Garcia, B., Ramirez, F. de M., Arteaga de Murphy, C., Pedraza-Lopez, M., Camacho-Lopez, M.A., Ferro-Flores, G., 2008. Gold nanoparticles conjugated to [Tyr3] octreotide peptide. *Biophys. Chem.*, 138, 83–90.

Templeton, A.C., Chen, S., Gross, S.M., Murray, R.W., 1999. Water-soluble, isolable gold clusters protected by tiopronin and coenzyme A monolayers. *Langmuir*, 15, 66–76.

Turkevich, J., Stevenson, P.C., Hillier, J., 1951. A study of the nucleation and growth processes in the synthesis of colloidal gold. *Discuss. Faraday Soc.*, 11, 55–75.

van de Wiele, C., Phonteyne, P., Pauwels, P., 2008. Gastrin-releasing peptide receptor imaging in human breast carcinoma versus immunohistochemistry. *J. Nucl. Med.*, 49, 260–264.

von Maltzahn, G., Ren, Y., Park, J.H., Min, D.H., Kotamraju, V.R., Jayakumar, J., Fogal, V., Sailor, M.J, Ruoslahti, E., Bhatia, S.N., 2008. *In vivo* tumor cell targeting with "Click" nanoparticles. *Bioconjug. Chem.*, 19, 1570–1578.

Wang, Z., Levy, R., Fernig, D.G., Brust, M., 2005. The peptide route to multifunctional gold nanoparticles. *Bioconjug. Chem.*, 16, 497–500.

Weissleder, R., Mahmood, U., 2001. Molecular imaging. *Radiology*, 219, 316–333.

West, J.L., Halas, N.J., 2003. Engineered nanomaterials for biophotonics applications: Improving sensing, imaging, and therapeutics. *Annu. Rev. Biomed. Eng.*, 5, 285–292.

Wild, D., Macke, H., Christ, E., Gloor, B., Reubi, J.C., 2008. Glucagon-like peptide 1-receptor scans to localize occult insulinomas. *N. Engl. J. Med.*, 359, 766–768.

Wuelfing, W.P., Gross, S.M., Miles, D.T., Murray, R.W., 1998. Nanometer gold clusters protected by surface-bound monolayers of thiolatedpoly(ethylene glycol) polymer electrolyte. *J. Am. Chem. Soc.*, 120, 12696–12697.

Zhang, J., Chi, Q., Nielsen, J.U., Friis, E.P., Anderson J.E.T., Ulstrup, J., 2000. Two-dimensional cysteine and cystine cluster networks on Au(111) disclosed by voltammetry and in situ scanning tunneling microscopy. *Langmuir*, 16, 7229–7237.

Zhang, L.X., Sun, X.P., Song, Y.H., Jiang, X., Dong, S.J., Wang, E.A., 2006. Didodecyldimethylammonium bromide lipid bilayer-protected gold nanoparticles: Synthesis, characterization, and selfassembly. *Langmuir*, 22, 2838–2843.

Zwanziger, D., Khan, I.U., Neundorf, I., 2008. Novel chemically modified analogues of neuropeptide Y for tumor targeting. *Bioconjug. Chem.*, 19, 1430–1438.

12 Bioadhesive Microspheres and Their Biotechnological and Pharmaceutical Applications

Jayvadan K. Patel

CONTENTS

12.1 INTRODUCTION

The term "bioadhesion" refers to any bond formed between two biological surfaces, or a bond between a biological and a synthetic surface. In the case of bioadhesive drug delivery systems, the term bioadhesion is typically used to describe the adhesion between polymers, ether synthetic, or natural and soft tissue (i.e., gastrointestinal mucosa). Although the target of many bioadhesive delivery systems may be a soft tissue cell layer (i.e., epithelial cells), the actual adhesive bond may form with the cell layer, a mucous layer, or a combination of the two. In instances in which bonds form between mucus and polymer, the term "mucoadhesion" is used synonymously with bioadhesion. In general, bioadhesion is an all-inclusive term to describe adhesive interactions with any biological or biologically derived substance, mucoadhesion is used only when describing a bond involving mucus or a mucosal surface, and cytoadhesion is the cell-specific bioadhesion.

The adhesion of pharmaceutical formulations to the mucosal tissue offers the possibility of creating an intimate and prolonged contact at the site of administration. This prolonged residence time can result in enhanced absorption and, in combination with a controlled release of drug, also improved patient compliance by reducing the frequency of administration. Carrier technology offers an intelligent approach for drug delivery by coupling the drug to a carrier particle such as microspheres, nanospheres, liposomes, nanoparticles, etc., which modulates the release and absorption of the drug. Microspheres constitute an important part of these particulate drug delivery systems by virtue of their small size and efficient carrier capacity. However, the success of these microspheres is limited due to their short residence time at the site of absorption. It would, therefore, be advantageous to have means for providing an intimate contact of the drug delivery system with the absorbing membranes. This can be achieved by coupling bioadhesion characteristics to microspheres and developing bioadhesive microspheres (Nagai et al. 1984, Illum et al. 1988, Ikeda et al. 1992, Schaefer and Singh 2000). Bioadhesive microspheres include microparticles and microcapsules (having a core of the drug) of 1–1000 μm in diameter and consisting either entirely of a bioadhesive polymer or having an outer coating of it, respectively (Mathiowitz et al. 2001). Microspheres, in general, have the potential to be used for targeted and controlled release drug delivery; but coupling of bioadhesive properties to microspheres has additional advantages, e.g., efficient absorption and enhanced bioavailability of the drugs due to a high surface-to-volume ratio, a much more intimate contact with the mucus layer, specific targeting of drugs to the absorption site achieved by anchoring plant lectins, bacterial adhesins, antibodies, etc., on the surface of the microspheres.

Bioadhesive microspheres can be tailored to adhere to any mucosal tissue including those found in eye, nasal cavity, urinary, colon, and gastrointestinal tract, thus offering the possibilities of localized as well as systemic controlled release of drugs. Bioadhesive microspheres can be prepared using different techniques. The application of bioadhesive microspheres to the mucosal tissues of ocular cavity, gastric, and colonic epithelium is used for the administration of drugs for localized action. The prolonged release of drugs and a reduction in frequency of drug administration to the ocular cavity can highly improve patient compliance (Vasir et al. 2003). The latter advantage can also be obtained for the drugs administered intranasally due to the reduction in mucociliary clearance of drugs adhering to nasal mucosa. Microspheres prepared with bioadhesive and bioerodible polymers undergo selective uptake by the M cells of Peyer's patches (PP) in gastrointestinal (GI) mucosa. This uptake mechanism has been used for the delivery of protein and peptide drugs, antigens for vaccination, and plasmid DNA for gene therapy. Moreover, by keeping the drugs in close proximity to their absorption window in the GI mucosa, the bioadhesive microspheres improve the absorption and oral bioavailability of drugs like furosemide and riboflavin (Vasir et al. 2003).

12.2 BIOADHESION MECHANISM

The mechanisms responsible for the formation of bioadhesive bonds are not completely clear. In order to develop ideal bioadhesive drug delivery systems, it is important to describe and understand the forces that are responsible for adhesive bond formation. Most research has focused on analyzing bioadhesive interactions between polymer hydrogel and soft tissue. The process involved in the formation of such bioadhesive bonds has been described in four stages.

12.2.1 Intimate Contact between Bioadhesive and Receptor Tissues

Bioadhesive material has to penetrate into the crevices of the tissue on to which it is applied. Hence, the roughness of the tissue surface is an important factor for bioadhesion. Roughness is defined as the ratio of maximum depth (d) to maximum/minimum width (h) i.e., (d/h). For adhesion to occur, the ratio must have the value of $d/h = 1/20$ (Figure 12.1 illustrates a roughness of the tissue surface). Insignificant roughness for bioadhesion occurs when the ratio has the value less than 1/20. For higher value, only highly fluid material can penetrate. So, their viscosity and wetting power are the factors for satisfactory adhesion when the material is solid there its constitute bonds (Helfand and Tagami 1972).

12.2.2 Wetting of Bioadhesive Surface or the Swelling of the Bioadhesive

The theory is predominantly applicable to liquid bioadhesive system and contact behavior in terms of the ability of liquid or paste to spread over biological tissue surface (Figure 12.2). Swelling theory is applied only to the solid bioadhesive system. Here water plays an important role in the bioadhesion of solid material. This theory is based on the hydration of colloids, which results in the relaxation of stretched entangled or twisted molecules, now the molecules are able to liberate their adhesive sites to create bonds. The hydration of colloids causes the dissociation of the already existing hydrogen bonding of the polymers (Helfand and Tagami 1972). Thus, polymer–water interaction will overcome the polymer–polymer interaction and cause the chain inters diffusion. The rupture of any inters chain or intra chain associations will increase the mobility of macromolecules and facilitate their penetration in the tissue surface crevice.

12.2.3 Penetration of the Bioadhesive into Grooves of Tissue Surface/ Interpenetrating of the Bioadhesive Chains with Those of Mucosa

During chain interpenetration, the molecules of the bioadhesive and glycoprotein networks are brought into intimate contact. So, due to the concentration gradient, the bioadhesive polymer chains will penetrate into the network. The rate of penetration depends upon the diffusion coefficient of the macromolecules through cross-linked network. Larger cross-linked polymer network faces more difficulty while the penetration will be more favored with smaller chains and chain ends. The solid

FIGURE 12.1 Roughness of the tissue surface.

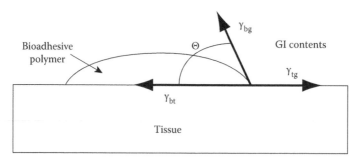

FIGURE 12.2 Schematic diagram showing the interfacial tensions involved in spreading a bioadhesive polymer over GI mucosa.

uncross-linked (swollen) and liquid systems can easily penetrate into crevices. Diffusion coefficient depends upon molecular weight and it decreases rapidly as the cross-linking increases (Kaelble 1977, Wake 1982). Exact interpenetration depth necessary to achieve bioadhesion is not known but hypothetical value ranges from 0.2–0.5.

12.2.4 Formation of the Low Chemical Bonds

There are two types of chemical bonds, first the primary bonds, in this permanent covalent bonds having high strength, they are undesirable in bioadhesion. Second is the secondary bonds, in this electrostatic attractions are the coulomb force between molecules of opposite charge. Van der Waals forces are the interactions between uncharged molecules. Hydrogen bonding occurs when specific hydrogen atom from one molecule is associated with another atom from second molecules. A hydrophobic bond occurs when nonpolar groups associate with each other in aqueous solution due to the tendency of water to exclude nonpolar molecules (Kaelble, 1977). This type of force is most important in bioadhesion.

12.3 THEORIES ON BIOADHESION

The study of adhesion is not a new science. The same theories of adhesion that were developed to explain and predict the performance of glues, adhesives, and paint can be and have been allied to bioadhesive systems (Figure 12.3). In general, five theories have been adapted to the study of bioadhesion.

12.3.1 Electrical Theory

The adhesive polymer and mucus typically have different electronic characteristics. When these two surfaces come in contact, a double layer of electrical charge forms at the interface and then adhesion develops due to attractive force from electron transfer across the electrical double layer.

12.3.2 Adsorption Theory

In the adsorption theory, a bioadhesive polymer adheres to mucus because of secondary surface forces such as van der Waals forces, hydrogen bonds, or hydrophobic interactions. For a bioadhesive polymer with a carboxyl group, hydrogen bonding is considered to be the dominant force at the interface. On the other hand, hydrophobic interactions can explain the fact that a bioadhesive may bind to a hydrophobic subtract more tightly than to a hydrophilic surface.

12.3.3 Wetting Theory

Primarily applicable to liquid bioadhesive systems, the wetting theory emphasizes the intimate contact between the adhesive and mucus. Thus, a wetted surface is controlled by structural similarity, degree of cross-linking of the adhesive polymer, or uses a surfactant.

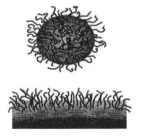

FIGURE 12.3   Mechanical bonding through interpenetration of bioadhesive and mucus polymer chains.

12.3.4 DIFFUSION THEORY

The essence of this theory is that chains of the adhesive and the substrate penetrate one another to a sufficient depth to create a semipermanent adhesive bond. The penetration rate depends on the diffusion coefficients of both interacting polymers, and the diffusion coefficient is known to depend on molecular weight and cross-linking density. In addition, segment mobility coprotein and the expanded nature of both networks are important parameters that need to be considered.

12.3.5 FRACTURE THEORY

Perhaps the most applicable theory for studying bioadhesion through mechanical measurements has been fracture theory. This theory analyzes the forces required to separate two surfaces after adhesion. The maximum tensile stress (S_m) produced during detachment can be determined by dividing the maximum force of detachment (F_m) by the total surface area (A_o) involved in the adhesive interaction:

$$S_m = \frac{F_m}{A_o}$$

12.4 FACTORS INFLUENCING BIOADHESIVE PROPERTIES

The adhesiveness of a bioadhesive polymer is determined by its intrinsic polymeric properties and the environment in which it is placed, and is influenced by many factors mentioned below.

12.4.1 BIOADHESIVE POLYMER-RELATED FACTORS

12.4.1.1 Molecular Weight

The optimum molecular weight for maximum bioadhesion depends on the type of bioadhesive polymer at tissue. It is generally understood that the threshold required for successful bioadhesion is the molecular weight at least 100,000. For example, PEG with a molecular weight of 20,000 has little adhesive character whereas PEG with 200,000 molecular weight has improved, and PEG with 400,000 molecular weight has superior adhesive properties. The fact that bioadhesiveness improves with increase in molecular weight for linear polymer implies two things: (1) interpenetration is more critical for lower molecular weight polymers to be a good bioadhesive, and (2) entanglement is important for high molecular weight polymers. Adhesiveness of nonlinear structure, by comparison, follows a quite different trend (Duchene et al. 1988). The adhesive strength of dextrin, with a very high molecular weight of 19,500,000, is similar to that of PEG, with a molecular weight of 200,000. The reason for this similarity may be that the helical conformation of dextran may shield many of the adhesive groups, which are primarily responsible for adhesion, unlike the conformation of PEG.

12.4.1.2 Concentration of a Bioadhesive Polymer

There is optimum concentration of a bioadhesive polymer to produce maximum bioadhesion. In highly concentrated systems, beyond the optimum level however, the adhesive strength drops significantly because the coiled molecules become separated from the medium so that the chains available for interpenetration become limited (Gurney et al. 1984).

12.4.1.3 Chain Flexibility

Chain flexibility is critical for interpenetration and entanglement as water soluble polymers become cross-linked, mobility of individual polymer chain decreases, and thus the effective length of the chain that can penetrate into the mucus layer decreases, which leads to reduction in bioadhesive strength (Barrer et al. 1968, Park 1986).

12.4.1.4   Type of Functional Groups

Mucin molecules are negatively charged at neutral pH due to the presence of carboxylate and sulfate groups. Because of the strong interactions between oppositely charged electrolytes poly-cations can be excellent mucoadhesive at neutral pH. At low pH values, however, where mucin is not charged, polycations are far less effective. At acidic pH, polyanions have excellent mucoadhesive properties due to their ability of forming hydrogen bonds with numerous carbohydrate hydroxyl groups of mucin molecules. Weak polyanions such as polyacrylic acid exist as protonated species at low pH value and behave as proton donors for hydrogen bonding. At neutral pH, even weak polyanions are fully ionized, and the interaction with negatively charged mucin is always repulsive.

12.4.2   Environment-Related Factors

12.4.2.1   pH

pH can influence the formal charge on the surface of mucus as well as certain ionizable bioadhesive polymers. Mucus will have a different charge density depending on pH due to difference in the dissociation of functional groups on the carbohydrate moiety and the amino acids of the polypeptide backbone. Some studies have shown that the pH of the medium is important for the degree of hydration of cross-linked polyacrylic acid, showing consistently increased hydration from pH 4 through pH 7 and then a decrease as alkalinity and ionic strength increase. For example, polycarbophil does not show a strong bioadhesive property above pH 5 because uncharged, rather than ionized, carboxyl group react with mucin molecules, presumably through numerous hydrogen bonds. However, at higher pH, the chains are fully extended due to electrostatic repulsion of the carboxylate anions (Park and Robinson 1985, Ch'ng et al. 1985).

12.4.2.2   Initial Contact Time

Contact time between the bioadhesive and mucus layer determines the extent of swelling and interpenetration of the bioadhesive polymer chains. Moreover, bioadhesive strength increases as the initial contact time increases (Kamath and Park 1994). However, longer initial contact time should be based on the tissue viability.

12.4.2.3   Swelling

Swelling characteristics are related to the bioadhesive itself and its environment. Swelling depends on the polymer concentration, ionic strength, as well as the presence of water. During the dynamic process of bioadhesion, maximum bioadhesion *in vitro* occurs with optimum water content. Overhydration results in the formation of a wet slippery mucilage without adhesion (Chen and Cyr 1970, Gurny and Peppas 1990, Mortazavi and Smart 1993).

12.4.2.4   Applied Strength

To place a solid bioadhesive system, it is necessary to apply a defined strength. Whatever the polymer, poly [acrylic acid/divinyl benzene poly (HENA)] or carbopol-934P, the adhesion strength increases with the applied strength or with the duration of its application, up to an optimum. The pressure initially applied to the bioadhesive tissue contact site can affect the depth of interpenetration. If high pressure is applied for a sufficiently long period of time, polymers become bioadhesive even though they do not have attractive interaction with mucin.

12.4.2.5   Selection of the Model Substrate Surface

The handling and treatment of biological substrates during the testing of bioadhesive is an important factor, since physical and biological changes may occur in the mucus gels or tissues under the experimental conditions. The viability of the biological substrate should be confirmed by examining properties such as permeability, electrophysiology, or histology. Such studies may be necessary before and after performing the *in vitro* tests using tissues.

12.4.3 PHYSIOLOGICAL VARIABLES

12.4.3.1 Mucin Turnover

The extent of interaction between the polymer and the mucus depends on mucus viscosity, degree of entanglement, and water content. How long the bioadhesive remains at the site depends on whether the polymer is soluble or insoluble in water, and the associate turnover rate of mucin. The natural turnover of mucin molecules from the mucus layer is important for at least two reasons, first the mucin turnover is expected to limit the residence time of the bioadhesive strength, bioadhesives are detached from the surface due to mucin turnover. The turnover rate may be different in the presence of bioadhesive, but no information is available on this aspect. Second, mucin turnover results in a substantial amount of soluble mucin molecules. These molecules interact with bioadhesive before they have a chance to interact with the mucin layer. Surface fouling is unfavorable for bioadhesive to the tissue surface. The physicochemical properties of the mucus are known to change during disease conditions if mucoadhesives are to be used in the diseased states. The bioadhesive property needs to be evaluated under the same conditions. Mucin turnover may depend on other factors as such as presence of food (Lehr et al. 1991). The gastric mucosa accumulates secreted mucin on the luminal surface of the tissue during the early stages of fasting. The accumulated mucin is subsequently released by freshly secreted acid or simply by the passage of ingested food, the exact turnover rate of the mucus layer remains to be determined. Lehr et al. (1991) calculated a mucin turnover time of 47–270 min. The ciliated cells in the nasal cavity are known to transport the mucus to the throat at a rate of 5 mm/min. The mucociliary clearance in tracheal region has been found to be in the range of 4–10 mm/min.

12.4.3.2 Disease States

The physiology properties of the mucus are known to change during disease conditions such as the common cold, gastric ulcers, ulcerative colitis, cystic fibrosis, bacterial and fungal infections of the female reproductive tract, and inflammation of the eye. The exact structural changes taking place in the mucus under these conditions are not clearly understood. If bioadhesives are to be used in the disease states, the bioadhesive property needs to be evaluated under the same condition.

12.5 BIOADHESIVE POLYMERS

A polymer is a substance formed by the linkage of a large number of small molecules known as monomers. A bioadhesive polymer is a synthetic or natural polymer that binds to biological substrates such as mucosal membranes. Such polymers are sometimes referred to as biological "glues" because they are incorporated into drugs to enable the drugs to bind to their target tissues.

The properties of the bioadhesive microspheres, e.g., their surface characteristics, force of bioadhesion, release pattern of the drug, and clearance, are influenced by the type of polymers used to prepare them. Suitable polymers that can be used to form bioadhesive microspheres include soluble and insoluble, nonbiodegradable, and biodegradable polymers. These can be hydrogels or thermoplastics, homopolymers, copolymers or blends, natural or synthetic polymers. Many types of forces can be used to anchor a polymer to mucus and/or a tissue surface. Covalent forces are suitable, provided the polymeric material is not toxic to the tissue. More likely polymer candidates will be those that are capable of either weak polar or electrostatic interactions. Undoubtedly, the ultimate force for any polymeric material attached to a tissue will be a combination of forces including hydrophilic and hydrophobic. It is also clear that strong interactions between chemical groups on the polymer and mucus/tissue are needed to keep the dosage form in contact with the tissue for an extended period of time. Polymers that can adhere to either soft or hard tissues are known as bioadhesive polymers. They are used to improve localization for oral-controlled or sustained release drug delivery system (Vasir et al. 2003). When such polymers hydrate with water, they can adhere to the mucosa and withstand salivation, tongue movement, and also swelling for a significant period of time.

12.5.1 BIOADHESIVE POLYMERS SHOULD HAVE THE FOLLOWING IDEAL PROPERTIES

1. Should be nontoxic.
2. Preferably form strong noncovalent bond with mucin–epithelial cell surface.
3. Adhere quickly to the moist tissue.
4. Allow easy incorporation of the drug and offer no hindrance as to its release.
5. Show specificity for attachment to an area or cellular site.
6. Show specificity for attachment and stimulate endocytosis.
7. Possess a wide margin of safety, both locally and systemically.
8. Show bioadhesive properties in both the dry and liquid state.
9. Should be able to accommodate both oil-and water-soluble drugs for purpose of controlled drug delivery.
10. Show specificity for attachment and stimulate release of intracellular cytokines.

12.5.2 CLASSIFICATION OF BIOADHESIVE POLYMER

Bioadhesive polymers can be classified on basis of its origin and charge (Vasir et al. 2003; Table 12.1). In the more functional types of classification, bioadhesive polymers can be grouped into water-soluble polymers that are typically linear or random. The water-soluble polymers that swell indefinitely in contact with water and eventually undergo complete dissolution, e.g., methylcellulose, hydroxyethyl cellulose, hydroxy propyl methyl cellulose, sodium carboxy methyl cellulose, carbomers, chitosan, and

---

TABLE 12.1
Classification of Bioadhesive Polymer

1. Based on the origin

a. Synthetic polymers

i. Polyacrylic acid derivatives	ii. Cellulose derivatives
Carbopol	Carboxymethyl cellulose
Polycarbophil	Hydroxyethyl cellulose
Polyacrylate	Hydroxypropylmethyl cellulose
Poly(methylvinyl ether-*co*-methacrylic) acid	
Poly(2-hydroxyethyl methacrylate)	
Poly(methacrylate)	
Poly(alkyl cyanoacrylate)	

b. Natural polymers

Pectin, alginate, gellan, carrageenan, chitosan, guar-gum, xanthan-gum

2. Based on the charge

i. Cationic	ii. Anionic	iii. Neutral
Polysine	Carboxymethyl cellulose	Dextran
Polyvinyl methyl	Polyvinyl sulfate	Gelatin
Polybrene	Polymeric acid	Albumin
	Polyacrylic acid	PVP

plant gums etc. Water insoluble or hydrogel, these are water swellable materials, usually a cross-link polymer with limited swelling capacity, e.g., poly(acrylic acid-*co*-acrylamide) copolymers, carrageenan, sodium alginate, guar gum, and modified guar gum etc. Thermoplastic polymers include the nonerodible neutral polystyrene and semicrystalline bioerodible polymers, which generate the carboxylic acid groups as they degrade, e.g., polyanhydrides and polylactic acid. Various synthetic polymers used in bioadhesive formulations include polyvinyl alcohol, polyamides, polycarbonates, polyalkylene glycols, polyvinyl ethers, esters and halides, polymethacrylic acid, polymethylmethacrylic acid, methylcellulose, hydroxypropyl cellulose, hydroxypropyl methylcellulose, and sodium carboxymethylcellulose. Various biocompatible polymers used in bioadhesive formulations include cellulose-based polymers, ethylene glycol polymers and its copolymers, oxyethylene polymers, polyvinyl alcohol, polyvinyl acetate, and esters of haluronic acid. Various biodegradable polymers used in bioadhesive formulations are poly(lactides), poly(glycolides), poly(lactide-*co*-glycolides), polycaprolactones, and polyalkyl cyanoacrylates. Polyorthoesters, polyphosphoesters, polyanhydrides, polyphosphazenes are the recent additions to the polymers. In case of water-soluble polymers, the duration of residence time on tissue surface is based on the dissolution rate of the polymers, given their lack of solubility in common solvents, have a residence time based on the rate of mucus/tissue turnover. Robinson and coworkers (1985) using the fluorescence technique concluded that, cationic and anionic polymers bind more effectively then neutral polymers. Polyanions are better than polycations in terms of binding and that water-insoluble polymers give greater flexibility in dosage form design compared to rapidly or slowly dissolving polymers. Anionic polymers with sulfate group bind more effectively than those with carboxylic groups. Degree of binding is proportional to the charged density on the polymer. Highly binding polymers include CMC, gelatin, hyaluronic acid, carbopol, and polycarbophil. Some important bioadhesive polymers and their properties and rank order of bioadhesive force for various polymers are described in Tables 12.2 and 12.3, respectively (Wallace 1990, Kumar and Banker 1993, Ahuja et al. 1997).

12.5.3 NEXT-GENERATION BIOADHESIVE POLYMERS

With the disappointment in the merger of bioadhesive systems into pharmaceuticals in the site-specific drug delivery area, there has been an increasing interest from researchers in targeting the regions of the gastrointestinal tract (GIT) using more selective compounds capable of distinguishing between the types of cells found in different areas of the GIT. Loosely termed "cytoadhesion," this concept is specifically based on certain materials that can reversibly bind to cell surfaces in the GIT (Shah and Rocca 2004). These next generation of bioadhesives function with greater specificity because they are based on receptor-ligand-like interactions in which the molecules bind strongly and rapidly directly onto the mucosal cell surface rather than the mucus itself (Lehr 2004). One such class of compounds that has these unique requirements is called lectins. Lectins are proteins or glycoproteins and share the common ability to bind specifically and reversibly to carbohydrates. They exist in either soluble or cell-associated forms and possess carbohydrate-selective and recognizing parts. They are found mostly in plants, to a lesser extent, in some vertebrates (referred to as endogenous lectins), and can also be produced from bacteria or invertebrates (Haltner et al. 1997). Lectin-based drug delivery systems have applicability in targeting epithelial cells, intestinal M cells, and enterocytes. The intestinal epithelial cells possess a cell surface composed of membrane-anchored glycoconjugates. It is these surfaces that could be targeted by lectins, thus enabling an intestinal delivery concept. One lectin that has been studied to considerable extent in *in vitro* binding and uptake is tomato lectin (TL), which has been shown to bind selectively to the small intestine epithelium. In one study, using the everted gut sac model, this lectin was bound to polystyrene microspheres. Uptake of TL into the serosal fluid was reported as eightfold higher than the control Bovine serum albumin (BSA) (Carreno-Gomez et al. 1999). Furthermore, BSA-coupled microspheres were shown to have slower uptake than TL-coupled microspheres by a factor of two. In another study, specific binding by TL-coated polystyrene microspheres (0.98 mm) to enterocytes *in vitro* was examined (Lehr et al. 1992). Fluorescently labeled

TABLE 12.2
Bioadhesive Polymers and Their Properties

Bioadhesive	Properties	Characteristics	References
Polycarbophil (polyacrylic acid cross-linked with divinyl glycol)	Mw 2.2×10^5 η 2,000–22,500 cps (1% aqueous solution) κ 15–35 mL/g in acidic media (pH 1–3) 100 mL/g in neutral and basic media φ viscous colloid in cold water	Synthesized by lightly cross-linking of divinyl glycol Swellable depending on pH, but insoluble in water Entangle the polymer with mucus on the surface of the tissue Hydrogen bonding between the nonionized carboxylic acid and mucin	Ch'ng et al. (1985)
Carbopol/carbomer (carboxy polymethylene)	Mw $1 \times 10^6 – 4 \times 10^6$ η 29,400–39,400 cps at 25°C with 0.5% aqueous solution ρ 5 g/cm³ in bulk pH 2.5–3.0 φ water, alcohol, glycerin	Synthesized by cross-linker of allyl sucrose or allyl pentaerythritol Excellent thickening, emulsifying, suspending, gelling agent Common component in bioadhesive dosage forms	Ahuja et al. (1997)
Sodium carboxymethyl cellulose (cellulose carboxymethyl ether sodium salt)	Mw $9 \times 10^4 – 7 \times 10^5$ η 1200 cps with 1% solution ρ 0.75 g/cm³ in bulk pH 6.5–8.5 φ water	Sodium salt of a polycarboxymethyl ether of cellulose Emulsifying, gelling, binding agent Good bioadhesive strength	Wallace (1990)
Hydroxypropyl cellulose (cellulose 2-hydroxypropyl ether)	Mw $6 \times 10^4 – 1 \times 10^6$ η 4000–6500 cps with 2% aqueous solution ρ 0.5 g/cm³ in bulk pH 5.0–8.0 φ soluble in water below 30°C, ethanol	Partially substituted polyhydroxypropyl ether of cellulose Granulating and film coating agent for tablet Thickening agent, emulsion stabilizer, suspending agent in oral and topical liquid solution or suspension formulation	Kumar and Banker (1993)
Hydroxypropyl methyl cellulose (cellulose 2-hydroxypropyl methyl cellulose	Mw 8.6×10^4 η 15–4000 cps (2% aqueous solution) φ cold water	Mixed alkyl hydroxyalkyl cellulosic ether Suspending, viscosity-increasing and film-forming agent Tablet binder and adhesive ointment ingredient	Ahuja et al. (1997)
Hydroxyethyl cellulose	ρ 0.6 g/mL pH 6.0–8.5	Used as suspending or viscosity-increasing agent Binder, film former, thickener	Ahuja et al. (1997)
Alginate	pH 7.2 η 20–400 cps (1%) φ water	Stabilizer in emulsion, suspending agent, tablet disintegrant, tablet binder	Wallace (1990)

polystyrene microspheres were coated with TL, and incubated in a CaCo-2 cell line. It was observed that the lectin-coated microspheres were resistant to repeat washings compared to the control (Bovine serum albumin-microsphere).

For optimal buccal bioadhesion, Shojaei and Li (1997) have designed, synthesized, and character-ized a copolymer of PAA and PEG monoethylether monomethacrylate(PAA-*co*-PEG) (PEGMM). By adding PEG to these polymers, many of the shortcomings of PAA for bioadhesion, outlined earlier, were eliminated. Hydration studies, glass transition temperature, bioadhesive force, surface

TABLE 12.3
Rank Order of Bioadhesive Force for Various Polymers

Test Polymer	Mean % Adhesive Force	Standard Deviation
Poly(acrylic acid)	185.0	10.3
Tragacanth	154.4	7.5
Poly(methylvinylether-*co*-maleic anhydride)	147.7	9.7
Poly(ethyleneoxide)	128.6	4.0
Methylcellulose	128.0	2.4
Sodium alginate	126.2	12.0
HPMC	125.2	16.7
Karaya gum	125.2	4.8
MEC	117.4	4.2
Soluble starch	117.2	3.1
Gelatin	115.8	5.6
Pectin	100.0	2.4
PVP	97.6	3.9
PEG	96.0	7.6
PVA	94.8	4.4
Poly(hydroxyethylmethacrylate)	88.4	2.3
Hydroxypropylcellulose	87.1	13.3

energy analysis, and effect of chain length and molecular weight on bioadhesive force were studied. The resulting polymer has a lower glass transition temperature than PAA and exists as a rubbery polymer at room temperature. Copolymers of 12- and 16-mol% PEGMM showed higher bioadhesion than PAA. The effects of hydration on bioadhesion seen by the copolymers revealed that film containing lower PEGMM content, which had higher hydration levels, had lower bioadhesive strengths. The 16-mol% PEGMM had the most favorable thermodynamic profile and the highest bioadhesive forces. Polymers investigated in this study also showed that the molecular weight and chain length had little or no effect on the bioadhesive force (Nagai and Kinishi 1987). Lele and Hoffman (2000) investigated novel polymers of PAA complexed with PEGylated drug conjugate. Only a carboxyl group containing drugs such as indomethacin could be loaded into the devices made from these polymers. An increase in the molecular weight of PEG in these copolymers resulted in a decrease in the release of free indomethacin, indicating that drug release can be manipulated by choosing different molecular weights of PEG.

A new class of hydrophilic pressure-sensitive adhesives (PSAs) that share the properties of both hydrophobic PSAs and bioadhesives has been developed by CoriumTechnologies (Cleary et al. 2003). These Corplex® adhesive hydrogels have been prepared by noncovalent (hydrogen bond) cross-linking of a film-forming hydrophilic polymer(e.g., PVP), with a short-chain plasticizer (typically PEG) bearing complementary reactive hydroxyl groups at its chain ends. Owing to the appreciable length and flexibility of PEG chains, a relatively large space can be provided for a stoichiometric complex and a "carcass-like" structure. The specific balance between enhanced cohesive strength and large free volume in PVP–PEG miscible blends influences their PSA behavior. Properties of these hydrophilic PSA hydrogels prepared by the "carcass-like" cross-linking method can be modified using a polymer with complementary reactive groups to form "ladder-like" cross-links with PVP. Thus, these Corplex PSA hydrogels have a broad range of unique adhesive/cohesive properties that enable topical and drug delivery systems to be applied to either skin or mucosa.

An AB block copolymer of oligo(methyl methacrylate) and PAA has been synthesized for prolonged mucosal drug delivery of hydrophilic drugs (Inoue et al. 1998). These block copolymers form micelles in an aqueous medium, which was confirmed by fluorescence probe technique using pyrene. A model drug, doxorubicin hydrochloride, when incorporated into these micelles, results in its release being prolonged at a slower rate. Polymers with thiol groups were also investigated as a new generation of bioadhesive polymers. A study conducted by Bernkop-Schnurch et al. (1999) demonstrated that the introduction of a sulfahydryl group increased the adhesive properties of bioadhesive polymers. In this study, cysteine was attached covalently to polycarbophil by using carbodiimide as a mediator, forming amide bonds between the primary amino group of the amino acid and the carboxylic acid moieties of the polymer. The results showed that there was considerable improvement in the overall behavior of adhesion and adhesive properties when tested on porcine intestinal mucosa at a pH level above five. Langoth et al. (2003) investigated the benefit of thiolated polymers (thiomers) for the development of buccal drug delivery systems. The matrix tablet based on this thiomer showed good stability, bioadhesion, and controlled drug release (for leuenkephalin over 24 h).

In addition, bioadhesive microspheres were studied recently by Bogataj et al. (1999) for application in the urinary bladder. The microspheres were prepared by a solvent evaporation method using eudragit RL or hydroxypropylcellulose as matrix polymers. In another study, microspheres with a eudragit RS matrix polymer and different bioadhesive polymers, i.e., chitosan hydrogen chloride, sodium salt of carboxymethyl cellulose and polycarbophil were prepared and found to be useful as platforms for oral peptide delivery, with a high capacity of binding to bivalent cations, which are essential cofactors for intestinal proteolytic enzymes (Chein and Langer 1998). Alur et al. (1999) studied the transmucosal sustained delivery of chlorpheniramine maleate in rabbits using a novel natural bioadhesive gum (from Hakea), as an excipient in buccal tablets. It was concluded that the gum not only sustained the release of drug but also provided sufficient mucoadhesion to tablets for clinical application.

12.6   PREPARATION OF BIOADHESIVE MICROSPHERES

Bioadhesive microspheres can be prepared using any of the following techniques.

12.6.1   SOLVENT EVAPORATION

It is the most extensively used method of microencapsulation. A buffered or plain aqueous solution of the drug (may contain a viscosity building or stabilizing agent) is added to an organic phase consisting of the polymer solution in solvents like dichloromethane (or ethyl acetate or chloroform) with vigorous stirring to form the primary water in oil emulsion. This emulsion is then added to a large volume of water containing an emulsifier like PVA or PVP to form the multiple emulsion (w/o/w). The double emulsion, so formed, is then subjected to stirring until most of the organic solvent evaporates, leaving solid microspheres. The microspheres can then be washed, centrifuged, and lyophilized to obtain the free flowing and dried microspheres.

12.6.2   HOT MELT MICROENCAPSULATION

In this method, the polymer is first melted and then mixed with solid particles of the drug that have been sieved to less than 50 μm. The mixture is suspended in a nonmiscible solvent (like silicone oil), continuously stirred, and heated to 5°C above the melting point of the polymer. Once the emulsion is stabilized, it is cooled until the polymer particles solidify. The resulting microspheres are washed by decantation with petroleum ether. The only disadvantage of this method is the moderate temperature to which the drug is exposed.

12.6.3 SOLVENT REMOVAL

In this method, drug is dispersed or dissolved in a solution of the selected polymer in a volatile organic solvent like methylene chloride. This mixture is then suspended in silicone oil containing Span 85 and methylene chloride (Carino et al. 1999). After pouring the polymer solution into silicone oil, petroleum ether is added and stirred until solvent is extracted into the oil solution. The resulting microspheres can then be dried in vacuum.

12.6.4 HYDROGEL MICROSPHERES

Microspheres made of gel-type polymers, such as alginate, are produced by dissolving the polymer in an aqueous solution, suspending the active ingredient in the mixture and extruding through a precision device, producing microdroplets that fall into a hardening bath that is slowly stirred. The hardening bath usually contains calcium chloride solution, whereby the divalent calcium ions cross-link the polymer forming gelled microspheres. The method involves an "all-aqueous" system and avoids residual solvents in microspheres. The particle size of microspheres can be controlled by using various size extruders or by varying the polymer solution flow rates.

12.6.5 SPRAY DRYING

In this process, the drug may be dissolved or dispersed in the polymer solution and spray dried. The quality of spray-dried microspheres can be improved by the addition of plasticizers, which promote polymer coalescence on the drug particles and hence promote the formation of spherical and smooth-surfaced microspheres. The size of microspheres can be controlled by the rate of spraying, the feed rate of polymer drug solution, nozzle size, and the drying temperature.

12.6.6 PHASE INVERSION MICROENCAPSULATION

The process involves addition of drug to a dilute solution of the polymer. The mixture is poured into an unstirred bath of a strong nonsolvent (petroleum ether) in a solvent to nonsolvent ratio of 1:100, resulting in the spontaneous production of microspheres through phase inversion. The microsphere then be filtered, washed with petroleum ether, and dried at room temperature. This simple and fast process of microencapsulation involves relatively little loss of polymer and drug.

12.6.7 SUPERCRITICAL FLUID TECHNOLOGY

Drug and polymeric bioadhesive microparticles have been prepared using SCFs as solvents and antisolvents. The advantages of SCF technology include use of mild conditions for pharmaceutical processing (which is advantageous for labile proteins and peptides), use of environmentally benign nontoxic materials (such as CO_2), minimization of organic solvent use, and production of particles with controllable morphology, narrow size distribution, and low static charge.

12.7 EVALUATION OF THE BIOADHESIVE MICROSPHERES

The best approach to evaluate bioadhesive microspheres is to evaluate the effectiveness of bioadhesive polymer to prolong the residence time of drug at the site of absorption, thereby increasing the absorption and bioavailability of the drug. The methods used to evaluate bioadhesive microspheres include the following.

12.7.1 Production Yield

The percentage of production yield (wt/wt) can be calculated from the weight of dried microspheres (w_1) recovered from each of the batches and the sum of the initial dry weight of starting materials (w_2) as the following equation:

$$\text{Percentage of production yield} = \frac{w_1}{w_2} \times 100$$

12.7.2 Drug Content and Loading Efficiency

The bioadhesive microspheres of each formulation can be extracted in dissolution medium and assayed by suitable analytical methods. The actual amount of drug loaded relative to the theoretical amount in the microspheres was calculated as a percentage and expressed as the loading efficiency.

12.7.3 Particle Size Measurement

The prepared bioadhesive microspheres are suspended in suitable solvents and sized by using a laser particle size distribution analyzer or microscopic method.

12.7.4 Surface Characterization of the Bioadhesive Microspheres

Surface morphology of microspheres and the morphological changes produced through polymer degradation can be investigated and documented using scanning electron microscopy (SEM), electron microscopy, and scanning tunneling microscopy (STM). To assess the effect of surface morphology on the bioadhesive properties, the microsphere samples are lyophilized and analyzed under SEM at 150× and 1000×. The smooth texture of the microsphere surface leads to weak bioadhesive properties, while the coarser surface texture improves the adhesion through stronger mechanical interactions. The morphological surface changes occurring due to the hydrolytic degradation of the polymers, e.g., polyanhydrides can be studied after incubating the microspheres in the PBS buffer for different intervals of time.

12.7.5 Determination of Bulk Density

Accurate weights of microspheres (w_m) are transferred into a 100 mL graduated cylinder to obtain the apparent volumes (v) between 50 and 100 mL. The bulk density is calculated in gram per milliliter by the following formula:

$$\text{Bulk density} = \frac{w_m}{v}$$

12.7.6 Angle of Repose

The angle of repose can be measured by heap carefully built up by dropping the microsphere samples through a glass funnel to the horizontal plate of a powder characteristic tester.

12.7.7 Zeta Potential Study

The zeta potential of bioadhesive microspheres dispersed in dissolution medium can be determined by a zeta meter. The directional movement of 200 microspheres from each formulation are observed and averaged from three determinations.

12.7.8 SWELLING PROPERTY

The swelling of bioadhesive microspheres can occur in dissolution medium. Their diameters are periodically measured by using a laser particle size distribution analyzer or microscope until they are decreased by erosion and dissolution. The percentage of swelling is determined at different time intervals by the difference between the diameter of microspheres at time t (D_t) and initial time ($t=0$ [D_0]) as calculated from the following equation:

$$\text{Percentage of swelling} = \frac{D_t - D_0}{D_0} \times 100$$

12.7.9 INFRARED ABSORPTION STUDY

The infrared (IR) spectra of drug and additives in the bioadhesive microspheres are examined using the potassium bromide disc method by an IR spectrophotometer in the required range.

12.7.10 MEASUREMENT OF ADHESIVE STRENGTH/*IN VITRO* TESTS

The quantification of the bioadhesive forces between polymeric microspheres and the mucosal tissue is a useful indicator for evaluating the bioadhesive strength of microspheres. *In vitro* techniques have been used to test the polymeric microspheres against a variety of synthetic and biological tissue samples, such as synthetic and natural mucus, frozen and freshly excised tissue etc. The different *in vitro* methods include the following and are summarized in Table 12.4.

12.7.10.1 Tensile Stress Measurement

12.7.10.1.1 Wilhelmy Plate Technique
The Wilhelmy plate technique is traditionally used for the measurement of dynamic contact angles and involves the use of a microtensiometer or a microbalance. The CAHN dynamic contact angle

TABLE 12.4
Measurement of Adhesive Strength/*In Vitro* Tests

Method	Comment
Wilhelmy plate technique	The measurement of dynamic contact angles and involves the use of a microtensiometer or a microbalance
Novel EMFT	The EMFT measures tissue adhesive forces by monitoring the magnetic force required to exactly oppose the bioadhesive force
Shear stress measurement	The shear stress measures the force that causes a mucoadhesive to slide with respect to the mucus layer in a direction parallel to their plane of contact
Adhesion number	Determined as the ratio of the number of particles attached to the substrate to the total number of applied particles, expressed as a percentage
Falling liquid film method	Quantitative *in situ* method, wherein an excised intestinal segment cut lengthwise is spread on a plastic flute and positioned at an incline and suspension of microspheres is allowed to flow down the intestinal strip. Particle concentrations entering the segment from the dilute suspension reservoir and leaving the intestinal segment can be determined with the help of the Coulter counter
Everted sac technique	A passive test for bioadhesion and involves polymeric microspheres and a section of the everted intestinal tissue
Novel rheological approach	The rheological properties of the mucoadhesive interface (i.e., of the hydrated gel) are influenced by the occurrence of interpenetration step in the process of bioadhesion

analyzer (model DCA 322, CAHN instruments, Cerritos) has been modified to perform adhesive microforce measurements. The DCA 322 system consists of an IBM compatible computer and a microbalance assembly (Chickering et al. 1999). The microbalance unit consists of stationary sample and tare loops and a motor-powered translation stage. The instrument measures the bioadhesive force between mucosal tissue and a single microsphere mounted on a small diameter metal wire suspended from the sample loop in microtensiometer (Santos et al. 1999). The tissue, usually rat jejunum, is mounted within the tissue chamber containing Dulbecco's phosphate buffered saline containing 100 mg/dL glucose and maintained at the physiologic temperature. The chamber rests on a mobile platform, which is raised until the tissue comes in contact with the suspended microsphere. The contact is held for 7 min, at which time the mobile stage is lowered and the resulting force of adhesion between the polymer and mucosal tissue is recorded as a plot of the load on microsphere versus mobile stage distance or deformation. The plot of output of the instrument is unique in that it displays both the compressive and the tensile portions of the experiment. By using the CAHN software system, three essential bioadhesive parameters can be analyzed. These include the fracture strength, deformation to failure, and work of adhesion:

- Fracture strength: It is the maximum force per unit surface area required to break the adhesive bond.
- Deformation to failure: It is the distance required to move the stage before complete separation occurs. This parameter is dependent on the material stiffness and the intensity of strength of adhesion.
- Work of adhesion: It is a function of both the fracture strength and the deformation to failure. It tends to be the strongest indicator of the bioadhesive potential.

This technique allows the measurement of bioadhesive properties of a candidate material in the exact geometry of the proposed microsphere delivery device, and the use of a physiological tissue chamber mimics the *in vivo* conditions. From a single tensile experiment, 11 bioadhesive parameters can be analyzed, out of which 3 are direct predictors of the bioadhesive potential (Chickering and Mathiowitz 1995). The CAHN instrument, although a powerful tool has inherent limitations in its measurement technique, makes it better suited for large microspheres (with a diameter of more than 300 μm) adhered to tissue *in vitro*. Therefore, many new techniques have been developed to provide the quantitative information of the bioadhesive interactions of the smaller microspheres.

12.7.10.1.2 Novel Electromagnetic Force Transducer

The electromagnetic force transducer (EMFT) is a remote sensing instrument that uses a calibrated electromagnet to detach a magnetic-loaded polymer microsphere from a tissue sample (Hertzog and Mathiowitz 1999). It has the unique ability to record, remotely and simultaneously, the tensile force information as well as the high magnification video images of bioadhesive interactions at near physiological conditions. The EMFT measures tissue adhesive forces by monitoring the magnetic force required to exactly oppose the bioadhesive force. To test a microsphere, it must first be attached to the sample of tissue; magnetic force is then generated by an electromagnet mounted on the microscope vertically above the tissue chamber. After the computer has calculated the position of microsphere, the tissue chamber is slowly moved down, away from the magnet tip. As the tissue slowly descends away from the magnet, the video analysis continuously calculates the position of microsphere until the latter is completely pulled free of the tissue. The computer can display the results either as raw data or convert it to a force versus displacement graph. The primary advantage of the EMFT is that no physical attachment is required between the force transducer and the microsphere. This makes it possible to perform accurate bioadhesive measurements on the small microspheres, which have been implanted *in vivo* and then excised (along with the host tissue) for measurement. This technique can also be used to evaluate the bioadhesion of polymers to specific cell types and hence, can be used to develop BDDS to target-specific tissues.

12.7.10.2 Shear Stress Measurement

The shear stress measures the force that causes a mucoadhesive to slide with respect to the mucus layer in a direction parallel to their plane of contact (Kamath and Park 1994). Adhesion tests based on the shear stress measurement involve two glass slides coated with polymer and a film of mucus. Mucus forms a thin film between the two polymer-coated slides, and the test measures the force required to separate the two surfaces. Mikos and Peppas (1990) designed the *in vitro* method of flow chamber. The flow chamber made of Plexiglass is surrounded by a water jacket to maintain a constant temperature. A polymeric microsphere placed on the surface of a layer of natural mucus is placed in a chamber. A simulated physiologic flow of fluid is introduced in the chamber and the movement of microsphere is monitored using video equipment attached to a goniometer, which also monitors the static and dynamic behavior of the microparticle (Chickering et al. 1996).

12.7.10.3 Other Tests to Measure the Adhesive Strength

12.7.10.3.1 Adhesion Number

Adhesion number for bioadhesive microspheres is determined as the ratio of the number of particles attached to the substrate to the total number of applied particles, expressed as a percentage. The adhesion strength increases with an increase in the adhesion number.

12.7.10.3.2 Falling Liquid Film Method

It is a simple, quantitative *in situ* method, wherein an excised intestinal segment cut lengthwise, is spread on a plastic flute and positioned at an incline. The suspension of microspheres is allowed to flow down the intestinal strip. Particle concentrations entering the segment from the dilute suspension reservoir and leaving the intestinal segment can be determined with the help of Coulter counter to quantify the steady-state fraction of particles adhered to the intestinal mucosa. The percent of particles retained on the tissue is calculated as an index of bioadhesion (Teng and Ho 1987).

12.7.10.3.3 Everted Sac Technique

The everted intestinal sac technique is a passive test for bioadhesion and involves polymeric microspheres and a section of the everted intestinal tissue. It is performed using a segment of intestinal tissue excised from the rat, everted, ligated at the ends, and filled with saline. It is then introduced into a tube containing a known amount of the microspheres and saline, and agitated while incubating for 30 min. Sac is then removed, microspheres are washed and lyophilized, and the percentage of binding to the sac is calculated from the difference in the weight of the residual spheres from the original weight of the microspheres.

The advantage of the technique is that no external force is applied to the microspheres being tested; microspheres are freely suspended in buffer solution and made to come in contact with the everted intestinal tissue randomly. The CAHN technique and the everted intestinal sac technique, both predict the strength of bioadhesion in a very similar manner. Santos and group (1999) established a correlation between the two *in vitro* bioadhesion assay methods, which thereby allows one to confidentially utilize a single bioadhesion assay to scan a variety of bioadhesive polymers.

12.7.10.4 Novel Rheological Approach

The rheological properties of the mucoadhesive interface (i.e., of the hydrated gel) are influenced by the occurrence of interpenetration step in the process of bioadhesion. Chain interlocking, conformational changes, and the chemical interaction, which occur between bioadhesive polymer and mucin chains, produce changes in the rheological behavior of the two macromolecular species. The rheological studies provide an acceptable *in vitro* model representative of the *in vivo* behavior of bioadhesive polymers (Riley et al. 2001). Due to intermolecular interactions between the two polymers (mucin and the bioadhesive polymer), the experimentally measured viscosity of the mixture is generally higher than the viscosity calculated as a weighted average of the viscosities of the individual components. Thus, the magnitude of the intermolecular interactions can be quantitated by the relative change of the solution viscosity. A synergistic increase in the viscosity of the gastric mucus glycoprotein has been observed with polyacrylates, which thereby reinforce the gastroduodenal mucus. It has been reported that an optimum polymer concentration is required for rheological synergy to be evident, above which any synergy is masked by the rheological properties of the polymer alone. The effect of pH on the mucus/polymer rheological synergism of polyacrylates has been examined using dynamic oscillatory rheology (Madsen et al. 1998). It has been shown that an optimum mucus polymer interaction occurs not only at the pK_a value but also at the pH regimes unique to each polymer type, being influenced by the hydrogen-bonded interactions.

12.7.11   MEASUREMENT OF THE RESIDENCE TIME/*IN VIVO* TECHNIQUES

The measurements of the residence time of bioadhesives at the application site provide quantitative information on their mucoadhesive properties. The GI transit times of many bioadhesive preparations have been examined using radioisotopes and the fluorescent-labeling techniques. The different *in vivo* methods include the following and are summarized in Table 12.5.

12.7.11.1   GI Transit Using Radio-Opaque Microspheres

It is a simple procedure involving the use of radio-opaque markers, e.g., barium sulfate, encapsulated in bioadhesive microspheres to determine the effects of bioadhesive polymers on GI transit time. Faeces collection (using an automated faeces collection machine) and x-ray inspection provide a noninvasive method of monitoring total GI residence time without affecting normal GI motility. Mucoadhesives labeled with Cr-51, Tc-99m, In-113m, or I-123 have been used to study the transit of the microspheres in the GI tract (Mathiowitz et al. 1999).

12.7.11.2   Gamma Scintigraphy Technique

The distribution and retention time of the bioadhesive intravaginal microspheres can be studied using the gamma scintigraphy technique. A study has reported the intensity and distribution of radioactivity in the genital tract after the administration of technetium-labeled Hyaluronic acid esters (HYAFF) microspheres. The dimensions of the vaginal cavity of the sheep can be outlined and imaged using labeled gellan gum, and the data collected is subsequently used to compare the distribution of radio-labeled HYAFF formulations. The retention of bioadhesive radio-labeled microspheres based on HYAFF polymer was found to be more for the dry powder formulation than for the pessary formulation after 12 h of administration to vaginal epithelium (Richardson et al. 1996). The combination of sheep model and gamma scintigraphy method has been proved to be an extremely useful tool for evaluating the distribution, spreading, and clearance of vaginally administered bioadhesive drug delivery system, including microbicides.

TABLE 12.5
Measurement of the Residence Time/*In Vivo* Techniques

Method	Comment
GI transit using radio-opaque microspheres	The use of radio-opaque markers, e.g., barium sulfate, encapsulated in bioadhesive microspheres to determine the effects of bioadhesive polymers on GI transit time
Gamma scintigraphy technique	The distribution and retention time of the bioadhesive intravaginal microspheres can be studied using the gamma scintigraphy technique

---

12.7.12 IN VITRO DRUG RELEASE STUDY

The *in vitro* drug release test of the bioadhesive microspheres can be performed on Franz diffusion cell with dialysis membrane or USP dissolution test apparatus.

12.7.13 DRUG PERMEATION STUDY

Cell cultures techniques are mostly used for drug permeation study.

12.8 BIOTECHNOLOGICAL APPLICATIONS

Many researchers have sought to enhance the mucoadhesion of particles in order to improve their retention at mucosal surfaces. To maximize association with mucus, a variety of mucoadhesive drug delivery systems have been engineered, driven by various interaction forces between mucus and nanoparticles, including hydrogen bonding, van der Waals interactions, polymer chain interpenetration, hydrophobic forces, and electrostatic/ionic interactions (Ponchel and Irache 1998, Woodley 2001). Electrostatic interaction is one of the most exploited forms of mucoadhesion, as exemplified by chitosan, a cationic polymer obtained from deacetylation of chitin, for a variety of oral and nasal drug delivery applications (Bernkop-Schnurch 2005, Prego et al. 2005). Jubeh et al. similarly concluded that cationic liposomes adhered to healthy colonic rat mucosa at rates threefold greater than neutral or anionic liposomes (Jubeh et al. 2004). Particles synthesized from common biomaterials, such as poly(ethylene glycol) (Peppas et al. 1999, Lele and Hoffman 2000, Yoncheva et al. 2005), polycarbophil and carbopol (derivatives of poly(acrylic acid)) (Takeuchi et al. 2001, Woodley 2001), polymethacrylates (Quintanar-Guerrero et al. 2001, Keely et al. 2005), and poly(sebacic acid) (Chickering et al. 1996) may achieve mucoadhesion via hydrogen bonding, polymer entanglements with mucins, hydrophobic interactions, or a combination of these mechanisms. The association of the aforementioned biomaterials with mucus is typically considered nonspecific, as they lack biological molecules that afford recognition and binding to specific target chemical structures on mucus or the surfaces of mucosal epithelia (Ponchel and Irache 1998, Galindo-Rodriguez et al. 2005).

Alternatively, mucoadhesion can also be achieved via ligands that recognize particular mucin glycoproteins. For example, TL that specifically recognize and bind to *N*-acetyl glucosamine-containing complexes on cell surfaces, were initially tested for specific interactions with intestinal tissues (Woodley 2001). However, there was little improvement in intestinal transit time of TL compared to other mucoadhesive systems (Naisbett and Woodley 1995), an observation later justified by the strong association of lectins to the mucus gel instead of the epithelia (Lehr et al. 1992, Irache et al. 1996, Montisci et al. 2001). Lectins with different sugar specificities have also been investigated, including those derived from *Canavalia ensiformis* (Haltner et al. 1996) and *Lotus tetragonolobus* (Irache et al. 1994).

Mucoadhesive particles usually improve the pharmacokinetics of therapeutics in the GI tract compared to free drug alone. For example, mucoadhesive polyanhydride copolymers of fumaric and sebacic acids were used to encapsulate insulin and dicumarol (an anticoagulant drug with poor water solubility and erratic intestinal adsorption), and in both cases, improved the bioavailability of both loaded drugs compared to their native form (Chickering et al. 1996). Chitosan-based or chitosan-coated nanoparticles represent another common group of mucoadhesive systems (Janes et al. 2001, Bernkop-Schnurch 2005, Prego et al. 2005), and several groups have studied nanoparticles made of these polymers to deliver proteins (Fernandez-Urrusuno et al. 1999, Pan et al. 2002), peptides (Prego et al. 2006), and DNA (Roy et al. 1999) to mucosal tissues. In addition to prolonged residence times, improved oral bioavailability of drugs observed *in vivo* may be partially attributed to improved drug stability in encapsulated form in the GI tract that is rich in proteolytic enzymes such as pepsin, trypsin, and chymotrypsin (Galindo-Rodriguez et al. 2005). For example, the stability of insulin and calcitonin exposed to protease degradation is significantly improved when loaded in polymeric nanoparticles (Lowe and Temple 1994, Sarmento et al. 2007, Damge et al. 1997).

A range of reports have suggested that nanoparticles are capable of entry into intestinal epithelia via M cells on the domes of the PP in small animals. Not surprisingly, the PP are the intestinal surfaces least protected by mucus and most exposed to chyme. Since no mucus is secreted in the region surrounding these cells, which protrude relatively unprotected into the lumen, the mucus barrier is minimal (Neutra and Forstner 1987). Indeed, M cells are positioned as sensory outposts for cellular immune functions, transcytosing particles that impinge on their surface into the interior of the patch. The dominant opinion is that particulate uptake in GIT of mammals is principally via the M cells of PP, and that uptake by enterocytes plays a minor role (Lavelle et al. 1995, Galindo-Rodriguez et al. 2005). Polystyrene nanopsheres and microspheres were found in the PP of the rat colon 12 h following oral administration (Jani et al. 1992). Similarly, Jani et al. found microspheres up to a diameter of 3 µm in PP and in the gut-associated lymphoid tissue 12 h after oral administration (Jani et al. 1989). An *ex vivo* model using a diffusion chamber with pig intestinal tissue shows that radiolabeled poly(isobutyl cyanoacrylate) nanoparticles (211 nm) cross pig PP, whereas tissue deprived of PP was impermeable to particles (Scherer et al. 1993). The extent of nanoparticle uptake in rabbits appeared at least an order of magnitude greater than that in mice, a finding attributed to the significantly greater abundance of M cells in rabbit PP (O'Hagan 1996). Although the above reports widely support the notion that M cells account for the bulk of particles crossing the intestinal epithelium, M cells occupy a relatively small portion of the total surface area of the GIT. The potential of drug delivery to the systemic circulation via M cells instead of enterocytes has thus remained debatable.

A number of studies have suggested that uptake of nanoparticles may occur through enterocytes as well as M cells (Rieux et al. 2006); however, as recently pointed out, no satisfying explanation has been advanced to elucidate conflicting reports (Delie 1998). A direct consequence of greater uptake into intestinal PP is the application of nanoparticles to enhance oral immunization (Clark et al. 2001), because the low intestinal permeability and high presystemic clearance of vaccines from the GIT requires the use of prohibitively large and repetitive doses to elicit even a modest immune response (Czerkinsky et al. 1993). In general, antigens incorporated into particles are more effective for oral immunization than soluble antigens (Langer et al. 1997, Galindo-Rodriguez et al. 2005). For example, ovalbumin incorporated into PLGA particles induced potent serum and mucosal immune response, including both IgA and IgG antibodies (Challacombe et al. 1992). Similar results have been achieved with a variety of toxins, including staphylococcal enterotoxin B toxoid (Eldridge et al. 1990), tetanus toxoid (Jung et al. 2001), as well as DNA vaccines (Roy et al. 1999). The use of muco- or bioadhesive agents in particulate DNA delivery systems for administration by the nasal or pulmonary routes is not a strategy that has attracted a large volume of research. The advent of DNA vaccination, however, is still relatively new, and the development of new and effective vectors may be the key event in realizing the potential of this strategy. From a vaccination point of view, vectors that elicit transient expression and are rapidly cleared from the body with a concomitant induction of strong immunological responses are highly desirable. Strategies that employ such facets without utilizing replication-competent live viruses have already demonstrated proof of principle that such vectors can achieve appreciable immune responses by the utilization of self-replicating RNA and induction of apoptosis (Ying et al. 1999, Restifo et al. 2000). In terms of peptide and protein delivery, some agents offer distinct potential for use in vaccine delivery systems for pulmonary and intranasal delivery. One such agent is chitosan. In addition to absorption-enhancing properties, interactions with the immune system may include the activation of macrophages and complement (Tokura et al. 1999). It is unlikely that immunostimulatory agents that engender responses by tissue destruction and stress would be suitable for lung or nasal delivery. The wide area of research gleaned mostly outside the realms of vaccinology, along with increased knowledge of the mechanisms of immunopotentiation, has provided the platforms of knowledge and ideas that are waiting to be exploited.

As an example, *Escherichia coli* have been reported to specifically adhere to the lymphoid follicle epithelium of the ileal Peyer's patch in rabbits (Inman and Cantey 1983). Additionally, different staphylococci possess the ability to adhere to the surface of mucus gel layers and not to the mucus-free surface (Sanford et al. 1989). Thus, it appears that drug delivery based on bacterial adhesion could be an efficient method to improve the delivery of particular drugs or carrier systems. Bernkop-Schnürch et al. covalently attached a fimbrial protein (antigen K99 from *E. coli*) to poly(acrylic acid) polymer and substantially improved the adhesion of the drug delivery system to the GI epithelium using a system. In this study, the function of the fimbrial protein was tested using a hemagglutination assay, along with equine erythrocytes expressing the same K99-receptor structures as those of GI epithelial cells. A 10-fold slower migration of the equine erythrocytes through the K99-poly(acrylic acid) gel, compared to the control gel without the fimbriae, was demonstrated, indicating the strong affinity of the K99 fimbriae to their receptors on the erythrocytes (Bernkop-Schnürch et al. 1995).

The biotechnological applications of muco adhesive microspheres by various routes are given under the section pharmaceutical applications.

12.9 PHARMACEUTICAL APPLICATIONS

Bioadhesive microspheres have been extensively studied for a number of applications. Majority of these can be understood by classifying these applications on the basis of route of administration. All these applications have been reviewed in the subsequent sections and listed in Table 12.6.

12.9.1 ORAL APPLICATIONS

12.9.1.1 Buccal

The small total surface area of ~50 cm^2, the relatively low permeability of buccal tissues, and typically short residence time of <5–10 min are considered as significant disadvantages to using the oral cavity for drug delivery. The oral cavity is used both for systemic delivery and local treatment. The systemic delivery of drugs is either sublingual, through the mucosal membranes lining the floor of the mouth or buccal, through the mucosal membranes lining the cheeks. Furthermore, oral transmucosal drug delivery bypasses the first-pass effect and avoids presystemic elimination in the GI tract. These factors make the oro-mucosal cavity a very attractive and feasible site for systemic drug delivery (Harris and Robinson 1992). The composition of the oral epithelium varies depending on the site in the oral cavity. The areas exposed to mechanical stress (the gingivae and hard palate) are keratinized similar to the epidermis. The mucosa of the soft palate, the sublingual, and the buccal regions, however, are not keratinized. The keratinized epithelia contain neutral lipids like ceramides and acylceramides, which have been associated with the barrier function. It is estimated that the permeability of the buccal mucosa is 4–4000 times greater than that of the skin. In general, the permeabilities of the oral mucosa decrease in the order of sublingual greater than buccal and buccal greater than palatal. The daily salivary volume secreted in humans is between 0.5 and 2 l, which is sufficient to hydrate oral mucosal dosage forms. This water-rich environment of the oral cavity is the main reason behind the selection of hydrophilic polymeric matrices as vehicles for oral transmucosal drug delivery systems. Vyas and Jain (1992) prepared polymer-grafted starch microspheres bearing Isosorbide dinitrate and evaluated their potential as sustained release buccal bioadhesive system both by *in vitro* release studies and *in vivo* absorption studies. Starch microspheres grafted with polymethyl methacrylate (PMMA) exhibited relatively slow drug release as compared to polyacrylate (PAA)-grafted microspheres. Moreover, the C_{max} and AUC recorded for the acrylic acid-grafted starch microspheres were found to be more than that for PMMA-grafted starch microspheres. It has been revealed by *in vivo* absorption studies that steady-state plasma levels can be maintained above the minimum effective concentration (MEC) over a period of 12 h after the buccal administration of the grafted microspheres.

TABLE 12.6
Pharmaceutical Applications of Bioadhesive Microspheres

Drug	Route of Administration	Bioadhesive Polymers Use	Comments/Results	References
Acyclovir	Ocular	Chitosan	Slow release rate increased AUC	Genta et al. (1997)
Methyl prednisolone	Ocular	Hyaluronic acid	Slow release rates sustained drug concentration in tear fluids	Kyyronen et al. (1992)
Gentamicin	Nasal	DSM+LPC	Increased nasal absorption	Farraj et al. (1990)
Insulin	Nasal	DSM+LPC	Efficient delivery of insulin into the systemic circulation via nasal route	Farraj et al. (1990)
Human growth hormone (hGH)	Nasal	DSM+LPC	Rapid and increased absorption	Illum et al. (1990)
Desmopressin	Nasal	Starch	Addition of LPC causes a fivefold increase in C_{max} and twofold increase in bioavailability	Critchley et al. (1994)
Hemagglutinin (HA) obtained from influenza A virus	Nasal	HYAFF	With mucosal adjuvant serum IgG antibody response as compared to i.m. immunization	Singh et al. (2001)
Furosemide	GI	AD-MMS (PGEFs)	Increased bioavailability Higher AUC effective absorption from the absorption window	Akiyama and Nagahara (1999)
Amoxicillin	GI	Ethyl cellulose-Carbopol-934P	Greater anti *H. pylori* activity	Liu et al. (2005)
Delapril HCL	GI	AD-MMS (PGEFs)	MRT of drug is increased	Akiyama et al. (1994)
Glipizide	GI	Chitosan	Prolonged blood glucose reduction	Patel et al. (2005)
Glipizide	GI	Chitosan-alginate	Prolonged blood glucose reduction	Patel et al. (2004)
Vancomycin	Colonic	PGEF coated with Eudragit S 100	Well absorbed even without absorption enhancers	Geary and Schlameus (1993)
Insulin	Colonic	PGEF coated with Eudragit S 100	Absorbed only in the presence of absorption enhancers, e.g., EDTA salts	Geary and Schlameus (1994)
Nerve growth factor (nGF)	Vaginal	HYAFF	Increased absorption from HYAFF microspheres as compared to aqueous solution of the drugs	Ghezzo et al. (1992)
Insulin	Vaginal	HYAFF	Increased absorption for HYAFF microspheres as compared to aqueous solution of the drugs	Illum et al. (1994)
Salmon calcitonin	Vaginal	HYAFF	Increased absorption from HYAFF microspheres as compared to aqueous solution of the drugs	Richardson and Armstrong (1999)

(continued)

TABLE 12.6 (continued)
Pharmaceutical Applications of Bioadhesive Microspheres

Drug	Route of Administration	Bioadhesive Polymers Use	Comments/Results	References
Pipedimic acid	Vesical	CMC as mucopolysaccharide + Eduragit RL as matrix polymer	—	Bogataj et al. (1999)
Amoxicillin	GI	Chitosan	Enhanced amoxicillin stability and compete *H. pylori* eradication	Patel and Patel (2007)
Amoxicillin	GI	Carbopol-934P as bioadhesive + ethyl cellulose as matrix polymer	Compete *H. pylori* eradication	Patel and Chavda (2008)

Kockisch et al. (2005) developed mucoadhesive microspheres that could be utilized for the controlled release of triclosan in oral care formulations, specifically dental pastes. Using a double emulsion-solvent evaporation technique, triclosan was incorporated into microspheres that were prepared from gantreztrade mark MS-955, carbopol-974P, polycarbophil, or chitosan, and the profiles for its release were established under simulated "in use" conditions. Triclosan was rapidly released into a sodium lauryl sulfate-containing buffer from all but the chitosan microspheres. The release of triclosan from microspheres suspended in a nonaqueous paste was found to be sustained over considerable time periods, which were influenced strongly by the nature of the polymeric carrier. For microspheres that were fabricated from gantrez, carbopol, or polycarbophil, the release appeared to obey zero-order kinetics, whereas in the case of chitosan-derived vehicles, the release profile fitted the Baker and Lonsdale model. The work has demonstrated that these polymeric microspheres, particularly those of chitosan, are promising candidates for the sustained release of triclosan in the oral cavity.

12.9.1.2 Gastrointestinal

There is no doubt that the oral route is the most favored and probably the most complex route of drug delivery. Critical barriers, such as mucus covering the GI epithelia, high turnover rate mucus, variable range of pH, transit time with broad spectrum, absorption barrier, degradation during absorption, hepatic first-pass metabolism, rapid luminal enzymatic degradation, longer time to achieve therapeutic blood levels, and inter and intra subject variability, are all possible issues with the oral route. The development of peroral-controlled release drug delivery system has been hindered by the inability to restrain and localize the drug delivery system in selected regions of the GIT. Bioadhesive drug delivery system forms an important approach to decrease the GI transit of drugs. Drug properties especially amenable to bioadhesive formulations include a relatively short biological half-life of about 2–8 h, a specific window for the absorption of drug by an active, saturable absorption process and small absorption rate constants (Longer et al. 1985). The GI epithelium consists of a single layer of simple, columnar epithelium lying above a collection of cells called the lamina propria and supported by a layer of smooth muscle known as the *Muscularis mucosae*. Tight junctions or the zona occludens holds the cells together. A special type of GI epithelium, the PP of the gut-associated lymphoid tissue (GALT) is also present. Polymeric microspheres can also be phagocytized by these microfold cells and hence can be used for vaccination purposes (Carino et al. 1999).

Specially engineered polymeric bioadhesive microspheres can traverse both the mucosal absorptive epithelium and follicle-associated epithelium covering the lymphoid tissues of PP depending on the particle size, polymer composition, and the surface charge of bioadhesive microspheres. Bioerodible bioadhesive microspheres have been reported to increase the peroral bioavailability of dicumarol, insulin and have been investigated for peroral gene delivery. The increased bioactivity of insulin and the plasmid DNA can be accounted to the uptake of microspheres by cells lining the GI epithelium. Thus, these uptake pathways can be used as a platform for the systemic delivery of a variety of therapeutic agents showing poor absorption through GI epithelium. Bioadhesive microspheres, by keeping the drug in the region proximal to its absorption window, allow targeting and localization of the drug at a specific site in the GIT. An adhesive micromatrix system (AD-MMS), a novel formulation approach, reported by Akiyama and Nagahara (1999) consists of the drug and an adhesive polymer dispersed in a spherical matrix of the PGEFs, with a diameter of 177–500 µm. This formulation showed strong adherence to the stomach mucosa. Drug release from this system could be regulated by the appropriate selection of HLB value of the PGEFs. Various channeling agents were reported to regulate drug release through the micromatrix systems, e.g., mannitol, acrylic acid, and lactose. In experiments using rats, the prolongation of GI transit time and improvement in the bioavailability of furosemide (with a narrow absorption window) has been shown. The MRT values after PGEF microspheres and the AD-MMS administration were found to be 6.1 ± 0.6 and 6.7 ± 0.7 h, respectively. While the AUC (0–24 h) after AD-MMS administration (11.57 ± 1.84 µg h/mL) was 1.8 times that of the PGEF microsphere (6.56 ± 0.93 µg/mL). The results could be explained to be due to the adherence of the AD-MMS to a more proximal area of the GIT rather than the absorption window, and furosemide was thereby effectively absorbed from the absorption window (Akiyama et al. 1998). AD-MMS containing amoxicillin has been evaluated against the amoxicillin suspension for *H. pylori* clearance *in vivo* using Mongolian gerbils as the animal model. A 10-times greater anti *H. pylori* activity after oral administration of AD-MMS as compared to the amoxicillin suspension has been reported, which could be due to the difference in gastric residence provided by the two dosage forms (Nagahara et al. 1998). Amoxicillin AD-MMS adheres to the infected mucosa and thereby provides a higher *H. pylori* eradication or clearance rate. Patel et al. (2005) prepared glipizide microspheres containing chitosan by simple emulsification phase separation technique using glutaraldehyde as a cross-linking agent. *In vivo* testing of the mucoadhesive microspheres to albino Wistar rats demonstrated the significant hypoglycemic effect of glipizide. Investigation by Patel and Patel (2007) and Patel and Chavda (2008) on *H. pylori* clearance effect showed that there was tendency for a more-effective *H. pylori* activity of bioadhesive amoxicillin microspheres prepared using chitosan and carbopol-934P as bioadhesive polymers. Chitosan microspheres (CMs) prepared by chemical cross-linking provide a longer residence time in the fasted gerbil stomach than either tetracycline solution or microspheres prepared by ionic precipitation (Hejazi and Amiji 2004). Majithiya et al. (2005) compared the bioavailability of clarithromycin from microsphere formulation and plain drug suspension *in vivo*, with AUC $0 \rightarrow$ alpha being 91.7 (µh/mL) and 24.9 (µh/mL), respectively. The results of the study demonstrated good mucoadhesion of the microspheres with the stomach mucosa as well as higher accumulation of drug in the stomach membrane. Microspheres also exhibited sustained release of drug. Thus, CMs appear, technically, promising mucoadhesive drug delivery systems for delivering clarithromycin to treat stomach ulcers. Govender et al. (2005) prepared tetracycline microsphere, and antimicrobial studies showed that the drug concentrations in the *in vitro* release samples were above the minimum concentration of drug required for inhibition of *Staphylococcus aureus* growth. Novel inherently fluorescent microspheres composed of a luminescent polyanhydride, poly[*p*-(carboxyethylformamido)-benzoic anhydride] (PCEFB), and poly(lactide-*co*-glycolide) (PLGA) (2:1, weight ratio) by the GIT was evaluated by fluorescent microscopy. Oral efficiency of the incorporated insulin also was determined by measuring the reduction of plasma glucose levels after feeding diabetic rats with a single dose of the microspheres. They found that PCEFB/PLGA microspheres could adhere to the intestinal epithelium and traverse the absorptive cells. A large number of the spheres were observed in spleen, whereas few were detected in liver within the evaluated period of time. The apparent reduction of the plasma glucose levels was observed over a span of 6 h postfeeding. The unique properties of the delivery system such as biodegradability, bioadhesivity, and inherently luminescent characteristics render it an ideal "visible" tracer for monitoring the oral fate of polymeric microspheres (Li et al. 2004). Diltiazem-loaded mucoadhesive microspheres were successfully prepared by emulsification/internal gelation technique with a maximum incorporation efficiency of $93.29 \pm 0.26\%$. The *in vitro* wash-off test indicated that the microspheres had good mucoadhesive properties. The wash-off was faster at simulated intestinal fluid (phosphate buffer, pH 7.4) than that at simulated gastric fluid (0.1 M HCl, pH 1.2). The *in vitro* drug release mechanism was nonfickian type, controlled by the swelling and relaxation of polymer. There was no significant change in drug content and cumulative drug release of drug-loaded microspheres stored at different storage conditions after 8 weeks of study (Das and Maurya 2008).

Zhou et al. (2007) prepared chitosan/cellulose acetate multimicrospheres (CCAM) with or without ranitidine by the method of w/o/w emulsion with no toxic reagents and had the size interval of 200–280 μm. The mucoadhesive tests showed that CCAM could retain in GIT for an extended period of time. There were 53.7% of CCAM that remained in stomach after administered for 2(1/2) h and 98.9% of CCAM remained in stomach and small intestine after administered for 3(1/2) h. These results suggest that CCAM is a useful dosage form targeting the gastric mucosa or prolonging gastric residence time as a multiple-unit mucoadhesive system.

12.9.1.3 Colon

Colon drug delivery has been used for molecules aimed at the local treatment of colonic diseases and for the delivery of molecules susceptible to enzymatic degradation such as peptides. The mucosal surface of colon resembles that of the small intestine at birth but changes with age causing the loss of villi, leaving a flat mucosa with deep crypt cells. Therefore, the absorptive capacity of the colon is much less as compared to small intestine. The mucus layer provides not only a stable pH environment but also acts as a diffusion barrier for the absorption of drugs. Mucus production is more in the elderly as the number of mucus-secreting goblet cells increase with age. Colonic mucosal environment is also effected by the colonic microflora as they degrade the mucins. Bioadhesive microspheres can be used during the early stages of colonic cancer for enhancing the absorption of peptide drugs and vaccines, for the localized action of steroids, and drugs with a high hepatic clearance, e.g., budesonide, and for the immunosuppressive agents such as cyclosporine. Colon-specific bioadhesive microspheres can be used for the protection of peptide drugs from the enzyme rich part of the GIT and to release the biologically active drug at the desired site for its maximum absorption. Insulin was found to be absorbed well in the colon only in the presence of absorption enhancers, e.g., EDTA salts, which cause chelation of calcium ions present in the tight junctions and hence opening of water channels in the cell membranes. Geary and Schlameus (1993) formulated targeted release utilizing Eudragit® S100 enteric coating and decreased GI transit time comparing two bioadhesive polymers (chitosan and Carbopol® 934). Salicylate microspheres incorporated in enteric and nonenteric formulations with chitosan or carbomer (Carbopol 934) were tested for sustained delivery in the lower GIT using rat and dog animal models. Optimized formulations were designed for porcine insulin, which were then tested in the nondiabetic dog model for bioavailability and pharmacological response. Microspheres of triglyceride mixtures and low-melting fat composition were prepared using a spinning disk method and subsequently incorporated in two-part gelatin capsules containing Carbopol 934 or incorporated in chitosan film prior to placement in two-part gelatin capsules. Improved bioavailability for the enteric-coated oral insulin formulation was observed in the dog model with prolonged decrease in plasma glucose.

Anande et al. (2008) developed cyst-targeted novel concanavalin-A (Con-A) conjugated mucoadhesive microspheres of diloxanide furoate (DF) for the effective treatment of amoebiasis. Eudragit microspheres of DF were prepared using emulsification-solvent evaporation method. Formulations were characterized for particle size and size distribution, % drug entrapment, surface morphology, and *in vitro* drug release in simulated GI fluids. Eudragit microspheres of DF were conjugated with Con-A. IR spectroscopy and DSC were used to confirm the successful conjugation of

Con-A to eudragit microspheres, while Con-A conjugated microspheres were further characterized using the parameters of zeta potential, mucoadhesiveness to colonic mucosa, and Con-A conjugation efficiency with microspheres. IR studies confirmed the attachment of Con-A with eudragit microspheres. All the microsphere formulations showed good % drug entrapment (78 ± 5%). Zeta potential of eudragit microspheres and Con-A conjugated eudragit microspheres were found to be 3.12 ± 0.7 and 16.12 ± 0.5 mV, respectively. The attachment of lectin to the eudragit microspheres significantly increases the mucoadhesiveness and also controls the release of DF in simulated GI fluids. Gamma scintigraphy study suggested that eudragit S100-coated gelatin capsule retarded the release of Con-A conjugated microspheres at low pH and released microspheres slowly at pH 7.4 in the colon. Wittaya-areekul et al. (2006) prepared mucoadhesive alginate/CMs containing prednisolone intended for colon-specific delivery. Two methods have been used for the preparation of the microspheres and the results for both preparation methods showed that the particle size and drug content were mainly depending on the amount of the drug concentration and not the amount of chitosan and calcium chloride. The *in vitro* mucoadhesive tests for microspheres prepared from both methods were carried out using the freshly excised gut of pigs. The microspheres prepared by the one-step method exhibited excellent mucoadhesive properties after 1 h test. Increased chitosan concentrations from 0%, 0.5%, 1.0% to 1.5% (w/v) resulted in 43%, 55%, 82%, and 88% of the particles remaining attached on the gut surface after 1 h, respectively. However, the particles prepared by the two-step method showed significant less mucoadhesion under the same experimental conditions. At chitosan concentrations of 0%, 0.5%, 1.0%, and 1.5% (w/v), the amount of particles remaining attached to the mucosal surface of the pig gut after 1 h was 43%, 3%, 11%, and 11%, respectively. The prednisolone release at a pH of 6.8 after 4 h was between 63% and 79% for the microspheres prepared by the one-step method and between 57% and 88% for the microspheres prepared by the two-step method with a prednisolone drug load of 5% and 10% (w/v), respectively. The results show that depending on the preparation method, these chitosan-coated alginate microspheres show different mucoadhesiveness, whereas their other properties are not statistically significantly different.

12.9.2 Topical Applications

12.9.2.1 Nasal

Histologically, the nasal mucosa provides a potentially good route for systemic drug delivery. With a surface area of 150 cm^2, a highly dense vascular network, and a relatively permeable membrane structure, the nasal route has good absorption potential. Drawbacks are potential local toxicity/irritation, relatively lower permeability for large macromolecules, rapid mucociliary clearance of ~4–6 mm/min, presence of proteolytic enzymes causing drug degradation in the nasal cavity, limited formulation for changing drug delivery profiles, and the influence of pathological conditions (i.e., colds or allergies). The basic function of the nose, in addition to functioning as a sensory organ, is the pretreatment of inspired air. The air is heated and humidified and its passage through the nose will help clear particles and bacteria from the air before it reaches the lung. The nasal cavity offers a large, highly vascularized subepithelial layer for efficient absorption. Also, blood is drained directly from nose into the systemic circulation, thereby avoiding first-pass effect (Soane et al. 1999). However, the nasal delivery of drugs has certain limitations due to the mucociliary clearance of therapeutic agents from the site of deposition, resulting in a short residence time for absorption. The use of bioadhesive drug delivery system increases the residence time of formulations in nasal cavity, thereby improving the absorption of drugs. It has been shown by gamma scintigraphy study (Illum et al. 1987) that radiolabeled microspheres made from diethyl amino ethyl dextran (DEAE–dextran), starch, and albumin are cleared significantly more slowly than solutions after nasal administration in human volunteers. Hence, it was suggested by Illum (1999) that the intranasal application of bioadhesive microspheres (in powder form) causes them to swell on coming in contact with the nasal mucosa to form a gel and decrease their rate of clearance from the nasal cavity, thereby providing poorly absorbed drugs a longer time for absorption.

Bioadhesive microspheres have also been investigated as vehicles for peptides and proteins. An increased bioavailability of FITC-dextran was observed in rats when microspheres made from carbopol were used as a vehicle in comparison with reference lactose microspheres (Ady El-Shafy et al. 2000). A significantly greater hypocalcemic effect was observed after the administration of salmon calcitonin in gelatin microspheres in comparison with salmon calcitonin in buffer (Morimoto et al. 2001). The bioavailability of salmon calcitonin was greater when using positively charged spheres than when using negatively charged spheres of the same size. Long-term use of bioadhesive microspheres however may reduce bioavailability. For example, after 8 days of nasal delivery of insulin from microspheres made from starch and carbopol, a reduced bioavailability and lower decrease of blood glucose levels was noticed (Callens et al. 2003). The excellent absorption-enhancing properties of bioadhesive microspheres are now being used extensively for both low molecular weight as well as macromolecular drugs like proteins. Nasal cavity as a site for systemic drug delivery has been investigated extensively and many nasal formulations have already reached commercial status, including leutinizing hormone-releasing hormone (LHRH) and calcitonin (Illum 1999).

Chitosan and starch are the two most widely employed bioadhesive polymers for nasal drug delivery. It has been reported that the clearance half-life was 25% greater for CMs than for starch microspheres. This may be due to the differences in the surface charge, molecular contact, and the flexibility of two polymers. Chitosan exerts a transient inhibitory effect on the mucociliary clearance of the bioadhesive formulations. The concept of using a bioadhesive delivery system in the form of degradable starch microspheres (DSM) for the nasal delivery of drugs was introduced in 1988. DSM system when combined with absorption enhancers, such as lysophosphatidylcholine (LPC), successfully improved the nasal absorption of gentamicin. The bioavailability of gentamicin was increased to 10% with the use of bioadhesive microspheres and was further increased to 57% by the addition of LPC to microsphere formulation. The DSM/LPC system has also been proposed as an efficient method for the delivery of insulin into the systemic circulation via nasal route (Farraj et al. 1990). A rapid and much higher absorption of the human growth hormone (hGH) has been observed when hGH was administered in the form of DSM/LPC system of microspheres (Illum et al. 1987). Critchley et al. (1994) evaluated bioadhesive starch microspheres as a nasal delivery system for desmopressin, and observed significant improvement in the absorption of drug, both in terms of peak plasma levels and bioavailability. A fivefold increase in maximum plasma concentration (C_{max}) and a doubling of bioavailability was observed on addition of LPC in a concentration of 0.2% to the starch microspheres. Other bioadhesive microspheres used for the nasal administration of peptides and proteins include the cross-linked dextran microspheres, which are water insoluble and water absorbable. Sephadex and DEAE–Sephadex were found to improve the nasal absorption of insulin, but to a lesser extent than the starch microspheres (Edman et al. 1992, Callens et al. 2003). Hyaluronic acid ester microspheres were used for the nasal delivery of insulin in sheep and the increase in nasal absorption was found to be independent of the dose of microspheres in the range of 0.5–2.0 mg/kg (Illum et al. 1994).

Preda and Leucuta (2003) studied *in vitro* and *in vivo* experiments in rats and showed good adhesive characteristics of gelatin/poly(acrylic acid) microspheres, which were greater if the poly(acrylic acid) content was greater. A significant retardation in the gastric and intestinal emptying time of the beads was observed. This was also suggested by the bioavailability of the model drug after intragastric and intranasal administration of the microspheres. The pharmacokinetic parameters after microsphere administration were more appropriate to a slow release drug delivery system. The *in vivo* performance of mucoadhesive microspheres formulations showed prolonged and controlled release of salbutamol as compared with oral administration of conventional dosage form (Jain et al. 2004). Gavini et al. (2005) prepared microspheres for the nasal administration of an antiemetic drug, metoclopramide hydrochloride and found that alginate/chitosan spray-dried microspheres have promising properties for use as mucoadhesive nasal carriers of an antiemetic drug.

12.9.2.2 Ocular

There are two major surface tissues of the eye, the conjunctiva and cornea that interface with the outside world. The conjunctiva is a thin tissue that extends from the edge of the eyelid and across the globe of the eye to the cornea. The cornea is the window of the eye and is the transparent tissue in the center of the globe surface. Mucin is secreted by conjunctival goblet cells but there are no goblet cells on the cornea. On this basis, a bioadhesive polymer will firmly attach to conjunctival mucus, but only loosely, if at all, to corneal mucus. The turnover rate of the mucin is ~15–20 h, whereas normal tear turnover time is ~16%/min in humans except during sleeping or anesthesia. Obviously, attaching a polymer to mucin will considerably slow the removal of ocular drug delivery systems from the front of the eye. Topical administration is the route of choice for the treatment of ophthalmic diseases because of the blood–ocular barrier. Achieving therapeutic concentrations in the eye by systemic administration necessitates the usage of such high systemic concentration that, in many cases, systemic side effects and toxicity result. Traditional ophthalmic formulations such as aqueous solutions and ointments have low (typically 2%–10%) bioavailability of drugs due to the small surface area available for penetration, the presence of absorption barriers, and a number of precorneal elimination factors (Saettone et al. 1999). These elimination factors include drainage of instilled solutions; lacrimation and tear turnover, drug metabolism, tear evaporation, and possible binding to lachrymal proteins. To prolong the residence time of drugs in the preocular area, bioadhesive drug delivery system has been developed taking advantage of the presence of a mucin–glycocalyx domain in the external portion of the eye.

Various bioadhesive drug delivery systems employed for the ocular delivery of drugs include the semisolids, viscous liquids, solids/inserts, and the particulate drug delivery system including bioadhesive microspheres and liposomes. The advantages of microspheres, i.e., increased residence time and decreased frequency of administration were quite evident with CMs of acyclovir (Genta et al. 1997) and methylprednisolone-loaded hyaluronic acid microspheres (Chickering et al. 1999). Acyclovir-loaded chitosan microparticles showed an increased drug bioavailability in the eye as compared to the drug administered alone. Genta et al. (1997) reported an approximately four times increase in the aqueous humor concentration of suspension (39.37 μg/mL min) after a single instillation into rabbit's eye. Increase in levels and the prolonged release of acyclovir from bioadhesive microspheres can be used to overcome the inconvenience caused by frequently applied ointments. The release of methylprednisolone from hyaluronic acid ester films and microspheres has been investigated *in vitro* and *in vivo* (in tear fluid of rabbits) (Kyyronen et al. 1992). Methylprednisolone was either physically dispersed in the polymeric matrix or covalently linked to hyaluronic acid. Microspheres containing methylprednisolone chemically bonded to the polymeric backbone of hyaluronic acid showed slower release of drug *in vitro* and produced sustained drug concentrations in the tear fluids of rabbits.

Durrani et al. (1995) investigated the effect of these parameters on the precorneal clearance of In[111]-labeled microspheres prepared using carbopol 907. The clearance of microspheres administered in dry form was faster than in the hydrated form, probably due to incomplete hydration in the tear fluid. The *in vivo* slow basal phase clearance constants were found to be 0.007 and 0.034 min^{-1} for the suspension of microspheres at a pH of 5.0 and 7.4, respectively. At pH 5, the presence of protonated carboxyl groups permits enhanced adhesion due to hydrogen bonding between the polymer and mucin strands, resulting in reduced clearance values. The clearance of microspheres, which significantly limits their residence time in the ocular cavity is a direct function of the pH and hydration state of microspheres and follows a biphasic process with an initial rapid clearance followed by a much slower basal phase. Initial clearance phase is independent of pH and hydration state, while the basal phase clearance value varies with these factors. Coating the microspheres with bioadhesive polymers has also been evaluated as a potential way of increasing the bioavailability. Chitosan-coated microspheres, which increased the bioavailability of indomethacin in rabbits (Calvo et al. 1997) and PEG-coated microspheres, resulted in an increase in the bioavailability of acyclovir (Fresta et al. 2001).

12.9.2.3 Vaginal

The vagina is a fibrovascular tube connecting the uterus to the exterior of the body. In adults, the length of the vagina varies from 6 to 10 cm, with the posterior wall a ~1.5–2.0 cm longer than the anterior wall. The vaginal epithelium is a stratified squamous epithelium resting on a lamina propria. The surface area of the vagina is increased by numerous folds in the epithelium and by microridges covering the epithelium cell surface. The permeability of the vaginal epithelium may vary during estrus or menstrual cycle. Traditionally, this route of administration is used for the delivery of therapeutic and contraceptive agents to exert a local effect (antifungal, spermicidal) and for the systemic delivery of drugs (Richardson et al. 1996). It has been used for the delivery of drugs that are susceptible to GI degradation or hepatic metabolism following peroral delivery, e.g., oestrogens and progestogens for the treatment of postmenopausal symptoms and for contraception. This route has also been explored for the delivery of therapeutic peptides, e.g., calcitonin and for microbicidal agents to help prevent the transmission of human immunodeficiency virus and other sexually transmitted diseases (STDs). Using absorption enhancers, e.g., surfactants and bile salts, can increase the absorption of peptides from the vagina. The adverse effects of absorption enhancers on the mucosal integrity can however be bypassed by employing bioadhesive microspheres within the vaginal cavity.

The retention of microspheres made from a benzyl ester of hyaluronic acid (HYAFF) was studied using gamma scintigraphy in sheep (Richardson et al. 1996). The microspheres were either delivered as a dry powder or included in Suppocire BS_2X pessaries. A substantial percentage of the radiolabeled spheres remained in the vagina till 12 h after administration. The retention was greater for the dry powder formulation than for the pessary formulation, which the authors suggested might be caused by the loss of microspheres on the leakage of the molten base. HYAFF microspheres have been successfully used for the incorporation of peptides such as nerve growth factor (Ghezzo et al. 1992) and salmon calcitonin. HYAFF microspheres have demonstrated good bioadhesive properties both *in vitro* and *in vivo*. In an unconscious rat model, these microspheres maintained contact with the vaginal epithelium for at least 6 h after administration. Hypocalcemic effects in the rat and sheep confirmed that absorption of salmon calcitonin was increased after the administration of bioadhesive (HYAFF) microspheres as compared with an aqueous solution of calcitonin (Richardson and Armstrong 1999). HYAFF microspheres due to their high biocompatibility and controllable degradation rate have been used for the localized drug delivery of steroids, analgesics, anti-inflammatory, and anti-infectives. This has led to a great deal of enthusiasm in the development of safe and effective bioadhesive vaginal contraceptive and anti-infective formulations to control pregnancy and help prevent the spread of STDs (Richardson et al. 1996).

Insulin was administered vaginally to sheep as an aqueous solution and as a lyophilized powder with bioadhesive starch microspheres. The effect of lysophosphatidylcholine (LPC) on the vaginal absorption of insulin from both formulations was studied. While the vaginal absorption of insulin from insulin solution was minimal, the addition of LPC resulted in a rapid rise in plasma insulin and a pronounced fall in plasma glucose levels. The absolute bioavailability of the peptide from the latter solution was 13%. The hypoglycemic response to vaginally administered insulin was also improved using the microspheres delivery system, compared to insulin solution alone, and was further enhanced by LPC. Vaginal absorption of insulin from each formulation appeared to be influenced by the oestrus cycle and was thought to correlate with changes in vaginal histology (Richardson et al. 1992).

12.9.3 MISCELLANEOUS APPLICATIONS

12.9.3.1 Vesicular Delivery

The mucosal layers in the urinary bladder are different from those present in both small as well as large intestine with regards to their structure and thickness. The vesical mucus contains oligosaccharides–glycosaminoglycans (GAG) that carry a large number of sulfate groups and thus a high negative charge density. Despite these differences, there are certain similarities between the mucus layers in urinary bladder and intestine as they both contain sugar chains completely or partly attached to proteins (Bogataj et al. 1999). Therefore, it is expected that polymers that show good mucoadhesive strength on the intestinal mucosa will exhibit some mucoadhesiveness on the vesical mucosa as well. Bogataj et al. (1999) evaluated various polymers for the mucoadhesion strength, swelling, and drug release from bioadhesive microspheres applied into the urinary bladder. It has been reported that the microspheres containing carboxymethyl cellulose (CMC) as mucoadhesive agent and eudragit RL as matrix polymer provided the longest release time from microspheres and showed high strength of mucoadhesion.

12.9.3.2 Mucosal Immunization

The majority of pathogens initially infect their hosts through mucosal surfaces; induction of mucosal immunity is therefore likely to make an important contribution to the protective immunity. Moreover, the mucosal administration of vaccine avoids the use of needles and is thus an attractive approach for the development of new generation vaccines. Current research in vaccine development has focused on treatment requiring a single administration, since the major disadvantage of many currently available vaccines is that repeated administrations are required. The ability to provide controlled release of antigens through bioadhesive microspheres has given an impetus to research in the area of mucosal immunization. Intravaginal immunization has been tried in sheep using DSM and LPC for the influenza virus hemagglutinin (TOPS) (O'Hagan et al. 1993). The highest levels of antibodies were detected after intramuscular injection than after intravaginal immunization since the vagina unlike intestine, lungs, and nasal cavity has no aggregates of lymphoid tissue within the epithelium. The HYAFF bioadhesive microspheres in the presence of a mucosal adjuvant-LTK 63 administered intranasally are reported to induce a significantly enhanced serum IgG antibody response in comparison to intramuscular immunization with hemagglutinin obtained from influenza A virus (Singh et al. 2001). Antigen–microsphere formulations prepared by adsorbing the antigen onto preformed polymeric hydrogel microspheres can be used to provide enhanced immune responses in animals. Polyphosphazene microspheres with adsorbed influenza antigen and tetanus toxoid can be administered intranasally to have increased immune responses (Payne et al. 2001).

12.9.3.3 Protein and Peptide Drug Delivery

Protein and peptide drugs offer formidable challenges for peroral delivery due to their relatively large size, enzymatic degradation, and very low permeability across the absorptive epithelial cells. Bioadhesive microspheres provide an interesting noninvasive patient compliant approach to improve the absorption of these drugs. The luminal enzymatic degradation of proteins and peptides can be effectively minimized by direct contact with the absorptive mucosa and avoiding exposition to body fluids and enzymes. Specific enzyme inhibitors can be attached to the surface of bioadhesive microspheres (Borchard et al. 1996). Moreover, certain polymers, e.g., chitosan have been reported to possess permeability-enhancing properties. Senel et al. (2000) observed a six- to sevenfold enhancement of permeability by chitosan for the bioactive peptide TGF-ß to which the oral mucosa was reported to be relatively impermeable. This permeability-enhancing effect can be attributed to the transient opening of the tight junctions in the cell membranes or due to an increase in the thermodynamic activity of penetrant or due to the ability of chitosan to disrupt the lipid organization of the cellular membranes. Microspheres prepared with polyacrylic acid derivatives can chelate the extracellular calcium ions *in vivo* and hence reduce the integrity of tight junctions, which results in a permeability-enhancing effect (Borchard et al. 1996). Polyacrylates can also inhibit the proteolytic enzymes present in the GIT by binding to the essential enzyme cofactors, such as calcium and zinc ions, resulting in a conformational change of enzyme and loss of its activity (Luessen et al. 1995). Quan et al. (2008) developed eudragit–cysteine conjugate to coat on CMs for an oral protein drug delivery system, having mucoadhesive and pH-sensitive property. BSA as a protein model drug was loaded in thiolated eudragit-coated CMs (TECMs) to study the release character of the delivery system. After thiolated eudragit coating, it was found that the release rate of BSA from BSA-loaded TECMs was observably suppressed at pH 2.0 PBS solution, while at pH 7.4 PBS solution, the BSA can be sustainingly released for several hours. The structural integrity of BSA released from BSA-loaded TECMs was guaranteed by sodium dodecylsulfate–polyacrylamide gel electrophoresis (SDS-PAGE) and circular dichroism (CD) spectroscopy. The mucoadhesive property of TECMs was evaluated and compared with CMs and eudragit-coated chitosan microspheres (ECMs). It was confirmed that after coating thiolated eudragit, the percentage of TECMs that remained on the isolated porcine intestinal mucosa surface was significantly higher than those of CMs and ECMs. Likewise, gamma camera imaging of Tc-99m labeled microsphere distribution in rats after oral administration also suggested that TECMs had comparatively stronger mucoadhesive characters. Results indicated that TECMs have potential to be an oral protein drug carrier.

12.10   CONCLUSION

Bioadhesive microspheres offer a unique carrier system for many pharmaceuticals and can be tailored to adhere to any mucosal tissue, including those found in eyes, oral cavity, and throughout the respiratory, urinary, and GIT. The bioadhesive microspheres can be used not only for controlled release but also for the targeted delivery of the drugs to specific sites in body. Recent advances in medicine have envisaged the development of polymeric drug delivery systems for protein/peptide drugs and gene therapy. These challenges put forward by the medicinal advances can be successfully met by using increasingly accepted polymers, e.g., polyacrylates, HYAFF, chitosan and its derivatives, polyphosphazenes, etc. Many studies have already been undertaken for exploring the prospects of bioadhesive microspheres in gene therapy, delivery of peptides, localized and targeted release of antitumor agents, and mucosal vaccination. Although significant advances have been made in the field of bioadhesives, there are still many challenges ahead in this field. Of particular importance is the development of universally acceptable standard evaluation methods and the development of newer site-directed polymers. Polymeric science needs to be explored to find newer bioadhesive polymers with the added attributes of being biodegradable, biocompatible, bioadhesive for specific cells or mucosa, and which could also function as enzyme inhibitors for the successful delivery of proteins and peptides. A multidisciplinary approach will therefore be required to overcome these challenges and to employ bioadhesive microspheres as a cutting edge technology for the site-targeted controlled release drug delivery of new as well as existing drugs.

ABBREVIATIONS

Area under curve	AUC
Absorption measured at water	κ
Bovine serum albumin	BSA
Bioadhesive drug delivery systems	BDDS
Centimeter	cm
Carboxymethyl cellulose	CMC
Concentration maximum	C_{max}
Degree Celsius	°C
Degradable starch microspheres	DSM
Depth	d
Deoxynucleic acid	DNA
Density	ρ
Deciliter	dL
Ethylenediaminetetraacetic acid	EDTA
Gastrointestinal	GI

Gastrointestinal tract	GIT
Hour	h
Helicobacter pylori	*H. pylori*
Hyaluronic acid esters	HYAFF
Immunoglobulin G	IgG
Kilogram	kg
Lysophosphatidylcholine	LPC
Maximum tensile stress	S_m
Maximum force of detachment	F_m
Minute (time)	min
Millimeter	mm
Milligram	mg
Micron	μm
Molar concentration	M
Millivolt	mV
Molecular weight	Mw
Microfold cells	M cells
Minimum effective concentration	MEC
Total surface area	A_o
Polyethylene glycol	PEG
Pressure-sensitive adhesives	PSAs
Polyvinyl pyrolin	PVP
Polyglycerol esters of fatty acids	PGEFs
Super critical fluid	SCF
Soluble solvent	ϕ
Tomato lectin	TL
Volumes	v
Viscosity	η
Weight	wt

REFERENCES

Ady El-Shafy, M., Kellaway, I.W., Taylor, G., and Dickinson, P.A. 2000. Improved nasal bioavailability of FITC-dextran from mucoadhesive microspheres in rabbits. *J. Drug Target* 7:355–361.

Ahuja, A., Khar, R.K., and Ali, J. 1997. Mucoadhesive drug delivery systems. *Drug Dev. Ind. Pharm.* 23:489–515.

Akiyama, Y. and Nagahara, N. 1999. Novel formulation approaches to oral mucoadhesive drug delivery system. In: *Bioadhesive Drug Delivery Systems-Fundamentals, Novel Approaches and Development*, Mathiowitz, E., Chickering, D.E., and Lehr, C.M. (eds.), pp. 477–505. Marcel Dekker, New York.

Akiyama, Y., Yoshioka, M., Horibe, H., Inada, Y., Hirai, S., Kitamori, N., and Toguchi, H. 1994. Antihypertensive effect of oral controlled release microspheres containing an ACE inhibitor (delapril hydrochloride) in rats. *J. Pharm. Pharmacol.* 46:661–665.

Akiyama, Y., Nagahara, N., Nara, E., Kitano, M., Iwasa, S., Yamamoto, I., Azuma, J., and Ogawa, Y. 1998. Evaluation of oral mucoadhesive microspheres in man on the basis of the pharmacokinetics of furosemide and riboflavin, compounds with limited gastrointestinal absorption sites. *J. Pharm. Pharmacol.* 50:159–166.

Alur, H.H., Pather, S.I., Mitra, A.K., and Johnston, T.P. 1999. Transmucosal sustained-delivery of chlorpheniramine maleate in rabbits using a novel, natural mucoadhesive gum as an excipient in buccal tablets. *Int. J. Pharm.* 88(1):1–10.

Anande, N.M., Jain, S.K., and Jain, N.K. 2008. Con-A conjugated mucoadhesive microspheres for the colonic delivery of diloxanide furoate. *Int. J. Pharm.* 359(1–2):182–189.

Barrer, R.M., Barrie, J.A., and Wong, P.S.L. 1968. The diffusion and solution of gases in highly cross-linked copolymers. *Polymers* 9:609–627.

Bernkop-Schnürch, A. 2005. Thiomers: A new generation of mucoadhesive polymers. *Adv. Drug Deliv. Rev.* 57:1569–1582.

Bernkop-Schnürch, A., Gabor, M.P., Szostak, F., and Lubitz, W. 1995. An adhesive drug delivery system based on K99-fimbriae. *Eur. J. Pharm. Sci.* 3:293–299.

Bernkop-Schnürch, A., Schwarch, V., and Steininger, S. 1999. Polymers with thiol groups: A new generation of mucoadhesive polymers. *Pharm. Res.* 16(6):876–881.

Bogataj, M., Mrhar, A., and Korosec, L. 1999. Influence of physicochemical and biological parameters on drug release from microspheres adhered on vesical and intestinal mucosa. *Int. J. Pharm.* 177:211–220.

Borchard, G., Luessen, H.L., deBoer, A.G., Verhoef, J.C., Lehr, C.M., and Junginger, H.E. 1996. The potential of mucoadhesive polymers in enhancing intestinal peptide drug absorption. III. Effects of chitosan-glutamate and carbomer on epithelial tight junctions *in vitro. J. Control. Release* 39:31–38.

Callens, C., Ceulemans, J., Ludwig, A., Foreman, P., and Remon, J.P. 2003. Influence of multiple nasal administrations of bioadhesive powder on the insulin bioavailability. *Int. J. Pharm.* 250:415–422.

Calvo, P., VilaJato, J.L., and Alonso, M.J. 1997. Evaluation of cationic polymer-coated nanocapsules as ocular drug carriers. *Int. J. Pharm.* 153:41–50.

Carino, P.G., Jacob, J.S., Chen, C.J., Santos, C.A., Hertzog, B.A., and Mathiowitz, E. 1999. Bioadhesive, bioerodible polymers for increased intestinal uptake. In: *Bioadhesive Drug Delivery Systems— Fundamentals, Novel Approaches and Development*, Mathiowitz, E., Chickering, D.E., and Lehr, C.M. (eds.), pp. 459–475. Marcel Dekker, New York.

Carreno-Gomez, B., Woodley, J.F., and Florence, A.T. 1999. Studies on the uptake of tomato lectin nanoparticles in everted gut sacs. *Int. J. Pharm.* 183:7–11.

Challacombe, S.J., Rahman, D., Jeffery, H., Davis, S.S., and O'Hagan, D.T. 1992. Enhanced secretory IgA and systemic IgG antibody responses after oral immunization with biodegradable microparticles containing antigen. *Immunology* 76:164–168.

Chein, H. and Langer, R. 1998. Oral particulate delivery: Status and future trends. *Adv. Drug Deliv. Rev.* 34:339–350.

Chen, J.L. and Cyr, G.N. 1970. Composition producing adhesion through hydration. In: *Adhesion in Biological Systems*, Manly, R.S. (ed.), pp. 163–181. Academic Press, New York.

Ch'ng, H.S., Park, H., Kelly, P., and Robinson, J.R. 1985. Bioadhesive polymers as platforms for oral controlled drug delivery II: Synthesis and evaluation of some swelling water-insoluble bioadhesive polymers. *J. Pharm. Sci.* 74:399–405.

Chickering, D.E. and Mathiowitz, E. 1995. Bioadhesive microspheres: A novel electrobalance-based method to study adhesive interactions between individual microspheres and intestinal mucosa. *J. Control. Release* 34:251–261.

Chickering, D., Jacob, J., and Mathiowitz, E. 1996. Poly(fumaric-*co*-sebacic) microspheres as oral drug delivery systems. *Biotechnol. Bioeng.* 52:96–101.

Chickering, D.E., Santos, C.A., and Mathiowitz, E. 1999. Adaptation of a microbalance to measure bioadhesive properties of microspheres. In: *Bioadhesive Drug Delivery System—Fundamentals, Novel Approaches and Development*, Mathiowitz, E., Chickering, D.E., and Lehr, C.M. (eds.), pp. 131–145. Marcel Dekker, New York.

Clark, M.A., Jepson, M.A., and Hirst, B.H. 2001. Exploiting M cells for drug and vaccine delivery. *Adv. Drug Deliv. Rev.* 50:81–106.

Cleary, G.W., Feldstein, M.M., Singh, P., and Plate, N.A. 2003. A new polymer blend adhesive with combined properties to adhere to either skin or mucosa for drug delivery, Podium Abstract, 30th Annual Meeting and Exposition of the Controlled Release Society, Glagsow, Scotland, July, 19–23.

Critchley, H., Davis, S.S., Farraj, N.F., and Illum, L. 1994. Nasal absorption of desmopressin in rats and sheep-effect of a bioadhesive microsphere delivery system. *J. Pharm. Pharmacol.* 46:651–656.

Czerkinsky, C., Svennerholm, A.M., and Holmgren, J. 1993. Induction and assessment of immunity at enteromucosal surfaces in humans: Implications for vaccine development. *Clin. Infect. Dis.* 16:S106–S116.

Damge, C., Vranckx, H., Balschmidt, P., and Couvreur, P. 1997. Poly(alkyl cyanoacrylate) nanospheres for oral administration of insulin. *J. Pharm. Sci.* 86:1403–1409.

Das, M.K. and Maurya, D.P. 2008. Evaluation of diltiazem hydrochloride-loaded mucoadhesive microspheres prepared by emulsification-internal gelation technique. *Acta Pol. Pharm.* 65(2):249–259.

Delie, F. 1998. Evaluation of nano- and microparticle uptake by the gastrointestinal tract. *Adv. Drug Deliv. Rev.* 34(2–3):221–233.

Duchene, D., Touchard, F., and Peppas, N.A. 1988. Pharmaceutical and medical aspects of bioadhesive systems for drug administration. *Drug Dev. Ind. Pharm.* 14:283–318.

Durrani, A.M., Farr, S.J., and Kellaway, I.W. 1995. Precorneal clearance of mucoadhesive microspheres from the rabbit eye. *J. Pharm. Pharmacol.* 47:581–584.

Edman, P., Bjork, E., and Ryden, L. 1992. Microspheres as a nasal delivery system for peptide drugs. *J. Control. Release* 21:165–172.

Eldridge, J.H., Hammond, C.J., Meulbroek, J.A., Staas, J.K., Gilley, R.M., and Tice, T.R. 1990. Controlled vaccine release in the gut-associated lymphoid-tissues orally administered biodegradable microspheres target the Peyers patches. *J. Control. Release* 11:205–214.

Farraj, N.F., Johansen, B.R., Davis, S.S., and Illum, L. 1990. Nasal administration of insulin using bioadhesive microspheres as a delivery system. *J. Control. Release* 13:253–261.

Fernandez-Urrusuno, R., Calvo, P., Remunan-Lopez, C., Vila-Jato, J.L., and Alonso, M.J. 1999. Enhancement of nasal absorption of insulin using chitosan nanoparticles. *Pharm. Res.* 16:1576–1581.

Fresta, M., Fontana, G., Bucolo, C., Cavallaro, G., Giammona, G., and Puglisi, G. 2001. Ocular tolerability and *in vivo* bioavailability of poly(ethylene glycol) (PEG)-coated polymethyl-2-cyanoacrylate nanospheres-encapsulated acyclovir. *J. Pharm. Sci.* 90:288–297.

Galindo-Rodriguez, S.A., Allemann, E., Fessi, H., and Doelker, E. 2005. Polymeric nanoparticles for oral delivery of drugs and vaccines: A critical evaluation of *in vivo* studies. *Crit. Rev. Ther. Drug Carr. Syst.* 22:419–464.

Gavini, E., Rassu, G., Sanna, V., Cossu, M., and Giunchedi, P. 2005. Mucoadhesive microspheres for nasal administration of an antiemetic drug, metoclopramide: *In-vitro/ex-vivo* studies. *J. Pharm. Pharmacol.* 57(3):287–294.

Geary, S. and Schlameus, H.W. 1993. Vancomycin and insulin used as models for oral delivery of peptides. *J. Control. Release* 23:65–74.

Genta, I., Conti, B., Perugini, P., Pavanetto, F., Spadaro, A., and Puglisi, G. 1997. Bioadhesive microspheres for ophthalmic administration of Acyclovir. *J. Pharm. Pharmacol.* 49:737–742.

Ghezzo, E., Benedetti, L., Rochira, N., Biviano, F., and Callegaro, L. 1992. Hyaluronane derivative microsphere as NGF delivery device: Preparation methods and *in vitro* release characterization. *Int. J. Pharm.* 87:21–29.

Govender, S., Pillay, V., Chetty, D.J., Essack, S.Y., Dangor, C.M., and Govender, T. 2005. Optimisation and characterisation of bioadhesive controlled release tetracycline microspheres. *Int. J. Pharm.* 306(1–2):24–40.

Gurney, R., Meyer, J.M., and Peppas, J.M. 1984. Bioadhesive intra-oral release systems: Design, testing and analysis. *Biomaterials* 5:336–340.

Gurny, R. and Peppas, N.A. 1990. Semisolid dosage forms as buccal bioadhesive. In: *Bioadhesive Drug Delivery System*, Lenaerts, V. and Gurny, R. (eds.), pp. 153–168. CRC Press, Boca Raton, FL.

Haltner, E., Easson, J.H., Russel-Jones, G., and Lehr, C.M. 1996. A rapid assay for bioadhesion of lectin-functionalized nano-particles to Caco-2 cell monolayers. *J. Control. Release* 41:S3.

Haltner, E., Easson, J.H., and Lehr, C.M. 1997. Lectins and bacterial invasion factors for controlling endotransytosis of bioadhesive drug carrier systems. *Eur. J. Pharm. Biopharm.* 44:3–13.

Harris, D. and Robinson, J.R. 1992. Drug delivery via the mucous membranes of the oral cavity. *J. Pharm. Sci.* 81:1–10.

Hejazi, R. and Amiji, M. 2004. Stomach-specific anti-*H. pylori* therapy; part III: Effect of chitosan microspheres crosslinking on the gastric residence and local tetracycline concentrations in fasted gerbils. *Int. J. Pharm.* 19:99–108.

Helfand, E. and Tagami, Y. 1972. Theory of the interface between immiscible polymers. *J. Chem. Phys.* 57:1812–1813.

Hertzog, B.A. and Mathiowitz, E. 1999. Novel magnetic technique to measure bioadhesion. In: *Bioadhesive Drug Delivery Systems: Fundamentals, Novel Approaches and Development*, Mathiowitz, E., Chickering, D.E., and Lehr, C.M. (eds.), pp. 147–171. Marcel Dekker, New York.

Ikeda, K., Murata, K., Kobayashi, M., and Noda, K. 1992. Enhancement of bioavailability of dopamine via nasal route in beagle dogs. *Chem. Pharm. Bull.* 40:2155–2158.

Illum, L. 1999. Bioadhesive formulations for nasal delivery. In: *Bioadhesive Drug Delivery Systems: Fundamentals, Novel Approaches and Development*, Mathiowitz, E., Chickering, D.E., and Lehr, C.M. (eds.), pp. 519–539. Marcel Dekker, New York.

Illum, L., Jorgensen, H., Bisgaard, H., Krogsgaard, O., and Rossing, N. 1987. Bioadhesive microspheres as a potential nasal drug delivery system. *Int. J. Pharm.* 39:189–199.

Illum, L., Furraj, N.F., Critcheley, H., and Davis, S.S. 1988. Nasal administration of gentamycin using a novel microsphere delivery system. *Int. J. Pharm.* 46:261–265.

Illum, L., Farraj, N.F., Davis, S.S., Johansen, B.R., and O'Hagan, D.T. 1990. Investigation of the nasal absorption of biosynthetic human growth hormone in sheep-use of a bioadhesive microsphere delivery system. *Int. J. Pharm.* 63:207–211.

Illum, L., Farraj, N.F., Fisher, A.N., Gill, J., Miglietta, M., and Benedetti, L.M. 1994. Hyaluronic acid ester microsphere as a nasal delivery system for insulin. *J. Control. Release* 29:133–141.

Inman, L.R. and Cantey, J.R. 1983. Specific adherence of *Escherichia coli* (strain RDEC-1) to membranous (M) cells of the Peyer's patch in *Escherichia coli* diarrhea in the rabbit. *J. Clin. Invest.* 71:1–8.

Inoue, T., Chen, G., Nakame, K., and Hoffman, A.S. 1998. An AB copolymer of oligo (methyl methyl-methacrylate) and poly (acrylic acid) for micellar delivery of hydrophobic drugs. *J. Control. Release* 51(2–3):221–229.

Irache, J.M., Durrer, C., Duchene, D., and Ponchel, G. 1994. Preparation and characterization of lectin–latex conjugates for specific bioadhesion. *Biomaterials* 15:899–904.

Irache, J.M., Durrer, C., Duchene, D., and Ponchel, G. 1996. Bioadhesion of lectin–latex conjugates to rat intestinal mucosa. *Pharm. Res.* 13:1716–1719.

Jain, S.K., Chourasia, M.K., Jain, A.K., Jain, R.K., and Shrivastava, A.K. 2004. Development and characteriza-tion of mucoadhesive microspheres bearing salbutamol for nasal delivery. *Drug Deliv.* 1(2):113–122.

Janes, K.A., Calvo, P., and Alonso, M.J. 2001. Polysaccharide colloidal particles as delivery systems for mac-romolecules. *Adv. Drug Deliv. Rev.* 47:83–97.

Jani, P.U., Halbert, G.W., Langridge, J., and Florence, A.T. 1989. The uptake and translocation of latex nano-spheres and microspheres after oral administration to rats. *J. Pharm. Pharmacol.* 41:809–812.

Jani, P.U., Mccarthy, D.E., and Florence, A.T. 1992. Nanosphere and microsphere uptake via Peyer patches-observation of the rate of uptake in the rat after a single oral dose. *Int. J. Pharm.* 86:239–246.

Jubeh, T.T., Barenholz, Y., and Rubinstein, A. 2004. Differential adhesion of normal and inflamed rat colonic mucosa by charged liposomes. *Pharm. Res.* 21:447–453.

Jung, T., Kamm, W., Breitenbach, A., Hungerer, K.D., Hundt, E., and Kissel, T. 2001. Tetanus toxoid loaded nanoparticles from sulfobutylated poly(vinyl alcohol)-graft-poly (lactide-*co*-glycolide): Evaluation of antibody response after oral and nasal application in mice. *Pharm. Res.* 18:352–360.

Kaelble, D.H. 1977. A surface energy analysis of bioadhesion. *Polymers* 18:475–482.

Kamath, K.R. and Park, K. 1994. Mucosal adhesive preparations. In: *Encyclopedia of Pharmaceutical Technology*, Swarbrick, J. and Boylan, J.C. (eds.), pp. 133–163. Marcel Dekker, New York.

Keely, S., Rullay, A., Wilson, C., Carmichael, A., Carrington, S., Corfield, A., Haddleton, D.M., and Brayden, D.J. 2005. *In vitro* and *ex vivo* intestinal tissue models to measure mucoadhesion of poly (methacrylate) and *N*-trimethylated chitosan polymers. *Pharm. Res.* 22:38–49.

Kockisch, S., Rees, G.D., Tsibouklis, J., and Smart, J.D. 2005. Mucoadhesive, triclosan-loaded polymer microspheres for application to the oral cavity: Preparation and controlled release characteristics. *Eur. J. Pharm. Biopharm.* 59(1):207–216.

Kumar, V. and Banker, G.S. 1993. Chemically-modified cellulose polymers. *Drug Dev. Ind. Pharm.* 19:1–13.

Kyyronen, K., Hume, L., Benedetti, L., Urtti, A., Topp, E., and Stella, V. 1992. Methylprednisolone esters of hyaluronic acid in ophthalmic drug delivery: *In vitro* and *in vivo* release studies. *Int. J. Pharm.* 80:161–169.

Langer, R., Cleland, J.L., and Hanes, J. 1997. New advances in microsphere-based single-dose vaccines. *Adv. Drug Deliv. Rev.* 28:97–119.

Langoth, N., Kalbe, J., and Bernkop-Schnurch, A. 2003. Development of buccal drug delivery systems based on a thiolated polymer. *Int. J. Pharm.* 252:141–148.

Lavelle, E.C., Sharif, S., Thomas, N.W., Holland, J., and Davis S.S. 1995. The importance of gastrointestinal uptake of particles in the design of oral delivery systems. *Adv. Drug Deliv. Rev.* 18:5–22.

Lehr, C.M. 2004. Lectins and glycoconjugates in drug delivery and targeting. *Adv. Drug Del. Rev.* 56:419–420.

Lehr, C.M., Poelma, F.G.J., Junginger, H.E., and Tukker, J.J. 1991. An estimate of turnover time of intestinal mucus gel layer in the rat in situ loop. *Int. J. Pharm.* 70:235–240.

Lehr, C.M., Bouwstra, J.A., Kok, W., Noach, A.B., de Boer, A.G., and Junginger, H.E. 1992. Bioadhesion by means of specific binding of tomato lectin. *Pharm. Res.* 9:547–553.

Lele, B.S. and Hoffman, A.S. 2000. Mucoadhesive drug carriers based on complexes of poly(acrylic acid) and PEGylated drugs having hydrolysable PEG-anhydride-drug linkages. *J. Control. Release* 69:237–248.

Li, Y., Jiang, H.L., Jin, J.F., and Zhu, K.J. 2004. Bioadhesive fluorescent microspheres as visible carriers for local delivery of drugs. II: Uptake of insulin-loaded PCEFB/PLGA microspheres by the gastrointestinal tract. *Drug Deliv.* 1(6):335–340.

Liu, Z., Lu, W., Qian, L., Zhang, X., Zeng, P., and Pan, J. 2005. *In vitro* and *in vivo* studies on mucoadhesive microspheres of amoxicillin *J. Control. Release* 102:135–144.

Longer, M.A., Ch'ng, H.S., and Robinson, J.R. 1985. Bioadhesive polymers as platforms for oral controlled drug delivery. III. Oral delivery of chlorthiazide using a bioadhesive polymer. *J. Pharm. Sci.* 74:406–411.

Lowe, P.J. and Temple, C.S. 1994. Calcitonin and insulin in isobutylcyanoacrylate nanocapsules: Protection against proteases and effect on intestinal absorption in rats. *J. Pharm. Pharmacol.* 46:547–552.

Luessen, H.L., Verhoef, J.C., Borchard, G., Lehr, C.M., deBoer, A.G., and Junginger, H.E. 1995. Mucoadhesive polymers in peroral peptide drug delivery. II. Carbomer and polycarbophil are potent inhibitors of the intestinal proteolytic enzyme trypsin. *Pharm. Res.* 12:1293–1298.

Madsen, F., Eberth, K., and Smart, J.D. 1998. A rheological examination of the mucoadhesive/mucus interaction: The effect of mucoadhesive type and concentration. *J. Control. Release* 50:167–178.

Majithiya, R.J. and Murthy, R.S. 2005. Chitosan-based mucoadhesive microspheres of clarithromycin as a delivery system for antibiotic to stomach. *Curr. Drug Deliv.* 2(3):235–242.

Mathiowitz, E., Chickering, D., Jacob, J.S., and Santos, C. 1999. Bioadhesive drug delivery systems. In: *Encyclopedia of Controlled Drug Delivery*, Mathiowitz, E. (ed.), pp. 9–44. John Wiley & Sons, New York.

Mathiowitz, E., Chickering, D.E., and Jacob, J.S. 2001. U.S. Patent No. 6,197,346.

Mikos, A.G. and Peppas, N.A. 1990. Bioadhesive analysis of controlled release systems. IV. An experimental method for testing the adhesion of microparticles with mucus. *J. Control. Release* 12:31–37.

Montisci, M.J., Dembri, A., Giovannuci, G., Chacun, H., Duchene, D., and Ponchel, G. 2001. Gastrointestinal transit and mucoadhesion of colloidal suspensions of *Lycopersicon esculentum* L. and *Lotus tetragonolobus* lectin-PLA microsphere conjugates in rats. *Pharm. Res.* 18:829–837.

Morimoto, K., Katsumata, H., Yabuta, T., Iwanaga, K., Kakemi, M., Tabuta, Y., and Ikada, Y. 2001. Evaluation of gelatin microspheres for nasal intramuscular administrations of salmon calcitonin. *Eur. J. Pharm.* 124:173–182.

Mortazavi, S.A. and Smart, J.D. 1993. An investigation into the role of water movement and mucus gel dehydration in mucoadhesion. *J. Control. Release* 25:197–203.

Nagahara, N., Akiyama, Y., Nako, M., Tada, M., Kitano, M., and Ogawa, Y. 1998. Mucoadhesive microspheres containing amoxicillin for clearance of *Helicobacter pylori*. *Antimicrob. Agent Chemother.* 42:2492–2494.

Nagai, T. and Kinishi, R. 1987. Buccal/gingival drug delivery systems, *J. Control. Release* 6:353–356.

Nagai, T., Nishimoto, Y., Nambu, N., Suzuki, Y., and Sekine, K. 1984. Powder dosage form of insulin for nasal administration. *J. Control. Release* 1:15–22.

Naisbett, B. and Woodley, J. 1995. The potential use of tomato lectin for oral-drug delivery bioadhesion *in-vivo*. *Int. J. Pharm.* 114:227–236.

Neutra, M.R. and Forstner, J.F. 1987. Gastrointestinal mucus: Synthesis, secretion and function. In: *Physiology of the Gastrointestinal Tract*, Johnson, L.R. (ed.), pp. 975–1009. Raven, New York.

O'Hagan, D.T. 1996. The intestinal uptake of particles and the implications for drug and antigen delivery. *J. Anat.* 189:477–482.

O'Hagan, D.T., Rafferty, D., Wharton, S., and Illum, L. 1993. Intravaginal immunization in sheep using a bioadhesive microsphere antigen delivery system. *Vaccine* 11:660–664.

Pan, Y., Li, Y.J., Zhao, H.Y., Zheng, J.M., Xu, H., Wei, G., Hao, J.S., and Cui, F.D. 2002. Bioadhesive polysaccharide in protein delivery system: Chitosan nanoparticles improve the intestinal absorption of insulin *in vivo*. *Int. J. Pharm.* 249:139–147.

Park, H. 1986. On the mechanism of bioadhesion. PhD dissertation, School of Pharmacy, University of Wisconsin-Madison, Madison, WI.

Park, K. and Robinson, J.R. 1984. Bioadhesive polymers as platforms for oral controlled drug delivery: Method to study bioadhesion. *Int. J. Pharm.* 19:107–127.

Park, H. and Robison, J.R. 1985. Physico-chemical properties of water insoluble polymers important to mucin epithelial adhesion. *J. Control. Release* 2:47–57.

Patel, J.K. and Chavda, J.R. 2008. Formulation and evaluation of stomach specific amoxicillin-loaded carbopol-934P mucoadhesive microspheres for anti-*Helicobacter pylori* therapy. *J. Microencapsul.* 26:365–376. 21 August: 1–12. DOI: 10.1080/02652040802373012.

Patel, J.K. and Patel, M.M. 2007. Stomach specific anti-*Helicobacter pylori* therapy: Preparation and evaluation of amoxicillin loaded chitosan mucoadhesive microspheres. *Curr. Drug Deliv.* 4:41–51.

Patel, J.K., Amin, A.F., and Patel, M.M. 2004. Formulation optimization and evaluation of controlled release mucoadhesive microspheres of glipizide for oral drug delivery using factorial design. *Drug Deliv. Technol.* 4:48–53.

Patel, J.K., Patel, R.P., Amin, A.F., and Patel, M.M. 2005. Formulation and evaluation of mucoadhesive glipizide microspheres. *AAPS PharmSciTech.* 6:49–55.

Payne, L.G., Woods, A.L., and Jenkins, S.A. 2001. U.S. Patent No. 6,207,171.

Peppas, N.A., Keys, K.B., Torres-Lugo, M., and Lowman, A.M. 1999. Poly(ethylene glycol)-containing hydrogels in drug delivery. *J. Control. Release* 62:81–87.

Ponchel, G. and Irache, J. 1998. Specific and non-specific bioadhesive particulate systems for oral delivery to the gastrointestinal tract. *Adv. Drug Deliv. Rev.* 34:191–219.

Preda, M. and Leucuta, S.E. 2003. Oxprenolol-loaded bioadhesive microspheres: Preparation and *in vitro/in vivo* characterization. *J. Microencapsul.* 20:777–789.

Prego, C., Torres, D., and Alonso, M.J. 2005. The potential of chitosan for the oral administration of peptides. *Expert Opin. Drug Deliv.* 2:843–854.

Prego, C., Torres, D., and Alonso, M.J. 2006. Chitosan nanocapsules as carriers for oral peptide delivery: Effect of chitosan molecular weight and type of salt on the *in vitro* behaviour and *in vivo* effectiveness. *J. Nanosci. Nanotechnol.* 6:2921–2928.

Quan, J.S., Jiang, H.L., Kim, E.M., Jeong, H.J., Choi, Y.J., Guo, D.D., Yoo, M.K., Lee, H.G., and Cho, C.S. 2008. pH-sensitive and mucoadhesive thiolated Eudragit-coated chitosan microspheres. *Int. J. Pharm.* 359(1–2):205–210.

Quintanar-Guerrero, D., Villalobos-Garcia, R., Alvarez-Colin, E., and Cornejo-Bravo, J.M. 2001. *In vitro* evaluation of the bioadhesive properties of hydrophobic polybasic gels containing *N,N*-dimethylaminoethyl methacrylate-*co*-methyl methacrylate. *Biomaterials* 22:957–961.

Restifo, N.P. 2000. Building better vaccines: How apoptotic cell death can induce inflammation and activate innate and adaptive immunity. *Curr. Opin. Immunol.* 12:597–603.

Richardson, J.L. and Armstrong, T.I. 1999. Vaginal delivery of calcitonin by hyaluronic acid formulations. In: *Bioadhesive Drug Delivery Systems: Fundamentals, Novel Approaches and Development*, Mathiowitz, E., Chickering, D.E., and Lehr, C.M. (eds.), pp. 563–599. Marcel Dekker, New York.

Richardson, J.L., Farraj, N.F., and Illum, L. 1992. Enhanced vaginal absorption of insulin in sheep using lysophosphatidylcholine and a bioadhesive microspheres delivery system. *Int. J. Pharm.* 88:319–325.

Richardson, J.L., Whetstone, J., Fisher, N.F., Watts, P., Farraj, N.F., Hinchcliffe, M., Benedetti, L., and Illum, L. 1996. Gamma scintigraphy as a novel method to study the distribution and retention of a bioadhesive vaginal delivery system in sheep. *J. Control. Release* 42:133–142.

Rieux, A., Fievez, V., Garinot, M., Schneider, Y.-J., and Preat, V. 2006. Nanoparticles as potential oral delivery systems of proteins and vaccines: A mechanistic approach. *J. Control. Release* 116:1–27.

Riley, R.G., Smart, J.D., Tsibouklis, J., Dettmar, P.W., Hampson, F., Davis, J.A., Kelly, G., and Wilber, R.W. 2001. An investigation of mucus/polymer rheological synergism using synthesised and characterised poly(acrylic acid)s. *Int. J. Pharm.* 217:87–100.

Roy, K., Mao, H.Q., Huang, S.K., and Leong, K.W. 1999. Oral gene delivery with chitosan–DNA nanoparticles generates immunologic protection in a murine model of peanut allergy. *Nat. Med.* 5:387–391.

Saettone, M.F., Burgalassi, S., and Chetoni, P. 1999. Ocular bioadhesive drug delivery systems. In: *Bioadhesive Drug Delivery Systems-Fundamentals, Novel Approaches, and Development*, Mathiowitz, E., Chickering, D.E., and Lehr C.M. (eds.), pp. 629–630. Marcel Dekker, New York.

Sanford, B.A., Thomas, V.L., and Ramsay, M.A. 1989. Binding of staphylococci to mucus *in vivo* and *in vitro*. *Infect. Immun.* 57:3735–3742.

Santos, C.A., Jacob, J.S., Hertzog, B.A., Freedman, B.D., Press, D.L., Harnpicharnchai, P., and Mathiowitz, E. 1999. Correlation of two bioadhesion assays: The everted sac technique and the CAHN microbalance. *J. Control. Release* 61:113–122.

Sarmento, B., Ribeiro, A., Veiga, F., Sampaio, P., Neufeld R., and Ferreira, D. 2007. Alginate/chitosan nanoparticles are effective for oral insulin delivery. *Pharm. Res.* 24:2198–2206.

Schaefer, M.J. and Singh, J. 2000. Effect of isopropyl myristic acid ester on the physical characteristics and *in-vitro* release of etoposide from PLGA microspheres. *AAPS PharmSciTech.* 1(4):article 32.

Scherer, D., Mooren, F.C., Kinne, R.K., and Kreuter, J. 1993. *In vitro* permeability of PBCA nanoparticles through porcine small intestine. *J. Drug Target.* 1:21–27.

Senel, S., Kremer, M.J., Kas, S., Wertz, P.W., Hincal, A.A., and Squier, C.A. 2000. Enhancing effect of chitosan on peptide drug delivery across buccal mucosa. *Biomaterials* 21:2067–2071.

Shah, K. and Rocca, J. 2004. Lectins as next-generation mucoadhesives for specific targeting of the gastrointestinal tract. *Drug Deliv. Tech.* 4(5):1.

Shojaei, A.M. and Li, X. 1997. Mechanism of buccal mucoadhesion of novel copolymers of acrylic acid and polyethylene glycol monomethylether monomethacrylate. *J. Control. Release* 47:151–161.

Singh, M., Briones, M., and O'Hagan, D.T. 2001. A novel bioadhesive intranasal delivery system for inactivated influenza vaccines. *J. Control. Release* 70:267–276.

Soane, R.J., Perkins, A.C., Jones, N.S., Davis, S.S., and Illum, L. 1999. Evaluation of the clearance characteristics of bioadhesive systems in humans. *Int. J. Pharm.* 178:55–65.

Takeuchi, H., Yamamoto, H., and Kawashima, Y. 2001. Mucoadhesive nanoparticulate systems for peptide drug delivery. *Adv. Drug Deliv. Rev.* 47:39–54.

Teng, C.L.C. and Ho, N.F.H. 1987. Mechanistic studies in the simultaneous flow and adsorption of poly coated latex particles on intestinal mucus. I. Methods and physical model development. *J. Control. Release* 6:133–149.

Tokura, S., Tamura, H., and Azuma, I. 1999. Immunological aspects of chitin and chitin derivatives administered to animals. *Experientia* Suppl. 87:279–292.

Vasir, J.K., Tambwekar, K., and Garg, S. 2003. Bioadhesive microspheres as a controlled drug delivery system. *Int. J. Pharm.* 255:13–32.

Vyas, S.P. and Jain, C.P. 1992. Bioadhesive polymer grafted starch microspheres bearing isosorbide dinitrate for buccal administration. *J. Microencapsul.* 9:457–464.

Wake, W.C. 1982. *Adhesion and the Formulation of Adhesives.* Applied Science, London, U.K., pp. 67–119.

Wallace, J.W. 1990. Cellulose derivatives and natural products utilized. In: *Encyclopedia of Pharmaceutical Technology*, Swarbrick, J. and Boylan, J.C. (eds.), pp. 319–337. Marcel Dekker, New York.

Wittaya-areekul, S., Kruenate, J., and Prahsarn, C. 2006. Preparation and *in vitro* evaluation of mucoadhesive properties of alginate/chitosan microparticles containing prednisolone. *Int. J. Pharm.* 312(1–2):113–118.

Woodley, J. 2001. Bioadhesion: New possibilities for drug administration? *Clin. Pharmacokinet.* 40:77–84.

Ying, H., Zaks, T.Z., Wang, R.F., Irvine, K.R., Kammula, U.S., Marincola, F.M., Leitner, W., and Restifo, N.P. 1999. Cancer therapy using a self-replicating RNA vaccine, *Nat. Med.* 5:823–827.

Yoncheva, K. Gomez, S., Campanero, M.A., Gamazo, C., and Irache, J.M. 2005. Bioadhesive properties of pegylated nanoparticles. *Exp. Opin. Drug Deliv.* 2:205–218.

Zhou, H.Y., Chen, X.G., and Zhang, W.F. 2007. *In vitro* and *in vivo* evaluation of mucoadhesiveness of chitosan/cellulose acetate multimicrospheres. *Biomed. Mater. Res. A* 83(4):1146–1153.

13 Microencapsulation of Probiotic Cells

*Thomas Heidebach, Elena Leeb,
Petra Först, and Ulrich Kulozik*

CONTENTS

13.1 INTRODUCTION

The ingestion of viable probiotic cells is associated with various well-documented health benefits, generated by providing a positive impact on the ecosystem in the human intestinal tract (Jia et al., 2008; Naidu et al., 1999; Tamime, 2005). Thus, the addition of living probiotic microorganisms to mostly fermented foods is a frequent strategy to meet the increasing demand for healthy "functional foods" (Rodgers, 2008; Stanton et al., 2001). The WHO recommends a description of probiotics as "live microorganisms which, when administered in adequate numbers, confer a health benefit on the host" (Ross et al., 2005). Therefore, maintenance of the viability and functionality of the probiotics until they reach their site of colonization in the human gut is one of the key requirements for an effective application (Mattila-Sandholm et al., 2002; Ross et al., 2005). However, a decline of bioactivity, that is the loss of living probiotic cell numbers during processing, storage, and gastro-intestinal transit has been found to be crucial for successful and effective application in functional foods (Mattila-Sandholm et al., 2002; Shah, 2000; Siuta-Cruce and Goulet, 2001). In this regard, the delivery of probiotic cells in microencapsulated form offers a promising way to enhance the survival of these sensitive bacteria by reducing the impact of detrimental factors from the outside environment, and hence protecting the cells against adverse external conditions (Augustin, 2003; Champagne et al., 2005; Ross et al., 2005).

13.2   GENERAL REQUIREMENTS REGARDING THE MICROENCAPSULATION OF PROBIOTIC CELLS

Microencapsulation is a defined technology, involving the coating or entrapment of a core material into capsules in the size-range between a few micrometers and a few millimeters (Kirby, 1991). In most cases of probiotic encapsulation, microcapsules are generated from precursors in the form of liquid droplets that undergo a liquid–solid phase transformation, caused by dehydration, gelation, or crystallization. The capsules are therefore mostly of spherical shape and the resulting capsule size can be controlled by manipulating the droplet generation process. Almost all attempts within probibtic microencapsulation involve the basic processing steps and are shown in Figure 13.1.

Capsules generated in this manner are the so-called matrix capsules, because the living cells are embedded as core material and immobilized randomly in a continuous matrix of wall material, which is often a hydrogel (Figure 13.2a) (Desai and Park, 2005). Matrix capsules therefore differ from the classic type of microcapsule, consisting of a continuous core, surrounded by a continuous layer of shell material (Figure 13.2b) (Kirby, 1991). On this account, the terms immobilization and encapsulation were used interchangeably in most reported literature about microencapsulation of probiotics (Anal and Singh, 2007; Krasaekoopt et al., 2003).

Since the probiotic cells are randomly immobilized within the capsule matrix, it must be considered that a small amount can be found on the surface of the capsules and therefore exposed to the surrounding area. On this account, a frequent strategy to modify the matrix capsules is coating, by which the capsule surface is covered by a continuous membrane layer (Figure 13.2c). This is typically achieved by an additional processing step where the polymer capsules are immersed in a solution of coating polymer. Poly-*l*-lysine and chitosan are commonly used as coating

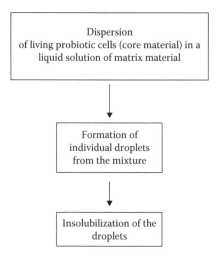

FIGURE 13.1 Basic processing steps for the production of matrix capsules in the field of probiotic microencapsulation.

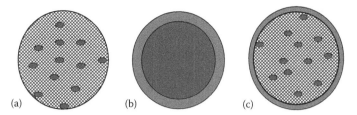

FIGURE 13.2 Different types of microcapsules: (a) matrix capsule, (b) classic microcapsule, (c) coated matrix capsule.

materials, because they form strong complexes with alginate (Krasaekoopt et al., 2004). This process serves to increase the mechanical strength of the capsule and creates a more pronounced barrier function.

In food technology there are numerous applications for microencapsulation with a multitude of technological and biofunctional benefits (Jackson and Lee, 1991). A prevalent objective is to protect the core material from degradation by reducing its reactivity to the outside environment (Gibbs et al., 1999; Schrooyen et al., 2001). This is mainly achieved by the control of mass transfer between the core and the surrounding by employing the matrix material as a physical barrier. Therefore, microencapsulation can serve as a technological method to protect living probiotic cells against adverse external conditions when introduced for the creation of functional foods (Champagne and Fustier, 2007; Desai and Park, 2005; Kailasapathy, 2002; Lopez-Rubio et al., 2006).

Compared to many other bioactive substances intended for microencapsulation, living probiotic cells are relatively sensitive toward adverse environmental conditions. On this account, many of the already available microencapsulation technologies are not suitable for probiotic encapsulation, because those techniques often comprise drastic processing steps, such as heating or acidification. In this context, particularly the insolubilization step during encapsulation has been shown to be the most critical for the successful encapsulation of probiotic bacteria without high losses of living cell numbers. Next to the insolubilization, the resulting capsule size is also crucial for an application in food. On the one hand, the capsules must be sufficiently small to avoid a negative sensorial impact on the product (Champagne and Fustier, 2007). On the other hand, larger capsule diameters, and hence a higher volume-to-surface ratio, increase the likeliness of a protecting effect (Anal and Singh, 2007).

However, the optimal size for microcapsules depends on the intended purpose and on the type of food. While probiotic microcapsules with diameters between 2 and 3 mm could not be sensorial detected in fermented sausage (Muthukumarasamy and Holley, 2006), capsules of that size-range give an adverse sensorial effect in yogurt, milk, and sour cream. For dairy foods a size-range of approximately 100 µm diameter was generally seen as a threshold to avoid negative sensorial impacts of microcapsules in most foods (Hansen et al., 2002).

13.3 METHODS OF ENCAPSULATION AND AVAILABLE ENCAPSULATION MATERIALS

The vast majority of probiotic microcapsules are produced by three methods: the extrusion technique, the emulsion method, and spray-drying. In the case of the emulsion- and extrusion technique, biopolymers, such as alginate, gellan-gum, xanthan, κ-carrageenan, locust bean gum, or mixtures thereof, are commonly used as gelation material, since aqueous solutions of these polymers can undergo mild ionotrophic and/or thermal gelation (Champagne et al., 1994; Doleyres and Lacroix, 2005; Krasaekoopt et al., 2003). Furthermore, the resulting gels are water-insoluble, which is important to avoid a premature release of cells in an aqueous environment. Recently the use of proteins as gelling agent has also gained considerable attraction (Chen et al., 2006).

13.3.1 Extrusion Technique

The extrusion technique is the oldest and the most common approach to produce matrix capsules from hydrocolloid solutions. Due to its easy and gentle formulation, this method allows a virtually loss-free entrapment of living microbial cells, such as probiotics (Kailasapathy, 2002).

To form microcapsules by the extrusion method, a probiotic cell suspension is mixed with a solution of one of the above-mentioned hydrocolloids, and the mixture is dripped into a solution containing multivalent cations (usually Ca^{2+} in the form of $CaCl_2$). On addition of sodium algi-nate droplets to a calcium solution, interfacial polymerization appears by precipitation of calcium alginate followed by a more gradual gelation of the interior as calcium ions permeate through the alginate systems. Figure 13.3 shows the schematic process flow of the extrusion technique.

The size and the shape of the capsules depend on the diameter of the orifice, the distance of free-fall into the hardening solution, the concentration of the calcium chloride in the hardening solution, duration of hardening of the capsules, and the viscosity of the hydrocolloid–probiotic cell-mixture. Limited by the geometry of the nozzle, predominantly relatively large capsules in size-ranges above 0.5 mm up to 5 mm are produced (Krasaekoopt et al., 2003).

13.3.2 Emulsion Technique

Next to the extrusion technique, the emulsion technique is the most commonly used method for the microencapsulation of probiotic cells (Krasaekoopt et al., 2003). With this technique, gener-ally smaller capsules can be produced, as compared to the extrusion technique. While the extru-sion method leads to a very narrow particle size distribution, given by the geometry of the nozzle, a wider distribution of particle sizes is a general limitation of the emulsion technique (Poncelet et al., 1992).

In this technique a small volume of the aqueous hydrocolloid–cell mixture (discontinuous phase) is emulsified into a larger volume of vegetable oil (continuous phase). Once a water-in-oil emul-sion is formed, the hydrocolloid–cell mixture must be insolubilized to form small capsules within the oil phase (Krasaekoopt et al., 2003). When alginate capsules are produced, the microcapsules are hardened by slowly adding aqueous $CaCl_2$ solution to the emulsion while stirring. By the time

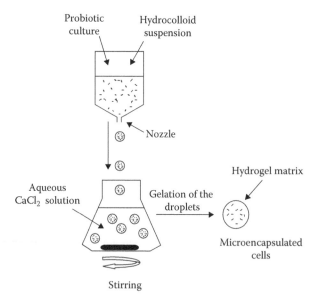

FIGURE 13.3 Schematic process flow of the extrusion technique.

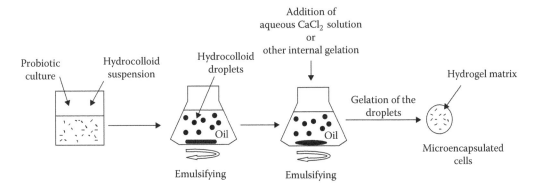

FIGURE 13.4   Schematic process flow of the emulsion technique.

the calcium solution contacts the dispersed alginate phase, gelling occurs. Figure 13.4 shows the schematic process flow of the emulsion technique. Other internal gelation mechanisms, driven by temperature, enzymatic, or chemical cross-linking are also possible.

The technique was first developed by Nilsson et al. (1983) as a gentle method for the immobilization of living cells. The authors stated that by adjusting the speed of a magnetic stirrer during the emulsifying process capsules with average size-ranges between 0.1 and 5 mm could be produced. For probiotic encapsulation, the emulsifying step is often carried out by means of a magnetic stirring bar or a mechanic stirrer. However, to generate small microcapsules in a size-range below 100 μm from highly viscous hydrocolloid–cell mixtures, sometimes high shear systems such as ultra-turrax or a high-pressure homogenizer must be applied. As a drawback, these processes can lead to significant losses of living cell numbers during the process caused by high shearing stress (Capela et al., 2007).

13.3.3   SPRAY-DRYING

Spray-drying is known as a routine process in food technology, converting liquids into dry powders. The process can also be employed to microencapsulate probiotic cells. In this case, mixtures of probiotic cell concentrates with aqueous solutions of various polymers, such as modified starch (O'Riordan et al., 2001), gum acacia (Desmond et al., 2002), gum arabic, gelatin (Lian et al., 2003), whey protein isolate (Picot and Lacroix, 2004), maltodextrin mixed with gum arabic (Su et al., 2007), and β-cyclodextrin mixed with gum acacia (Zhao et al., 2008), have been spray-dried, and their ability to protect the probiotic cells against subsequent adverse conditions has been investigated. The advantage of spray-drying is that sufficient small capsules with average diameters below 100 μm are usually generated at comparably low costs and the process is widely established in food technology.

However, it must be considered that microcapsules prepared by this method are water soluble in most cases and therefore not suitable to protect probiotic cells in liquid formulations and in the human gastrointestinal tract (Krasaekoopt et al., 2003). However, some innovative approaches already showed the feasibility to produce water-insoluble probiotic microcapsules from spray-drying, by using pre-denatured whey protein solutions (Picot and Lacroix, 2004) or protein–sugar mixtures (Crittenden et al., 2006) as precursors.

Furthermore, it must be considered that the high temperatures and rapid dehydration during spray-drying generally leads to a deterioration of the cells, resulting in a significant loss of living cells and a diminished resistance against following adverse environmental conditions (Meng et al., 2008). It has been shown that the probiotic cell survival during spray-drying increases with decreasing outlet air temperature (Ananta et al., 2005; Gardiner et al., 2000; Lian et al., 2002; To and Etzel,

1997), and is often as low as 1%–10% in sufficiently dried powders (Desmond et al., 2002; Gardiner et al., 2000; Lian et al., 2002; Picot and Lacroix, 2004; Wang et al., 2004; Zhao et al., 2008).

As a conclusion, the technique of spray-drying could thus be advantageous over the classical encapsulation techniques if water-insoluble microcapsules are produced and higher survival rates are achieved by selection of heat-resistant strains (Meng et al., 2008) or the use of protective solutes (Santivarangkna et al., 2008).

13.4 EFFECT OF MICROENCAPSULATION ON ADVERSE ENVIRONMENTAL CONDITIONS

13.4.1 PROTECTION DURING STORAGE IN FOOD

The vast majority of foods that were supplemented with encapsulated probiotics are fermented products, in which lactic acid bacteria have already been used since ancient times (Naidu et al., 1999). Just recently the application of probiotics in non-fermented foods has also been suggested (Prado et al., 2008).

The purpose of using microencapsulated probiotics is to decrease the unavoidable drop of living cell numbers from the first addition of the probiotic concentrate to the food, until colonization in the gut after ingestion. This can be achieved by entrapment of probiotic cells in a protective microenvironment of the capsule matrix, and thus diminish negative influences during product storage.

13.4.1.1 Yogurt

Yogurt is the most extensively investigated product so far, since the remaining probiotic activity at the end of the determined expiration date was often found to be rather low (Kailasapathy and Rybka, 1997; Shah, 2000).

However, microencapsulation of various probiotic strains in hydrogel microcapsules based on κ-carrageenan (Adhikari et al., 2003), alginate (Hussein and Kebary, 1999), and gellan-xanthan (Sun and Griffiths, 2000) has been shown to increase the survival of probiotic cells in yogurt after refrigerated storage of about 1 log cycle, compared to the respective free cells. Besides the use of simple matrix capsules based on the above-mentioned hydrogels, some modifications were applied toward an optimization of the protective effect.

In this context, a possible strategy is to co-encapsulate prebiotic substances, such as insoluble resistant starch granules, together with the microorganisms. It has been shown that the simultaneous application of pro- and prebiotics, so-called synbiotics, can enhance the survival of probiotic cells during storage in fermented dairy products (Capela et al., 2006; Desai et al., 2004). By using such synbiotic microcapsule systems, a 1-log-cycle (Kailasapathy, 2006) and 0.5-log-cycle (Sultana et al., 2000) higher survival was found after 7–8 weeks of refrigerated storage in yogurt, compared to the respective free cells.

Another strategy being discussed to improve the protective properties of probiotic microcapsules is coating with chitosan. By using this modification, an increase of about 3 log cycles (Iyer and Kailasapathy, 2005) and 1 log cycle (Krasaekoopt et al., 2006) was found for encapsulated probiotics after 4–6 weeks of storage in yogurt.

From the available results, it can be deduced that encapsulation of various probiotic strains in hydrocolloid gels generally enhances their survival during storage in yogurt, but with various success. Along with the rather low pH of 4.5 and below (Lourens-Hattingh and Viljoen, 2001), cell death from oxygen has been discussed as one of the major factors for the low survival rates of probiotics in yogurt. Talwalkar and Kailasapathy (2003) have shown that encapsulation in alginate hydrogels offers substantial protection for probiotics under aerobic conditions for several probiotic strains and could therefore be responsible for higher survival rates of encapsulated cells during storage in yogurt.

However, it must be considered that some authors found inferior sensory attributes as a result of the addition of probiotic microcapsules, compared with yogurt containing the respective probiotic cells in free form. In some cases this could be explained by a rather simple relationship between large capsule sizes and graininess in yogurt (Adhikari et al., 2003; Kailasapathy, 2006). Further changes in flavor, due to more complex reactions between hydrocolloid and yogurt matrix, as well as altered metabolic profiles of microorganisms caused by encapsulation, must also be considered (Adhikari et al., 2000, 2003; Maragkoudakis et al., 2006).

13.4.1.2 Cheese

Besides yogurt, rennet cheese was also investigated several times as a matrix for encapsulated probiotics.

While Gobbetti et al. (1998) found no protective effect after storage in fresh cheese for 2 weeks, Ozer et al. (2009) report about 2-log-cycle higher survival of encapsulated cells after storage in white-brined cheese for 90 days.

However, in the case of rennet cheese, microencapsulation seems not always useful, since in some cases an even higher probiotic cell count was found in cheeses containing free cells at the end of storage, compared to those containing encapsulated cells (Dinakar and Mistry, 1994; Godward and Kailasapathy, 2003; Kailasapathy and Masondole, 2005). It therefore seems that the physiological conditions in a hydrocolloid matrix, such as κ-carrageenan or alginate, are less favorable for the probiotic cells compared with the cheese matrix.

Furthermore, several authors have shown that the dense protein matrix of rennet cheese offers outstanding protection effects to probiotic bacteria during passage through the gastrointestinal tract (Boylston et al., 2004). Since the rennet cheese matrix itself offers good protection, the application of encapsulated probiotics seems to be dispensable in most of the cases. However, besides the protection, encapsulation could also help to avoid high losses of probiotic cells together with the whey during the renneting process (Godward and Kailasapathy, 2003).

13.4.1.3 Other Types of Food

Along with yogurt and cheese, various other foods were already supplemented with encapsulated probiotic cells. Improved survival of probiotic cells, compared to the respective free cells at the end of the storage time, was found in fermented frozen dairy dessert with a pH of 4.5 (Shah and Ravula, 2000); ice cream (Homayouni et al., 2008), frozen ice milk (Sheu et al., 1993), refrigerated, non-acidified milk (Hansen et al., 2002), fermented maize beverage with a pH of 3.5 (McMaster et al., 2005a); dry fermented sausages with a pH of about 4.8 (Muthukumarasamy and Holley, 2006); and mayonnaise with a pH of about 4.4 (Khalil and Mansour, 1998).

Hence, many studies have shown the potential of probiotic encapsulation to increase the survival rates during storage in various foods. However, the achievable protective effects partly depend on the surrounding food matrix and the used probiotic strain, and should be therefore individually evaluated for each application. Furthermore, it must be stated that in most of the cases no explanation for the protective effects was given. Further research should therefore target toward the molecular mechanisms, responsible for the observed protective effects during storage in food matrices.

13.4.2 PROTECTION DURING DRYING AND SUBSEQUENT DRIED STORAGE

13.4.2.1 Drying of Microencapsulated Probiotics

Processing and application of probiotic cell concentrates in foods often require long-term storage. Liquid cell concentrates must generally be stored in frozen form to preserve high living cell numbers. For long-time storage at elevated temperatures drying of the cells is required, also in case of microencapsulated probiotic cells. With regard to drying and subsequent storage of microencapsulated probiotics, two major strategies are available. While spray-drying provides capsule building and drying in one step (see Section 13.3.3), hydrogel-based microcapsules, generated from an emulsion-or-extrusion process can subsequently be freeze-dried in an additional processing step. Conventional fluidized bed air drying at room temperature was also sometimes applied to dry hydrogel-based probiotic microcapsules, but with at least 10% residual water (Champagne and Gardner, 2001; Selmer-Olsen et al., 1999), the achievable grade of dryness is not sufficient. Such high amounts of residual water lead to unacceptable high losses of living probiotic cells during subsequent storage as a powder (Zayed and Roos, 2004). Independent of microencapsulation, freeze-drying has been shown to be generally less detrimental for probiotic cells, compared to spray-drying under direct comparison (To and Etzel, 1997; Wang et al., 2004). As a disadvantage, the cost of freeze-drying has been estimated to be six times higher per kilogram of removed water, compared to spray-drying (Knorr, 1998).

Freeze-drying is frequently used as a method to dry hydrogel-based microcapsules, containing probiotic cells. However, the influence of the encapsulated state of the probiotics on the drying process and vice versa was not investigated in most of the studies, and the dried capsules were subsequently rehydrated by addition to simulated gastric juice (Cui et al., 2000; Lian et al., 2003) or foods containing high amounts of water (Godward and Kailasapathy, 2003; Shah and Ravula, 2000).

The rehydration rate and fluid volume after drying influences cell survival (Meng et al., 2008). It can be manipulated by encapsulation of the cells, since rehydration is influenced by the microenvironment of the capsule. When protective substances were used for free and encapsulated samples, higher survival after freeze-drying and subsequent rehydration was found in the encapsulated samples, compared to free samples. This was explained by a possible protection against an osmotic shock, caused by a delayed rehydration (Kearney et al., 1990; Kushal et al., 2006).

In contrast to this, in a study of Champagne et al. (1992), no difference in survival for free and alginate-encapsulated cells after freeze-drying was found when protective substances were added to both free and encapsulated cells. Furthermore, Champagne et al. (1996) encapsulated various probiotic strains in alginate microcapsules. The freshly prepared capsules were soaked in a protectant solution and were subsequently freeze-dried. The authors even found a detrimental effect of encapsulation, which was highly strain-dependent. Also, when whey-protein-based microcapsules were used, no increase in survival during freeze-drying was found, due to microencapsulation (Reid et al., 2005).

Incorporation of cryoprotectants into the probiotic microcapsules before drying is a frequent strategy, because it has been shown that alginate alone offers no particular protection during freezing or freeze-drying (Champagne et al., 1992; Champagne and Gardner, 2001). This is commonly done by soaking the capsules in cryoprotectant solution as an additional technological step. However, it is suggested that protective substances with high molecular weight, such as casein micelles, cannot enter the gel network in the case of alginate (Champagne et al., 1992, 1996).

13.4.2.2 Storage of Dried Microcapsules

A loss of viability can be generally found during storage of probiotic microorganisms in the dried state. The adverse effect of oxygen plays a major role, leading to detrimental oxidative changes on the cell membrane (Castro et al., 1995). Furthermore, storage temperature and relative humidity (RH) are of importance regarding survival during storage. A temperature closely above 0°C generally leads to higher survival compared to more elevated storage temperatures (Gardiner et al., 2000; Higl et al., 2007; Teixeira et al., 1995), because lower temperatures result in reduced rates of detrimental chemical reactions, such as fatty acid oxidation (Castro et al., 1995). Therefore, to achieve high survival rates refrigeration is necessary for long-time storage of probiotic powders.

The residual water content of the sample correlates with the RH in the ambient air during storage, expressed in the sorption isotherms. Storage at high RH leads to an increase of the water content in the sample and is therefore detrimental. Optimal survival was found to be in the intermediate moisture range. At very low RH removal of structural water from the cells lead to increased damage. At higher RH levels oxidation is accelerated (Castro et al., 1995; Higl et al., 2007; Teixeira et al., 1995).

In this context, it must be considered that alginate, because of its good water binding properties, has shown to have higher water content at a certain level of RH, compared to the nonencapsulated cell sample stored under the same conditions (Champagne et al., 1996).

During an accelerated storage test at 45°C, Champagne et al. (1992) found higher survival rates for encapsulated cells. Furthermore, various authors found higher survival for encapsulated cells, compared to free cells during storage for up to 6 weeks (Capela et al., 2006; Song et al., 2003). However, no protective effect was found in a study of Champagne et al. (1996).

Next to freeze-dried hydrogel capsules, sometimes microcapsules generated from spray-drying processes were investigated regarding storage as powder.

Using modified starch as matrix material, O'Riordan et al. (2001) found no enhanced survival after 20 days of storage in dry muesli mix as well as dry malted beverage powder, stored at ambient temperature when encapsulated cells were compared to free cells. In contrast to this, by spray-drying in gum acacia (Desmond et al., 2002), protein-polysaccharide-oil-emulsions (Crittenden et al., 2006), and β-cyclodextrin with gum acacia mixture (Zhao et al., 2008) higher cell numbers, compared with free cells, were achieved after storage up to 8 weeks. The higher survival was explained by the oxygen-scavenging properties of the microcapsule matrix (Crittenden et al., 2006).

It is likely that the use of protective substances has a far more significant impact on probiotic survival during drying and storage, compared to microencapsulation itself. However, microencapsulation could help to apply these protective substances more effectively, or the capsule matrix itself could (partly) consist of protective substances. In other words, microencapsulation cannot replace the use of protective substances but can render it more effectively.

13.4.3 PROTECTION OF ENCAPSULATED PROBIOTICS DURING GASTRIC TRANSIT

To obtain a positive probiotic effect, the microorganisms must survive the transit through the low-pH gastric environment. The strong acidic conditions in the human stomach, as a natural barrier of the host, lead to a considerable reduction of living probiotic cell numbers (Naidu et al., 1999; Ross et al., 2005). Therefore, the gastric transit is still considered as the most critical hurdle regarding probiotic survival in food applications (Agrawal, 2005).

In this context, microencapsulation can help to protect the core material from degradation by reducing its reactivity to the outside environment in providing a physical barrier of a gel matrix. On this account, it can be presumed that the protective effect of encapsulated probiotics mainly depends on the physical characteristics of the capsule matrix and the capsule size.

Investigation of the protective effect by comparing the survival rates of encapsulated and free probiotic cells during incubation under simulated gastric conditions at low pH values has been carried out by many researchers. However, it must be considered that many different factors influence the survival rates of encapsulated probiotic cells during simulated gastric transit, especially the microbial strain, and the composition of the simulated gastric fluid, which widely differs.

13.4.3.1 Alginate-Based Microcapsules

Many researchers investigated whether a protective effect due to microencapsulation of probiotic cells in an alginate gel matrix, generated by ionotrophic gelation of alginate solutions of various concentrations between 1% and 3% during gastric transit can be achieved. In this context, an important prerequisite is the structural integrity of the microcapsules during gastric transit. By now several *in vitro* studies have shown that alginate microcapsules remain physically stable at low pH values in gastric fluid (Allan-Wojtas et al., 2008; Annan et al., 2008; Cui et al., 2000; Hansen et al., 2002; Iyer et al., 2004; Martoni et al., 2007). However, regarding the achievable protective effect, the results from the scientific literature available so far must be considered inconsistent, as outlined in this chapter. Several authors found an 2–4 log cycles increased survival of various probiotic strains after incubation under acidic simulated gastric conditions, due to encapsulation in alginate microcapsules (Cui et al., 2000; Ding and Shah, 2007; Song et al., 2003).

In contrast to these results, in another approach no protective effect under such conditions was found (Liserre et al., 2007). Hansen et al. (2002) suggested the capsule size as a relevant factor for a protective effect at low pH. In their study alginate capsules with an average diameter of 20 or 70 μm were produced. Capsules of both size-ranges were not able to protect probiotic cells during incubation at simulated gastric conditions. The authors concluded that alginate capsules with average sizes below 100 μm, likely to not degrade the textural quality of the delivery food, are not able to protect probiotic cells from the harsh environment in the human stomach. However, in some cases also large alginate capsules in millimeter size-range have been shown to be unable in protecting viable cells under such conditions (Leverrier et al., 2005; Trindade and Grosso, 2000).

Some authors particularly studied the influence of the capsule size on the protective effect. In this case the protective effect was found to be increased with increasing capsule size in a range between 0.2 and 3 mm (Chandramouli et al., 2004; Lee and Heo, 2000; Muthukumarasamy et al., 2006).

Regarding the influence of alginate concentration and therefore the resulting network density on the protective effect, contradictory results were found. Le-Tien et al. (2004) report a protective effect due to microencapsulation during simulated gastric conditions, which was not significantly different for 1.5% or 2.5% alginate as matrix material. Chen et al. (2005) found an application of 1% sodium alginate leading to an even higher protective effect compared to 2% or 3% alginate.

In contrast to this, some authors found an increasing protective effect under simulated gastric conditions with an increasing alginate concentration (Chandramouli et al., 2004; Mandal et al., 2006).

From the above displayed results it becomes clear that a general statement about the suitability of an alginate gel matrix as a protective barrier against a strongly acidic environment is not possible. Furthermore, the influence of capsule size and gel network density remains unclear. A possible explanation for this phenomenon could be that the variation of experimental simulated gastric conditions and the strain-dependent acid sensitivity dominates over the general improvement of alginate encapsulated cells in comparison to free cells.

The failure regarding a protective effect of alginate during incubation under simulated gastric conditions was explained by the porosity of the alginate gel, allowing the diffusion of H^+-ions into the gel, thus affecting the cells (Trindade and Grosso, 2000). Le-Tien et al. (2004) encapsulated a pH-sensitive color indicator in 3 mm alginate beads from 1.5% or 2.5% alginate solutions. The authors found that after incubation in simulated gastric juice with a pH of 1.5 the internal capsule-pH went below a value of 2 after approximately 8 min incubation time, independent of the used alginate concentration. The authors thus concluded that alginate gels have only a limited buffering capacity, and the gel structure can be seen as a large-porous hydrogel, allowing the diffusion of H^+-ions into the gel almost freely. As a result of these sometimes unsatisfactory results, several strategies have been applied to increase the protective effect of alginate encapsulation.

13.4.3.2 Additional Coating of the Capsules

The highly porous alginate network found in bacteria-loaded microcapsules may explain the often limited protective effects found under simulated gastric conditions, and it is therefore suggested that manipulating the alginate gel toward a less porous structure would lead to an increased protective effect (Allan-Wojtas et al., 2008). This can be achieved by coating, which includes the building of an additional membrane layer on the capsule surface. Poly-*l*-lysin and chitosan are commonly used as coating materials, because they form strong complexes with alginates.

However, in the case of poly-*l*-lysine no significant difference regarding the protective effect was found by comparison of capsules with and without additional coating (Cui et al., 2000; Krasaekoopt et al., 2004).

In contrast to this, coating with chitosan led to an enhancement of the protective effect. An improved survival over uncoated or poly-*l*-lysin-coated microcapsules during simulated gastric treatment was found in several investigations (Iyer and Kailasapathy, 2005; Krasaekoopt et al., 2004; Lee et al., 2004).

13.4.3.3 Other Polysaccharide-Based Matrix Materials

Until today alginate is the most frequently used polymer for probiotic encapsulation. However, in some cases alternatives, such as gellan-xanthan were used instead. In this case a mixture of

0.75% gellan-gum with 1% xanthan-gum was mixed with the probiotic cell concentrate and gelled by contact with a $CaCl_2$ solution during the encapsulation process. Some authors found a protective effect in such capsules under simulated gastric conditions (McMaster et al., 2005a; Sun and Griffiths, 2000), but in a study of Leverrier et al. (2005) no increase in probiotic cell survival was found. Muthukumarasamy et al. (2006) investigated microcapsules from 0.5% gellan with 1% xanthan solutions, 1.75% κ-carageenan with 0.75% locust bean gum solutions or solely 3% alginate solutions. The authors found that alginate capsules generally offer a more enhanced protective effect during incubation at simulated gastric conditions compared to the other polymers used.

The use of other polysaccharides besides alginate is only sporadic, but from the available results it seems rather unlikely that superior protective characteristics are achievable.

13.4.3.4 Protein-Based Microcapsules

The application of food-protein-based matrix materials instead of polysaccharide polymers is a relatively new strategy and can be seen as an alternative approach to the primarily used hydrocolloids. In this case, the implementation of alternative gelling mechanisms during the encapsulation process was often necessary, as outlined below.

By using protein-based microcapsules such as Ca^{2+}-induced gelation of alginate–pectin–whey protein mixtures (Guerin et al., 2003), transglutaminase-induced gels from concentrated 15% caseinate solutions (Heidebach et al., 2009b), alginate-coated gelatine microcapsules generated by using the natural cross-linker genipin from plants (Annan et al., 2008), or rennet-induced gelation of a 35% skim-milk concentrate (Heidebach et al., 2009a), an increased survival under simulated gastric conditions was found.

In all the above-mentioned studies, the higher survival of encapsulated cells compared to free cells after incubation under simulated gastric conditions was explained by the buffering capacity of the proteinaceous gel matrix. Hebrard et al. (2006), who encapsulated a recombinant yeast in whey protein-based microcapsules, came to a similar conclusion that a protein matrix could exert a buffering effect toward low pH or attack of pepsin during simulated gastric digestion. In contrast to the commonly used hydrocolloids, even highly concentrated aqueous solutions from many proteins still have a relatively low viscosity and can therefore be used as precursor for microencapsulation. This enables the formation of microcapsules with a high-density gel network, supporting the idea of a protective barrier between the sensitive core material and the surrounding.

13.5 SURVIVAL AND CELL RELEASE DURING SIMULATED INTESTINAL CONDITIONS

Simulated intestinal conditions have shown to be generally less detrimental to probiotic cells, compared to simulated gastric conditions at low pH values. Sometimes even an increase in CFU was found during incubation under simulated intestinal conditions of encapsulated and respective free probiotic cells (Annan et al., 2008; Guerin et al., 2003; Picot and Lacroix, 2004). Also, in some cases no decrease of living cell numbers was found, regardless of whether free or encapsulated cells were used (Favaro-Trindade and Grosso, 2002; Muthukumarasamy et al., 2006; Reid et al., 2005; Trindade and Grosso, 2000). However, in cases where the cell count was reduced, most authors found a lesser reduction for microencapsulated cells (Chandramouli et al., 2004; Ding and Shah, 2007; Iyer and Kailasapathy, 2005; Kim et al., 2008; Krasaekoopt et al., 2004; Lee and Heo, 2000; Mandal et al., 2006; McMaster et al., 2005a; Song et al., 2003). Only a few authors found no protective effect due to microencapsulation (Leverrier et al., 2005; Reid et al., 2005; Sultana et al., 2000).

However, to effectively use probiotic microcapsules in final food products, it must be ensured that the matrix materials on the one hand not only provide the desired barrier effect during storage in the food and under acidic conditions in the stomach but on the other hand are also digestible whereby the probiotic cells are released in the human gut. This release is an essential step for colonization of the probiotic cells in the human intestinal tract, and therefore a mandatory prerequisite for the successful application of probiotics in functional foods.

Since the pore size of alginate gels is small enough to retain microorganisms within the gel network (Allan-Wojtas et al., 2008), a quantitative release is only achieved due to complete degradation of the gel matrix during digestion. On this account some authors studied the cell release from alginate-based microcapsules under various *in vitro* simulated intestinal conditions. Several authors found a complete release of probiotic cells during incubation in simulated intestinal juice (Iyer et al., 2004; Mandal et al., 2006; Shah and Ravula, 2000). A more delayed release of probiotic cells was found with chitosan-coated alginate microcapsules, due to the additional coating layer (Iyer et al., 2005). In contrast to this, Urbanska et al. (2007) found that chitosan-coated alginate capsules, containing *Lactobacillus acidophilus*, were not dissolved after incubation in simulated intestinal fluid for 24 h. Cui et al. (2000) assessed that poly-*l*-lysine-coated alginate microcapsules were completely dissolved within 12 h in simulated intestinal juice, accompanied by a complete cell release. In contrast to this, Martoni et al. (2007) found poly-*l*-lysine-coated alginate capsules to be physically stable after incubation in simulated intestinal fluid for 10 h.

However, *in vivo* trials regarding the digestibility of alginate capsules reveal a higher stability as expected from the *in vitro* trials. Hoad et al. (2008) investigated the fate of alginate microcapsules from 1.5% alginate solutions in the human intestinal tract after ingestion *in vivo,* by means of magnetic resonance imaging. The physically intact capsules were still visible in the human intestine 4 h after ingestion, while they were dissolved in a parallel *in vitro* trial (Rayment et al., 2009). The authors therefore concluded that *in vitro* trials do not ideally reflect the conditions in the human intestinal tract. Also, chitosan- or poly-*l*-lysine-coated alginate microcapsules fed to mice were found intact in the intestine of the mouse 6 h after ingestion, with the chitosan-coated capsules showing a higher mechanical strength (Lin et al., 2008). Furthermore, Van Venrooy (2004) used six different alginate preparations, varying in molecular weight and viscosity, to produce probiotic microcapsules. After feeding the microcapsules to pigs, it was found that the capsules were generally excreted undigested, independent of the type of alginate used.

Alginate is extracted from the cell walls of marine algae and serves as a carbohydrate source for various marine molluscs (Gacesa, 1992). However, it is indigestible for humans, showing several characteristics of a dietary fiber (Brownlee et al., 2005). From these results it becomes clear that further investigations regarding the release characteristic of alginate capsules *in vivo* are necessary to ensure an effective application of microencapsulated probiotic cells.

13.6 DEVELOPMENT OF NOVEL APPROACHES FOR PROBIOTIC MICROENCAPSULATION

The available technologies for probiotic microencapsulation are generally limited to food grade materials. Furthermore, even within these limitations many factors must be considered to achieve a successful application. Figure 13.5 shows some of the most important tasks that should be considered as commented below.

I: The amount of living probiotic cells required to achieve a health effect due to ingestion of probiotic food is different for each application and probiotic strain. However, several authors declare 10^6–10^7 CFU of probiotic cells per gram functional food product as a recommended target value (Anal and Singh, 2007; Champagne et al., 2005). In most of the studies regarding the addition of microencapsulated probiotics to several foods, addition levels of 1%–2% capsules were applied (Gobbetti et al., 1998; Lahtinen et al., 2007; Lemay et al., 2002; Muthukumarasamy and Holley, 2006; Reid et al., 2007). Therefore, the capsules should contain at least 10^8–10^9 CFU of probiotic cells per gram to match the desired amount, if no cells are lost during storage.

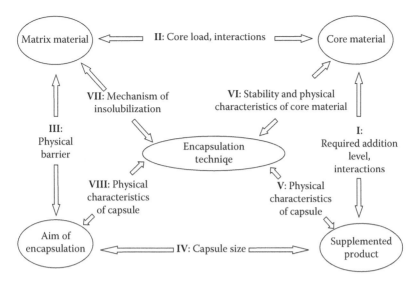

FIGURE 13.5   Important influencing factors regarding the microencapsulation of probiotic cells.

II: The core load, i.e., the concentration of culturable cells, of the resulting capsules must be kept as high as possible to reduce the required addition level, which should be as low as possible, to reduce the likeliness of a negative sensorial impact on the food (McMaster et al., 2005b). Since probiotic-containing microcapsules for food applications are generally not intended to be exposed to a fermentation process accompanied by cell growth within the capsules in most of the cases, a high core load can only be achieved by using a highly concentrated probiotic cell suspension as core material precursor. In this context it must also be considered that dilution steps, as well as a loss in living cell numbers during the encapsulation process itself, can significantly reduce the concentration of living probiotic cells in the final capsules.

   The choice of a suitable matrix material can be driven by favorable interactions between the core and the matrix. In case of probiotic cells, addition of protectants, prebiotics, or even a prebiotic matrix could lead to advanced capsule systems. Such applications could be the use of porous starch granules, filled with probiotics (Puupponen-Pimia et al., 2002), or a dense matrix, based on skim-milk solids (Heidebach et al., 2009a), as well as whey protein-based microcapsules with a high nutritional and nutraceutical value (Reid et al., 2007).

III: The generation of a protective effect mainly depends on the ability of the matrix material to provide a physical barrier against adverse conditions. From this point of view, a dense matrix with a small pore size, containing functional groups that provide oxygen-scavenging (Crittenden et al., 2006; Talwalkar and Kailasapathy, 2003) or buffering properties could be advantageous (Le-Tien et al., 2004). Recently, coating of microcapsules with hydrophobic stearic acid was successfully performed by Sabikhi et al. (2008), to render the capsule more resistant toward penetration of acid.

IV: For food applications, the average diameter of the resulting microcapsules is one of the most significant bottlenecks. On the one side, larger capsule diameters and hence a higher volume-to-surface ratio generally increase the likeliness of a protecting effect (Anal and Singh, 2007). Furthermore an additional coating of microcapsules significantly increases the capsules size (Krasaekoopt et al., 2004). On the other side, the capsules must be sufficiently small to avoid a negative sensorial impact on the food product (Champagne and Fustier, 2007). This conflict of targets causes the need to examine the acceptable capsule size for each food individually. In this context, foods exhibiting a structure which is naturally associated with coarseness or even gelled foods, such as cheese, are capable of larger capsule sizes, compared to yogurt, sour cream, or milk, where capsules in size-ranges between 1 and 3 mm have been shown to adversely affect the mouth feel (Hansen et al., 2002).

V: A reasonable strategy is to adjust the applied matrix material and encapsulation technology toward the food the capsules are added to.

For example, most of the probiotic foods available today are dairy products. A higher consumer acceptance for dairy-based microcapsules compared to those of nondairy origin in such products seems likely (Chen et al., 2006). Furthermore, possible separation problems caused by inappropriate capsule characteristics must be considered, as mentioned by Lahtinen et al. (2007), who used crystallized fat-based matrices to create microcapsules, subsequently stored in oat drinks. As well, components of the food matrix, such as high salt concentrations, can support an undesired disintegration of the microcapsules during storage (Kailasapathy and Masondole, 2005).

VI: Several authors have shown that the encapsulation process itself can be detrimental to the cells (Picot and Lacroix, 2004; Reid et al., 2005). In this context, the nature of the gelling mechanism during the encapsulating process seems to be one of the key factors for a successful encapsulation of living probiotic bacteria. Thus, novel approaches should target on gentle gelation mechanisms, in order to avoid detrimental effects on the sensitive core material. A useful gelling mechanism recently suggested could be the cold-set gelation of whey-protein solutions (Chen et al., 2006; Hebrard et al., 2006). Also the use of enzymes as natural biocatalysts (Heidebach et al., 2009a,b), nontoxic cross-linking agents from plants, such as genipin (Annan et al., 2008), or pre-denatured protein–saccharide mixtures (Crittenden et al., 2006) enable the formation of gels under mild circumstances.

VII: When alginate capsules are produced by the emulsion technology the capsules are hardened by slowly adding calcium chloride to the alginate solution/oil emulsion during stirring, in most of the cases. By the time the calcium solution contacts the dispersed alginate phase, gelling occurs. Thus, the gelation kinetic is inhomogeneous, which often leads to capsules of irregular shape (Cui et al., 2000; Muthukumarasamy et al., 2006; Sheu and Marshall, 1993). In contrast to this, internal gelation processes triggered by thermal or enzymatic reaction lead to spherical capsules (Heidebach et al., 2009b), which is generally favored because of the reduced exposure of surface compared to irregular-shaped particles.

VIII: In addition to the matrix material, the chosen encapsulation technique determines the physical characteristics of the resulting capsules. While spray-drying processes are relatively cheap and comparably small capsules are created, as a disadvantage they are mostly water soluble. In contrast to this, predominantly large, water-insoluble capsules are generated by the extrusion method. In comparison, the emulsion technique has the advantage that smaller capsules can be created. However, the resulting capsules must be separated from oil which is more laborious, instead of separation from an aqueous hardening solution (Muthukumarasamy et al., 2006). Future approaches should therefore target toward encapsulation processes that selectively combine the advantages of different encapsulation techniques. In this context, a promising strategy could be the application of water-insoluble microcapsules from spray-drying, based on proteins (Picot and Lacroix, 2004) or even alginate and carageenan (Burey et al., 2009). It would also be advantageous if the emulsion technique could be brought forward to water-in-water emulsions, a phenomena caused by thermodynamic incompatibility of protein–polysaccharide mixtures. Syrbe et al. (1998) already demonstrated the feasibility to create whey-protein-based microgel particles by means of this method.

ABBREVIATIONS

CFU   Colony forming units
RH   Relative humidity

REFERENCES

Adhikari, K., Mustapha, A., Grun, I. U., and Fernando, L. 2000. Viability of microencapsulated bifidobacteria in set yogurt during refrigerated storage. *Journal of Dairy Science* 83(9):1946–1951.

Adhikari, K., Mustapha, A., and Grun, I. U. 2003. Survival and metabolic activity of microencapsulated *Bifidobacterium longum* in stirred yogurt. *Journal of Food Science* 68(1):275–280.

Agrawal, R. 2005. Probiotics: An emerging food supplement with health benefits. *Food Biotechnology* 19(3):227–246.

Allan-Wojtas, P., Hansen, L. T., and Paulson, A. T. 2008. Microstructural studies of probiotic bacteria-loaded alginate microcapsules using standard electron microscopy techniques and anhydrous fixation. *LWT— Food Science and Technology* 41(1):101–108.

Anal, A. K. and Singh, H. 2007. Recent advances in microencapsulation of probiotics for industrial applications and targeted delivery. *Trends in Food Science & Technology* 18(5):240–251.

Ananta, E., Volkert, M., and Knorr, D. 2005. Cellular injuries and storage stability of spray-dried *Lactobacillus rhamnosus* GG. *International Dairy Journal* 15(4):399–409.

Annan, N. T., Borza, A. D., and Hansen, L. T. 2008. Encapsulation in alginate-coated gelatin microspheres improves survival of the probiotic *Bifidobacterium adolescentis* 15703T during exposure to simulated gastro-intestinal conditions. *Food Research International* 41(2):184–193.

Augustin, M. A. 2003. The role of microencapsulation in the development of functional dairy foods. *Australian Journal of Dairy Technology* 58(2):156–160.

Boylston, T. D., Vinderola, C. G., Ghoddusi, H. B., and Reinheimer, J. A. 2004. Incorporation of bifidobacteria into cheeses: Challenges and rewards. *International Dairy Journal* 14(5):375–387.

Brownlee, I. A., Allen, A., Pearson, J. P., Dettmar, P. W., Havler, M. E., Atherton, M. R., and Onsoyen, E. 2005. Alginate as a source of dietary fiber. *Critical Reviews in Food Science and Nutrition* 45(6):497–510.

Burey, P., Bhandari, B., Howes, T., and Gidley, M. J. 2009. Gel particles from spray-dried disordered polysaccharides. *Carbohydrate Polymers*, 76(2):206–213.

Capela, P., Hay, T. K. C., and Shah, N. P. 2006. Effect of cryoprotectants, prebiotics and microencapsulation on survival of probiotic organisms in yoghurt and freeze-dried yoghurt. *Food Research International* 39(2):203–211.

Capela, P., Hay, T. K. C., and Shah, N. P. 2007. Effect of homogenisation on bead size and survival of encapsulated probiotic bacteria. *Food Research International* 40(10):1261–1269.

Castro, H. P., Teixeira, P. M., and Kirby, R. 1995. Storage of lyophilized cultures of *Lactobacillus bulgaricus* under different relative humidities and atmospheres. *Applied Microbiology and Biotechnology* 44(1–2):172–176.

Champagne, C. P. and Fustier, P. 2007. Microencapsulation for the improved delivery of bioactive compounds into foods. *Current Opinion in Biotechnology* 18(2):184–190.

Champagne, C. P. and Gardner, N. 2001. The effect of protective ingredients on the survival of immobilized cells of *Streptococcus thermophilus* to air and freeze-drying. *Electronic Journal of Biotechnology* 4(3):146–152.

Champagne, C. P., Morin, N., Couture, R., Gagnon, C., Jelen, P., and Lacroix, C. 1992. The potential of immobilized cell technology to produce freeze-dried, phage-protected cultures of *Lactococcus lactis*. *Food Research International* 25(6):419–427.

Champagne, C. P., Lacroix, C., and Sodinigallot, I. 1994. Immobilized cell technologies for the dairy-industry. *Critical Reviews in Biotechnology* 14(2):109–134.

Champagne, C. P., Mondou, F., Raymond, Y., and Brochu, E. 1996. Effect of Immobilization in alginate on the stability of freeze-dried *Bifidobacterium longum*. *Bioscience Microflora* 15(1):9–15.

Champagne, C. P., Gardner, N. J., and Roy, D. 2005. Challenges in the addition of probiotic cultures to foods. *Critical Reviews in Food Science and Nutrition* 45(1):61–84.

Chandramouli, V., Kailasapathy, K., Peiris, P., and Jones, M. 2004. An improved method of microencapsulation and its evaluation to protect *Lactobacillus* spp. in simulated gastric conditions. *Journal of Microbiological Methods* 56(1):27–35.

Chen, K. N., Chen, M. J., Liu, J. R., Lin, C. W., and Chiu, H. Y. 2005. Optimization of incorporated prebiotics as coating materials for probiotic microencapsulation. *Journal of Food Science* 70(5):M260–M266.

Chen, L. Y., Remondetto, G. E., and Subirade, M. 2006. Food protein-based materials as nutraceutical delivery systems. *Trends in Food Science & Technology* 17(5):272–283.

Crittenden, R., Weerakkody, R., Sanguansri, L., and Augustin, M. 2006. Synbiotic microcapsules that enhance microbial viability during nonrefrigerated storage and gastrointestinal transit. *Applied and Environmental Microbiology* 72(3):2280–2282.

Cui, J. H., Goh, J. S., Kim, P. H., Choi, S. H., and Lee, B. J. 2000. Survival and stability of bifidobacteria loaded in alginate poly-*l*-lysine microparticles. *International Journal of Pharmaceutics* 210(1–2):51–59.

Desai, A. R., Powell, I. B., and Shah, N. P. 2004. Survival and activity of probiotic lactobacilli in skim milk containing prebiotics. *Journal of Food Science* 69(3):M57–M60.

Desai, K. G. H. and Park, H. J. 2005. Recent developments in microencapsulation of food ingredients. *Drying Technology* 23(7):1361–1394.

Desmond, C., Ross, R. P., O'Callaghan, E., Fitzgerald, G., and Stanton, C. 2002. Improved survival of *Lactobacillus paracasei* NFBC 338 in spray-dried powders containing gum acacia. *Journal of Applied Microbiology* 93(6):1003–1011.

Dinakar, P. and Mistry, V. V. 1994. Growth and viability of *Bifidobacterium bifidum* in cheddar cheese. *Journal of Dairy Science* 77(10):2854–2864.

Ding, W. K. and Shah, N. P. 2007. Acid, bile, and heat tolerance of free and microencapsulated probiotic bacteria. *Journal of Food Science* 72(9):M446–M450.

Doleyres, Y. and Lacroix, C. 2005. Technologies with free and immobilised cells for probiotic bifidobacteria production and protection. *International Dairy Journal* 15(10):973–988.

Favaro-Trindade, C. S. and Grosso, C. R. F. 2002. Microencapsulation of *L. acidophilus* (La-05) and *B. lactis* (Bb-12) and evaluation of their survival at the pH values of the stomach and in bile. *Journal of Microencapsulation* 19(4):485–494.

Gacesa, P. 1992. Enzymatic degradation of alginates. *International Journal of Biochemistry* 24(4):545–552.

Gardiner, G. E., O'Sullivan, E., Kelly, J., Auty, M. A. E., Fitzgerald, G. F., Collins, J. K., Ross, R. P., and Stanton, C. 2000. Comparative survival rates of human-derived probiotic *Lactobacillus paracasei* and *L. salivarius* strains during heat treatment and spray drying. *Applied and Environmental Microbiology* 66(6):2605–2612.

Gibbs, B. F., Kermasha, S., Alli, I., and Mulligan, C. N. 1999. Encapsulation in the food industry: A review. *International Journal of Food Sciences and Nutrition* 50(3):213–224.

Gobbetti, M., Corsetti, A., Smacchi, E., Zocchetti, A., and De Angelis, M. 1998. Production of Crescenza cheese by incorporation of bifidobacteria. *Journal of Dairy Science* 81(1):37–47.

Godward, G. and Kailasapathy, K. 2003. Viability and survival of free and encapsulated probiotic bacteria in cheddar cheese. *Milchwissenschaft—Milk Science International* 58(11–12):624–627.

Guerin, D., Vuillemard, J. C., and Subirade, M. 2003. Protection of bifidobacteria encapsulated in polysaccharide-protein gel beads against gastric juice and bile. *Journal of Food Protection* 66(11):2076–2084.

Hansen, L. T., Ian-Wojtas, P. M., Jin, Y. L., and Paulson, A. T. 2002. Survival of Ca-alginate microencapsulated *Bifidobacterium* spp. in milk and simulated gastrointestinal conditions. *Food Microbiology* 19(1):35–45.

Hebrard, G., Blanquet, S., Beyssac, E., Remondetto, G., Subirade, M., and Alric, M. 2006. Use of whey protein beads as a new carrier system for recombinant yeasts in human digestive tract. *Journal of Biotechnology* 127(1):151–160.

Heidebach, T., Först, P., and Kulozik, U. 2009a. Microencapsulation of probiotic cells by means of rennet-gelation of milk proteins. *Food Hydrocolloids*, 23(7):1670–1677.

Heidebach, T., Först, P., and Kulozik, U. 2009b. Transglutaminase-induced caseinate gelation for the microencapsulation of probiotic cells. *International Dairy Journal* 19(2):77–84.

Higl, B., Kurtmann, L., Carlsen, C. U., Ratjen, J., Forst, P., Skibsted, L. H., Kulozik, U., and Risbo, J. 2007. Impact of water activity, temperature, and physical state on the storage stability of *Lactobacillus paracasei* ssp paracasei freeze-dried in a lactose matrix. *Biotechnology Progress* 23(4):794–800.

Hoad, C., Rayment, P., Cox, E., Wright, P., Butler, M., Spiller, R., and Gowland, P. 2008. Investigation of alginate beads for gastro-intestinal functionality, Part 2: *In vivo* characterisation. *Food Hydrocolloids* 23:833–839.

Homayouni, A., Azizi, A., Ehsani, M. R., Yarmand, M. S., and Razavi, S. H. 2008. Effect of microencapsulation and resistant starch on the probiotic survival and sensory properties of synbiotic ice cream. *Food Chemistry* 111(1):50–55.

Hussein, S. A. and Kebary, K. M. K. 1999. Improving viability of bifidobacteria by microentrapment and their effect on some pathogenic bacteria in stirred yoghurt. *Acta Alimentaria* 28(2):113–131.

Iyer, C. and Kailasapathy, K. 2005. Effect of co-encapsulation of probiotics with prebiotics on increasing the viability of encapsulated bacteria under *in vitro* acidic and bile salt conditions and in yogurt. *Journal of Food Science* 70(1):M18–M23.

Iyer, C., Kailasapathy, K., and Peiris, P. 2004. Evaluation of survival and release of encapsulated bacteria in *ex vivo* porcine gastrointestinal contents using a green fluorescent protein gene-labelled *E. coli*. *LWT—Food Science and Technology* 37(6):639–642.

Iyer, C., Phillips, M., and Kailasapathy, K. 2005. Release studies of *Lactobacillus casei* strain Shirota from chitosan-coated alginate-starch microcapsules in *ex vivo* porcine gastrointestinal contents. *Letters in Applied Microbiology* 41(6):493–497.

Jackson, L. S. and Lee, K. 1991. Microencapsulation and the food-industry. *LWT—Food Science and Technology* 24(4):289–297.

Jia, W., Li, H. K., Zhao, L. P., and Nicholson, J. K. 2008. Gut microbiota: A potential new territory for drug targeting. *Nature Reviews Drug Discovery* 7(2):123–129.

Kailasapathy, K. 2002. Microencapsulation of probiotic bacteria: Technology and potential applications. *Current Issues in Intestinal Microbiology* 3(2):39–48.

Kailasapathy, K. 2006. Survival of free and encapsulated probiotic bacteria and their effect on the sensory properties of yoghurt. *LWT—Food Science and Technology* 39(10):1221–1227.

Kailasapathy, K. and Masondole, L. 2005. Survival of free and microencapsulated *Lactobacillus acidophilus* and *Bifidobacterium lactis* and their effect on texture of feta cheese. *Australian Journal of Dairy Technology* 60(3):252–258.

Kailasapathy, K. and Rybka, S. 1997. *L. acidophilus* and *Bifidobacterium* spp.—Their therapeutic potential and survival in yogurt. *Australian Journal of Dairy Technology* 52(1):28–35.

Kearney, L., Upton, M., and Mcloughlin, A. 1990. Enhancing the viability of *Lactobacillus plantarum* inoculum by immobilizing the cells in calcium-alginate beads incorporating cryoprotectants. *Applied and Environmental Microbiology* 56(10):3112–3116.

Khalil, A. H. and Mansour, E. H. 1998. Alginate encapsulated bifidobacteria survival in mayonnaise. *Journal of Food Science* 63(4):702–705.

Kim, S. J., Cho, S. Y., Kim, S. H., Song, O. J., Shin, I. S., Cha, D. S., and Park, H. J. 2008. Effect of microencapsulation on viability and other characteristics in *Lactobacillus acidophilus* ATCC 43121. *LWT—Food Science and Technology* 41(3):493–500.

Kirby, C. 1991. Microencapsulation and controlled delivery of food ingredients. *Food Science and Technology Today* 5(2):74–78.

Knorr, D. 1998. Technology aspects related to microorganisms in functional foods. *Trends in Food Science & Technology* 9(8–9):295–306.

Krasaekoopt, W., Bhandari, B., and Deeth, H. 2003. Evaluation of encapsulation techniques of probiotics for yoghurt. *International Dairy Journal* 13(1):3–13.

Krasaekoopt, W., Bhandari, B., and Deeth, H. 2004. The influence of coating materials on some properties of alginate beads and survivability of microencapsulated probiotic bacteria. *International Dairy Journal* 14(8):737–743.

Krasaekoopt, W., Bhandari, B., and Deeth, H. C. 2006. Survival of probiotics encapsulated in chitosan-coated alginate beads in yoghurt from UHT- and conventionally treated milk during storage. *LWT—Food Science and Technology* 39(2):177–183.

Kushal, R., Anand, S. K., and Chander, H. 2006. *In vivo* demonstration of enhanced probiotic effect of co-immobilized *Lactobacillus acidophilus* and *Bifidobacterium bifidum*. *International Journal of Dairy Technology* 59(4):265–271.

Lahtinen, S. J., Ouwehand, A. C., Salminen, S. J., Forssell, P., and Myllarinen, P. 2007. Effect of starch- and lipid-based encapsulation on the culturability of two *Bifidobacterium longum* strains. *Letters in Applied Microbiology* 44(5):500–505.

Le-Tien, C., Millette, M., Mateescu, M. A., and Lacroix, M. 2004. Modified alginate and chitosan for lactic acid bacteria immobilization. *Biotechnology and Applied Biochemistry* 39:347–354.

Lee, J. S., Cha, D. S., and Park, H. J. 2004. Survival of freeze-dried *Lactobacillus bulgaricus* KFRI 673 in chitosan-coated calcium alginate microparticles. *Journal of Agricultural and Food Chemistry* 52(24):7300–7305.

Lee, K. Y. and Heo, T. R. 2000. Survival of *Bifidobacterium longum* immobilized in calcium alginate beads in simulated gastric juices and bile salt solution. *Applied and Environmental Microbiology* 66(2):869–873.

Lemay, M. J., Champagne, C. P., Gariepy, C., and Saucier, L. 2002. A comparison of the effect of meat formulation on the heat resistance of free or encapsulated cultures of *Lactobacillus sakei*. *Journal of Food Science* 67(9):3428–3434.

Leverrier, P., Fremont, Y., Rouault, A., Boyaval, P., and Jan, G. 2005. *In vitro* tolerance to digestive stresses of propionibacteria: Influence of food matrices. *Food Microbiology* 22(1):11–18.

Lian, W. C., Hsiao, H. C., and Chou, C. C. 2002. Survival of bifidobacteria after spray-drying. *International Journal of Food Microbiology* 74(1–2):79–86.

Lian, W. C., Hsiao, H. C., and Chou, C. C. 2003. Viability of microencapsulated bifidobacteria in simulated gastric juice and bile solution. *International Journal of Food Microbiology* 86(3):293–301.

Lin, J. Z., Yu, W. T., Liu, X. D., Xie, H. G., Wang, W., and Ma, X. J. 2008. *In vitro* and *in vivo* characterization of alginate-chitosan-alginate artificial microcapsules for therapeutic oral delivery of live bacterial cells. *Journal of Bioscience and Bioengineering* 105(6):660–665.

Liserre, A. M., Re, M. I., and Franco, B. D. G. M. 2007. Microencapsulation of *Bifidobacterium animalis* subsp *lactis* in modified alginate-chitosan beads and evaluation of survival in simulated gastrointestinal conditions. *Food Biotechnology* 21(1–2):1–16.

Lopez-Rubio, A., Gavara, R., and Lagaron, J. A. 2006. Bioactive packaging: Turning foods into healthier foods through biomaterials. *Trends in Food Science & Technology* 17(10):567–575.

Lourens-Hattingh, A. and Viljoen, B. C. 2001. Yogurt as probiotic carrier food. *International Dairy Journal* 11(1–2):1–17.

Mandal, S., Puniya, A. K., and Singh, K. 2006. Effect of alginate concentrations on survival of microencapsulated *Lactobacillus casei* NCDC-298. *International Dairy Journal* 16(10):1190–1195.

Maragkoudakis, P. A., Miaris, C., Rojez, P., Manalis, N., Magkanari, F., Kalantzopoulos, G., and Tsakalidou, E. 2006. Production of traditional Greek yoghurt using *Lactobacillus* strains with probiotic potential as starter adjuncts. *International Dairy Journal* 16(1):52–60.

Martoni, C., Bhathena, J., Jones, M. L., Urbanska, A. M., Chen, H., and Prakash, S. 2007. Investigation of microencapsulated BSH active *Lactobacillus* in the simulated human GI tract. *Journal of Biomedicine and Biotechnology*, doi: 10.1155/2007/13684.

Mattila-Sandholm, T., Myllarinen, P., Crittenden, R., Mogensen, G., Fonden, R., and Saarela, M. 2002. Technological challenges for future probiotic foods. *International Dairy Journal* 12(2–3):173–182.

McMaster, L. D., Kokott, S. A., Reid, S. J., and Abratt, V. 2005a. Use of traditional African fermented beverages as delivery vehicles for *Bifidobacterium lactis* DSM 10140. *International Journal of Food Microbiology* 102(2):231–237.

McMaster, L. D., Kokott, S. A., and Slatter, P. 2005b. Micro-encapsulation of *Bifidobacterium lactis* for incorporation into soft foods. *World Journal of Microbiology & Biotechnology* 21(5):723–728.

Meng, X. C., Stanton, C., Fitzgerald, G. F., Daly, C., and Ross, R. P. 2008. Anhydrobiotics: The challenges of drying probiotic cultures. *Food Chemistry* 106(4):1406–1416.

Muthukumarasamy, P. and Holley, R. A. 2006. Microbiological and sensory quality of dry fermented sausages containing alginate-microencapsulated *Lactobacillus reuteri*. *International Journal of Food Microbiology* 111(2):164–169.

Muthukumarasamy, P., Ian-Wojtas, P., and Holley, R. A. 2006. Stability of *Lactobacillus reuteri* in different types of microcapsules. *Journal of Food Science* 71(1):M20–M24.

Naidu, A. S., Bidlack, W. R., and Clemens, R. A. 1999. Probiotic spectra of lactic acid bacteria (LAB). *Critical Reviews in Food Science and Nutrition* 39(1):13–126.

Nilsson, K., Birnbaum, S., Flygare, S., Linse, L., Schroder, U., Jeppsson, U., Larsson, P. O., Mosbach, K., and Brodelius, P. 1983. A general-method for the immobilization of cells with preserved viability. *European Journal of Applied Microbiology and Biotechnology* 17(6):319–326.

O'Riordan, K., Andrews, D., Buckle, K., and Conway, P. 2001. Evaluation of microencapsulation of a *Bifidobacterium* strain with starch as an approach to prolonging viability during storage. *Journal of Applied Microbiology* 91(6):1059–1066.

Ozer, B., Kirmaci, H. A., Senel, E., Atamer, M., and Hayaloglu, A. 2009. Improving the viability of *Bifidobacterium bifidum* BB-12 and *Lactobacillus acidophilus* LA-5 in white-brined cheese by microencapsulation. *International Dairy Journal* 19(1):22–29.

Picot, A. and Lacroix, C. 2004. Encapsulation of bifidobacteria in whey protein-based microcapsules and survival in simulated gastrointestinal conditions and in yoghurt. *International Dairy Journal* 14(6):505–515.

Poncelet, D., Lencki, R., Beaulieu, C., Halle, J. P., Neufeld, R. J., and Fournier, A. 1992. Production of alginate beads by emulsification internal gelation. 1. Methodology. *Applied Microbiology and Biotechnology* 38(1):39–45.

Prado, F. C., Parada, J. L., Pandey, A., and Soccol, C. R. 2008. Trends in non-dairy probiotic beverages. *Food Research International* 41(2):111–123.

Puupponen-Pimia, R., Aura, A. M., Oksman-Caldentey, K. M., Myllarinen, P., Saarela, M., Mattila-Sandholm, T., and Poutanen, K. 2002. Development of functional ingredients for gut health. *Trends in Food Science & Technology* 13(1):3–11.

Rayment, P., Wright, P., Hoad, C., Ciampi, E., Haydock, D., Gowland, P., and Butler, M. F. 2009. Investigation of alginate beads for gastro-intestinal functionality, Part 1: *In vitro* characterisation. *Food Hydrocolloids* 23(3):816–822.

Reid, A. A., Vuillemard, J. C., Britten, M., Arcand, Y., Farnworth, E., and Champagne, C. P. 2005. Microentrapment of probiotic bacteria in a Ca2+-induced whey protein gel and effects on their viability in a dynamic gastro-intestinal model. *Journal of Microencapsulation* 22(6):603–619.

Reid, A. A., Champagne, C. P., Gardner, N., Fustier, P., and Vuillemard, J. C. 2007. Survival in food systems of *Lactobacillus rhamnosus* R011 microentrapped in whey protein gel particles. *Journal of Food Science* 72(1):M31–M37.

Rodgers, S. 2008. Novel applications of live bacteria in food services: Probiotics and protective cultures. *Trends in Food Science & Technology* 19(4):188–197.

Ross, R. P., Desmond, C., Fitzgerald, G. F., and Stanton, C. 2005. Overcoming the technological hurdles in the development of probiotic foods. *Journal of Applied Microbiology* 98(6):1410–1417.

Sabikhi, L., Babu, R., Thompkinson, D. K., and Kapila, S. 2008. Resistance of microencapsulated *Lactobacillus acidophilus* LA1 to processing treatments and simulated gut conditions. *Food Bioprocess Technology*, doi:10.1007/s11947-008-0135-1.

Santivarangkna, C., Higl, B., and Foerst, P. 2008. Protection mechanisms of sugars during different stages of preparation process of dried lactic acid starter cultures. *Food Microbiology* 25(3):429–441.

Schrooyen, P. M. M., van der Meer, R., and De Kruif, C. G. 2001. Microencapsulation: Its application in nutrition. *Proceedings of the Nutrition Society* 60(4):475–479.

Selmer-Olsen, E., Sorhaug, T., Birkeland, S. E., and Pehrson, R. 1999. Survival of *Lactobacillus helveticus* entrapped in Ca-alginate in relation to water content, storage and rehydration. *Journal of Industrial Microbiology & Biotechnology* 23(2):79–85.

Shah, N. P. 2000. Probiotic bacteria: Selective enumeration and survival in dairy foods. *Journal of Dairy Science* 83(4):894–907.

Shah, N. P. and Ravula, R. R. 2000. Microencapsulation of probiotic bacteria and their survival in frozen fermented dairy desserts. *Australian Journal of Dairy Technology* 55(3):139–144.

Sheu, T. Y. and Marshall, R. T. 1993. Microentrapment of *Lactobacilli* in calcium alginate gels. *Journal of Food Science* 58(3):557–561.

Sheu, T. Y., Marshall, R. T., and Heymann, H. 1993. Improving survival of culture bacteria in frozen desserts by microentrapment. *Journal of Dairy Science* 76(7):1902–1907.

Siuta-Cruce, P. and Goulet, J. 2001. Improving probiotic survival rates. *Food Technology* 55(10):36–42.

Song, S. H., Cho, Y. H., and Park, J. 2003. Microencapsulation of *Lactobacillus casei* YIT 9018 using a microporous glass membrane emulsification system. *Journal of Food Science* 68(1):195–200.

Stanton, C., Gardiner, G., Meehan, H., Collins, K., Fitzgerald, G., Lynch, P. B., and Ross, R. P. 2001. Market potential for probiotics. *American Journal of Clinical Nutrition* 73(2):476S–483S.

Su, L. C., Lin, C. W., and Chen, M. J. 2007. Development of an oriental-style dairy product coagulated by microcapsules containing probiotics and filtrates from fermented rice. *International Journal of Dairy Technology* 60(1):49–54.

Sultana, K., Godward, G., Reynolds, N., Arumugaswamy, R., Peiris, P., and Kailasapathy, K. 2000. Encapsulation of probiotic bacteria with alginate-starch and evaluation of survival in simulated gastrointestinal conditions and in yoghurt. *International Journal of Food Microbiology* 62(1–2):47–55.

Sun, W. R. and Griffiths, M. W. 2000. Survival of bifidobacteria in yogurt and simulated gastric juice following immobilization in gellan-xanthan beads. *International Journal of Food Microbiology* 61(1):17–25.

Syrbe, A., Bauer, W. J., and Klostermeyer, N. 1998. Polymer science concepts in dairy systems—An overview of milk protein and food hydrocolloid interaction. *International Dairy Journal* 8(3):179–193.

Talwalkar, A. and Kailasapathy, K. 2003. Effect of microencapsulation on oxygen toxicity in probiotic bacteria. *Australian Journal of Dairy Technology* 58(1):36–39.

Tamime, A. Y., Saarela, M., Korslund Sondergaard, A., Mistry, V. V., and Shah, N. P. 2005. Production and maintenance of viability of probiotic micro-organisms in dairy products. In *Probiotic Dairy Products*, ed. A. Y. Tamime, pp. 39–71, Oxford, U.K.: Blackwell publishing.

Teixeira, P. C., Castro, M. H., Malcata, F. X., and Kirby, R. M. 1995. Survival of *Lactobacillus delbrueckii* ssp *bulgaricus* following spray-drying. *Journal of Dairy Science* 78(5):1025–1031.

To, B. C. S. and Etzel, M. R. 1997. Survival of *Brevibacterium linens* (ATCC 9174) after spray drying, freeze drying, or freezing. *Journal of Food Science* 62(1):167–170.

Trindade, C. S. F. and Grosso, C. R. F. 2000. The effect of the immobilisation of *Lactobacillus acidophilus* and *Bifidobacterium lactis* in alginate on their tolerance to gastrointestinal secretions. *Milchwissenschaft— Milk Science International* 55(9):496–499.

Urbanska, A. M., Bhathena, J., and Prakash, S. 2007. Live encapsulated *Lactobacillus acidophilus* cells in yogurt for therapeutic oral delivery: Preparation and *in vitro* analysis of alginate-chitosan microcapsules. *Canadian Journal of Physiology and Pharmacology* 85(9):884–893.

Van Venrooy, I. 2004. Mikroverkapselung von probiotischen Bakterien mit Hydokolloiden zur Stabilitäts-verbesserung und zum Colontargeting; Dissertation, Christian-Albrechts-Universität, Kiel, Germany.

Wang, Y. C., Yu, R. C., and Chou, C. C. 2004. Viability of lactic acid bacteria and bifidobacteria in fermented soymilk after drying, subsequent rehydration and storage. *International Journal of Food Microbiology* 93(2):209–217.

Zayed, G. and Roos, Y. H. 2004. Influence of trehalose and moisture content on survival of *Lactobacillus salivarius* subjected to freeze-drying and storage. *Process Biochemistry* 39(9):1081–1086.

Zhao, R. X., Sun, J. L., Torley, P., Wang, D. H., and Niu, S. Y. 2008. Measurement of particle diameter of *Lactobacillus acidophilus* microcapsule by spray drying and analysis on its microstructure. *World Journal of Microbiology & Biotechnology* 24(8):1349–1354.

14 Structuring of Emulsions by Tailoring the Composition of Crystalline Emulsifier

Adam Macierzanka and Halina Szeląg

CONTENTS

14.1 INTRODUCTION

The engineering of emulsion structures has attracted the interest of many research laboratories over the last few years due to increasing industrial demand for colloidal matrices that could be used as bio-microreactors in biotechnology, vehicles for delivering and controlled release of lipophilic (or hydrophilic) active compounds in pharmaceutical and cosmetic formulations, tunable systems for controlling texture, stability and rheological properties of food products, and many other well-defined and profitable applications. In this chapter, we have shown an example of how an overall structure of emulsion can be controlled by tailoring the emulsifier composition. The emulsifiers described here are fatty acid ester derivatives of glycerol and propylene glycol that have been synthesized in the presence of various fatty acid carboxylates. The emulsion preparation conditions and the fatty acid profile of the amphiphiles, the emulsifiers were composed of, enabled the preparation of the systems where a dispersed phase was stabilized by crystalline interfacial films. The type and the structure of emulsions were largely dependent on the type and the concentration of the carboxylates, as well as the type of the polyol used to synthesize emulsifiers.

14.2 PREPARATION OF ACYLGLYCEROL AND ACYLPROPYLENEGLYCOL EMULSIFIERS

Glycerol can be converted to many commodity chemicals by using methods such as oxidation, reduction, dehydration, halogenation, etherification, and esterification (Pagliaro and Rossi 2008, Zheng et al. 2008, Zhou et al. 2008). The direct esterification of this polyhydric alcohol with fatty acids can be used to synthesize mixtures of mono- and diacylglycerols (mono-diglycerides,

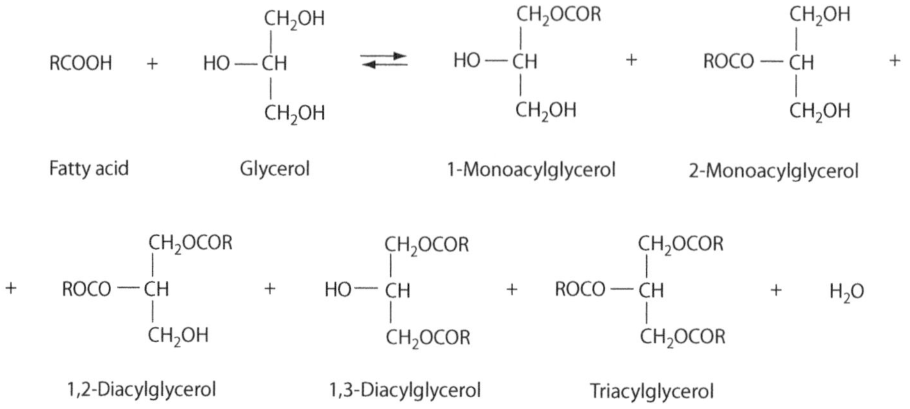

FIGURE 14.1 Reaction scheme for esterification of glycerol with fatty acid.

MAG-DAG; typically 40%–55% of MAG), Figure 14.1. The commercial mono-diglycerides always contains some amounts of triacylglycerols (TAG) and residues of fatty acids and glycerol. Upon further purification by molecular distillation, over 90% MAG product can be prepared (Szeląg and Zwierzykowski 1983, 1995, Zielinski 1997, Stauffer 2002). Both, mono-diglycerides and distilled MAGs are generally recognized as safe and can be used as emulsifiers and viscosity modifiers in pharmaceutical, food, cosmetic, and biotechnological industries to stabilize various dispersed systems. In many functional emulsions, such as some cosmetic creams and colloidal food products, MAGs are used in combination with another lipophilic emulsifier, propylene glycol monostearate (PGMS), in order to make stable α-crystalline blends of emulsifiers (Krog, 1999). Like mono-diglycerides, fatty acid mono-diesters of propylene glycol can be obtained by means of its direct esterification with fatty acids (Figure 14.2).

The type and the functional properties of dispersions are very much dependent on the amphiphilic properties of the emulsifiers. It has recently been shown that these can be modified be selective use of the polyol and the fatty-acid substrates of the esterification reaction as well as the type and concentration of other compounds in the reaction mixture (Szeląg and Zwierzykowski 1998, Szeląg and Macierzanka 2001, Macierzanka and Szeląg 2004, 2006). Fatty acid salts of sodium, potassium, and zinc were used to effectively emulsify the substrates. Those carboxylates have an ability to adsorb onto the polyol–fatty acid interface under the conditions of the esterification process and convert the reaction mixture into a microemulsion. As a result of the creation of huge

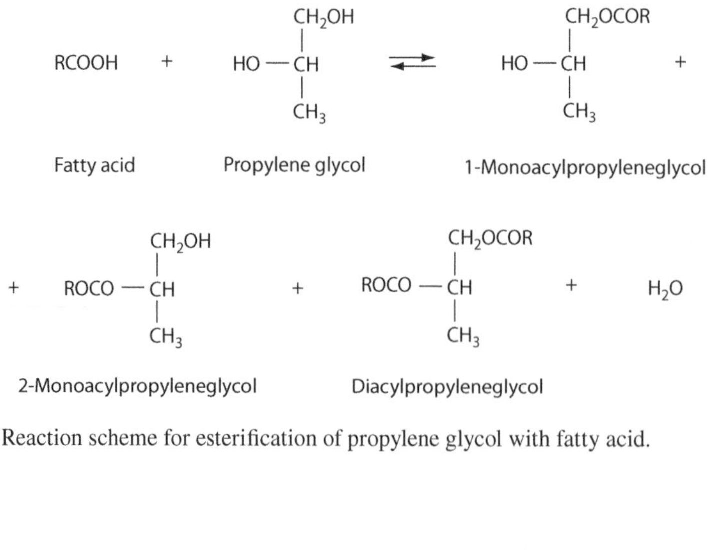

FIGURE 14.2 Reaction scheme for esterification of propylene glycol with fatty acid.

contact area between the substrates, their conversion to mono- and diglycerides was much faster than in uncatalyzed reactions. The final products of the esterification processes, carried out in the presence of various carboxylates, were used directly as emulsifiers. Since they contained contrasting carboxylates, hydrophobic zinc salts, or hydrophilic sodium (or potassium) salts, the production of vast range of emulsifiers with various, well-defined hydrophobic properties was possible (Table 14.1).

14.3 FORMATION AND PROPERTIES OF EMULSIONS

14.3.1 EFFECT OF THE EMULSIFIER NATURE

In this section, we have summarized our recent studies on the effect of acylglycerol (AG; mono-diglycerides) and acylpropyleneglycol (APG; fatty acid mono-diesters of propylene glycol) lipophilic emulsifiers, as well as their mixtures, on the structural transformations of emulsions, as observed during emulsion formation and storage (Macierzanka et al. 2009). The aim of the work was to identify the phenomena that can be encountered during the manufacturing and shelf life of emulsions prepared with these commonly used emulsifiers. Pathways of the structural transitions of emulsions are described below in detail for the example of the use of emulsifiers, which were produced in the absence of fatty acid carboxylates. The influence of carboxylates on the changes in emulsion microstructure and rheological properties will be presented in Section 14.3.2.

The phase diagrams formulated for emulsions stabilized by AG and APG emulsifiers as well as for the equivalent amounts of both emulsifiers (AG/APG mixed emulsifier) have been depicted in Figure 14.3. They show the dependence of emulsion type/structure on the temperature of emulsification, decreasing during the process, and the water:oil phase ratio. As one may expect for the use of strongly hydrophobic emulsifiers, water-in-oil (W/O) emulsions were produced at the beginning of emulsification. For AG and AG/APG, after cooling to, respectively, 59°C–58°C or 56°C–54.5°C, a sudden significant increase in fluidity of the systems was observed, associated with rapid destabilization of the emulsions (i.e., a massive separation of the water phase). The transition is thought to be caused by the spontaneous formation of the MAG–water or MAG/MAPG–water hydrophilic lamellar gel phase, which by entrapping of oil droplets led to the local phase inversion to the oil-in-water (O/W) type system. This process and subsequent changes of emulsion microstructure have been schematically shown in Figure 14.4.

The lamellar liquid-crystalline phases, produced with MAG, water and usually an admixture of other surfactants, have been found to be very effective stabilizers of O/W emulsions (Friberg et al. 1969, Friberg and Mandell 1970, Krog 1997, Batte et al. 2007a, Marangoni et al. 2007) due to their specific architecture comprising the bimolecular layers of emulsifier separated by layers of water. The lamellar structures are capable of encapsulating oil droplets. The MAG-based gels can be produced from the lamellar mesophase when the MAG–water systems are cooled below the Krafft temperature (the malting temperature of the hydrocarbon chains; Krog 1997), as explained in more detail further in the text. With this in mind, one may assume that the formation of the gel phase, caused by a gradual cooling of the emulsion systems, was a driving mechanism of the destabilization of the W/O emulsion and its inversion to the O/W system (Figure 14.3).

The emulsification processes were also repeated directly in a rheometer set-up, which gave an opportunity to measure *in situ* the apparent viscosity of the emulsions as they were being formed. The aforementioned destabilization of the emulsions based on AG and AG/APG emulsifiers was observed as a rapid decrease in viscosity (Figure 14.5b). This was in turn followed by an increase, attributed to the dispersion of the formed O/W emulsion gel in the continuous oil phase, for the systems with up to 55 wt.% water contents. The change was consistent with the onset of the crystallization of both AG and AG/APG emulsifiers (Figure 14.5a), so one may assume it was promoted by the interfacial solidification of the emulsifiers. The transition led to the formation of huge agglomerates of the O/W emulsion gel suspended in the continuous oil ("O/W+O" system). As the emulsification

TABLE 14.1
Composition of the Emulsifiers Synthesized by Means of the Esterification of Glycerol or Propylene Glycol with Fatty Acids in the Presence of Various Fatty Acid Carboxylates

| Components of the Esterification Reaction Mixture | | | Emulsifier Composition (wt.%) | | | | | | | | |
Polyol	Fatty Acid	Carboxylate (Metal/Fatty Acid Moiety)	Carboxylate	MAG	DAG+TAG	MAPG	DAPG	FA	G	PG	HLB
G	$C_{18:0}$	Without	N/A	43.0	31.5	N/A	N/A	15.3	10.2	N/A	4.0
G	$C_{18:0}$	$Zn/(C_{18:0})_2$	4.0	44.8	33.5	N/A	N/A	9.5	8.2	N/A	4.7
G	$C_{16:0}$	$Zn/(C_{16:0})_2$	4.0	45.3	33.0	N/A	N/A	9.5	8.2	N/A	4.9
G	$C_{14:0}$	$Zn/(C_{14:0})_2$	3.9	45.5	33.4	N/A	N/A	8.7	8.5	N/A	5.3
G	$C_{12:0}$	$Zn/(C_{12:0})_2$	3.8	46.8	31.9	N/A	N/A	9.3	8.2	N/A	5.7
G	$C_{16:0}-C_{18:0}$	$Na/C_{16:0}-C_{18:0}$	5.3	47.2	38.3	N/A	N/A	2.5	6.7	N/A	5.9
G	$C_{16:0}-C_{18:0}$	$Na/C_{16:0}-C_{18:0}$	8.0	47.5	34.0	N/A	N/A	4.4	6.1	N/A	6.3
G	$C_{16:0}-C_{18:0}$	$Na/C_{16:0}-C_{18:0}$	10.6	47.4	32.2	N/A	N/A	4.0	5.8	N/A	6.7
G	$C_{16:0}-C_{18:0}$	$K/C_{16:0}-C_{18:0}$	5.9	45.3	24.7	N/A	N/A	14.7	9.4	N/A	5.7
G	$C_{16:0}-C_{18:0}$	$K/C_{16:0}-C_{18:0}$	7.4	46.0	25.3	N/A	N/A	13.3	8.0	N/A	6.0
G	$C_{16:0}-C_{18:0}$	$K/C_{16:0}-C_{18:0}$	12.5	45.7	20.3	N/A	N/A	13.0	8.5	N/A	6.9
PG	$C_{18:0}$	Without	N/A	N/A	N/A	48.7	14.3	25.1	N/A	11.9	3.4
PG	$C_{18:0}$	$Zn/(C_{18:0})_2$	4.0	N/A	N/A	55.9	24.7	8.8	N/A	6.6	3.4

Notes: G, glycerol; PG, propylene glycol; MAG, monoacylglycerol; DAG, diacylglycerol; TAG, triacylglycerol; MAPG, monoacylpropyleneglycol (propylene glycol monoester); DAPG, diacylpropyleneglycol (propylene glycol diester); FA, fatty acid; HLB, hydrophile-lipophile balance.

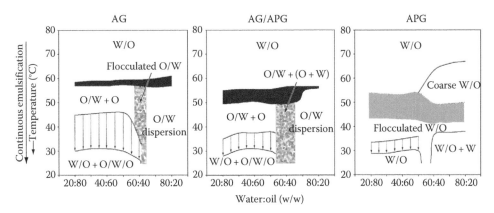

FIGURE 14.3 The phase diagrams of the water–oil systems prepared with AG, AG/APG, and APG emulsifiers. For AG and AG/APG, the black-colored areas show the temperature ranges at which a sudden decrease of the viscosity and a phase separation were observed during the emulsification. The dashed arrows correspond to the temperature ranges of the gradual shear-induced transition of "O/W+O" system to "W/O+O/W/O" emulsion (both systems coexisted at that stage). In the diagram obtained for APG emulsifier, the gray-colored area represents the temperature range of the progressive decrease in the fluidity of emulsion before the interfacial crystallization of the emulsifier and the flocculation of droplets took place. The dashed arrows correspond to the temperature range of the gradual shear-induced deflocculation of W/O emulsion. (Adapted from Macierzanka, A. et al., *Colloids Surf. A Physicochem. Eng. Aspects*, 334, 40, 2009. With permission from Elsevier.)

was continued under decreasing temperature, the emulsion gel was gradually dispersed in the oil. Ongoing decrease in a particle size as well as interfacial and bulk crystallizations of the components of the emulsifiers led to the formation of viscous emulsion of the complex "W/O+O/W/O" type; the O/W emulsion gel dispersed in the external oil phase formed droplets of the oil-in-water-in-oil (O/W/O) type, which coexisted with the water droplets, both covered by the crystallized films of emulsifier formed *in situ* at the interface (Figures 14.4a and 14.6). In the O/W/O droplets, the internal oil droplets were stabilized by the lamellar liquid crystals that formed layers oriented at the surface of the oil droplets. The droplets were immobilized by the gel matrix located between them (Figure 14.6b). The existence and location of lamellar liquid crystals were revealed in cross-polarized light as a birefringent pattern of the droplets (adsorbed lamellar phase; Macierzanka et al. 2009). For high water:oil proportions, the phase inversion led to the formation of an O/W dispersion, which had very poor creaming stability.

Different pathway of emulsification was observed when the APG emulsifier was applied solely (Figure 14.3). For the water contents up to 50 wt.%, the structural transitions of emulsions identified during the emulsification from high to low temperatures were as follows: (1) the initial formation of W/O emulsion for the temperature range above the level of the crystallization of APG emulsifier, then (2) the progressive decrease in the fluidity of emulsion, (3) the formation of huge flocs of water droplets, due to simultaneous interfacial crystallization of the emulsifier and shear-induced collisions of droplets, leading to the formation of crystalline links between droplets (this phenomenon was observed as a substantial increase in the apparent viscosity at 42°C–40°C, Figure 14.5b). The flocculation was followed by a deflocculation caused by the ongoing emulsification at the temperature below the range of the emulsifier crystallization. The final emulsification product was a viscous, ointment-like and smooth W/O emulsion. As revealed by confocal and differential interference contrast (DIC) microscopy, the water droplets were covered by thick, rigid membrane of the crystallized emulsifier (Figure 14.7). From the above, the most pronounced difference in the behavior of the emulsions, compared to those produced with AG and AG/APG emulsifiers, was a lack of the phase inversion step. As reported by Krog (1977), in contrast to monoglycerides, propylene glycol esters cannot associate with water into mesomorphic phases; the characteristic that seems

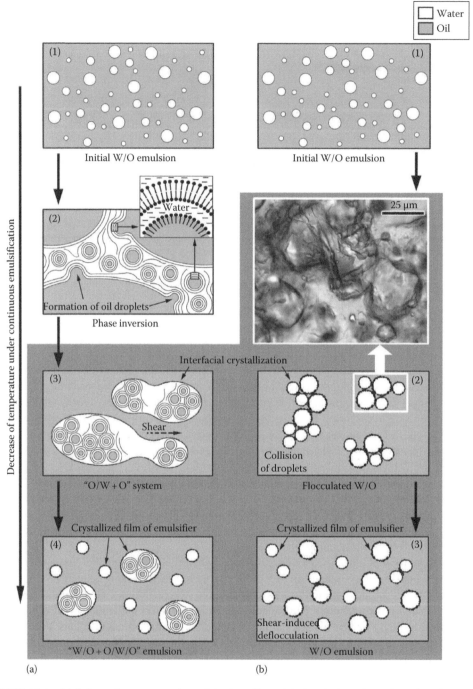

FIGURE 14.4 (a) Schematic representation of the lamellar gel-induced phase inversion of W/O emulsion and the formation of the O/W emulsion gel-loaded O/W/O double droplets, in the systems produced with AG or AG/APG emulsifiers (for water contents ≤55 wt.%). (b) Schematic representation of the orthokinetic flocculation and deflocculation of water droplets in the W/O emulsions produced with APG emulsifier (for water contents ≤50 wt.%). The micrograph shows an example of flocculated droplets. The transitions in emulsion structure were observed during the emulsification at decreasing temperature, as shown in Figure 14.3. (Adapted from Macierzanka, A. et al., *Colloids Surf. A Physicochem. Eng. Aspects*, 334, 40, 2009. With permission from Elsevier.)

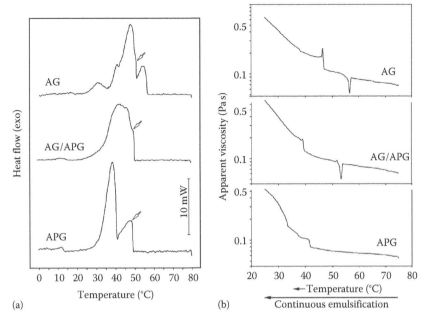

(a)   (b)

FIGURE 14.5 (a) DSC crystallization profiles of AG, AG/APG, and APG emulsifiers in their mixtures with paraffin wax (emulsifier:wax weight ratio = 5:3). The arrows indicate the peak crystallization temperature of paraffin wax, which was used in the emulsion preparation as a component of the oil phase. (b) Evolution of the viscosity of emulsions prepared with the three emulsifiers, as observed during the emulsification processes. Both, the emulsifier:wax ratio and the viscosity profiles characterized in (a) and (b), respectively, correspond to the emulsifications for the W:O proportions of 40:60 (w/w). (From Macierzanka, A. et al., *Colloids Surf. A Physicochem. Eng. Aspects*, 334, 40, 2009. With permission from Elsevier.)

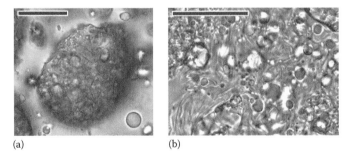

(a)   (b)

FIGURE 14.6 (a) DIC micrographs of the surface and (b) the internal structure of representative O/W/O droplet from the "W/O + O/W/O" emulsion produced with AG emulsifier. The droplet was gently compressed by a slow pressing down the cover slip with the microscope lens; this way the crystallized film of emulsifier was destroyed and the core of the droplet leaked out into the oil phase and was easily observed (b). The interior comprised small oil droplets surrounded by and suspended in the liquid-crystalline gel matrix. The scale bars corresponds to 10 μm.

to be responsible for the different overall structure of the oil-continuous emulsions formed. For the water contents ≥60 wt.%, the W/O emulsion collapsed after the crystallization of the emulsifier and the flocculation took place (Figure 14.3), due to the shear-induced destruction of the interfacial film of many water droplets. At the end of emulsification, some remaining W/O emulsion and the separated water were observed ("W/O + W" system).

The relative AG–APG proportions determined the structure of freshly prepared emulsions and the type of the structural changes observed during their storage. The mean photon transport length

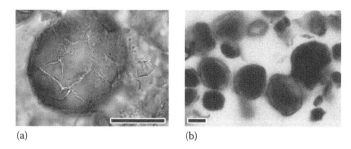

(a)   (b)

FIGURE 14.7 DIC (a) and confocal (b) micrographs of W/O emulsion (40:60, w/w) stabilized by APG emulsifier. Image (a) shows surface of the representative water droplet covered by rigid, crystallized film of the emulsifier. Image (b) shows the cross-section of the water droplets; the droplets are encapsulated by distinct, thick membrane of the emulsifier. The scale bars correspond to 10 μm.

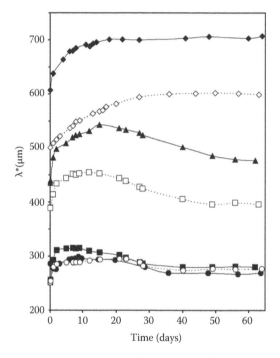

FIGURE 14.8 Time-dependent evolution of the emulsion structure (W:O = 40:60, w/w) expressed as changes in the mean photon transport length through emulsion (λ^*). Effect of the relative proportions of AG and APG emulsifiers (w/w): (●) 100:0, (○) 70:30, (■) 50:50, (□) 30:70, (▲) 20:80, (◇) 10:90, (◆) 0:100. Emulsions were stored at 23 ± 1°C. (Adapted from Macierzanka, A. et al., *Colloids Surf. A Physicochem. Eng. Aspects*, 334, 40, 2009. With permission from Elsevier.)

through emulsion (λ^*, measured by using a multiple light scattering method; Mengual et al. 1999a,b) recorded for fresh emulsions was dependent mainly on the degree of dispersity, increasing with the increase of droplet size for the APG-rich systems (Figure 14.8). The formation of network of droplets, observed during the first days of storage, allowed photons of incident light to penetrate deeper into the examined samples, so the parameter λ^* increased. It was probably caused by a restoration of the crystalline links between droplets destroyed during the emulsification. When the AG content was more than 10 wt.% in AG–APG mixture, the increase was followed by a step-by-step decrease, which might correspond to the sintering of the continuous-oil crystals (the crystalline material, which was not involved in the formation of the interfacial layer) as well as the new crystals growth in the oil phase. In the emulsions stabilized by the emulsifiers containing at least 90 wt.% of the APG component, the dominant post-crystallization process was the formation of solid crystal bridges between crystalline membranes of droplets, so after the aforementioned increase of λ^* the parameter remained almost constant suggesting only little subsequent changes in the emulsion structure (Figure 14.8). All emulsions were kinetically stable for at least 2 months at room temperature. After this time, their structures varied from a meshed network of coherent continuous-oil crystals and water droplets in the AG-rich systems to huge droplets linked by crystal bridges in the APG-stabilized emulsions.

14.3.2 TAILORING OF THE EMULSION STRUCTURE BY CHANGING THE EMULSIFIER COMPOSITION

From the previous section, it is clear that a range of oil-continuous emulsions that differ significantly in microstructure can be created by using only two crystalline emulsifiers in various proportions. However, under the conditions applied, it was not possible to formulate stable water-continuous systems for high water contents, due to the hydrophobic nature of the emulsifiers.

The O/W dispersions, which contained over 60 wt.% water and were stable for considerable time, have been produced with the acylglycerol emulsifier containing some amount of zinc carboxylate (AG/ZnC; Macierzanka et al. 2006). The dispersion was stabilized by liquid-crystalline α-gel phase formed during the emulsification from high to low temperatures.

When distilled, saturated MAGs are mixed with water, and heated to the Krafft temperature, so-called lamellar mesophase will be formed. In this phase, bimolecular lipid layers are separated by water, which penetrates into the planes of polar head groups of the MAG molecules (Figure 14.9). The swelling properties of lamellar mesophase depend on pH and the presence of ionic cosurfactants in the system. Addition of cosurfactants influences the increase of electrostatic repulsion between the layers, resulting in a higher degree of swelling. Usually DATEM (diacetyl tartaric acid ester of MAG) or SSL (sodium stearoyl lactylate) are used to facilitate this effect (Cassin et al. 1998, Heertje et al. 1998, Borné et al. 2001, Sein et al. 2002). As reported by Krog (2001), saturated C16/C18 MAGs are dispersible in water between 55°C and 70°C in the form of lamellar aggregates or vesicles.

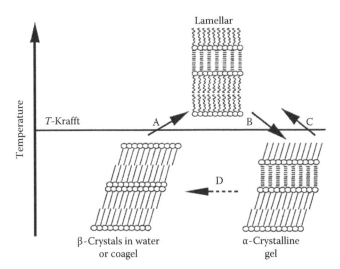

FIGURE 14.9 Phase transitions of monoglyceride systems at Krafft temperature (A, B, and C) and below Krafft temperature (D). (Adopted from Heertje, I. et al., *Lebensm. Wiss. Technol.*, 31, 387, 1998. With permission from Elsevier.) The Krafft temperature is defined as the melting point of the hydrocarbon chains (Krog 1997).

Below the Krafft temperature the mesophase transforms to the α-gel phase, in which MAG molecules are still aligned in individual bilayers, separated by water (Figure 14.9). However, the hydrocarbon chains are not in a liquid-like state any more. Since the α-gel is not thermodynamically stable, it converts into the coagel phase (a network of plate-like β-crystals in water). Formation of hydrogen bonds between the glycerol moieties causes the release of water into the bulk medium. A possible mechanism of the α-gel → coagel conversion has been reported by Sein et al. (2002) and van Duynhoven et al. (2005). The rate of this transition depends on a number of factors, such as the composition of the MAGs, the water/MAG ratio, the presence of other surfactants, the storage temperature, the time, and the gel processing conditions (e.g., shear rate; Heertje et al. 1998). Larsson (1967) have reported that pure 1-MAGs with fatty acid chains shorter than C14 transform directly from the lamellar mesophase to the mixture of β-crystals and water. Both the α-gel and coagel are used in the processing of some food products in order to obtain a desired texture (Doerfert 1962, Heertje et al. 1991). Since these phases exhibit a different functionality (e.g., with respect to firmness or foaming), control of the preferential formation of one of the two phases has been a subject of scientific investigations (Heertje et al. 1998). The amount of β-crystalline coagel can be determined by deferential scanning calorimetry (DSC). For binary MAG–water systems, the enthalpy of melting of pure coagel phase is about twice the value obtained for the melting of the liquid–crystalline α-gel phase (Cassin et al. 1998, Heertje et al. 1998, Sein et al. 2002). Coagel Index (CI), defined as the ratio between the enthalpies of the heating (ΔH_h) and reheating (ΔH_{rh}) scans (Cassin et al. 1998, Sein et al. 2002), indicates the amount of α-gel that converted to coagel. When the CI = 1, the examined material is in the α-gel state. For a pure coagel this parameter is about 2. Both phases convert into the lamellar phase at melting temperature, but when cooled down, the lamellar phase converts only to the α-gel phase (Figure 14.9; Heertje et al. 1998). Since the transition of α-gel to coagel usually takes days or weeks, the immediate reheating scan exhibits the melting of the α-gel.

In our experiments (Macierzanka et al. 2006), we examined the kinetics of the similar phase transition in the O/W dispersion prepared with AG/ZnC emulsifier. The ratio $\Delta H_h/\Delta H_{rh}$, obtained from the DSC measurements, has been referred as to the Dispersion Coagel Index (DCI). As shown in Figure 14.10A, almost complete conversion to coagel (DCI ≈ 2) took about 90 days at room temperature. In this case (Figure 14.10B), the endothermic peak in the first heating curve, corresponding to the melting of coagel phase (curve a), reveals about twice the heat content of the sample compared to the melting of α-gel phase, represented by the peak in the second heating curve (c). Reheating was performed just after controlled cooling of the dispersion sample (curve b). After storage at room temperature for several days after preparation, the O/W dispersions developed a firmer consistency. It was observed that the penetration depth of the measuring punch into the samples, dropped to less than 1 mm. The change of consistency, from an initial semi-fluid to a final semi-solid state, took 4–135 days, depending on the water content (Figure 14.11). It appears that such an increase in hardness was caused by the transition in the MAG–crystalline phase, as proposed above. Formation of the MAG coagel phase was previously studied mainly for the binary MAG/cosurfactants–water gels (Krog 1997, Cassin et al. 1998, Heertje et al. 1998, Chupin et al. 2001, Sein et al. 2002), whereas our work shows it can also take place in a MAG-based, multicomponent (complex emulsifier–water–oil) dispersed system. Recently, it has been shown that the addition of sodium soap stabilized the lamellar dispersion of a commercial distilled monoglyceride based on fully hydrogenated palmstearin in function of temperature (Van de Walle et al. 2008). The presence of that anionic cosurfactant also retarded the transition from the metastable α-gel to the stable coagel phase, thus the use of anionic cosurfactant provided an opportunity to control the transition kinetics. The transition to the coagel phase was also studied by Marangoni et al. for ternary systems composed of vegetable oil, water, and monoglyceride/cosurfactants (Batte et al. 2007a, Marangoni et al. 2007). Encapsulation of oil by the MAG liquid–crystalline lamellar phase, followed by droplet wall crystallization, made it possible to create complex colloidal systems with the functionality and properties of semisolid fats for various proportions of the components used and different preparation conditions (Batte et al. 2007a,b). The authors reported on an increase in the storage modulus and peak melting temperature

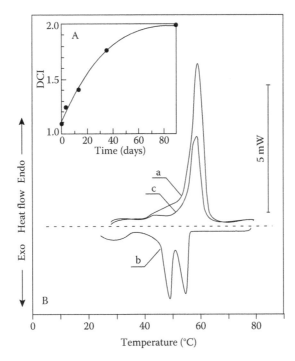

FIGURE 14.10 Thermal behavior of O/W dispersion (O:W weight ratio = 20:80) prepared with AG emulsifier containing zinc stearate. Section A shows the storage time-dependent formation of coagel in the O/W dispersion (at room temperature), displayed as an increase in the Dispersion Coagel Index (DCI). Section B presents the DSC thermograms of heating (a), cooling (b), and reheating (c) of the O/W dispersion, stored 90 days at room temperature prior to examination. (From Macierzanka, A. et al., *Langmuir*, 22, 2487, 2006. With permission from the American Chemical Society.)

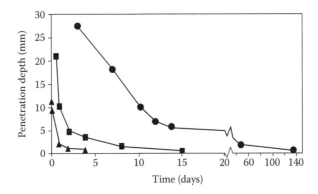

FIGURE 14.11 Influence of storage time on the consistency of O/W dispersions obtained with AG emulsifier containing zinc stearate. Water content in the dispersion (wt.%): 60 (▲), 70 (■), 80 (●). (From Macierzanka, A. et al., *Langmuir*, 22, 2487, 2006. With permission from the American Chemical Society.)

over time, showing that the viscoelastic behavior of that material was dependent on the microstructural transitions of the liquid-crystalline phase.

In our recent investigations, very promising results in terms of a satisfactory stability and an ability to control rheological properties of O/W emulsions were also achieved with the use of AG and APG emulsifiers containing strongly hydrophilic surfactants such as sodium fatty acid carboxylates (NaC). The emulsification carried out from high to low temperature led to the phase inversion

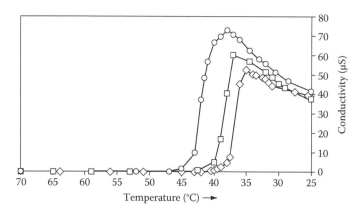

FIGURE 14.12 Phase inversion of emulsion from the W/O to the O/W type observed as an increase of the electrical conductivity of emulsion during the emulsification under decreasing temperature. Effect of the concentration of sodium fatty acid carboxylate (NaC) in the APG emulsifiers used to produce the emulsions (paraffin oil/water, 40:60, w/w): (◊) 5.3, (□) 8.1, (○) 10.8 wt.% NaC.

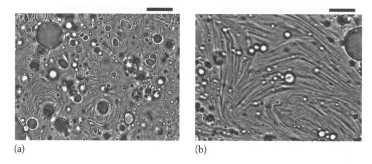

FIGURE 14.13 DIC micrographs of O/W emulsion (40:60, w/w) stabilized by APG emulsifier modified with sodium stearate. The gel phase was located around the oil droplets (a; encapsulation) and between them (b; embedment of oil droplets). The scale bars correspond to 10 μm.

from the initial W/O to the final O/W system when the crystallization temperature of emulsifier was reached. The inversion was clearly seen from the rapid increase in the electrical conductivity of emulsion (Figure 14.12). The resulting O/W emulsions were consisted of droplets stabilized by the liquid–crystalline multilamellar gel phase presumably composed of nonionic MAG (or MAPG) and anionic NaC, forming bilayers separated by the layers of water (Figure 14.13). The presence of sufficient amounts of the charged cosurfactant (NaC) in the bilayer structures enabled incorporation of water into the gel phase. As mentioned above, such a phase can encapsulate oil droplets, imparting both the steric hindrance (an onion-like lamellar shell swollen with water) and the ionic barrier (repulsion of charged encapsulated oil droplets) against coagulation. Novales and coworkers reported on the formulation of O/W emulsions with monopalmitin and palmitic acid, whereas no stable emulsions could be obtained with pure constituents (Novales et al. 2005). The preparation of the vesicle dispersions of both emulsifying components in an aqueous phase was found to be a necessary initial step in emulsion formulation.

In the emulsion produced with APG emulsifier modified with NaC, some oil droplets were seen to be embedded in the gel phase also located in the continuous water phase (Figure 14.13b). Existence of such a continuous-phase gel structures was likely to largely account for a viscoelastic behavior of emulsions. The rheological properties could be controlled by the concentration of the anionic surfactant present in the emulsifier composition. For example, as shown in Figure 14.14a, the emulsions exhibited a distinct linear viscoelastic region as long as the yield stress has not been

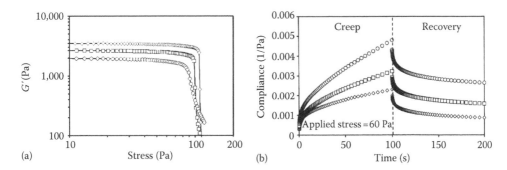

FIGURE 14.14   (a) Storage modulus (G') vs. shear stress and (b) a creep recovery test for the O/W emulsions (40 wt.% paraffin oil) prepared with APG emulsifiers modified with sodium fatty acid carboxylates. Effect of the concentration of the carboxylate in the emulsifier: (◊) 5.3, (□) 8.1, (○) 10.8 wt.%.

achieved, however both the level of the storage modulus (G') in that region as well as the stress required to break the emulsion structure (yielding) decreased with an increase in the concentration of sodium carboxylate in the APG emulsifier. At lower shear stress (up to 50–80 Pa, depending on the emulsifier used) the G' values were only slightly dependent on the stress applied. This indicates the structure of emulsions to be elastic in that shear stress range. As a consequence, a linear response of the emulsion samples was recorded. The transition from elastic to viscous domain, at a certain critical stress (yield stress), observed as a sharp reduction of G' value (Pal 1999), indicates the moment at which the increasing stress became sufficiently large to allow the droplets to move past one another. The data were consistent with the results of the creep recovery measurements (Figure 14.14b), showing the percent of recoverable deformation to be highest for the emulsion produced with the emulsifier containing the lowest amount of sodium carboxylate. This direct measurement of the emulsion elasticity implies that the amount of the energy stored during the creep phase was dependent on the structure of the gel phase in emulsion, which in turn seems to be governed by the carboxylate concentration.

The concentration and the type of carboxylate could also have an impact on the structure of W/O emulsions, as shown in Figure 14.15 for the example of the use of APG emulsifiers modified with zinc fatty acid carboxylates (ZnC) (Macierzanka and Szeląg 2006). Among the emulsifiers studied, the largest, and in many cases, irregular (i.e., not spherical) droplets were produced with the emulsifier containing the least amount of ZnC (Figure 14.15a). When using more ZnC-rich emulsifiers, water droplets became smaller and spherical. The finest emulsion was obtained with the emulsifier containing the highest concentration of ZnC (Figure 14.15b). An effective adsorption of ZnC crystals in the studied systems was only limited to the sub-micron particles (Macierzanka and Szeląg 2006). The aggregates formed in the continuous oil phase by larger ZnC crystals also were observed in the emulsions prepared with the emulsifiers containing the highest ZnC concentrations. Because the sub-micron ZnC crystals coadsorbed with MAPG at the water–oil interface, the differences in the size and shape of droplets might derive from changes in the elasticity of the interfacial films produced for various MAPG/ZnC proportions, depending of the emulsifier type. The structure of emulsions had a profound effect on their rheological properties. For small and spherical droplets in the emulsion shown in Figure 14.15b, the easiest flow was observed, compared to the emulsion shown in Figure 14.15a, for which much higher shear stress was required to align large, irregular droplets of dispersed water in the direction of the flow (Figure 14.15c). The differences in the yield stress values might also be due to the aforementioned presence of the unadsorbed ZnC crystals in the oil phase. Zinc stearate is used as an anti-caking agent in applications in which a free flowing final product is desired, such as cosmetic powders and fire extinguishers (Markley 1961). Thus one may expect that the unadsorbed ZnC crystals can act as shear-thinning agents in the W/O emulsions obtained with the emulsifiers containing considerable amounts of the carboxylate.

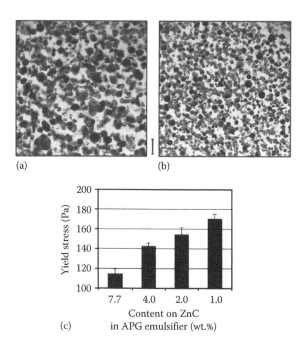

FIGURE 14.15 Confocal micrographs of W/O emulsions (40:60, w/w) prepared with APG emulsifiers containing (a) 1.0 or (b) 7.7 wt.% zinc carboxylate (ZnC; zinc stearate was used in this study). The scale bar corresponds to 80 μm. The graph (c) shows an effect of the ZnC content on the yield stress of emulsions. (Adapted from Macierzanka A. and Szeląg H., *Colloids Surf. A Physicochem. Eng. Aspects*, 281, 125, 2006. With permission).

14.3.3 POTENTIAL BIOTECHNOLOGICAL AND OTHER APPLICATIONS

Over the last 10–20 years, there has been increasing interest in *in vitro* compartmentalization, a strategy based on partitioning of biochemical reactions in dispersed water droplets to form bio-microreactors in W/O emulsions (Tawfik and Griffiths 1998, Griffiths and Tawfik 2006, He 2008, Huebner et al. 2008). In such systems, each droplet can function as an independent reactor. The small volumes of compartments, their large number per emulsion volume unit, the ease of preparing emulsions, and their high stability to changes in temperature, pH, and salt concentrations have rendered W/O emulsions an ideal tool for both miniaturizing and parallelizing biochemical reactions in areas such as genomics, genetics, and proteomics. Physical and chemical manipulations of the content of droplets can be achieved by different approaches such as the controlled delivery of hydrophobic substrates through the oil phase (Griffiths and Tawfik 2003) and hydrophilic components through nanodroplets or micelles (Bernath et al. 2005), altering pH, for example, by delivery of acetic acid (Griffiths and Tawfik 2003), changing the temperature (Ghadessy et al. 2001, Nakano et al. 2003, 2005, Musyanovych et al. 2005) and fusion of droplets derived from different emulsions to combine reagents (Pietrini and Luisi 2004). The above techniques have been recently reviewed by Griffiths and Tawfik (2006).

In this view, the oil-continuous emulsions presented in the previous sections seem to be promising candidates as a source of bio-microreactors. Water droplets stabilized by various AG and APG emulsifiers can vary in size, shape, spatial organization, and even internal structure (e.g., O/W/O-type droplets containing an O/W emulsion gel) depending on the relative proportions of the emulsifiers and the presence, the type, and the concentration of fatty acid carboxylates. A fatty acid profile of the emulsifiers can be selected during their manufacturing (Macierzanka 2004). This can provide desired melting and crystallization characteristics of the emulsifiers. Hence, for example, the interfacial crystallization and thus the encapsulation of hydrophilic substrates for the reaction in water droplets can be performed at relatively low temperatures. It would be of great importance if the substrates, which are introduced into the droplets during the emulsification process, might be inactivated by high temperature.

On the other hand, the O/W emulsions produced with AG and APG emulsifiers modified with sodium or potassium carboxylates can serve as vehicles of poorly water-soluble substances that might be incorporated either into the oil droplets or the liquid-crystalline structures of the emulsifier, forming a gel matrix around and between the droplets. Self-assembled structures of polar lipids have been utilized as reaction media (Vauthey et al. 2000, Leser et al. 2006) or for delivery and controlled release of active compounds (Leser et al. 2006, Sagalowicz et al. 2006, Amar-Yuli et al. 2009).

14.4 SUMMARY

We have shown an example of how a composition of crystalline emulsifiers can affect the pathways of emulsification, the microstructural evolution of the produced emulsions, and their rheological properties. The above can be influenced either by changing the relative proportions of two emulsifiers crystallizing during the emulsification process or by using the emulsifiers that were synthesized in the presence of various, contrasting ionic surfactants, which modified their hydrophilic properties. Both approaches seem to be promising ways for the controlled manufacturing of complex water- or oil-continuous colloidal structures with desired structure and functional properties.

ACKNOWLEDGMENTS

The work was financially supported by the Polish Ministry of Science and Information Society Technologies (Research Projects 3T09B08927 and 7T09B10820). Part of the work was funded by the Biotechnology and Biological Sciences Research Council (BBSRC, United Kingdom) through the core grant to the Institute of Food Research.

ABBREVIATIONS

AG	Acylglycerol emulsifier (consisting predominantly MAG and DAG)
AG/APG	Mixed emulsifier consisting of equivalent amounts of AG and APG emulsifiers
APG	Acylpropyleneglycol emulsifier (consisting predominantly fatty acid mono-diesters of propylene glycol)
CI	Coagel Index
DAG	Diacylglycerol
DAPG	Diacylpropyleneglycol (propylene glycol diester)
DATEM	Diacetyl tartaric acid ester of MAG
DCI	Dispersion Coagel Index
DIC	Differential interference contrast (microscopy)
DSC	Differential scanning calorimetry
FA	Fatty acid
G	Glycerol
G'	Storage modulus
HLB	Hydrophile–lipophile balance
MAG	Monoacylglycerol
MAPG	Monoacylpropyleneglycol (propylene glycol monoester)
NaC	Sodium fatty acid carboxylates
O/W	Oil-in-water (emulsion)
O/W+O	Unstable emulsion system consisting of agglomerates of the O/W emulsion suspended in the continuous oil phase
O/W/O	Oil-in-water-in-oil (double emulsion)

PG   Propylene glycol
PGMS   Propylene glycol monostearate
SSL   Sodium stearoyl lactylate
TAG   Triacylglycerol
W/O   Water-in-oil (emulsion)
W/O+O/W/O   Emulsion system consisting of the water and the O/W/O-type droplets both
                      suspended in the external oil phase
W/O+W   Unstable system consisting of W/O emulsion and separated water phase
ZnC   Zinc fatty acid carboxylate
λ^* Mean photon transport length through emulsion

REFERENCES

Amar-Yuli, I., Libster, D., Aserin, A., and N. Garti, 2009. Solubilization of food bioactives within lyotropic liquid crystalline mesophases. *Curr. Opin. Colloid Interface Sci.* 14: 21–32.

Batte, H. D., Wright, A. J., Rush, J. W., Idziak, S. H. J., and A. G. Marangoni, 2007a. Phase behavior, stability, and mesomorphism of monostearin-oil-water gels. *Food Biophys.* 2: 29–37.

Batte, H. D., Wright, A. J., Rush, J. W., Idziak, S. H. J., and A. G. Marangoni, 2007b. Effect of processing conditions on the structure of monostearin–oil–water gels. *Food Res. Int.* 40: 982–988.

Bernath, K., Magdassi, S., and D. S. Tawfik, 2005. Directed evolution of protein inhibitors of DNA-nucleases by *in vitro* compartmentalization (IVC) and nano-droplet delivery. *J. Mol. Biol.* 345: 1015–1026.

Borné, J., Nylander, T., and A. Khan, 2001. Phase behavior and aggregate formation for the aqueous monoolein system mixed with sodium oleate and oleic acid. *Langmuir* 17: 7742–7751.

Cassin, G., de Costa, C., van Duynhoven, J. P. M., and W. G. M. Agterof, 1998. Investigation of the gel to coagel phase transition in monoglyceride–water systems. *Langmuir* 14: 5757–5763.

Chupin, V., Boots, J.-W. P., Killian, J. A., Demel, R. A., and B. de Kruijff, 2001. Lipid organization and dynamics of the monostearoylglycerol–water system. A ^{2}H NMR study. *Chem. Phys. Lipids* 109: 15–28.

Doerfert, G. H., 1962. Distilled monoglycerides, *Food Eng.* 34: 97–100.

Friberg, S. and L. Mandell, 1970. Phase equilibria and their influence on the properties of emulsions. *J. Am. Oil Chem. Soc.* 47: 149–152.

Friberg, S., Mandell, L., and M. Larsson, 1969. Mesomorphous phases, a factor of importance for the properties of emulsions. *J. Colloid Interface Sci.* 29: 155–156.

Ghadessy, F. J., Ong, J. L., and P. Holliger, 2001. Directed evolution of polymerase function by compartmentalized self-replication. *Proc. Natl. Acad. Sci. USA* 98: 4552–4557.

Griffiths, A. D. and D. S. Tawfik, 2003. Directed evolution of an extremely fast phosphotriesterase by *in vitro* compartmentalization. *EMBO J.* 22: 24–35.

Griffiths, A. D. and D. S. Tawfik, 2006. Miniaturising the laboratory in emulsion droplets. *Trends Biotechnol.* 24: 395–402.

He, M., 2008. Cell-free protein synthesis: Applications in proteomics and biotechnology. *New Biotechnol.* 25: 126–132.

Heertje, I., Hendrickx, H. A. C. M., Knoops, A. J., Roijers, E. C., and H. Turksma, 1991. Use of mesomorphic phases in food products. European Patent 0,558,523,B1.

Heertje, I., Roijers, E. C., and H. A. C. M. Hendrickx, 1998. Liquid crystalline phases in the structuring of food products. *Lebensm. Wiss. Technol.* 31: 387–396.

Huebner, A., Sharma, S., Srisa-Art, M., Hollfelder, F., Edel, J. B., and A. J. DeMello, 2008. Microdroplets: A sea of applications? *Lab Chip Miniat. Chem. Biol.* 8: 1244–1254.

Krog, N., 1977. Functions of emulsifiers in food systems. *J. Am. Oil Chem. Soc.* 54: 124–131.

Krog, N. J., 1997. Food emulsifiers and their chemical and physical properties. In *Food Emulsions* (3rd edn.), eds. S. E. Friberg and K. Larsson, pp. 141–188 (and references therein). New York: Marcel Dekker.

Krog, N., 1999. Food emulsifiers. In *Lipid Technologies and Applications* (2nd edn.), eds. F. D. Guneston and F. B. Padley, pp. 521–534 (and references therein). New York: Wiley-VCH.

Krog, N., 2001. Crystallization properties and lyotropic phase behavior of food emulsifiers: Relation to technical applications. In *Crystallization Processes in Fats and Lipid Systems*, eds. N. Garti and K. Sato, pp. 505–528. New York: Marcel Dekker.

Larsson, K., 1967. The structure of mesomorphic phases and micelles in aqueous glyceride systems. *Z. Phys. Chem. Neue Folge.* 56: 173–198.

Leser, M. E., Sagalowicz, L., Michel, M., and H. J. Watzke, 2006. Self-assembly of polar food lipids. *Adv. Colloid Interface Sci.* 123–126: 125–136.

Macierzanka A., 2004, Synthesis and properties of emulsifiers: Ester derivatives of glycerol and propylene glycol, synthesized in the presence of selected carboxylates. PhD thesis, Gdansk University of Technology, Gdansk, Poland.

Macierzanka, A. and H. Szeląg, 2004. Esterification kinetics of glycerol with fatty acids in the presence of zinc carboxylates: Preparation of modified acylglycerol emulsifiers. *Ind. Eng. Chem. Res.* 43: 7744–7753.

Macierzanka, A. and H. Szeląg, 2006. Microstructural behavior of water-in-oil emulsions stabilized by fatty acid esters of propylene glycol and zinc fatty acid salts. *Colloids Surf. A Physicochem. Eng. Aspects* 281: 125–137.

Macierzanka, A., Szeląg, H., Moschakis, T., and B. S. Murray, 2006. Phase transitions and microstructure of emulsion systems prepared with acylglycerols/zinc stearate emulsifier. *Langmuir* 22: 2487–2497.

Macierzanka, A., Szeląg, H., Szumała, P., Pawłowicz, R., Mackie, A. R., and M. J. Ridout, 2009. Effect of crystalline emulsifiers composition on structural transformation of water-in-oil emulsions: Emulsification and quiescent conditions. *Colloids Surf. A Physicochem. Eng. Aspects* 334: 40–52.

Marangoni, A. G., Idziak, S. H. J., Vega, C., Batte, H., Ollivon, M., Jantzi, P. S., and J. W. E. Rush, 2007. Encapsulation-structuring of edible oil attenuates acute elevation of blood lipids and insulin in humans. *Soft Matter* 3: 183–188.

Markley, K. S., 1961. Salts of fatty acids. In *Fatty Acids* (2nd edn.), ed. K. S. Markley, pp. 715–756. London, U.K.: Interscience Publishers Ltd.

Mengual, O., Meunier, G., Cayré, I., Puech, K., and P. Snabre, 1999a. TURBISCAN MA 2000: Multiple light scattering measurement for concentrated emulsions and suspensions instability analysis. *Talanta* 50: 445–456.

Mengual, O., Meunier, G., Cayre, I., Puech, K., and P. Snabre, 1999b. Characterisation of instability of concentrated dispersions by a new optical analyser: The TURBISCAN MA 1000. *Colloids Surf. A Physicochem. Eng. Aspects* 152: 111–123.

Musyanovych, A., Mailander, V., and K. Landfester, 2005. Miniemulsion droplets as single molecule nanoreactors for polymerase chain reaction. *Biomacromolecules* 6: 1824–1828.

Nakano, M., Komatsu, J., Matsuura, S., Takashima, K., Katsura, S., and A. Mizuno, 2003. Single-molecule PCR using water-in-oil emulsion. *J. Biotechnol.* 102: 117–124.

Nakano, M., Nakai, N., Kurita, H., Komatsu, J., Takashima, K., Katsura, S., and A. Mizuno, 2005. Single-molecule reverse transcription polymerase chain reaction using water-in-oil emulsion. *J. Biosci. Bioeng.* 99: 293–295.

Novales, B., Ropers, M. H., and J.-P. Douliez, 2005. Use of fatty acid/monoglyceride vesicle dispersions for stabilizing O/W emulsions. *Colloids Surf. A Physicochem. Eng. Aspects* 269: 80–86.

Pagliaro, M. and M. Rossi, 2008. *The Future of Glycerol: New Usages for a Versatile Raw Material.* Cambridge, U.K.: RSC Publishing.

Pal, R., 1999. Yield stress and viscoelastic properties of high internal phase ratio emulsions. *Colloid Polym. Sci.* 277: 583–588.

Pietrini, A. V. and P. L. Luisi, 2004. Cell-free protein synthesis through solubilisate exchange in water/oil emulsion compartments. *ChemBioChem* 5: 1055–1062.

Sagalowicz, L., Leser, M. E., Watzke, H. J., and M. Michel, 2006. Monoglyceride self-assembly structures as delivery vehicles. *Trends Food Sci. Technol.* 17: 204–214.

Sein, A., Verheij, J. A., and G. M. Agterof, 2002. Rheological characterization, crystallization, and gelation behavior of monoglyceride gels. *J. Colloid Interface Sci.* 249: 412–422.

Stauffer, C. E., 2002. Emulsifiers and stabilizers. In *Fats in Food Technology*, ed. K. K. Rajah, pp. 228–274. Sheffield, U.K.: Sheffield Academic Press.

Szeląg, H. and A. Macierzanka, 2001. Synthesis of modified emulsifiers in the presence of Na, K and Zn fatty acid carboxylates. *Tenside Surf. Det.* 38: 377–380.

Szeląg, H. and W. Zwierzykowski, 1983. The application of molecular distillation to obtain high concentration of monoglycerides. *Fette Seifen Anstrichmittel* 85: 443–446.

Szeląg, H. and W. Zwierzykowski, 1995. Molecular distillation of selected fatty acid derivatives. *S&OUML;FW J.* 121: 444–448.

Szeląg, H. and W. Zwierzykowski, 1998. Esterification kinetics of glycerol with fatty acids in the presence of sodium and potassium soaps. *Fett/Lipid* 100: 302–307.

Tawfik, D. S. and A. D. Griffiths, 1998. Man-made cell-like compartments for molecular evolution. *Nat. Biotechnol.* 16: 652–656.

Van de Walle, D., Goossens, P., and K. Dewettinck, 2008. Influence of sodium soap and ionic strength on the mesomorphic behavior and the α-gel stability of a commercial distilled monoglyceride. *Food Res. Int.* 41: 247–254.

van Duynhoven, J. P. M., Broekmann, I., Sein, A., van Kempen, G. M. P., Goudappel, G.-J. W., and W. S. Veeman, 2005. Microstructural investigation of monoglyceride–water coagel systems by NMR and CryoSEM. *J. Colloid Interface Sci.* 285: 703–710.

Vauthey, S., Milo, C., Frossard, P., Garti, N., Leser, M. E., and H. J. Watzke, 2000. Structured fluids as micro-reactors for flavor formation by the Maillard reaction. *J. Agric. Food Chem.* 48: 4808–4816.

Zheng, Y., Chen, X., and Y. Shen, 2008. Commodity chemicals derived from glycerol, an important biorefinery feedstock. *Chem. Rev.* 108: 5253–5277.

Zhou, C.-H., Beltramini, J. N., Fan, Y.-X., and G. Q. Lu, 2008. Chemoselective catalytic conversion of glycerol as a biorenewable source to valuable commodity chemicals. *Chem. Soc. Rev.* 37: 527–549.

Zielinski, R. J., 1997. Synthesis and composition of food-grade emulsifiers. In *Food Emulsifiers and their Applications*, eds. G. L. Hasenhuettl and R. Hartel, pp. 11–38. New York: Chapman & Hall.

15 Multiple Emulsions in Biomedical and Biotechnological Applications

Ozgen Ozer and Gulten Kantarci

CONTENTS

15.1 INTRODUCTION

Multiple emulsions were first described in 1925 by Seifriz (1925); in 1965, Herbert described these systems in detail as a vaccine delivery system (Herbert 1965); and in 1968, Engel investigated the intestinal absorption of insulin in water-in-oil-in-water (W/O/W) multiple emulsions (Engel et al. 1968). In the last 10–20 years, besides different research groups, detailed investigations on these systems were performed by many researchers.

Multiple emulsions are the systems wherein droplets of the dispersed phase contain additional but smaller droplets, identical to or different from the continuous phase. Multiple emulsions are complex polydispersed systems and are also called "emulsion of emulsions" or, most recently, "emulsion liquid membranes" (Figure 15.1). The liquid phase separates the internal droplets from the external phase and acts as a semipermeable membrane, allowing solute to diffuse between the two phases. Multiple emulsions have shown promise in pharmaceuticals, cosmetics, food technology, and separation sciences (Davis et al. 1985, Buszello and Müller 2000, Kovács et al. 2005). Recently, there are many interesting fields of applications, like vaccine formulations, enzyme immobilization (May and Landgraff 1976, Iso et al. 1989, Giovagnoli et al. 2004), and drug overdose treatment (Frankenfeld et al. 1976,

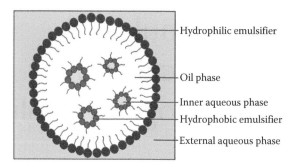

FIGURE 15.1 Schematic representation of W/O/W multiple emulsions; structure of a multiple emulsion droplet showing the role of the various components. (From Kovács, A. et al., *J. Therm. Anal. Calorim.*, 82, 491, 2005. With permission from Springer.)

Chiang et al. 1978). They also serve as blood substitute (Zheng et al. 1991, 1992, 1993). On the other hand, multiple emulsions are presented as a suitable vehicle for oral administration of insulin (Engel et al. 1968, Shichiri et al. 1974, Silva-Cunha et al. 1998). The potential applications of these systems in food technology were studied more recently to protect the reactive food nutrients and volatile flavors, and to develop low calorie, reduced fat food products (Raynal 1996).

These systems have some other advantages, such as the protection of the entrapped substances in the inner aqueous phase and modulation of their release rate to improve product efficacy, and the incorporation of active substances in the different compartments. The encapsulation effect of these systems is of primary importance as a prolonged release system. They are more suitable systems than water-in-oil (W/O) emulsions for parenteral formulations due to their low viscosity. In addition to these well-known properties, they lead to the development of cosmetic formulations characterized by their efficient, light, nongreasy, and nonsticky textures (Ferrero and Doucet 1996). The disadvantages of these systems are the difficulties on their stability, characterization, and production at industrial scale.

15.2 FORMULATION DEVELOPMENT

15.2.1 Formulation Methods

Several methods are used for the formulation of multiple emulsions. The equipment used has an important effect on the viscosity and stability. Temperature, agitation speed, quantity of primary emulsion in the multiple emulsion, and ratios of surfactant are the essential parameters for the formulation (Prybilski et al. 1991). The most common method for their preparation is the two-step emulsification method with a high yield and reproducibility. The primary emulsion would be re-emulsified using a secondary emulsifier. Phase inversion, the usage of pseudo-ternary diagram of water-emulsifier-oil for improvement of the one-step emulsification, membrane emulsification method, and microchannel emulsification methods are also used in some cases. The phase inversion method can be preferred for the production of multiple emulsions with fine droplet size, but it has not been widely used because of the stability and reproducibility problems (Raynal et al. 1993, Yazan et al. 1993, Silva-Cunha et al. 1997a, Khan et al. 2006).

A new thermally reversible polymer was synthesized and used as a thickener for the external aqueous phase of the multiple emulsions. The polymer consisted of poly(acrylic acid) blocks covalently grafted to Pluronic segments. In this case, the emulsion was free flowing at ambient temperature and thickened in human body. The ease of fabrication with high entrapment yield was reported as the main advantage of this system (Olivieri et al. 2003).

In another study, membrane emulsification technology was applied to the W/O/W multiple emulsions. First, a submicron W/O emulsion containing an anticancer drug and water and Lipidol as the dispersion and continuous phase, respectively, was prepared. Polyoxyethylene (40) hydrogenerated caster oil was used as the hydrophobic surfactant. Then, a W/O/W multiple emulsion was prepared using the W/O emulsion as the dispersion phase. It was reported that the formulation prepared by double-membrane emulsification method was a medically safe emulsion with the precise design of the particle size and with a high production yield (Nakashima et al. 2000).

The selection of the proper emulsifiers is of primary importance for the formulation of multiple emulsions. The concentration of lipophilic primary emulsifier markedly affected the rheological behavior, droplet size, and *in vitro* drug release of a W/O/W multiple emulsion. The more lipophilic surfactant is increased, the more the oil globule swelling capacity is increased, and the more the release is delayed (Jager-Lezer et al. 1997, Ozer et al. 2000a). W/O/W multiple emulsions were formulated with high content of inner phase and relatively low concentrations of lipophilic polymeric primary emulsifier, PEG 30-dipolyhydroxystearate. W/O/W multiple emulsion was obtained with a good long-term stability (Vasiljevic et al. 2005, 2006).

Generally, a high amount of classical surfactants was used to stabilize the simple and multiple emulsions. Cyclodextrins were used in the formulation of oil-in-water-in-oil (O/W/O) multiple emulsions to decrease the high amount of classical surfactant. They have a cavity in the center of the molecule, and a hydrophobic molecule, which can be included in the cyclodextrin cavity, can be introduced into the emulsion formulation without modification of emulsion stability. O/W/O multiple emulsion was formulated in the absence of classical surfactants and by using appropriate cyclodextrins (Duchêne et al. 2003). Cyclodextrins can form *in situ* surface-active agents by including a fatty acid chain of the glycerides of the oily phase. The stability of the emulsion that contains an active agent was found to be dependent on the possible competition between the additive and the fatty acid to enter the cyclodextrin cavity. Small cyclodextrins were found more suitable than the ones that have a larger cavity (Yu et al. 2003).

In another study, the effect of surfactant structure, hydrophilic–lipophilic balance, and the volume fraction of water and oil phases on the development of one-step method were evaluated. The results suggested that the formation of multiple droplets could involve a combination of transitional and catastrophic phase inversions, which are the two types of phase inversion. The results have established the conditions under which multiple emulsions were formed using the nonionic surfactants. The pair Span 60 and Tween 60 series gave more stable emulsions. Furthermore, the results revealed the wide range of water/oil ratios to form stable double emulsions and gave some information about the potential mechanism for the formation of double emulsions using the one-step method (Morais et al. 2008).

Recently, spontaneous multiple emulsion was developed by using two oil systems (Labrafil M 1944CS and Labrafac Lipophile WL 1349) to suggest a potential route for large-scale or *in vivo* production of multiple emulsions via one-step process with minimal agitation. The study indicated the possibility of producing a rather stable multiple emulsion (Devani et al. 2005).

15.2.2 STABILITY OF MULTIPLE EMULSIONS

The potential applications of multiple emulsions are so promising; therefore, their stability has lead many researchers to overcome this problem. There are a number of factors affecting the stability of multiple emulsions. The mechanism of breakdown and the methods that have been developed to assess stability were investigated as well (Florence and Whitehill 1980, 1982). Among the factors affecting the stability, the type of oil is of particular interest. The type of oil affects the entrapment yield of the multiple emulsion and characteristics of interfacial film, and these two factors are of crucial importance in determining the globule size and the stability. For this reason, in the study of Ozer et al., the oils from hydrocarbon, triglyceride, and ester groups were used for the formulation of multiple emulsions. Different methods have been proposed for the stabilization of multiple emulsions. Multiple emulsions prepared with hydrocarbon oils suggested the ones with best stability (Ozer et al. 2006).

The effect of viscosifying agents on the stability of multiple emulsions were studied, and stable multiple emulsions were formulated with a small portion of primary emulsion containing different viscosifying agents in the outer phase. It was observed that the compactibility, thickness, and stability of multiple emulsion containing cellulose derivatives and carbomer as viscosifying agents in the outer phase were increased not only at room temperature but also at 40°C (Ozer et al. 2000b). Related to this study, multiple emulsions that were able to break up and release smaller or larger proportion of the inner aqueous phase under moderate shear rates were formulated. Viscosification of the outer aqueous phase with a viscosifying agent or with a thickening agent was found necessary to obtain suitable release, to increase the stability, and to have a pleasant texture (Muguet et al. 2001).

Hino et al. stabilized the multiple emulsion system by making inner aqueous phase hypertonic, addition of chitosan to the inner phase, and techniques of phase inversion with porous membrane. All these methods were found effective, and lipiodol W/O/W emulsion for transcatheter arterial embolization (TAE) therapy was prepared with epirubicin hydrochloride. The formulation was suggested as an effective and stabile system (Hino et al. 2000).

Multiple emulsions ($W_1/O/W_2$) consist of water droplets (W_1) dispersed into oil globules (O), which are redispersed in an external aqueous phase (W_2). A small-molecule surfactant and an amphiphilic polymer were used to stabilize the inverse emulsion (W_1 in oil globules) and the inverse emulsion (oil globules in W_2), respectively. The composition of the droplet interface was determined by combining mechanical and interfacial tension measurements, and it was showed that due to the absence of polymeric surfactant diffusion, the stability of multiple emulsion enhanced (Michaut et al. 2004).

Emulsifiers that affect the external water/oil interface and the osmotic pressure may affect the stability of W/O/W multiple emulsions. So, the effects of interfacial film strength and pressure balance were investigated, and a microscopic method was described for accelerated stability test of multiple emulsions. The type and the concentration of Span were found important for emulsion stability with respect to film strength, viscosity, and surfactant coverage of the emulsion interfacial film (Jiao et al. 2002). In another study, the rheological properties of W/O/W multiple emulsions containing Span 83 and Tween 80 were determined, and long-term stability was found to be dependent upon the Span 83 concentration (Jiao and Burgess 2003).

Some other techniques were applied to the formulation of multiple emulsions for improved stability. For example, a water-soluble dye was added to the inner phase of W/O/W multiple emulsion, and the emulsion was dried using different freeze-drying procedures with different cryoprotectants in order to use this technique to produce a multiple emulsion with enhanced stability. Trehalose was found to be the most effective cryoprotectant for protecting the stability of multiple emulsion (Choi et al. 2007).

As it was indicated before, thickening the outer phase of multiple emulsion resulted in an increased stability. For this purpose, biopolymeric complexes of protein and polysaccharide were applied to W/O/W multiple emulsions for the entrapment of glucose and vitamin B1. Stability of the system was improved by using whey protein isolate–xanthan gum complex at the external oil/water interface (Benichou et al. 2007a). The same authors prepared an O/W/O multiple emulsions for the entrapment of flumethrin with the whey protein isolat–xanthan gum complex. This complex provided a thick and efficient barrier against release of entrapped material for both of the formulations (Benichou et al. 2007b).

15.3 MULTIPLE EMULSION APPLICATIONS

15.3.1 MULTIPLE EMULSIONS IN IMMUNIZATION

Since the 1960s, it has been known that multiple emulsions are effective adjuvants (Herbert 1965). Adjuvants are compounds or preparations that nonspecifically stimulate the immune response to an immunogen such as aluminum compounds, aluminum phosphate, and aluminum hydroxide. In 1937, Freund reported the development of a potent adjuvant consisting of killed tubercle bacilli in a water-in-mineral oil emulsion. This preparation is known as Freund's complete adjuvant (FCA), and the W/O emulsion alone is termed as Freund's incomplete adjuvant (FIA). Freund's adjuvants are used only in animals because they cause unacceptable morbidity in humans. Even mineral oil alone can produce granulomata and sterile abscesses (Klegerman and Groves 1992). Multiple emulsions had first been suggested as a method of producing prolonged antibody response instead of Freund's adjuvant, because of their notable advantage of low viscosity and ease of injection (Herbert 1965). Cholera toxin is an effective mucosal adjuvant but causes significant intestinal secretion, which limits its usefulness. By loading cholera toxin into the multiple emulsion with a soluble antigen for oral immunization, intestinal secretion was inhibited because of the presence of cholera toxin in the inner aqueous phase (Tomasi et al. 1997). A W/O/W type oil adjuvant vaccine containing an outer aqueous phase with a polyethylene glycol derivative and an inner aqueous phase containing an antigen was developed, and was granted patent right. This oil adjuvant vaccine formulation shows a high adjuvant effect, reduced side effects such as topical response, superior preparation stability, and superior convenience to allow a person to perform an injection easily due to the lowered viscosity (Saito et al. 2001).

Multiple emulsions were used also as vehicles for the parenteral administration of certain vaccines. Multiple emulsion was preferable compared to simple emulsion systems since they were easier to inject and the resultant antibody titer was much higher (Herbert 1965). Multiple W/O/W emulsions contain both W/O and O/W simple emulsions, because of the presence of the reservoir phase inside the droplets of another phase that can be used for prolonged release of active ingredients, multiple emulsions can avoid the booster dose of the vaccine (Bozkır and Saka 2007).

Verma and Jaiswal, in 1997, prepared a multiple emulsion vaccine using *Pasteurella multocida* as seed. This vaccine was easily administered and offered protection for up to one year in calves vaccinated against hemorrhagic septicemia, which is an economically important disease of cattle and buffaloes. Many countries use oil adjuvant vaccine for this infection, which provides both a higher degree and a long duration of immunity up to one year, but it has disadvantages such as high viscosity resulting in its poor injectability; moreover, swelling or irritation can occur at the inoculation site. This multiple emulsion vaccine against hemorrhagic septicemia provided an immunity parallel to oil adjuvant vaccine. In addition, several advantages were provided, such as stability, low viscosity, easy administration, and absence of undesirable effects at the inoculation site (Verma and Jaiswal 1997).

Singh et al., in 1997, examined induction of cytotoxic T lymphocyte (CTL) with peptide antigen in the form of W/O/W multiple emulsion in mice. In addition, the multiple emulsion–loaded peptide antigen was compared with a simple emulsion (O/W) for the CTL response induced. The results of this study showed that the induction of CTL response with a multiple emulsion was as potent as a simple emulsion (Singh et al. 1997). In another study, it was explained that, the induction of antibody response was also stimulated by multiple emulsion as antigen delivery vehicle as well as simple emulsion. In addition, it was stated that multiple emulsion had greater stability, easier application, and has produced fewer nodules at the injection site than simple emulsion (Silva-Cunha et al. 1996).

In 2004, Bozkır and Hayta developed stable W/O/W multiple formulations for intramuscular administration, which contains influenza hemagglutinin (HA) in the internal aqueous phase. Immune responses of formulations have been investigated in Wistar albino rats and compared with immune response raised against the conventional vaccine. The results of this study demonstrated that HA was well entrapped in the multiple emulsion formulation, and a single administration of entrapped HA was shown to stimulate a more effective immune response than conventional vaccine (Bozkır and Hayta 2004). These multiple emulsion formulations were compared with HA-entrapped nanoparticles, and multiple emulsions were shown to be more effective in immune response than nanoparticles. Nanoparticles are taken up by macrophages more easily than multiple emulsion formulations, so more nanoparticles were phagocytized than emulsions. Antigen entrapment was higher in multiple emulsions (99%) than in nanoparticles (74%), so emulsion formulations have given higher immunization than nanoparticles (Bozkır et al. 2004).

In another study, recombinant hepatitis B surface antigen (HBsAg) was incorporated into W/O/W multiple emulsion for examining antibody response in mice. The primary W/O emulsion was added into the outer water phase, containing hydrophilic emulsifier and recombinant HBsAg.

This multiple emulsion was compared with niosome and an aluminum-based suspension for antibody response to rHBsAg by immunizing mice intramuscularly. It was reported that responses to antigen in the form of multiple emulsion was higher than responses of niosomes and aluminum suspension, and the hemolytic effect of multiple emulsion was lower than the hemolytic effect of niosomes and aluminum suspension (Eraltay et al. 2004).

In 2008, Shahiwala and Amiji examined the potential of squalene oil-containing W/O/W multiple emulsions to induce systemic and nasal immunity following intranasal and oral administration in mice. The multiple emulsion formulations resulted in higher immunoglobulin G (IgG) and immunoglobulin A (IgA) responses as compared with aqueous solutions. In this study, immune response was evaluated by the administration of the first dose followed by a second "booster" dose. After first immunization, low IgG and IgA responses were observed with multiple emulsions; after the second dose, IgG and IgA response increased significantly with both routes of administration. It was shown that squalene oil-containing W/O/W multiple emulsion–based formulations can significantly enhance the immune response, especially after oral administration, and as a conclusion, they can be successfully adopted as a better formulation for administration of prophylactic and therapeutic vaccines (Shahiwala and Amiji 2008).

15.3.2 Multiple Emulsions for Peptide and Protein Delivery

Peptides and structurally-similar non-peptides have poor intestinal permeability and low oral bioavailability. Most peptides are poorly absorbed via the intestinal membrane because of extensive proteolytic degradation by intestinal enzymes and insufficient membrane permeability due to high molecular weight and low lipophilicity (Aungst et al. 1996). Engel et al. have investigated the possible use of W/O/W emulsions for the oral administration of insulin. A W/O/W multiple emulsion of insulin was delivered to normal rats and significant decrease in blood glucose level was observed (Engel et al. 1968).

The W/O/W emulsions also have been proposed to enhance the enteral bioavailability of insulin (Shichiri et al. 1974, Matsuzawa et al. 1995, Suzuki et al. 1998, Onuki et al. 2004). Shichiri et al. have reported that W/O/W insulin emulsions were quite resistant to the action of pepsin, chymotrypsin, and trypsin *in vitro*. Administration of multiple emulsion of insulin into the jejunum of normal rabbits was three to four times more than the administration of aqueous insulin. On the other hand, in the same study, multiple emulsion of insulin was administered orally by gastric gavage to rabbits, and large variations of the blood glucose and plasma insulin responses were observed (Shichiri et al. 1974). Matsuzawa et al., in 1995, also developed W/O/W emulsions as a carrier for insulin via the enteral route. Lecithin, Span 80, and soybean oil were used as oily phase, and gelatin was added in the inner water phase for improving stability and avoiding insulin escape from inner aqueous phase. In order to obtain smaller emulsion particles, emulsions were filtered with a membrane filter (0.45 μm). A significant hypoglycemic effect of insulin was observed using an *in situ* loop method in rats (Matsuzawa et al. 1995). This W/O/W emulsion formulation was slightly modified by incorporating unsaturated fatty acids to oily phase, and was evaluated for enteral delivery of insulin. Enhanced absorption of insulin without tissue damage was observed after ileal and colonic administration (Morishita et al. 1998). Silva-Cunha et al. examined W/O/W multiple emulsions of insulin for oral delivery. It has been shown that W/O/W multiple emulsions of medium-chain triglycerides containing sodium insulin are able to protect insulin against enzymatic degradation *in vitro*. The stability of encapsulated insulin in the presence of digestive enzymes such as pepsine, trypsin, or α-chymotrypsin was determined. It was reported that almost all the entrapped insulin was recovered after treatment with enzyme solutions while the free insulin in control preparations was degraded (Silva-Cunha et al. 1997b). The biological effects of W/O/W multiple emulsions of medium-chain triglycerides after oral administration were also examined. It was shown that, multiple emulsions containing sodium insulin are able to reduce glycemia of diabetic rats after oral administration (Silva-Cunha et al. 1998).

In order to optimize, insulin-loaded W/O/W multiple emulsions for oral delivery with different formulations were examined (Cournarie et al. 2004a). A stable fish oil–containing multiple emulsion was developed and compared with multiple emulsion composed of medium-chain triglycerides. It was explained that medium-chain triglycerides allow formation of more stable multiple emulsion of insulin than fish oil (Cournarie et al. 2004b). Cournarie et al. developed a multiple emulsion with reduced oil concentration (20%) by using two polymeric surfactants for insulin delivery. They reported that insulin surface properties favored drug-loaded multiple emulsion formation; both the characteristics and stability of the multiple emulsion were influenced by the conformation of the protein in solution (Cournarie et al. 2004a).

Dogru et al. prepared a W/O/W multiple emulsion for oral use of a peptide salmon calcitonin (sCT) for the treatment of osteoporosis. The peptide sCT was incorporated in the inner aqueous phase, and a protease inhibitor, aprotinin, was included in the outer water phase of the system. *In vitro* diffusion studies indicated that W/O/W multiple emulsion seem to protect sCT against proteolytic degradation. This formulation produced effects on serum calcium in rats similar to the commercial formulation (Doğru et al. 2000).

15.3.3 MULTIPLE EMULSIONS AS TARGETING DELIVERY SYSTEMS

The conventional W/O/W emulsions are readily taken up by the reticuloendothelial system (RES), so stealth-type or tissue-specific multiple emulsions for targeting were developed. Therefore, drugs can be targeted to the RES by incorporating them into W/O/W emulsion. The W/O/W emulsion is a dispersion of oil drops containing smaller water droplets that allow the delivery of drugs preferentially to the RES (Khopade et al. 1996, Uno et al. 1997).

For example, Tacrolimus, which is a very powerful immunosuppressive drug, was incorporated into W/O/W emulsion and given intravenously to the rats for targeting RES. It was shown that tacrolimus levels in the liver and spleen, which are primary components of the RES, were significantly higher in the animal group injected with multiple emulsion relative to the control group injected with Tacrolimus solution. This study suggests that the W/O/W emulsion is applicable as an intravenous drug carrier for local immunosuppression (Uno et al. 1997).

In another study, a W/O/W multiple emulsion system containing rifampicin was prepared with small droplet size to study targeting to the lungs. Polysaccharide-coated multiple emulsion was given intravenously to the rats and the tissue distribution of the drug was investigated. A significant enhancement in lung uptake and a decreased internalization by spleen was observed (Khopade et al. 1996).

Multiple emulsions were also used for targeting liver and TAE therapy. After hepatic arterial administration of epirubicin hydrochloride–loaded W/O/W emulsion to rats, it was observed that the concentration of the drug in liver was increased and the toxic effects of it on the normal hepatic cells were reduced. In the same study, Lipiodol W/O/W emulsion was suggested to be a suitable formulation for TAE therapy (Hino et al. 2000).

Talegaonkar and Vyas prepared a sphere-in-oil-in-water (S/O/W) multiple emulsions containing diclofenac sodium (DS) by gelatinization of inner aqueous phase. They aimed to target DS to non-RES organs such as lungs and inflammatory tissue (synovial fluid). It is concluded that the amount of drug in RES-rich organs (spleen, liver) were significantly lower than the values in non-RES organs, with poloxamer containing multiple emulsion (S/O/Wp) (Talegaonkar and Vyas 2005).

15.3.4 PROLONGED/CONTROLLED DRUG DELIVERY

Multiple emulsions can be used as an effective carrier for both lipophilic and hydrophilic drugs through oral, parenteral, and topical routes of administration. They have been used as controlled/prolonged drug delivery systems via both oral and parenteral routes. The main objective is to locate the best absorption region of drug after its encapsulation in multiple emulsions.

There are two types of multiple emulsions, W/O/W and O/W/O emulsions, and W/O/W emulsions have been studied in a wide area of application. They are a less dispersive system than W/O emulsions, which makes them more convenient for use in parenteral formulations.

The W/O/W multiple emulsions of vancomycin were prepared by using different formulations, emulsification methods, and conditions, and the properties of the formulations were evaluated. Stable W/O/W multiple emulsions with a particle size at about 3 μm and an entrapment efficiency at about 70% were prepared. The formulations showed higher plasma concentrations when compared with vancomycin solutions alone, indicating the potential clinical application of these systems (Okachi and Nakano 2000).

Due to their structure, multiple emulsions can offer promising pharmaceutical applications for controlled, prolonged drug delivery. Pandit et al. described a study by using indomethacin in the inner oily phase of an O/W/O multiple emulsion (Pandit et al. 1988). In another work of Mishra and Pandit, a prolonged release of an analgesic, pentazocine, was also obtained with O/W/O multiple emulsion. When the results were compared with the aqueous solutions of drugs, a significantly prolonged effect was observed with both multiple emulsions (Mishra and Pandit 1989).

The W/O/W multiple emulsion system of DS was prepared for prolonged delivery. The encapsulation efficiency of multiple emulsion was determined. The particle size of oil droplets was found to be influenced by the pressure and temperature. It was indicated that the inner water droplets have to be much smaller than the oil droplets in the multiple emulsion for a better encapsulation, thus a better prolonged effect was obtained (Lindenstruth and Müller 2004).

Multiple emulsions could have two compartments with different pH, and by using efficient ingredients such as octadecylamine and nonpolar oil (Parleam), the diffusion process between internal acidic aqueous phase and alkaline external phase could be reduced to maintain the two aqueous phases at two different pH values (Silva-Cunha et al. 1996, Tedajo et al. 2001). Related to this work, the antimicrobial properties of a vaginal multiple emulsion containing benzalkonium chloride in the outer phase of W/O/W multiple emulsion was studied (Tedajo et al. 2002). Then, the authors decided to incorporate chlorhexidine digluconate, which is a widely used antiseptic agent for vaginal applications, to the inner phase of the system. Assessment of the minimal bactericidal concentration for the active substances under hypoosmotic conditions demonstrated the synergistic activity and the applicability of the carrier system (Tedajo et al. 2005).

Stable multiple emulsions containing two nitroimidazole derivates, metronidazole and ornidazole for vaginal therapy, were formulated. Metronidazole and ornidazole were located in the internal and external phases of multiple emulsion separately, and the release was determined in acidic and alkaline medium to investigate better the effect of pH and location of active substance on the release. The *in vivo* release of ^{99m}Tc (99m Technetium)-labeled metronidazole and ornidazole from the multiple emulsions through rabbit vagina was also investigated. A prolonged release was observed in alkaline medium and the multiple emulsions were found locally effective in vagina (Ozer et al. 2007).

There are some studies about the effectiveness of multiple emulsions in topical applications. O/W/O multiple emulsion was used for the prolonged release of topical formulations of hydrocortisone. By this formulation, hydrocortisone was kept longer in the epidermis and dermis than simple emulsions. The entrapped substance can be transferred from the internal phase to the external phase through the middle, membrane phase, which provides a prolonged release. The topical use of O/W/O multiple emulsion with hydrocortisone presented as a challenge to decrease the side effects (Laugel et al. 1998).

Multiple emulsions were used to enhance the absorption of drugs as a bioavailability enhancer. A stable multiple emulsion system containing isoniazid in the inner phase was prepared, and the effect of drug concentration and additives in oleaginous phase on the release was evaluated. The release was enhanced with the increasing concentration of active substance and was followed by a steady-state release at high concentrations. Egg lecithin and oleic acid in the oily phase also produced an increased release. It was reported that the bioavailability was increased with multiple emulsion (Khopade and Jain 1998).

A W/O/W multiple emulsion was developed to protect bioactive substances from degradation by pancreatic enzymes. The W/O/W multiple emulsion was prepared with octanoic acid triacylglycerol as oil phase, which produces octanoic acid monoacylglycerol with the enzymatic hydrolysis. The multiple emulsion was found to be stable in the artificial stomach solution, and it was shown to be possible to control the delivery of the system to the intestine without deformation. The deformation was found to be dependent upon the diameter of the oil droplets (Shima et al. 2004).

The activated form of N-glutaryl-phosphatidylethanolamine (aNGPE) has been incorporated into $W_1/O/W_2$ double-emulsion globules to determine the rate of external coalescence in a $W_1/O/W_2$ emulsion. The reported findings indicated that the presence of aNGPE extended the rate of external coalescence in a $W_1/O/W_2$ emulsion. The incorporation of this molecule into these globules would not only enable protein conjugation to the globule surface but would also allow control over the rate of release of solubilized components from the globule. This would promote prolonged release of a therapeutic agent or any water-soluble compound from the inner aqueous phase (Lawson and Papadopoulos 2004).

15.3.5 MULTIPLE EMULSIONS AS ARTIFICIAL CELLS

W/O emulsions can be used to compartmentalize and select large gene libraries for a predetermined function. Genes can be transcribed and translated in the aqueous droplets of a W/O emulsion (Figure 15.2) (Tawfik and Griffiths 1998, Aharoni et al. 2005a, Rothe et al. 2006). Typically, an aqueous solution of genes and an *in vitro* transcription–translation system is stirred (or homogenized) into

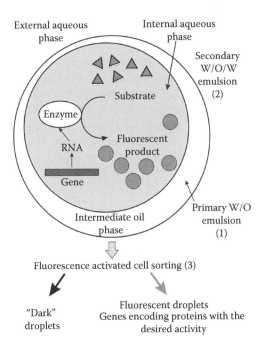

FIGURE 15.2   Gene selection in W/O/W emulsions: (1) Genes are compartmentalized in a W/O emulsion and translated *in vitro* in the presence of a fluorogenic substrate. Compartments in which the gene encodes an active enzyme become fluorescently labeled. (2) A second W/O/W emulsion is formed with an external aqueous phase. (3) Fluorescent compartments are isolated by FACS together with the genes encoding the protein of interest. (From Bernath, K. et al., *Anal. Biochem.*, 325, 151, 2004. With permission from Elsevier.)

an oil-surfactant mixture to create a W/O emulsion with ~10^{10} aqueous droplets per ml of emulsion. The majority of droplets contain no more than a single gene along with all of the molecular machinery needed to express that gene (Miller et al. 2006). The oil phase remains largely inert and restricts the diffusion of genes and proteins between compartments (Tawfik and Griffiths 1998, Aharoni et al. 2005a). The expressed proteins and the products of their catalytic activities cannot leave the droplets, and so genotype is coupled to phenotype *in vitro*, making it possible to select very large libraries of genes (Miller et al. 2006).

Subsequent conversion of the primary W/O emulsions into a W/O/W emulsion makes the emulsion amenable to sorting by flow cytometry without compromising the integrity of the inner aqueous droplets within the oil phase. So, this W/O/W emulsion can be sorted by fluorescence-activated cell sorting (FACS) while their droplets remain intact. Subsequently, genes embedded in the aqueous droplets of the primary W/O emulsion together with a fluorescent marker can be isolated and enriched from a large excess of genes embedded in W/O emulsion droplets that do not contain a fluorescent marker. The ability to create miniature aqueous compartments a few microns in diameter (by double emulsions) and then sort these compartments by FACS provides a powerful tool for the *in vitro* evolution of enzymes (Bernarth et al. 2004, Mastrobattista et al. 2005, Miller et al. 2006).

W/O/W emulsions have also been used to screen enzyme variants expressed in *Escherichia coli*. This approach was found useful when it is difficult to express an enzyme *in vitro* or when the enzyme is not very efficient (Aharoni et al. 2005b, Miller et al. 2006).

15.3.6 OXYGEN-CARRYING MULTIPLE EMULSIONS

Multiple emulsions had been tried out as a substitute to blood where hemoglobin has been incorporated in the inner phase of emulsion (Khan et al. 2007). Zheng et al. presented a prototype Hb multiple emulsion as a stable oxygen delivery system. A concentrated solution of hemoglobin (Hb) was encapsulated in the form of a Hb-in-oil-in-water (Hb/O/W) multiple emulsion. Hb multiple emulsions showed several important characteristics such as satisfactory rheological properties and good hydrodynamic stability, high encapsulation concentration of Hb, and satisfactory oxygen affinity. It was reported that Hb/O/W multiple emulsion system may provide utility as an oxygen-carrying red blood cell substitute or organ perfusion media (Zheng et al. 1991, 1992, 1993).

Borwanker et al. have developed a multiple emulsion encapsulating hemoglobin solution as a viable blood substitute. The encapsulation efficiency of the formulation was found excellent (greater than 90%), giving high oxygen-carrying capacity and was maintained over a 4 day storage period at 4°C (Borwanker et al. 1988).

15.3.7 ENZYME IMMOBILIZATION

In order to stabilize enzymes to facilitate their recovery and recycle, they are frequently immobilized (Brady et al. 2008). Multiple emulsions have become a new tool in the field of biotechnology for the immobilization of various enzymes, proteins, amino acids, etc., in biotechnology (Khan et al. 2006). Purified enzymes encapsulated in liquid surfactant membranes have been shown to retain their catalytic activity. Liquid membrane system is basically known as stabilized W/O/W multiple emulsion system, in which two miscible phases are separated by an organic phase. Liquid membrane acts as a thin semipermeable film through which solute must diffuse moving from one phase to another (Khan et al. 2007). Reports of using multiple emulsions for enzyme immobilization goes back to 1972, where hydrocarbon-based multiple emulsions were used to entrap urease enzyme for kidney diseases (May and Landgraff 1976, Khan et al. 2007). May and Landgraff prepared multiple emulsion for immobilization of yeast alcohol dehydrogenase (ADH)–nicotinamide adenine dinucleotide (NAD+) system. Enzymes or cofactors were placed in the inner aqueous phase and the external solution periodically assayed for the leakage of them. It was determined that, the extent of leakage was very small during experiments, and no significant increase was observed with time. Besides enzyme utilization, the other advantage of enzyme immobilization is reversibility; the emulsion can be broken and the enzymes recovered at the completion of process (May and Landgraff 1976).

Iso et al. prepared microcapsules by using W/O/W two-phase emulsion technique in order to provide immobilization of lipase (Iso et al. 1989).

Giovagnoli et al. have encapsulated superoxide dismutase (SOD) and catalase (CAT) in biodegradable microspheres for protein delivery. A modified W/O/W double-emulsion method was used for poly(D,L-lactide-co-glycolide) (PLGA) and poly(D,L-lactide) (PLA) microsphere preparation by co-encapsulating mannitol, trehalose, and PEG400 for protein stabilization (Giovagnoli et al. 2004).

In 2008, Shum et al. presented a novel approach for fabricating monodisperse phospholipid vesicles with high encapsulation efficiency using controlled double emulsions as templates. Glass-capillary microfluidics was used to generate monodisperse double-emulsion templates. They showed that the high uniformity in size and shape of the templates were maintained in the final phospholipid vesicles after a solvent removal step (Shum et al. 2008).

15.3.8 DRUG OVERDOSE TREATMENT

W/O/W emulsions have been used for removal of toxins from body fluids and were also valuable as alternative methods for the emergency treatment of drug overdose. An aqueous base-in-oil system was proposed to remove acidic drugs such as barbiturates and salicylates. The unionized drug, having significant oil solubility, permeates from the aqueous donor phase through the oil membrane into the basic inner phase where it is trapped in the form of an oil-insoluble anion. It was concluded that the emulsion can absorb the drug from either the stomach or small intestine (Frankenfeld et al. 1976). When emulsion is administered orally, the acidic pH of stomach becomes the external aqueous phase where an acidic drug exists in unionized form. The unionized drug would easily transport through the oil phase into the inner aqueous phase across the concentration gradient. Basic buffer in the inner phase ionizes the drug and prevents its backward transport. Thus, entrapping excess drug in multiple emulsions treats over dosage (Khan et al. 2006, 2007).

Frankenfeld et al. reported that liquid membranes are capable of rapid uptake of phenobarbital and acetyl salicylic acid as model drugs from various donor systems. It was determined that under favorable *in vitro* conditions, up to 95% of phenobarbital was removed in 5 min and acetyl salicylic acid was extracted slightly faster (Frankenfeld et al. 1976).

Chiang et al. also suggested liquid membranes for the drug overdose treatment. The *in vitro* removal of six barbiturates from pH 2 donor solutions by liquid membranes with pH control was examined and evaluated. They reported that more than 90% of barbiturates were removed within 10 min by the liquid membranes (Chiang et al. 1978).

15.3.9 MULTIPLE EMULSIONS IN COSMETICS

Common properties of multiple emulsions, especially their excellence on product presentation and skin feel, are of great interest for cosmetic industry. Multiple emulsions can be used widely in cosmetics and dermatological products depending on their many advantages such as high capacity of entrapment; protection of fragile substances; possibility to introduce incompatible substances in same formulation; and a pleasant skin feel with their efficient, light, nongreasy texture. There is no need to disperse them into another vehicle to obtain a convenient cosmetic form (Seiller et al. 1994).

Multiple emulsions have been recommended for lots of cosmetic formulations such as sun creams, moisturizing creams, makeup cleansers, perfume preparations, and so on. Skin moisturizing is the main effect claimed for the W/O/W products. Some companies have been able to carry these products to the market since 1991. The processes used in these products are often patented (Ferrero and Doucet 1996). In past years, different cosmetic active substances have been entrapped in multiple emulsion systems. W/O/W multiple emulsion systems were used to improve the stability of ascorbic acid (Gallarate et al. 1999), calcium thioglycolate (Gallarate et al. 2001), and mandelic acid (Carlotti et al. 2004). Recently, stable W/OW multiple emulsions with ethylene oxide free emulsifiers, which were more compatible with skin, were prepared. It was shown that these systems could encapsulate glycolic acid in their inner phase, which did not destabilize them (Carlotti et al. 2005).

In spite of the many useful and important roles on skin and mucous membranes, retinol is very unstable against light, heat, pH, and oxygen. Thus, to develop a gel microstructure by using amorphous silica, a microparticle production method was reported by combining sol-gel technology with O/W/O emulsion. Slower release of retinol from silica microspheres was obtained (Lee et al. 2001). Following this work, a O/W/O multiple emulsion was prepared with retinol entrapped spherical silica particles. To overcome the stability problems, different polymers were used as a stabilizer of the water phase. Hydoxypropyl cellulose was used as a stabilizer in the external oil phase. The encapsulation efficiency was found higher with Pluronic P123 (Hwang et al. 2005).

Ascorbic acid is a labile molecule and is very sensitive to hydrolysis. To avoid degradation and to improve stability, it was introduced into the inner phase of O/W/O multiple emulsions. Furthermore, some more advantages such as occlusivity, low viscosity, and nonoily texture have also been observed (Farahmand et al. 2006).

In another study, to achieve the deposition of water-soluble benefit agents onto skin, a multiple emulsion system was developed with low-HLB emulsifiers and three types of oil (mineral oil, vegetable oil, and silicone oil). A higher yield value was obtained with mineral oils compared to the vegetable oils. Stable multiple emulsion for personal cleansing systems could not be prepared when either the oil or the emulsifier or both contained unsaturated chains (Vasudevan and Naser 2002).

O/W microemulsions, O/W and W/O emulsions, and W/O/W multiple emulsion of ascorbic acid were prepared by using nonionic, nonethoxylated, glucose- or saccharose-derived, and skin-compatible emulsifiers to evaluate the oxidative stability of the vitamin. Ascorbic acid was incorporated into the inner phase of multiple emulsion at a pH value in which the vitamin is known to be stable. W/O/W multiple emulsion systems were found to be a more stable vehicle for ascorbic acid when compared with other emulsions (Gallarate et al. 1999).

15.4 CONCLUSION

Multiple emulsions are interesting forms with their protection, sustained and controlled release possibilities, and textures. This has facilitated the application of these systems in treatment, especially in areas such as targeting and prolonged/controlled drug delivery systems, for administration of prophylactic and therapeutic vaccines, peptide and protein delivery, the immobilization of enzyme systems, and for compartmentalizing and selecting gene libraries. W/O/W emulsions can also be used as a blood substitute by encapsulating hemoglobin solution in the inner phase. These systems are suggested for the removal of toxins from body fluids and are also valuable as alternative methods for the emergency treatment of drug overdose.

The main factors that limited their use are the long-term stability of these systems and their scaled-up production. By the way, the cosmetic industry is highly interested in multiple emulsions with a large number of patents by offering many advantages to cosmetic formulators. Thus, multiple emulsions are promising carriers as a delivery system, and remarkable advances have been made about their potential applications. Novel and more efficient multiple emulsion systems might be promising for the future.

ABBREVIATIONS

aNGPE   *N*-Glutaryl-phosphatidylethanolamine
ADH   Alcohol dehydrogenase
CAT   Catalase

CTL	Cytotoxic T lymphocyte
DS	Diclofenac sodium
FACS	Fluorescence activated cell sorting
FCA	Freund's complete adjuvant
FIA	Freund's incomplete adjuvant
HA	Hemagglutinin
Hb	Hemoglobin
Hb/O/W	Hb-in-oil-in-water
HBsAg	Hepatitis B surface antigen
IgA	Immunoglobulin A
IgG	Immunoglobulin G
ml	Milliliter
NAD+	Nicotinamide adenine dinucleotide
O	Oil globules
O/W	Oil-in-water
O/W/O	Oil-in-water-in-oil
PLA	Poly(D,L-lactide)
PLGA	Poly(D,L-lactide-*co*-glycolide)
rHBsAg	Recombinant HBsAg
RES	Reticuloendothelial system
sCT	Salmon calcitonin
SOD	Superoxide dismutase
S/O/W	Sphere-in-oil-in-water
TAE	Transcatheter arterial embolization
W	Aqueous phase
W/O	Water-in-oil
W/O/W	Water-in-oil-in-water
^{99m}Tc	99m Technetium
°C	Celsius temperature
μm	Micron

REFERENCES

Aharoni, A., Amitai, G., Bernath, K., Magdassi, S., and Tawfik, D.S. 2005a. High-throughput screening of enzyme libraries: Thiolactonases evolved by fluorescence-activated sorting of single cells in emulsion compartments. *Chemistry and Biology* 12:1281–1289.

Aharoni, A., Griffiths, A.D., and Tawfik, D.S. 2005b. High-throughput screens and selections of enzyme encoding genes. *Current Opinion in Chemical Biology* 9:210–216.

Aungst, J.B., Saitoh, H., Burcham, D.L., Huang, S.-M., Mousa, S.A., and Hussain, M.A. 1996. Enhancement of the intestinal absorption of peptides and nonpeptides. *Journal of Controlled Release* 41:19–31.

Benichou, A., Aserin, A., and Garti, N. 2007a. O/W/O double emulsions stabilized with WPI-polysaccharide conjugates. *Colloids and Surfaces A: Physicochemical and Engineering Aspects* 297:211–220.

Benichou, A., Aserin, A., and Garti, N. 2007b. O/W/O double emulsions stabilized with WPI-polysaccharide complex. *Colloids and Surfaces A: Physicochemical and Engineering Aspects* 294:20–32.

Bernath, K., Hai, M., Mastrobattista, E., Griffiths, A.D., Magdassi, S., and Tawfik, D.S. 2004. *In vitro* compartmentalization by double emulsions: Sorting and gene enrichment by fluorescence activated cell sorting. *Analytical Biochemistry* 325:151–157.

Borwanker, C.M., Pfeiffer, S.B., Zheng, S. et al. 1988. Formulation and characterization of a multiple emulsion for use as a red blood cell substitute. *Biotechnology Progress* 4:210–217.

Bozkır, A. and Hayta, G. 2004. Preparation and evaluation of multiple emulsions water-in-oil-in-water (W/O/W) as delivery system for Influenza virus antigens. *Journal of Drug Targeting* 12:157–164.

Bozkır, A. and Saka, O.M. 2007. Multiple emulsions: Delivery system for antigens. In *Multiple Emulsions: Technology and Applications*, A. Aserin (Ed.), pp. 293–306. Hoboken, NJ: John Wiley and Sons, Inc.

Bozkır, A., Hayta, G., and Saka, O.M. 2004. Comparison of biodegradable nanoparticles and multiple emulsions (water-in-oil-in-water) containing influenza virus antigen on the *in vivo* immune response in rats. *Pharmazie* 59:723–725.

Brady, D., Jordaan, J., Simpson, C., Chetty, A., Arumugam, C., and Moolman, F.S. 2008. Sperezymes: A novel structured self-immobilisation enzyme technology. *BMC Biotechnology* 8:8.

Buszello, K. and Müller, B.W. 2000. Emulsions as drug delivery systems. In *Pharmaceutical Emulsions and Suspensions*, F. Nielloud and M. Gilberte (Eds.), pp. 191–228. New York/Basel, Switzerland: Marcel Dekker.

Carlotti, M.E., Sapino, S., Morel, S., and Gallarate, M. 2004. W/O/W multiple emulsions with mandelic. *Journal of Drug Delivery Science and Technology* 14:409–417.

Carlotti, M.E., Gallarate, M., Sapino, S., Ugazio, E., and Morel, S. 2005. W/O/W multiple emulsions for dermatological and cosmetic use, obtained with ethylene oxide free emulsifier. *Journal of Dispersion Science and Technology* 26:183–192.

Chiang, C.W., Fuller, G.C., Frankenfeld, J.W., and Rhodes, C.T. 1978. Potential of liquid membranes for overdose treatment: *In vitro* studies. *Journal of Pharmaceutical Sciences* 67:63–66.

Choi, M.J., Briançon, S., Bazile, D., Royere, A., Min, S.G., and Fessi, H. 2007. Effect of cryoprotectant and freeze-drying process on stability of W/O/W emulsions. *Drying Technology* 25:809–819.

Cournarie, F., Rosilio, V., Chèron, M. et al. 2004a. Improved formulation of W/O/W multiple emulsion for insulin encapsulation. Influence of the chemical structure of insulin. *Colloid Polymer Science* 282:562–568.

Cournarie, F., Savelli, M.-P., Rosilio, V. et al. 2004b. Insulin-loaded W/O/W multiple emulsions: Comparison of the performances of systems prepared with medium-chain-triglycerides and fish oil. *European Journal of Pharmaceutics and Biopharmaceutics* 58:477–482.

Davis, S.S., Hadgraft, J., and Palin, K.J. 1985. Medical and pharmaceutical applications of emulsions. In *Encyclopedia of Emulsion Technology: Applications*, P. Becher (Ed.), pp. 159–238. New York/Basel, Switzerland: Marcel Dekker.

Devani, M.J., Ashford, M., and Craig, D.Q.M. 2005. The development and characterisation of triglyceride-based 'spontaneous' multiple emulsions. *International Journal of Pharmaceutics* 300:76–88.

Doğru, S.T., Çalış, S., and Öner, F. 2000. Oral multiple W/O/W emulsion formulation of a peptide salmon calcitonin: *In vitro-in vivo* evaluation. *Journal of Clinical Pharmacy and Therapeutics* 25:435–443.

Duchêne, D., Bochot, A., Yu, S.-C., Pépin, C., and Seiller, M. 2003. Cyclodextrin and emulsions. *International Journal of Pharmaceutics* 266:85–90.

Engel, R.H., Riggi, S.J., and Fahrenbach, M.J. 1968. Insulin: Intestinal absorption as water-in-oil-in-water emulsions. *Nature* 219: 856–857.

Eraltay, A., Öner, F., Özcengiz, E., and Alpar, R. 2004. Adjuvant effects of niosome and water/oil/water multiple emulsion carrier systems for recombinant Hepatitis B surface antigen. *Hacettepe University, Journal of the Pharmacy Faculty* 24:8–94.

Farahmand, S., Tejerzadeh, H., and Farboud, E.S. 2006. Formulation and evaluation of a vitamin C multiple emulsion. *Pharmaceutical Development and Technology* 11:255–261.

Ferrero, L. and Doucet, O. 1996. Cosmetic applications. In *Multiple Emulsions: Structure, Properties and Applications*, J.L. Grossiord and M. Seiller (Eds.), pp. 313–348. Paris, France: Editions de Santé Paris.

Florence, A.T. and Whitehill, D. 1980. Some features of breakdown in water-in-oil-in-water multiple emulsions. *Journal of Colloid and Interface Science* 79:243–256.

Florence, A.T. and Whitehill, D. 1982. The formulation and stability of multiple emulsions. *International Journal of Pharmaceutics* 11:277–308.

Frankenfeld, J.W., Fuller, G.C., and Rhodes, C.T. 1976. Potential use of liquid membranes for emergency treatment of drug overdose. *Drug Development and Industrial Pharmacy* 2:405–419.

Gallarate, M., Carlotti, M.E., Trotta, M., and Aimaretti, M. 2001. Formulation and characterization of w/o/w multiple emulsions of calcium thioglycolate. *Journal of Dispersion Science and Technology* 22:13–20.

Giovagnoli, S., Blasi, P., Ricci, M., and Rossi, C. 2004. Biodegradable micropheres as carriers for native superoxide dismutase and catalase delivery. *AAPS PharmSciTech* 5:1–9.

Herbert, W.J. 1965. Multiple emulsions: A new form of mineral-oil antigen adjuvant. *Lancet* 16:771–775.

Hino, T., Kawashima, Y., and Shimabayashi, S. 2000. Basic study for stabilization of w/o/w emulsion and its application to transcatheter arterial embolization therapy. *Advanced Drug Delivery Reviews* 45:27–45.

Hwang, Y.J., Oh, C., and Oh, S.G. 2005. Controlled release of retinol from silica particles prepared in o/w/o emulsion: The effects of surfactants and polymers. *Journal of Controlled Release* 106:339–349.

Iso, M., Shirahaze, T., Hanamura, S., Urishiyama, S., and Omi, S. 1989. Immobilization of enzyme by micro-encapsulation and application of the encapsulated enzyme in the catalysis. *Journal Microencapsulation* 6:165–176.

Jager-Lezer, N., Terrisse, I., Bruneau, F., Tokgoz, S., Ferreira, L., Clausse, D., Seiller, M., and Grossiord, J.L. 1997. Influence of lipophilic surfactant on the release kinetics of water-soluble molecules entrapped in a W/O/W multiple emulsion. *Journal of Controlled Release* 45:1–13.

Jiao, J. and Burgess, D.J. 2003. Rheology and stability of water-in-oil-in-water multiple emulsions containing Span 83 and Tween 80. *AAPS PharmSci* 5:1–12.

Jiao, J., Rhodes, D.G., and Burgess, D.J. 2002. Multiple emulsion stability: Pressure balance and interfacial film strength. *Journal of Colloid and Interface Science* 250:444–450.

Khan, A.Y., Talegeonkar, S., Iqbal, Z., Ahmed, F.J., and Khar, R.K. 2006. Multiple emulsions: An overview. *Current Drug Delivery* 3:429–433.

Khan, A.Y., Talegeonkar, S., Khar, R.K., Ahmed, F.J., and Iqbal, Z. 2007. Potentials of liquid membrane system: An overview. In latest reviews. http://www.pharmainfo.net/reviews/potentials-liquid-membrane-system-overview

Khopade, A.J. and Jain, N.K. 1998. Effects of drug concentration in inner aqueous phase and additives in oleaginous phase on release and bioavailability of Isoniazid from multiple emulsion. *Drug Development and Industrial Pharmacy* 24:677–680.

Khopade, A.J., Mahadik, K.R., and Jain, N.K. 1996. Targeting multiple emulsions to the lungs. *Pharmazie* 51:558–562.

Klegerman, E.M. and Groves, M.J. 1992. Vaccines. In *Pharmaceutical Biotechnology: Fundamentals and Essentials*, E.M. Klegerman and M.J. Groves (Eds.), pp. 64–69. Buffalo Grove, IL: Interpharm Press, Inc.

Kovács, A., Csòka, I., Kònya, M., Csányi, E., Fehér, A., and Erős, I. 2005. Structural analysis of w/o/w multiple emulsions by means of DSC. *Journal of Thermal Analysis and Calorimetry* 82:491–497.

Laugel, C., Baillet, A., Youenang Piemi, M.P., Marty, J.P., and Ferrier, D. 1998. Oil-water-oil multiple emulsions for prolonged delivery of hydrocortisone after topical applications: Comparison with simple emulsions. *International Journal of Pharmaceutics* 160:109–117.

Lawson, L.B. and Papadopoulos, K.D. 2004. Effects of phospholipid cosurfactant on external coalescence in water-in-oil-water double-emulsion globules. *Colloids and Surface A: Physicochemical and Engineering Aspects* 250:337–342.

Lee, M.-H., Oh, S.-G., Moon, S.-K., and Bae, S.-Y. 2001. Preparation of silica particles encapsulating retinol using o/w/o multiple emulsions. *Journal of Colloid and Interface Science* 240:83–89.

Lindenstruth, K. and Müller, B.W. 2004. W/O/W multiple emulsions with diclofenac sodium. *European Journal of Pharmaceutics and Biopharmaceutics* 58:621–627.

Mastrobattista, E., Taly, V., Chanudet, E., Treacy, P., Kelly, B.T., and Griffiths, A.D. 2005. High-throughput screening of enzyme libraries: *In vitro* evolution of a β-galactosidase by fluorescence-activated sorting of double emulsions. *Chemistry and Biology* 12:1291–1300.

Matsuzawa, A., Morishita, M., Takayama, K., and Nagai, T. 1995. Absorption of insulin using water-in-oil-in-water emulsion from an enteral loop in rats. *Biological and Pharmaceutical Bulletin* 18:1718–1723.

May, S.W. and Landgraff, L.M. 1976. Cofactor recycling in liquid membrane-enzyme systems. *Biochemical and Biophysical Research Communications* 68:786–792.

Michaut, F., Perrin, P., and Hbraud, P. 2004. Interface composition of multiple emulsions: Rheology as a probe. *Langmuir* 20:8576–8581.

Miller, O.J., Bernath, K., Agresti, J.J. et al. 2006. Directed evolution by *in vitro* compartmentalization. *Nature Methods* 3:561–570.

Mishra, B. and Pandit, J.K. 1989. Prolonged release of pentozocine from multiple O/W/O emulsions. *Drug Development and Industrial Pharmacy* 15:1217–1230.

Morais, J.M., Santos, O.D.H., Nunes, J.R.L., Zanatta, C.F., and Rocha-Filho, P.A. 2008. W/O/W multiple emulsions obtained by one-step emulsification method and evaluation of the involved variables. *Journal of Dispersion Science and Technology* 29:63–69.

Morishita, M., Matsuzawa, A., Takayama, K., Isowa, K., and Nagai, T. 1998. Improving insulin enteral absorption using water-in-oil-in-water emulsion. *International Journal of Pharmaceutics* 172:189–198.

Muguet, V., Seiller, M., Barratt, G., Ozer, O., Marty, J.P., and Grossiord, J.L. 2001. Formulation of shear rate sensitive multiple emulsions. *Journal of Controlled Release* 70:37–49.

Nakashima, T., Shimizu, M., and Kukizaki, M. 2000. Particle control of emulsion by membrane emulsification and its applications. *Advanced Drug Delivery Reviews* 45:47–56.

Okachi, H. and Nakano, M. 2000. Preparation and evaluation of w/o/w type emulsions containing vancomycin. *Advanced Drug Delivery Reviews* 45:5–26.

Olivieri, L., Seiller, M., Bromberg, L., Besnard, M., Duong, T.-N.-L., and Grossiord, J.-L. 2003. Optimization of a thermally reversible W/O/W multiple emulsion for shear-induced drug release. *Journal of Controlled Release* 88:401–412.

Onuki, Y., Morishita, M., and Takayama, K. 2004. Formulation optimization of water-in-oil-water multiple emulsion for intestinal insulin delivery. *Journal of Controlled Release* 97:91–99.

Ozer, O., Baloğlu, E., Ertan, G., Muguet, V., and Yazan Y. 2000a. The effect of the type and the concentration of the lipophilic surfactant on the stability and release kinetics of the w/o/w multiple emulsions. *International Journal of Cosmetic Science* 22:459–470.

Ozer, O., Muguet, V., Roy, E., and Grossiord, J.L. 2000b. Stability study of w/o/w viscosified multiple emulsions. *Drug Development and Industrial Pharmacy* 26:1185–1189.

Ozer, O., Aydın, B., and Yazan, Y. 2006. Effect of oil type on stability of w/o/w emulsions. *Cosmetics and Toiletries* 121:57–64.

Ozer, O., Ozyazıcı, M., Tedajo, M., Taner, M.S., and Köseoglu, K. 2007. W/O/W multiple emulsions containing nitroimidazole derivates for vaginal delivery. *Drug Delivery* 14:139–145.

Pandit, J.K., Mishra, B., Krishnaswamy, Y., and Mishra, D.N. 1988. Prolonged plasma and brain levels of indomethacin from O/W/O emulsions. *Indian Journal of Pharmaceutical Sciences* 50:274–275.

Prybilski, C., Luca, M., Grossiord, J.-L., and Vaution, C. 1991. W/O/W multiple emulsions manufacturing and formulation considerations. *Cosmetics and Toiletries* 106:97–100.

Raynal, S. 1996. Applications of multiple emulsions in the food industry. In *Multiple Emulsions: Structure, Properties and Applications*, J.L. Grossiord and M. Seiller (Eds.), pp. 349–372. Paris, France: Editions de Santé Paris.

Raynal, S., Grossiord, J.L., Seiller, M., and Clausse, A. 1993. A topical W/O/W multiple emulsion containing several active substances: Formation, characterization and study of release. *Journal of Controlled Release* 26:129–140.

Rothe, A., Surjadi, N.R., and Power, B.E. 2006. Novel proteins in emulsions using *in vitro* compartmentalization. *Trends in Biotechnology* 24:587–592.

Saito, K., Kishimoto, Y., Miyahara, T., and Takase, K. 2001. Oil adjuvant vaccine. http://www.freshpatents. com/Oil-adjuvant-vaccine-dt20050721ptan20050158330.php

Seifriz W. 1925. Studies on emulsions. *Journal of Physical Chemistry* 29:738.

Seiller, M., Orecchioni, A.M., and Vaution, C. 1994. Vesicular systems and multiple emulsions in cosmetology. In *Cosmetic Dermatology*, H.I. Maibach, and R. Baran, (Eds.), pp. 27–35. Singapore: Kyoda printing Co.

Shahiwala, A. and Amiji, M.M. 2008. Enhanced mucosal and systemic immune response with squalene oil-containing multiple emulsions upon intranasal and oral administration in mice. *Journal of Drug Targeting* 16(4):302–310.

Shichiri, M., Shimizu, Y., Yoshida, Y. et al. 1974. Enteral absorption of water-in-oil-in-water insulin emulsions in rabbits. *Diabetologica* 10:317–321.

Shima, M., Tanaka, M., Kimura, Y., Adachi, S., and Matsuno, R. 2004. Hydrolysis of the oil phase of a w/o/w emulsion by pancreatic lipase. *Journal of Controlled Release* 94:53–61.

Shum, H.C., Lee, D., Yoon, I., Kodger, T., and Weitz, D.A. 2008. Double emulsion templated monodisperse phospholipid vesicles. *Langmuir* 24:7651–7653.

Silva-Cunha, A., Grossiord, J.L., and Seiller, M. 1996. Pharmaceutical applications. In *Multiple Emulsions: Structure, Properties and Applications*, J.L. Grossiord and M. Seiller (Eds.), pp. 279–312. Paris, France: Editions de Santé Paris.

Silva-Cunha, A., Grossiord, J.L., Puisieux, F., and Seiller, M. 1997a. Insulin in W/O/W multiple emulsions: Preparation characterization and determination of stability towards proteases *in vitro*. *Journal of Microencapsulation* 14:311–319.

Silva-Cunha, A., Grossiord, J.L., Puisieux, F., and Seiller, M. 1997b. W/O/W multiple emulsions of insulin containing a protease inhibitor and an absorption enhancer: Preparation, characterization and determination of stability towards proteases *in vitro*. *International Journal of Pharmaceutics* 158:79–89.

Silva-Cunha, A., Chèron, M., Grossiord, J.L., Puisieux, F., and Seiller, M. 1998. W/O/W multiple emulsions of insulin containing a protease inhibitor and an absorption enhancer: Biological activity after oral administration to normal and diabetic rats. *International Journal of Pharmaceutics* 169:33–34.

Singh, M., Hioe, C., Qiu, H. et al. 1997. CTL induction using synthetic peptides delivered in emulsions-critical role of the formulation procedure, *Vaccine* 15:1173–1778.

Suzuki, A., Morishita, M., Kajita M. et al. 1998. Enhanced colonic and rectal absorption of insulin using a multiple emulsion containing eicosapentaenoic acid and docosahexaenoic acid. *Journal of Pharmaceutical Sciences* 87:1196–1202.

Talegaonkar, S. and Vyas, S.P. 2005. Inverse targeting of diclofenac sodium to reticuloendothelial system-rich organs by sphere-in-oil-in-water (s/o/w) multiple emulsion containing poloxamer 403. *Journal of Drug Targeting* 13:173–178.

Tawfik, D.S. and Griffiths, A.D. 1998. Man-made cell like compartments for molecular evolution. *Nature Biotechnology* 16:652–656.

Tedajo, G.M., Seiller, M., Prognon, P., and Grossiord, J.L. 2001. pH compartmented W/O/W Multiple emulsions: A diffusion study. *Journal of Controlled Release* 75:45–53.

Tedajo, G.M., Bouttier, S., Grossiord, J.L., Marty, J.P., Seiller, M., and Fourniat, J. 2002. *In vitro* microbicidal activity of W/O/W multiple emulsion for vaginal administration. *International Journal of Antimicrobial Agents* 20:50–56.

Tedajo, G.M., Bouttier, S., Fourniat, J., Grossiord, J.L., Marty, J.P., and Seiller, M. 2005. Release of antiseptics from the aqueous compartments of a W/O/W multiple emulsion. *International Journal of Pharmaceutics* 288:63–72.

Tomasi, M., Dertzbaugh, M.T., Hearn, T., Hunter, L., and Elson, C.O. 1997. Strong mucosal adjuvanticity of cholera toxin within lipid particles of a new multiple emulsion delivery system for oral immunization. *European Journal of Immunology* 27:2720–2725.

Uno, T., Yamaguchi, T., Li, X.K. et al. 1997. The pharmacokinetics of water-in-oil-in water-type multiple emulsion of a new tacrolimus formulation. *Lipids* 32:543–548.

Vasiljevic, D, Vuleta, G., and Primorac, M. 2005. The characterization of the semi-solid W/O/W emulsions with low concentrations of the primary polymeric emulsifier. *International Journal of Cosmetic Science* 27:81–87.

Vasiljevic D., Parojcic, J., Primorac, M., and Vuleta, G. 2006. An investigation into the characteristics and drug release properties of multiple w/o/w emulsion systems containing low concentration of lipophilic polymeric emulsifier. *International Journal of Pharmaceutics* 309:171–177.

Vasudevan, T.V. and Naser, M.S. 2002. Some aspects of stability of multiple emulsions in personal cleansing systems. *Journal of Colloid and Interface Science* 256:208–215.

Verma, R. and Jaiswal, T.N. 1997. Protection, humoral and cell mediated immune responses in calves immunized with multiple emulsion haemorrhagic septicaemia vaccine, *Vaccine* 15:1254–1260.

Yazan, Y., Seiller, M., and Puisieux, F. 1993. Multiple emulsions. *Bollettino Chimico Farmaceutico* 132:187–196.

Yu, S.-C., Bochot, A., Bas, G.L. et al. 2003. Effect of camphor/cyclodextrin complexation on the stability of O/W/O multiple emulsions. *International Journal of Pharmaceutics* 261:1–8.

Zheng, S., Beissinger, R.L., and Wasan, D.T. 1991. The stabilization of hemoglobin multiple emulsion for use as a red blood cell substitute, *Journal of Colloid and Interface Science* 144:72–85.

Zheng, S., Beissinger, R.L., and Wasan, D.T. 1992. Measurement of yield of hemoglobin (Hb)-in-oil-in-water multiple emulsion based on Hb encapsulation efficiency. *Journal of Dispersion Science and Technology* 13:33–44.

Zheng, S., Zheng, Y., Beissinger, R.L., Wasan, D.T., and Mc Cormick, D.L. 1993. Hemoglobin multiple emulsion as an oxygen delivery system. *Biochimica et Biophysica Acta* 20(1158):65–74.

16 Multiple Emulsion–Solvent Evaporation Technique

Tivadar Feczkó

CONTENTS

16.1 INTRODUCTION

Microencapsulation by solvent evaporation is widely applied in pharmaceutical investigations in order to control drug release. Forming polymer nano- or microspheres can degrade and release the encapsulated drug slowly with a specific release profile. The sustained drug release has clinical benefits: reducing the dosing frequency, more convenience and acceptance for patients, and drug targeting to specific locations resulting in a higher efficiency. The appropriate choice of the encapsulation method to achieve an efficient drug incorporation depends on the hydrophilicity or the hydrophobicity of the drug.

The water-in-oil-in-water (W/O/W) method is the multiple emulsion procedure most often applied, which is primarily used for encapsulating water-soluble drugs. It involves dissolving the drug in distilled water or in a buffer solution (inner water phase) and the polymer in a volatile organic solvent that is not miscible in water (organic phase). The inner water phase (containing a surfactant or not) is then poured into the organic phase. This mixture is generally emulsified forming the first inner emulsion or the primary emulsion (W/O). Stability of the primary emulsion is a prerequisite for the successful encapsulation of multiple emulsions. This W/O emulsion is then poured into the outer aqueous phase that contains an emulsifier, and homogenized while forming the double W/O/W emulsion. The multiple emulsion is continuously stirred and the solvent diffuses into the aqueous phase and evaporates at the water/air interface inducing polymer precipitation and, thereby, the formation of solid drug-loaded microspheres. Solvent evaporation can take place under atmospheric or reduced pressure with an evaporator (Li et al., 2001, Olivier et al., 2002), at various temperatures. The solid microspheres are isolated by filtration or centrifugation, washed several times to eliminate the emulsifier agent, and dried under vacuum or freeze-dried. Besides the W/O/W method, other multiple emulsion–solvent evaporation procedures exist to produce sustained delivery devices for pharmaceutical purposes mostly, although their prevalence is rather low. At the end of this chapter, these processes and some examples of their applications are demonstrated.

16.2 PARAMETERS OF EMULSION PREPARATION INFLUENCING THE PROPERTIES OF FORMING PARTICLES

The most important process parameters that affect the properties (size, encapsulation efficiency and release kinetics, morphology) of the forming particles are summarized in this section. Increasing the volume fraction of the internal aqueous phase in the primary W/O emulsion resulted in lower encapsulation efficiencies according to Herrmann and Bodmeier (1998). However, Yang et al. (2000) stated that the volume ratio of oil to internal water has less effect on the encapsulation efficiency, but has a significant impact on surface morphology of the resulting microspheres. To obtain high encapsulation efficiencies, the volumes of the inner and outer phases had to be decreased and the osmotic pressure had to be equilibrated in order to efficiently limit the diffusion of water from and to the inner water phase and consequently to achieve a less porous structure (Al haushey et al., 2007). On the contrary, Lamprecht et al. (1999) found that an increase in the volume of external aqueous phase led to an increase in both the entrapment efficiency and the size of nanoparticles. The increase in particle size is probably attributed to a reduction of the shear during the homogenization process, because of a decrease in mixing efficiency associated with larger volume, involving an increase of the size of the emulsion droplets and consequently of the nanoparticles. Depending on the inner aqueous phase content of the emulsion droplet and on the size of the inner microdroplet relative to the emulsion droplet, the outcome will be a mixture of microparticles with honeycomb, capsule, and plain structure (Rosca et al., 2004). When the polymer does not precipitate rapidly at the outer surface, the diffusion of protein into the aqueous phase can take place before the formation of particles. In spite of its low solubility in the organic polymer solution, this phenomenon could involve a part of the loaded protein to relocate at the particle surface (Lamprecht et al., 1999). During solvent elimination and shrinkage, the inner microdroplets gradually coalesce under the pressure of the precipitating polymer. The polymer wall around them may break forming holes on the microparticle surface, through which the inner aqueous phase is partly expelled, affecting the loading efficiency. After particle hardening, these holes will contribute to the encapsulated substances leakage trough partitioning with the external aqueous phase and to the initial burst release.

The microsphere size tends to decrease with increasing mixing speed to form the double emulsion (Herrmann and Bodmeier, 1998). Particle mean diameter also depends upon the mixing methods used to prepare the first inner W/O emulsion and the second W/O/W emulsion. When both are fabricated by vortex-mixing, the obtained microspheres are large. However, when the inner W/O emulsion is prepared by sonication and the second W/O/W emulsion by vortex-mixing, a microfine inner emulsion is formed. The resulting microspheres become smaller and very homogenous. Finally, when the inner emulsion is formed with a homogenizer and the second emulsion with a vortex mixer, the particle diameter is intermediate between the sizes of the microspheres obtained by both methods (Couvreur et al., 1997). The stirring rate during the second emulsion should be high to obtain small particles, but this could have a negative effect on the protein stability and a compromise could be needed (Al haushey et al., 2007). Nevertheless, the interface between water and a solvent is the most significant destabilizing factor of proteins (Yang et al., 2000; Bilati et al., 2005). Typically, proteins become especially prone to aggregation from the moment they migrate and adsorb at the interface. They undergo unfolding when their hydrophobic area gets to the organic solvent. The formation of the primary W/O emulsion, and to a lesser extent of the secondary W/O/W emulsion, is mainly responsible for protein denaturation. Protein degradation might also take place upon mechanical shearing and exposure to ultrasound.

Particle size can be influenced also by the preparation temperature, as increasing it might lead to a decrease in microsphere size (Yang et al., 2000). This is due to the fact that the W/O/W double emulsion at high temperatures is less viscous; thus, it is much easier for the emulsion to be broken up into smaller droplets at the same input power of mixing. However, higher preparation temperatures (above 29°C) yield larger sizes of microspheres, probably due to rapid solvent evaporation. An interesting phenomenon found was that the microspheres fabricated at 4°C and 38°C yielded the highest encapsulation efficiency (52.0%–48.0%) and the lowest initial protein release (18.8%–20.0%), and microspheres produced at 22°C had the lowest encapsulation efficiency and the highest initial burst. Decreasing the preparation temperature results in an increase in encapsulation efficiency and a decrease in the initial burst. This interesting result is due to many factors. Since solubility parameter, mass transfer coefficient, and diffusivity of protein decrease with reducing temperature, a low preparation temperature leads to less dissolution loss of protein in the external water phase. In addition, at low temperatures, the coagulation rate during the phase inversion for the W/O/W emulsion becomes faster because of a higher solubility parameter difference of solvent between the polymer and water at lower temperatures. As a consequence, the lower the temperature, the faster and the tighter the outer skin formation. This creates more difficulties for proteins to migrate outward the surface of microspheres and reduces the initial burst. Nucleation growth and spinodal decomposition dominates the skin formation at low preparation temperatures. At high preparation temperatures, the skin formation is mainly dominated by evaporation process, which may yield microspheres with a dense skin and a highly packed substructure (underneath the skin), and hence results in slower release profiles.

Among these stages, solvent removal and solidification process may be the dominant stage to determine morphology and release characteristics of microspheres.

16.2.1 Materials Used and Their Effects

Beside process parameters the compounds such as polymers, solvents, surfactants, and other additives applied during the particle manufacturing impact the particle characteristics substantially.

16.2.1.1 Polymers

Polymer concentration is a key factor to influence the features and release profiles of microspheres (Yang et al., 2000). Enhancing polymer concentration results in microspheres with large size and low initial burst. This arises from the fact that a viscous polymer solution is more difficult to be broken up into smaller droplets at the same input power of mixing. An increase in polymer concentration can lead to an increase of the viscosity of the first emulsion and consequently to a reduction in the partitioning of the protein into the external aqueous phase, resulting in an increase in protein entrapment efficiency (Lamprecht et al., 1999). However, beyond a level of polymer concentration, particles aggregated due to the collapse of the particles.

Size, encapsulation efficiency, and *in vitro* release properties (enzymatic activity retention and protein quantification) were affected by the polymer molecular weight (Gaspar et al., 1998). Nanoparticles made of high-molecular-weight poly(lactic-*co*-glycolic acid) (PLGA) had a larger size, a higher loading, and a slower release rate than those made of low-molecular-weight PLGA.

The most relevant factor affecting the entrapment and release of protein from PLGA nanoparticles was the presence of free carboxyl-end groups in the PLGA chain. The nanoparticles made of PLGA with free carboxyl-end groups had a high protein loading and provided a continuous delivery. A substantial increase in the molecular weight of polymer required a dilution of the polymer solution to prevent an exceedingly high viscosity, and led to less stable primary emulsions and more porous solid microspheres (Yang et al., 2000). A semicrystalline polymer was found to be poorly suited for the preparation of sustained-release microparticles. The crystallinity of polymers affected the stability of the primary emulsion by exclusion of the internal aqueous droplets from the polymer matrix. This exclusion adversely impacted the encapsulation efficiency, morphology, and porosity of microparticles prepared from a double emulsion. An increase in the amount of polymer used appeared to give rise to microcapsules with a dense, less-porous polymeric phase, thereby inhibiting the burst effect. Porosity of the microspheres was also related to the initial burst effect and release patterns of the microspheres. The particle size increases when the polymer concentration in the organic phase increases (Couvreur et al., 1997). Increasing the concentration of the dissolved polymer increased the viscosity of the organic phase, which resulted in turn in a reduction of the stirring efficiency. The higher the viscosity of polymer solution, the more difficult was the formation of small emulsion droplets.

The degradation rate depends on polymer molecular weight, crystallinity, and ratio of homopolymers in a copolymer (Couvreur et al., 1997), as well as tacticity and presence of additives (Panyam et al., 2003). For example, L- and D-lactic acids have crystalline form, and the D,L-polymers are amorphous and more rapidly degradable (Couvreur et al., 1997). On the other hand, because of methyl group, the lactic acid polymer is more hydrophobic than the glycolic one, and more hydrophilic polymers degrade quicker than that with hydrophobic nature (Bai et al., 2001). The release of proteins from PLGA spheres is not only governed by the degradation rate of the polymer, but also by the affinity of the protein versus the polymer. PLGA with free terminal carboxyl groups can be used to encapsulate proteins very efficiently, and it slows down the protein release (Blanco and Alonso, 1997).

PLGA is the most preferably used polymer in the W/O/W double emulsion method. Active agents are also often encapsulated by polylactic acid (PLA) (Conway et al., 1997; Leo et al., 2006). The structure of PLA and PLGA are sometimes modified with a hydrophilic, flexible, and nonionic polymer, polyethylene glycol (PEG) in the so-called pegylation to prepare stealth, or long-circulating nanoparticles (Li et al., 2001; Dorati et al., 2007). The next polymer in the preference order may be poly(ε-caprolactone) (PCL) (Benoit et al., 1999; Al haushey et al., 2007; Coccoli et al., 2008). Poly(hydroxybutyrate) (PHB) and poly(hydroxybutyratehydroxyvalerate) (PHBHV) were also applied in some multiple emulsion processes (Atkins, 1997; Conway et al., 1997; Eligio et al., 1999). More rarely, encapsulating polymers are poly(ethylene adipate) (PEAD) (Atkins, 1997), polyphosphazene (Caliceti et al., 2000), poly(methylidene malonate) (Le Visage et al., 2001), polyanhydrides (Tabata et al., 1993), poly(anhydride-*co*-imides) (Chiba et al., 1997), polymethylmethacrylate (PMMA) (Zydowicz et al., 2002), ethylcellulose (Ubrich et al., 1996), and poly(ethylene carbonate); the latter one is exclusively surface bioerodible (Lambert et al., 2000).

16.2.1.2 Solvents

The volatile organic solvent used in the preparation of the microspheres by the multiple emulsion–solvent evaporation technique needs to possess a low boiling point to overcome the problems associated with residual solvent. Many solvents can be used for the preparation of microspheres: acetonitrile (ACN), ethyl acetate, chloroform, benzene, and dichloromethane (DCM). Methylene chloride is the most common solvent for the encapsulation using solvent evaporation technique (Li et al., 2008) because of its high immiscibility with water, low boiling point, and high volatility that facilitates easy removal by evaporation, and also a range of encapsulating polymers shows good solubility in it. Its high saturated vapor pressure compared to other solvents (at least two times higher) promises a high solvent evaporation rate, which shortens the duration of fabrication of microspheres. However, this solvent is confirmed carcinogenic according to Environmental Protection Agency data, and the researchers are making great efforts to find less toxic and non-chlorinated replacements. One type of endeavor is to decrease the amount of DCM, e.g., by using mixture of it and acetone (Ricci et al., 2004). Ethyl acetate shows promising potential as a less toxic substitute of DCM, and it induces less emulsification-induced denaturation of proteins than methylene chloride (Wolf et al., 2003). But due to the partial miscibility of ethyl acetate in water (4.5 times higher than that of methylene chloride), microspheres cannot form if the dispersed phase is introduced directly into the continuous phase. The sudden extraction of a large quantity of ethyl acetate from the dispersed phase makes the polymer precipitate into fiber-like agglomerates. To resolve this problem created by the miscibility of solvent with water, three methods can be used (Li et al., 2008):

1. The aqueous solution is presaturated with solvent.
2. The dispersed phase is first emulsified in a little quantity of aqueous solution. After the formation of drops, this emulsion is poured into a large quantity of aqueous solution.
3. The dispersed phase is emulsified in a little quantity of aqueous solution, the solution is agitated, and the solvent evaporates leading to solidification of microspheres.

After using the aforementioned methods, the microspheres are manufactured successfully with ethyl acetate. However, the microspheres prepared by methylene chloride are spherical and more uniform, while the use of ethyl acetate results in particles that appear to be partly collapsed and more porous (Herrmann and Bodmeier, 1998). The drug encapsulation efficiency reduces significantly compared to the microspheres made by methylene chloride. It is assumed that the high solubility of ethyl acetate in water leads to the loss of drug. More drug is entrained into the continuous phase by the higher mass flux of solvent, which is driven by the diffusion from the dispersed phase into the continuous phase; further, the big quantity of solvent present in the continuous phase increases the solubility of drug in the continuous phase, facilitating the diffusion of drug into the continuous phase.

Ethyl formate, as another organic solvent instead of DCM, shows also interesting results; the evaporation rate of it in water was 2.1 times faster than that of methylene chloride although ethyl formate possesses a lower vapor pressure and a higher boiling point (Li et al., 2008). This phenomenon is explained by the fact that more molecules of ethyl formate are exposed to the air/liquid interface because of its higher water solubility. In summary, some less toxic solvents have been tested and show a promising future, but there are not enough results to compare the quality of microspheres prepared by different solvents.

16.2.1.3 Surfactants

The surfactant, also called tensioactive agent, is frequently employed for the dispersion of one phase in another immiscible phase and for the stabilization of the obtained emulsion (Bilati et al., 2005). The presence of an emulsifying agent in the external water phase is critical for the successful formation of individual spherical microspheres. The role of the emulsifier is to prevent the coagulation of microspheres during solvent removal. When the concentration of the surfactant in the outer water phase increases, a significant reduction in particle size is achieved (Couvreur et al., 1997). At high surfactant concentration, the rate of emulsion stabilizer molecules that diffuse to the emulsion droplets/aqueous phase interface may increase. This would provide an improvement in the protection of droplets against coalescence, resulting in the formation of smaller emulsion droplets. As the solvent evaporates from the system, these droplets harden to form the nano- or microspheres. Therefore, the size of the finally obtained microspheres is dependent upon the size and stability of the emulsion droplets formed during the agitation process. At low surfactant concentration, small emulsion droplets are not stable, and the resulting microspheres are larger in size than those prepared with higher surfactant concentrations. When adding a hydrosoluble stabilizing agent [gelatin, ovalbumin (OVA)] to the inner water phase, particle size increases due to the rise in the diameter of the internal globules. The addition of surfactant lowers the surface tension of the continuous phase and the diminution of the latter one decreases the particles size and prevents protein adsorption and/ or aggregation at hydrophobic surfaces (Bilati et al., 2005). However, due to the critical micelle concentration (CMC), the surface tension cannot decrease infinitively. When surfactant concentration reaches a certain level, the solution surface is completely loaded. Any further additions of surfactant will arrange as micelles, and the surface tension of the aqueous phase will not decrease any more.

Before choosing the type of surfactant and its concentration, it is important to know the polarity of the two immiscible phases, the desired size of microspheres, and the demand on the sphericity of microspheres. Surfactants for emulsions are amphiphilic: That means one part of the molecule has more affinity to polar solutes such as water (hydrophilic) and the other part has more affinity to nonpolar solutes such as hydrocarbons (hydrophobic). When it is present in an emulsion, the surfactant covers the surface of drops with its hydrophobic part in the drop and its hydrophilic part in the water. There are four different types of surfactants classified by the nature of the hydrophilic part of a molecule: anionic, cationic, amphoteric, and nonionic. The anionic surfactants release a negative charge in the aqueous solution. They have a relatively high hydrophile–lipophile balance level because they are prone to be hydrophilic. The cationic surfactants, on the contrary, release a positive charge in aqueous solution. The amphoteric surfactants behave as anionic in alkali pH and as cationic in acid pH. Nonionic surfactants have no charge.

In the W/O/W multiple emulsion system, anionic stabilizers such as sodium dodecyl sulfate (SDS) and cationic stabilizers such as cetyltrimethylammonium bromide (Basarkar et al., 2007) are used very rarely, since ionic surfactants might bind to groups in proteins and cause denaturation. Among nonionic surfactants preferred, partially hydrolyzed polyvinyl alcohol (PVA) is mostly used due to its good emulsifying properties (Feczko et al., 2008). Panyam et al. (2003) showed that the surface-associated PVA rather than the particle size play a dominant role in controlling the degradation of nano- and microparticles. Pluronic® F68 (also called as poloxamer 188) was used both in the external water phase as an emulsifier (Rosa et al., 2000; Feczko et al., 2008) and in the internal phase as a lyoprotectant, which saved the entrapped enzyme activity (Wolf et al., 2003). It was also particularly successful in minimizing protein denaturation during encapsulation and/ or release. Poloxamer with higher molecular weight, called Pluronic F127, was more rarely applied (e.g., Leo et al., 2006). Polyvinyl pyrrolidones (PVP) are also appropriate surfactants (Coombes et al., 1998; Feczko et al., 2008) in the double emulsion method. Some proteins can behave as surface-active agents, e.g., bovine serum albumin (BSA) and human serum albumin have been effective in improving the encapsulation efficiency and stability of the encapsulated protein (Sanchez et al., 2003). Polysorbate 20 can be applied in the external phase (Rosa et al., 2000); however, it was found not to be a good stabilizer against the unfolding effect of the water/methylene chloride interface (Bilati et al., 2005). Both the hydrophilic (PEG chains) and the hydrophobic part (fatty acid chain) of the polysorbate molecules were preferentially partitioned in the methylene chloride phase, leading to low protection efficacy. Sodium cholate was also used for emulsifying (Sanchez et al., 2003) as it did not interfere with blood coagulation factors (Olivier et al., 2002). Monomethoxypolyoxyethylene–PLA hydrophilic block copolymer surfactant was used to coat PLA or PLGA hydrophobic particles, and to prevent the protein adsorption or to increase the blood circulation half-lives of the carriers (Bouillot et al., 1999). It can be applied instead of PVA when a biocompatible and bioeliminable surfactant is required for biological or medical applications in order to synthesize micro- or nanoparticles (Chognot et al., 2006).

16.2.1.4 Sugars

Sugars (e.g., sucrose, trehalose, and mannitol) are known to stabilize proteins in an aqueous solution; hence, during a W/O/W procedure, sugars are often added to the inner aqueous phase (Bilati et al., 2005). They are depleted in the surface layer of a protein, because they increase the surface tension of water by forming cooperative structures with water. This surface exclusion of sugars results in the preferential hydration of the protein and increases the free energy of the system. Because protein denaturation would augment this thermodynamically unfavorable effect, proteins in a solution state can be stabilized by the presence of sugars (Sah, 1999). Protecting effect of sugars against stress factors was not established at every protein. Some examples are demonstrated here both for effective and ineffective protection.

Trehalose was shown to partially improve the BSA secondary structure protection within PLGA microspheres and to facilitate BSA monomer release (Bai et al., 2001; Bilati et al., 2005). Trehalose and mannitol had a significant effect on the recovery of soluble non-aggregated proteins, e.g., interferon-γ after emulsification and ultrasonication, whereas no or very little influence on insulin-like growth factor was observed. Trehalose also improved asparaginase biological activity. No effect of trehalose, mannitol, and sucrose was observed against O/W interface-induced degradation of lysozyme, whereas lactose and lactulose significantly improved its structural stability and activity, mostly if these additives were also added to the second aqueous phase. Sugars could affect the encapsulation efficiency in special cases. Wang et al. (2004) described the significant reduction in initial burst release of a highly water-soluble model peptide, octreotide acetate, from PLGA microspheres by the coencapsulation of a small amount of glucose. The effect of glucose on initial burst are determined by two factors: (1) increased initial burst due to increased osmotic pressure during encapsulation and drug release, and (2) decreased initial burst due to decreased permeability of microspheres.

The biopolymer chitosan (CS) was used as an emulsifier in food double emulsions (Benichou et al., 2004). CS has surface activity and seems to stabilize W/O/W emulsions. CS reacts with anionic emulsifiers such as sodium dodecylsulfate at certain ratios to form a water-insoluble complex that has strong emulsification capabilities. Benichou et al. (2004) studied also the stabilization of double emulsions with some so-called molecular-recognition hybrids of whey protein isolate with different charged and uncharged polysaccharides. Double emulsions were prepared with different whey protein–xanthan ratios. By increasing the protein-to-polysaccharide ratio, the double emulsion droplets were smaller and the stability was increased. One can conclude that the binodal boundary line corresponds to the limit of complexation of the two biopolymers that will give optimum stabilization. At biopolymer content higher than the binodal composition, the addition of one of the two biopolymers will only slightly affect the double emulsion droplets stability, and will only add small depletion stabilizing effect to the emulsion droplets. More detailed presentation of sugar additives can be found in the work of Bilati et al. (2005).

16.2.1.5 Protein Additives

Gelatin, whey proteins, BSA, human serum albumin, caseins, and other proteins were used usually in combination with other nonionic monomeric emulsifiers (Benichou et al., 2004). Proteins or other macromolecular stabilizers are unlikely to completely replace lipophilic monomeric emulsifiers in double emulsions. However, proteins in combination with stabilizers do have the capacity to confer some enhanced degree of stability on a double emulsion system, and therefore the lipophilic emulsifier concentration is substantially reduced. A significant improvement in the stability of the emulsions was shown when these macromolecules were encapsulated onto the external interface (Sanchez et al., 2003).

Albumins and gelatins are proteins mainly used for protection purposes in W/O/W double emulsions. The protective effect of albumins against protein unfolding and aggregation has been extensively documented (Bilati et al., 2005) and is likely due to their surface-active properties. Albumins (i.e., bovine, human, or rat serum albumins) are thought to stabilize the emulsion and occupy the interfaces and shield the therapeutic proteins from contact with the solvents or hydrophobic surfaces. BSA together with small amounts of monomeric emulsifiers (or hydrocolloids) serves as good steric stabilizers and improves stability, shelf-life, and slows down the release of the markers. BSA plays double role in the emulsions: film former and barrier to the release of small molecules from the internal interface, and steric stabilizer at the external interface. The release mechanisms involving reverse micellar transport were also examined, and it was concluded that BSA reduces the chance of reverse micelle formation in the oil phase, and thus reduces the release rate. Since the particle size is related to a great extent to the stability of the first emulsion, serum albumin can act as a surfactant by stabilizing the first emulsion and consequently hampering the fast coalescence of the droplets (Lamprecht et al., 1999). However, when a hydrosoluble stabilizing agent (gelatin, OVA) is added to the inner water phase, particle size increases due to the rise in the diameter of the internal globules. OVA added to the internal water phase was shown to improve the encapsulation rate (Blanco-Prieto et al., 1997).

16.3 AGENTS ENCAPSULATED

W/O/W double emulsion procedure is a suitable method for encapsulating hydrophilic substances. Water soluble peptides and proteins have been incorporated most times with this method, since there are a lot of rather sensitive and valuable pharmaceutical agents in this material group.

16.3.1 PEPTIDES AND PROTEINS

Any type of water-soluble peptides and proteins could be encapsulated practically by W/O/W method. Agents most frequently investigated for the sake of developing sustained delivery devices can be grouped as model materials, hormones, enzymes, antigens, antibodies, cytokines, and growth factors. At the end of this section, some other entrapped substances, which cannot be divided into these categories, are also displayed.

16.3.1.1 Model Compounds

Model compounds have been usually chosen to study the influence of process parameters on the properties of forming micro- or nanoparticles. Some examples are summarized here to show interesting aspects of examinations carried out by encapsulating most important model proteins and a particular peptide.

BSA has been often used as a model antigen in the incorporation studies of vaccine adjuvants. *In vivo* experiments demonstrated that continuous release of a model antigen BSA from PLGA particles evoked high-titered immune responses in mice, which persisted for more than 142 days (Sah et al., 1995). The adjuvanticity of the microcapsules was found to be superior related to that of aluminum hydroxide and comparable to that of Freund's incomplete adjuvant. Control experiments substantiated that the mixture of antigen and blank microcapsules did not possess adjuvanticity. Immunization of animals with BSA-containing microcapsules was more effective in stimulating its immunogenicity than that with one prime and two booster injections of BSA in saline solution. Therefore, the microcapsule providing continuous release of antigen can be an effective alternative to multiple injections of antigen and have a potential for use as vaccine adjuvants. Microspheres were formed using a new family of anhydride polymers: tyrosine-containing poly(anhydride-*co*-imides), specifically poly[trimellitylimido-L-tyrosineco-sebacic acid-*co*-1,3-bis(carboxyphenoxy)propane] anhydrides [poly(TMA-Tyr:SA:CPP)] (Chiba et al., 1997). These polymers may be of particular interest for controlled delivery of vaccine antigens due to the incorporation of the immunological adjuvant, L-tyrosine modeled by BSA in the study of Chiba et al. (1997). Protein release rates from polymers of identical composition could be varied from 0.3 to over 125 pg/mg spheres/month by changing the amount of protein encapsulated. This effect could be magnified by using polymers with various monomer ratios. A close correlation between protein release and polymer weight loss was observed, suggesting a release mechanism controlled mainly by polymer erosion. BSA release from poly(TMA-Tyr:SA:CPP) microspheres was pH sensitive, being enhanced at high pH and depressed under acidic conditions. Preliminary tissue response and toxicological studies with poly(TMA-Tyr:SA:CPP) are encouraging. In addition, *in vivo* degradation studies in mice showed that the polymer eroded with time and the monomers completely disappeared from the implantation site. The gross appearance of tissue around the injection site following complete polymer erosion was indistinguishable from the normal tissue, suggesting tissue healing.

The process of endocytosis, exocytosis, and intracellular retention of PLGA nanoparticles including BSA as a model protein and 6-coumarin as a fluorescent marker were studied *in vitro* using human arterial vascular smooth muscle cells (Panyam and Labhasetwar, 2003). Cellular uptake of nanoparticles (mean particle size 97 nm) was a concentration-, time-, and energy-dependent endocytic process. Confocal microscopy demonstrated that nanoparticles were internalized rapidly, with nanoparticles seen inside the cells as early as within 1 min after incubation. The nanoparticle uptake increased with incubation time; however, once the extracellular nanoparticle concentration gradient was removed, exocytosis of nanoparticles occurred. Exocytosis of nanoparticles was slower than the exocytosis of the fluid phase marker, Lucifer yellow. Furthermore, the exocytosis of nanoparticles was reduced after the treatment of cells with the combination of sodium azide and deoxyglucose, suggesting that exocytosis of nanoparticles is an energy-dependent process. The nanoparticle retention increased with enhancing nanoparticle dose in the medium, but the effect was relatively less significant with the increase in incubation time. Interestingly, the exocytosis of nanoparticles was almost completely inhibited when the medium was depleted of serum. The above result suggested that the protein in the medium was either adsorbed onto nanoparticles and/or carried along with nanoparticles inside the cells, which probably interacts with the exocytic pathway and led to greater exocytosis of nanoparticles.

The biological fate of BSA-loaded nanoparticles following intravenous administration was determined over 24 h in rats (Li et al., 2001). Pegylated PLGA nanoparticles could extend half-life of BSA from 13.6 min that was found for PLGA nanoparticles to 4.5 h, and obviously change the protein biodistribution in rats compared with that of PLGA nanoparticles.

A screening design methodology was used to evaluate the effects of the process and formulation variables on PCL-BSA microsphere properties (Al haushey et al., 2007). Twelve operating factors were retained, and the particle properties considered were the mean size, the encapsulation efficiency, and the surface state. The statistical analysis of the results allowed determining the most influent factors. It appeared that the polymer concentration, the osmotic pressure equilibrium, and the volume of the inner, outer, and organic phases were the most important parameters. Following this screening study, it was possible to produce particles of small size (4 μm) with high entrapment efficiency (near to 80%) and smooth surface. A good batch-to-batch reproducibility was obtained. Table 16.1 shows the properties of polymer-BSA particles mostly investigated (polymer type, encapsulation efficiency, initial burst, release duration, and released protein ratio) in the studies of W/O/W

TABLE 16.1

Type of Polymer, Encapsulation Efficiency, Initial Burst, Release Duration Investigated, and Ratio of Protein Released by Polymer-BSA Particles

Polymer	Encapsulation Efficiency (%)	Initial Burst (%)	Release Duration (day)	Release Protein Amount (%)	References
PLGA	—	12	13	92	Sah et al. (1995)
		22	23	85	
PLGA	90	5	90	7, 20, 32	Panyam et al. (2003)
PLGA	24	25	20	60	Song et al. (1997)
	74	30	20	95	
PEG-PLGA	49	18	7	71	Li et al. (2001)
PLGA	64	8	7	50	
PLGA	80	20	60	90	Lamprecht et al. (1999)
PCL	70	10	60	70	
PCL	7–80	—	—	—	Al haushey et al. (2007)

systems. It can be observed in the data, PLGA is the encapsulation polymer most frequently examined. Further, the encapsulation efficiency, the initial burst, and the release properties of a particular agent can be modified in a wide range. Their variation depends on the process parameters as shown in Section 16.2. The frequency of PLGA use and the considerable variety of particle characteristic are typical not only for BSA but also for most of the peptides and proteins. Although, other compounds have not been investigated so thoroughly as BSA.

CS is a very interesting biomaterial for drug delivery; however, its use in oral administration is restricted by its fast dissolution in the stomach and limited capacity for controlling the release of drugs (Remunan-Lopez et al., 1998). To address this limitation, a new microparticulate CS controlled-release system, consisting of hydrophilic CS microcores entrapped in a hydrophobic cellulosic polymer, such as cellulose acetate butyrate or ethyl cellulose (EC), was proposed. Using sodium diclofenac (SD) and fluorescein isothiocyanate-labeled bovine serum albumin (FITC-BSA) as model compounds, the properties of these new microparticles for the entrapment and controlled release of drugs and proteins were investigated. Results showed that the entrapment efficiency of SD was very high irrespective of the processing conditions. Furthermore, for both model compounds (SD and FITC-BSA), it was possible to modulate the *in vitro* release of the encapsulated molecules by changing the core properties (molecular weight of polymer, core:coat ratio) or the coating polymer. The microparticles were stable at low pH, and thus suitable for oral delivery without requiring any harmful cross-linkage treatment.

OVA is another well-characterized model protein used preferentially in encapsulation experiments. PVP-stabilized microparticles exhibited higher protein loading and increased core loading of protein (Coombes et al., 1998). Initial burst release of OVA decreased by PVP. The changes in protein loading and delivery characteristics were considered to arise in part from an increase in the viscosity of the droplets of polymer solution, constituting the primary W/O emulsion, by diffusion of PVP from the external aqueous phase.

Formulation of poly(methylidene malonate 2.1.2) (PMM 2.1.2) microparticles entrapping OVA as a model protein was achieved (Le Visage et al., 2001). Parameters such as the nature of the solvent, polymer concentration, and polymer molecular weight were investigated. Preparation process led to the formation of spherical and smooth particles with a mean diameter of 5 μm, and an encapsulation efficiency and protein loading level of up to 16% and 2.9% w/w, respectively. After an initial burst of approximately 10%, the protein was released at a rate of less than 1% per day. Degradation of PMM 2.1.2 microparticles in the presence of esterases indicated that side-chain hydrolysis of the polymer was the rate-determining step in bioerosion; cleavage of the ester side chain, which was further hydrolyzed to glycolic acid and ethanol, led to an acrylic acid and subsequent solubilization of the polymer. However, slow polymer backbone solubilization after degradation was observed.

PLGA microspheres containing β-lactoglobulin (β-LG), the major whey protein of cow's milk, as a model protein, were manufactured (Rojas et al., 1999). Tween 20 emulsifier was shown to increase 2.8 fold the encapsulation efficiency of β-LG and to be responsible for removing the β-LG molecules that were adsorbed on the particle surface or very close to the surface as shown by confocal microscopy and zeta potential measurements. Tween 20 reduced the number of aqueous channels between the internal aqueous droplets as well as those communications with the external medium. Thus, the more dense structure of β-LG microspheres could explain the decrease of the burst release.

In order to study the mechanism of initial burst release from drug-loaded PLGA microspheres, a model peptide, octreotide acetate was encapsulated (Wang et al., 2002). A simple and accurate continuous monitoring system was developed to obtain a detailed release profile. Both the external and internal morphology of the microspheres changed substantially during release of 50% of the peptide over the first 24 h into an acetate buffer. After 5 h, a 1–3 μm "skin" layer with decreased porosity was observed forming around the microsphere surface. The density of the "skin" appeared to increase after 24 h with negligible surface pores present, suggesting the formation of a diffusion barrier. Similar morphological changes also occurred at pH 7.4, but more slowly.

It is noted that the encapsulation efficiency, *in vitro* release properties, and size of particles are important features of pharmaceutical agents encapsulated; however, in this study, the application and *in vivo* kinetics of sustained-release formulates if given are emphasized principally.

16.3.1.2 Hormones

The aim of encapsulation of therapeutic agents is generally to change unpleasant administration route (e.g., injection) to a more convenient one. Sustained delivery of insulin is one of the biggest challenges, as incidence of diabetes is substantially high and regrettably growing in the consumer societies. The acylation of peptides during the erosion of PLA PLGA microspheres is a potential risk to inactivate pharmaceutical agents. To investigate whether insulin is prone to the covalent attachment of lactic or glycolic acid, insulin-loaded PLA and PLGA microspheres containing 5% bovine insulin were manufactured (Ibrahim et al., 2005). The total ion chromatogram of insulin samples revealed that deamidation was the major mechanism of instability. No acylation products were found; however, control experiments in concentrated lactic acid solutions confirmed a minimal reactivity of the peptide. Further, bovine insulin did not undergo any chemical modification in PLGA microparticles within 6 months of storage (Rosa et al., 2000).

Uniform-sized biodegradable PLA/PLGA microcapsules loaded by recombinant human insulin were successfully prepared by combining a Shirasu porous glass membrane emulsification technique and a double emulsion/evaporation method (Liu et al., 2006) The unique advantage of this method is that the size of microcapsules can be controlled accurately, and the drug release profile can be adjusted just by changing the size of microcapsules. Large porous particles made of PLGA were produced for the pulmonary delivery of insulin (Ungaro et al., 2006). Hydroxypropyl-β-cyclodextrin (HP-β-CD), also known as absorption enhancer for pulmonary protein delivery, was tested as aid excipient to optimize the aerodynamic behavior of the microparticles. HP-β-CD-containing large porous particles with flow properties and dimensions suitable for aerosolization and deposition in deep regions of the lung following inhalation were manufactured. A contemporary release of insulin and HP-β-CD from the system could be achieved by selecting appropriate formulation conditions. Insulin-loaded PLGA microcapsules exhibited marked rapid release of insulin within several hours in both *in vivo* and *in vitro* experiments (Yamaguchi et al., 2002). On the other hand, the addition of glycerol or water in the primary DCM dispersion resulted in drastically suppressed initial release. As an additional effect of glycerol, the initial burst was further suppressed due to the decrease of the glass transition temperature of PLGA from 42.5°C to 36.7°C. Since the annealing of PLGA molecules took place at around 37°C, the porous structure of microspheres immediately disappeared after immersion in phosphate-buffered saline or subcutaneous administration. The insulin diffusion through the water-filled pores would be effectively prevented. The strict controlled initial release of insulin from the PLGA microsphere suggested the possibility of utilization in insulin therapy for type I diabetic patients who need construction of a basal insulin profile. Polyphosphazene-insulin microspheres induced a remarkable decrease in glucose levels and the activity was maintained throughout 1000 h (Caliceti et al., 2000). It stimulated anti-insulin antibody production that constantly increased over a period of 8 weeks. The glucose concentration was maintained in the range of 80% and 50% of the starting value for about 500 h and then the glycemia slowly increased to achieve the starting hyperglycemic value in 1000 h.

Martin et al. (2000) entrapped follicle stimulating hormone (FSH) into PHB microparticles. This biomaterial, dissolved in the inner aqueous phase, was able to stabilize the primary emulsion without using a surfactant. Similarly, Eligio et al. (1999) incorporated FSH as well as luteinizing hormone (LH) into PHB as well as PHBHV. In spite of the fact that release profiles obtained *in vitro* showed an adequate qualitative trend, the release rates are too slow to attain a hormone level to be used for the superovulation of cattle. Primary emulsion was found to be critical factor, and it was not able to be stabilized appropriately. More dramatic changes in the polymer material were suggested.

Triptorelin is a decapeptide analog of LH releasing hormone, currently used for the treatment of diseases dependent on sex hormones. Triptorelin-loaded nanospheres useful for transdermal iontophoretic administration were incorporated into PLGA (Nicoli et al., 2001). The release profiles obtained with this copolymer were characterized by the absence of burst effect. This behavior as well as the high encapsulation efficiency was explained by an ionic interaction occurring between the peptide and the copolymer. This supports the already expressed theory that the release of peptides and proteins from PLGA nanospheres is also governed by the affinity of the encapsulated molecule versus the polymer. The obtained nanoparticles, regarding their size, amount encapsulated, and zeta potential, were shown to be suitable for transdermal iontophoretic administration. Synthetic analogues of gonadotropin-releasing hormone (GnRH) are recognized as potent drugs for sex hormone–dependent diseases, most commonly for prostate cancer, but they are also indicated for endometriosis, uterine fibroids, *in vitro* fertilization, and precocious puberty (Schwach et al., 2003). A new GnRH antagonist, degarelix in PLGA microparticles, was prepared for 3-month sustained release in order to treat prostate cancer.

Parathyroid hormone (PTH) bioactivity was maintained largely during incorporating it into PLGA microspheres (Wei et al., 2004). These particles released detectable hormone in the initial 24 h, and it was biologically active as evidenced by the stimulated release of cAMP from osteosarcoma cells as well as increased serum calcium levels when injected subcutaneously into mice.

Recombinant human erythropoietin (EPO) and fluorescein isothiocyanate-labeled dextran (FITC-dextran) loaded biodegradable microspheres were prepared from PLGA (Bittner et al., 1998). EPO is a glycoprotein hormone. The protein is mainly produced in the kidneys and plays an important role in stimulating the red cell proliferation and their differentiation in the bone marrow. EPO is clinically used for the treatment of anemia associated with chronic renal failure. The protein is administered in cases of neoplastic diseases and chronic inflammation. The study of Bittner et al. (1998) highlights the importance of protein stability and compatibility with biodegradable parenteral delivery systems.

Recombinant human growth hormone (rhGH) was encapsulated within semicrystalline PLA and amorphous PLGA (Kim and Park, 2004). Both of the microspheres similarly exhibited rugged surface and porous internal structures, but their inner pore wall morphologies were quite different. The slowly degrading PLA microspheres had many nanoscale reticulated pores on the wall, while

TABLE 16.2
Hormones Encapsulated, Polymers Used, and Application or Main Result of the Products

Hormone	Polymer(s)	Main Result/Application	References
Bovine insulin	PLA, PLGA	Instability due to deamidation	Ibrahim et al. (2005)
Bovine insulin	PLGA	No chemical modification within 6 months of storage	Rosa et al. (2000)
Insulin	PLA, PLGA	Accurately controllable size	Liu et al. (2006)
Insulin	PLGA	Contemporary release, pulmonary delivery	Ungaro et al. (2006)
Insulin	PLGA	Diabetes type I	Yamaguchi et al. (2002)
Insulin	Polyphosphazene	Anti-insulin antibody production over 8 weeks	Caliceti et al. (2000)
FSH	PHB	FSH stabilized the emulsion without surfactant	Martin et al. (2000)
FSH, LH	PHB, PHBHV	Release rate was too slow for the superovulation of cattle	Eligio et al. (1999)
Triptorelin	PLGA	Transdermal iontophoretic administration	Nicoli et al. (2001)
Degarelix	PLGA	Prostate cancer	Schwach et al. (2003)
PTH	PLGA	Increased serum calcium level in mice subcutaneously	Wei et al. (2004)
EPO	PLGA	Neoplastic diseases and chronic inflammation	Bittner et al. (1998)
rhGH	PLA, PLGA	Different release profiles of the polymers	Kim and Park (2004)
Somatostatin	PLGA	Endocrine system regulation	Herrmann and Bodmeier (1998)

the relatively fast degrading PLGA microspheres had a nonporous and smooth wall structure. From the PLA microspheres, rhGH was released out in a sustained manner with an initial 20% burst, followed by constant release, and almost 100% complete release after a 1-month period. In contrast, the PLGA microspheres showed a similar burst level of 20%, followed by much slower release, but incomplete release of 50% after the same period. The different rhGH release profiles between PLA and PLGA microspheres were attributed to different morphological characters of the pore wall structure. The interconnected nanoporous structure of PLA microspheres was likely to be formed due to the preferable crystallization of PLA during the solvent evaporation process.

Encapsulation of somatostatin into PLGA was thoroughly investigated with four solvent evaporation methods involving W/O/W (Herrmann and Bodmeier, 1998). Somatostatin (also known as growth hormone-inhibiting hormone or somatotropin release-inhibiting factor) is a peptide hormone that regulates the endocrine system, affects neurotransmission, cell proliferation, and inhibition of the release of numerous secondary hormones. Most important data of hormones encapsulated (polymers used, application) are summarized in Table 16.2.

16.3.1.3 Antigens

Antigen encapsulation is of similar importance to development of hormone sustained release, as vaccines could be made more efficient by this method. Previously, Nakaoka et al. (1996) investigated the incorporation of OVA antigen into PLA microspheres. Following the intraperitoneal (i.p.) and subcutaneous (s.c.) injection of the microspheres to mice, the titer of anti-OVA antibody in the serum was measured to assess the size effect on the profile of antibody production. OVA was released from the microspheres for 80 days, irrespective of the microsphere size. In both the s.c. and i.p. immunization, the serum level of anti-OVA IgG antibody in the mice induced by the microspheres containing OVA was higher than that of free OVA when compared at the same dose. The serum level of antibody in the mice i.p. injected with the microspheres tended to increase with the decreasing size. On the other hand, in the s.c. immunization, the microsphere size had little influence on the antibody production. It is possible that the injected microspheres tend to aggregate in the s.c. tissue, disappearing the size effect on the antibody production.

Helicobacter pylori infection causes gastritis and peptic ulcer disease, and is associated with the appearance of gastric cancers and lymphomas. Kim et al. (1999) found that lysate of *H. pylori* in PLGA microparticles could stimulate the *H. pylori*-specific mucosal and systemic response in mice, and might be useful adjuvant in future *H. pylori* vaccine development. Due to the important drawbacks of the *Brucella melitensis* vaccine, a safer vaccine based on an outer membrane complex from *Brucella ovis* encapsulated in PCL microparticles was developed and tested in rams (Munoz et al., 2006). *B. ovis* causes a clinical or subclinical chronic disease in ovine that is characterized by genital lesions and low fertility in rams and placentitis and abortions in ewes. Homogeneous batches of microparticles were prepared by a new double emulsion–solvent evaporation method called "total recirculation one-machine system." One single dose of the microparticles conferred similar protection against *B. ovis* than the reference vaccine, and a dose-response effect existed in the rams immunized with microparticles. When using the higher dose of microparticles only two animals were found to be infected, in contrast with the lower dose of vaccine that resulted in infection levels similar to that found in unvaccinated controls. The lack of interference in *B. melitensis* diagnostic tests and the intrinsic avirulence and innocuousness of microparticles make this formulation an attractive anti-Brucella vaccine candidate.

Parenteral administration of vaccines often does not lead to optimal or long-lasting protection against disease-causing organisms, particularly that are inhaled, ingested, or sexually transmitted (Saraf et al., 2006). For optimal mucosal protection induction of immune response via mucosal routes is therefore highly desirable. Hepatitis B surface antigen (HBsAg) loaded lipid microparticles were manufactured for intranasal mucosal administration. To prepare the delivery system, biocompatible and least irritable, soyalecithin (phospholipid) was taken instead of polymer, because phosphatidylcholine is the major component of endogenous lung surfactant. Lipid microparticles (lipid-based novel particulate system) were designed for effective immunization through the naso-pulmonary route using protein antigen entrapped within it. These systems via better presentation, controlled release, immunoadjuvant properties, cationic nature, and subsequent mucoadhesive property can be exploited for efficient mucosal vaccination. Mucosal immunization with lipid microparticles through nasal administration may be effective in prophylaxis of diseases transmitted through mucosal routes as well as systemic infections, e.g., hepatitis B (HB). The strategy can be made more appropriate by direct determination of T-cell responses, paracellular transport, nasal mucociliary clearance, and mucosal toxicity assessment. Oral hepatitis B vaccine formulation was prepared by successful encapsulation of immunogenic peptide representing residues 127–145 of the immunodominant B-cell epitope (BCEP) of HBsAg in PLGA microparticles (Rajkannan et al., 2006). Single oral immunization of mice with the microparticles led to the significant induction of specific serum IgG and IgM anti-HB antibodies. After the termination of antibody induction, the orally immunized mice were infected with HBsAg, which resulted in the rapid production of antibodies against HBsAg as a result of secondary immune response. PLGA microparticles formulation approach may have potential in increasing the efficacy of microparticulate systems for the oral administration of hepatitis B vaccine (HBV). From the above results, it is very clear that the levels of anti-HB antibodies generated by BCEP against the HbsAg infection was more appreciable in biodegradable PLGA microparticles than in its native form. This microencapsulation of BCEP leading to slow release of the antigen *in vivo* would negate the need for the booster dose, which may lead, to adjuvant effect and thereby enhancing the immune response against the HBsAg of HBV. All these preliminary results find potential implication for the development of single oral vaccination strategies for the induction of sustained and prolonged anti-HBs antibody response against HBV.

PCL, PLGA, PLGA-PCL blend, and their copolymer nanoparticles encapsulating diphtheria toxoid were investigated for their potential as a mucosal vaccine delivery system (Singh et al., 2006). An *in vitro* experiment using Caco-2 cells showed significantly higher uptake of PCL nanoparticles in comparison to PLGA, PLGA-PCL blend, and their copolymer nanoparticles. The highest uptake mediated by the most hydrophobic nanoparticles using Caco-2 cells was mirrored in the *in vivo* studies following nasal administration. PCL nanoparticles induced DT serum specific IgG antibody responses significantly higher than PLGA. A significant positive correlation between hydrophobicity of the nanoparticles and the immune response was observed following intranasal administration.

A challenging target is to encapsulate particulate antigens such as viruses. Two small-scale double emulsion techniques for incorporation of formaldehyde-inactivated rotavirus particles (FRRV) into PLGA microspheres were developed and optimized. Rotavirus was released *in vitro* in a triphasic manner with both techniques (Sturesson et al., 1999). The more robust vortex technique was selected for preparation of PLGA microspheres containing rotavirus for *in vivo* studies. After immunization of mice with a single intramuscular injection, the PLGA-FRRV microspheres elicited an IgG antibody response in serum detected by ELISA equally high as that elicited with FRRV alone. These results indicate that the antigenicity of FFRV was retained after incorporation into PLGA microspheres.

Erythrocyte-ghost-loaded ethylcellulose microparticles could be a promising affinity support for antibody removal (Ubrich et al., 1996). Large volume plasma exchanges are used for the removal of anti-A or anti-B antibodies from the plasma of patients undergoing transplantation from donors with major ABO incompatibility. Studies were undertaken to determine the feasibility of an original immunoadsorbent based on porous microparticles within which erythrocyte ghosts carrying blood-group antigens were entrapped. The decrease of the antibody hemagglutinating titer after adsorption onto encapsulated ghosts suggests that antibodies can cross the polymeric membrane and bind to the antigens. This original approach of using encapsulated antigens for the batchwise removal of antibodies could be extended to affinity chromatography and immunoadsorption therapy. Table 16.3 exhibits antigens that were encapsulated, the polymers used, and the main result or application of the formulates.

TABLE 16.3
Antigens Encapsulated, Polymers Used, and Application or Main Result of the Products

Antigen	Polymer(s)	Main Result/Application	References
Ovalbumin	PLA	Size dependence of antibody level i.p.: yes, s.c.: no	Nakaoka et al. (1996)
Lysate of *H. pylori*	PLGA	Stimulate mucosal and systemic response	Kim et al. (1999)
Complex of *B. ovis*	PCL	Dose–response effect in rams	Munoz et al. (2006)
HBsAg	Soyalecithin[a]	Mucosal immunization against HB nasally	Saraf et al. (2006)
BCEP of HBsAg	PLGA	Oral vaccination against HB	Rajkannan et al. (2006)
Diphtheria toxoid	PCL, PLGA, PLGA-PCL	Correlation between hydrophobicity and immune response intranasally	Singh et al. (2006)
FRRV	PLGA	Antigenicity was retained intramuscularly	Sturesson et al. (1999)
Erythrocyte-ghost	Ethylcellulose	Removal of antibody	Ubrich et al. (1996)

[a] Phospholipid, not polymer.

16.3.1.4 Enzymes

L-Asparaginase is a preferred enzyme of incorporation studies. It exhibits a short *in vivo* half-life, and is active against acute lymphoblastic leukemia only in its tetrameric form. It was encapsulated in PLGA nanospheres (Gaspar et al., 1998; Wolf et al., 2003). The nanoparticles made of PLGA with free carboxyl-end groups had a high protein loading and provided a continuous delivery of the active enzyme for 20 days (Gaspar et al., 1998). Lysozyme is a family of enzymes that damage bacterial cell walls. Its encapsulation into PLGA microspheres showed that the release profile is highly dependent on the media (Jiang et al., 2002). Adsorption severely limited the amount of lysozyme available for release from microspheres. Protein stability and adsorption are crucial factors controlling protein release kinetics.

Prolidase is a naturally occurring enzyme involved in the final stage of protein catabolism. Deficient enzyme activity causes prolidase deficiency, a rare autosomal recessive inherited disorder whose main manifestation is chronic, intractable ulcerations of the skin, particularly of lower limbs. Although several attempts have been made toward the treatment of this pathology, a cure for this disease has yet to be found. Prolidase-loaded PLGA microparticulate systems were prepared to evaluate the possibility of enzyme replacement therapy (Genta et al., 2001). The results of prolidase from pig kidney (available on the market) demonstrated the positive role of microencapsulation as a process of enzymatic activity stabilization inside PLGA microspheres achieving both *in vitro* and *ex vivo* active enzyme release. This formulation can be proposed as a parenteral depot drug delivery system.

Glucose oxidase (GOD) was encapsulated as a model protein within PLA-PEG microspheres to evaluate the activity retention during microencapsulation process (Li et al., 2000). The obtained results showed that the solvent extraction/evaporation method based on the formation of double emulsion W/O/W benefited the activity retention compared with the phase separation method based on the formation of W/O/W. In the emulsion-evaporation system, a significant part of the protein activity was lost during the first emulsification procedure to form primary emulsion W/O (ca. 28%) and the second emulsification procedure to form the double emulsion W/O/W (ca. 20%), in contrast to other processes occurring during microsphere preparation. The matrix polymer and the solvent system in the oil phase had an impressive impact on the activity retention. The encapsulation of GOD in PLA-PEG microspheres was effective in reducing its specific activity loss. Sixty-seven per cent of the initial specific activity retention was detected for the released GOD from microspheres formulation during 1 week of incubation, but nearly all the activity was lost for GOD in solution incubated under the same condition. SDS-PAGE results showed that, although the activity loss was

TABLE 16.4
Enzymes Encapsulated, Polymers Used, and Application or Main Result of the Products

Enzyme	Polymer(s)	Main Result/Application	References
L-Asparaginase	PLGA	Continuous active enzyme delivery for 20 days	Gaspar et al. (1998)
Lysozyme	PLGA	Adsorption limited the amount of lysozyme available	Jiang et al. (2002)
Prolidase	PLGA	Parenteral depot delivery	Genta et al. (2001)
GOD	PLA-PEG	Specific activity loss reduced	Li et al. (2000)

detected, no rough changes of molecular weight of GOD was observed during encapsulation procedure and the initial days of incubation into the *in vitro* release medium. Table 16.4 summarizes the most important findings of studies about enzymes.

16.3.1.5 Growth Factors and Cytokines

Biodegradable microspheres containing recombinant human epidermal growth factor (rhEGF) were prepared using PLA (Han et al., 2001). Epidermal growth factor plays an important role in the regulation of cell growth, proliferation, and differentiation by binding to its receptor. The blood concentration of rhEGF was maintained at constant levels for 9–11 days after a single s.c. administration of rhEGF microspheres. The gastric ulcer healing effect of a single s.c. administration of rhEGF microspheres was increased 1.44-fold compared with twice-a-day s.c. administration of rhEGF saline solution after 11 days. The enhanced curative ratio of rhEGF loaded microspheres may be due to the optimized osmotic pressure, high encapsulation efficiency, and sustained-release pattern. This system might be applicable to treat chronic gastric ulcers by a single s.c. depot injection. This microsphere system can also help to improve other types of therapy for rhEGF treatment such as skin and corneal injury. Kim et al. (2002) conjugated the same growth factor with PEG to improve its physical stability during microencapsulation in PLGA microspheres. rhEGF was conjugated with *N*-hydroxysuccimide-derivatized methoxy-PEG. Pegylated rhEGF showed much enhanced physical stability against homogenization. Pegylated rhEGF was encapsulated in PLGA microspheres, and exhibited a triphasic release profile with a reduced initial burst, compared with unpegylated rhEGF.

PLGA biodegradable microparticles, which allow the local delivery of a precise amount of a drug by stereotactic injection in the brain, were prepared (Pean et al., 1998). Nerve growth factor (NGF) is a small secreted protein that induces the differentiation and survival of particular target neurons (nerve cells). A honeycomb-like structure characterized the internal morphology of the microspheres. It appeared that microparticles made from a hydrophilic uncapped PLGA in the absence of salt allowed the release of intact NGF at least during the first 24 h as determined by both ELISA and a PC12 cell-based bioassay.

Cytokines are proteins of high value secreted by the immune system, and are effective against several viruses. Due to their high sensitivity to enzymatic digestion, their encapsulation could be an appropriate procedure to elongate their lifetime in the body. Sanchez et al. (2003) synthesized PLGA nano- and microparticles containing interferon (IFN)-α coencapsulated with poloxamer having an average diameter of 280 nm and 10 μm, respectively. The particles exhibited a similar pattern of release characterized by an initial burst followed by small pulses of immunoenzymatically detected IFN-α for up to 1 month. Pegylated interferon showed better stability when exposed to DCM as compared to native IFN. Native IFN and pegylated IFN released 16% and 72%, respectively, after 3 weeks, as pegylated IFN largely retained its native aqueous solubility after being exposed to detrimental conditions of microencapsulation (Diwan and Park, 2003).

TABLE 16.5
Growth Factors/Cytokines Encapsulated, Polymers Used, and Application or Main Result of the Products

Growth Factor/ Cytokine	Polymer(s)	Main Result/Application	References
rhEGF	PLA	To treat chronic gastric ulcers by a single s.c. depot injection	Han et al. (2001)
rhEGF	PEG, PLGA	Enhanced stability of pegylated rhEGF	Kim et al. (2002)
NGF	PLGA	Stereotactic injection in the brain	Pean et al. (1998)
IFN-α	PLGA	Elongation of IFN-α lifetime	Sanchez et al. (2003)
IFN	PLGA	Pegylated IFN showed better stability	Diwan and Park (2003)
GM-CSF	PEC	Homogeneous distribution of GM-CSF in spheres	Lambert et al. (2000)

The granulocyte-macrophage colony-stimulating factor (GM-CSF), a cytokine that functions as a white blood cell growth factor, was encapsulated in poly(ethylene carbonate) microspheres (Lambert et al., 2000). The above described formulation and process allowed to achieve high encapsulation efficiency and very low *in vitro* initial drug release (<5.6%). It has also been shown that the internal structure of the microspheres is related both to the polymer's molecular weight and to the polymer concentration in the organic phase. By setting the formulation parameters to the appropriate values, it is therefore possible to obtain a homogeneous distribution of the drug substance in the microspheres, which is a prerequisite to *in vivo* zero-order drug release profiles. Table 16.5 displays parameters and results of growth factor and cytokine encapsulations.

16.3.1.6 Other Peptides and Proteins

Couvreur et al. (1997) incorporated the V3 BRU, which is a synthetic peptide of human immunodeficiency virus (HIV), and the pBC 264 derived from cholecystokinin into PLGA microspheres. The pBC 264 neuropeptide is a very potent agonist for CCK-B receptors with a dopaminergic activity and could be of great interest for the treatment of Parkinson's diseases. However, chronic local stimulation of CCK-B receptors is hindered by the rapid elimination of pBC 264 from the brain after central administration. In order to study the long-term effects resulting from local stimulation of CCK-B receptors, the possibility of slowly delivering pBC 264 into a particular brain structure for a long period of time has been considered using its encapsulation into PLGA microspheres. These microspheres were shown to very efficiently encapsulate a 33-amino-acid peptide (V3 BRU), and *in vitro* release kinetics studies showed that such microspheres could be employed for both oral vaccination and controlled release. The encapsulation of a seven-amino-acid peptide (pBC 264) led on the contrary to a very low encapsulation efficiency. In order to increase the encapsulation of pBC 264, two strategies were adopted: taking into account the solubility of pBC 264 at different pH, the inner aqueous phase was maintained at a basic pH where the peptide was soluble, while the external aqueous phase was acidic; OVA was added to stabilize the inner emulsion. These two strategies allowed to increase significantly the encapsulation rate of pBC 264. Nevertheless, the *in vitro* release kinetics of the peptide were strongly influenced by the presence of OVA which seems to form pores in the microsphere structure. By contrast, when OVA was replaced by Pluronic F68, microspheres did not have pores, thus the release profile and the extent of the burst were much smaller. When microspheres were stereotactically implanted in the rat brain, *in vivo* release profiles were in good agreement with the release observed *in vitro*. In conclusion, these microspheres are well suited for the slow delivery of neuropeptides in the brain, a feature expected to facilitate the study of long-term effects of these compounds. Blanco-Prieto et al. (1997) studied also incorporation of pBC 264 into PLGA microspheres. The encapsulation efficiency was very low when the inner emulsion contained no stabilizing agent. Thus, the encapsulation rate was improved by the addition of OVA used as stabilizer of the inner emulsion. In addition, the presence of a pH gradient between the inner and the outer aqueous phases of the multiple emulsion led to an increase of the retention of the peptide within the microspheres. Release kinetics were characterized by a dramatic burst effect corresponding to the release of the major part of the entrapped pBC 264. There was no interaction between OVA and the polymer suggesting that the protein is not dissolved in the polymer network but mainly present at the interfaces of the inner aqueous phase/polymer and microsphere surface/outer aqueous phase.

Leucinostatin-A (Leu-A) is a nonapeptide exerting a remarkable activity especially against Gram-positive bacteria, *Candida albicans*, and *Cryptococcus neoformans*, although its use is limited due to its toxicity. It is especially important if considering the fact that *C. neoformans* infections are often associated with HIV insurgences. Leu-A loaded PLGA nanospheres represent a new effective therapeutic system for candida infection (Ricci et al., 2004). *In vivo* experiments showed a drastic reduction of Leu-A toxicity, that is, the LD50 was increased more than 18-fold, and the study on systemic candidiasis models revealed high effectiveness of the nanospheres in reducing either the growth of fungal colonies in infected mice liver or in the mortality index. Therefore, this formulation is able to maintain a drug hematic level sufficiently high to ensure activity against candida without determining lethal side effects. It is suggested in this work that nanosphere surface should be coated to reduce the uptake by the immune system in order to prolong blood half-life and make possible the intravenous administration.

Melittin, the predominant peptide isolated from honeybee venom, is a typical representative of biologically active peptide drugs with high therapeutic potential. Bee venom has a marked effect on the immune system, cardiovascular system, and blood, and it also exhibits antitumor activity. Melittin is commonly used in the treatment of arthritic disorders, such as rheumatoid arthritis and osteoarthritis. The water-soluble peptide, melittin, was encapsulated in PLGA microspheres to deliver it over a period of about 1 month (Cui et al., 2005). The drug release profiles *in vitro* exhibited a significant burst release, followed by a lag phase of little or no release, and then a phase of constant melittin release. The type of polymer used was a critical factor in controlling the release of melittin from the microspheres. The rate of peptide release from the microspheres correlated well with the rate of polymer degradation. Moreover, melittin was released completely during the study period of 30 days, which agreed well with the polymer degradation rate.

Large porous microspheres potentially useful for capreomycin sulfate entrapment in PLGA were prepared for pulmonary delivery (Giovagnoli et al., 2007). Capreomycin is a peptide antibiotic, commonly grouped with the aminoglycosides, which is given in combination with other antibiotics for tuberculosis. The double emulsion method enabled the preparation of capreomycin sulfate–loaded large porous microspheres having suitable characteristics to match respirability requirements. Using a simple preparation method with a consistent time, cost, and material saving, the microspheres formed could be useful for capreomycin sulfate pulmonary delivery.

Recombinant human bone morphogenetic protein-2 (rhBMP-2) is a homodimeric protein currently being tested for its use in bone healing. The protein's osteoinductive property of causing mesenchymal differentiation into chondrocytes, with subsequent calcification of the cartilaginous matrix, can be enhanced by prolonging its presence at the site of healing. Clinical use of rhBMP-2 has been hampered by a lack of suitable systems for its delivery. Such systems should be capable of maintaining the protein *in situ* for sufficient time to interact with target cells, release the protein at effective concentrations during bone formation, cause no unnecessary tissue distress, and be resorbed. RhBMP-2 was encapsulated in PLGA, and binding capacity, kinetics, and total incorporation of protein were determined (Schrier and DeLuca, 1999), although, biological activity of released protein and effect on bone healing were not examined.

NC-1900, an active fragment analog of arginine vasopressin, was proved to be capable of improving the spatial memory deficits and the impairments in passive avoidance test. A novel drug carrier for brain delivery, cationic bovine serum albumin conjugated pegylated (CBSA-PEG) nanoparticles

TABLE 16.6
Other Peptide(s)/Protein Encapsulated, Polymers Used, and Application or Main Result of the Products

Peptide(s)/Protein	Polymer(s)	Main Result/Application	References
V3 BRU, pBC264	PLGA	Slow delivery in the brain	Couvreur et al. (1997)
pBC264	PLGA	Encapsulation rate improved by OVA	Blanco-Prieto et al. (1997)
Leu-A	PLGA	Drastic reduction of Leu-A toxicity	Ricci et al., (2004)
Melittin	PLGA	Completely released during 30 days	Cui et al. (2005)
Capreomycin sulfate	PLGA	Pulmonary delivery against tuberculosis	Giovagnoli et al. (2007)
rhBMP-2	PLGA	Bone healing	Schrier and DeLuca (1999)
NC-1900	BSA-PEG	Treatment of memory deficits intravenously	Xie et al. (2006)
Thymopentin	WGA-PLGA	Improved intestinal absorption	Yin et al. (2006)

holding NC-1900, was developed, and its improvement on scopolamine-induced memory deficits was investigated in mice using the platform-jumping avoidance test (Xie et al., 2006). The half-life of NC-1900 loaded in CBSA-PEG nanoparticles in plasma was about 78 h, which was fourfold longer than that of free NC-1900 (19 h). The active avoidance behavioral results showed that the s.c. administration of NC-1900 tended to improve memory deficits, but the difference did not present any statistical significance, whereas this peptide failed to produce any positive effects by i.v. administration. However, the i.v. injection of CBSA-PEG nanoparticles loaded with NC-1900 greatly improved memory impairments to a normal level, but the efficacy was slight if the loaded nanoparticles (NPs) were exclusive of the conjugation of CBSA, indicating that CBSA-PEG nanoparticles was a promising brain delivery carrier for NC-1900 with CBSA as a potent brain targetor. It was concluded that CBSA-PEG nanoparticles loaded with NC-1900 was potentially efficacious in the treatment of memory deficits via i.v. administration.

Thymopentin (TP5) is a synthetic pentapeptide, which constitutes the segment 32–36 of thymopoietin, a well-defined thymic hormone. As an immunomodulating agent, TP5 is clinically used in the treatments of autoimmune diseases, such as, e.g., atopic dermatitis, chronic lymphocytic leukemia, Sezary's syndrome, rheumatoid arthritis, as well as decreased immune competency in elderly surgical patients due to poor membrane permeability, extensive metabolism in the gastrointestinal tract, and extremely short half-life of 30 s, repeated injections or intravenous infusions of TP5 are necessary which greatly restrict its clinical applications. Lectin-conjugated PLGA nanoparticles for oral delivery of thymopentin were designed (Yin et al., 2006). Novel WGA-PLGA conjugates were synthesized by coupling the amino groups of wheat germ agglutinin (WGA) to the carbodiimide-activated carboxylic groups of PLGA, and were incorporated into nanoparticle preparation to take mucoadhesive properties. The retention of biorecognitive activity of WGA after covalent coupling was confirmed by hemagglutination test. *In vitro* experiments with pig mucin demonstrated that the conjugation of WGA enhanced the interaction about 1.8–4.2 fold compared with that of the nonconjugated nanoparticles, and still exhibited sugar specificity. The pharmacodynamical studies on oral administration confirmed that the conjugation of WGA onto PLGA nanoparticles effectively improved the intestinal absorption of TP5 due to specific bioadhesion on gastrointestinal cell membrane. Table 16.6 shows properties of the encapsulated peptides and proteins presented in this section.

16.3.2 NUCLEOSIDE, OLIGONUCLEOTIDES, DNA

There is no doubt HIV treatment is one of the highest challenges of medical sciences. Microencapsulation could develop new methods to achieve this goal. Zidovudine is a nucleoside analog reverse transcriptase inhibitor, a type of antiretroviral drug. It was the first approved treatment for HIV. PLGA–zidovudine microcapsules were prepared to study the effect of several formulation and processing factors on the efficiency of encapsulation, surface morphology, and drug release profile in a previous article (Mandal et al., 1996). Akhtar and Lewis (1997) improved the delivery of anti-HIV oligonucleotides (ODNs) with PLGA microspheres to macrophage cells in culture. Phosphodiester or phosphorothioate sequences, including those antisense to the *tat* gene in HIV, were incorporated into PLGA microspheres. For a given polymer molecular weight and ODN chemistry, entrapment efficiencies and *in vitro* release rates were dependent on microsphere size. Smaller microspheres (1–2 μm) released 70% of the entrapped ODN within 4 days compared with 40 days for larger microspheres (10–20 μm), suggesting that this delivery system also offers the potential for sustained release of ODNs. The cellular association of ODNs entrapped within small microspheres was improved 10-fold in murine macrophages compared with free ODNs. Uptake was enhanced when macrophages were activated with interferon-γ and lipopolysaccharide treatment but decreased significantly in the presence of metabolic and phagocytosis inhibitors. Fluorescence microscopy studies with macrophages showed that a more diffuse subcellular distribution of ODNs was observed when delivered as a microsphere formulation compared with free ODNs, which exhibited the characteristic punctate periplasmic distribution. These results indicated that polymer microspheres represent an attractive strategy for the improved cellular delivery of ODNs.

Microspheres allowing the controlled release of the model oligonucleotide oligodeoxythymidilate (pdT16) were designed (Rosa et al., 2002). The oligonucleotide, alone or associated with polyethyl-enimine (PEI) at different nitrogen/phosphate ratios, was encapsulated within PLGA microspheres. The introduction of PEI in the internal aqueous phase resulted in a strong increase of the oligonucle-otide encapsulation efficiency. PEI also affected microsphere morphology inducing the formation of very porous particles, yielding an accelerated release of pdT16. However, when incubated with HeLa cells, microspheres encapsulating pdT16/PEI complexes allowed an improvement of the intracellu-lar penetration of the released oligonucleotide. The developed strategy appears to be a very interest-ing tool to obtain a sustained-release system for ODNs with an efficient cellular delivery. Nanosized complexes of antisense transforming growth factor-β2 (TGF-β2) phosphorothioate oligonucleotides (PS-ODN) with PEI and naked PS-ODN were encapsulated into PLGA microspheres (Rosa et al., 2003; Gomes dos Santos et al., 2006). The PS-ODN was introduced either naked or complexed in the inner aqueous phase of the first emulsion. The presence of PEI induced the formation of large pores observed onto microsphere surface. The introduction of NaCl in the outer aqueous phase increased the encapsulation efficiency and reduced microsphere porosity. *In vitro* release kinetics of PS-ODN showed that the early phase of PS-ODN and PS-ODN-PEI complex release was primarily controlled by pure diffusion, irrespectively of the type of microsphere. Microspheres containing antisense TGF-β2 nanosized complexes significantly increased intracellular penetration of ODN in conjunctival cells after subconjunctival administration to rabbit, and subsequently improved bleb survival in a rabbit experimental model of glaucoma filtration surgery. These results open up inter-esting prospective for the local controlled delivery of genetic material into the eye.

Successful gene delivery requires efficient DNA encapsulation during formulation and intracel-lular gene delivery once the carriers have been endocytosed (Pfeifer et al., 2005). Plasmid DNA encoding luciferase protein has been mostly investigated for DNA encapsulation studies (Hsu et al., 1999; Prabha et al., 2002; Howard et al., 2004; Pfeifer et al., 2005). Oral induction of a dissemi-nated mucosal immune response with polyplex-based DNA vaccines requires the delivery of intact polyplexes (polyelectrolyte complexes formed by self-assembly of plasmid DNA with a cationic polymer) to subepithelial lymphoid tissue (e.g., Peyer's patches) within the gastrointestinal tract (Howard et al., 2004). PEGylated PEI/DNA polyplexes were stable to salt-induced aggregation. Polyplexes containing 1:1 PEG/PEI ratio (mass/mass) gave similar levels of luciferase gene expres-sion in B16F10 cells compared to non-PEG complexes. PLGA microparticles containing PEGylated polyplexes were formulated. Microparticles containing PEGylated polyplexes were given orally to Wistar rats. Significant transgene expression (compared to background) was found in periph-eral tissue (spleen) 72 h after administration. Nanoparticles formulated from PLA and PLGA were extensively investigated as nonviral gene delivery systems containing also plasmid DNA encoding luciferase protein (Prabha et al., 2002). Smaller-sized nanoparticles (mean diameter of 70 nm) showed a 27-fold higher transfection than the larger-sized nanoparticles (mean diameter of 202 nm) in COS-7 cell line and a four-fold higher transfection in HEK-293 cell line. The greater transfection of the smaller-sized fraction as compared to the larger-sized fraction of nanoparticles did not seem to be related to their surface properties (zeta potential), cellular uptake, or the rate and extent of release of DNA. The results of the study were important because they signified the importance of particle size in nanoparticle-mediated gene transfection. The results suggest that formulating nanoparticles of smaller diameter is critical to improve the efficiency of nanoparticle-mediated transfection. In another study of them (Prabha and Labhasetwar, 2004) gene expression of nanoparticles was determined in breast cancer (MCF-7) and prostate cancer (PC-3) cell lines. Nanoparticles formulated using PLGA polymer demonstrated greater gene transfection than those formulated using PLA polymer, and this was attributed to the higher DNA release from PLGA nanoparticles. Higher-molecular-weight PLGA resulted in the formation of nanoparticles with higher DNA loading, which demonstrated higher gene expression than those formulated with lower molecular-weight PLGA. In addition, the nanoparticles with lower amount of surface-associated PVA demonstrated higher gene transfection in both the cell lines. Higher gene transfection with these nanoparticles was attributed to their higher intracellular uptake and cytoplasmic levels. Results thus demonstrate that the DNA loading in nanoparticles and its release, and the surface-associated PVA influencing the intracellular uptake and endolysosomal escape of nanoparticles, were some of the critical determinants in nanoparticle-mediated gene transfection.

Cationic surfactant, cetyltrimethylammonium bromide, was used to provide positive charge on the surface of nanoparticles (Basarkar et al., 2007). Reporter plasmid gWIZ™ Beta-gal was loaded on the surface of nanoparticles by incubation. *In vitro* cytotoxicity study showed a low toxicity. Structural integrity of the pDNA released from nanoparticles was maintained. Transfecting human embryonic kidney (HEK293) cells with nanoparticles prepared from low-molecular-weight PLGA and PLA resulted in an increased expression of beta-galactosidase as compared to those prepared from high-molecular-weight polymer. The results demonstrated that the PLGA and PLA cationic nanoparticles can be used to achieve prolonged release of pDNA. Nucleoside, ODNs, and plasmid DNAs encapsulated by polymers and application or main result of the products are summarized in Table 16.7.

TABLE 16.7
Nucleoside, Oligonucleotides, and DNA Encapsulated by Polymers, and Application or Main Result of the Products

Material Encapsulated	Polymer(s)	Main Result/Application	References
Zidovudine	PLGA	Reverse transcriptase inhibitor against HIV	Mandal et al. (1996)
Anti-HIV ODNs	PLGA	Improved cellular delivery	Akhtar and Lewis (1997)
Oligodeoxythymidilate	PLGA	Improvement of intracellular penetration	Rosa et al. (2002)
PS-ODN	PLGA	Genetic material into eye	Gomes dos Santos et al. (2006)
Luciferase encoding pDNA	PEG-PEI, PLGA	Orally, significant transgene expression in spleen tissue	Howard et al. (2004)
Luciferase encoding pDNA	PLA, PLGA	Transfection improvement by size decrease	Prabha et al. (2002)
Luciferase encoding pDNA	PLA, PLGA	High gene transfection	Prabha and Labhasetwar (2004)
pDNA gWIZ Beta-gal	PLA, PLGA	Prolonged release with structural integrity maintained	Basarkar et al. (2007)

16.3.3 STEROID HORMONES AND ANTIBIOTICS

A new formulation of levonorgestrel (LNG) and ethinylestradiol (EE) drugs that are hydrophobic contraceptive hormones was developed, using PCL polymer (Dhanaraju et al., 2003). The result of this work demonstrates the feasibility of formulating both LNG and EE (both are hydrophobic in nature) in the same biodegradable polymeric microspheres. Because of less-pronounced burst effect, microspheres prepared by such method may present a promising approach for achieving contraception for longer period after single administration. Furthermore, the present microspheres are attractive for parenteral application because of their optimum micron size structure and their biodegradability (Dhanaraju et al., 2004). Histological examination of LNG/EE steroid-loaded PLGA microspheres injected intramuscularly into the thigh muscle of Wistar rats showed minimal inflammatory reaction, demonstrating that contraceptive-steroid-loaded microspheres were biocompatible (Dhanaraju et al., 2006). The LNG/EE loaded PLGA microspheres can be used as an intramuscularly injectable drug delivery carrier, in consideration of their biodegradation, biocompatibility, and particle size. Betamethasone, a steroid hormone; and sulfasalazine, a sulfa drug (both are effective anti-inflammatory agents in the treatment of inflammatory bowel disease) were encapsulated into PLA, PLGA, or PCL polymer by W/O/W and S/O/W methods (Lamprecht et al., 2000). Although both of the process was suitable for synthesizing sustained drug delivery devices, the encapsulation efficiency and the *in vitro* release of particles prepared by S/O/W were preferable.

More recently, antibiotics encapsulation have been achieved by some groups using double emulsion method. Gentamicin could be an effective antibiotic against Brucella *in vitro*; however, during infection the bacterium is localized intracellularly, making the treatment difficult. The therapeutics requires the association of more than one antimicrobial for weeks and that often leads to poor patient compliance, contributing to low therapy efficiency. Gentamicin-loaded PLGA microspheres were prepared as sustained delivery formulation. *In vitro* studies indicated an initial burst and a continuous drug release for up to 4 weeks (Blanco-Prieto et al., 2002). Virto et al. (2007) found similar release pattern for PLGA-gentamicin microspheres. However, their aim was to treat chronic osteomyelitis in orthopedic surgery.

Econazole (ECZ) and moxifloxacin (MOX) had been applied individually against tuberculosis (TB) caused by multidrug-resistant and latent *Mycobacterium tuberculosis*. PLGA nanoparticle-encapsulated ECZ and MOX were evaluated against murine TB (drug susceptible) in order to develop a more potent regimen for TB (Ahmad et al., 2008). A single oral dose resulted in therapeutic drug concentrations in plasma for up to 5 days (ECZ) or 4 days (MOX), while in the organs (lungs, liver, and spleen) it was up to 6 days. In comparison, free drugs were cleared from the same organs within 12–24 h. In *M. tuberculosis*-infected mice, 8 oral doses of the formulation administered weekly were found to be equipotent to 56 doses (MOX administered daily) or 112 doses (ECZ administered twice daily) of free drugs. Furthermore, the combination of MOX + ECZ proved to be significantly efficacious compared with individual drugs. Addition of rifampicin to this combination resulted in total bacterial clearance from the organs of mice in 8 weeks.

Ciprofloxacin is a synthetic antibiotic belonging to a group called fluoroquinolones. Its mode of action depends upon blocking bacterial DNA replication by binding itself to an enzyme called DNA gyrase, thereby inhibiting the unwinding of bacterial chromosomal DNA during and after the replication. Ciprofloxacin HCl (CIP)-PLGA nanoparticles and its antibacterial potential was evaluated with pathogenic bacteria, *Escherichia coli*, *in vitro* and *in vivo* (Jeong et al., 2008). In the *in vitro* antibacterial activity test, CIP-encapsulated nanoparticles showed relatively lower antibacterial activity compared to free CIP. However, CIP-encapsulated PLGA nanoparticles effectively inhibited the growth of bacteria due to the sustained-release characteristics of nanoparticles, while free CIP was less effective in the inhibition of bacterial growth. These results indicated that CIP-encapsulated PLGA nanoparticles have superior effectiveness to inhibit the growth of

TABLE 16.8
Antibiotics and Steroid Hormones Encapsulated by Polymers and Application or Main Result of the Products

Material Encapsulated	Polymer(s)	Main Result/Application	References
LNG, EE	PCL	Biocompatible intramuscularly	Dhanaraju et al. (2003)
Betamethasone	PLA, PLGA, PCL	Anti-inflammatory	Lamprecht et al. (2000)
Gentamicin	PLGA	Continuous release for 4 weeks	Virto et al. (2007)
Econazole, moxifloxacin	PLGA	TB clearance from mice combined with rifampicin	Ahmad et al. (2008)
Ciprofloxacin	PLGA	Effective inhibition of *E. coli*	Jeong et al. (2008)

bacteria *in vivo*. Table 16.8 illustrates the main characteristics of antibiotics and steroid hormones encapsulated in the studies represented.

16.3.4 OTHER AGENTS

Some other medical uses of particles, synthesized by double emulsion–solvent evaporation, that cannot be divided into one of the former groups are shown here. Most of these pharmaceutical agents are encapsulated in PLGA; e.g., lidocaine, the first amino amide-type local anesthetic antiarrhythmic drug (Klose et al., 2006) or iron chelator desferrioxamine that demonstrates antimalarial activity upon multiple or continuous parenteral administration was encapsulated to PLGA microparticles (Schlicher et al., 1997). 5-iodo-2′-deoxyuridine (IDU) is a potent inhibitor of replication and DNA synthesis of vaccinia virus (VV) in a dose-dependent way, and a well-known inhibitor of herpesvirus replication. Sustained local release systems were developed for radioiodinated iodo-2′-deoxyuridine ((125)IUdR) from biodegradable polymeric microspheres to facilitate the controlled delivery of (125)IUdR to brain tumors (Reza and Whateley, 1998).

Highly open porous biodegradable polymeric microspheres were fabricated for use as injectable scaffold microcarriers for cell delivery (Kim et al., 2006). The incorporation of an effervescent salt, ammonium bicarbonate, in the primary water droplets spontaneously produced carbon dioxide and ammonia gas bubbles during the solvent evaporation process, which not only stabilized the primary emulsion, but also created well-interconnected pores in the resultant microspheres. The surface pores became as large as 20 μm in diameter with increasing the concentration of ammonium bicarbonate, being sufficient enough for cell infiltration and seeding. Cells were attached to the particles. These porous scaffold microspheres could be potentially utilized for cultivating cells in a suspension manner and for delivering the seeded cells to the tissue defect site in an injectable manner. The feasibility of large porous PLGA particles as long-acting carriers for pulmonary delivery of low-molecular-weight heparin (LMWH) was tested by Rawat et al. (2008). LMWHs are negatively charged oligosaccharides used in the treatment of deep vein thrombosis and pulmonary embolism. Because of their relatively high molecular weight, negative surface charge, and short half-lives, LMWHs are administered by s.c. injection multiple times a day. PLGA was used to encapsulate it in a sustained delivery form. The drug entrapment efficiencies of the microspheres were increased by modifying them with three different additives: polyethyleneimine (PEI), Span 60, and stearylamine. After pulmonary administration, the half-life of the drug from the PEI- and stearylamine-modified microspheres was increased by five- to sixfold compared to the drug entrapped in plain microspheres. The viability of Calu-3 cells was not adversely affected when incubated with the microspheres. The developed porous microspheres of LMWH had the potential to be used in a form deliverable by dry-powder inhaler as an alternative to multiple parenteral administrations of LMWH.

16.4   W/O/O, W/O/O/O, S/O/O/O, AND O/W/O

It might seem to be arbitrary to classify the procedures mentioned in this section title into the category of multiple emulsion–solvent evaporation methods, although, they also match to this category
name. These four methods are discussed as a matter of curiosity, as they are very rarely applied for
encapsulation.

The water-in-oil-in-oil (W/O/O) emulsion–solvent evaporation method was used recently to
entrap α-cobrotoxin (Li et al., 2005), which is a primary postsynaptic neurotoxic protein isolated
from the venom of Chinese cobra (*Naja naja atra*). Based on its effect of blocking nerve transmission, α-cobrotoxin has demonstrated a wide range of pharmacological properties including
analgesic effects against any pain. In addition, α-cobrotoxin also shows the potential for stop-
ping drug addiction due to its characteristics of high effectivity, no endurance, and no addiction.
Currently, α-cobrotoxin is typically administered orally or intravenously. Such administration
route always results in uneven distribution of α-cobrotoxin to the whole body and brain, and may
cause potential toxicity and adverse effects. α-Cobrotoxin was incorporated into the microspheres
composed of poly(lactide-*co*-glycolide) (PLGA) and poly[1,3-bis(*p*-carboxy-phenoxy) propane-
co-*p*-(carboxyethylformamido) benzoic anhydride] (P(CPP:CEFB)) and intranasally delivered to
model rats in order to improve its analgesic activity. PLGA and P(CPP:CEFB) were dissolved in
ACN/DMC (3:2 in volume ratio) cosolvent. α-Cobrotoxin was dissolved in double-distilled water
containing certain amount of alginate. The aqueous solution was mixed with the above organic
solution and emulsified with a probe sonicator to form W/O emulsion. The primary W/O emulsion was then emulsified into liquid paraffin containing Span 80 emulsifier with a homogenizer
to form the W/O/O emulsion. The resulting secondary emulsion was immediately transferred into
liquid paraffin containing Span 80. The mixture was magnetically stirred to evaporate the organic
solvents. The microspheres with high entrapment efficiency (>80%) and average diameter of about
25 μm could be prepared by a modified W/O/O emulsion–solvent evaporation method. Scanning
electron micrograph (SEM) study indicated that P(CPP:CEFB) content played a considerable role
on the morphology and degradation of the microspheres. The presence of P(CPP:CEFB) in the
microspheres increased their residence time at the surface of the nasal rat mucosa. The toxicity of
the composite microspheres to nasal mucosa was proved to be mild and reversible. A tail flick assay
was used to evaluate the antinociceptive activity of the microspheres after nasal administration.
Compared with the free α-cobrotoxin and PLGA microspheres, PLGA/P(CPP:CEFB) microspheres
showed an apparent increase in the strength and duration of the antinociceptive effect at the same
dose of α-cobrotoxin.

W/O/O/O multiple emulsion system is achieved by homogenizing aqueous phase in a protective
oil phase (e.g., soybean oil) to form a primary W/O emulsion. The primary emulsion was dispersed
in a polymer solution (O), and the emulsion was then emulsified in a hardening solution (O). The
W/O/O emulsion was dispersed in the hardening solution by forcing it at a constant rate of infusion
through a narrow electrically conductive stainless steel tube (O'Donnell and McGinity, 1998). The
hardening solution was continuously agitated by means of a magnetic stirrer for 24 h to enable evaporation of the organic solvent from the microspheres. The W/O/O/O type microspheres successfully
prevented encapsulated thioridazine HCl drug (O'Donnell and McGinity, 1998) from direct contact
with the polymer/solvent system and can control the release of reactive compounds. Thioridazine is
a piperidine antipsychotic drug that was previously widely used in the treatment of schizophrenia
and psychosis.

A very special suspension-multiple emulsion system (S/O/O/O) was developed by Iwata et al.
(1998). Tumor necrosis factor (TNF, cachexin, or cachectin) is a cytokine involved in systemic
inflammation, and is a member of a group of cytokines that stimulate the acute phase reaction. The
primary role of TNF is in the regulation of immune cells. TNF is also able to induce apoptotic cell
death, to induce inflammation, and to inhibit tumorigenesis and viral replication. In order to select
a suitable solvent system for the preparation of PLGA microspheres containing TNF, the stability of

TNF when mixed with DCM and ACN under various phase conditions was investigated. When the TNF solution was emulsified into DCM to form a W/O emulsion prior to solvent evaporation using the W/O/W technique, a significant loss in activity of TNF was found. When the TNF was dispersed as a dry powder in the DCM phase, the protein was inactivated due to immediate hydration under the conditions of the S/O/W system. Since TNF also was inactivated in a buffered saline containing ACN, the stability of the protein in microspheres prepared from an anhydrous solvent system was studied using ACN as the polymer solvent. Multiphase microspheres prepared by an anhydrous multiple emulsion process had a significantly higher loading efficiency of intact TNF than conventional matrix-type microspheres prepared by an anhydrous method using TNF powder and ACN. Soybean oil in which aluminum monostearate was dispersed to dissolve the dispersed solid in the oil. Dried gelatin powder containing TNF was dispersed by an ultrasonic homogenizer into oil phase containing Span 80 to obtain a fine oily suspension of the solid-in-oil (S/O) type. The oily suspension was poured into PLGA-ACN solution and dispersed gently to form a solid-in-oil-in-oil (S/O/O) type emulsion. The S/O/O emulsion was poured through a narrow nozzle into mineral oil containing Span 80, and then agitated. The resulting S/O/O/O multiple emulsion was agitated using a three-blade propeller to evaporate the ACN. The hardened microspheres were filtered. The anhydrous solvent evaporation system (S/O/O/O) utilized to produce multiphase PLGA microspheres was the most suitable encapsulation process for TNF.

Finally, a completely different multiple emulsion method is mentioned that is called O/W/O or dry-in-oil multiple emulsion. Its main steps are as follows: (1) preparation of a primary O/W emulsion in which the oily dispersed phase was constituted of DCM and drug and the aqueous continuous phase is a mixture of acetic acid solution (2%) and methanol containing the polymer (CS) and an emulsifier (Tween 80); (2) oily outer phase, mineral oil containing Span 20 surfactant; (3) evaporation of aqueous solvents at reduced pressure according to Genta et al. (1997). Ketoprofen, a nonsteroidal anti-inflammatory drug with analgesic and antipyretic effects (acts by inhibiting the body's production of prostaglandin) was entrapped in CS microspheres with good morphological characteristics and narrow size distribution, however, with a low encapsulation efficiency. Ketoprofen release was modulated using chemical polymer cross-linking with glutaraldehyde.

Astaxanthin (3,3'-dihydroxy-β-β'-carotene-4–4'-dione) is the main ketocarotenoid responsible for the red-orange color in salmonids and crustacean. In human nutrition, astaxanthin has been gaining widespread popularity as a dietary supplement due to its powerful antioxidant properties. Currently, several astaxanthin products derived from microalgae are available in the marketplace, and being promoted as anticancer and anti-inflammatory agents as well as immunostimulants. As most carotenoids, astaxanthin is a highly unsaturated molecule, and thus can easily be degraded by thermal or oxidative processes during the manufacture and storage of foods. This can cause the loss of their nutritive and biological desirable properties as well as the production of undesirable flavor or aroma compounds. Synthetic astaxanthin was microencapsulated in a CS matrix cross-linked with glutaraldehyde by using the method of multiple emulsion O/W/O solvent evaporation (Higuera-Ciapara et al., 2004). The stability of the pigment in the microcapsules was studied under storage at 25°C, 35°C, and 45°C for 8 weeks by measuring isomerization and loss of concentration of pigment. Results showed that the microencapsulated pigment did not suffer isomerization nor chemical degradation under the investigated storage conditions.

Paclitaxel is a mitotic inhibitor used in cancer chemotherapy. Recently, Trickler et al. (2008) provided conceptual proof that CS/glyceryl monooleate can form polycationic nanosized particles (400–700 nm) in multiple emulsion (O/W/O) solvent evaporation methods. The nanoparticles have a hydrophobic inner core with a hydrophilic coating that exhibits a significant positive charge and sustained-release characteristics. This novel nanoparticle formulation shows evidence of mucoadhesive properties; a fourfold increased cellular uptake and a 1000-fold reduction in the half maximal inhibitory concentration of paclitaxel. These advantages allow lower doses of paclitaxel to achieve a therapeutic effect, thus presumably minimizing the adverse side effects. Some data of different multiple emulsion–solvent evaporation methods used in some special cases can be found in Table 16.9.

TABLE 16.9
Pharmaceutical Compounds Encapsulated in Polymers by Different Multiple Emulsion–Solvent Evaporation Methods

Material Encapsulated	Polymer(s)	Method	References
α-Cobrotoxin	PLGA, P(CPP:CEFB)	W/O/O	Li et al. (2005)
Thioridazine HCl	PLGA	W/O/O/O	O'Donnell and McGinity (1998)
TNF	PLGA	S/O/O/O	Iwata et al. (1998)
Ketoprofen	Chitosan	O/W/O	Genta et al. (1997)
Astaxanthin	Chitosan	O/W/O	Higuera-Ciapara et al. (2004)
Paclitaxel	Chitosan	O/W/O	Trickler et al. (2008)

16.5   SUMMARY

In the 1990s, the aim of most studies was to find relationships exhaustively between process parameters and properties of nano- and microparticles formed by multiple emulsion–solvent evaporation method. Then, more and more potential applications of sustained delivery devices became emphasized, which was expressed in the experimental work in such a way the *in vitro* release investigations were supplemented or substituted by *in vivo* ones. The results of several products are encouraging for the future. Some of the systems fabricated possess the possibility of becoming an effective drug release formulation. However, toxicological and clinical examinations of huge expense must precede commercial applications. That is why most of the promising improvements have not reached yet or can never get to the market at all. Double emulsion methods were used for the encapsulation of water-soluble peptides and proteins primarily. Hormones, antigens, enzymes, growth factors, and cytokines were investigated most frequently with this procedure. However, important sustained release forms of DNA, antibiotics, and steroid hormones were also fabricated and studied. Other aspects of multiple emulsion–solvent evaporation are W/O/O, W/O/O/O, S/O/O/O, and O/W/O; nevertheless, these processes are rarely applied. It can also be concluded considering all types of experiments, PLGA is kept to be the most appropriate encapsulating polymer.

ABBREVIATIONS

ACN	Acetonitrile
BCEP	B-cell epitope
βLG	β-Lactoglobulin
B. melitensis	*Brucella melitensis*
BSA	Bovine serum albumin
cAMP	Cyclic adenosine monophosphate
CBSA-PEG	Cationic bovine serum albumin conjugated pegylated
CIP	Ciprofloxacin
CS	Chitosan
DCM	Dichloromethane
ECZ	Econazole
EE	Ethinylestradiol
ELISA	Enzyme-linked immunosorbent assay
EPO	Erythropoietin
FITC-BSA	Fluorescein isothiocyanate-labeled bovine serum albumin
FRRV	Formaldehyde-inactivated rotavirus
FSH	Follicle stimulating hormone
GM-CSF	Granulocyte-macrophage colony-stimulating factor

GnRH	Gonadotropin-releasing hormone
GOD	Glucose oxidase
HB	Hepatitis B
HBsAg	Hepatitis B surface antigen
HIV	Human immunodeficiency virus
HPβCD	Hydroxypropyl-β-cyclodextrin
H. pylori	*Helicobacter pylori*
IFN	Interferon
Ig	Immunoglobulin
i.p.	Intraperitoneal
Leu-A	Leucinostatin-A
LH	Luteinizing hormone
LNG	Levonorgestrel
MOX	Moxifloxacin
M. tuberculosis	*Mycobacterium tuberculosis*
NGF	Nerve growth factor
OVA	Ovalbumin
O/W/O	Oil-in-water-in-oil
PC12	Pheochromocytoma 12 cell line
PCL	Poly(ε-caprolactone)
pDNA	Plasmid deoxyribonucleic acid
P(CPP:CEFB)	Poly[1,3-bis(*p*-carboxy-phenoxy) propane-*co*-*p*-(carboxyethylformamido) benzoic anhydride]
PEAD	Poly(ethylene adipate)
PEG	Polyethylene glycol
PEI	Polyethylenimine
PHB	Poly(hydroxybutyrate)
PHBHV	Poly(hydroxybutyratehydroxyvalerate)
PLA	Polylactic acid
PLGA	Poly(lactic-*co*-glycolic acid) or poly(lactide-*co*-glycolide)
PMMA	Polymethylmethacrylate
PMM 2.1.2	Poly(methylidene malonate 2.1.2
poly(TMA-Tyr:SA:CPP)	Poly[trimellitylimido-L-tyrosineco-sebacic acid-*co*-1,3-bis(carboxyphenoxy)propane] anhydrides
PS-ODN	Phosphorothioate oligonucleotide
PTH	Parathyroid hormone
PVA	Polyvinyl alcohol
PVP	Polyvinyl pyrrolidone
rhBMP-2	Recombinant human bone morphogenetic protein-2
rhEGF	Recombinant human epidermal growth factor
rhGH	Recombinant human growth hormone
s.c.	Subcutaneous
SD	Sodium diclofenac
SDS-PAGE	Sodium dodecyl sulfate polyacrylamide gel electrophoresis
S/O	Solid-in-oil
S/O/O	Solid-in-oil-in-oil
S/O/O/O	Solid-in-oil-in-oil-in-oil
TB	Tuberculosis
TGF-β2	Transforming growth factor-β2
TNF	Tumor necrosis factor
W/O	Water-in-oil

W/O/O   Water-in-oil-in-oil
W/O/O/O   Water-in-oil-in-oil-in-oil
W/O/W   Water-in-oil-in-water

REFERENCES

Ahmad, Z., Pandey, R., Sharma, S., and Khuller, G. K. 2008. Novel chemotherapy for tuberculosis: Chemotherapeutic potential of econazole- and moxifloxacin-loaded PLG nanoparticles. *Int. J. Antimicrob. Agents* 31:142–146.

Akhtar, S. and Lewis, K. J. 1997. Antisense oligonucleotide delivery to cultured macrophages is improved by incorporation into sustained-release biodegradable polymer microspheres. *Int. J. Pharm.* 151:57–67.

Al haushey, L., Bolzinger, M. A., Bordes, C., Gauvrit, J. Y., and Briancon, S. 2007. Improvement of a bovine serum albumin microencapsulation process by screening design. *Int. J. Pharm.* 344:16–25.

Atkins, T. W. 1997. Fabrication of microcapsules using poly(ethylene adipate) and a blend of poly(ethylene adipate) with poly(hydroxybutyratehydroxyvalerate): Incorporation and release of bovine serum albumin. *Biomaterials* 18:173–180.

Bai, X.-L., Yang, Y.-Y., Chung, T.-S., Ng, S., and Heller, J. 2001. Effect of polymer compositions on the fabrication of poly(ortho-ester) microspheres for controlled release of protein. *J. Appl. Polymer Sci.* 80:1630–1642.

Basarkar, A., Devineni, D., Palaniappan, R., and Singh, J. 2007. Preparation, characterization, cytotoxicity and transfection efficiency of poly(DL-lactide-*co*-glycolide) and poly(DL-lactic acid) cationic nanoparticles for controlled delivery of plasmid DNA. *Int. J. Pharm.* 343:247–254.

Benichou, A., Aserin, A., and Garti, N. 2004. Double emulsions stabilized with hybrids of natural polymers for entrapment and slow release of active matters. *Adv. Colloid Interface Sci.* 108–109:29–41.

Benoit, M.-A., Baras, B., and Gillard, J. 1999. Preparation and characterization of protein-loaded poly(ε-caprolactone) microparticles for oral vaccine delivery. *Int. J. Pharm.* 184:73–84.

Bilati, U., Allemann, E., and Doelker, E. 2005. Strategic approaches for overcoming peptide and protein instability within biodegradable nano- and microparticles. *Eur. J. Pharm. Biopharm.* 59:375–388.

Bittner, B., Morlock, M., Koll, H., Winter, G., and Kissel, T. 1998. Recombinant human erythropoietin (rhEPO) loaded poly(lactide-*co*-glycolide) microspheres: Influence of the encapsulation technique and polymer purity on microsphere characteristics. *Eur. J. Pharm. Biopharm.* 45:295–305.

Blanco, M. D. and Alonso, M. J. 1997. Development and characterization of protein-loaded poly(lactide-*co*-glycolide) nanospheres. *Eur. J. Pharm. Biopharm.* 43:287–294.

Blanco-Prieto, M. J., Fattal, E., Gulik, A., Dedieu, J. C., Roquest, B. P., and Couvreur, P. 1997. Characterization and morphological analysis of a cholecystokinin derivative peptide-loaded poly(lactide-*co*-glycolide) microspheres prepared by a water-in-oil-in-water emulsion solvent evaporation method. *J. Control. Release* 43:81–87.

Blanco-Prieto, M. J., Lecaroz, C., Renedo, M. J., Kunkova, J., and Gamazo, C. 2002. *In vitro* evaluation of gentamicin released from microparticles. *Int. J. Pharm.* 242:203–206.

Bouillot, P., Babak, V., and Dellacherie, E. 1999. Novel bioresorbable and bioeliminable surfactants for microsphere preparation. *Pharm. Res.* 16:148–154.

Caliceti, P., Veronese, F. M., and Lora, S. 2000. Polyphosphazene microspheres for insulin delivery. *Int J. Pharm.* 211:57–65.

Chiba, M., Hanes, J., and Langer, R. 1997. Controlled protein delivery from biodegradable tyrosine-containing poly(anhydride-*co*-imide) microspheres. *Biomaterials* 18:893–901.

Chognot, D., Leonard, M., Six, J.-L., and Dellacherie, E. 2006. Surfactive water-soluble copolymers for the preparation of controlled surface nanoparticles by double emulsion/solvent evaporation. *Colloid. Surf. B* 51:86–92.

Coccoli, V., Luciani, A., Orsi, S., Guarino, V., Causa, F., and Netti, P. A. 2008. Engineering of poly(ε-caprolactone) microcarriers to modulate protein encapsulation capability and release kinetic. *J. Mater. Sci.* 19:1703–1711.

Conway, B. R., Eyles, J. E., and Alpar, H. O. 1997. A comparative study on the immune responses to antigens in PLA and PHB microspheres. *J. Control. Release* 49:1–9.

Coombes, A. G. A., Yeh, M.-K., Lavelle, E. C., and Davis, S. S. 1998. The control of protein release from poly(DL-lactide *co*-glycolide) microparticles by variation of the external aqueous phase surfactant in the water-in oil-in water method. *J. Control Release* 52:311–320.

Couvreur, P., Blanco-Prieto, M. J., Puisieux, F., and Roques, B. 1997. Multiple emulsion technology for the design of microspheres containing peptides and oligopeptides. *Adv. Drug Del. Rev.* 28:85–96.

Cui, F., Cun, D., Tao, A., Yang, M., Shi, K., Zhao, M., and Guan, Y. 2005. Preparation and characterization of melittin-loaded poly(DL-lactic acid) or poly(DL-lactic-*co*-glycolic acid) microspheres made by the double emulsion method. *J. Control. Release* 107:310–319.

Dhanaraju, M. D., Vema, K., Jayakumar, R., and Vamsadhara, C. 2003. Preparation and characterization of injectable microspheres of contraceptive hormones. *Int. J. Pharm.* 268:23–29.

Dhanaraju, M. D., Jayakumar, R., and Vamsadhara, C. 2004. Influence of manufacturing parameters on development of contraceptive steroid loaded injectable microspheres. *Chem. Pharm. Bull.* 52:976–979.

Dhanaraju, M. D., RajKannan, R., Selvaraj, D., Jayakumar, R., and Vamsadhara, C. 2006. Biodegradation and biocompatibility of contraceptive-steroid-loaded poly (DL-lactide-*co*-glycolide) injectable microspheres: *In vitro* and *in vivo* study. *Contraception* 74:148–156.

Diwan, M. and Park, T. G. 2003. Stabilization of recombinant interferon-α by pegylation for encapsulation in PLGA microspheres. *Int. J. Pharm.* 111–122.

Dorati, R., Genta, I., Colonna, C., Modena, T., Pavanetto, F., Perugini, P., and Conti, B. 2007. Investigation of the degradation behaviour of poly(ethylene glycol-*co*-D,L-lactide) copolymer. *Polym. Degr. Stab.* 92:1660–1668.

Eligio, T., Rieumont, J., Sanchez, R., and Silva, J. F. S. 1999. Characterization of chemically modified poly(3-hydroxy-alkanoates) and their performance as matrix for hormone release. *Die Angew. Makromol. Chem.* 270:69–75.

Feczko, T., Toth, J., and Gyenis, J. 2008. Comparison of the preparation of PLGA-BSA nano- and microparticles by PVA, poloxamer and PVP. *Colloid. Surf. A* 319:188–195.

Gaspar, M. M., Blanco, D., Cruz M. E. M., and Alonso, M. J. 1998. Formulation of L-asparaginase-loaded poly(lactide-*co*-glycolide) nanoparticles: Influence of polymer properties on enzyme loading, activity and *in vitro* release. *J. Control. Release* 52:53–62.

Genta, I., Perugini, P., Conti, B., and Pavanetto, F. 1997. A multiple emulsion method to entrap a lipophilic compound into chitosan microspheres. *Int. J. Pharm* 152:237–246.

Genta, I., Perugini, P., Pavanetto, F., Maculotti, K., Modena, T., Casado, B., Lupi, A., Iadarola, P., and Conti, B. 2001. Enzyme loaded biodegradable microspheres *in vitro ex vivo* evaluation. *J. Control. Release* 77:287–295.

Giovagnoli, S., Blasi, P., Schoubben, A., Rossi, C., and Ricci, M. 2007. Preparation of large porous biodegradable microspheres by using a simple double-emulsion method for capreomycin sulfate pulmonary delivery. *Int. J. Pharm.* 333:103–111.

Gomes dos Santos, A. L., Bochot, A., Doyle, A., Tsapis, N., Siepmann, J., Siepmann, F., Schmaler, J., Besnard, M., Behar-Cohen, F., and Fattal, E. 2006. Sustained release of nanosized complexes of polyethyleneimine and anti-TGF-β2 oligonucleotide improves the outcome of glaucoma surgery. *J. Control. Release* 112:369–381.

Han, K., Lee, K.-D., Gao, Z.-G., and Park, J.-S. 2001. Preparation and evaluation of poly(L-lactic acid) microspheres containing rhEGF for chronic gastric ulcer healing. *J. Control. Release* 75:259–269.

Herrmann, J. and Bodmeier, R. 1998. Biodegradable, somatostatin acetate containing microspheres prepared by various aqueous and non-aqueous solvent evaporation methods. *Eur. J. Pharm. Biopharm.* 45:75–82.

Higuera-Ciapara, I., Felix-Valenzuela, L., Goycoolea, F. M., and Argüelles-Monal, W. 2004. Microencapsulation of astaxanthin in a chitosan matrix. *Carbohydr. Polym.* 56:41–45.

Howard, K. A., Li, X. W., Somavarapu, S., Singh, J., Green, N., Atuah, K. N., Ozsoy, Y., Seymour, L. W., and Alpar, O. 2004. Formulation of a microparticle carrier for oral polyplex-based DNA vaccines. *Biochim. Biophys. Acta* 1674:149–157.

Hsu, Y.-Y., Hao, T., and Hedley, M. L. 1999. Comparison of process parameters for microencapsulation of plasmid DNA in poly(D,L-lactic-*co*-glycolic) acid microspheres. *Drug Target.* 7:313–323.

Ibrahim, M. A., Ismail, A., Fetouh, M. I., and Gopferich, A. 2005. Stability of insulin during the erosion of poly(lactic acid) and poly(lactic-*co*-glycolic acid) microspheres. *J. Control. Release* 106:241–252.

Iwata, M., Tanaka, T., Nakamura, Y., and McGinity, J. W. 1998. Selection of the solvent system for the preparation of poly(D,L-lactic-*co*-glycolic acid) microspheres containing tumor necrosis factor-alpha (TNF-α). *Int. J. Pharm.* 160:145–156.

Jeong, Y.-I., Na, H.-S., Seo, D.-H., Kim, D.-G., Lee, H.-C., Jang, M.-K., Na, S.-K., Roh, S.-H., Kim, S.-I., and Nah, J.-W. 2008. Ciprofloxacin-encapsulated poly(DL-lactide-*co*-glycolide) nanoparticles and its antibacterial activity. *Int. J. Pharm.* 352:317–323.

Jiang, G., Woo, B. H., Kang, F., Singh, J., and DeLuca, P. P. 2002. Assessment of protein release kinetics, stability and protein polymer interaction of lysozyme encapsulated poly(D,L-lactide-*co*-glycolide) microspheres. *J. Control. Release* 79:137–145.

Kim, H. K. and Park, T. G. 2004. Comparative study on sustained release of human growth hormone from semi-crystalline poly(L-lactic acid) and amorphous poly(D,L-lactic-*co*-glycolic acid) microspheres: Morphological effect on protein release. *J. Control. Release* 98:115–125.

Kim, S. Y., Doh, H. J., Ahn, J. S., Ha, Y. J., Jang, M. H., Chung, S. I., and Park, H. J. 1999. Induction of mucosal and systemic immune response by oral immunization with *H. pylori* lysates encapsulated in poly(D,L-lactide-*co*-glycolide) microparticles. *Vaccine* 17:607–616.

Kim, T. H., Lee, H., and Park, T. G. 2002. Pegylated recombinant human epidermal growth factor (rhEGF) for sustained release from biodegradable PLGA microspheres. *Biomaterials* 23:2311–2317.

Kim, T. K., Yoon, J. J., Lee, D. S., and Park, T. G. 2006. Gas foamed open porous biodegradable polymeric microspheres. *Biomaterials* 27:152–159.

Klose, D., Siepmann, F., Elkharraz, K., Krenzlin, S., and Siepmann, J. 2006. How porosity and size affect the drug release mechanisms from PLGA-based microparticles. *Int. J. Pharm.* 314:198–206.

Lambert, O., Nagele, O., Loux, V., Bonny, J.-D., and Marchal-Heussler, L. 2000. Poly(ethylene carbonate) microspheres: Manufacturing process and internal structure characterization. *J. Control. Release* 67:89–99.

Lamprecht, A., Ubrich, N., Hombreiro Pérez, M., Lehr, C.-M., Hoffman, M., and Maincent, P. 1999. Biodegradable monodispersed nanoparticles prepared by pressure homogenization-emulsification, *Int. J. Pharm.* 184:97–105.

Lamprecht, A., Torres, H. R., Shafer, U., and Lehr, C.-M. 2000. Biodegradable microparticles as a two-drug controlled release formulation: A potential treatment of inflammatory bowel disease. *J. Control Release* 69:445–454.

Le Visage, C., Quaglia, F., Dreux, M., Ounnar, S., Breton, P., Bru, N., Couvreur, P., and Fattal, E. 2001. Novel microparticulate system made of poly(methylidene malonate 2.1.2). *Biomaterials* 22:2229–2238.

Leo, E., Ruozi, B., Tosi, G., and Vandelli, M. A. 2006. PLA-microparticles formulated by means a thermoreversible gel able to modify protein encapsulation and release without being co-encapsulated. *Int. J. Pharm.* 323:131–138.

Li, M., Rouaud, O., and Poncelet, D. 2008. Microencapsulation by solvent evaporation: State of the art for process engineering approaches. *Int. J. Pharm.* 363:26–39.

Li, X., Zhang, Y., Yan, R., Jia, W., Yuan, M., Deng, X., and Huang, Z. 2000. Influence of process parameters on the protein stability encapsulated in poly-DL-lactide–poly(ethylene glycol) microspheres. *J. Control. Release* 68:41–52.

Li, Y.-P., Pei, Y.-Y., Zhang, X.-Y., Gu, Z.-H., Zhou, Z.-H., Yuan, W.-F., Zhou, J.-J., Zhu, J.-H., and Gao, X.-J. 2001. PEGylated PLGA nanoparticles as protein carriers: Synthesis, preparation and biodistribution in rats. *J. Control. Release* 71:203–211.

Li, Y., Jiang, H. L., Zhu, K. J., Liu, J. H., and Hao, Y. L. 2005. Preparation, characterization and nasal delivery of α-cobrotoxin-loaded poly(lactide-*co*-glycolide)/polyanhydride microspheres. *J. Control. Release* 108: 10–20.

Liu, R., Huang, S.-S., Wan, Y.-H., Ma, G.-H., and Su, Z.-G. 2006. Preparation of insulin-loaded PLA/PLGA microcapsules by a novel membrane emulsification method and its release *in vitro*. *Colloid. Surf. B* 51:30–38.

Mandal, T. K., Shekleton, M., Onyebueke, E., Washington, L., and Penson, T. 1996. Effect of formulation and processing factors on the characteristics of biodegradable microcapsules of zidovudine. *J. Microencapsul.* 13:545–557.

Martin, M. A., Miguens, F. C., Rieumont, J., and Sanchez, R. 2000. Tailoring of the external and internal morphology of poly-3-hydroxy butyrate microparticles. *Colloid. Surf. B* 17:111–116.

Munoz, P. M., Estevan, M., Marin, C. M., De Miguel, M. J., and Grillo, M. J. 2006. Brucella outer membrane complex-loaded microparticles as a vaccine against *Brucella ovis* in rams. *Vaccine* 24:1897–1905.

Nakaoka, R., Inoue, Y., Tabata, Y., and Ikada, Y. 1996. Size effect on the antibody production induced by biodegradable microspheres containing antigen. *Vaccine* 14:1251–1256.

Nicoli, S., Santi, P., Couvreur, P., Couarraze, G., Colombo, P., and Fattal, E. 2001. Design of triptorelin loaded nanospheres for transdermal iontophoretic administration. *Int. J. Pharm.* 214:31–35.

O'Donnell, P. B. and McGinity, J. W. 1998. Influence of processing on the stability and release properties of biodegradable microspheres containing thioridazine hydrochloride. *Eur. J. Pharm. Biopharm.* 45:83–94.

Olivier, J.-C., Huertas, R., Lee, H. J., Calon, F., and Pardridge, W. M. 2002. Synthesis of pegylated immunonanoparticles. *Pharm. Res.* 19:1137–1143.

Panyam, J. and Labhasetwar, V. 2003. Dynamics of endocytosis and exocytosis of poly(D,L-lactide-*co*-glycolide) nanoparticles in vascular smooth muscle cells. *Pharm. Res.* 20:212–220.

Panyam, J., Dali, M. M., Sahoo, S. K., Ma, W., Chakravarthi, S. S., Amidon, G. L., Levy, R. J., and Labhasetwar, V. 2003. Polymer degradation and *in vitro* release of a model protein from poly(D,L-lactide-*co*-glycolide) nano- and microparticles. *J. Control. Release* 92:173–187.

Pean, J.-M., Venier-Julienne, M.-C., Boury, F., Menei, P., Denizot, B., and Benoit, J.-P. 1998. NGF release from poly(D,L-lactide-*co*-glycolide) microspheres. Effect of some formulation parameters on encapsulated NGF stability. *J. Control. Release* 56:175–187.

Pfeifer, B. A., Burdick, J. A., Little, S. R., and Langer, R. 2005. Poly(ester-anhydride):poly(β-amino ester) micro- and nanospheres: DNA encapsulation and cellular transfection. *Int. J. Pharm.* 304:210–219.

Prabha, S. and Labhasetwar, V. 2004. Critical determinants in PLGA/PLA nanoparticle-mediated gene expression. *Pharm. Res.* 21:354–364.

Prabha, S., Zhou, W.-Z., and Panyam, J. 2002. Size-dependency of nanoparticle-mediated gene transfection: Studies with fractionated nanoparticles. *Int. J. Pharm.* 244:105–115.

Rajkannan, R., Dhanaraju, M. D., Gopinath, D., Selvaraj, D., and Jayakumar, R. 2006. Development of hepatitis B oral vaccine using B-cell epitope loaded PLG microparticles. *Vaccine* 24:5149–5157.

Rawat, A., Majumder, Q. H., and Ahsan, F. 2008. Inhalable large porous microspheres of low molecular weight heparin: *In vitro* and *in vivo* evaluation. *J. Control. Release* 128:224–232.

Remunan-Lopez, C., Lorenzo-Lamosa, M. L., Vila-Jato, J. L., and Alonso, M. J. 1998. Development of new chitosan–cellulose multicore microparticles for controlled drug delivery. *Eur. J. Pharm. Biopharm.* 45:49–56.

Reza, M. S. and Whateley, T. L. 1998. Iodo-2′-deoxyuridine (IUdR) and (125)IUdR loaded biodegradable microspheres for controlled delivery to the brain. *J. Microencapsul.* 15:789–801.

Ricci, M., Blasi, P., Giovagnoli, S., Perioli, L., Vescovi, C., and Rossi, C. 2004. Leucinostatin-A loaded nanospheres: Characterization and *in vivo* toxicity and efficacy evaluation. *Int. J. Pharm.* 275:61–72.

Rojas, J., Pinto-Alphandary, H., Leo, E., Pecquet, S., Couvreur, P., Gulik, A., and Fattal, E. 1999. A polysorbate-based non-ionic surfactant can modulate loading and release of β-lactoglobulin entrapped in multiphase poly(DL-lactide-*co*-glycolide) microspheres. *Pharm. Res.* 16:255–260.

Rosa, G. D., Iommelli, R., La Rotonda, M. I., Miro, A., and Quaglia, F. 2000. Influence of the co-encapsulation of different non-ionic surfactants on the properties of PLGA insulin-loaded microspheres. *J. Control. Release* 69:283–295.

Rosa, G. D., Quaglia, F., La Rotonda, M. I., Besnard, M., and Fattal, E. 2002. Biodegradable microparticles for the controlled delivery of oligonucleotides. *Int. J. Pharm.* 242:225–228.

Rosa, G. D., Bochot, A., Quaglia, F., Besnard, M., and Fattal, E. 2003. A new delivery system for antisense therapy: PLGA microspheres encapsulating oligonucleotide/polyethyleneimine solid complexes. *Int. J. Pharm.* 254:89–93.

Rosca, I. D., Watari, F., and Uo, M. 2004. Microparticle formation and its mechanism in single and double emulsion solvent evaporation. *J. Control. Release* 99:271–280.

Sah, H. 1999. Stabilization of proteins against methylene chloride/water interface-induced denaturation and aggregation. *J. Control. Release* 58:143–151.

Sah, H., Toddywala, R., and Chien, Y. W. 1995. Continuous release of proteins from biodegradable microcapsules and *in vivo* evaluation of their potential as a vaccine adjuvant. *J. Control. Release* 35:137–144.

Sanchez, A., Tobio, M., Gonzalez, L., Fabra, A., and Alonso, M. J. 2003. Biodegradable micro- and nanoparticles as long-term delivery vehicles for interferon-alpha. *Eur. J. Pharm. Sci.* 18:221–229.

Saraf, S., Mishra, D., Asthana, A., Jain, R., Singh, S., and Jain, N. K. 2006. Lipid microparticles for mucosal immunization against hepatitis B. *Vaccine* 24:45–56.

Schlicher, E. J. A. M., Postma, N. S., Zuidema, J., Talsma, H., and Hennink, W. E. 1997. Preparation and characterization of poly (D,L-lactic-*co*-glycolic acid) microspheres containing desferrioxamine. *Int. J. Pharm.* 153:235–245.

Schrier, J. A. and DeLuca, P. P. 1999. Recombinant human bone morphogenetic protein-2 binding and incorporation in PLGA microsphere delivery systems. *Pharm. Dev. Technol.* 4:611–621.

Schwach, G., Oudry, N., Delhomme, S., Luck, M., Lindner, H., and Gurny, R. 2003. Biodegradable microparticles for sustained release of a new GnRH antagonist—Part I: Screening commercial PLGA and formulation technologies. *Eur. J. Pharm. Biopharm.* 56:327–336.

Singh, J., Pandit, S., Bramwell, V. W., and Alpar, H. O. 2006. Diphtheria toxoid loaded poly-(ε-caprolactone) nanoparticles as mucosal vaccine delivery systems. *Methods* 38:96–105.

Song, C. X., Labhasetwar, V., Murphy, H., Qu, X., Humphrey, W. R., Shebuski, R. J., and Levy, R. J. 1997. Formulation and characterization of biodegradable nanoparticles for intravascular local drug delivery. *J. Control. Release* 43:197–212.

Sturesson, C., Artursson, P., Ghaderi, R., Johansen, K., Mirazimi, A., Uhnoo, I., Svensson, L., Albertsson, A.-C., and Carlfor, J. 1999. Encapsulation of rotavirus into poly(lactide-*co*-glycolide) microspheres. *J. Control. Release* 59:377–389.

Tabata, Y., Gutta, S., and Langer, R. 1993. Controlled delivery systems for proteins using polyanhydride microspheres. *Pharm. Res.* 10:487–496.

Trickler, W. J., Nagvekar, A. A., and Dash, A. K. 2008. A novel nanoparticle formulation for sustained paclitaxel delivery. *AAPS PharmSciTech* 9:486–493.

Ubrich, N., Ngondi, J., Rivat, C., Pfister, M., Vigneron, C., and Maincent, P. 1996. Selective *in vitro* removal of anti-A antibodies by adsorption on encapsulated erythrocyte-ghosts. *J. Biomed. Mater. Res.* 37:155–160.

Ungaro, F., De Rosa, G., Miro, A., Quaglia, F., and La Rotonda, M. I. 2006. Cyclodextrins in the production of large porous particles: Development of dry powders for the sustained release of insulin to the lungs. *Eur. J. Pharm. Sci.* 28:423–432.

Virto, M. R., Elorza, B., Torrado, S., Elorza, M. L. A., and Frutos, G. 2007. Improvement of gentamicin poly(D,L-lactic-*co*-glycolic acid) microspheres for treatment of osteomyelitis induced by orthopedic procedures. *Biomaterials* 28:877–885.

Wang, J., Wang, B. M., and Schwendeman, S. P. 2002. Characterization of the initial burst release of a model peptide from poly(D,L-lactide-*co*-glycolide) microspheres. *J. Control. Release* 82:289–307.

Wang, J., Wang, B. M., and Schwendeman, S. P. 2004. Mechanistic evaluation of the glucose-induced reduction in initial burst release of octreotide acetate from poly(D,L-lactide-*co*-glycolide) microspheres. *Biomaterials* 25:1919–1927.

Wei, G., Pettway, G. J., McCauley, L. K., and Ma, P. X. 2004. The release profiles and bioactivity of parathyroid hormone from poly(lactic-*co*-glycolic acid) microspheres. *Biomaterials* 25:345–352.

Wolf, M., Wirth, M., Pittner, F., and Gabor, F. 2003. Stabilization and determination of the biological activity of L-asparaginase in poly(D,L-lactide-*co*-glycolide) nanospheres. *Int. J. Pharm.* 256:141–152.

Xie, Y.-L., Lu, W., and Jiang, X.-G. 2006. Improvement of cationic albumin conjugated pegylated nanoparticles holding NC-1900, a vasopressin fragment analog, in memory deficits induced by scopolamine in mice. *Behavioural Brain Res.* 173:76–84.

Yamaguchi, Y., Takenaga, M., Kitagawa, A., Ogawa, Y., Mizushima, Y., and Igarashi, R. 2002. Insulin-loaded biodegradable PLGA microcapsules: Initial burst release controlled by hydrophilic additives. *J. Control. Release* 81:235–249.

Yang, Y.-Y., Chung, T.-S., Bai, X.-L., and Chan, W.-K. 2000. Effect of preparation conditions on morphology and release profiles of biodegradable polymeric microspheres containing protein fabricated by double-emulsion method. *Chem. Eng. Sci.* 55:2223–2236.

Yin, Y., Chen, D. W., Qiao, M. X., Lu, Z., and Hu, H. Y. 2006. Preparation and evaluation of lectin-conjugated PLGA nanoparticles for oral delivery of thymopentin. *J. Control. Release* 116:337–345.

Zydowicz, N., Nzimba-Ganyanad, E., and Zydowicz, N. 2002. PMMA microcapsules containing water-soluble dyes obtained by double emulsion/solvent evaporation technique. *Polym. Bull.* 47:457–463.

17 Microemulsions in Biotechnology and Pharmacy: An Overview

Ambikanandan Misra, Kiruba Florence,
Manisha Lalan, and Tapan Shah

CONTENTS

17.1 INTRODUCTION

Microemulsions or micellar emulsions are defined as single optically isotropic and thermodynami-cally stable multicomponent fluids composed of oil, water, and surfactant (usually in conjunction with a cosurfactant). The droplets in a microemulsion are in the range of 1–100 nm in diameter. It is well established that dispersed particles having a diameter less than one-fourth of the wavelength of visible light, that is, less than approximately 120 nm, do not refract light and therefore microemul-sions appear transparent to the eye. The basic difference between emulsions and microemulsions are that emulsions exhibit excellent kinetic stability but are thermodynamically unstable when com-pared with microemulsions. The concept of microemulsion was introduced as early as 1943 by Hoar and Schulman (Hoar and Schulman, 1943). They generated clear single-phase solution by titrating a milky emulsion with hexanol. Schulman et al. introduced the term "microemulsion" for this system in 1959. In recent years, microemulsions have attracted a great deal of attention because of their biocompatibility, biodegradability, ease of handling and preparation, and appreciable solubilization capacity for both water- and oil-soluble drugs. The differences between emulsions and microemul-sions are enlisted in Table 17.1.

Microemulsions have various textures such as oil in water (O/W) (Figure 17.1), water in oil (W/O) (Figure 17.2), bicontinuous mixtures (Figure 17.3), ordered droplets, or lamellar mixtures with a wide range of phase equilibria among them and with excess oil and/or water phases. This great variety of textures is governed by variations in the composition of the whole system and in the structure of the interfacial layers. The rationale for developing and using medicated microemulsions are listed in Table 17.2.

17.2 THEORIES AND THERMODYNAMICS OF MICROEMULSION FORMULATION

Three different approaches have been proposed to explain microemulsion formation and the sta-bility aspects. The important features of the microemulsion are thermodynamic stability, optical transparency, large overall interfacial area (about $100 \, m^2/mL$), variety of structures, low interfacial tension, and increased solubilization of oil/water dispersed phase. Microemulsion requires more surfactant than emulsion to stabilize a large overall interfacial area.

The interfacial tension between the oil and the water can be lowered by the addition and adsorption of surfactant. When the surfactant concentration is increased further, it lowers the

TABLE 17.1
Differences between Emulsions and Microemulsions

Characteristics	Emulsion	Microemulsion
Droplet size	100–100,000 nm	10–100 nm
Phase	Two	One
Appearance	Opaque	Transparent
Proportion of dispersed phase	30%–60%	23%–40% without corresponding increase in viscosity
Energy requirement	Requires large energy input at the time of preparation	Forms spontaneously, so no energy requirement
Stability	Theoretically stable but thermodynamically unstable	Kinetically unstable but thermodynamically stable
Surfactant concentration	2–3 wt%	>6 wt%

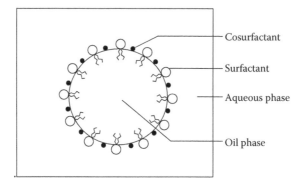

FIGURE 17.1 Oil-in-water microemulsion.

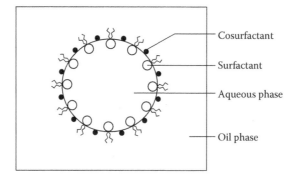

FIGURE 17.2 Water-in-oil microemulsion.

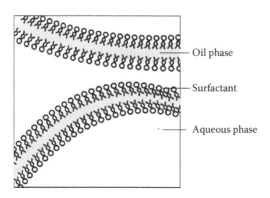

FIGURE 17.3 Bicontinuous microemulsion.

interfacial tension till CMC (critical micelle concentration). The micellar formation commences beyond this concentration of the surfactant. This negative interfacial tension leads to a simultaneous and spontaneous increase in the area of the interface. The large interfacial area formed may divide itself into a large number of closed shells around small droplets of either O/W or W/O and further decreases the free energy of the system. In many cases, the interfacial tension is not yet ultralow when the CMC is reached. It has been studied and observed by Schulman et al. that the addition of a cosurfactant (medium-sized alcohol or amine) to the system results in virtually zero interfacial tension (Bowcott and Schulman, 1955). The further addition of a surfactant (where interfacial tension (γ) is zero) leads to negative interfacial tension.

TABLE 17.2
Rationale for Developing and Using Medicated Microemulsions

Reason	Drug Examples
Solubilization of poorly water-soluble drugs	Diazepam, vitamin A and E, and dexamethasone palmitate
Solubilization of hydrolytically susceptible compounds	Lomustine and physostigmine salicylate
Reduction of irritation, pain, or toxicity of intravenously administered drugs	Diazepam
Potential for sustained release dosage forms	Barbiturates
Site specific drug delivery to various organs	Cytotoxic drugs

17.2.1 MIXED FILM THEORIES

The relatively large entropy of mixing of droplets and continuous medium explains the spontaneous formation of microemulsion. Schulman and coworkers (Hoar and Schulman, 1943; Schulman et al., 1959) emphasized the importance of the interfacial film. They considered that the spontaneous formation of microemulsion droplets was due to the formation of a complex film at the oil–water interface by the surfactant and the cosurfactant. This caused a reduction in oil–water interfacial tension to very low values (from close to zero to negative), which is represented by following equation.

$$\gamma_i = \gamma_{O/W}^{-\pi_i}$$

where
$\gamma_{O/W}$ = oil–water interfacial tension without the film present
π_i = spreading pressure
γ_i = interfacial tension

17.2.1.1 Mechanism of Curvature of a Duplex Film

The interfacial film should be curved to form small droplets to explain both the stability of the system and the bending of the interface. A flat duplex film would be under stress because of the difference in tension and spreading of pressure on either side of it. Reduction of this tension gradient by equalizing the two surface tensions is the driving force for the film curvature. Both sides of the interface expand spontaneously with penetration of oil and cosurfactant until the pressures become equal. The side with higher tension would be concave and would envelop the liquid on that side, making it an internal phase. It is generally easier to expand the oil side of an interface than the waterside and hence W/O microemulsion can be formed easily than O/W microemulsion (Tadros, 1984).

17.2.2 SOLUBILIZATION THEORIES

Shinoda et al. (Shinoda and Kuneida, 1973; Friberg et al., 1976, 1977; Rance and Friberg, 1977; Friberg and Venable, 1983; Shinoda and Lindman, 1987) considered microemulsion to be thermodynamically stable monophasic solution of water-swollen or oil-swollen spherical micelles. Rance and Friberg (1977) illustrated the relationship between reverse micelles and W/O microemulsion with the help of phase diagrams. The inverse micelle region of ternary system, that is, water, pentanol, and sodium dodecyl sulfate (SDS) is composed of water-solubilized reverse micelles of

SDS in pentanol. Addition of up to 50% O-xylene gives rise to transparent W/O region containing a maximum of 28% water with 5% pentanol and 6% surfactant (i.e., microemulsions). The quaternary phase diagram constructed on adding p-xylene shows a relationship of these areas to the isotropic inverse micellar phase. These four component systems could be prepared by adding hydrocarbon directly to the inverse micellar phase by titration. Thus, the system mainly consists of swollen inverse micelle rather than small emulsion droplets (Shinoda and Kuneida, 1973; Rance and Friberg, 1977).

17.2.3 Thermodynamic Theories

This theory explains the formation of microemulsion even in the absence of cosurfactant. For microemulsion to form spontaneously, the free energy is

$$\Delta G = \Gamma \Delta A$$

where
  ΔG is the free energy
  Γ is the interfacial tension
  ΔA is the increase in surface area

Thermodynamic theory takes into account entropy of droplets and thermal fluctuations at the interface as important parameters leading to interfacial bending instability. Ruckenstein and coworkers (Ruckenstein and Chi, 1975; Ruckenstein and Krishnan, 1979, 1980; Ruckenstein, 1985) and Overbeek (1978) considered that microemulsion formation is entropically driven. The dispersion of droplets in the continuous phase increases the entropy of the system and produces a negative free energy change that is significantly important for very small droplets as in microemulsions. The free energy (ΔG_m) for microemulsion formation can be calculated using the following equation.

$$\Delta G_m = \Delta G_1 + \Delta G_2 + \Delta G_3$$

where
  ΔG_m is the free energy
  ΔG_1 is the interfacial energy
  ΔG_2 is the free energy of inter droplet interactions
  ΔG_3 is the entropy for dispersion of droplets in continuous medium

Later it was shown that accumulation of the surfactant and cosurfactant at the interface results in a decrease in chemical potential generating an additional negative free energy change called as dilution effect. This theory explained the role of cosurfactant and salt in a microemulsion formed with ionic surfactants. The cosurfactant produces an additional dilution effect and decreases interfacial tension further. The addition of salts to system containing ionic surfactants causes similar effects by shielding the electric field produced by the adsorbed ionic surfactant in the adsorption of large amount of surfactant.

17.3 FORMATION AND PHASE BEHAVIOR OF MICROEMULSION

17.3.1 Types of Microemulsion

The formation of oil- or water-swollen microemulsion depends on the packing ratio, property of surfactant, oil phase, temperature, chain length, type, and nature of cosurfactant.

17.3.1.1 Packing Ratio

The hydrophilic lipophilic balance (HLB) of surfactant determines the type of microemulsion through its influence on molecular packing and film curvature. The analysis of film curvature for surfactant associations leading to microemulsion formation has been explained by Israelachvili et al. (1976) and Mitchell and Ninham (1977) in terms of packing ratio, also called as critical packing parameter.

$$\text{Critical packing parameter (CPP)} = \frac{V}{a*l}$$

where
   V is the volume of surfactant molecule
   a is the head group surface area
   l is the length

If CPP has value between 0 and 1, interface curves toward water (positive curvature) and O/W systems are favored, but when CPP is greater than 1, interface curves spontaneously toward oil (negative curvature) so W/O microemulsions are favored. At zero curvature, when the HLB is balanced (p is equivalent to 1), then either bicontinuous or lamellar structures may be formed according to the rigidity of the film (zero curvature).

17.3.1.2 Properties of Surfactant, Oil Phase, and Temperature

The type of emulsion, to a large extent, depends on the nature of surfactant; Gerbacia and Rosano (1973) observed that the interfacial tension could be temporarily reduced due to diffusion of cosurfactant through the interface. Microemulsion is formed by the combination of dispersion and stabilization processes. The dispersion process involves a transient reduction of interfacial tension to nearly zero or negative value, at which the interface expands to form fine dispersed droplets. Subsequently, droplets absorb more surfactant until the bulk phase is depleted enough to bring the value of interfacial tension positive. The interfacial film of alcohol and surfactant initiates the stabilization process. Stability of O/W emulsion system can be controlled by the interfacial charge. If the diffused double layer at the interface is compressed by high concentration of counterions, W/O microemulsions are formed.

Type of surfactant also determines type of microemulsion formed. Surfactant contains hydrophilic head group and lipophilic tail group. The oil component influences curvature by its ability to penetrate and hence swells the tail group region of the surfactant monolayer. Short-chain oils such as alkanes penetrate the lipophilic group region largely than long-chain alkanes, and swelling of this region to a great extent results in an increased negative curvature.

Temperature is extremely important in determining the effective head group size of nonionic surfactants. Winsor studied the effect of temperature on the type of microemulsion formed. For the given amount of components in ternary system with nonionic surfactant, oil, and water, at relatively low temperatures, type I system (an O/W with excess oil) is formed. At intermediate temperature, type III system (microemulsion with excess of both oil and water) is present. At relatively higher temperature, type II (W/O microemulsion with excess water) system exists (Winsor, 1954, 1968).

17.3.1.3 Chain Length, Type, and Nature of Cosurfactant

Alcohols are widely used as cosurfactants in microemulsions. Addition of shorter chain cosurfactant (e.g., ethyl alcohol) gives positive curvature effect, as alcohol swells the head region more than the tail region favoring the formation of O/W type of system, while longer chain cosurfactant (e.g., cetyl alcohol) favors W/O type by alcohol swelling more in tail region than head region.

17.3.2 PHASE BEHAVIOR OF MICROEMULSION

17.3.2.1 Salinity

At low salinity, the droplet size of O/W microemulsion increases. This corresponds to an increase in the solubilization of oil. As salinity further increases, the system becomes bicontinuous over an intermediate salinity range. Increase in salinity leads to the formation of continuous microemulsion with reduction in globule size. Further increase in salinity ultimately results in complete phase transition.

17.3.2.2 Alcohol Concentration

Increasing the concentration of low molecular weight alcohol as a cosurfactant leads to the phase transition from W/O to bicontinuous and ultimately to O/W type microemulsion. Exactly opposite phase transition is noticed in case of high molecular weight alcohol.

17.3.2.3 Surfactant Hydrophobic Chain Length

The increase in length of hydrophobic chain of the surfactant shows the change of O/W microemulsion to W/O via bicontinuous phase.

17.3.2.4 pH

Change in pH influences the microemulsions containing pH-sensitive surfactants. This effect is more pronounced in case of acidic or alkaline surfactants. Carboxylic acids and amines change the phase behavior from W/O to O/W by increasing the pH.

17.3.2.5 Nature of Oil

Increase in the aromaticity of oil leads to phase transition from O/W to W/O and is opposite to that of increase in the oil alkane carbon number.

17.3.2.6 Ionic Strength

As the ionic strength increases, the system passes from O/W microemulsion in equilibrium with excess oil to the middle phase and finally to W/O microemulsion in equilibrium with excess water.

17.4 CONSTRUCTION OF PHASE DIAGRAMS

When water, oil, and surfactants are mixed, microemulsions are only one of the association structures. Preparation of a stable, isotropic homogeneous, transparent, nontoxic microemulsion requires consideration of a number of variables. Construction of phase diagrams reduces a number of trials and labor. Phase diagrams help to find the microemulsion region in ternary or quaternary system and also help to determine the minimum amount of surfactant required for microemulsion formation.

17.4.1 TERNARY SYSTEMS

The phase behavior of surfactant–oil–water (S/O/W) is best reported by using ternary diagram. Here, two independent composition variables are sufficient, since third one is complement to 100%. Figure 17.4 shows the phase diagram that allows one to determine the ratio of oil:water:surfactant: cosurfactant at the boundary of microemulsion region. To plot the composition of four-component systems, a regular tetrahedron is composed by fixing and varying the other three or by using a constant ratio of two components (surfactant and cosurfactant or cosolvent). Figure 17.5 shows the pseudoternary diagram at constant surfactant to cosurfactant ratio. It also shows that single-phase or multiphase regions of microemulsion domains are near the center of diagram in areas containing

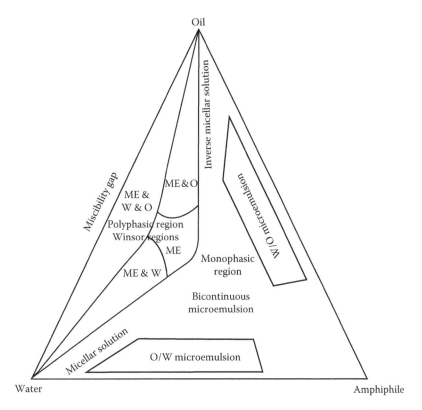

FIGURE 17.4 Different regions of phase diagram (ME, microemulsion; W, water; and O, oil).

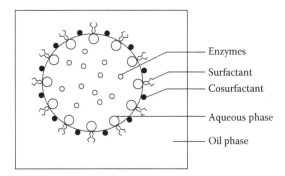

FIGURE 17.5 Enzymes entrapped in microemulsion.

large amounts of surfactant that is toxic. The phase behavior of surfactants, which form microemulsions in absence of cosurfactant, can be completely represented by ternary diagram.

17.4.2 Winsor's Phase Diagram

Winsor (1954) reported the relationship between the phase behavior of amphiphiles–oil–water and nature of the different components of ternary system. Different regions of a phase diagram are shown in Figure 17.4.

Winsor I

The microemulsion composition corresponding to Winsor I is characterized by two phase, the lower O/W microemulsion phase in equilibrium with excess oil.

Winsor II

The microemulsion composition corresponding to Winsor II is characterized by very low interfacial tension and maximal solubilization of oil and water for a given quantity of surfactant. Since, in this phase, microemulsion coexists with both excess phases, no one can distinguish the dispersed phase from the continuous phase.

Winsor III

This phase comprises of three phases, middle microemulsion phase (O/W plus W/O, called bicontinuous) in equilibrium with upper excess oil and lower water.

Winsor IV

Microemulsions can be distinguished from the micelles by its inner core swollen with oil. The microemulsion structure depends on the chemical composition, temperature, and concentration of the constituents. Different surfactants stabilize different microstructures due to aggregation. This aggregation phenomenon leads to a system with minimum free energy and thermodynamic stability. Even though the spherical micelles are considered to have minimal water–hydrocarbon contact area for a given volume, the inter micellar free energy and the impossibility of the existence of voids in the hydrophobic region lead to other amphiphilic assemblies like cylinders and planes. They are organized in the form of liquid crystalline phases or liquid isotropic phases. A wide variety of surfactant molecules obeys the geometric rules embodied in the packing parameter.

In concentrated aqueous solutions, amphiphiles exist as lamellar phase with two configuration (planar and continuous lamellar phase), hexagonal phase (surfactant molecules aggregate into circular cylinder micelles that pack onto the hexagonal lattice), cubic phases, and nematic phases. While in dilute solution, they exist as worm- or thread-like micelles.

17.4.3 QUATERNARY PHASE DIAGRAMS

Microemulsion is generally a quaternary system. To study their phase behavior, pseudoternary phase diagram consisting of the oil–water amphiphiles is usually drawn. Optimization done by pseudoternary diagram is, however, not an accurate method. Hence, it is better to use quaternary phase diagram for such systems.

17.4.4 METHODS FOR CONSTRUCTING PHASE DIAGRAM

Quaternary phase diagrams should be constructed to define the extent and nature of the microemulsion regions and surrounding regions. Several methods can be used to achieve the same. In one of the methods, a large number of samples of different composition must be prepared. The microemulsion region is identified by its isotropic nature and low viscosity. Other regions can be identified by their characteristic optical structure (Shinoda and Friberg, 1986). These diagrams are complicated and time consuming to prepare. In another method, microemulsion region can be located by titration method. At a constant ratio of SAA/CoS, various combinations of oil and SAA/CoS are produced. The water is added dropwise. After the addition of each drop, the mixture is stirred and examined through a polarized filter. The appearance (transparency, opalescence, and isotropy) is recorded along with the number of phases. Thus, an appropriate delineation of the boundaries can be obtained, in which it is possible to refine through the production of compositions point by point beginning with the four basic components.

The original method for the construction of phase diagram developed by Hoar and Schulman (1943) can be used for the preparation of microemulsion. In this method, adding the oil–surfactant mixture to some of the aqueous phase in a temperature-controlled container with agitation makes a coarse macroemulsion as a first step, which is then titrated with cosurfactant until clarity is obtained and then diluted with water to give a microemulsion of the desired concentration.

Rosano et al. (1988) suggested a simple routine test for rapid evaluation of components for their stability in microemulsions without construction of phase diagram. In this test, coarse emulsion is prepared and titrated to clarity with the chosen cosurfactant. The minimum concentration of surfactant required to cover the interface is calculated. If the system does not clarify after adding the cosurfactant in an amount equivalent to the primary surfactant, the system is considered to be unacceptable and first the cosurfactant and finally the oil is changed in a logical manner.

17.5 FORMULATION OF MICROEMULSIONS

Microemulsions are isotropic systems that are difficult to formulate, than ordinary emulsions, because of involvement of spontaneous interactions among the constituent molecules. Generally, the microemulsion formulation requires following components:

1. Oil phase: Toluene, cyclohexane, mineral oil or vegetable oils, silicone oils, or esters of fatty acids, etc., have been widely investigated as oil components.
2. Aqueous phase: Aqueous phase may contain hydrophilic active ingredients and preservatives. Some workers have utilized buffer solutions as aqueous phase.
3. Primary surfactant: The surfactants are generally ionic, nonionic, or amphoteric. The surfactants chosen are generally nonionic because of their good tolerance on topical application. Amphoteric surfactants are used only for specialized purposes.
4. Secondary surfactant (cosurfactant): Cosurfactants originally used were short-chain fatty alcohols (pentanol, hexanol, benzyl alcohol). These are most often polyols, esters of polyols, derivatives of glycerol, and organic acids. Their main purpose is to render interfacial film fluid by wedging themselves between the surfactant molecules.

17.6 CHARACTERIZATION OF MICROEMULSIONS

The determination of microemulsion structure is difficult, although it is important for the successful commercial exploitation of microemulsions as a drug delivery system.

17.6.1 PHASE BEHAVIOR STUDIES

Visual observations, phase contrast microscopy, and freeze fracture transmission electron microscopy can differentiate microemulsions from liquid crystals and coarse emulsions. Clear isotropic single-phase systems are identified as microemulsions, whereas opaque systems showing bifringence when viewed by cross-polarized microscopy may be taken as liquid crystalline system.

Phase behavior studies provide information about the boundaries of different phases as a function of composition variables and temperature. They also allow comparison of the efficiency of different surfactants for a given application.

17.6.2 SCATTERING TECHNIQUES FOR MICROEMULSIONS CHARACTERIZATION

Small-angle x-ray scattering (SAXS), small-angle neuron scattering (SANS), as well as static and dynamic light scattering are widely applied techniques in the study of microemulsions. In the static scattering techniques, the intensity of scattered radiation (q) is measured as a function of the scattering vector (q)

$$Q = \left(4\frac{\pi}{\lambda} \right) \sin \frac{\theta}{2}$$

where
    θ is the scattering
    λ is the wavelength of the radiation

The lower limit of size that can be measured with these techniques is about 2 nm. The upper limit is about 100 nm for SANS and SAXS and up to a few micrometers for light scattering. These methods are very valuable for obtaining quantitative information on the size, shape, and dynamics of the components. The major drawback of these techniques is that the dilution of the sample is required for the reduction of interparticulate interaction. This can modify the structure and the composition of the pseudophases. Nevertheless, successful determinations are carried out using a dilution technique that maintains the integrity of droplets.

17.6.2.1 Static Light Scattering Techniques
These have also been widely used to determine microemulsion droplet size and shape. Here, the intensity of scattered light is generally measured at various angles and for different concentrations of microemulsion droplets.

17.6.2.2 Dynamic Light Scattering Techniques
It is also referred to as photon correlation spectroscopy (PCS) and can analyze the fluctuations in the intensity of scattering by the droplets due to Brownian motion. This technique allows the determination of z-average diffusion coefficients, D. In the absence of interparticle interactions, the hydrodynamic radius of the particles R_H can be determined from the diffusion coefficient using the Stokes-Einstein equation

$$D = \frac{kT}{6\pi\acute{\eta}R_H}$$

where
    k is Boltzmann constant
    T is the absolute temperature
    $\acute{\eta}$ is the viscosity of the medium

17.6.3 Nuclear Magnetic Resonance Studies

The structure and dynamics of microemulsions can be studied by using nuclear magnetic techniques. Self-diffusion measurements using different tracer techniques, generally radio labeling, supply information of the mobility of the components. The Fourier transform pulsed-gradient spin-echo (FT-PGSE) technique uses the magnetic gradient on the samples and allows simultaneous and rapid determination of the self-diffusion coefficients (in the range of 10^4–10^{12} m²/s) of many components.

17.6.4 Electron Microscopic Studies

The microemulsions can be characterized by electron microscopic techniques. In spite of high liability of the samples and the possibility of artifacts, electron microscopy is used to study the microstructures. The microemulsion systems are observed under microscope followed by either chemical or thermal fixation methods. But the freeze fracture electron microscopy has also been used to study microemulsion structure, in which extremely rapid cooling of the sample is required in order to maintain structure and minimize the possibility of artifacts. It has been reported that other than CRYO-TEM, the direct observation of the microemulsion over the grid followed by normal air drying is also a useful tool in the study of microstructure and its size analysis.

17.6.5 INTERFACIAL TENSION AND ELECTRICAL CONDUCTIVITY MEASUREMENTS

The formation and the properties of microemulsion can be studied by measuring the interfacial tension. Ultralow values of interfacial tension are correlated with phase behavior, particularly the existence of surfactant phase or middle phase microemulsion in equilibrium with aqueous and oil phases. Spinning drop apparatus can measure the ultralow interfacial tension. Interfacial tension is derived from the measurement of the shape of a drop of the low-density phase, rotating it in cylindrical capillary filled with the high-density phase. To determine the nature of the continuous phase and to detect the phase inversion phenomenon, the electrical conductivity measurements are highly useful. A sharp increase in conductivity in certain W/O microemulsion systems was observed at low-volume fractions and such behavior was interpreted as an indication of a "percolative behavior" or exchange of ions between droplets before the formation of bicontinuous structures. Dielectric measurements are a powerful means of probing both structural and dynamic features of microemulsion systems.

17.6.6 RHEOLOGICAL PROPERTIES AND VISCOSITY MEASUREMENTS

In general, microemulsions have low viscosity and exhibit Newtonian flow behavior. At very high shear rates, shear thinning is observed. Viscosity data are helpful in determining the shape of the corresponding aggregates or extract information regarding the interaction potential between the droplets. Even though microemulsions of bicontinuous structure possess highly interconnected structure, they show Newtonian flow with low viscosity because of their very short structural relaxation time (less than 1 ms). When there is transition from a droplet structure to a bicontinuous structure, viscosity of the system increases. Viscosity measurements can indicate the presence of rod-like or work-like reverse micelle. Viscosity measurements as a function of volume fraction have been used to determine the hydrodynamic radius of droplets, interactions between the droplets, and deviations from spherical shape.

17.7 STABILITY STUDIES

The stability of the microemulsion has been assessed by conducting long-term stability study and accelerated stability studies. In long-term stability study, the system is kept at room temperature and refrigeration temperature. Over the time period, microemulsion systems are evaluated for their size, zeta potential, assay, pH, viscosity, and conductivity. On long-term study, the activation energy for the system and shelf life of the system may be calculated as like other conventional delivery system (Nonoo and Chow, 2008).

Accelerated stability studies are the essential tools to study the thermodynamic stability of microemulsions. It can be done by centrifugation, heating/cooling cycle, and freeze/thaw cycles.

1. In the centrifugation, the system is subjected to centrifugation at 822 g for 30 min and followed by the observation for phase separation.
2. The heating/cooling cycle of keeping the system at 4°C and 45°C for not less than 48 h at each stage.
3. The freeze/thaw cycle of microemulsion can be done between −21°C and 25°C or between 5°C and 10°C (Sheikh and Faiyaz, 2007).

17.8 MICROEMULSIONS IN BIOTECHNOLOGY

Recent years have witnessed an increase in the use of microemulsions in biotechnological applications in general, with establishment of microemulsions as inevitable means for protein-based reactions in particular. The reason underlying the marked increase in popularity and need of microemulsions for protein-based materials is mainly enhanced solubilization of proteins in nonpolar and lipophilic solvents and retention of their activity in microemulsions, which is otherwise negligible.

Wells and coworkers were the pioneers in the field and extensively worked out the role of enzymes in catalyzing a plethora of biochemical reactions. This pioneering contribution provided a major impetus to the study of enzyme-based catalysis. The field has grown up since then and numerous other applications in enzyme immobilization as well as bioseparations have come up. Behavioral profile of enzymes in hydrophobic media serves to mimic the natural phenomenon since the biocatalysts are also working in similar condition in cellular microenvironments.

Research in the microemulsion-based enzymatic reactions has yielded varying results as far as reactivity is concerned. Some enzymes show maximal reactivity in organic media, while others exhibit opposing characteristics.

A large number of endogenous and exogenous factors can affect enzymatic activity. Amount and rate of addition of water, conformational and steric factors, and surface charge on enzyme and substrate are crucial parameters to decide the course of enzymatic activity. In W/O microemulsions, initial addition of water leads to rigidization of protein molecule due to the interaction of water with ionizable groups in protein. Further addition results in hydration of hydrogen bonding sites and finally shielding of nonpolar region by aqueous monolayer.

Conformational and steric changes were earlier thought to play a key role in enzymatic activity. Earlier it was believed that anomalous sudden rise in enzymatic activity was due to conformational changes taking place in enzymes solubilized in microemulsion. However, recent theories regarding the repulsive interactions between surface charge of the droplet and charged substrate particles have been considered to be responsible for the anomalous rise in the activity of solubilized enzymes.

The presence of nonionic surfactant bilayer causes an increased repulsion leading to restriction of enzyme from penetration into the interfacial bilayer causing to erratic enzymatic activity.

A diverse array of enzymatic reactions has been investigated. To name a few, synthesis of esters, peptides, sugar fatty acid esters transesterification, and antibiotic and steroid transformations have been widely studied. The various enzymes that have been explored include lipases, phospholipases, alkaline phosphatase, pyrophosphates, trypsin lysozymes, α-chymotrypsin, peptidases, glucosidase, and oxidases.

17.8.1 Enzymes in Biocatalysis

Apart from enzyme immobilization in microemulsions, use of enzymes solubilized in microemulsions or hydrated reverse micelles has been under active research. In addition, microemulsion-based organogels (MBGs), that is, gelled microemulsion systems based on biopolymers like gelatin, agar, etc., have recently attracted attention as solid phase catalysts in organic solvents due to problems in product isolation and enzyme reuse in microemulsion systems. Specialized enzymes mostly catalyze reactions in biological systems. Catalysis by enzymes is necessary, as noncatalyzed chemical reactions are by far too slow to be effective under the prevalent physiological conditions in a biological system at neutral to slightly alkaline pH and 20°C–40°C temperatures. High stereoselectivity and specificity as well as highly specific catalytic activity on substrates render them far superior over synthetic catalysts, which lack above attributes of enzymes as biocatalysts. Microemulsion-based enzymes and use of enzymes in organic synthesis and nonconventional solvents have been perceived as powerful substitute for conventional chemical methods. Microemulsions can be successfully used as media for hosting various biocatalytic reactions. Enzymes can be incorporated or entrapped in the aqueous core of reverse micelles catalyzing reactions of both hydrophilic and hydrophobic substrates (Figure 17.5). Microemulsion-based systems employed in industrial biotechnology mainly compose of (a) ternary surfactantless microemulsions and (b) MBGs.

There are two basic advantages in using enzymes as catalysts in organic media instead of aqueous solutions. First, organic solvents favor the solubility of hydrophobic substrates and, second, the presence of such solvents shifts the thermodynamic equilibrium of condensation/hydrolysis reactions in favor of the desired product. Hence, microemulsions can serve as potential reaction media for hosting enzyme-based catalytic reactions. Owing to the presence of surfactants, they are capable of solubilizing both hydrophilic or polar substrates as well as lipophilic reactants. Microemulsions can communicate and exchange their content rapidly through elastic collisions. In contrast to emulsion-based medium, microemulsions are more thermodynamically stable that provide stable microenvironment for the enzymatic activity of the trapped enzymes. Hayes (2000) reviewed that microemulsions are well-suited reaction medium for hosting lipid modification reactions involving lipases and other enzymes. Enzymes of almost of all classes and structures have been solubilized in the microemulsions and used for the reactions (Sheild et al., 1986).

17.8.1.1 Hydrolysis

A number of investigations have been done on characterizing the hydrolysis reactions in microemulsion media. Burrier and Brecher (1983) showed that synthetically prepared lecithin-containing microemulsions are stable and effective substrates for the acid lipase purified from rat liver lysosomes and carried out the hydrolysis of triolein in both lipid vesicle and microemulsion preparations. They concluded that V_{max} for the hydrolysis kinetics is much higher than that carried out in lipid vesicle. Similarly, Hui et al. (1995) reported that lipase from *Candida cylindracea* possessed higher catalytic activity, both in the hydrolysis of (D,L)-ethyl mandelate in SDS/*n*-butanol/*n*-octane O/W microemulsion and in the esterification of α-bromopropionic acid with *n*-butanol in SDS/*n*-butanol/*n*-octane W/O microemulsion, than in traditional water and oil biphasic solutions.

The activity of lipase from *C. cylindracea* has been investigated by Oconnor et al. (1991) in the detergentless microemulsion media of *n*-hexane/propan-2-ol/water and in paraffin/water biphasic mixtures and it was also found that the enzyme-catalyzed hydrolysis of methyl palmitate was very slow in a complete nonaqueous system, but addition of small amounts of water in microemulsion medium led to an increased rate of reaction at atmospheric pressure and 37°C.

Not only hydrolysis but even reverse hydrolysis has also been studied in microemulsion media. Sathish et al. (2008) synthesized *n*-hexyl-β-D-glucopyranoside (HGP) by β-glucosidase-catalyzed condensation of glucose and hexanol in an aqueous–organic bicontinuous microemulsion system with both water and oil as continuous phases and studied the reverse hydrolysis reaction. They revealed that bicontinuous microemulsion system is more applicable to the enzymatic hydrolysis with minimal degradation.

17.8.1.2 Esterification

Esterification is one of the important reactions catalyzed by enzymes and hence, these reactions have also been explored and optimized using microemulsions for improved yield and stability. Peter et al. (1997) synthesized sugar fatty acid ester (SFAE) surfactant, ethyl 6-*O*-decanoyl glucoside, in microemulsions by lipase component B from *C. antarctica* with higher yield. Natural gelling agents such as gelatin, agar, and κ-carrageenan have been used in the form of hydrogels for enzyme immobilization. Incorporation of microemulsions in the hydrogels showed enhanced catalytic activity of lipases. Stamatis and Xenakis (1999) reported that high yields (80%) were obtained with agar and κ-carrageenan organogels in isooctane for esterification reaction of propanol with lauric acid in various hydrocarbons at room temperature by lipase from *Pseudomonas cepacia*. Investigations reported by Giuliani et al. (2001) show that high yield of pentylferulate (50%–60%) were obtained in microemulsion composed of *n*-hexane, 1-pentanol, water, and cetyl trimethyl ammonium bromide as surfactant using feruloyl esterases.

Yuchun and Ying (2005) carried out a model esterification reaction of hexanol and hexanoic acid by *C. cylindracea* lipase in the cyclohexane/dodecylbenzenesulfonic acid (DDBSA)/water microemulsion and found that there was significant increase in the rate of reaction. Comparison of conversion in several acid-catalyzed reaction systems was also performed and the results showed that the conversion in DBSA system showed the highest reaction rate.

On the same line, Stamatis et al. (2006) reported the role of composition of ternary systems in the catalytic efficiency, and stability of lipases and high conversions (up to 95%) were obtained in catalyzing the esterification of fatty acids or natural phenolic acids including cinnamic acid derivatives by lipases from *Rhizomucor miehei* (RmL) and *C. antarctica* (CaL-B) entrapped in surfactant-free microemulsions.

Blattner et al. (2006) showed that biocatalysis using MBGs in super crtitical CO_2 is the effective alternative to conventional bioconversion processes. They reported that lipases from *C. antarctica* and *Mucor miehei* encapsulated in lecithin W/O MBGs with either hydroxypropylmethyl cellulose or gelatin had been successfully used for the esterification of lauric acid and 1-propanol. Esterification of oleic acid with hexanol by lipase in sodium bis(2-ethylhexyl)sulfosuccinate (AOT)/isooctane/water (anionic) CTAB/decane/hexanol (cationic) nonionic ($C_{12}EO_4$)/decane system in trioelin and tributyrin substrates was carried out by Xenakis et al. (2006) and found that the rate of hydrolytic reaction was slower in nonionic surfactant-based system and more enzymatic stability was in cationic system.

17.8.1.3 Transesterification

Holmberg and Österberg (1987) used microemulsions based on aliphatic hydrocarbon, surfactant, and aqueous buffer as reaction medium for the lipase-catalyzed transesterification of a triglyceride and a fatty acid with composition similar to that of natural cocoa butter from a palm oil distillation fraction. Osterberg et al. (1988) also studied the lipase catalyzed transesterification of unsaturated lipids with stearic acid in the microemulsion medium.

17.8.1.4 Glycerolysis

Holmberg et al. (1989) performed glycerolysis of palm oil by 1,3-specific lipase in the microemulsion medium containing isooctane, AOT, palm oil, and a combination of water and glycerol as the polar component and showed that the reaction was very slow in a completely nonaqueous system, while the reaction was markedly faster accompanied with improved yield in microemulsion medium.

Pahn et al. (1991) showed continuous glycerolysis of olive oil by *Chromobacterium viscosum* lipase by adsorption on liposome and solubilization in microemulsion droplets of glycerol containing a little amount of water and found that the system had higher operational stability for the enzyme and the half-life of the enzyme.

Glycerolysis of olive oil by lipase in the quaternary system containing glycerol, olive oil, *n*-butane, and AOT was optimized by Bender et al. (2008).

17.8.1.5 Steroid Transformation

The $\Delta^{1,2}$-dehydrogenation of high concentrations of the steroid-methyl-Reichstein's compound S-21-acetate (16MRSA) in a microemulsion system was studied by Smolders et al. (1991) using heat-dried and thawed *Arthrobacter simplex* cells as biocatalyst. The investigation showed marked improvement in the yield and efficiency of the steroid biotransformation in presence of enzymes immobilized in microemulsions.

17.8.1.6 Oxidation/Reduction

Larsson et al. (1987) reported that horse-liver alcohol dehydrogenase (HLADH) solubilized in AOT/cyclohexane reverse micelles was used for the oxidation of ethanol and reduction of cyclohexanone in a coupled substrate–coenzyme recycling system and found that the reaction rate was approximately five times higher in a reverse micellar solution than in buffer. Catalytic activity and stability of cholesterol oxidase dissolved in ternary systems composed of *n*-hexane, isopropanol, and water were studied by Yuri et al. (1988). They studied preparative conversion of cholesterol to cholestenone catalyzed by cholesteroloxidase in *n*-hexane/isopropanol/water ternary systems and could get

100% yield. Larsson et al. (1991a,b) carried out the oxidation of racemic 3-methylcyclohexanone by HLADH using a coupled substrate–coenzyme regenerating cycle in an AOT–isooctane–buffer microemulsion. The oxidation of a range of substituted benzaldehydes by xanthine oxidase in three different micellar systems (DTBA/hexanol/heptane/water, Triton X-1000/hexanol/cycle hexane/water, and AOT/isooctane/water) were investigated by Bommarius et al. (1995) and showed that reversed micellar systems allowed the partitioning of the substrates between different polarity domains with retention of reactivity. Orlich et al. (2000) studied the phase behavior and compatibility of ternary system consisting of Marlipal O13–60 ($C_{13}EO_6$ in industrial quality)/cyclohexane/water and its effect on the activity and stability of alcohol dehydrogenase from yeast (YADH) and horse liver (HLADH) and the carbonyl reductase from *Candida parapsilosis*. They found that except formate dehydrogenase, the other enzymes show significant change in activity and stability by varying the water or surfactant concentration of the microemulsion. The catalytic behavior of mushroom polyphenol oxidase has been studied by Rojo et al. (2001) in AOT/cyclohexane reverse micelles and they optimized the stability conditions for the activity of the polyphenol oxidase toward several substrates like *p*-cresol and 4-methylcatechol. Rodakiewicz-Nowak et al. (2002) isolated *Agaricus bisporus* tyrosinase and evaluated its activity in oxidation of 4-*t*-butylcatechol to 4-*t*-butyl-*o*-quinone and found that the initial rate of tyrosinase reactions in W/O microemulsions was reduced and this reduction may be attributed to substrate partitioning between the water pool, interfacial layer, and isooctane phase. Hirofumi et al. (2005) investigated the suitability of bacterial glycerol dehydrogenase (GLD) and soluble transhydrogenase (STH) in nanostructural reverse micelles for the regeneration of redox cofactors NADH and NADPH and showed that GLD and STH entrapped in AOT/isooctane reverse micelles have potential for use in redox cofactor recycling in reverse micelle. Papadimitriou et al. (2005) developed biomimetic medium consisting of olive oil microemulsions for enzymatic oxidation of oleuropein by tyrosinase. They used microemulsions with olive oil as the nonpolar solvent, lecithin as surfactant, 1-propanol as cosurfactant, and water and found that addition of L-proline as a coupling reactor did not succeed in preventing enzyme inactivation in the microemulsions, probably owing to substrate localization and product accumulation around the entrapped enzyme molecules in the micellar interface. Moniruzzaman et al. (2009) found that horseradish peroxidase (HRP)-catalyzed oxidation of pyrogallol by hydrogen peroxide in water-in-ionic liquid (w/IL) microemulsions was much more effective than in a conventional AOT/water/isooctane microemulsion.

17.8.1.7 Antibiotics Production

Papapanagiotou et al. (2005) investigated the use of a rapeseed oil emulsion feed, produced by a phase inversion temperature (PIT) process and found increased production of biomass, with threefold increase in oil utilization and a higher oxytetracycline titer.

17.8.2 MICROEMULSIONS IN ENZYME/PROTEIN IMMOBILIZATION

Enzymes can be defined as organic, proteinous, water-soluble colloidal endogenous substances in the body, found intra/extracellularly that catalyze a series of biochemical reactions to maintain the physiological homeostasis in the body of an organism. Their role is mainly catalysis in a biological system. Their action as catalyst has attracted considerable attention of researchers for their commercial exploitation in industrial, agricultural, pharmaceutical, biotechnological, and food applications. High stereoselectivity and stereospecificity render enzymes extremely useful for their use as catalyst in industries. Ecofriendliness of enzyme further render them highly promising in industries due to serious issues like environmental pollution caused by synthetic catalysts. However, isolation of enzymes in highly purified form, avoidance of enzymatic contamination with end products, and lability of enzymes are major obstacles in commercial applications of enzymes. Hence, enzyme immobilization has been developed as a technique since last many decades to overcome most of the above problems.

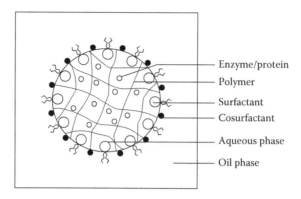

FIGURE 17.6 Enzyme immobilization in microemulsion-based gel.

Enzyme immobilization may be defined as attachment of an enzyme to a solid matrix or a rigid support by means of physical or chemical bonds, without loss of enzymatic activity (Figure 17.6). Commercially, enzyme immobilization has been largely exploited due to their attributes like convenience, economy, and stability. More than 5000 publications and patents have been published regarding the versatile techniques involving enzyme immobilization. Emulsions have been used for enzyme immobilization since 1972. Earlier, scientists have investigated the use of multiple emulsions to entrap urease enzyme for renal disorders. Literature is full of examples regarding the use of immobilized enzymes form, *M. dentrifricans* for waste water treatment, immobilization of alcohol dehydrogenase for conversion of alcohol to acetaldehyde for industrial purpose, conversion of ketoisocaproate to L-leucine by L-leucine dehydrogenase, enzymatic production of amino acid (L-phenyl alanine) from immobilized chymotrypsin, and immobilized lipase enzyme for the hydrolysis of fatty acids.

Microemulsions are clear, transparent, and isotropic systems that facilitate easy spectroscopic determination of encapsulated contents. Similarly, there is enhanced stability of enzymes in aqueous microenvironment of microemulsions. By virtue of higher transparency and improved shelf life of entrapped contents, microemulsions are considered to be very fascinating systems for enzyme immobilization. In case of water-insoluble or poorly water-soluble substrates that result in poor yield and purification problems, microemulsions serve as highly promising means solubilizing such water-insoluble or poorly water-soluble substrates. By virtue of enhancing their solubility, microemulsions appreciably enhance the availability of substrates. In case of very poorly water-soluble substrates, many efforts have been undertaken in searching for suitable organic media in order to increase the substrate concentrations and ultimately the reaction rates. For such challenging substrates, microemulsions can be used as troubleshooting tools to meet such challenges posed by poorly water-soluble or highly insoluble substrates. W/O microemulsions have attracted attention as media for carrying out enzyme-catalyzed reactions because of their interesting properties as hosts for biocatalysts. The unique properties of microemulsion systems allow hydrophilic enzymes to be incorporated into the nanometre-sized water core of droplets of microemulsions. Enzymes are confined to aqueous microdomain of the microemulsions and hence, are protected from detrimental effects of organic solvents with retention of their catalytic activity and integrity. However, as discussed earlier, product isolation, scale up at commercial level, is a major constraint in the path of commercial exploitation of enzymes at industrial level.

Schuleit and Pier (2000) investigated immobilization of *Ch. viscosum* lipase enzyme in silica-hardened organogels. Nesamony and Koling (2009) investigated the rheological properties of the microemulsions and found them to be suitable templates for enzyme immobilization and nanoparticle synthesis or as carriers for DNA, polypeptides, and protein drugs.

Daubresse et al. (1996) reported the immobilization of alkaline phosphatase by physical entrapment within colloidal carriers using inverse microemulsion polymerization method. Huang et al.

(1998) found microemulsions to be excellently useful tool for studying the kinetics of enzymes in an immobilized form. The investigators studied the human placental alkaline phosphatase embedded in a reverse micellar system prepared by dissolving the surfactant AOT in 2,2,4-trimethylpentane and concluded that the rate-limiting step of the hydrolytic reaction changes from phosphate releasing in aqueous solution to a covalent phosphorylation or dephosphorylation step in reverse micelles. Thus, basic kinetics of enzymes can be elucidated using microemulsions.

Zoumpanioti et al. (2006) studied the lipases from *R. miehei* (RmL) and *C. antarctica* (CaL-B) entrapped in surfactant-free microemulsion-like ternary systems consisting of *n*-hexane, short-chain alcohols (1-propanol or 2-methyl-2-propanol), and water, to catalyze esterification of fatty acids or natural phenolic acids including cinnamic acid derivatives. Lipase enzymes immobilized and entrapped in microemulsions and analogous ternary systems showed high conversions (up to 95%). They used hydroxypropylmethyl cellulose (HPMC) MBGs formulated with various surfactant-free microemulsions as a matrix for enzyme immobilization. These lipase-containing MBGs were found to serve as new solid phase catalyst for use with appreciable retention of catalytic activity of enzymes for 8 months. Surfactant-free MBGs still exhibited higher catalytic activity during storage.

Similarly, Xenakis and Stamatis (1999) reported the retention of enzymatic activity following immobilization in agar- or gelatin-based microemulsion gels of lipase enzyme from *P. cepacia*.

Famiglietti et al. (2004) investigated the photosynthetic activity of cyanobacteria in micro-emulsions. Parameters like HLB and other factors affecting the physical state and stability of the cells in the microemulsion were investigated. Microemulsion-based systems were found to display photosynthetic activity.

Microemulsions serve as excellent media for solubilizing hydrophobic substrates. Malaviya and Gomes (2008) reported the biotransformation of a hydrophobic substrate—sitosterol to androstene-dione using a microemulsion-based nutrient broth and found the latter to be superior in much respects as compared with conventional nutrient media. Reverse micellar systems serve as very good media for biocatalysis. Komives et al. (1994) described the use of reverse micelle-based organophosphorus hydrolase enzyme for the degradation of organophosphorus compound-based pesticides.

Solubilization of enzymes in hydrated reverse micelles or W/O microemulsions with retention of catalytic activity has been widely studied. Microemulsion-based systems serve as ideal means of solubilizing enzymes. The enzymes are encapsulated in aqueous core of the microemulsions, surfactants adsorbed at the interface, and organic phase mainly composed of hydrophobic substrates and products. Precise control over water concentration in the microemulsion also facilitates maintenance of thermodynamic equilibria in enzyme-catalyzed hydrolytic and condensation reactions. Lipases have been studied extensively due to their versatile biotechnological applications. Lipases have been found to be efficacious and stable in microemulsions and are capable of catalyzing synthetic reactions in case of fatty acids. However, problems associated with postsynthetic purification, isolation of products, and enzymes limit the usefulness of these methods. Hence, surfactant-free MBGs prepared using biopolymers like gelatin and agar have been worked out. These MBGs are discussed earlier. Rigidity and stability of MBGs render them useful as solid phase catalyst. Further, the matrix-like network formed by gel facilitates retention of enzymatic activity due to diffusion of substrates and thereby establishment of contact between enzymes and substrates. Apart from conventional microemulsion, surfactantless microemulsions have also been found to be favorable for enzymatic catalysis in oxidation and esterification reactions. Zoumpanioti has investigated esterification reactions in isooctane using lipase-containing surfactantless microemulsions immobilized on cellulose-based MBGs as solid phase catalysts.

Jenta et al. (1991) have investigated the use of MBGs to immobilize the enzyme *Ch. viscosum* lipase. MBGs have been found to augment the catalytic efficacy of lipase due to increased surface area to volume ratio of gel in contact with substrate in external organic phase. Parmar et al. (1996) have immobilized lipases from porcine pancreas (PPL) and *Curzdidu cylindruceu* in MBGs and studied the hydrolysis of perpropanoates of different classes of polyphenolics with them. In addition, the lipase from *Cundidu antrrrticu* has been used for selective propanoylation of the primary hydroxyl groups in

2-deoxy-D-ribose and D-ribose, for the synthesis of corresponding nucleosides. Phenolic compounds have been used as precursors for naturally occurring polyphenols like flavonoids, chalcones, coumarins, etc. Yield and stereoselectivity of functional groups in the precursor are major constraints. Protection/deprotection of various functional groups at various stages results in subsequent poor yield at the final stage. In this work, scientists have studied the hydrolysis of perpropanoates and peracetates of different classes of polyphenolics such as (1) catechol dipropanoate, (2) resorcinol dipropanoate, (3) hydroquinone dipropanoate, (4) 2,6-dimethoxyhydroquinone dipropanoate, (5) 2-*tert*-butylhydroquinone dipropanoate, (6) phloroglucinol tripropanoate, (7) 2-methylresorcinol dipropanoate, (8) 2,4-diacetoxyacetophenone, (9) ethyl 3,5-diacetoxybenzoate, (10) ethyl 3,4,5-triacetoxybenzoate, (11) methyl 2,4-diacetoxybenzoate, and (12) ethyl 2,4-diacetoxybenzoate in the presence of lipases immobilized in MBGs in benzene. The resultant products have exhibited good yield with high degree of selectivity. Second, lipases in MBGs were found to catalyze the selective deacetylation reactions at accelerated rates as compared with conventional organic solvents. The end products obtained as the outcome of the reactions displayed good yield accompanied by high degree of selectivity.

17.8.3 Microemulsions for Bioseparations

Salient features of microemulsion like difference in affinities of protein molecules toward different microemulsions render them unique and are used for selective separation of proteins and similar biotherapeutics. Owing to gentle solvent action of microemulsions over proteins, they can be used for selective extraction of proteins and natural molecules without significant alteration in their chemical structure, conformational stability, and minimal loss of physio/pharmacological activities. Hatton (1989) explored the use of oily microemulsions for extraction of proteins. Partition coefficient of proteins between aqueous phase and a microemulsion, ionic strength, pH, and formulation components of microemulsion need to be optimized prior to scale up of microemulsions for protein immobilization and enzymatic extraction. W/O microemulsion of Winsor II type have been used to separate biomolecules like amino acids.

17.8.3.1 Microemulsions in Protein Separation

Sun et al. (2008) reported the use of reverse micellar two-phase separation technique for protein separation. Selectivity and capacity of reversed micelles can be enhanced by an attachment of affinity ligand. Sun et al. (2008) used Cibacron Blue F3G-A (CB) as an affinity ligand. The affinity partitioning of lysozyme and bovine serum albumin (BSA) to the CB–lecithin micelles was studied. Formation of mixed micelles by additionally introducing a nonionic surfactant, Tween 85, to the CB–lecithin micelles was effective to increase the solubilization of lysozyme due to the increase of W_0 (water/surfactant molar ratio)/micellar size.

Bergstrome and Holmberg (1992) have investigated the protein immobilization to hydrophilized surfaces using microemulsions. They have investigated microemulsion as reaction medium for protein immobilization to polystyrene hydrophilized with polyethylene glycol (PEG) or polysaccharides and found significant rise in protein immobilization to such hydrophilized surfaces in case of albumin, anti-IgG, and collagen. Similarly, marked improvement in the stability of anti-IgG and interleukin-2 was also reported in microemulsion system.

Guadalupe et al. (1992) investigated the solubilization and activity of proteins in compressible fluid-based microemulsions. They have delineated the effect of the microemulsion and compressible propane along with the effect of pressure on protein transfer from aqueous to organic phase on the structure and activity of cytochrome C and subtilisin.

17.8.4 Miscellaneous Biotechnological Applications

Apart from the role of microemulsions in protein and enzyme stabilization and kinetics, some newer and relatively unexplored horizons of applications of microemulsion have also been currently expedited. Pfammater et al. (2004) have investigated innovative aspects of microbiology in W/O microemulsions using *C. pseudotropicalis*. Lower aggregation tendency of cellular contents in microemulsions is an added attribute of microemulsion. The implications of these findings support the fact that microemulsions can serve as an utmost useful tool in studying possible cellular metabolism, thereby authenticating their utility in the field of industrial and pharmaceutical biotechnology.

17.8.4.1 Microemulsions in Biodiesel

Ever increasing population and widespread industrialization have resulted in unprecedented demand of nonrenewable, conventional fossil fuels. Dwindling reserves of these fossil fuels have resulted in a precarious condition and it is the need of hour to find out some potential, renewable substitutes of these fossil fuels. Scarcity and nonrenewability of fossil fuels have generated serious concerns globally. Extensive research is being undertaken to find the potential substitutes for fossil fuels. Biodegradable fuels obtained from cash crops have received quite good attention in world market. Biodiesel may be defined as the monoalkyl esters of long-chain fatty acids derived from renewable vegetation matter such as vegetable oils and is synthesized from triglycerides in vegetable oils by transesterification with alcohol in presence/absence of catalysts like alkalies, acids, or enzymes. Microemulsions of vegetable oils like sunflower oil with alcohols apparently have lower viscosity and ecofriendliness, rendering them more acceptable. Ecofriendliness and easy renewability of biofuel have made them fascinating and superior to conventional fossil fuels. Researchers have prepared microemulsions of vegetable oils in diesel. Madras et al. (2004), has investigated the role of microemulsions as the biofuels. In current research work, they have investigated transesterification of sunflower oil in supercritical methanol and supercritical ethanol at various temperatures, synthesized in supercritical carbon dioxide. Blending of microemulsions with thermal cracking and transesterification of vegetable oils are seen as potential biofuel alternatives.

17.8.4.2 Microemulsions in Gene Library

The aqueous droplets of the W/O emulsion function as a cell-like compartment facilitating transcription of single gene followed by its subsequent translation to give multiple copies of the protein (e.g., an enzyme). Compartmental nature of aqueous droplets ensures the linkage of gene and the protein encoded by the corresponding gene.

17.9 PHARMACEUTICAL APPLICATIONS OF MICROEMULSIONS

17.9.1 ORAL DRUG DELIVERY

Ritschel et al. (1990) studied the absorption of cyclosporine—a drug having poor oral absorption and bioavailability. W/O microemulsions when administered orally in rats were found to be having higher absolute and relative bioavailability as compared with commercially available solutions. A W/O microemulsion-based cyclosporine preparation containing a sorbitan ester–polyoxyethylene glycol monoether mixture of surfactant, a low molecular weight alcohol, fatty ester, and water as the vehicle for the drug was administered. Tarr and Yalkowsky (1989). Silica was added to obtain a gel-like consistency and the microemulsions were filled and administered in hard gelatin capsules. Better absolute and relative bioavailabilities were observed with microemulsion-based preparation as compared with parenteral preparation. This augmentation in bioavailability may be attributed to the microemulsion droplets (Ritschel et al., 1990). Considering the various benefits and patient-friendly nature of oral route, it has always been the major attraction for the formulation chemists. However, conditions prevailing in gastrointestinal microenvironment like acidic pH, presence of proteolytic enzymes in gastrointestinal tract, and extensive first pass hepatic metabolism are major constraints that limit the usefulness of the oral drug delivery. Administration of peptide and protein drugs by oral route is a major challenge for a formulation scientist. Ritschel (1991) studied the gastrointestinal absorption of insulin, vasopressin, and cyclosporine by formulating them in a series of

O/W microemulsions. The investigators concluded that lipid phase of the microemulsion, types of surfactants, and biocompatibility of the lipids are major critical factors that need to be manipulated to produce a good microemulsion. Drewe et al. (1992) demonstrated improved oral absorption of cyclosporine administered by hard gelatin capsules containing two O/W microemulsions (fast and sustained release) and a solid micellar solution. A significant rise in oral absorption of cyclosporine may be attributed to greater solubility of lipophilic cyclosporine in microemulsion-based formulations. Microemulsions can be potential drug delivery systems for peptide drugs like insulin as they are capable of protecting peptide drug candidates from adverse microenvironment prevailing in the biological system. Cliek et al. (2006) have tried to develop a stable oral microemulsion system for insulin delivery using lecithin. Oral microemulsion of mometosane furoate has been found to be highly effective as compared with conventionally available and marketed preparations of the drug in the treatment of erosive-ulcerative oral *Lichen planus* (Aguire et al., 2004). Mucoadhesive polymers interact with glycoproteins in the mucus layer that covers mucosal epithelial surfaces in the body. The popular routes in which mucoadhesive materials are used include nasal, ocular, buccal, vaginal, rectal, and oral route. The rheology of the formulation might be more important in such cases. A second generation of bioadhesives, lectin-like cytoadhesives, is now in focus. These bioadhesives achieve more specific mucoadhesion that is independent of mucus turnover. This class of substances will probably be most useful for the oral route, rather than the nasal or ocular routes. Microemulsions have a great potential for oral delivery of sparingly soluble lipophilic drugs with poor bioavailability and drugs that are unstable at the conditions prevailing in the gastrointestinal tract. Ceccuti et al. (1989) used a microemulsion to formulate WR2721, which is employed in radiotherapy and chemotherapy of cancer and needs to be protected from hydrolysis in stomach so as to retain its biological activity. Using a W/O microemulsion of CTAB, isooctane, and butanol, scientists found that hydrolysis was reduced considerably in the microemulsion as compared with aqueous solution.

17.9.2 Topical Drug Delivery

Despite the number of advantages associated with topical/transdermal route over the oral route, drug efficacy is greatly curtailed by transdermal penetration route. Presence of stratum corneum—outermost layer of the skin comprising of keratin-rich dead cells embedded in a lipid matrix—acts as a rate-limiting barrier to drug penetration. Considering the solubilizing capacity of microemulsions, they are expected to significantly alter the structure of the stratum corneum lipid assemblies, with obvious consequences for drug penetration. Bhatnagar and Vyas (1994) investigated the bioavailability of transdermally administered propranolol, a β-receptor blocking drug, which undergoes extensive first pass hepatic metabolism. It was found that bioavailability of this drug could be extensively improved by transdermal application from a lecithin-based W/O microemulsions. Willimann et al. (1992) prepared lecithin-containing W/O microemulsions for the transdermal administration of scopolamine and broxaterol and found that the transport rate obtained with the lecithin microemulsion gel was much higher than that obtained with an aqueous solution at the same concentration. Percutaneous route of administration has been extensively studied where the drug transport from microemulsions was recorded usually better than that from other ointment, gels, and creams. Osborne et al. (1988) studied microemulsion consisting of AOT–octanol–water and reported sixfold increase in transdermal flux on increasing the water content of the microemulsion increased from 15% to 68%. Schmalfub et al. (1997) studied transdermal delivery of hydrophilic diphenhydramine hydrochloride from a W/O microemulsion formulation through excised human skin. The formulation was based on the combination of Tween 80 and Span 20 with isopropyl myristate (IPM). Cholesterol and oleic acid containing two other formulations increased the permeation, whereas oleic acid containing formulation had no measurable effect on permanent flux. Dalmora and Oliveira (1999) described a formulation of an inclusion complex of the anti-inflammatory piroxicam with β-cyclodextrin in an O/W microemulsion system for topical use. A gelatin MBG has been used in the iontophoretic transdermal delivery of a model hydrophilic drug.

The MBGs were prepared using surfactants and oils including Tween 80 and IPM (Kantaria et al., 1999). Opaque and thermoreversible organogels can be regarded as innovative delivery systems for drugs and antigens (Murdan et al., 1999). Microemulsions appear to have the ability to deliver large amounts of water and topically applied agents into the skin as they provide a better reservoir for a poorly water-soluble drug. Percutaneous absorption of drug is increased due to retention of micro-emulsion-based dosage form. Further, incorporation of charged moieties to microemulsions may also result in changed release and retention profile of a drug on topical administration. Cationic microemulsions exhibited enhanced drug retention to the skin that might be ascribed to interaction between positively charged microemulsion systems and negatively charged skin sites. The results further endorsed the fact that charged microemulsions can be used to optimize drug targeting without a concomitant increase in systemic absorption (Peira et al., 2008). Microemulsion-based system for delivery of Quercetin—a potent antioxidant—has also been studied. Study demonstrated the nonirritancy of the topically applied microemulsion as well as significant reduction of the UV irradiation-induced glutathione depletion and secretion/activity of metalloproteinases (Vincentini et al., 2008). Microemulsion showed potent enhancement effect for sodium nonivamide—a capsacin derivative—by 3.7–7.1-folds as compared with control group, justifying the use of micro-emulsion as topical drug delivery system (Huang et al., 2007). Microemulsions can also enhance the stability, retention, and absorption of highly labile materials like nucleic acids. DNA delivery is always a great challenge for the formulation scientists. Various delivery systems like liposomes and nanoparticles are under extensive screening to evaluate their role in site-specific delivery of DNA. Topical administration of plasmid DNA in ethanol-fluorocarbon microemulsion resulted in increased luciferase expression justified by 45- and 1000-fold increase in serum levels of IgG and IgA, respectively (Cui et al., 2003). Apart from improving the stability, shelf life of a drug and protection of drug or biomolecule from adverse biological microenvironment, microemulsions serve as excellent carrier systems for solubilizing the poorly water-soluble drugs like nimesulide as well as facilitate in retarding the drug release, thus providing a sustained release profile for a water-insoluble or poorly soluble drug. MBG of poorly water-soluble nimesulide was successfully developed with *in vitro* release rates comparable with marketed formulations (Derle et al., 2006). Similar studies were undertaken to enhance the solubility and thereby percutaneous absorption of poorly water-soluble drug like Aceclofenac. Lee et al. (2005) used Labrafil 1944 CS, Cremophor ELP, and ethanol as a cosurfactant and reported fivefold increase in percutaneous absorption of aceclofenac indicating microemulsion as a promising tool for percutaneous delivery of aceclofenac.

17.9.2.1 Microemulsions for Transdermal Delivery

Scientists reported the potential of W/O microemulsions to achieve deeper and faster penetration into human skin and hairless mouse skin. They considered that the microemulsion might be able to lower the interfacial energy between the skin and the vehicle simultaneously on its intimate contact with skin lipids and water, further enhancing the drug delivery.

17.9.2.2 Microemulsion-Based Formulation in Vaginal Drug Delivery

D'Cruz et al. (2001) developed a lipophilic, spermicidal gel-microemulsion containing GM-144 as a vaginal contraceptive that was as effective as the commercially available N-9 gel, and GM-144 was found to have the potential to become a clinically useful safe vaginal contraceptive and a vehicle for formulating lipophilic drugs used in reducing the risk of heterosexual transmission of sexually transmitted diseases using microemulsion-based drug delivery systems.

17.9.3 Ocular Delivery

O/W microemulsions were investigated mainly to achieve solubilization of poorly soluble drugs, to enhance absorption, and to accomplish sustained release profile in ophthalmic drug delivery. Siebenbordt and Keipert (1991) developed and studied lecithin-Tween 80-based microemulsion for dissolution of some poorly soluble drugs such as atropine, chloramphenicol, and indomethacin. The formulation exhibited low physiological irritation and showed improved and sustained *in vitro* release. A lecithin-based microemulsion containing timolol, a β-blocker, used in glaucoma therapy could prolong the time of drug absorption. It is probably due to the retention of tiny nanodroplets on the cornea for some time and serving as a reservoir of timolol, thus prolonging the absorption time of the drug. Hasse and Keipert (1997) formulated pilocarpine microemulsions of low viscosity using lecithin, propylene glycol, and PEG 200 as cosurfactant and IPM as the oil phase to provide the sustained release of pilocarpine.

17.9.4 PARENTERAL ADMINISTRATION

Microemulsions have been successfully used for parenteral administration of molecules with very poor affinity for water and are highly lipophilic in nature and also for achieving the sustained release profile. Kakutani et al. (1991) reported the role of O/W microemulsions in targeting very lipophilic drugs to reticuloendothelial system (RES) tissues like liver and spleen. Results indicated that higher partition coefficient of the drug favors drug targeting to various tissues. W/O microemulsions are expected to prolong the release rate of hydrophilic drugs by parenteral route. Date and Nagarsenker (2008) investigated microemulsion-based delivery of propofol, an anesthetic, and found the efficacy of the anesthetic comparable with marketed preparation by conducting *in vivo* studies in rats and also found the microemulsion system to be having good stability with reduced toxicity potential.

17.9.5 INTRANASAL DRUG DELIVERY

The nasal route is one of the most permeable and highly vascularized sites for drug administration, ensuring rapid absorption and onset of therapeutic action. Microemulsion has generated considerable amount of interest as a potential drug delivery system for nasal delivery of drugs. The viscosity of microemulsions can be tailored as per the need of application or in some instances through incorporation of specific gelling agents such as a carbopol or gelatin. The dispersal of drug as a solution in nanosized droplets enhances the rate of dissolution into contacting aqueous phase and *in vivo* generally results in an increase in drug bioavailability. In addition, the presence of surfactant and in some cases cosurfactant, for example, medium-chain triglycerides in many cases, serves to increase membrane permeability, thereby increasing the drug uptake. Improved solubilization and subsequent enhanced brain uptake of Nimodipine following intranasal administration of O/W microemulsion was observed by Zhang et al. (2004). The results suggested microemulsions to be a promising approach for intranasal delivery of Nimodipine in the treatment, prophylaxis, and management of progressive neurodegenerative disorders.

Microemulsion formulations of clonazepam incorporated without and with mucoadhesive agents exhibited faster onset of action followed by prolonged duration of action in treatment of status epilepticus. Brain–blood uptake ratio of clonazepam at 0.5 h of administration of clonazepam microemulsion, clonazepam mucoadhesive microemulsion, clonazepam solution, and intravenous administration of clonazepam microemulsion exhibited 0.5, 0.67, 0.48, and 0.13, respectively, indicating more effective brain targeting following intranasal administration and best brain targeting with Clonazepam microemulsion (Vyas et al., 2006). Microemulsion-based systems are currently under investigation for their possible role in brain targeting in neurodegenerative disorders like Alzheimer's disease and Parkinsonism. Jogani et al. (2008) studied microemulsion/mucoadhesive microemulsion of tacrine, and assessed its pharmacokinetic and pharmacodynamic performances for brain targeting and for improvement in memory in scopolamine-induced amnesic mice. The results demonstrated rapid and larger extent of transport of tacrine into the mice brain and faster regain of memory loss in scopolamine-induced amnesic mice after intranasal microemulsion administration.

Microemulsions have also been under active research for the formulation of various popular and relevant herbal medicines. Bhanushali and Bajaj (2007) developed intranasally administered microemulsion-based formulation of essential oils like eucalyptus oil and peppermint oil for the treatment of migraine. Intranasal administration of microemulsion loaded with antiepileptic drugs circumvent the BBB, thus providing a better and safer alternative for brain targeting of drugs and thereby timely and effective management of chronic and serious disorders like epilepsy. Microemulsion-based formulations of Lamotrigine have been studied for enhancing solubilization of Lamotrigine, thereby achieving faster drug uptake in the brain, facilitating rapid and effective management of emergency conditions arising during epilepsy (Shinde et al., 2007). Intranasal administration of microemulsion-based systems help in overcoming the first pass hepatic metabolism of labile drugs, thereby assist in improving the therapeutic levels of drug at the site of action and enhance bioavailability of drugs. Bhanushali and Bajaj (2007) reported improved brain targeting and higher brain uptake of Sumatriptan following intranasal administration of mucoadhesive microemulsion formulations of the drug.

17.9.6 LONG-ACTING DELIVERY SYSTEM

Long-acting delivery of luteinizing hormone-releasing hormone (LHRH) using biocompatible W/O microemulsion has been reported (Gasco). In this study, $55\,\mu g/mL$ of LHRH microemulsion was injected (IM) to rats and the testosterone levels of the rats were found to be significantly lower following the drug administration. The formulation also resulted in sustained release profile and effective plasma levels were maintained for 10–20 days. There was significant increase in rate, extent, and duration of action as compared to conventional IV. Drug was accompanied with prolonged effect of drug (between 10 and 20 days) as compared with $100\,mcg/kg$ intravenously.

17.9.7 SUSTAINED RELEASE FROM MICROEMULSIONS

In an O/W microemulsion, hydrophobic drugs, solubilized mainly in oil droplets, experience hindered diffusion and hence its release rate gets retarded, providing a sustained release profile (Shinoda et al., 1991; Olssen and Wennerstrom, 1994). Possibility of controlling the drug release rate by microemulsion structure and composition as well as by drug partitioning makes microemulsions very promising for controlled release applications. Bello et al. (1994) compared the release of pertechnate from a lecithin-based microemulsion with an aqueous solution on subcutaneous administration in rabbits. He found that pertechnate carried by W/O microemulsions was released faster from an aqueous solution.

17.9.8 COLLOIDAL SYSTEMS BASED ON MICROEMULSIONS

Organic nanoparticles of cholesterol and retinol were synthesized in different microemulsion systems (AOT/heptane/water, CTABr/hexanol/water, and Triton X-100/decanol/water) by direct precipitation of the active principle in the aqueous cores of the microemulsion. The size of the nanoparticles can be influenced by the concentration of the organic molecules and the diameter of the water cores that is related to the ratio $R = [H_2O]/[surfactant]$. The particles remained stable for several months. The average diameter of cholesterol nanoparticles is comprised between 5.0 and 7.0 nm, while that of retinol is smaller, being 2.5 nm. The average size of the cholesterol nanoparticles do not change much as a function either of the ratio R or of the concentration of cholesterol. The constant size of the nanoparticles can be explained by the thermodynamic stabilization of a preferential size of the particles (Shinoda et al., 1993).

17.9.9 SOLUBILIZATION OF DRUGS IN MICROEMULSIONS

Solubilizing capacity of a W/O microemulsion for water-soluble drugs is significantly lesser than that of an O/W microemulsion and vice versa. Researchers studied the solubilization of hydrocortisone by

W/O microemulsion and found that the amount that could be incorporated in the microemulsion was dependent on the concentration of both the surfactant (Brij 35/Arlacel 186) and the cosurfactant (short-chain alcohols). The solubility of a drug in microemulsion also depends on the molecular weight of the drug and oil. For high molecular weight drugs like proteins and peptides, loss of entropy due to phase separation limits the solubilization of drugs. Fleer et al. (1993) and Carlfors et al. (1991) prepared the microemulsions using water, IPM, and nonionic surfactant mixtures for solubilization of a local anesthetic—lidocaine. NMR self-diffusion measurements indicated that surface-active but lipophilic lidocaine lowered the phase inversion temperature (PIT) due to simple packing measurement.

17.9.10 MICROEMULSIONS FOR PROTEINS AND PEPTIDE DELIVERY

Recent advances in pharmaceutical sciences and biotechnology permit rather high amounts of peptides and proteins to be delivered, Short half-life, conformational stability, and biodegradability of these molecules cause considerable difficulties in their formulation for oral administration. Microemulsions extensively studied for the protection of biodegradable drugs like proteins and peptides from biological environment of peroral route.

Cilek et al. (2006) prepared and characterized a stable microemulsion formulation for oral administration of a peptide, for example, rh-insulin. The microemulsions were prepared using Labrafil M 1944 CS, phospholipon 90G (lecithin), absolute alcohol, and bidistilled water. Wang et al. (2008) have shown that the insulin-loaded W/O microemulsion possesses eminent sustained release efficacy, and the cytostatic as well as cytotoxic assays illustrate that the microemulsion can be used as drug delivery in small doses.

17.9.11 SITE-SPECIFIC DRUG DELIVERY AND DRUG TARGETING BY MICROEMULSIONS

Pharmacokinetics study performed by Li Zhang showed that Norcantharidin microemulsion had relatively longer circulating time in mice than Norcantharidin injection. Moreover, the overall drug-targeting efficiency of liver was enhanced from 3.66% to 6.10%. These results suggest that a norcantharidin microemulsion system is a promising candidate for the treatment of hepatogenic diseases (Zhang et al., 2005).

17.9.11.1 Site-Specific Drug Delivery from W/O Microemulsions

Colloidal drug carriers like nanoparticles can be used to achieve targeted drug delivery. Nanoparticles prepared by emulsion polymerization can be regarded as promising delivery candidates for drugs. Couvreur et al. (1977) reported the preparation method based on emulsification of the insoluble monomers in aqueous phase. Literature citations report the synthesis and use of polycyanoacrylate nanocapsules prepared by dissolving the monomer alkylcyano acrylate and a lipophilic drug in an oily phase. This oily phase is then injected into an aqueous solution of a nonionic surfactant (Al Khouri Fallouh et al., 1986). Polyalkylcyanoacrylate (PACA) nanocapsules have also been prepared from W/O microemulsions (Gasco and Trotta, 1986). The principle underlying PACA nanocapsule preparation by this method (Gasco and Trotta, 1986) is *in situ* nucleophilic polymerization taking place on gradually adding the lipophilic monomer to the microemulsion. The payload for hydrophilic drug (doxorubicin) can be improved by entrapping it by dissolving it in aqueous phase (Carpigno et al., 1992).

17.10 NONPHARMACEUTICAL APPLICATIONS

17.10.1 PERFLUORO MICROEMULSIONS

Fluorocarbon-based microemulsions were prepared by using fluorinated surfactants. Chabert et al. (1976) reported preparation of microemulsion consisting of mixtures of fluorinated polyoxyethylene derivatives but were found to be toxic. Ceccuti studied these aspects of microemulsions and found that microemulsion prepared from fluorinated oil, water, and nonionic surfactant displayed oxygen absorption similar to that of blood and was also found to be patient friendly, well tolerated, and relatively free from toxicity.

17.10.2 MICROEMULSIONS IN COSMETICS

Since the development of modern cosmetology, emulsions hold an important position in the field of cosmetics particularly skin care products. Properties of microemulsions like transparency and "clean" appearance, small globule size, ease of application, excellent stability, appreciable shelf life, solution-like behavioral characteristics make them not only very appealing to customers but render them more customer friendly also. Microemulsion formulations are believed to enhance the topical drug uptake in terms of rate and extent. Selection of suitable oil phase, surfactants and cosurfactants (qualitatively and quantitatively), safety, toxicity and irritation potential, and cost effectiveness are the major issues to be attended and are the most critical parameters that a cosmetic chemist or formulation chemist should be aware of, in order to optimize the microemulsions. Sodium alkyl sulfates, tetraethylene glycol monododecyl ether, lecithin, dodecyl ligoglucoside, alkyl dimethyl amine oxide, propanol, hexadecane, and IPM have been used as surfactants, cosurfactants, and oils. Hair care products containing an amino functional polyorganosiloxane (a nonionic surfactant) and an acid and/or a metal salt are successfully commercialized. Microemulsions can further stabilize the cosmetics and add to their aesthetic appeal due to their capacity to solubilize fragrance and flavored oils. Nonionic surfactants and oils are well established for their role in cosmetic formulations for skin care products. Currently, ionic surfactants have also been thoroughly studied for their use in cosmetics. Large numbers of patent filings support the fact.

17.10.2.1 Hair Products

Microemulsions have been used for hair treatment formulations that contain hydrolyzing agents (NaOH and KOH) and reducing agents (ammonium thioglycolates, metal, or ammonium sulfates or bisulfates). Combined amount of these agents used for hair relaxers is 0.4%–5% for permanent wave solutions 2%–10% for depilatories 3%–15%. It is claimed that the high level of surfactant does not make this product any more irritant to skin than a conventional emulsion. Dimethicone is effective as a hair conditioner and is under trial for being used in shampoos. Toray Silicon Company (Japan) has patented compositions for clear hair conditioners and conditioning shampoos using a microemulsion of 20% dimethicone with an average particle size of 0.05 μm prepared by emulsion polymerization of dimethylsiloxane tetramer stabilized with dicocoyldimethyl ammonium chloride and tallow trimethylammonium chloride or dodecylbenzene sulfonic acid and heated to 85°C. Microemulsions of functionalized silicone oils are also used to make clear conditioning shampoos in patents by General Electric, Dow Corning, and Unilever (Riccio and Merifield, 1995; Graiver and Tanaka, 1992; Birtwistle, 1989). A microemulsified silicone oil solution with a particle size less than 0.15 μm is prepared by emulsion polymerization of the cyclomethicone precursor. Latex of polystyrene is prepared using a microemulsion stabilized with nonionic surfactants and this is then incorporated as an ingredient in hairspray and setting compositions. The very fine particulate size (0.01–0.1 μm) improves the style forming and retaining capability of the product as well as the feel of the resin on the hair. Microemulsion liquid crystalline gel formulations from Croda, Inc. based on nonionic surfactants and ethoxylated phosphates for clear hair gels comprising of mineral oil, volatile silicone 345, DEA leth-3 phosphate, DEA oleth-10 phosphate, oleth-3, oleth-5, acetamide MEA, water, propylene glycol, sorbitol, and hexylene glycol have been used for reduced oily hair feeling and the other one containing a higher level of polyol for humectant effect, classifying it as a curl activator.

17.10.2.2 Fragrances and Perfumes

Solubilization of perfume O/W produces a clear solution without the use of volatile or drying solvents such as ethanol. Generally, a fatty alcohol or sorbitan ester ethoxylates are the most suitable surfactants. A 1992 patent by Yves Saint Laurent Perfumes (France) describes the solubilization of perfume oils in combination with emollient using PEG ester surfactants and polyglycerol ester cosurfactants (Dartel and Breda, 1993).

17.10.2.3 Gels and Antiperspirants

Microemulsion gels are usually based on nonionic ethoxylated ether surfactants based on oleyl, lauryl, or isoceryl alcohol either alone or in combination with ethoxylated phosphate esters. Long hexagonal liquid crystalline structure present in the gel can vibrate when the container is tapped, resulting in a ringing gel. Clear antiperspirants that do not leave behind any visible residue are gaining popularity. Clear sticks can be based on dibenzyl sorbitol acetal (DBSA) gelling agent and zirconium glycine-complexed aluminum chlorhydrate antiperspirant actively soluble in propylene glycol without water. Refractive indices of oil and aqueous phases in close proximity can also be sought as an alternative approach to obtain clear, transparent, and nonsticky antiperspirant gel (Abrutyn, 1993; Hoffmann and Ebert, 1988).

17.10.3 INDUSTRIAL AND MISCELLANEOUS APPLICATIONS

17.10.3.1 Chemical Reactions in Microemulsion

Chemical reactions inside microemulsion and other molecular aggregates containing micelles and vesicles proceed differently from those in free organic environment. Microemulsion alters ionization potential, oxidation–reduction properties, dissociation constant, chemical pathway, and photochemistry. They stabilize reactants, intermediate, and products.

17.10.4 MICROEMULSION AS A MICROELECTRODE

Since microemulsions are acting as a microreservior for the encapsulated components, they tend to possess certain characteristic features and act as microcatalyst and microelectrode. Yang et al. (2006) demonstrated the microelectrode electrochemistry of a simple electroactive probe (ferrocene) in SDS/n-C_4H_9OH/H_2O microemulsion systems. The oxidation of ferrocene within the microemulsion environment was carried out. Excellent Nernstian electrochemical responses were observed.

17.10.5 MODEL FOR BIOLOGICAL MEMBRANE

Microemulsions are also used as model in membrane mimetic chemistry as reactions in such systems resemble highly organized biochemical processes in a number of ways and components of metabolic pathways are compartmentalized into membranes or intercellular particles.

17.10.6 MICROEMULSIONS IN FOODS

Microemulsions due to lipoidal nature are used in the preparation of foods. Microemulsions form in the intestine during the digestion and absorption of fat. The major differences between food and other microemulsions are in the composition of the oil component and food grade surfactants. Food grade surfactants, viz., phosphatidylcholine (lecithin), AOT, and sorbatin monostearate/monolaurate (Tweens) have been extensively studied with regard to the formation of O/W and W/O microemulsions (Staufer, 1992). Larsson et al. (1991a,b) have elucidated the role of cereal and edible lipid systems in the formation of microemulsions. Literature survey has shown that microemulsions of carnauba wax form better protective coatings on citrus fruit as compared with shellac, wood resin, oxidized polyoxyethylene, etc. It has been found that carnauba wax-based microemulsions offer better protection to citrus fruits, minimize the oxidative degradation, and thereby help to prolong the shelf life of fruits and similar foodstuffs along with maintaining intact their cosmetic appeal. Microemulsions have also been used to produce glycerides for application in food products. Microemulsions are hypothesized to provide enhanced antioxidant protection due to synergistic effect between hydrophilic and lipophilic antioxidants. Despite the sound theoretical conceptualization and noteworthy research, microemulsions have failed to invoke interest of the food technologists and the area largely remains unexplored.

17.11 CONCLUSION

Microemulsions or micellar emulsions are optically isotropic and thermodynamically stable multicomponent liquid composed of oil, water, and surfactant (usually in conjunction with a cosurfactant). It has the unique ability to solubilize pharmaceuticals/biotechnological/cosmetic/food/paints/refinery products, etc., within it and hence has wide applications in these fields. Microemulsion-based drug delivery systems have shown important features like incorporation of high quantity of poorly water-soluble drugs and their improved bioavailability, and site-specific, sustained, and controlled delivery of drugs once administered in different routes. Microemulsion systems are the precursors, intermediates, or catalysts and are used in the formulation of highly fascinating drug delivery systems like nanoparticles. In biotechnological products, microemulsion is the accepted technology in enzyme reactions, immobilization, and bioseparation of biomolecules. Still, there is enough scope for the improvement and development of technologies to make microemulsions more efficacious and user friendly by modifying its stability aspects and rheological properties and methods of preparation on a small scale and even for commercial purposes on a larger scale.

ABBREVIATIONS

16MRSA	16-Methyl-Reichstein's compound S-21-acetate
AOT	Aerosol OT
AOT	Sodium bis(2-ethylhexyl)sulfosuccinate
BBB	Blood brain barrier
CB	Cibacron Blue F3G-A
CMC	Critical micelle concentration
CPCR	Carbonyl reductase from *Candida parapsilosis*
CoS	Cosurfacatant
CPP	Critical packing parameter
CRYO TEM	Freeze fracture transmission electron microscopy
CTAB	Cetyl trimethyl ammonium bromide
DBSA	Dibenzyl sorbitol acetal
DDBSA	Dodecylbenzenesulfonic acid
DNA	Deoxy ribonucleic acid
FT-PGSE	Fourier transform pulsed-gradient spin-echo
GLD	Glycerol dehydrogenase
HGP	*n*-Hexyl-β-D-glucopyranoside
HLB	Hydrophilic lipophilic balance
HLADH	Horse-liver alcohol dehydrogenase
HRP	Horseradish peroxidase
Ig G, Ig A	Immunoglobulins of G and A
IM Injection	Intramuscular injection
IPM	Isopropyl myristate
IV	Intravenous
LHRH	Leuteinizing hormone releasing hormone
MBG	Microemulsion-based organogel

O/W	Oil-in-water type
PACA	Polyalkylcyanoacrylate
PCS	Photon correlation spectroscopy
PEG	Polyethylene glycol
PIT	Phase inversion temperature
PPL	Porcine pancreatic lipase
RES	Reticuloendothelial system
SAA	Surface active agents (surfactants)
SANS	Small-angle neutron scattering
SAXS	Small-angle x-ray scattering
SDS	Sodium dodecyl sulfate
SFAE	Sugar fatty acid ester
S/O/W	Surfactant, oil, and water mixture
STH	Soluble transhydrogenase
TEM	Transmission electron microscopy
w/IL	Water-in-ionic liquid
W/O	Water-in-oil type
YADH	Alcohol dehydrogenase from yeast

SYMBOLS

Γ	Interfacial tension
ΔG	Free energy
ΔA	Increase in area
q	Intensity of scattered light
λ	Wavelength of the radiation
θ	Scattering angle
k	Boltzmann constant
T	Absolute temperature
$\acute{\eta}$	Viscosity of the medium
D	Diffusion coefficients
R_H	Hydrodynamic radius of the particles

REFERENCES

Abrutyn E., 1993, SCC continuing Education Program, Antiperspirant & Deodorant Technology, February 24.

Aguire J.M., Bagan J.V., Rodriguez C., Jimenez Y., martinez C.R., Diaz de Rojas F., and Ponte A., 2004, Efficacy of Mometasone furoate microemulsion in the treatment of erosive, ulverative oral lichen planus: Pilot study, *J. Oral Pathol. Med.*, 33(7):381–385.

Al Khouri Fallouh N., Roblot-Treupel L., Fessi H., Devissaguet J.P., and Puisiex F., 1986, Development of a new process for the manufacture of poluiobutylcyanoacrylate nanocapsules, *Int. J. Pharm.*, 28:125–132.

Bello M., Colangelo D., Gasco M.R., Maranetto F., Morel S., Podio V., Turco G.L., and Viano I., 1994, Pertechnate release from water oil microemulsion and aqueous solution after subcutaneous injection in rabbits, *J. Pharm. Pharmacol.*, 46:508–510.

Bender J P., Junges A., and Franceschi E., 2008, High pressure cloud point data for this system glycerol + olive oil + n-butane + AOT, *Braz. J. Chem. Eng.*, 25(3):563–570.

Bergstrome K. and Holmberg K., 1992, Microemulsion as reaction media for immobilization of proteins to hydrophilized surfaces, *Colloids Surf.*, 63(3–4):273–280.

Bhanushali R.S. and Bajaj A.N., 2007, Design and development of thermoreversible mucoadhesive microemulsion of intranasal delivery of sumatriptan succinate, *Ind. J. Pharm. Sci.*, 69(5):709–712.

Bhatnagar S. and Vyas S.P., 1994, Organogel based system for transdermal delivery of propranolol, *J. Microencapsul.*, 4(11):431–438.

Birtwistle D.H., 1998, Shampoo compositions and methods, International Patent WO/1998/005296, to Unilever.

Blattner C., Zoumpanioti M., Kröner J., Schmeer G., Xenakis A., and Kunz W., 2006, Biocatalysis using lipase encapsulated in microemulsion-based organogels in supercritical carbon dioxide, *J. Supercrit. Fluids*, 36(3):182–193.

Bommarius A.S., Hatton T.A., and Wang D.I.C., 1995, Xanthine oxidase reactivity in reversed micellar systems: A contribution to the prediction of enzymic activity in organized media, *J. Am. Chem. Soc.*, 117(16):4515–4523.

Bowcott J.E. and Schulman H.H.Z., 1955, Emulsion, control of droplet size and phase continuity in transparent oil water dispersions stabilised with soap and alcohol, *Elektrochemie*, 59:283–288.

Burrier R.E. and Brecher P., 1983, Hydrolysis of triolein in phospholipid vesicles and microemulsions by a purified rat liver acid lipase, *J. Biol. Chem.*, 258(19):12043–12050.

Carlfors J., Blute I., and Schmidt V., 1991, Lidocaine in microemulsions: A dermal delivery system, *J. Dispers. Sci. Technol.*, 12(5–6):467–482.

Carpigno R., Gasco M.R., and Morel S., 1992, Optimization of incorporation of deoxycotisone acetate in lipsoheres, *Eur. J. Pharm. Biopharm.*, 38:7–10.

Ceccuti C., Rico I., Lattes A., Novelli A., Rico G., Marion A., Graciaa A., and Lachaise J., 1989, New formulation of blood substitutes: Optimization of novel fluorinated microemulsions, *Eur. J. Med. Chem.*, 24(5):485–492.

Chabert P., Foulletier L., and Lantz A., 1976, Emulsification of fluorocarbon compounds for biological applications as oxygen transporters, U.S. patent 3989343.

Cilek A., Čelebi N., and Timaksiz F., 2006, Lecithin-based microemulsion of a peptide for oral administration: Preparation, characterization, and physical stability of the formulation, *Drug Deliv.*, 13(1):19–24.

Couvreur P., Roland M., and Speiser P., 1977, Nanoparticles: A new type of lysosomotropic carrier, *FEBS Lett.* 84(2):323–326.

Cui Z., Fountain W., Clark M., Jay M., and Mump R.J., 2003, Novel ethanol/fluorocarbon microemulsion for topical genetic Immunization, *Pharm. Res.*, 20(1):16–23.

D'Cruz O.J., Yiv S. H., and Uckun F.M., 2001, GM-144, a novel lipophilic contraceptive gel microemulsion, *AAPS PharmSciTech*, 2(2):Article 5.

Dalmora M.E.A. and Oliveira A.J., 1999, Inclusion complex of piroxicam with beta cyclodextrin and incorporation in hexadecyltrimethyl ammonium bromide based microemulsion, *Int. J. Pharm.*, 184(2): 157–164.

Dartel N. and Breda B., 1993, Microemulsion containing a perfuming concentrate and corresponding product, U.S. Patent 5,252,555 to Yves Saint Laurent Parfumes, France.

Date A.A. and Nagarsenker M.S., 2008, Design and evaluation of microemulsions for improved parenteral delivery of propofol, *AAPS PharmSciTech*, 9(1):138–145.

Daubresse C., Grandfils Ch., Jerome R., and Teyssie Ph., 1996, Enzyme immobilization in reactive nanoparticles produced by inverse microemulsion polymerization, *Colloids Polym. Sci.*, 274(5):482–489.

Derle D.V., Sagar B.S.H., and Pimpale S., 2006, Microemulsion as vehicle for transdermal permeation of nimesulide, *Ind. J. Pharm. Sci.*, 68(5):622–625.

Drewe J., Beglinger C., and Kisssel T., 1992, The absorption site of cyclosporine in human gastro intestinal tract, *Br. J. Clin. Pharmacol.*, 33(1):39–43.

Famiglietti M., Hochkoppler A., Wehrli E., and Luisi P.L., 2004, Photosynthetic activity of cyanobacteria in w/o microemulsions, *Biotechnol. Bioeng.*, 40(11):173–178.

Fleer G.J., Cohen Stuart M.A., Scheutjens J.M.H.M., Cosgrove T., and Vincent B., 1993, *Polymers at Interfaces*, Chapman & Hall, London, U.K.

Friberg S.E. and Venable R.L., 1983, A non aqueous microemulsion. In *Encyclopedia of Emulsion Technology*, vol. 1 (P. Becher, Ed.), Marcel Dekker, New York, pp. 287–336.

Friberg S., Lapczynska L., and Gillberg G., 1976, Microemulsion containing the non ionic surfactant: The importance of Pit value, *J. Colloid Interface Sci.*, 56:19–32.

Friberg S.E., Buraczewska I., and Ravery J.C., 1977, In: *Micellization, Solubilization and Microemulsion* (K.L. Mittal, Ed.), Plenum, New York, p. 901.

Gerbacia E. and Rosano H.L., 1973, Microemulsions: Formation and stabilization, *J. Colloid Interface Sci.* 233–244.

Giuliani S., Piana C., Setti L., Hochkoeppler A., Pifferi P.G., Williamson G., and Faulds C.B., 2001, Synthesis of pentylferulate by a feruloyl esterase from *Aspergillus niger* using microemulsions, *Biotechnol. Lett.* 23(4):325–330.

Graiver D. and Tanaka O., 1992, Methods for making polydiorganosiloxane microemulsions, U.S. Patent 4,999,398, to Dow Corning.

Guadalupe A., Kamat S., Komives C., Beckman E.J., and Russel A.J., 1992, Solubilization and activity of proteins in compressible fluid based microemulsions, *Nat. Biotechnol.* 10:1584–1588.

Halloran D.J., 1998, Optically clear hair conditioning compositions containing aminofunctional silicone micro-emulsions, U.S. Patent 6147038 to Dow Corning.

Hasse A. and Keipert S., 1997, Development and characterization of microemulsions for ocular application, *Eur. J. Pharm. Biopharm.*, 43:179–183.

Hatton T.A., 1989, *Surfactant Based Separation Processes* (Surfactant Series No. 30), *Reversed Micellar Extraction of Proteins*, Marcel Dekker, New York, pp. 55–90.

Hayes D.G., 2000, Lipid modification in water-in-oil microemulsion. In *Enzymes in Lipid Modification* (U.T. Bornscheuer, Ed.), Wiley-VCH Verlag GmbH, Weinheim, Germany.

Hirofumi I., Noriho K., and Masahiro G., 2005, Enzymatic redox cofactor regeneration in organic media: Functionalization and application of glycerol dehydrogenase and soluble transhydrogenase in reverse micelles, *Biotechnol. Prog.*, 21(4):1192–1197.

Hoar T.P. and Schulman J.H., 1943, Transparent oil in water dispersions: The oleopathic hydrophillic micelle, *Nature*, 152:102–103.

Hoffmann H. and Ebert G., 1988, Thermoreversible gelation of polymers and biopolymers, *Angew: Chem. Int. Ed. Engl.*, 27:902.

Holmberg K. and Österberg E., 1987, Enzymatic transesterification of a triglyceride in microemulsions. In *Surface Forces and Surfactant Systems*, Springer, Berlin/Heidelberg, Germany, pp. 74, 98–102.

Holmberg K., Bo L., and Maj-Britt S., 1989, Enzymatic glycerolysis of a triglyceride in aqueous and nonaqueous microemulsions, *J. Am. Oil Chem. Soc.*, 66(12):1796–1800.

Huang T.M., Huang H.C., Chang T.C., and Chang C.C., 1998, Solvent kinetic isotope effects of human alkaline phosphatase in reverse micelles, *Biochem. J.*, 330(Pt 1):267–275.

Huang Y.B., Lin Y.H., Ming L., Wang R.J., Yi Hung T., Pao W.C., 2007, Transdermal delivery of capsaicin derivative-sodium nonivamide acetate using microemulsion as vehicle, *Int. J. Pharm.*, 349(1–2):206–211.

Hui P.X., Zu Y.L., and Ward O.P., 1995, Hydrolysis of ethyl mandelate and esterification of 2-bromopropionic acid in micro-emulsions, *J. Ind. Microbiol. Biotechnol.*, 14(5):416–419.

Israelachvili J.N., Mitchell D.J., and Ninham B.W., 1976, Phase behaviour of amphiphile-water binary mixtures, *J. Chem. Soc. Faraday Trans.*, 1(72):1525.

Jenta T.R.J., Robinson B.H., Batts G., and Thomson A.R., 1991, Enzymatic kinetic studies using lipase immobilized in microemulsion based Organogel, *Trends Colloid Interface Sci. V*, 84:334–337.

Jogani V.V., Shah P.J., Mishra P., Mishra A.K., and Misra A.N., 2008, Intranasal mucoadhesive microemulsion of tacrine to improve brain targeting, *Alzheimer Dis. Assoc. Disord.*, 22(2):116–124.

Kakutani T., Nishiara Y., Takahashi K., and Kirano K., 1991, O/W lipid microemulsion for parenteral delivery, *Proc. Int. Symp. Control. Release Bioact. Mater.*, 18:359.

Kantaria S., Rees G.D., and Lawrence M.J., 1999, Gelatin-stabilised microemulsion-based organogels: Rheology and application in iontophoretic transdermal drug delivery, *J. Control. Release*, 60(2–3):355–365.

Komives C.F., Lilley E., and Russel A.J., 1994, Biodegradation of pesticides in non ionic w/o microemulsion of Tween 85: Relationship between micelle structure and activity, *Biotechnol. Bioeng.*, 43(10):946–959.

Larsson K.M., Adlercreutz P., and Mattiasson B., 1987, Activity and stability of horse-liver alcohol dehydrogenase in sodium dioctylsulfosuccinate/cyclohexane reverse micelles, *Eur. J. Biochem.*, 166(1):157–161.

Larsson K., Osborne D.W., Pesheck C.V., and Chipman R.J., 1991a, *Microemulsions and Emulsions in Foods* (El-Nokaly, M. and Cornell, D., Eds.), American Chemical Society, Washington, DC, pp. 44–50 and 62–79.

Larsson K.M., Adlercreutz P., Mattiasson B., and Olsson U., 1991b, Enzyme catalysis in uni- and bi-continuous microemulsions: Dependence of kinetics on substrate partitioning, *J. Chem. Soc. Faraday Trans.*, 87:465–471.

Lee J., Lee Y., Kim J., Yoon M., and Choi Y.W., 2005, Formulation of microemulsion systems for transdermal delivery of aceclofenac, *Arch. Pharm. Res. (N. Engl. J. Med)*, 28(9):1097–1102.

Madras G., Kolluru C., and Rajnish K., 2004, Synthesis of biodiesel in supercritical fluids, *Fuel*, 83:2029–2033.

Malaviya A. and Gomes J., 2008. Nutrient Broth/PEG 200/Triton X114/Tween 80/chloroform microemulsion as a reservoir of solubilized sitosterol for biotransformation to androstenedione, *J. Ind. Microbiiol. Biotechnol.*, 35(11):1435–1440.

Mitchell D.J. and Ninham B.W., 1977, Micelles, vesicles and microemulsions, *J. Chem. Soc., Faraday Trans.*, 21981:601–629.

Moniruzzaman M., Kamiya N., and Goto M., 2009, Biocatalysis in water-in-ionic liquid microemulsions: A case study with horseradish peroxidase, *Langmuir*, 25(2):977–982.

Murdan S., Gregoriadis G., and Florence A.T., 1999, Novel sorbitan monostearate organo gels, *J. Pharm. Sci.*, 88(6):608–614.

Nesamony J. and Koling W.M., 2009, Rheological properties and dynamic microstructures of polymeric microemulsions, From www.appsj.org/abstracts/Am-2003/AAPS2003-001625.pdf (Accessed on 22.04.2010).

Nonoo A.O. and Chow D.S.-L., 2008, Cremophor-free intravenous microemulsions for paclitaxel II. Stability, *in vitro* release and pharmacokinetics, *Int. J. Pharm.*, 49(1–2):117–123.

Oconnor J., Aggett A., Williams D.R., and Stanley R.A., 1991, *Candida cylindracea* lipase-catalyzed hydrolysis of methyl palmitate in detergentless microemulsion and paraffin/water biphasic media, *Aust. J. Chem.*, 44(1):53–60.

Olssen U. and Wennerstrom H., 1994, Globular and bicontinuous phases of non ionic surfactant films, *Adv. Colloid Interface Sci.*, 49:113–146.

Orlich B., Berger H., Lade M., and Schomäcker R., 2000, Stability and activity of alcohol dehydrogeinase W/O-microemulsions: Enantioselective reduction including cofactor regeneration, *Biotechnol. Bioeng.*, 70(6):638–646.

Osborne D.W., Ward A.J., and O'Neil K., 1988, Microemulsions as delivery of vehicles: 1. Characterization of a model system, *Drug Dev. Ind. Pharm.*, 14:1203.

Osterberg E., Blomstrom A.C., and Holmberg K., 1989, Lipase catalysed transesterification of unsaturated lipids in microemulsion, *J. Am. Oil Chem. Soc.*, 66:1330–1333.

Overbeek J.Th.G., 1978, Microemulsion, a field at the border between lyophobic and lyophilic colloids, 1st Rideal Lecture, *Faraday Discuss. Chem. Soc.*, 65:20.

Pahn S.C., Joon S.R., and Jae-Jin K., 1991, Continuous glycerolysis of olive oil by *Chromobacterium viscosum* lipase immobilized in liposome in reversed micelles, *Biotechnol. Bioeng.*, 38(10):1159–1165.

Papadimitriou A., Sotiroudis T.G., and Xenakis A., 2005, Olive oil microemulsions as a biomimetic medium for enzymatic studies: Oxidation of oleuropein, *J. Am. Oil Chem. Soc.*, 82(5):335–340.

Papapanagiotou P.A., Quinn H., Molitor J.P., Nienow A.W., and Hewitt C.J., 2005, The use of phase inversion temperature (PIT) microemulsion technology to enhance oil utilisation during *Streptomyces rimosus* fed-batch fermentations to produce oxytetracycline, *Biotechnol. Lett.*, 27(20):1579–1585.

Parmar V.S., Bisht K.S., Pati H.N., Sharma N.K., Ajay K., Naresh K., Malhotra S., Singh A., Prasad A.K., and Wengel J., 1996, Novel biotransformation on peracetylated polyphenolics by lipase immobilized in microemulsion based gels and on carbohydrates by *Candida antarctica* lipase, *Pure Appl. Chem.*, 68(6):1309–1314.

Peira E., Carlotti M.E., Trotta C., Cavalli R., and Trotta M., 2008, Positively charged microemulsion for topical application, *Int. J. Pharm.*, 346(1–2):119–123.

Peter S., Karin L., Martin B., and Karl H., 1997, Glucoside ester synthesis in microemulsions catalyzed by *Candida antarctica* component B lipase, *J. Am. Oil Chem. Soc.*, 74(1):39–42.

Pfammater N., Hochkoppler A., and Luisi P.L., 2004, Soubilization and growth of *Candida pseudotropicalis* in w/o microemulsions, *Biotechnol. Bioeng.*, 4000(1):167–172.

Rance D.G. and Friberg S., 1977, Micellar solutions versus microemulsions, *J. Colloid Interface Sci.* 60:207–209.

Riccio D. and Merrifield J.H., 1995, Method of preparing microemulsions of amino silicone fluids and MQ resin mixtures, U.S. Patent 6180117 to General Electric Company.

Ritschel W.A., 1991, Microemulsions for improved peptide absorption from the gastrointestinal tract, *Methods Find. Exp. Clin. Pharmacol.*, 13(3):205–220.

Ritschel W.A., Adolph S., Ritschel G.B., and Schroeder T., 1990, Improvement of peroral absorption of cyclosporin A by microemulsions, *Methods Find. Exp. Clin. Pharmacol.*, 12:127–134.

Rodakiewicz-Nowak J., Monkiewicz M., and Haber J., 2002, Enzymatic activity of the *A. bisporus* tyrosinase in AOT/isooctane water-in-oil microemulsions, *Colloids Surf. A: Physicochem. Eng. Aspects*, 208(1–3):347–356.

Rojo M., Gómez M., Isorna P., and Estrada P., 2001, Micellar catalysis of polyphenol oxidase in AOT/cyclohexane, *J. Mol. Catal. B Enzyme*, 11(4–6):857–865.

Ruckenstein E. and Chi J.C., 1975, Stability of microemulsion, *J. Chem. Soc. Faraday Trans.* 2:1690–1707.

Ruckenstein E. and Krishnan R., 1979, Swollen micellar models for solubilization, *J. Colloid Interface Sci.*, 71:321–335.

Ruckenstein E. and Krishnan R., 1980, The equilibrium radius of microemulsions formed with ionic surfactants, *J. Colloid Interface Sci.*, 75:476–492.

Sathish K., Jeong E.S., Yun S.E., Mun S.P., and Rusling, J.F., 2008, Bicontinuous microemulsion as reaction medium for the β-glucosidase-catalyzed synthesis of n-hexyl-β-D-glucopyranoside, *Enzyme Microb. Technol.*, 42(3):252–258.

Schmalfub U., Neubert W., and Wohlarb W., 1997, Modification of drug penetration into human skin using microemulsions, *J. Control. Release*, 46:279–285.

Schuleit M. and Pier L.L., 2000, Enzyme immobilization in silica hardened organogels, *Biotechnol. Bioeng.*, 72(2):249–253.

Schulman J.H., Stoekenius W., and Prince L.M., 1959, Physical chemistry of lipids from alkanes to phospholipids, *Phys. Chem.*, 63:1677.

Sheikh S. and Faiyaz S., 2007, Nano emulsions as vehicles for trans dermal delivery of Aceclofenac, *AAPS PharmSciTech*, 8(4):article 104, E1–E9.

Sheild W.J., Fergusan H.D., Bommarius A.S., and Hatton, T.A., 1986, Enzymes in reversed micelles as catalysts for organic phase reactions, *Ind. Eng. Chem. Found.*, 25:603–612.

Shinde A.J., Patil R.R., and Devarajan P.V., 2007, Microemulsion of Lamotrigine for nasal delivery, *Ind. J. Pharm. Sci.*, 69(5):721–722.

Shinoda K. and Friberg S.E., 1986, Stability of emulsion. In: *Emulsion and Solubilization*, Wiley, New York.

Shinoda K. and Kuneida H., 1973, Conditions to produce so called microemulsions: Factors to increase mutual solubility of oil and water by solubilizer, *J. Colloid Interface Sci.*, 42:381–387.

Shinoda K. and Lindman B., 1987, Self organizing structures of lecithin, organized surfactant systems: Microemulsions, *Langmuir*, 3:135–149.

Shinoda K., Araki M., Sadaghiani A., Khan A., and Lindman B., 1991, Lecithin based microemulsions: Phase behaviour and structure, *J. Phys. Chem.*, 95:989.

Shinoda, K., Shibata, Y., and Lindman, B., 1993, Interfacial tensions for lecithin microemulsions including the effect of surfactant and polymer addition, *Langmuir*, 9:1254.

Siebenbordt I. and Keipert S., 1991, Versuche zur Entwickelung und Characterisierung ophtalmologisch verwendbarer tensidhaltiger Mehrkomponentsysteme. *Pharmazie*, 46:435–439.

Smolders A.J.J., Pinheiro H.M., Noronha P., Cabral J.M.S., 1991, Steroid bioconversion in a microemulsion system, *Biotechnol. Bioeng.*, 38(10):1210–1217.

Stamatis H. and Xenakis A., 1999, Biocatalysis using microemulsion-based polymer gels containing lipase, *J. Mol. Catal. B*, 6(4):399–406.

Staufer C.E., 1992, Application of starch-based fat replacers, *Food Technol.*, 46:70.

Sun Y., Lei Gu., Xiao D.T., Shu B., Sokasu I., Shintaro F., 2008, Protein separation using affinity based reverse micelles, *Biotechnol. Prog.*, 25(3):506–512.

Tadros Th.F., 1984, *Surfactants and in Solution*, Vol. III (K.L. Mittal and B. Lindman, Eds.), Plenum Press, New York, pp. 1501–1532.

Tarr B.D. and Yalkowsky S.H., 1989, Enhanced intestinal absorption of cyclosorine in rats through reduction of emulsion droplet size, *Pharm. Res.*, 6(1):40–43.

Vincentini F.T., Simi T.R., Delciampo J.D., Wolga N.O., Pitol D.L., Iyomasa M.M., Bentley M.V., and Fonseca M.J., 2008, Quercetin in W/O microemulsions: *In Vitro* and *In Vivo* skin penetration and efficacy against UVB induced skin damages evaluated *in vivo*, *Eur. J. Biopharm.*, 69(3):948–957.

Vyas T.K., Babbar A.K., Sharma R.K., Singh S., and Misra A., 2006, Intranasal mucoadhesive microemulsions of clonazepam: Preliminary studies on brain targeting, *J. Pharm. Res.*, 95(3):570–580.

Wang J.L., Wang Z.W., Liu F., and Zhao D.Y., 2008, Preparation and *in vitro* release test of insulin loaded W/O microemulsion, *J. Disp. Sci. Technol.*, 29(5):756–762.

Willimann H., Walde J.I., Luisi A., Gazzaniga A., and Dtroppolo F., 1992, Lecithin organogel as matrix for transdermal transport of drugs, *J. Pharm. Sci.*, 81(9):871–874.

Winsor P.A., 1954, *Solvent Properties of Amphiphilic Compounds*, Butterworth, London, U.K.

Winsor P.A., 1968, Binary and multicomponent solutions of amphiphilic compounds. Solubilization and the formation, structure, and theoretical significance of liquid crystalline solutions, *Chem. Rev.*, 68:1.

Xenakis A. and Stamatis H., 1999, Lipase immobilization on microemulsion based polymer gels, *Prog. Colloid Polym. Sci.*, 112:132–135.

Xenakis, A., Valis TP., and Kolisis F.N., 2006, Lipase-catalyzed esterification of fatty acids in nonionic microemulsions, *Ann. NY Acad. Sci.*, 613(10):674–680.

Yang Z., Zhao J., Gao L., Wang T., Cao Q., Zhang N., and Yang Z., 2006, Microelectrode electrochemistry in microemulsion system, *Anal. Lett.*, 39(9):1801–1808.

Yuchun H. and Ying C., 2005, The catalytic properties and mechanism of cyclohexane/DBSA/water microemulsion system for esterification, *J. Mol. Catal.*, 237(1–2):232–237.

Yuri L.K., Riet H., and Cees V., 1988, Detergentless microemulsions as media for enzymatic reactions, choles-
terol oxidation catalyzed by cholesterol oxidase, *Eur. J. Biochem.*, 176(2):265–271.
Zhang Q., Jiang X., Jiang W., Wei C., and Zhengi S., 2004, Preparation of nimodipine loaded microemulsion
for Intranasal delivery and evaluation on targeting efficiency to brain, *Int. J. Pharm.* 275(1–2):85–96.
Zhang L., Sun X., and Zhang Z.R., 2005, An investigation on liver-targeting microemulsions of norcantharidin,
Drug Deliv., 12(5):289–295.
Zoumpanioti M., Karali M., Xenakis A., and Stamatis H., 2006, Lipase biocatalytic processes in surfactant free
microemulsions like ternary systems and related organogels, *Enzyme Microb. Technol.*, 39(4):531–539.

18 Biocompatible Microemulsions

Monzer Fanun

CONTENTS

18.1 INTRODUCTION

Microemulsions are transparent colloidal assemblies that have polar and nonpolar microdomains. They consist either of water droplets in oil or oil droplets in water where the droplets are surrounded by surfactant film, or they have a bicontinuous microstructure, i.e., continuous channels of oil and water separated by the surfactant. Due to their thermodynamic stability, these systems have good shelf life, large surface area, low viscosity, and ultralow surface tension (Hoar and Schulman 1943, Prince 1977, Danielsson and Lindman 1981, Bourrel and Schecter 1988, Sjoblom et al. 1996, Kunieda and Solans 1997, Paul and Moulik 1997, Moulik and Paul 1998, Salager and Anton 1999, Fanun 2009a). The formulation and characterization of microemulsions was extensively investigated (Kahlweit et al. 1996, Tlusty and Safran, 2000, Tlusty et al. 2000, Hellweg 2002, Salager et al. 2005). Due to the existence of polar, nonpolar, and interfacial microdomains, microemulsion systems operate as excellent solvents of active ingredients, including drugs that are relatively insoluble in both aqueous and organic solvents (Kreuter 1994, Paul and Moulik 2001, Flanagan and Singh 2006, Kogan and Garti 2006, Gupta and Moulik 2008, Fanun 2010). Due to their special physicochemical properties, such as in the nanometer range of controlled drop size, the almost monodisperse size distribution, a large specific inner surface, and a distinctive dissolving power for reactants, they are of interest for a number of industrial processes. Enhancing solubilization in microemulsions is directly related to the presence of a smooth, blurred, and expanded transition across the interfacial region from polar to apolar bulk phases (Salager et al. 2005). Improved solubilization in microemulsions could be achieved by the addition of polar oil through the so-called lipophilic linker effect. The best lipophilic linker was found to have a hydrophobe chain length as an average between that of the surfactant tail and the oil (Salager et al. 1998). Lipophilic linkers are sometimes added to oil–water–ionic surfactant microemulsions in order to increase the solubilization of hydrophobic oils (Szekeres et al. 2006). The development of novel microemulsions to be used as biocompatible nanomaterial for biomedical and pharmaceutical applications, to improve the quality of the already existing medical devices has assumed great importance. The relationship between the structural features of biocompatible microemulsions and drug solubilization and delivery was studied (Dalmora et al. 2001, Carlotti et al. 2003, Formariz et al. 2007, 2008). Biocompatible microemulsions are qualified to be prospective drug delivery systems, provided they are composed of biocompatible excipients (Das et al. 1991, Paul and Moulik 1991, Mitra et al. 1994, Constantinides

1995, Mitra et al. 1996, Tenjarla 1999, Acharya et al. 2001a,b, Mele et al. 2004, Acosta et al. 2005, Mitra and Paul 2005). Biocompatible microemulsions have also been found to improve the bioavailability of certain drugs, e.g., orally administered labile peptides and proteins (Sarciaux et al. 1995, Mitra et al. 1996, Majhi and Moulik 1999, Watnasirichaikul et al. 2000, 2002, Radwan and Aboul-Enein 2002, Krauel et al. 2005, Pinto-Reis et al. 2006, Graf et al. 2009). For example, contact lenses made of microemulsion-laden gels are expected to deliver drugs at therapeutic levels for a few days. The delivery rates can be tailored by controlling the particle and the drug loading (Le Bourlais et al. 1998, Arriagada and Osseo-Asare 1999, Gulsen and Chauhan 2005). Doxorubicin biocompatible oil-in-water (O/W) microemulsions stabilized by mixed surfactants containing soya phosphatidyl-choline were reported by Formariz et al. (2006). The characterization of caprylocaproyl macrogol-glycerides-based microemulsion drug delivery vehicles for an amphiphilic drug was studied by Djordjevic et al. (2004). Other authors reported (Monduzzi et al. 1997, Rohloff et al. 2003, La Mesa 2005, Shimek et al. 2005, Gochman-Hecht and Bianco-Peled 2006, Zimmerberg and Kozlov 2006, Kim and Dungan 2008) on the ability of water-soluble, globular proteins to tune surfactant/oil/water self-assemblies and their potential for the formation of microemulsions for biological applications. The use of microemulsions for the delivery of proteins and nutraceuticals was also reported (Pauletti et al. 1996, Watnasirichaikul et al. 2000, 2002, Eaimtrakarn et al. 2002, Pitaksuteepong et al. 2002, Radwan and Aboul-Enein 2002, Djordjevic et al. 2004, des Rieux et al. 2006, Pinto Reis et al. 2006, Spernath and Aserin 2006). Microemulsions composed of fluorinated oil with a biocompatible hydrogenated surfactant were investigated as blood substitutes (Cecutti et al. 1989, 1990). Microemulsions were also used for the preparation of poly(alkylcyanoacrylate) nanoparticles (Watnasirichaikul et al. 2000, Krauel et al. 2005, 2006). Methylene blue (MB)-doped silica nanoparticles (NPs) were prepared in a reverse microemulsion and used as a novel matrix for biochemical application (Watnasirichaikul et al. 2000, Krauel et al. 2005, 2006). A biocompatible microenvironment was provided for heme proteins retaining their native conformation and biological activity by the silica matrix with hydrophilic groups (Bagwe et al. 2004, Zhao et al. 2004, Xian et al. 2006). Studies on biocompatible nanocapsules formed in microemulsion-templated processes were also reported (Zielińska et al. 2008). Microemulsions along with other systems have been used as precursors for the synthesis of encapsulating agents (Cortesi and Nastruzzi 1999, Couvreur et al. 2002, Chávez et al. 2005). Biocompatible microemulsions were used for the formation of organogels (Luisi et al. 1990, Angelico et al. 2005), and in the design and production of skin care products, which are nowadays defined by terms like quality, safety, efficacy, exclusiveness, and consumer confidence (Magdassi and Touitov 1998, Gasperlin and Kristl 2000). A number of studies were conducted on the formulation and characterization of pseudoternary biocompatible systems containing surfactants (single or mixed), cosurfactant(s), oil (single or mixed), and water (Das et al. 1991, Paul and Moulik 1991, Mitra et al. 1994, 1996, Constantinides 1995, Tenjarla 1999, Acharya et al. 2001a,b, Shukla et al. 2002, Mele et al. 2004, Mitra and Paul 2004, 2005, Acosta et al. 2005). The influence of surfactant types along with their chain lengths, structure and mixing ratios, type of cosurfactants and oils on the extent of the one-phase microemulsions region in the phase diagrams has been investigated (Leung and Shah 1987, Kunieda and Aoki 1996, Kumar and Mittal 1999, Evans and Wennerstrom 1999). The formation of microemulsion using biocompatible sucrose alkanoate systems was reported (Kunieda and Shinoda 1985, Herrington and Sahi 1988, Pes et al. 1996, Aramaki et al. 1997a, Nakamura et al. 1997, 1999). Microemulsions are useful as reaction media as a way to overcome the reactant incompatibility problem that one frequently encounters in organic synthesis. The use of a microemulsion can be seen as an alternative to phase-transfer catalysis. Holmberg (2007) reported on the usefulness of microemulsions for overcoming reactant incompatibility, speeding up reactions of one polar and one apolar reactant, and inducing regiospecificity. The use of water-in-oil (W/O) microemulsions as medium for enzymatic reactions is a promising application. Microemulsions were also extensively tested for their potential for this application (Avramiotis et al. 1999, Klier et al. 2000, Garti 2003, Bauduin et al. 2005, Garti et al. 2005, Papadimitriou et al. 2005, 2007). The enzyme efficiency was affected by the chain length of the surfactant and the nature of the cosurfactant used in the formulation of the biocompatible micro-emulsions as reported (Stamatis and Xenakis 1999, Stamatis et al. 1999, Zhou et al. 2001, Fadnavis and Deshpande 2002, Matura et al. 2002, Orlich and Schomacker 2002, Garti 2003, Flanagan and Singh 2006, Leser et al. 2006, Papadimitriou et al. 2007, 2008). The solubilization of polar oils in surfactant self-organized structure was investigated by small-angle x-ray scattering (SAXS) and small-angle neutron scattering (SANS) (Barlow et al. 2000, Kunieda et al. 2001). The formation of middle-phase microemulsions of polar oils was reported (Nishimi 2008). This chapter will review on the development and characterization of biocompatible microemulsion systems and their evalu-ation as probable vehicles for encapsulation, stabilization, and delivery of bioactive natural products and prescription drugs. The review will be based on the classification of microemulsions according to the type of oil used in its formulation. The types of oils that will be reviewed are mainly the polar ones that include cyclic flavors, linear esters, and mono-, di-, and triglycerides.

18.2 MICROEMULSIONS BASED ON CYCLIC OILS

Microemulsions formulated with cyclic polar oils were extensively studied (Kahlweit et al. 1990, Burauer et al. 2000, Kogan and Garti 2006, Fanun 2007a, 2008, 2008b, 2008c, and 2008d, Nishimi 2008). Among these cyclic oils; flavors and perfumes were extensively used for the preparation of biocompatible microemulsions. Both volume- and temperature-induced percolation of the conduc-tance of W/O microemulsions formed with Aerosol-OT (AOT) in cyclic aliphatic and aromatic oils have been studied (Chakraborty and Moulik 2005). The phase behavior of the SDS (sodium dodecyl sulfate)/BA (benzyl alcohol)/water system and the structures of the aggregates determined by small-angle x-ray diffraction and cyclic voltammetry were investigated (Guo et al. 1999). The phase diagrams of ternary water–cyclic oil–C_iE_j (alkylpolyglycol ether) systems have been determined for cyclic oils such as cyclohexane, ethylbenzene and toluene, and mixtures of cyclohexane with tolu-ene. Nishimi (2008) reported on the monomeric solubility of the surfactant in oil obtained from precise volume measurements of the middle phase at the phase inversion temperature (PIT). Phase equilibria in systems composed of water, sodium chloride, cyclohexane, SDS, and n-pentanol was reported (Guo et al. 1999). Ai and Ishitobi (2006) examined the solubilization abilities of decaglyc-erol laurates for the oils m-xylene and (R)-(+)-limonene. The role of the alcohol in the interfacial layer of microemulsions supported by octyl monoglucoside and geraniol was reported (Stubenrauch et al. 1997). The average effective diameters of particles formed in the water–AOT–(n-decane + limonene) microemulsion were measured by the light scattering technique (Kuznetsova et al. 1996). Improved R (+)-limonene solubilization in food-grade nonionic microemulsions in the presence of polyols and ethanol was reported (Garti et al. 2001, Yaghmur et al. 2002a). The microstructure was studied by pulsed gradient spin-echo NMR, conductivity, and viscosity (Yaghmur et al. 2003, De Campo et al. 2004). Emulsified microemulsion (EME) was prepared in a W/O microemulsion for-mulated using glycerol monooleate (GMO), R (+)-limonene, ethanol, and glycerol dispersed in an aqueous phase containing Pluronic F127 as a steric stabilizer (Yaghmur et al. 2005b, Lutz et al. 2007). The impact of glycerol and ethanol on Tween 80 micelles in water was also shown. Middle-phase microemulsion formation was evaluated by varying the mole ratio of anionic and cationic surfactants in mixtures with R (+)-limonene. Mixed anionic–cationic surfactant systems solubi-lized more oil than the anionic surfactant alone under optimum middle-phase microemulsion con-ditions (Upadhyaya et al. 2006). Microemulsions were obtained using nonethoxylated surfactants derived from saccharose and glucose and R (+)-limonene (Carlotti et al. 1999). The phase behavior and microstructure of edible microemulsions of R (+)-limonene with concentrated aqueous sugar solutions using sucrose laurate and sucrose oleate as surfactants was studied (Dave et al. 2007). Nguyen and Sabatini (2009) reported on the alcohol-free microemulsions based on R (+)-limonene and rhamnolipid biosurfactant and rhamnolipid mixtures. The phase behavior and properties of biocompatible W/O microemulsions based on R-(+)-limonene and a mixture of lecithin and either 1-propanol or 1,2-propanediol as surfactants were reported (Papadimitriou et al. 2008). These systems were used for the enzymatic esterification of fatty acids with the short-chained alcohols used as cosurfactants for the formulation of the microemulsions. The characteristics of microemulsions based on sucrose monostearate (SMS), R-(+)-limonene, alcohols, and water were investigated using conductivity, viscosity, and differential scanning calorimetry (DSC) (Garti et al. 2000b). Fanun (2006b) reported on the temperature effect on the phase behavior of mixed nonionic surfactants and R (+)-limonene microemulsions. O/W microemulsion system based on oleic acid and terpenes was developed for the solubilization of ketoprofen (Rhee et al. 2001). Fanun (2007) reported on the formation of temperature-insensitive microemulsions using sugar ester surfactants. Propylene glycol and ethoxylated surfactant affects the phase behavior of water/sucrose stearate/R (+)-limonene. Water solubilization, microstructure, viscous and conductive flow parameters of microemulsions based on mixed nonionic surfactants and R (+)-limonene were reported (Fanun and Salah Al-Diyn 2006, 2007, Fanun 2008a,b,c,d, 2009a,b). It was found that mixed nonionic surfactants are able to solubilize water and R (+)-limonene in microemulsions better than the single surfactants. The effect of humidity on evaporation rates from aqueous and nonaqueous microemulsion with R (+)-limonene was also reported by (Hamdan et al. 1999). Strickley (2004) reported on solubilizing excipients in oral and injectable formulations done using plant oils in microemulsions and other colloidal systems. The incorporation of the cyclic mono-terpene linolene, a known skin penetration enhancer, as oil component on microemulsion formation both in water- and propylene glycol-containing systems was reported (Yotsawimonwat et al. 2006). The incorporation of terpenes as penetration enhancers into a topical microemulsion affects its physical characteristics. This in turn may lead to the instability of the microemulsion and/or can influence the release patterns of drugs from these microemulsions when applied as topical formulations (Barry and Williams 1989, Okabe et al. 1989, Obata et al. 1990, Williams and Barry 1991, Yotsawimonwat et al. 2006). R (+)-limonene-based microemulsions were prepared to improve the transdermal permeation of sodium diclofenac (Escribano et al. 2003, Fanun 2009d). Celecoxib solubilized in U-type R (+)-limonene-based nonionic microemulsions was reported (Garti et al. 2006, Fanun 2009a). The effect of R (+)-limonene on the transdermal permeation of nifedipine and domperidone was also reported (Calpena et al. 1994). Microemulsions containing 5% w/w tea tree oil (terpene) were applied for the treatment of acne vulgaris (Biju et al. 2005). O/W microemulsion based on oleic acid and terpenes added at the level of 5% showed a good solubilizing capacity and excellent skin permeation rate of ketoprofen. R (+)-limonene resulted in a powerful enhancing activity (threefold increase over control) (Yaghmur et al. 2005a). The enhanced solubilization of nutraceuticals (i.e., cholesterol and phytosterols, lycopene, lutein, and lutein esters) was performed in food-grade microemulsions based on nonionic surfactants and R (+)-limonene (Garti et al. 2004b, 2005). The influence of sterols on structural transitions was studied using NMR (Spernath et al. 2003, Garti et al. 2004a, Rozner et al. 2008). Solubilized lutein- and lutein esters-induced microstructure transitions were studied (Amar et al. 2004, Yaghmur et al. 2004). Microemulsion catalysis was used in the furfural–cysteine model reaction studied in oil/water microemulsions based on nonionic surfactants and R (+)-limonene for selective flavor formation. The chemical reaction was found to occur preferably at the interfacial film (Yaghmur et al. 2002b). The microstructure of the O/W microemulsion formulated using nonionic surfactants and R (+)-limonene served as a microreactor or a solubilization vehicle for food applications, has been studied using SAXS and SANS techniques (Gochman-Hecht and Bianco-Peled 2006). Microemulsion formulated using R (+)-limonene was used for aquifer washing (Martel et al. 1998a,b). Fanun (2009c,e) studied the water/nonionic surfactant/peppermint oil systems in order to determine the one-phase microemulsion regions. The surfactants were of two types: hydroxylated (sugar esters) and ethoxylated (ethoxylated mono-di-glyceride and polyoxyethylene sorbitan monooleate). These surfactants were able to form microemulsions without the use of cosurfactants or cosolvents. The ternary phase behavior at 25°C, 37°C, and 45°C was explored to determine the effect of temperature on the amount of solubilized water in the microemulsions. For each system investigated, the total area of the one-phase microemulsion region, A_T, was estimated. Temperature-insensitive microemulsions were observed in the sugar ester-based systems. It was found that minor changes in the surfactant structure suffice to provoke a considerable change in the total monophasic area of the system. Hydrophilic sugar esters were able to solubilize the higher quantities of water in peppermint oil compared to the polyoxyethylene-type surfactants. Alcohol-free microemulsions formulated using peppermint oil and sugar esters, ethoxylated mono-di-glyceride, and polyoxyethylene sorbitan monooleate for food, cosmetics, and pharmaceutical applications were also reported (Fanun 2009c). The effects of four essential oils (rosemary, ylang, lilacin, and peppermint oils), and three plant oils (jojoba oil, corn germ oil, and olive oil) on the permeation of aminophylline were studied using human skin. The permeation effects of these oils were compared with those of three chemical penetration enhancers. Microemulsions containing 10% jojoba oil and 30% corn germ oil were found to be superior vehicles for the percutaneous absorption of aminophylline (Sinha and Pal Kaur 2000, Monti et al. 2002, Wang et al. 2007).

18.3   MICROEMULSIONS BASED ON LINEAR OILS

Isopropylmyristate was widely used in the formulation of biocompatible microemulsions for biological applications (Keipert and Schulz 1994, Von Corswant et al. 1998a, Acharya et al. 2001a,b, Lucangioli et al. 2003, Gupta et al. 2005, Liu et al. 2009). Water solubilization, microstructure, diffusion, viscous and conductive flow parameters, and diclofenac solubilization studies were conducted on systems based on mixed nonionic biocompatible surfactants and isopropylmyristate (Fanun and Salah Al-Diyn 2006, 2007, Fanun 2008a,b,c,d). The formation and characterization of a pharmaceutically useful microemulsion based on isopropylmyristate, polyoxyethylene (4) lauryl ethers (Brij-30, and Brij-92), isopropyl alcohol, and water were reported (Acharya et al. 2001a,b). The effect of adding hydrophilic surfactants (n-alkyl α-D-maltosides (C_nG_2), sucrose monododecanoate, and sodium taurocholate) to microemulsions based on soybean phosphatidylcholine (SbPC) and isopropylmyristate was investigated (Von Corswant et al. 1998a). Microemulsions based on AOT or egg lecithin and isopropylmyristate as potential drug delivery systems were studied by microcalorimetry (Fubini et al. 1989). The nonsteroidal anti-inflammatory naproxen was incorporated into cationic O/W microemulsions based on hexadecyltrimethylammonium bromide, ethanol, isopropylmyristate, or butylstearate as oil (Correa et al. 2005). Microemulsions made from phosphatidylcholine and isopropylmyristate were used in the electrokinetic chromatography to better simulate the biopartitioning of some steroidal drugs (Lucangioli et al. 2003). Pilocarpine was incorporated in microemulsion based on isopropylmyristate, propylene glycol, and a mixture of two sucrose fatty acid ester surfactans with hydrophilic–lipophilic balance (HLB) values of 5 and 16 (Keipert and Schulz, 1994). Microemulsion systems based on isopropylmyristate and the mixed AOT and Tween85 were tested for the topical delivery of Cyclosporin A (Liu et al. 2009). W/O microemulsions based on AOT and isopropylmyristate as a penetration enhancer were used in the transdermal drug delivery of 5-fluorouracil (Gupta et al. 2005). Biocompatible microemulsions based on lecithin and sugar-based surfactants and isopropylmyristate were investigated for their potential as templates to produce NPs for the delivery of proteins and peptides (Alany et al. 2001, Fiedler 2002, Krauel et al. 2005, 2006, Boonme et al. 2006a,b, Graf et al. 2008a). Delivering insulin orally, entrapped in NPs and dispersed in a biocompatible microemulsion based on isopropylmyristate, caprylocaproyl macrogolglycerides, polyglyceryl oleate was studied by Graf et al. (2008b, 2009). Biocompatible microemulsions based on isopropylmyristate as the oil phase were discussed as potential substitutes for chlorinated solvents in dry cleaning applications and as solvent delivery systems for pharmaceutical applications. It was found that the proposed linker-based formulations were able to form alcohol-free microemulsions while achieving higher solubilization capacity than similar systems reported in the literature (Graciaa et al. 1993a,b, Kunieda and Solans 1997, Salager et al. 1998, Kumar and Mittal 1999, Uchiyama et al. 2000, Acosta et al. 2002, 2003a,b, 2005, Sabatini et al. 2003, Tongcumpou et al. 2003a,b, Fanun 2009a). Microemulsions formed in a water/ABA block copolymer [poly(hydroxystearic acid)-poly(ethylene oxide)-poly(hydroxystearic acid)]/1,2-hexanediol/isopropylmyristate system were characterized using SAXS and conductivity (Graf et al. 2008a). It was found that the isopropylmyristate resulted in a smaller microemulsion phase than with nonpolar oils for the nonionic ethoxylated/alkyl polyglycoside/glyceryl monooleate system. It has been observed that microemulsions containing alkyl polyglycoside (APG) and isopropylmyristate are much less temperature sensitive than microemulsions containing just nonionic ethoxylates (Alany et al. 2001).

18.4 MICROEMULSIONS BASED ON MONO-, DI-, AND TRIGLYCERIDES

Triglyceride-based oil microemulsification formulation is challenging due to the creation of unwanted phases such as macroemulsions, liquid crystals, or sponge phases. Microemulsions based on triglycerides exhibit a smaller stability region than microemulsions based on hydrocarbons or fatty acid esters. The formulation of microemulsions through the hydrophilic–lipophilic deviation (HLD) concept based on phase behavior observations, has mostly been applied to triglyceride-based oils/surfactant systems (Van Hecke et al. 2003). The Winsor IV microemulsion systems composed of water, sucrose monostearate, 1-butanol, and caprylic capric triglyceride have been investigated using SAXS, pulsed gradient spin-echo (PGSE), NMR, and viscosity measurements (Fanun et al. 2001). It was reported that monoglycerides are able to form water-in-triglyceride oil microemulsions (Gulik-Krzywicki and Larsson 1984). Linear alkyl polypropoxylated sulfate (LAPS) surfactants and linear alkyl polypropoxylated ethoxylated sulfate (LAPES) surfactants were used for the formulation of environmentally friendly vegetable oil microemulsions (Do et al. 2008). Some authors explained the role of cosurfactant of different lipophilicities vis-a-vis structures for microemulsions using plant oils (Hoffmann et al. 1992, Aboofazeli et al. 1995, Schubert and Kaler 1996, Guo et al. 1999). Other authors explored the effect of temperature and additives on the stability as well as on the extent of one-phase microemulsions region of single and mixed microemulsion systems (Oh et al. 1995, Kunieda and Aoki 1996, Pes et al. 1996, Aramaki et al. 1997a,b, Binks et al. 1997, Aarra et al. 1999, Evans and Wennerstrom 1999, Ryan and Kaler 2000, Garti et al. 2000b). N-Methyl pyrrolidone (NMP) was used as a cosurfactant for the development of biocompatible microemulsions using triglycerides and isopropylmyristate (Debuigne et al. 2001, Kreilgaard 2002, Lee et al. 2005, Bachhav et al. 2006). The biocompatible microemulsions of soybean oil in systems made of anionic surfactant, oleic acid, water, and several glycols were prepared by Comelles et al. (2005). Aqueous and nonaqueous microemulsion systems with a palm oil-based emollient, i.e., medium-chain triglyceride (MCT) stabilized by two oppositely charged ionic surfactants and a medium-chain alcohol, were investigated (Hamdan et al. 1995). Phase behavior, diffusion, conductive and viscous flow parameters of mixed nonionic surfactants and caprylic–capric triglyceride microemulsions were reported (Fanun and Salah Al-Diyn 2006, Fanun 2007, 2008a,b,c,d). The effect of n-butanol on the microstructure of tripalmitin system was studied (Caboi et al. 2005). The effects of triglyceride oils on butanol distribution in W/O cationic microemulsions were also investigated (Yao and Romsted 1997). Extended surfactants containing one or more intermediate-polarity groups between the hydrophilic head and the hydrophobic tail have been proposed for the formulation of triglyceride oils microemulsions with high solubilization capacity and ultralow interfacial tension (Witthayapanyanon et al. 2006). Improved MCT solubilization in food-grade microemulsions in the presence of polyols and ethanol was reported (Garti et al. 2001). Triglyceride microemulsions were formulated using linker molecules, with the addition of co-oil to help enhance sebum solubilization. The effect of several types of co-oil on the phase behavior and the microstructure was studied; the co-oils evaluated here are squalene, squalane, isopropylmyristate, and ethyl laurate. Salt addition shifts the fish diagram toward more hydrophobic oil systems and higher surfactant/linker concentrations (Komesvarakul et al. 2006). The effect of adding isopropylmyristate to the microemulsions systems based on water, 1-propanol, SbPC, and two different triglycerides; a MCT (C_8–C_{10}) and a long-chain triglyceride (soybean oil) was described by Von Corswant and Söderman (1998). A triblock surfactant is reported, which allows for the efficient microemulsification of triglyceride without the use of cosurfactants or dilution with co-oils (Witthayapanyanon et al. 2006). Six polar oils including the alkanoic acids, octanoic and oleic, their corresponding ethyl esters and the medium and long chain triglycerides, Miglyol 812, and soybean oil were used in the formation of phospholipid microemulsions (Huang et al. 2004). The microstructure of microemulsions formulated with both MCT and long-chain triglycerides and SbPC was investigated by Von Corswant et al. (1997). The hydrodynamic size and surfactant aggregation number first decreased then increased in microemulsions formulated by $C_{18:1}E_{10}$ and containing one of the larger molecular volume oils (i.e., either a triglyceride, Miglyol 812, or soybean oil), suggesting that the asymmetric $C_{18:1}E_{10}$ micelles became spherical upon the addition of a small amount of oil and grew thereafter because of further oil being incorporated into the core of the spherical microemulsion droplet (Warisnoicharoen et al. 2000a and 2000b). W/O microemulsions regions formulated of soybean oil, polyoxyethylene (40) sorbitol hexaoleate, and water-ethanol were found to be strongly dependent on temperature and water/alcohol ratios (Joubran et al. 1993). The effects of temperature and cosurfactant on the formation of microemulsions with palm oil derivatives and the nonionic surfactant Imbentin coco $_{6.9}$EO, a commercial fatty alcohol ethoxylate was reported (Raman et al. 2005). The PIT increased with increased molecular weight for triglycerides in the systems formulated with nonionic surfactants. The water solubilization capacity of these systems was dependent on surfactant and oil types in analogy to ordinary hydrocarbon systems (Alander and Wärnheim 1989a,b). The effect of temperature on the phase behavior of the systems water/sucrose laurate/ethoxylated mono-di-glyceride/caprylic–capric triglyceride was reported by Fanun and Salah Al-Diyn (2006). Factors affecting water solubilization in sucrose esters nonionic microemulsion based on the caprylic–capric triglyceride have been investigated (Garti et al. 2000a). Microemulsion formulated using MCT and tocopheryl polyethylene glycol 1000 succinate (TPGS) as a surfactant for the oral delivery of protein drugs was investigated (Ke et al. 2005). The microstructure of a triolein oil microemulsion of, aqua (water-ethanol, 80/20 wt %), and polyoxyethylene (40) sorbitol hexaoleats has been studied with SANS (Trevino et al. 1994, 1998). The types of microstructures formed along the one-phase microemulsion channel water/sucrose laurate/ethoxylated mono-di-glycerides/caprylic–capric triglyceride system were examined using electrical conductivity and self-diffusion NMR by Fanun (2007c). A synergetic effect of sucrose and ethanol on the formation of triglyceride microemulsions was reported (Joubran et al. 1994). The microemulsification of vegetable oils used as biodiesel fuel was tested to reduce their viscosity in order to improve the diesel engine performance (Freedman et al. 1990a,b, Demirbas 2005, Balat 2008). W/O microemulsions consisting of long- or medium-chain triglycerides, a blend of a low and high HLB surfactants and an aqueous phase, have been developed by Constantinides and Scalart (1997) using commercially available and pharmaceutically acceptable components. The absorption and bioavailability of seocalcitol solubilized in self-microemulsifying drug delivery systems (SMEDDS) containing either MCT or long chain triglycerides, with the same ratio between lipid, surfactant and cosurfactant was tested (Grove et al. 2006). The characterization of water-in-triglyceride microemulsion for the oral delivery of earthworm fibrinolytic enzyme clinically was used for the management of cardiovascular diseases (Cheng et al. 2008). Chitosan-based triglyceride microemulsions were used for the study of the distribution of nobiletin in mice brain following i.v. administration (Yao et al. 2008). Microemulsions containing the triglyceride oils exhibited a significant increase in the testosterone propionate solubilization over the corresponding micellar solution in the absence of oil (Malcolmson et al. 1998). A mixture of BA and MCT (3/1) was chosen as the oil phase to develop an aqueous parenteral formulation containing itraconazole (ITZ) using an O/W microemulsion system (Rhee et al. 2007). SMEDDS were formulated using MCT, and a long-chain triglyceride to improve the oral bioavailability of a poorly absorbed, antimalarial drug (Halofantrine, Hf) (Khoo et al. 1998). O/W microemulsions made up of triglycerides, a mixture of lecithin and *n*-butanol, and an aqueous solution were developed as topical drug carrier systems for the percutaneous delivery of anti-inflammatory drugs, i.e., ketoprofen (Paolino et al. 2002). MCT-based O/W microemulsions were formulated and characterized for intravenous administration (Hu et al. 2008). Triglyceride-based microemulsions were also investigated for the intravenous administration of sparingly soluble substances (Von Corswant 1998b). MCT microemulsions were used as transdermal drug delivery vehicles (Kogan and Garti 2006). The percutaneous delivery of anti-inflammatory drugs using biocompatible O/W microemulsions based on triglycerides as oil phase as topical drug carrier systems was reported. These systems showed good human skin tolerability when tested on human volunteers (Osborne et al. 1991, Shinoda et al. 1991, Schurtenberger et al. 1993, Jakobsson and Sivik 1994, Barlow et al. 2000, Paolino et al. 2002). Tricaprylin-based microemulsions were evaluated for the oral delivery of low molecular weight heparin conjugates (Sang et al. 2005). Microemulsions prepared using caprylic–capric triglyceride with varying weight ratios of surfactant to cosurfactant were used to improve the solubility and enhance the bioavailability of poor water-soluble cyclosporin A (Gao et al. 1998). El-Laithy (2003) constructed microemulsion using AOT, and MCT with oleic acid/glycerol monooleate and water were characterized for oral drug delivery. Biodegradable insulin nanocapsules were prepared from biocompatible microemulsions based on caprylic–capric triglycerides and mono/diglycerides (Watnasirichaikul et al. 2000). Quantitative solubility relationships between small molecule/water-uptake in triglyceride/monoglyceride microemulsions was reported (Rane et al. 2008). The enzymatic hydrolysis of palm oil was carried out with lipase from Rhizopus sp. in microemulsions with varying water content (Stark et al. 1990). Microemulsions made of triglycerides were used for the reactions of interesterification and synthesis catalyzed by *Candida cylindracea* lipase (Bello et al. 1987). A tricaprylin microemulsion was also used as microreactor for the hydrolysis of phosphatidylcholine by phospholipase A_2 (Garti et al. 1997). Garti (2003) reported on the use of triglyceride microemulsions as microreactors for Maillard thermal degradation between sugars and amino acids for the preparation of food flavors. Food-grade-based surfactants and various triglycerides were investigated to prepare O/W microemulsions for food applications (Parris et al. 1994). Soybean oil was solubilized in microemulsions using food-grade and nonfood-grade surfactants (Flanagan et al. 2006). Biocompatible microemulsions formulated using mixtures of natural lipids (glycerol trioleate, glycerol monooleate, diglycerol monooleate, and lecithin) and water for use in the food, cosmetics, and pharmaceuticals (Mele et al. 2004). W/O microemulsions based on olive oil, either extravirgin (EVOO) or refined (ROO), were tested for the catalytic activities of two oxidizing enzymes that have been detected in virgin olive oil, namely, tyrosinase and peroxidase, and the activity of a proteolytic enzyme such as trypsin was studied in olive oil microemulsions (Papadimitriou et al. 2005, 2007). Olive oil microemulsions were used as a biomimetic medium for the enzymatic oxidation of oleuropein, the most abundant olive phenolic compound (Papadimitriou et al. 2005, 2007). Two polar lipids, Sn-1/3 and Sn-2 monopalmitin effect on the activity of lipase in microemulsions demonstrate that even if the lipase is expelled from the interface, it can catalyze esterification of the Sn-2 monoglyceride with fatty acids in microemulsions, leading to formation of di- and triglycerides (Reis et al. 2008). The oxidation of fish oil was decreased by its transformation into a microemulsion with ascorbic acid and α±-tocopherol used as antioxidants. Tocopherol was localized in the hydrocarbon chain region whereas ascorbic acid was accumulated in the polar parts of the microemulsion (Jakobsson and Sivik 1994). It was found that systems with polar oils such as ethyl butyrate or with cationic surfactants such as stearyl trimethylammonium chloride are more efficient in removing phenol than systems with normal alkanes or anionic surfactants. It was also shown that microemulsion formed using polar oil performs better than using only the polar oil as the extraction solvent (López-Montilla et al. 2005).

18.5 SUMMARY

Biocompatible microemulsions based on flavor oils, isopropylmyristate, triglycerides were extensively used for the enhancement of the solubilization and delivery of pharmaceutical active ingredients and nutraceuticals. Some of these systems were used as blood substitutes. These systems were also characterized as microreactors for enzymatic and chemical reactions that generate a variety of bioingredients and food flavors. These systems have also other important applications that include the preparation of nanoparticles for drug delivery, in dry cleaning applications, and in bioseparations. The characterized properties of these systems were found to highly influence the intended applications.

SYMBOLS AND TERMINOLOGY

AOT	Aerosol-OT
APG	Alkyl polyglycoside
BA	Benzyl alcohol
Brij	Polyoxyethylene (4) lauryl ether
C_iE_j	Alkylpolyglycol ether
C_nG_2	n-Alkyl α-D-maltosides
DSC	Differential scanning calorimetry
EME	Emulsified microemulsion
EVOO	Extravirgin olive oil
GMO	Glycerol-monooleate
Hf	Halofantrine
HLB	Hydrophilic-lipophilic balance
HLD	Hydrophilic lipophilic deviation
ITZ	Itraconazole
LAPS	Linear alkyl polypropoxylated sulfate
LAPES	Linear alkyl polypropoxylated ethoxylated sulfate
MB	Methylene blue
MCT	Medium-chain triglycerides
NMP	N-Methyl pyrrolidone
NMR	Nuclear magnetic resonance
NPs	Nanoparticles
PIT	Phase inversion temperature
PGSE	Pulsed gradient spin echo
ROO	Refined olive oil
SAXS	Small angle x-ray scattering
SANS	Small angle neutron scattering
SbPC	soybean phosphatidylcholine
SDS	Sodium dodecyl sulfate
SMEDDS	Self-microemulsifying drug delivery systems
SMS	Sucrose mono stearate
TPGS	Tocopheryl polyethylene glycol 1000 succinate

REFERENCES

Aarra, M.G., Hailand, H., and Skauge, A. (1999) Phase behavior and salt partitioning in two- and three-phase anionic surfactant microemulsion systems: Part I, phase behavior as a function of temperature, *Journal of Colloid and Interface Science*, 215:201–215.

Aboofazeli, R., Pate, N., Thomas, M., and Lawrence, M.J. (1995) Investigations into the formation and characterization of phospholipid microemulsions, IV. Pseudo-ternary phase diagrams of systems containing water-lecithin-alcohol and oil; the influence of oil, *International Journal of Pharmaceutics*, 125:107–116.

Acharya, A., Sanyal, S.K., and Moulik, S.P. (2001a) Formation and characterization of a useful biological microemulsion system using mixed oil (ricebran and isopropyl myristate), polyoxyethylene (2)oleyl ether (Brij 92), isopropyl alcohol, and water, *Journal of Dispersion Science and Technology*, 22:551–561.

Acharya, A., Sanyal, S.K., and Moulik, S.P. (2001b) Formation and characterization of a pharmaceutically useful microemulsion derived from isopropylmyristate, polyoxyethylene (4) lauryl ether (Brij-30), isopropyl alcohol and water, *Current Science*, 81:362–370.

Acosta, E., Tran, S., Uchiyama, H., Sabatini, D.A., and Harwell, J.H. (2002) Formulating chlorinated hydro-carbon microemulsions using linker molecules, *Environmental Science and Technology*, 36:4618–4624.

Acosta, E., Do Mai, P., Harwell, J.H., and Sabatini, D.A. (2003a) Linker-modified microemulsions for a variety of oils and surfactants, *Journal of Surfactants and Detergents*, 6:353–363.

Acosta, E.J., Le, M.A., Harwell, J.H., and Sabatini, D.A. (2003b) Coalescence and solubilization kinetics in linker-modified microemulsions and related systems, *Langmuir*, 19:566–574.

Acosta, E.J., Nguyen, T., Witthayapanyanon, A., Harwell, J.H., and Sabatini, D.A. (2005) Linker-based bio-compatible microemulsions, *Environmental Science and Technology*, 39:1275–1282.

Ai, S. and Ishitobi, M. (2006) Effects of the number of fatty acid residues on the phase behaviors of decaglyc-erol fatty acid esters, *Journal of Colloid and Interface Science*, 296:685–689.

Alander, J. and Wärnheim, T. (1989a) Model microemulsions containing vegetable oils. Part 1. Nonionic sur-factant systems, *JAOCS, Journal of the American Oil Chemists Society*, 66:1656–1660.

Alander, J. and Wärnheim, T. (1989b) Model microemulsions containing vegetable oils. 2. Ionic surfactant systems, *Journal of the American Oil Chemists Society*, 66:1661–1665.

Alany, R.G., Tucker, I.G., Davies, N.M., and Rades, T. (2001) Characterizing colloidal structures of pseu-doternary phase diagrams formed by oil/water/amphiphile systems, *Drug Development and Industrial Pharmacy*, 27:31–38.

Amar, I., Aserin, A., and Garti, N. (2004) Microstructure transitions derived from solubilization of lutein and lutein esters in food microemulsions, *Colloids and Surfaces B: Biointerfaces*, 33:143–150.

Angelico, R., Ceglie, A., Colafemmina, G., Lopez, F., Murgia, S., Olsson, U., and Palazzo, G. (2005) Biocompatible lecithin organogels: Structure and phase equilibria, *Langmuir*, 21:140–148.

Aramaki, K., Kunieda, H., Ishitobi, M., and Tagawa, T. (1997a) Effect of added salt on three-phase behavior in a sucrose monoalkanoate system, *Langmuir*, 13:2266–2270.

Aramaki, K., Ozawa, K., and Kunieda, H. (1997b) Effect of temperature on the phase behavior of ionic-non-ionic microemulsions, *Journal of Colloid and Interface Science*, 196:74–78.

Arriagada, F.J. and Osseo-Asare, K. (1999) Synthesis of nanosize silica in a nonionic water-in-oil microemul-sion: Effects of the water/surfactant molar ratio and ammonia concentration, *Journal of Colloid and Interface Science*, 211:210–220.

Avramiotis, S., Cazianis, C.T., and Xenakis, A. (1999) Interfacial properties of lecithin microemulsions in the presence of lipase. A membrane spin-probe study, *Langmuir*, 15:2375–2379.

Bachhav, Y.G., Date, A.A., and Patravale, V.B. (2006) Exploring the potential of *N*-methyl pyrrolidone as a cosurfactant in the microemulsion systems, *International Journal of Pharmaceutics*, 326:186–189.

Bagwe, R.P., Yang, C., Hilliard, L.R., and Tan, W. (2004) Optimization of dye-doped silica nanoparticles pre-pared using a reverse microemulsion method, *Langmuir*, 20:8336–8342.

Balat, M. (2008) Biodiesel fuel production from vegetable oils via supercritical ethanol transesterification, Energy Sources, Part A: Recovery, *Utilization and Environmental Effects*, 30:429–440.

Barlow, D.J., Lawrence, M.J., Zuberi, T., Zuberi, S., and Heenan, R.K. (2000) Small-angle neutron-scattering studies on the nature of the incorporation of polar oils into aggregates of *N,N*-dimethyldodecylamine-*N*-oxide, *Langmuir*, 16:10398–10403.

Barry, B.W. and Williams, A.C. (1989) Human skin penetration enhancement: The synergy of propylene glycol with terpenes, *Proceedings of the International Symposium On Controlled Release of Bioactive Material*, 16:33–34.

Bauduin, P., Touraud, D., Kunz, W., Savelli, M.-P., Pulvin, S., and Ninham, B.W. (2005) The influence of struc-ture and composition of a reverse SDS microemulsion on enzymatic activities and electrical conductivi-ties, *Journal of Colloid and Interface Science*, 292:244–254.

Bello, M., Thomas, D., and Legoy, M.D. (1987) Interesterification and synthesis by *Candida cylindracea* lipase in microemulsions, *Biochemical and Biophysical Research Communications*, 146:361–367.

Biju, S.S., Ahuja, A., and Khar, R.K. (2005) Tea tree oil concentration in follicular casts after topical delivery: Determination by high-performance thin layer chromatography using a perfused bovine udder model, *Journal of Pharmaceutical Sciences*, 94:240–245.

Binks, B.P., Fletcher, P.D.I., and Taylor, D.J.F. (1997) Temperature insensitive microemulsions, *Langmuir*, 13:7030–7038.

Boonme, P., Krauel, K., Graf, A., Rades, T., and Junyaprasert, V.B. (2006a) Characterisation of microstructures formed in isopropyl palmitate/water/AerosolÂ®OT: 1-butanol (2:1) system, *Pharmazie*, 61:927–932.

Boonme, P., Krauel, K., Graf, A., Rades, T., Junyaprasert, V.B. (2006b) Characterization of microemulsion structures in the pseudoternary phase diagram of isopropyl palmitate/water/Brij 97:1-butanol, *AAPS PharmSciTech*, art. no. 45, 7(2):E1–E6.

Bourrel, M. and Schecter, R. (1988) *Microemulsions and Related Systems*, Marcel Dekker Inc.: New York.

Burauer, S., Sottmann, T., and Strey, R. (2000) Nonionic microemulsions with cyclic oils: Oil penetration, efficiency and monomeric solubility, *Tenside, Surfactants, Detergents*, 37:8–16.

Caboi, F., Lazzari, P., Pani, L., and Monduzzi, M. (2005) Effect of 1-butanol on the microstructure of lecithin/water/tripalmitin system, *Chemistry and Physics of Lipids*, 135:147–156.

Calpena, A.C., Lauroba, J., Suriol, M., Obach, R., and Domenech, J. (1994) Effect of D-limonene on the transdermal permeation of nifedipine and domperidone, *International Journal of Pharmaceutics*, 103:179–186.

Carlotti, M.E., Gallarate, M., Morel, S., and Ugazio, E. (1999) Micellar solutions and microemulsions of odorous molecules, *Journal of Cosmetic Science*, 50:281–295.

Carlotti, M.E., Gallarate, M., and Rossatto, V. (2003) O/W microemulsion as a vehicle for sunscreens, *Journal of Cosmetic Science*, 54:451–462.

Cecutti, C., Rico, I., Lattes, A., Novelli, A., Rico, A., Marion, G., Gracia, A., Lachaise, J. (1989) New formulation of blood substitutes: Optimization of novel fluorinated microemulsions, *European Journal of Medicinal Chemistry*, 24:485–492.

Cecutti, C., Novelli, A., Rico, I., and Lattes, A. (1990) New formulation for blood substitutes, *Journal of Dispersion Science and Technology*, 11:115–123.

Chakraborty, I. and Moulik, S.P. (2005) Physicochemical studies on microemulsions: 9. Conductance percolation of AOT-derived W/O microemulsion with aliphatic and aromatic hydrocarbon oils, *Journal of Colloid and Interface Science*, 289:530–541.

Chávez, J.L., Wong, J.L., Jovanovic, A.V., Sinner, E.K., and Duran, R.S. (2005) Encapsulation in sub-micron species: A short review and alternate strategy for dye encapsulation, *IEE Proceedings Nanobiotechnology*, 152:73–84.

Cheng, M.-B., Wang, J.-C., Li, Y.-H., Liu, X.-Y., Zhang, X., Chen, D.-W., Zhou, S.-F., and Zhang, Q. (2008) Characterization of water-in-oil microemulsion for oral delivery of earthworm fibrinolytic enzyme, *Journal of Controlled Release*, 129:41–48.

Comelles, F., Sánchez-Leal, J., and González, J.J. (2005) Soybean oil microemulsions with oleic acid/glycols as cosurfactants, *Journal of Surfactants and Detergents*, 8:257–262.

Constantinides, P.P. (1995) Lipid microemulsions for improving drug dissolution and oral absorption: Physical and biopharmaceutical aspects, *Pharmaceutical Research*, 12:1561–1572.

Constantinides, P.P. and Scalart, J.-P. (1997) Formulation and physical characterization of water-in-oil microemulsions containing long- versus medium-chain glycerides, *International Journal of Pharmaceutics*, 158:57–68.

Correa, M.A., Scarpa, M.V., Franzini, M.C., and Oliveira, A.G. (2005) On the incorporation of the non-steroidal anti-inflammatory naproxen into cationic O/W microemulsions, *Colloids and Surfaces B: Biointerfaces*, 43:108–114.

Cortesi, R. and Nastruzzi, C. (1999) Liposomes, micelles and microemulsions as new delivery systems for cytotoxic alkaloids, *Pharmaceutical Science and Technology Today*, 2:288–298.

Couvreur, P., Barratt, G., Fattal, E., Legrand, P., and Vauthier, C. (2002) Nanocapsule technology: A review, *Critical Reviews in Therapeutic Drug Carrier Systems*, 19:99–134.

Dalmora, M.E., Dalmora, S.L., and Oliveira, A.G. (2001) Inclusion complex of piroxicam with Î²-cyclodextrin and incorporation in cationic microemulsion. *In vitro* drug release and *in vivo* topical anti-inflammatory effect, *International Journal of Pharmaceutics*, 222:45–55.

Danielsson, I. and Lindman, B. (1981) The definition of microemulsion. *Colloids and Surfaces*, 3:391–392.

Das, M.L., Bhattacharya, P.K., and Moulik, S.P. (1991) Model biological microemulsions. 2. Water-(cholesteryl benzoate + heptane)-Triton X-100-butanol microemulsions containing dextran, gelatin, bovine serum albumin, and NaCl, *Langmuir*, 7:636–642.

Dave, H., Gao, F., Schultz, M., and Co, C.C. (2007) Phase behavior and SANS investigations of edible sugar-limonene microemulsions, *Colloids and Surfaces A: Physicochemical and Engineering Aspects*, 296:45–50.

De Campo, L., Yaghmur, A., Garti, N., Leser, M.E., Folmer, B., and Glatter, O. (2004) Five-component food-grade microemulsions: Structural characterization by SANS, *Journal of Colloid and Interface Science*, 274:251–267.

Debuigne, F., Cuisenaire, J., Jeunieau, L., Masereel, B., and Nagy, J.B. (2001) Synthesis of nimesulide nanoparticles in the microemulsion epikuron/isopropyl myristate/water/*n*-butanol (or isopropanol), *Journal of Colloid and Interface Science*, 243:90–101.

Demirbas, A. (2005) Biodiesel production from vegetable oils via catalytic and non-catalytic supercritical methanol transesterification methods, *Progress in Energy and Combustion Science*, 31:466–487.

des Rieux, A., Fievez, V., Garinot, M., Schneider, Y.-J., and Préat, V. (2006) Nanoparticles as potential oral delivery systems of proteins and vaccines: A mechanistic approach, *Journal of Controlled Release*, 116:1–27.

Djordjevic, L., Primorac, M., Stupar, M., and Krajisnik, D. (2004) Characterization of caprylocaproyl macro-golglycerides based microemulsion drug delivery vehicles for an amphiphilic drug, *International Journal of Pharmaceutics*, 271:11–19.

Do, L.D., Withayyapayanon, A., Harwell, J.H., and Sabatini, D.A. (2008) Environmentally friendly vegetable oil microemulsions using extended surfactants and linkers, *Journal of Surfactants and Detergents*, 12:91–99.

Eaimtrakarn, S., Rama Prasad, Y.V., Ohno, T., Konishi, T., Yoshikawa, Y., Shibata, N., and Takada, K. (2002) Absorption enhancing effect of Labrasol on the intestinal absorption of insulin in rats, *Journal of Drug Targeting*, 10:255–260.

El-Laithy, H.M. (2003) Preparation and physicochemical characterization of dioctyl sodium sulfusuccinate (aerosol OT) microemulsion for oral drug delivery, *AAPS PharmSciTech*, 4:2–12.

Escribano, E., Calpena, A.C., Queralt, J., Obach, R., and Doménech, J. (2003) Assessment of diclofenac permeation with different formulations: Anti-inflammatory study of a selected formula, *European Journal of Pharmaceutical Sciences*, 19:203–210.

Evans, D.F. and Wennerstrom, H. (1999) *The Colloidal Domain Where Physics, Chemistry, Mathematics and Technology Meet*, Wiley: New York.

Fadnavis, N.W. and Deshpande, A. (2002) Synthetic applications of enzymes entrapped in reverse micelles & organo-gels, *Current Organic Chemistry*, 6:393–410.

Fanun, M. (2007a) Propylene glycol and ethoxylated surfactant effects on the phase behavior of water/sucrose stearate/oil system, *Journal of Dispersion Science and Technology*, 28:1244–1253.

Fanun, M. (2007b) Conductivity, viscosity, NMR and diclofenac solubilization capacity studies of mixed nonionic surfactants microemulsions, *Journal of Molecular Liquids*, 135:5–13.

Fanun, M. (2007c) Structure probing of water/mixed nonionic surfactants/caprylic-capric triglyceride system using conductivity and NMR, *Journal of Molecular Liquids*, 133:22–27.

Fanun, M. (2008a) Water solubilization in mixed nonionic surfactants microemulsions, *Journal of Dispersion Science and Technology*, 29:1043–1052.

Fanun, M. (2008b) Viscous flow parameters of mixed nonionic surfactants microemulsions, *Journal of Dispersion Science and Technology*, 29:1257–1265.

Fanun, M. (2008c) Conductive flow parameters of mixed nonionic surfactants microemulsions, *Journal of Dispersion Science and Technology*, 29:1426–1434.

Fanun, M. (2008d) A study of the properties of mixed nonionic surfactants microemulsions by NMR, SAXS, viscosity and conductivity, *Journal of Molecular Liquids*, 142:103–110.

Fanun, M. (2009a) *Microemulsions: Properties and Applications*, Surfactant Science Series, Vol. 144, Taylor & Francis/CRC press: Boca Raton, FL.

Fanun, M. (2009b) Microstructure of mixed nonionic surfactants microemulsions studied by SAXS and DLS, *Journal of Dispersion Science and Technology*, 30:115–123.

Fanun, M. (2009c) Microemulsions formation on water/nonionic surfactant/peppermint oil mixtures, *Journal of Dispersion Science and Technology*, 30:399–405.

Fanun, M. (2009d) Oil type effect on the solubilization of diclofenac in mixed nonionic surfactants microemulsions, *Colloids and Surfaces A*, 343:75–82.

Fanun, M. (2009e) Properties of microemulsions based on sugar surfactants and peppermint oil, *Colloid and Polymer Science*, 287:899–910.

Fanun, M. (2010) *Colloids in Drug Delivery*, Surfactant Science Series Vol., Taylor & Francis/CRC Press: Boca Raton, FL.

Fanun, M. and Al-Diyn, W.S. (2006a) Electrical conductivity and self diffusion-NMR studies of the system: Water/sucrose laurate/ethoxylated mono-di-glyceride/isopropylmyristate, *Colloids and Surfaces A: Physicochemical and Engineering Aspects*, 277:83–89.

Fanun, M. and Al-Diyn, W.S. (2006b) Temperature effect on the phase behavior of the systems water/sucrose laurate/ethoxylated-mono-di-glyceride/oil, *Journal of Dispersion Science and Technology*, 27:1119–1127.

Fanun, M. and Al-Diyn, W.S. (2007) Structural transitions in the system water/mixed nonionic surfactants/R(+)-limonene studied by electrical conductivity and self-diffusion-NMR, *Journal of Dispersion Science and Technology*, 28:165–174.

Fanun, M., Wachtel, E., Antalek, B., Aserin, A., and Garti, N. (2001) A study of the microstructure of four-component sucrose ester microemulsions by SAXS and NMR, *Colloids and Surfaces A: Physicochemical and Engineering Aspects*, 180:173–186.

Fiedler, H.P. (2002) *Encyclopedia of Excipients for Pharmaceuticals, Cosmetics and Related Areas*, 5th edn., Editio Cantor Verlag: Aulendorf, Germany.

Flanagan, J. and Singh, H. (2006) Microemulsions: A potential delivery system for bioactives in food, *Critical Reviews in Food Science and Nutrition*, 46:221–237.

Flanagan, J., Kortegaard, K., Neil Pinder, D., Rades, T., and Singh, H. (2006) Solubilisation of soybean oil in microemulsions using various surfactants, *Food Hydrocolloids*, 20:253–260.

Formariz, T.P., Sarmento, V.H.V., Silva-Junior, A.A., Scarpa, M.V., Santilli, C.V., and Oliveira, A.G. (2006) Doxorubicin biocompatible O/W microemulsion stabilized by mixed surfactant containing soya phosphatidylcholine, *Colloids and Surfaces B: Biointerfaces*, 51:54–61.

Formariz, T.P., Chiavacci, L.A., Sarmento, V.H.V., Santilli, C.V., Tabosa do Egito, E.S., and Oliveira, A.G. (2007) Relationship between structural features and *in vitro* release of doxorubicin from biocompatible anionic microemulsion, *Colloids and Surfaces B: Biointerfaces*, 60:28–35.

Formariz, T.P., Chiavacci, L.A., Sarmento, V.H.V., Franzini, C.M., Silva, A.A. Jr., Scarpa, M.V., Santilli, C.V., Egito, E.S.T., and Oliveira, A.G. (2008) Structural changes of biocompatible neutral microemulsions stabilized by mixed surfactant containing soya phosphatidylcholine and their relationship with doxorubicin release, *Colloids and Surfaces B: Biointerfaces*, 63:287–295.

Freedman, B., Bagby, M.O., Callahan, T.J., and Ryan T.W. III. (1990a) Cetane numbers of fatty esters, fatty alcohols and triglycerides determined in a constant volume combustion bomb, in *SAE Technical Paper Series*, p. 900343. Society of Automotive Engineers: Warrendale, PA.

Freedman, B., Bagby, M.O., Callahan, T.J., and Ryan, T.W. (1990b) Cetane numbers of fatty esters, fatty alcohols and triglycerides determined in a constant volume combustion bomb, *SAE (Society of Automotive Engineers) Transactions*, 99:153–161.

Fubini, B., Gasco, M.R., and Gallarate, M. (1989) Microcalorimetric study of microemulsions as potential drug delivery systems. II. Evaluation of enthalpy in the presence of drugs, *International Journal of Pharmaceutics*, 50:213–217.

Gao, Z.-G., Choi, H.-G., Shin, H.-J., Park, K.-M., Lim, S.-J., Hwang, K.-J., and Kim, C.-K. (1998) Physicochemical characterization and evaluation of a microemulsion system for oral delivery of cyclosporin A, *International Journal of Pharmaceutics*, 161:75–86.

Garti, N. (2003) Microemulsions as microreactors for food applications, *Current Opinion in Colloid and Interface Science*, 8:197–211.

Garti, N., Lichtenberg, D., and Silberstein, T. (1997) The hydrolysis of phosphatidylcholine by phospholipase A2 in microemulsion as microreactor, *Colloids and Surfaces A: Physicochemical and Engineering Aspects*, 128:17–25.

Garti, N., Aserin, A., and Fanun, M. (2000a) Non-ionic sucrose esters microemulsions for food applications. Part 1. Water solubilization, *Colloids and Surfaces A: Physicochemical and Engineering Aspects*, 164:27–38.

Garti, N., Clement, V., Fanun, M., and Leser, M.E. (2000b) Some characteristics of sugar ester nonionic microemulsions in view of possible food applications, *Journal of Agricultural and Food Chemistry*, 48:3945–3956.

Garti, N., Yaghmur, A., Leser, M.E., Clement, V., and Watzke, H.J. (2001) Improved oil solubilization in oil/water food grade microemulsions in the presence of polyols and ethanol, *Journal of Agricultural and Food Chemistry*, 49:2552–2562.

Garti, N., Amar-Yuli, I., Spernath, A., and Hoffman, R.E. (2004a) Transitions and loci of solubilization of nutraceuticals in U-type nonionic microemulsions studied by self-diffusion NMR, *Physical Chemistry Chemical Physics*, 6:2968–2976.

Garti, N., Zakharia, I., Spernath, A., Yaghmur, A., Aserin, A., Hoffman, R.E., Jacobs, L. (2004b) Solubilization of water-insoluble nutraceuticals in nonionic microemulsions for water-based use, *Progress in Colloid and Polymer Science*, 126:184–189.

Garti, N., Spernath, A., Aserin, A., and Lutz, R. (2005) Nano-sized self-assemblies of nonionic surfactants as solubilization reservoirs and microreactors for food systems, *Soft Matter*, 1:206–218.

Garti, N., Avrahami, M., and Aserin, A. (2006) Improved solubilization of Celecoxib in U-type nonionic microemulsions and their structural transitions with progressive aqueous dilution, *Journal of Colloid and Interface Science*, 299:352–365.

Gasperlin, M. and Kristl, J. (2000) Innovation concepts in skin care product design, *Farmacevtski Vestnik*, 51:269–276.

Gochman-Hecht, H. and Bianco-Peled, H. (2006) Structure modifications of AOT reverse micelles due to protein incorporation, *Journal of Colloid and Interface Science*, 297:276–283.

Graciaa, A., Lachaise, J., Cucuphat, C., Bourrel, M., and Salager, J.L. (1993a) Improving solubilization in microemulsions with additives. 1. The lipophilic linker role, *Langmuir*, 9:669–672.

Graciaa, A., Lachaise, J., Cucuphat, C., Bourrel, M., and Salager, J.L. (1993b) Improving solubilization in microemulsions with additives. 2. Long chain alcohols as lipophilic linkers, *Langmuir*, 9:3371–3374.

Graf, A., Ablinger, E., Peters, S., Zimmer, A., Hook, S., and Rades, T. (2008a) Microemulsions containing lecithin and sugar-based surfactants: Nanoparticle templates for delivery of proteins and peptides, *International Journal of Pharmaceutics*, 350:351–360.

Graf, A., Jack, K.S., Whittaker, A.K., Hook, S.M., Rades, T. (2008b) Protein delivery using nanoparticles based on microemulsions with different structure-types, *European Journal of Pharmaceutical Sciences*, 33:434–444.

Graf, A., Rades, T., and Hook, S.M. (2009) Oral insulin delivery using nanoparticles based on microemulsions with different structure-types: Optimisation and *in vivo* evaluation, *European Journal of Pharmaceutical Sciences*, 37:53–61.

Grove, M., Müllertz, A., Nielsen, J.L., and Pedersen, G.P. (2006) Bioavailability of seocalcitol. II: Development and characterisation of self-microemulsifying drug delivery systems (SMEDDS) for oral administration containing medium and long chain triglycerides, *European Journal of Pharmaceutical Sciences*, 28:233–242.

Gulik-Krzywicki, T. and Larsson, K. (1984) An electron microscopy study of the L2-phase (microemulsion) in a ternary system: Triglyceride/monoglyceride/water, *Chemistry and Physics of Lipids*, 35:127–132.

Gulsen, D. and Chauhan, A. (2005) Dispersion of microemulsion drops in HEMA hydrogel: A potential ophthalmic drug delivery vehicle, *International Journal of Pharmaceutics*, 292:95–117.

Guo, R., Tianqing, L., and Weili, Y. (1999) Phase behavior and structure of the sodium dodecyl sulfate/benzyl alcohol/water system, *Langmuir*, 15:624–630.

Gupta, S. and Moulik, S.P. (2008) Biocompatible microemulsions and their prospective uses in drug delivery, *Journal of Pharmaceutical Sciences*, 97:22–45.

Gupta, R.R., Jain, S.K., and Varshney, M. (2005) AOT water-in-oil microemulsions as a penetration enhancer in transdermal drug delivery of 5-fluorouracil, *Colloids and Surfaces B: Biointerfaces*, 41:25–32.

Hamdan, S., Lizana, R., and Laili, C.R. (1995) Aqueous and nonaqueous microemulsion systems with a palm oil-base emollient, *Journal of the American Oil Chemists' Society*, 72:151–155.

Hamdan, S., Ahmad, F.B.H., Dai, Y.Y., Dzulkefly, K., and Ku Bulat, K.H. (1999) Effect of humidity on evaporation from aqueous and nonaqueous microemulsion with perfume, *Journal of Dispersion Science and Technology*, 20:415–423.

Hellweg, T. (2002) Phase structures of microemulsions, *Current Opinion in Colloid and Interface Science*, 7:50–56.

Herrington, T.M. and Sahi, S.S. (1988) Phase behavior of some sucrose surfactants with water and *n*-decane, *Journal of the American Oil Chemists' Society*, 65:1677–1681.

Hoar, T.P. and Schulman, J.H. (1943) Transparent water-in-oil dispersions: The oleopathic hydro-micelle, *Nature*, 152(3847):102–103.

Hoffmann, H., Thunig, C., and Valiente, M. (1992) The different phases and their macroscopic properties in ternary surfactant systems of alkyldimethylamine oxides, intermediate chain *n*-alcohols and water, *Colloids and Surfaces*, 67:223–237.

Holmberg, K. (2007) Organic reactions in microemulsions, *European Journal of Organic Chemistry*, 5:731–742.

Hu, H.Y., Huang, Y., Liu, J., Xu, X.L., Gong, T., Xiang, D., and Zhang, Z.R. (2008) Medium-chain triglycerides based oil-in-water microemulsions for intravenous administration: Formulation, characterization and *in vitro* hemolytic activities, *Journal of Drug Delivery Science and Technology*, 18:101–107.

Huang, L., Lips, A., and Co, C.C. (2004) Microemulsification of triglyceride sebum and the role of interfacial structure on bicontinuous phase behavior, *Langmuir*, 20:3559–3563.

Jakobsson, M. and Sivik, B. (1994) Oxidative stability of fish oil included in a microemulsion, *Journal of Dispersion Science and Technology*, 15:611–619.

Joubran, R.F., Cornell, D.G., and Parris, N. (1993) Microemulsions of triglyceride and non-ionic surfactant: Effect of temperature and aqueous phase composition, *Colloids and Surfaces A: Physicochemical and Engineering Aspects*, 80:153–160.

Joubran, R., Parris, N., Lu, D., and Trevino, S. (1994) Synergetic effect of sucrose and ethanol on formation of triglyceride microemulsions, *Journal of Dispersion Science and Technology*, 15:687–704.

Kahlweit, M., Strey, R., and Busse, G. (1990) Microemulsions: A qualitative thermodynamic approach, *Journal of Physical Chemistry*, 94:3881–3894.

Kahlweit, M., Busse, G., and Faulhaber, B. (1996) Preparing nontoxic microemulsions with alkyl monoglucosides and the role of alkanediols as cosolvents, *Langmuir*, 12:861–862.

Ke, W.T., Lin, S.Y., Ho, H.O., and Sheu, M.T. (2005). Physical characterizations of microemulsion systems using tocopheryl polyethylene glycol 1000 succinate (TPGS) as a surfactant for the oral delivery of protein drugs, *Journal of Controlled Release*, 102:489–507.

Keipert, S. and Schulz, G. (1994) Microemulsions with sucrose fatty ester surfactants. Part 1, *Pharmazie*, 49:195–197.

Khoo, S.-M., Humberstone, A.J., Porter, C.J.H., Edwards, G.A., and Charman, W.N. (1998) Formulation design and bioavailability assessment of lipidic self-emulsifying formulations of halofantrine, *International Journal of Pharmaceutics*, 167:155–164.

Kim, J.Y. and Dungan, S.R. (2008) Effect of α-lactalbumin on aerosol-OT phase structures in oil/water mixtures, *Journal of Physical Chemistry B*, 112(17):5381–5392.

Klier, J., Tucker, C.J., Kalantar, T.H., and Green, D.P. (2000) Properties and applications of microemulsions, *Advanced Materials*, 12:1751–1757.

Kogan, A. and Garti, N. (2006) Microemulsions as transdermal drug delivery vehicles, *Advances in Colloid and Interface Science*, 123–126:369–385.

Komesvarakul, N., Sanders, M.D., Szekeres, E., Acosta, E.J., Faller, J.F., Mentlik, T., Fisher, L.B., (…), and Scamehorn, J.F. (2006) Microemulsions of triglyceride-based oils: The effect of co-oil and salinity on phase diagrams, *Journal of Cosmetic Science*, 57:309–325.

Krauel, K., Davies, N.M., Hook, S., and Rades, T. (2005) Using different structure types of microemulsions for the preparation of poly(alkylcyanoacrylate) nanoparticles by interfacial polymerization, *Journal of Controlled Release*, 106:76–87.

Krauel, K., Graf, A., Hook, S.M., Davies, N.M., and Rades, T. (2006) Preparation of poly (alkylcyanoacrylate) nanoparticles by polymerization of water-free microemulsions, *Journal of Microencapsulation*, 23:499–512.

Kreilgaard, M. (2002) Influence of microemulsions on cutaneous drug delivery, *Advanced Drug Delivery Reviews*, 54(Suppl.):S77–S98.

Kreuter J. (Ed.) (1994) *Colloidal Drug Delivery Systems*, Marcel Dekker: New York.

Kumar, P. and Mittal, K.L. (1999) *Handbook of Microemulsion Science and Technology*, Marcel Dekker: New York.

Kunieda, H. and Aoki, R. (1996) Effect of added salt on the maximum solubilization in an ionic-surfactant microemulsion, *Langmuir*, 12:5796–5799.

Kunieda, H. and Shinoda, K. (1985) Evaluation of the hydrophile-lipophile balance (HLB) of nonionic surfactants. I. Multisurfactant systems, *Journal of Colloid And Interface Science*, 107:107–121.

Kunieda, H. and Solans, C. (1997) How to prepare microemulsions: Temperature-insensitive microemulsions, *Industrial Applications of Microemulsions*, edited by C. Solans and H. Kunieda, pp. 21–45, Marcel Dekker: New York.

Kunieda, H., Horii, M., Koyama, M., and Sakamoto, K. (2001) Solubilization of polar oils in surfactant self-organized structures, *Journal of Colloid and Interface Science*, 236:78–84.

Kuznetsova, G.M., Kartasheva, Z.S., and Kasaikina, O.T. (1996) Kinetics of limonene autooxidation, *Russian Chemical Bulletin*, 45:1592–1595.

La Mesa, C. (2005) Polymer-surfactant and protein-surfactant interactions, *Journal of Colloid and Interface Science*, 286:148–157.

Le Bourlais, C., Acar, L., Zia, H., Sado, P.A., Needham, T., and Leverge, R. (1998) Ophthalmic drug delivery systems: Recent advances, *Progress in Retinal and Eye Research*, 17:33–58.

Lee, P.J., Langer, R., and Shastri, V.P. (2005) Role of *n*-methyl pyrrolidone in the enhancement of aqueous phase transdermal transport, *Journal of Pharmaceutical Sciences*, 94:912–917.

Leser, M.E., Sagalowicz, L., Michel, M., and Watzke, H.J. (2006) Self-assembly of polar food lipids, *Advances in Colloid and Interface Science* (SPEC. ISS.), 123–126:125–136.

Leung, R. and Shah, D.O. (1987) Solubilization and phase equilibria of water-in-oil microemulsions. I. Effects of spontaneous curvature and elasticity of interfacial films, *Journal of Colloid and Interface Science*, 120:320–329.

Liu, H., Wang, Y., Lang, Y., Yao, H., Dong, Y., and Li, S. (2009) Bicontinuous cyclosporin A loaded water-AOT/tween 85-isopropylmyristate microemulsion: Structural characterization and dermal pharmacokinetics *in vivo*, *Journal of Pharmaceutical Sciences*, 98:1167–1176.

López-Montilla, J.C., Pandey, S., Shah, D.O., and Crisalle, O.D. (2005) Removal of non-ionic organic pollutants from water via liquid-liquid extraction, *Water Research*, 39:1907–1913.

Lucangioli, S.E., Carducci, C.N., Scioscia, S.L., Carlucci, A., Bregni, C., and Kenndler, E. (2003) Comparison of the retention characteristics of different pseudostationary phases for microemulsion and micellar electrokinetic chromatography of betamethasone and derivatives, *Electrophoresis*, 24:984–991.

Luisi, P.L., Scartazzini, R., Haering, G., and Schurtenberger, P. (1990). Organogels from water-in-oil microemulsions, *Colloid & Polymer Science*, 268:356–374.

Lutz, R., Aserin, A., Wachtel, E.J., Ben-Shoshan, E., Danino, D., and Garti, N. (2007) A study of the emulsified microemulsion by SAXS, Cryo-TEM, SD-NMR, and electrical conductivity, *Journal of Dispersion Science and Technology*, 28:1149–1157.

Magdassi, S. and Touitov E. (1998) *Novel Cosmetic Delivery Systems*, Marcel Dekker Inc.: New York.

Majhi, P.R. and Moulik, S.P. (1999) Physicochemical studies on biological macro-and microemulsions VI: Mixing behaviors of Eucalyptus oil, water and polyoxyethylene sorbitan monolaurate (Tween 20) assisted by n-Butanol or cinnamic alcohol, *Journal of Dispersion Science and Technology*, 20:1407–1427.

Malcolmson, C., Satra, C., Kantaria, S., Sidhu, A., and Lawrence, M.J. (1998) Effect of oil on the level of solubilization of testosterone propionate into nonionic oil-in-water microemulsions, *Journal of Pharmaceutical Sciences*, 87:109–116.

Martel, R., Gélinas, P.J., and Desnoyers, J.E. (1998a) Aquifer washing by micellar solutions: 1 optimization of alcohol-surfactant-solvent solutions, *Journal of Contaminant Hydrology*, 29:319–346.

Martel, R., Lefebvre, R., and Gélinas, P.J. (1998b) Aquifer washing by micellar solutions: 2. DNAPL recovery mechanisms for an optimized alcohol-surfactant-solvent solution, *Journal of Contaminant Hydrology*, 30:1–31.

Matura, M., Goossens, A., Bordalo, O., Garcia-Bravo, B., Magnusson, K., Wrangsja, K., and Karlberg, A.-T. (2002) Oxidized citrus oil (R-limonene): A frequent skin sensitizer in Europe, *Journal of the American Academy of Dermatology*, 47:709–714.

Mele, S., Murgia, S., Caboi, F., and Monduzzi, M. (2004) Biocompatible lipidic formulations: Phase behavior and microstructure, *Langmuir*, 20:5241–5246.

Mitra, R.K. and Paul, B.K. (2004) Physicochemical studies of some single and mixed microemulsion systems. I: Phase behaviour of surfactant/cosurfactant/oil/water systems and the effect of temperature, additives and oil, *Journal of Surface Science and Technology*, 20:105–135.

Mitra, R.K. and Paul, B.K. (2005) Physicochemical investigations of microemulsification of eucalyptus oil and water using mixed surfactants (AOT + Brij-35) and butanol, *Journal of Colloid and Interface Science*, 283:565–577.

Mitra, N., Mukhopadhyay, L., Bhattacharya, P.K., and Moulik, S.P. (1994) Biological microemulsions: Part IV—Phase behaviour and dynamics of microemulsions prepared with vegetable oils mixed with aerosol-OT, cinnamic alcohol and water, *Indian Journal of Biochemistry and Biophysics*, 31:115–120.

Mitra, N., Mukherjee, L., Bhattacharya, P.K., and Moulik, S.P. (1996) Biological microemulsions V: Mutual mixing of oils, amphiphiles and water in ternary and quaternary combinations, *Indian Journal of Biochemistry and Biophysics*, 33:206–212.

Monduzzi, M., Caboi, F., and Moriconi, C. (1997) On the bicontinuous microstructure induced by a guest protein in a typical AOT microemulsion, *Colloids and Surfaces A: Physicochemical and Engineering Aspects*, 129–130:327–338.

Monti, D., Chetoni, P., Burgalassi, S., Najarro, M., Saettone, M.F., and Boldrini, E. (2002) Effect of different terpene-containing essential oils on permeation of estradiol through hairless mouse skin, *International Journal of Pharmaceutics*, 237:209–214.

Moulik, S.P. and Paul, B.K. (1998) Structure, dynamics and transport properties of micro emulsions, *Advances in Colloid and Interface Science*, 78:99–195.

Nakamura, N., Tagawa, T., Kihara, K., Tobita, I., and Kunieda, H. (1997) Phase transition between microemulsion and lamellar liquid crystal, *Langmuir*, 13:2001–2006.

Nakamura, N., Yamaguchi, Y., Hakansson, B., Olsson, U., Tagawa, T., and Kunieda, H. (1999) Formation of microemulsion and liquid crystal in biocompatible sucrose alkanoate systems, *Journal of Dispersion Science and Technology*, 20:535–557.

Nguyen, T.T. and Sabatini, D.A. (2009) Formulating alcohol-free microemulsions using rhamnolipid biosurfactant and rhamnolipid mixtures, *Journal of Surfactants and Detergents*, 12:109–115. DOI: 10.1007/s11743–008–1098-y.

Nishimi, T. (2008) The formation of middle-phase microemulsions of polar oils, *Macromolecular Symposia*, 270:48–57.

Obata, Y., Takayama, K., Okabe, H., and Nagai, T. (1990) Effect of cyclic monoterpenes on percutaneous absorption in the case of a water-soluble drug (diclofenac sodium), *Drug Design and Delivery*, 6:319–328.

Oh, K.-H., Baran J.R. Jr., Wade, W.H., Weerasooriya, V. (1995) Temperature insensitive microemulsion phase behavior with non-ionic surfactants, *Journal of Dispersion Science and Technology*, 16:165–188.

Okabe, H., Takayama, K., Ogura, A., and Nagai, T. (1989) Effects of limonene and related compounds on the percutaneous absorption of indomethacin, *Drug Design and Delivery*, 4:313–321.

Orlich, B. and Schomacker, R. (2002) Enzyme catalysis in reverse micelles, *Advances in Biochemical Engineering/Biotechnology*, 75:185–208.

Osborne, D.W., Ward, A.J.I., and O'Neill, K.J. (1991) Microemulsions as topical drug delivery vehicles: *In-vitro* transdermal studies of a model hydrophilic drug, *Journal of Pharmacy and Pharmacology*, 43:451–454.

Paolino, D., Ventura, C.A., Nisticò, S., Puglisi, G., and Fresta, M. (2002) Lecithin microemulsions for the topical administration of ketoprofen: Percutaneous adsorption through human skin and *in vivo* human skin tolerability, *International Journal of Pharmaceutics*, 244(1–2):21–31.

Papadimitriou, V., Sotiroudis, T.G., and Xenakis, A. (2005) Olive oil microemulsions as a biomimetic medium for enzymatic studies: Oxidation of oleuropein, *JAOCS, Journal of the American Oil Chemists' Society*, 82:335–340.

Papadimitriou, V., Sotiroudis, T.G., and Xenakis, A. (2007) Olive oil microemulsions: Enzymatic activities and structural characteristics, *Langmuir*, 23:2071–2077.

Papadimitriou, V., Pispas, S., Syriou, S., Pournara, A., Zoumpanioti, M., Sotiroudis, T.G., and Xenakis, A. (2008) Biocompatible microemulsions based on limonene: Formulation, structure, and applications, *Langmuir*, 24(7):3380–3386.

Parris, N., Joubran, R.F., and Lu, D.P. (1994) Triglyceride microemulsions: Effect of nonionic surfactants and the nature of the oil, *Journal of Agricultural and Food Chemistry*, 42:1295–1299.

Paul, B.K. and Moulik, S.P. (1991) Biological microemulsions: Part III—The formation characteristics and transport properties of saffola-aerosol OT-hexylamine-water system, *Indian Journal of Biochemistry and Biophysics*, 28:174–183.

Paul, B.K. and Moulik, S.P. (1997) Microemulsions: An overview, *Journal of Dispersion Science and Technology*, 18:301–367.

Paul, B.K. and Moulik, S.P. (2001) Uses and applications of microemulsions, *Current Science*, 80:990–1001.

Pauletti, G.M., Gangwar, S., Knipp, G.T., Nerurkar, M.M., Okumu, F.W., Tamura, K., Siahaan, T.J., and Borchardt, R.T. (1996) Structural requirements for intestinal absorption of peptide drugs, *Journal of Controlled Release*, 41:3–17.

Pes, Ma.A., Aramaki, K., Nakamura, N., and Kunieda, H. (1996) Temperature-insensitive microemulsions in a sucrose monoalkanoate system, *Journal of Colloid and Interface Science*, 178:666–672.

Pinto Reis, C., Neufeld, R.J., Ribeiro, A.J., and Veiga, F. (2006) Nanoencapsulation II. Biomedical applications and current status of peptide and protein nanoparticulate delivery systems, *Nanomedicine: Nanotechnology, Biology, and Medicine*, 2(2):53–65.

Pitaksuteepong, T., Davies, N.M., Tucker, I.G., and Rades, T. (2002) Factors influencing the entrapment of hydrophilic compounds in nanocapsules prepared by interfacial polymerisation of water-in-oil microemulsions, *European Journal of Pharmaceutics and Biopharmaceutics*, 53:335–342.

Prince, L.M. (1977) *Micellization, Solubilization and Microemulsions*, Academic Press: London, U.K.

Radwan, M.A. and Aboul-Enein, H.Y. (2002) The effect of oral absorption enhancers on the *in vivo* performance of insulin-loaded poly(ethylcyanoacrylate) nanospheres in diabetic rats, *Journal of Microencapsulation*, 19:225–235.

Raman, I. Ab., Suhaimi, H., and Tiddy, G.J.T. (2005) Microemulsions with palm oil derivatives stabilized by a nonionic surfactant: Effects of temperature and cosurfactant, *Journal of Dispersion Science and Technology*, 26:355–364.

Rane, S.S., Cao, Y., and Anderson, B.D. (2008) Quantitative solubility relationships and the effect of water uptake in triglyceride/monoglyceride microemulsions, *Pharmaceutical Research*, 25:1158–1174.

Reis, C., Miller, R., Leser, M., and Watzke, H. (2008) Lipase-catalyzed reactions at interfaces of two-phase systems and microemulsions, *Applied Biochemistry and Biotechnology*, 1–16.

Rhee, Y.-S., Choi, J.-G., Park, E.-S., and Chi, S.-C. (2001) Transdermal delivery of ketoprofen using microemulsions, *International Journal of Pharmaceutics*, 228:161–170.

Rhee, Y.-S., Park, C.-W., Nam, T.-Y., Shin, Y.-S., Chi, S.-C., and Park, E.-S. (2007) Formulation of parenteral microemulsion containing itraconazole, *Archives of Pharmacol Research*, 30:114–123.

Rohloff, C.M., Shimek, J.W., and Dungan, S.R. (2003) Effect of added α±-lactalbumin protein on the phase behavior of AOT-brine-isooctane systems, *Journal of Colloid and Interface Science*, 261:514–523.

Rozner, S., Aserin, A., and Garti, N. (2008) Competitive solubilization of cholesterol and phytosterols in nonionic microemulsions studied by pulse gradient spin-echo NMR, *Journal of Colloid and Interface Science*, 321:418–425.

Ryan, L.D. and Kaler, E.W. (2000) Alkyl polyglucoside microemulsion phase behavior, *Colloids and Surfaces A: Physicochemical and Engineering Aspects*, 176:69–83.

Sabatini, D.A., Acosta, E., and Harwell, J.H. (2003) Linker molecules in surfactant mixtures, *Current Opinion in Colloid and Interface Science*, 8:316–326.

Salager, J.L. and Anton, R.E. (1999) *Handbook of Microemulsion Science and Technology*, P. Kumar and K. L. Mittal (Eds.), Chap. 8, Marcel Dekker Inc.: New York.

Salager, J.-L., Graciaa, A., and Lachaise, J. (1998) Improving solubilization in microemulsions with additives. Part III: Lipophilic linker optimization, *Journal of Surfactants and Detergents*, 1:403–406.

Salager, J.-L., Antón, R.E., Sabatini, D.A., Harwell, J.H., Acosta, E.J., and Tolosa, L.I. (2005) Enhancing solubilization in microemulsions: State of the art and current trends, *Journal of Surfactants and Detergents*, 8:3–21.

Sang, K.K., Eun, H.L., Vaishali, B., Lee, S., Lee, Y.-K., Kim, C.-Y., Hyun, T.M., and Byun, Y. (2005) Tricaprylin microemulsion for oral delivery of low molecular weight heparin conjugates, *Journal of Controlled Release*, 105:32–42.

Sarciaux, J.M., Acar, L., and Sado, P.A. (1995) Using microemulsion formulations for oral drug delivery of therapeutic peptides, *International Journal of Pharmaceutics*, 120:127–136.

Schubert, K.-V. and Kaler, E.W. (1996) Nonionic microemulsions, *Berichte der Bunsengesellschaft/Physical Chemistry Chemical Physics*, 100:190–205.

Schurtenberger, P., Peng, Q., Leser, M.E., and Luisi, P.-L. (1993) Structure and phase behavior of lecithin-based microemulsions: A study of the chain length dependence, *Journal of Colloid and Interface Science*, 156:43–51.

Shimek, J.W., Rohloff, C.M., Goldberg, J., and Dungan, S.R. (2005) Effect of $\alpha\pm$-lactalbumin on the phase behavior of AOT-brine-isooctane mixtures: Role of charge interactions, *Langmuir*, 21:5931–5939.

Shinoda, K., Araki, M., Sadaghiani, A., Khan, A., and Lindman, B. (1991) Lecithin-based microemulsions: Phase behavior and microstructure, *Journal of Physical Chemistry*, 95:989–993.

Shukla, A., Janich, M., Jahn, K., Krause, A., Kiselev, M.A., and Neubert, R.H.H. (2002) Investigation of pharmaceutical oil/water microemulsions by small-angle scattering, *Pharmaceutical Research*, 19:881–886.

Sinha, V.R. and Pal Kaur, M. (2000) Permeation enhancers for transdermal drug delivery, *Drug Development and Industrial Pharmacy*, 26:1131–1140.

Sjoblom, J., Lindberg, R., and Friberg, S.E. (1996) Microemulsions-phase equilibria characterization, structure, applications and chemical reactions, *Advances in Colloid Interface Science*, 95:125–287.

Spernath, A. and Aserin, A. (2006) Microemulsions as carriers for drugs and nutraceuticals, *Advances in Colloid and Interface Science*, 128–130:47–64.

Spernath, A., Yaghmur, A., Aserin, A., Hoffman, R.E., and Garti, N. (2003) Self-diffusion nuclear magnetic resonance, microstructure transitions, and solubilization capacity of phytosterols and cholesterol in winsor IV food-grade microemulsions, *Journal of Agricultural and Food Chemistry*, 51:2359–2364.

Stamatis, H. and Xenakis, A. (1999) Biocatalysis using microemulsion-based polymer gels containing lipase, *Journal of Molecular Catalysis—B Enzymatic*, 6:399–406.

Stamatis, H., Xenakis, A., and Kolisis, F.N. (1999) Bioorganic reactions in microemulsions: The case of lipases, *Biotechnology Advances*, 17:293–318.

Stark, M.-B., Skagerlind, P., Holmberg, K., and Carlfors, J. (1990) Dependence of the activity of a rhizopus lipase on microemulsion composition, *Colloid & Polymer Science*, 268:384–388.

Strickley, R.G. (2004) Solubilizing excipients in oral and injectable formulations, *Pharmaceutical Research*, 21:201–230.

Stubenrauch, C., Paeplow, B., and Findenegg, G.H. (1997) Microemulsions supported by octylmonoglucoside and geraniol. 1. The role of the alcohol in the interfacial layer, *Langmuir*, 13:3652–3656.

Szekeres, E., Acosta, E., Sabatini, D.A., and Harwell, J.H. (2006) Modeling solubilization of oil mixtures in anionic microemulsions: II. Mixtures of polar and non-polar oils, *Journal of Colloid and Interface Science*, 294:222–233.

Tenjarla, S. (1999) Microemulsions: An overview and pharmaceutical applications, *Critical Reviews in Therapeutic Drug Carrier Systems*, 16:461–521.

Tlusty, T. and Safran, S.A. (2000) Microemulsion networks: The onset of bicontinuity, *Journal of Physics Condensed Matter*, 12:A253–A262.

Tlusty, T., Safran, S.A., and Strey, R. (2000). Topology, phase instabilities, and wetting of microemulsion networks. *Physical Review Letters*, 84:1244–1247.

Tongcumpou, C., Acosta, E.J., Quencer, L.B., Joseph, A.F., Scamehorn, J.F., Sabatini, D.A., Chavadej, S., and Yanumet, N. (2003a) Microemulsion formation and detergency with oily soils: I. Phase behavior and interfacial tension, *Journal of Surfactants and Detergents*, 6:191–203.

Tongcumpou, C., Acosta, E.J., Quencer, L.B., Joseph, A.F., Scamehorn, J.F., Sabatini, D.A., Chavadej, S., and Yanumet, N. (2003b) Microemulsion formation and detergency with oily soils: II. Detergency formulation and performance, *Journal of Surfactants and Detergents*, 6:205–214.

Trevino, S.F., Joubran, R., Parris, N., and Berk, N.F. (1994) Structure of a triglyceride microemulsion: A small angle neutron scattering study, *Langmuir*, 10:2547–2552.

Trevino, S.F., Joubran, R., Parris, N., and Berk, N.F. (1998) Structure of a triglyceride microemulsion: A small-angle neutron scattering study, *Journal of Physical Chemistry B*, 102:953–960.

Uchiyama, H., Acosta, E., Tran, S., Sabatini, D.A., and Harwell, J.H. (2000) Supersolubilization in chlorinated hydrocarbon microemulsions: Solubilization enhancement by lipophilic and hydrophilic linkers, *Industrial and Engineering Chemistry Research*, 39:2704–2708.

Upadhyaya, A., Acosta, E.J., Scamehorn, J.F., and Sabatini, D.A. (2006) Microemulsion phase behavior of anionic-cationic surfactant mixtures: Effect of tail branching, *Journal of Surfactants and Detergents*, 9:169–179.

Van Hecke, E., Catté, M., Poprawski, J., Aubry, J.-M., and Salager, J.-L. (2003) A novel criterion for studying the phase equilibria of non-ionic surfactant-triglyceride oil-water systems, *Polymer International*, 52:559–562.

Von Corswant, C. and Söderman, O. (1998) Effect of adding isopropyl myristate to microemulsions based on soybean phosphatidylcholine and triglycerides, *Langmuir*, 14:3506–3511.

Von Corswant, C., Engström, S., and Söderman, O. (1997) Microemulsions based on soybean phosphatidylcholine and triglycerides. Phase behavior and microstructure, *Langmuir*, 13:5061–5070.

Von Corswant, C., Olsson, C., and Söderman, O. (1998a) Microemulsions based on soybean phosphatidylcholine and isopropylmyristate-effect of addition of hydrophilic surfactants, *Langmuir*, 14: 6864–6870.

Von Corswant, C., Thorén, P., and Engström, S. (1998b) Triglyceride-based microemulsion for intravenous administration of sparingly soluble substances, *Journal of Pharmaceutical Sciences*, 87:200–208.

Wang, L.-H., Wang, C.-C., and Kuo, S.-C. (2007) Vehicle and enhancer effects on human skin penetration of aminophylline from cream formulations: Evaluation *in vivo*, *Journal of Cosmetic Science*, 58:245–254.

Warisnoicharoen, W., Lansley, A.B., and Lawrence, M.J. (2000a) Light scattering investigations on dilute nonionic oil-in-water microemulsions, *AAPS PharmSci*, [electronic resource] 2:E12.

Warisnoicharoen, W., Lansley, A.B., and Lawrence, M.J. (2000b) Nonionic oil-in-water microemulsions: The effect of oil type on phase behaviour, *International Journal of Pharmaceutics*, 198:7–27.

Watnasirichaikul, S., Davies, N.M., Rades, T., and Tucker, I.G. (2000) Preparation of biodegradable insulin nanocapsules from biocompatible microemulsions, *Pharmaceutical Research*, 17:684–689.

Watnasirichaikul, S., Rades, T., Tucker, I.G., and Davies, N.M. (2002) *In-vitro* release and oral bioactivity of insulin in diabetic rats using nanocapsules dispersed in biocompatible microemulsion, *Journal of Pharmacy and Pharmacology*, 54:473–480.

Williams, A.C. and Barry, B.W. (1991) The enhancement index concept applied to terpene penetration enhancers for human skin and model lipophilic (oestradiol) and hydrophilic (5-fluorouracil) drugs, *International Journal of Pharmaceutics*, 74:157–168.

Witthayapanyanon, A., Acosta, E.J., Harwell, J.H., and Sabatini, D.A. (2006) Formulation of ultralow interfacial tension systems using extended surfactants, *Journal of Surfactants and Detergents*, 9:331–339.

Xian, Y., Liu, F., Xian, Y., Zhou, Y., and Jin, L. (2006) Preparation of methylene blue-doped silica nanoparticle and its application to electroanalysis heme proteins, *Electrochimica Acta*, 51(28):6527–6532.

Yaghmur, A., Aserin, A., and Garti, N. (2002a) Phase behavior of microemulsions based on food-grade nonionic surfactants: Effect of polyols and short-chain alcohols, *Colloids and Surfaces A: Physicochemical and Engineering Aspects*, 209:71–81.

Yaghmur, A., Aserin, A., and Garti, N. (2002b) Furfural-cysteine model reaction in food grade nonionic oil/water microemulsions for selective flavor formation, *Journal of Agricultural and Food Chemistry*, 50:2878–2883.

Yaghmur, A., Aserin, A., Antalek, B., and Garti, N. (2003) Microstructure considerations of new five-component Winsor IV food-grade microemulsions studied by pulsed gradient spin-echo NMR, conductivity, and viscosity, *Langmuir*, 19:1063–1068.

Yaghmur, A., De Campo, L., Aserin, A., Garti, N., and Glatter, O. (2004) Structural characterization of five-component food grade oil-in-water nonionic microemulsions, *Physical Chemistry Chemical Physics*, 6:1524–1533.

Yaghmur, A., Aserin, A., Abbas, A., and Garti, N. (2005a) Reactivity of furfural-cysteine model reaction in food-grade five-component nonionic O/W microemulsions, *Colloids and Surfaces A: Physicochemical and Engineering Aspects*, 253:223–234.

Yaghmur, A., De Campo, L., Sagalowicz, L., Leser, M.E., and Glatter, O. (2005b) Emulsified microemulsions and oil-containing liquid crystalline phases, *Langmuir*, 21:569–577.

Yao, J. and Romsted, L.S. (1997) Effects of hydrocarbon and triglyceride oils on butanol distribution in water-in-oil cationic microemulsions, *Colloids and Surfaces A: Physicochemical and Engineering Aspects*, 123–124:89–105.

Yao, J., Zhou, J.P., Ping, Q.N., Lu, Y., and Chen, L. (2008) Distribution of nobiletin chitosan-based microemulsions in brain following i.v. injection in mice, *International Journal of Pharmaceutics*, 352:256–262.

Yotsawimonwat, S., Okonoki, S., Krauel, K., Sirithunyalug, J., Sirithunyalug, B., and Rades, T. (2006) Characterisation of microemulsions containing orange oil with water and propylene glycol as hydrophilic components, *Pharmazie*, 61:920–926.

Zhao, X., Bagwe, R.P., and Tan, W. (2004) Development of organic-dye-doped silica nanoparticles in a reverse microemulsion, *Advanced Materials*, 16:173–176.

Zhou, G.-W., Li, G.-Z., Xu, J., and Sheng, Q. (2001) Kinetic studies of lipase-catalyzed esterification in water-in-oil microemulsions and the catalytic behavior of immobilized lipase in MBGs, *Colloids and Surfaces A: Physicochemical and Engineering Aspects*, 194:41–47.

Zieliñska, K., Wilk, K.A., Seweryn, E., Pietkiewicz, J., and Saczko, J. (2008) Studies on biocompatible nano-capsules formed in microemulsion templated processes, *Materials Science-Poland*, 26(2):443–450.

Zimmerberg, J. and Kozlov, M.M. (2006) How proteins produce cellular membrane curvature, *Nature Reviews Molecular Cell Biology*, 7:9–19.

19 Hyaluronan-Based Nanofibers

Miloslav Pekař, Vladimír Velebný, and Helena Bilerová

CONTENTS

19.1 NANOFIBERS OVERVIEW

Nanofibers are the ultrafine solid fibers of very small diameters—preferably lower than 100 nm (Frenot and Chronakis, 2003), although fibers of hundreds of nanometers in diameter are also included among the nanofibers. Traditional polymer fibers made by spinning using pressure-driven flow through an extruder have diameters in the order of 10–100 μm (McKee et al., 2004). By shrinking the fiber diameter from micrometers to submicron or nanometer region, new amazing properties appear (Huang et al., 2003). Nanofibers have a large surface area per unit mass and very small pore size. Thus, 1 g of polymeric nanofibers of 50 nm diameter has an area about 100 m^2 (Vrieze et al., 2007). The surface area-to-volume ratio for a nanofiber can be up to 1000 times higher than that of a microfiber (Huang et al., 2003). Surface functionality of nanofibers can be controlled and tailored flexibly, and they possess superior mechanical performance like stiffness and tensile strength.

Nanofibers have been reported to be produced by a variety of techniques including (Huang et al., 2003)

- Drawing
- Template synthesis
- Phase separation
- Self-assembly
- Electrospinning

The latter seems to be the only suitable candidate for the large-scale production of one-by-one continuous nanofibers from various polymers or biopolymers. Nanofibers are usually prepared in the form of nonwoven fabrics intended for applications as, e.g., wound dressings or tissue engineering scaffolds, filtration, and implant coating films. Fibers spun from solution can, in principle, be easily modified controlling the solution composition. For example, core–shell nanofibers can be fabricated from a solution containing two polymers that will undergo a phase separation upon the solvent evaporation (Li and Xia, 2004). Various functional components like molecular species (drugs, dyes, enzymes, DNA, etc.) can be added directly to the spun solution, and nanofibers with a diverse composition and defined or tailored functionalities are thus obtained. Nanofibers can also be post-treated to obtain desired properties. Fluorescent molecules, conducting polymers, or various nanoparticles can be deposited on nanofibers by electrostatic interactions or liquid-phase attachment. Vapor deposition techniques can also be exploited to coat nanofibers with metals or ceramics. Modifications alter chemical, thermal, mechanical, absorption, or wetting properties of nanofibers without changing their original morphology. Much broader application expansion can be expected when technologies for the production of continuous single nanofibers or uniaxial fiber bundles will be developed.

Nanofibers prepared from biopolymers or biocompatible synthetic polymers are very attractive for medical and cosmetic applications. Especially, biopolymers nanofibers can be used in regenerative medicine as wound dressings or three-dimensional scaffolds (cell supports for tissue engineering).

19.2   NANOFIBERS BY ELECTROSPINNING: PRINCIPLES

Electrospinning is an abbreviated form of "electrostatic spinning" (Huang et al., 2003). The electrospinning uses electrostatic field of high voltage to create liquid jets from the tip of a capillary (Dietzel et al., 2001; Viswanathan et al., 2006). From this point of view, the electrospinning can be considered a variation of the well-known electrospraying process.

Liquid drop is held at the end of a capillary tube (spinneret) by its surface tension. Applying electrical field to the capillary, a charge on the surface of the liquid drop is induced (Frenot and Chronakis, 2003). The original hemispherical shape of the drop is distorted by a balancing of the charge repulsion forces with the surface tension. As the intensity of the field is increased, the drop elongates and forms a conical shape called Taylor cone. When a critical voltage is attained, the repulsive electrostatic forces overcome the surface tension, and charged jet of liquid is ejected from the cone tip and travels to a grounded plate—a collector. If the viscosity of ejected fluid is low, the jet breaks up into droplets as a result of surface tension, and this process is known as electrospraying. In the case of polymer solutions or melts of sufficiently high viscosity, the jet does not break and undergoes a whipping process wherein the solvent evaporates or the melt solidifies forming a fiber, which is further stretched and reduced in diameter. Jet breaking does not occur at fluid viscosities about 1–200 Pa (Dietzel et al., 2001; Viswanathan et al., 2006). Schematic illustration of the electrospinning process is shown in Figure 19.1.

Interesting modification of traditional capillary electrospinning was also developed (Jirsák et al., 2004, 2006) and now is known under the trade name Nanospider. The modification is based on the discovery of Taylor cone formation also from the thin liquid layer (polymer solution layer). One electrode can be made as a revolving cylinder partially immersed in the spun solution. Rotating cylinder brings the solution to its non-immersed surface in a thin layer. Electric field formed between the cylindrical electrode and (usually) a planar counter-electrode creates many sources of Taylor cone in the liquid film and a dense web of jets appears. Figure 19.2 shows the principle of Nanospider modification. Nanospider is a promising technology for the mass production of nanofibrous nonwoven products. However, most published literature still discusses results obtained on capillary-based apparatuses.

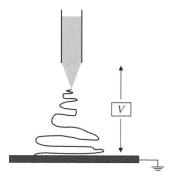

FIGURE 19.1 Principle of nanofiber electrospinning. The fiber is drawn from the top of Taylor cone located at the tip of capillary containing the polymer solution.

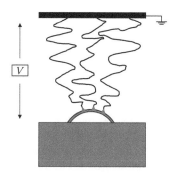

FIGURE 19.2 Principle of Nanospider modification of electrospinning process. The fibers are drawn from the thin layer of polymer solution on the top of rotating cylindrical electrode.

Formation of the drops is considered to be the main source of defects in nanofiber electrospinning. Drops appear in the final product as beads on or between the nanofibers, and the formation of beads is usually a pre-stage preceding the nanofibers formation before the optimal production parameters are achieved.

The electrospinning process is affected by many parameters and variables that can be specified as system (liquid) or process characteristics. The system parameters include

- Polymer molecular weight and its distribution
- Polymer chain architecture (linear, branched, etc.)
- Solution properties (viscosity, surface tension, conductivity)

The process parameters can be divided into two groups—processing and ambient conditions. The first group includes

- Solution concentration
- Electrical voltage
- Distance between the capillary and collector
- Solution flow rate to the tip of capillary

The ambient conditions are

- Solution temperature
- Humidity of ambient air
- Air velocity in the spinning chamber

The literature contains a lot of studies of influence of various parameters (Frenot and Chronakis, 2003; Huang et al., 2003; Li and Xia, 2004) and only brief notes on solution properties are given here because they are among crucial conditions of successful electrospinning.

It is generally accepted that the polymer solution must have an optimum concentration that is high enough to create polymer chain entanglements and to prevent collapsing into droplets yet not so high that the viscosity restricts chain movement and suppresses forcing of the electrical field. The optimum concentration is also a function of polymer molecular weight and architecture. Therefore, solution concentration is the principal parameter subjected to investigation in development of electrospinning of any polymer.

The surface tension of the solution should be low enough so that the applied voltage can overcome the surface forces and a "jetting cone" is formed. Although the surface tension can be a function of the polymer concentration, it is usually dependent much more on the solvent composition, which further affects also its evaporation rate during the fiber formation stage. Of course, the surface tension can be reduced by adding a small amount of surfactant.

McKee et al. (2004) and Shenoy et al. (2005) published detailed studies on rheological and entanglement effects on electrospinning concluded with some general rules. McKee et al. (2004) investigated electrospinning of a series of linear and branched polyesters—copolymers of poly(ethylene terephthalate) and poly(ethylene isophthalate). From the concentration dependence of the zero shear viscosity, different concentration regimes were found—especially the transition between the semidilute unentangled and entangled regimes that occurred at the entanglement concentration c_e. For the majority of tested copolymers, c_e was the minimum polymers concentration necessary for successful electrospinning of nanofibers but depreciated by co-formation of beads. To obtain uniform, defect-free fibers, the minimum required concentration was 2–2.5 times higher than c_e. Low molecular weight, linear copolymer required even higher concentration than c_e (higher in about 25%) to obtain (beaded) nanofibers. This was explained by the fact that the critical molecular weight for entanglements was only slightly below the molecular weight of the low molecular weight sample.

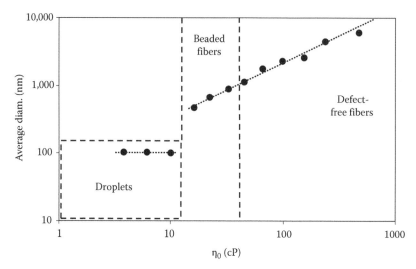

FIGURE 19.3 Dependence of electrospun morphology on the solution zero-shear viscosity. Based on data obtained for PET-*co*-PEI copolymers. (Adapted from McKee, M.G. et al., *Macromolecules*, 37, 1760, 2004. With permission from the American Chemical Society.)

Because of direct relationships between the zero shear viscosity, polymer concentration, and spinnability, some generalizations could be made. Figure 19.3 shows the dependence of electrospun morphology on the zero shear viscosity for linear and branched phthalate copolymers. When the nanofibers are formed, their diameter was found to scale either with the zero shear viscosity or with the normalized concentration as follows:

$$\text{diameter} \approx (\eta_0)^{0.8} \text{ or } \left(\frac{c}{c_e}\right)^{2.6}$$

Shenoy et al. (2005) based their analysis on the assumption that electrospinning is essentially comparable to conventional dry spinning. They stress that the conventional "spinnability" window, i.e., the regime where uniform fibers are obtained, is limited by instabilities arisen either from capillary wave breakup or due to the breakage of the fiber (brittle fracture). The presence of an elastically deformable entanglement network has been found to be essential for avoiding the fiber fracture in conventional spinning. In other words, viscoelastic properties are important for a proper fiber formation.

This common background formed a basis for the theoretical analysis of optimum polymer solution for electrospinning. It is based on a single parameter—the entanglement molecular weight of the undiluted polymer (M_e), i.e., the entanglement weight in the melt. Numerous experimental observations have validated a relationship between M_e and corresponding quantity in solution, which is more relevant for analysis of electrospinning: $M_{e,soln} = M_e/\varphi$, where φ is the polymer volume fraction in solution. Using $M_{e,soln}$, the solution entanglement number is defined as $n_{e,soln} = M_w/M_{e,soln}$, where M_w is the polymer weight-average molecular weight. Combining the two equations, the solution entanglement number can be calculated for polymer solution of a given molecular weight at a given concentration simply as $n_{e,soln} = \varphi M_w/M_e$. The zero shear viscosity of a polymer solution exhibits an upturn for $n_{e,soln} \sim 2$, indicating forming at least one entanglement per chain.

The authors analyzed published data for a number of polymers and found a simple general rule for predicting morphology of electrospun products. Solutions of polymers at concentrations corresponding to values of $n_{e,soln}$ lower than 2 leads to formation of beads only. Complete fiber formation is observed when $n_{e,soln}$ is about 3.5 or higher. In the intermediate interval, $n_{e,soln}$ between 2 and 3.5, beaded fibers are formed. Figure 19.4 illustrates the method for polystyrene in tetrahydrofurane

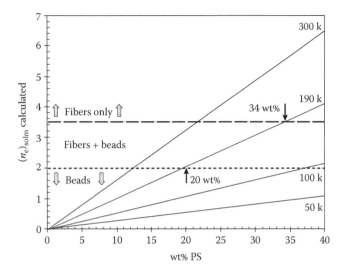

FIGURE 19.4 Relation of electrospun morphology to the entanglement number and polymer (PS) concentration in solution. (From Shenoy, S.L. et al., *Polymer*, 46, 3372, 2005. With permission from Elsevier.)

systems. The analysis is strictly valid only for noninteracting polymers in a good solvent and does not include the effect of other variables. When applicable, utilizing just the two molecular weights, the requisite polymer concentration for nanofiber formation may be estimated a priori.

19.3 HYALURONAN BIOCOLLOIDS: BASIC PROPERTIES

Hyaluronan is a naturally occurring polysaccharide, commonly found in connective tissues in the body such as vitreous, umbilical cord, joint fluid, etc. (Lowry and Beavers, 1994). As a component of the extracellular matrix, hyaluronan plays an important role in lubrication, water sorption, water retention, and a number of cellular functions such as attachment, migration, and proliferation (Bulpitt and Aeschlimann, 1999; Oerther et al., 1999). Hyaluronan is therefore an attractive building block for new biocompatible and biodegradable polymers that have applications in drug delivery, tissue engineering, and viscosupplementation (Balazs and Denlinger, 1989; Piacquadio et al., 1997; Abantangelo and Weigel, 2000; Pei et al., 2002).

It is a charged monotonic co-polymer with repeating disaccharide units composed of D-glucuronate and N-acetyl-D-glucosamine residues linked by β(1–4) and β(1–3) bonds (Lapčík et al., 1998):

Hyaluronan secondary structure is different for solid and liquid state, but in both cases, the form of helical structure stabilized by hydrogen bonds, can be found (Hascall and Laurent, 1997). In a physiological solution, the backbone of hyaluronan molecule is stiffened by a combination of the chemical structure of the disaccharide, internal hydrogen bonds, and interactions with solvent.

Molecular weight of hyaluronan can be in the range of a few thousands for oligosaccharides to several millions of grams per mole for high molecular weight material.

Due to its enormous water affinity, macromolecules of hyaluronan dissolve in water forming very viscous solutions or gel-like colloids. The high viscosity of hyaluronan aqueous solutions is caused not only by chain entanglements that are formed in solution above the critical overlap concentration but is also due to the huge hydration shell of hyaluronan chains. The viscosity and, consequently, also the conformation of hyaluronan macromolecules in the solution is affected by temperature, pH, ionic strength, and also by the presence of specific cations (Gómez-Alejandre et al., 2000). The hyaluronan conformation is flexible and very sensitive to water and cation arrangements around the chains (Sheenan and Almond, 2001).

Hyaluronan is a unique molecule not only from the view of physical properties but also from the view of its biological properties. It is of great interest that such a simple molecule that hyaluronan is, can exercise so many biological functions. As stated above, hyaluronan is composed of repeating disaccharide units consisting only of glucuronic acid and N-acetyl glucosamine, which are connected to unbranched chains. In tissues, hyaluronan is mostly found as a high-molecular compound. In addition to these macromolecules there are also products of hyaluronan degradation in tissues, fragments with much lower molecular weight. Just in different molecular weights of hyaluronan lies the diversity of its biological functions (Almond, 2007; Itano, 2008).

Hyaluronan with the molecular weight in millions of g/mol is an important component of different connective tissues and extracellular matrix in general. Due to its size, hyaluronan chain intertwines with the extracellular matrix and binds onto its structure different proteoglycane subunits with the structure similar to bottle brush. The core of such subunit is composed of protein to which chains of sulfated glykosaminoglycanes, sticking to different sides, are binded. By interaction of proteoglycane subunits with hyaluronan, proteoglycanes are formed as organized supramolecular units that are capable to interact with wide range of protein components of extracellular matrix and so create highly organized structure into which cells of particular tissues are sown. Hyaluronan in principle plays the role of organizer of extracellular matrix and most of connective tissues. Proteoglycanes, due to the high content of polysaccharides, are able to bind a huge amount of water in tissues. This is very important especially in extracellular matrix surrounding the cells. Environment saturated with water is very well permeable for low molecular substances that are used by cells as nutrients as well as an information molecules (e.g., different cytokines, growth factors) regulating behavior of cells, and also for oxygen needed by cells (Itano, 2008).

There are also small fragments of hyaluronan found in tissue. These fragments may, to a lesser extent, result from high molecular weight catabolism, in much greater extent in response to stressful conditions that cause hyaluronan splitting. Stressful conditions can occur, for example, at mechanical damage of tissue when immune cells, which are able to split high molecular weight hyaluronan, penetrate the wound. Similar stressful conditions can occur due to influence of different chemical agents on cells, due to UV light, heat stimulation, in the presence of microorganisms, etc. The generated fragments are usually significantly lower molecular weight, in units to tens of thousands, sometimes hundreds of thousands. For this reason, these fragments diffuse from extra cellular matrix and interact with different receptors on cell walls. The biological role of these fragments is no longer to organize the extracellular matrix as they are too small for that. These fragments therefore act as information messengers carrying information that something happened in extracellular matrix that led to its disintegration and it is therefore necessary to initiate steps to induce the original state (Tesar et al., 2006; Bastow et al., 2008).

Biological activity of hyaluronan fragments strongly depends on their molecular weight. Generally we can say that greater biological activity can be found at fragments with molecular weight around 500,000 g/mol and increases with decreasing molecular weight. Molecular weight also determines how the fragments will influence cells and which kind of cells it will influence (Stern et al., 2006).

Because of its unique rheological properties and complete biocompatibility, hyaluronan has been used quite extensively in many biomedical applications, including ophthalmology, drug delivery, dermatology, surgery, and medical implants.

19.4   ELECTROSPINNING OF HYALURONAN NANOFIBERS: PRACTICAL EXPERIENCE

Paper by Um et al. (2004) is perhaps the first report on electrospinning of hyaluronan. The authors state that the unusually high viscosity and the surface tension of hyaluronan solutions are major obstacles to the electrospinning process. They therefore modified the conventional electrospinning assembly by adding air blowing feature to overcome especially the high viscosity of hyaluronan solutions. Two hyaluronan samples, high ($M_w = 3,500,000$ g/mol) and low molecular weight ($M_w = 45,000$ g/mol), were tested. They were dissolved at unusually low pH (=1.5) in aqueous solutions. The strongly acidic solutions were explained by the subsequent crosslinking of the formed nanofibers. The authors still speak about polyelectrolytic character of hyaluronan, which should be the cause of intermolecular interactions and consequent high viscosity, but at such a low pH a majority of its carboxyl groups should be protonated.

The high molecular weight preparation did not show suitable concentration range, within the tested interval from 0.01% to 2% (w/v), where the initial liquid jet maintained its stability. Though some fibers were observed for concentrated solutions (above 1.3%), their number was low and suffered from droplet co-production and the jet stability was poor. Moreover, solutions above 1.5% (w/v) had so high viscosity that electrospinning could not be carried out. For the concentration range 1.3%–1.5% (w/v), the corresponding zero-shear viscosity range was 3–30 Pa·s.

The authors conclude the first part by stating that unusually high viscosity and high surface tension of hyaluronan aqueous solutions are the major obstacles in successful electrospinning. However, there are no data concerning the latter.

To decrease solution viscosity while keeping the hyaluronan concentration sufficiently high to entangle, the authors mixed the high and low molecular weight preparations. Some improvement of fiber formation was observed from solutions containing 1.3% (w/v) of high molecular weight polymer and 1%–2% (w/v) of the low molecular weight one. However, the result still was not satisfactory and drop defects were still observed. Note, that electrospinning of the low molecular weight hyaluronan was totally unsuccessful.

To further improve spinnability, the authors tested more volatile solvent, viz. ethanol. Ethanol was added up to 10%–20% (v/v) to the mixed high and low molecular weight hyaluronan solutions. Fiber content was increased slightly but the jet remained unstable.

The final solution was found in adding an air blowing system and constructing "electroblowing" apparatus. The air flow should enhance the solvent evaporation and introduce additional pulling force for the jet formation. The air stream was guided by ducts to pass closely around the spinneret and could also be preheated. Some success was achieved with 2.5% (w/v) hyaluronan solution when at the air blowing rate of 150 ft^3/h a stable jet stream was obtained. However, fiber formation was still unsatisfactory and drop defect still survived. This was attributed to the still insufficient solvent evaporation rate. The next improving step therefore was raising the temperature of the blown air.

At 47°C, the jet was stabilized and nanofibers formed with only a small fraction of beads. Temperature of 57°C was found satisfactory because a consistent and stable liquid jet stream was established and fine nanofibers were produced with fairly uniform diameters in the range of 49–74 nm. Whereas air blowing contributes to the stretching of the fluid jet and increases the evaporation of the solvent, the temperature rise leads to a reduction of viscosity of the spun hyaluronan solution.

Results of electroblowing confirm that parameters of the initial solution (concentration, viscosity, entanglements) are closely related to the behavior of streaming jet between the droplet at the spinneret and the collector, and that the space between spinneret and collector is a place where the formation of fibers and its stability is born. Though initial polymer solution parameters like sufficient concentration, not too high viscosity, or surface tension are the principal conditions of successful electrospinning, their realization is a matter of solution behavior during jet streaming, shortly, solvent evaporation rate.

Next contribution from the same laboratory (Wang et al., 2005) summarizes advantages of the electroblowing process:

- The combination of blowing and electric forces is better capable to overcome the high solution viscosity as well as surface tension.
- Elevated temperature of the blowing air can further decrease the solution viscosity and facilitate the jet formation.
- The blowing air accelerates the process of solvent evaporation before the jet reaches the collector, which is a necessary condition for the fiber formation and bead defect suppression.
- The fiber diameter can be controlled by the air flowing rate, the air temperature, and the direction of air flow.

The new contribution is devoted to the optimization of several key parameters in electroblowing: the air blowing rate, hyaluronan solution concentration, rate of solution feed, applied electric field, and type of collectors. Further, the unfeasible dissolving of prepared hyaluronan nanofibers in water is to be resolved. High molecular weight hyaluronan ($M_w = 3,500,000$ g/mol) was used again as well as highly acidic solutions (pH = 1.5) at concentration from 2% to 3% (w/v).

The optimum air blowing rate was found to be about 70 ft^3/h at which the bead formation was minimized. An increase of the blowing rate above this value deteriorated the nanofiber formation process. This was attributed to the rapid viscosity increase due to faster solvent evaporation at high blow rates. At lower blow rates, on contrary, the solvent evaporation rate is too low. The authors also observed some decrease in fiber diameter when increasing the blow rate from 40 to 100 ft^3/h, but the error bars at data points are so wide and overlapping that this effect seems not to be confirmed.

Hyaluronan concentration was investigated within the range 2%–3% (w/v) and an optimum 2.5%–2.7% (w/v) was found on the basis of uniformity of fibrous morphology as revealed by SEM. The authors related this optimum to optimum solution viscosity and surface tension. Corresponding zero-shear viscosity is in a range from 200 to 800 Pa·s. There are no data on surface tension. Both the concentration and viscosity optima are different from those found in the previous work (Um et al., 2004). Particularly the optimum viscosity is about two orders of magnitude higher. The authors claim that even the hyaluronan concentration about 2% (w/v) is too low for the classical electrospinning due to the low viscosity, low entanglement number, and high solvent content. However, electroblowing is capable to spin even such a solution, due to the additional force of the blowing air, and produces thinner fibers than electrospinning.

Dependence of the average fiber diameter on the hyaluronan solution concentration was also presented showing a direct linear proportionality. Once more, the error bars are so wide and overlapped that statistical (and also technological) significance of this conclusion is uncertain and it could be better stated that the fiber diameter is between 40 and 100 nm.

Interestingly, the authors state that hyaluronan concentration of around 2% (w/v) is still too low for successful electrospinning due to low viscosity, less chain entanglements, and high solvent amount. Electroblowing is capable of processing such low concentrated solutions just due to the additional blowing force and to produce thinner fibers than conventional electrospinning.

Feeding rate of hyaluronan solution to the spinneret affects the stability of the Taylor cone. Too low solution feeding rate can break the cone; at high feeding rates, the extra solution is dripped out and this may interfere with electrospinning. The optimal feeding rate of hyaluronan solution was found to be 20–60 μL/min. Optimum feeding should ensure a proper balance between solution delivery to the spinneret and its outflow to the collector. Points in the dependence of the average fiber diameter on the feeding rate have wide error bars once more; therefore, within the optimum feeding rate the diameter is essentially independent on feeding.

The voltage of electrical field was changed from 25 to 40 kV at the distance between the spinneret and the ground collector of 9.5 cm and showed no significant effect. Aluminum foil as one of the most popular electrospinning collectors was not suitable for hyaluronan spinning because it was difficult to separate the spun sheet from the collector. The wire screen allows much better separation and was preferred by the authors. Moreover, in electroblowing, the permeability of the wire collector to the air flow is another advantage.

As the membranes made from nanofibers prepared from native hyaluronan were soluble in water, the authors tried to make them water resistant by a subsequent treatment where the use of additional chemicals should be avoided. Two methods were used. The first one was treatment with HCl vapors for up to 10 min and then freezing at −20°C for 20–40 days. In the second procedure, the hyaluronan nanofibrous membranes were immersed in a mixture of varying ratios of ethanol, concentrated HCl, and water for several days at different temperatures (−20°C to 20°C). In principle, the two methods can be described as physical crosslinking.

Although the HCl-vapor post-treatment led to hyaluronan crosslinking, confirmed by swelling of prepared membranes in water, the desired mechanical strengths of nanofibrous materials were not achieved by this approach. The second method was much more successful. The membrane prepared at 4°C in an optimal mixture, consisting of about 24% (v/v) of water, 71% (v/v) of ethanol, and concentrated HCl solution forming the rest to 100%, retained its shape in neutral water for at least one week at 25°C. Crosslinking, process is supposed to include three different effects—the presence of high amount of ethanol prevents the dissolution of hyaluronan, the presence of HCl destabilizes hyaluronan and induces crosslinking, and the cooling slows down the reaction. Infrared spectra of crosslinked membranes indicated that the crosslinking was due to the formation of a hydrogen bonding network among the hyaluronan chains.

Li et al. (2006) mention that the strong hydration ability of hyaluronan may lead to the insufficient evaporation of aqueous solvents and to the fuse of nanofibers on the collector. To overcome this problem they used a "liquid" collector, viz. the ethanol bath because ethanol is a poor solvent for hyaluronan. The bath was grounded by a piece of aluminum foil immersed in a vessel containing ethanol. After electrospinning, ethanol was dried off in vacuum. Hyaluronan of average molecular weight $M_w = 2,000,000$ g/mol was dissolved in water and water-ethanol mixture (9:1 by volume) at concentrations of 1.3% and 1.5% (w/v). The electrospinning was not satisfactory due to large bead formation that could not have been suppressed by adjusting the applied voltage and spinneret-collector distance.

Next experiments were therefore made using dimethylformamide-water mixtures with volume ratios from 0.5:1 to 2:1. Dimethylformamide (DMF) was selected because it is a polar solvent with poor solubility for hyaluronan. The best results were obtained when the volume ratio of DMF to water was between 1:1 and 1.5:1 giving average fiber diameter 200 and 250 nm, respectively. Large beads occurred at lower ratios, whereas hyaluronan could not be dissolved completely at ratio 2:1. Surface tension and conductivity measurements revealed that both parameters decreased with increasing DMF content. The authors therefore concluded that the decrease in surface tension benefited the electrospinning process and made the beads disappear. The data on surface tension given in the paper show that the decrease was only moderate (in about 5–7 mN/m) and no data on its variance were given.

Rheological measurements showed that the flow curves of hyaluronan solutions changed very little after DMF was added. However, more distinct changes were observed on the normal force curves—the viscoelasticity of hyaluronan solutions in DMF-water mixtures seemed to be higher than that of hyaluronan solutions in water or water–ethanol mixtures. Rheological behavior (elasticity) of electrospinned solutions, especially upon stretching, thus may be very important for successful production of hyaluronan nanofibers.

The study further included electrospinning of hyaluronan-gelatin blends. Gelatin (average $M_n = 80,000$ g/mol) was selected to improve processability and to endow hyaluronan with protein characteristics to improve the cell adhesion. As a solvent, the DMF-water mixture at DMF:water ratio 1.5 was used and the concentration of hyaluronan was fixed at 1.5% (w/v). The hyaluronan:gelatin weight ratio ranged from 1:1 to 5:1. The blend spinning was successful giving average fiber diameters from 190 to 500 nm; the diameter increased upon increasing gelatin contents. Addition of gelatin increased the total polymer concentration, slightly decreased the conductivity, and decreased the surface tension. Rheological properties of solution containing 1:1 hyaluronan-gelatin blend were essentially equal to those of analogical hyaluronan solution.

The authors concluded that they succeeded in electrospinning of pure hyaluronan solutions without the assistance of air blowing. The main reason for the improved processability was attributed to using DMF–water mixture and (or) adding gelatin to the hyaluronan solution and corresponding decrease in surface tension. Although the beneficial effect of the surface tension is indisputable, the data also indicate important role of viscoelastic properties.

Ji et al. (2006) consider the temperature used in electroblowing by Um et al. (2004) and Wang et al. (2005) to be too high from the physiological point of view. They also prefer fabrication of three-dimensional hyaluronan nanofibrous structures with microporous channels facilitating the migration and proliferation of cells. Otherwise, the hydrophilic and anionic surfaces of hyaluronan based materials do not favor the cell attachment and subsequent tissue formation. Creating a microporous scaffold is one strategy to direct the cell growth and support the tissue formation. The authors therefore prepared a thiolated hyaluronan derivative by coupling dithiobis(propanoic dihydrazide) to the carboxylic groups of hyaluronan. Thiolated hyaluronan could be then crosslinked through poly(ethylene glycol)-diacrylate. Poly(ethylene oxide) of average molecular weight $M_w = 3400$ g/mol was used as a viscosity modifier in electrospinning.

Initial hyaluronan was of the average molecular weight $M_w = 1,500,000$ g/mol, after thiolation the molecular weight decreased to $M_w = 158,000$ g/mol with the index of polydispersity equal to 2.03. As a solvent for preparing the electrospinning solutions, Dulbecco's modified eagle's medium was utilized to keep the biocompatibility of electrospun products. The concentration of thiolated hyaluronan was fixed at 2% (w/v; no explanation or concentration optimization are given), poly(ethylene oxide) was applied at different weight ratios—thiolated hyaluronate:poly(ethylene oxide) ranged from 1:1 to 4:1.

At the lowest concentration of poly(ethylene oxide), the products of electrospinning showed a beads-on-string morphology with a high beads density. Increasing the amount of poly(ethylene oxide), beads density decreased dramatically and a uniform nanofibrous material was obtained at the ratio of 1:1, which was the selected as the optimal weight ratio for further fabrication. The authors claim that solution viscosity is among the most important parameters affecting the electrospinning process since it correlates with the entanglement number. They therefore reduced the viscosity by using a low molecular weight hyaluronan derivative and further by blending it with poly(ethylene oxide). However, they gave no data on viscosity and its change with changing solution composition.

The as-spun product was crosslinked as given above and subsequently soaked in deionized water to remove poly(ethylene oxide). The crosslinked scaffold maintained a three-dimensional structure after extraction though some fibers fused together, and the shape of fibers was not as uniform as before extraction. It was found that before extraction, more than 85% of fibers were within the diameter range between 70 and 110 nm, whereas after extraction, the diameter distribution became wider and 85% of fibers had the diameter within the range between 50 and 300 nm.

The electrospun scaffolds swelled in the phosphate-buffered saline solution (pH = 7.4). Scaffolds prepared without adding the acrylate crosslinker had an equilibrium swelling ratio much higher than the swelling ration of scaffolds crosslinked with the poly(ethylene glycol)-diacrylate. This was explained by an air-induced oxidation of thiols to disulfide linkages either during electrospinning or during subsequent exposition of electrospun products in air. Swelling ratios decreased with the addition of acrylate crosslinker and/or as a function of increasing exposing time in air. However, the disulfide-only crosslinked scaffolds had a poor mechanical behavior because of the low degree of crosslinking.

To improve the cell attachment, the crosslinked scaffolds were physically adsorbed with intact human plasma fibronectin. After 24 h of seeding with fibroblasts, the cells seemed to migrate inside the scaffold and developed a dendritic morphology, which is typical in three-dimensional fibrous matrices. These results suggested potential applications of prepared nanofibrous materials in cell encapsulation and tissue regeneration.

Different approach to nanofibrous hyaluronan-based scaffolds was described by Bhattacharyya et al. (2008). The nanofibrous structure was not achieved by manipulating on the hyaluronan but by dispersing carbon nanotubes within hyaluronan aqueous solutions. The objective was to apply the tubes as a nano-filler enhancing the mechanical properties of hyaluronan hydrogels without changing their water absorption capacity.

Commercial single-walled carbon nanotubes were carboxylated using nitric acid. Functionalized carbon nanotubes were mixed with hyaluronan in water at alkaline pH to facilitate subsequent crosslinking by divinyl sulfone. Finally, hyaluronan-carbon nanotube hydrogels were thus prepared. Carbon nanotubes amounted 2% (w) of the hyaluronan weight and their total concentration in resulting hydrogels was between 0.03% and 0.06% (w).

Incorporation of carbon nanotubes changed the hydrogel morphology considerably. Original "honeycomb" structure of pure hyaluronan hydrogel was transformed into the morphology of separate fibers or slivers. Nevertheless, the filled hydrogels almost retained the water sorption capacity of native gels (in the phosphate buffer at pH = 7.4)—it decreased from 190% to 170% after 24 h swelling. Mechanical properties, measured by frequency sweep rheometry, were strengthened as desired. Both the native and reinforced hydrogels showed frequency-independent storage (elastic) modulus and a considerably smaller (in three orders of magnitude) loss (viscous) modulus. However, about three to fourfold increase of especially the storage modulus was observed after reinforcing with carbon nanotubes.

Although only two concentrations of the carbon nanotubes were tested, the authors tried to explain the unexpected combination of a high water retention capacity and a large increase of storage modulus by the formation of carbon nanotube network in which the tubes are linked by segments of hyaluronan chains either covalently (divinyl sulfone crosslinking) or noncovalently (wrapping the tubes by hyaluronan). The authors also consider the combination of shear rigidity with the high water uptake as essential for biomedical applications of materials.

Rheological properties were mentioned several times as one of the most significant parameters in the electrospinning, especially in relationship to entanglements in spun solution. In conclusion we will therefore report on a study of hyaluronan rheological properties under physiological conditions (Krause et al., 2001). Hyaluronan of molecular weight 1,600,000 g/mol and a protein content less than 0.1% was dissolved in a phosphate buffered saline mix (0.138 mol/L NaCl, 0.0027 mol/L KCl; pH = 7.4) to prepare solutions in the concentration range from 0.094 to 8.7 g/L. Viscosity-shear rate curves showed a broad Newtonian plateau for low concentrated solutions, which was progressively shortened with increasing the hyaluronan concentration (Figure 19.5). All these curves could be extrapolated to determine the zero-shear viscosity that was then used to construct the plot of the specific viscosity dependence on hyaluronan concentration (Figure 19.6). In dilute solutions, specific viscosity was proportional to concentration ($\eta_{sp} \sim c^{1.1}$) as predicted by theory for polyelectrolytes in the high salt limit. The end of this proportionality is the overlap concentration (c^*), and this was determined to be 0.59 g/L. In following semidilute unentangled solution, the authors find $\eta_{sp} \sim c^{2.0}$; theoretical exponent prediction is 1.25. Semidilute entangled solution region started at the concentration $c_e = 2.4$ g/L and in this region $\eta_{sp} \sim c^{4.1}$. This exponent corresponds to those found for neutral polymers and is slightly higher than the theoretical prediction for polyelectrolytes (3.75).

From this study follows that hyaluronan solutions with concentration lower than about 0.6 g/L are so dilute that they are of no use in electrospinning, and minimum suitable concentration is about 2.4 g/L (entanglements start to be formed), i.e., the optimum concentration according to McKee et al. (2004) is about 5–6 g/L. Comparing this value with that used in published studies on hyaluronan electrospinning, it is found that concentrations used in practice are much higher. Theoretical prediction is thus probably not valid for polyelectrolyte nanofibers spinning.

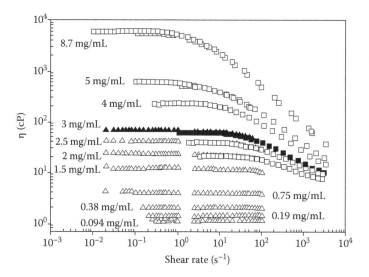

FIGURE 19.5 Apparent viscosity curves of hyaluronate at various concentration in phosphate-buffered saline solution at 25°C. (Adapted from Krause, W.E. et al., *Biomacromolecules*, 2, 65, 2001. With permission from the American Chemical Society.)

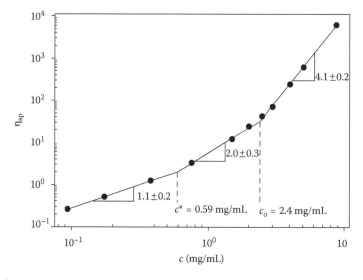

FIGURE 19.6 Concentration dependence of the zero-shear specific viscosity of hyaluronate in phosphate-buffered saline solution at 25°C. (Adapted from Krause, W.E. et al., *Biomacromolecules*, 2, 65, 2001. With permission from the American Chemical Society.)

19.5 PATENTED TECHNOLOGIES FOR HYALURONAN NANOFIBERS

Although the fabrication of hyaluronan-based nanofibers is not a trivial task, several patents have already appeared in this area. Chu et al. (2005) patented their electroblowing technology mentioned above to protect, directly and only, the fabrication of hyaluronan nanofibers and fibers. However, their patent embraces also fibers prepared from mixtures of hyaluronan with a variety of both synthetic and natural polymers. Comparing with the conventional electrospinning, the electroblowing should allow for higher throughput production and broader range of accessible operating conditions.

Patent by Chu et al. (2006) describes crosslinking of hyaluronan solutions and nanofibrous membranes made from them. Physical crosslinking is realized by contacting hyaluronan with an aqueous acidic solution (preferably HCl) for a period of time and a suitable temperature. The solvent mixture has a content of acidic component sufficient to prevent dissolution of the hyaluronan, whereas the acid is miscible with water, and sufficient to affect the hyaluronan crosslinking. Electrospun hyaluronan membranes can be crosslinked by treatment with HCl vapors. Crosslinked nonwoven hyaluronan is water resistant.

International patent (Knotková et al., 2008) covers preparation of nanofibers from polysaccharides, including hyaluronan of molecular weight between 10,000 and 3,500,000 g/mol, by electrospinning their mixture with polyvinyl alcohol as a fiber-formation supporting agent. Nanofibers are spun either from aqueous solution or from the water-ethanol mixture in the presence of suitable surfactant, e.g., Tween 20. Experimental work that originated this patent was done just on laboratory version of Nanospider technology. Example of hyaluronan nanofibers produced by the patented technology is shown in Figure 19.7.

American patent (Hashi and Li, 2008) describes, in fact, preparation of biomimetic nanofibrous scaffold linked with, e.g., an extracellular matrix component either covalently or non-covalently. Resulting material can be considered as nanofibrous scaffold modified by, e.g., hyaluronan. As

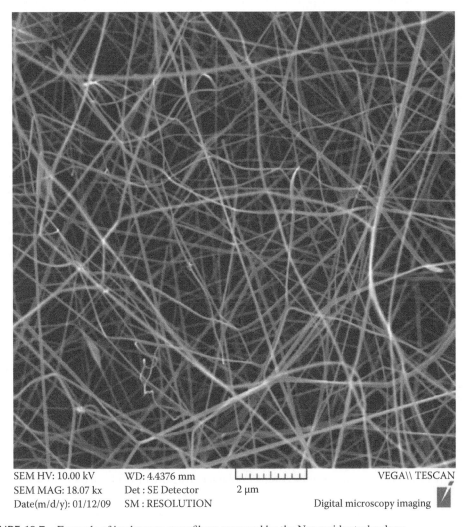

SEM HV: 10.00 kV	WD: 4.4376 mm		VEGA\\ TESCAN
SEM MAG: 18.07 kx	Det : SE Detector	2 μm	
Date(m/d/y): 01/12/09	SM : RESOLUTION		Digital microscopy imaging

FIGURE 19.7 Example of hyaluronan nanofibers prepared by the Nanospider technology.

an example of parental fiber-forming polymer, a copolymer of lactic and glycolic acids is given. Scaffolds are intended for use as a new tissue support.

19.6 SUMMARY AND FUTURE PROSPECTS

A lot of parameters affect nanofiber electrospinning, and finding optimal values for all of them need not be a simple task. Parameters were usually optimized by the one-variable-at-a-time approach and multiparametric optimization has not been made. Nevertheless, it seems to be confirmed that each parameter has some acceptable range of values, upper and lower limits outside which the electrospinning is not successful. For example, viscosity of spun solution must be within certain limits, otherwise no uniform nanofibers can be electrospun neither by changing any of the other parameters, e.g., electrical voltage. Some parameters seem to be really critical and essential, whereas deviations of others are more tolerable. Among the former belongs the concentration of spunned solution and related viscosity, or rheological behavior in general, and polymer molecular weight. The voltage is optimized relatively easily and usually the optimum is of tens of kilovolts. Surface tension or conductivity of spun solution is probably not so critical.

There is a general agreement on relevance of entanglements in the spun polymer solution, i.e., its concentration should be at least equal but preferably several times higher than the critical entanglement concentration. This conclusion is valid for noninteracting polymers in a good solvent and electrospinning behavior of charged polymers or polyelectrolytes or of hydrogen-bonding polymers may be different. Systematic studies on effects of interacting polymers, on relationships between (mechanical or sterical) entanglements and physical interactions are still lacking.

It is also interesting that little attention has been paid to the influence of viscoelastic properties of spun solution although they were mentioned in some studies. Just the elastic behavior could play an important role in fiber forming by jet straining and in fiber stabilization. At this point, elongation viscosity could be of greater significance than the usually measured and considered shear viscosity. Shenoy et al. (2005) note, however, that even the elongation viscosity is related to entanglements and the deductions based on them might be sufficient.

Despite of several papers and even patents, nanofibers production from hyaluronan still remains a technical and technological challenge. Problems in the hyaluronan electrospinning are usually ascribed to too high viscosity attained already for solutions of still low concentration. Um et al. (2004) give 0.135% (w/v) as the overlapping concentration for a typical hyaluronan from synovial fluids. This is much lower than found by Krause et al. (2001) (see above) and much lower than typical concentrations used in hyaluronan electrospinning. Critical concentrations can be influenced, besides the hyaluronan molecular weight, by the ionic strength, i.e., the composition of the aqueous solution. Hyaluronan is also well known for its huge hydration shell, and atypically high viscosity of hyaluronan aqueous solutions is often explained just by its high hydration. Viscosity of hyaluronan aqueous solutions is thus a result of not only the size of the hyaluronan chain but also of its hydration shell, which should move with the chain. Influence of hydration shell on electrospinning is unknown except consideration of hindered evaporation of water as a solvent. Ethanol is known as a dehydrating agent, thus water-ethanol mixtures can be advantageous not only for the good ethanol volatility (cf. also Section 19.5).

Optimum hyaluronan concentration for electrospinning calculated according to McKee et al. (2004) from the data by Krause et al. (2001) should be about 0.5%–0.6% (w/v). This is several times lower than the experimentally found acceptable value (see Section 19.4). An explanation can be given considering the non-Newtonian behavior of hyaluronan solutions. Analyses based on chains overlaps and entanglements work with the extrapolated, zero shear viscosity, i.e., a hypothetical value corresponding to an unperturbed, "still," polymer structure or conformation, to some "equilibrium" state in a given good solvent. Electrospinning processing conditions are definitely different, shear (and elongation) rates are really high, and the viscosity would be much lower the zero shear value. Concentration necessary for successful electrospinning thus could be higher than it follows from the zero shear extrapolation. Besides the zero shear viscosity also the concentration dependence of viscosities at selected shear (elongation) rates could be followed and analyzed.

Electric field used in electrospinning may affect structure in solutions of polyelectrolytes. Ions of the low molecular electrolyte and the counterions need not screen the polyelectrolyte charges as in the normal case of zero field. Chain conformations and inter-chain interactions can be changed. These can also be affected by the type of the low molecular background electrolyte, e.g., by its cosmotropic or chaotropic nature.

Unique biological properties of hyaluronan make it a very attractive biopolymer for producing nanofibrous materials. At the same time, hyaluronan belongs to polymers manifesting serious problems in electrospinning, to polymers that are very uneasily electrospinnable. Perfect optimization of hyaluronan electrospinning conditions leading to the nanofibers with well defined parameters is still in demand. It is necessary to thoroughly test all the parameters effecting the nanofiber producing mentioned above. Then it will be perhaps possible to change the size of fibers by changing only viscosity of solution or electrical voltage, for example. The future of hyaluronan nanofibers is particularly in the tissue engineering where one of the main tasks is a production of fabrics with oriented (aligned) hyaluronan fibers. Proper orientation of nanofibers and formation of an adequate three-dimensional structure is essential not only for proper cell penetration and seeding but also for maintaining the desired cell phenotype. Further, materials prepared from the native hyaluronan are soluble in aqueous media. The solubility can be suppressed or removed by a chemical modification of spun materials as seen above. Another approach that has not been tested yet, is the preparation of nanofibers from modified, i.e., hydrophobized hyaluronan. Modification should be made not only to improve resistance to aqueous media, i.e., by hyaluronan crosslinking or hydrophobizing, but also to simulate natural cell media. That is, e.g., by grafting peptides to the hyaluronan or hyaluronan-based nanofibrous products.

ACKNOWLEDGMENT

This work was supported by CPN, Nanomedic and Czech government—project No. MSM0021630501.

ABBREVIATIONS

c	Concentration
c_e	Entanglement concentration
M_w	Polymer weight-average molecular weight
M_e	Entanglement molecular weight
$n_{e,soln}$	Solution entanglement number
η	Apparent viscosity
η_0	Zero-shear viscosity
η_{sp}	Specific viscosity
φ	Polymer volume fraction in solution

REFERENCES

Abantangelo, G. and Weigel, P. 2000. *New Frontiers in Medical Science: Redefining Hyaluronan*. Amsterdam, the Netherlands: Elsevier.

Almond, A. 2007. Hyaluronan. *Cell Mol. Life Sci.* 64:1591–1596.

Balazs, E.A. and Denlinger, J.L. 1989. Clinical uses of hyaluronan: The biology of hyaluronan. In *Clinical Uses of Hyaluronan: The Biology of Hyaluronan*, D. Evered and J. Welan (Eds.), pp. 265–280. New York: Wiley.

Bastow, E.R., Byers, S., Golub, S.B., Clarkin, C.E., Pitsillides, A.A., and Fosang, A.J. 2008. Hyaluronan synthesis and degradation in cartilage and bone. *Cell Mol. Life Sci.* 65:395–413.

Bhattacharyya, S., Guillot, S., Dabboue, H., Tranchant, J.-F., and Salvetat, J.-P. 2008. Carbon Nanotubes as structural nanofibers for hyaluronic acid hydrogel scaffolds. *Biomacromolecules* 9:505–509.

Bulpitt, P. and Aeschlimann, D. 1999. New strategy for chemical modification of hyaluronic acid: Preparation of functionalized derivatives and their use in the formation of novel biocompatible hydrogels. *J. Biomed. Mater. Res.* 47:152–169.

Chu, B., Fang, D., Hsiao, B.S., and Okamoto, A. 2005. Electroblowing fibers to produce hyaluronan fibers/ nanofibers used for biomedical material, comprises forcing polymer fluid through spinneret towards collector, while simultaneously blowing gas through orifice. U.S. patent 2005073075, international patent WO-2005033381.

Chu, B., Hsiao, B.S., Fang, D., and Okamoto, A. 2006. Crosslinking of hyaluronan solutions and nanofibrous membranes made therefrom. U.S. patent 2006046590, international patent WO-2006026104.

Dietzel, J.M., Kleinmeyer, J., Harris, D., and Beck Tan, N.C. 2001. The effect of processing variables on the morphology of electrospun nanofibers and textiles. *Polymer* 42:261–272.

Frenot, A. and Chronakis, I.S. 2003. Polymer nanofibers assembled by electrospinning. *Curr. Opin. Colloid Interface Sci.* 8:64–75.

Gómez-Alejandre, S., Sánchez de la Blanca, E., Abradelo de Usera, C., Rey-Stolle, M.F., and Hernández-Fuentes, I. 2000. Partial specific volume of hyaluronic acid in different media and conditions. *Int. J. Biol. Macromol.* 27:287–290.

Hascall, V.C. and Laurent, T.C. 1997. Hyaluronan: Structure and physical properties. http://www.glycoforum. gr.jp/science/hyaluronan/HA01/HA01E.html (accessed January 30, 2009).

Hashi, C. and Li, S. 2008. Biomolecule-linked biomimetic scaffolds comprising nanofiber polymer and extracellular matrix component and/or growth factor. U.S. patent 2008220042.

Huang, Z.M., Zhang, Y.Z., Kotaki, M., and Ramakrishna, S. 2003. A review on polymer nanofibers by electrospinning and their applications in nanocomposites. *Compos. Sci. Technol.* 63:2223–2253.

Itano, N. 2008. Simple primary structure, complex turnover regulation and multiple roles of hyaluronan. *J. Biochem.* 144:131–137.

Ji, Y., Ghosh, K., Shu, X.Z., Li, B., Sokolov, J.C., Prestwich, G.D., Clark, R.A.F., and Rafailovich, M.H. 2006. Electrospun three-dimensional hyaluronic acid nanofibrous scaffolds. *Biomaterials* 27:3782–3792.

Jirsák, O., Sanetrník, F., Lukáš, D., Kotek, V., Martinová, L., and Chaloupek, J. 2004. Process and apparatus for producing nanofibers from polymer solution by electrostatic spinning. Czech patent 294274.

Jirsák, O., Sanetrník, F., Mareš, L., and Petráš, D. 2006. Textiles containing at least one layer of polymeric nanofibres and production of the layer of polymeric nanofibres from the polymer solution through electrostatic spinning. International patent WO-2006108364.

Knotková, K., Hrubá, J., and Velebný, V. 2008. Preparation of nanofibers from polysaccharides and polyvinyl alcohol by electrospinning for pharmaceutical and therapeutic applications. International patent WO-2008128484.

Krause, W.E., Bellomo, E.G., and Colby, R.H. 2001. Rheology of sodium hyaluronate under physiological conditions. *Biomacromolecules* 2:65–69.

Lapčík, L., Lapčík, L., Smedt, S., De Demeester, J., and Chabreček, P. 1998. Hyaluronan: Preparation, structure, properties, and applications. *Chem. Rev.* 98:2663–2684.

Li, D. and Xia, Y. 2004. Electrospinning of nanofibers: Reinventing the wheel? *Adv. Mater.* 16:1151–1170.

Li, J., He, A., Han, C.C., Fang, D., Hsiao, B.S., and Chu, B. 2006. Electrospinning of hyaluronic acid (HA) and HA/gelatin blends. *Macromol. Rapid Commun.* 27:114–120.

Lowry, K.M. and Beavers, E.M. 1994. Thermal stability of sodium hyaluronate in aqueous solution. *J. Biomed. Mater. Res.* 28:1239–1244.

McKee, M.G., Wilkes, G.L., Colby, R.H., and Long, T.E. 2004. Correlations of solution rheology with electrospun fiber formation of linear and branched polyesters. *Macromolecules* 37:1760–1767.

Oerther, S., Le Gall, H., Payan, E., Lapicque, F., Presle, N., Hubert, P., Dexheimer, J., and Netter, P. 1999. Hyaluronate-alginate gel as a novel biomaterial: Mechanical properties and formation mechanism. *Biotechnol. Bioeng.* 63:206–215.

Piacquadio, D., Jarcho, M., and Goltz, R. 1997. Evaluation of hylan b gel as a soft-tissue augmentation implant material. *J. Am. Acad. Dermatol.* 36:544–549.

Pei, M., Solchaga, L.A., Seidel, J., Zeng, L., Vunjak-Novakovic, G., Caplan, A.I., and Freed, L.E. 2002. Bioreactors mediate the effectiveness of tissue engineering scaffolds. *FASEB J.* 16:1691–1694.

Sheenan, J. and Almond, A. 2001. Hyaluronan: Statistic, hydrodynamic and molecular dynamic views. http:// www.glycoforum.gr.jp/science/hyaluronan/HA21/HA21E.html (accessed January 30, 2009).

Shenoy, S.L., Bates, W.D., Frisch, H.L., and Wnek, G.E. 2005. Role of chain entanglements on fiber formation during electrospinning of polymer solutions: Good solvent, non-specific polymer–polymer interaction limit. *Polymer* 46:3372–3384.

Stern, R., Asari, A.A., and Sugahara, K.N. 2006. Hyaluronan fragments: An information-rich system. *Eur. J. Cell Biol.* 85:699–715.

Tesar, B.M., Jiang, D., Liang, J., Palmer, S.M., Noble, P.W., and Goldstein, D.R. 2006. The role of hyaluronan degradation products as innate alloimmune agonists. *Am. J. Transplant.* 6:2622–2635.

Um, I.C., Fang, D., Hsiao, B.S., Okamoto, A., and Chu, B. 2004. Electro-spinning and electro-blowing of hyaluronic acid. *Biomacromolecules* 5:1428–1436.

Viswanathan, G., Murugesan, S., Pushparaj, V., Nalamasu, O., Ajayan, P.M., and Linhardt, R.J. 2006. Preparation of biopolymer fibers by electrospinning from room temperature ionic liquids. *Biomacromolecules* 7:415–418.

Vrieze, S.D., Westbroek, P., Van Camp, T., and Van Langenhove, L. 2007. Electrospinning of chitosan nanofibrous structures: Feasibility study. *J. Mater. Sci.* 42:8029–8034.

Wang, X., Um, I.C., Fang, D., Okamoto, A., Hsiao, B.S., and Chu, B. 2005. Formation of water-resistant hyaluronic acid nanofibers by blowing-assisted electro-spinning and non-toxic post treatments. *Polymer* 46:4853–4867.

Radha Gupta and Ashok Kumar

CONTENTS

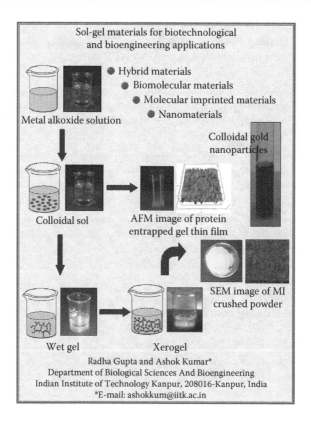

Radha Gupta and Ashok Kumar*
Department of Biological Sciences And Bioengineering
Indian Institute of Technology Kanpur, 208016-Kanpur, India
*E-mail: ashokkum@iitk.ac.in

20.1   INTRODUCTION

Sol-gel technology is a wonderful advancement in science and requires a multidisciplinary approach for its various applications. It is the process of making ceramic and glassy materials at a relatively low temperature that allows the doping of various inorganic, organic, and biomolecules during the formation of a glassy matrix. The sol-gel process was known as early as the 1800s, but in the last two decades sol-gel applications have increased manifold. It has been used for the fabrication of optical fibers, optical coatings, electro-optic materials, nanocrystalline semiconductor-doped xerogels, colloidal silica powders for chromatographic stationary phase and as catalytic support, nanoporous carbon xerogels and aerogels as hydrogen storage materials, luminescence concentrators, tunable lasers, active wave guides, semiconducting devices, sunscreen formulations (sol-gel pearls) and chemical sensors for detecting gases, heavy metals, and pH, and for many biosensor applications. Potential applications of sol-gel technology in the areas of defense, nanotechnology, environmental monitoring, and biomedical devices are now continuously emerging. In this chapter, we have focused on sol-gel materials for various biotechnological and bioengineering applications using sol-gel technology. The low processing temperature of sol-gel technology combined with the intrinsic biocompatibility (the ability of materials not to produce a significant rejection or immune response when they are inserted into the body) and environmental friendliness make it an ideal technology for the fabrication of bioactive materials. A bioactive material is defined as a material that elicits a specific response at the interface of the material, which results in the formation of a bond between the tissue and that material. In addition, the ability of sol-gel technology to manipulate the structure of materials at the molecular level as well as its ability to precisely control the nature of interfaces makes it an interesting approach for a wide range of practical applications. The sol-gel technology based on various alkoxides allows the production of conventional silica glasses as well as multicomponent materials, merging silicates with titanates, borates, and a variety of other oxides. The alkoxide gel method can also be used for the production of certain nonsilicate oxide glass-like materials (e.g., ZrO^{2-}). Using sol-gel technology, organic–inorganic hybrid materials can be prepared either by dissolution of organic molecules in a liquid sol-gel or by impregnation of a porous gel in the organic solution. Another way is to use inorganic precursor containing an organic group or carry out sol-gel reactions in a liquid solution to form chemical bonds in the hybrid gel (Gupta and Kumar, 2008a).

20.2 CONVENTIONAL SOL-GEL PROCESSING

The sol-gel process has been known since the late 1800s. The versatility of the technique was rediscovered in the early 1970s when glasses were produced without high-temperature melting processes. In general, the sol-gel process involves the transition of a system from a liquid "sol" (mostly colloidal) into a solid "gel" phase, and the understanding of various mechanisms underlying sol-gel processes have been the subject of several books and reviews (Klein, 1988; Livage et al., 1988; Brinker and Scherer, 1990; Hench and West, 1990). The schematic of sol-gel process and various products are shown in Figure 20.1. The starting materials used in the preparation of the sol are usually inorganic metal salts or metal organic compounds such as metal alkoxides [M(OR)n], where M represents a network forming element such as Si, Ti, Zr, Al, B, etc., and R is typically an alkyl group. The most commonly used precursors are tetramethyl-orthosilicate (TMOS) and tetraethyl-orthosilicate (TEOS) in the sol-gel process. The basic sol-gel reaction begins when metal alkoxide is mixed with water and a mutual solvent (mostly alcohol) in the presence of acid or base catalyst. Generally, both the hydrolysis and condensation reactions occur simultaneously once the hydrolysis reaction has been initiated. Hydrolysis leads to the formation of silanol groups ($\equiv$Si–OH) and condensation reactions produce siloxane bonds ($\equiv$Si–O–Si$\equiv$), resulting in the production of alcohol and water as by-products (Scheme 20.1). The chemical reactions occurring during the sol-gel process strongly influence the properties of the final material. An increased value of water to alkoxide molar ratio (R) is expected to promote hydrolysis reaction. In general, under stoichiometric addition of water ($R < 2$), the alcohol-producing condensation process is dominant; whereas at $R \geq 2$ water-forming condensation reaction is favored. The higher value of R causes more complete hydrolysis of the monomers before significant condensation can occur. The properties of sol-gel matrix are dependent on various physical and chemical properties of the composition of the sol-gel. One can tune several properties of sol-gel, viz., porosity, surface area, polarity, and rigidity by a selection of precursors, water-to-precursor molar ratios (R), solvent and cosolvent, pressure, temperature, aging, drying, and curing conditions. Further processing of the sol enables one to make sol-gel materials

Solution of precursors
Hydrolysis and condensation
Powder
Drying
Grinding
Dense thin film
Thermal treatment
Ambient press and temp
Sintering
Dip, spin and spray coating
Xerogel
Thin films on substrate
Sol (colloid)
Bulk
Spinning
Gelation
Dense ceramic
Supercritical drying
Sintering
Fibers
Gel
Bulk
Aerogel

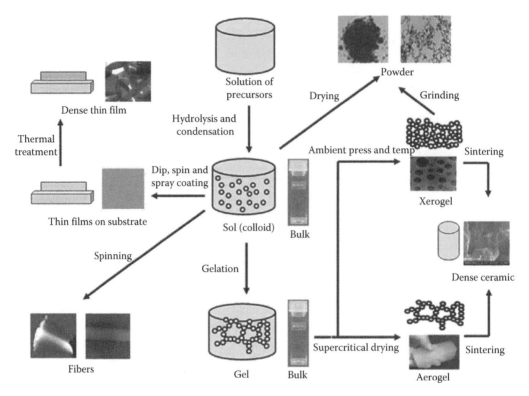

FIGURE 20.1 Schematic diagram showing sol-gel process and its various products (Reproduced from Gupta, R. and Kumar, A., *Biomed. Mater.*, 3, 034005, 2008a, and references cited therein. With permission.)

≡Si—OR + H2O
Hydrolysis
Esterification
HO—Si≡ + ROH
(i)
≡Si—OR + HO—Si≡
Alcohol condensation
Alcoholysis
≡Si–O—Si≡ + ROH
(ii)
≡Si–OH + HO —Si≡
Water condensation
Hydrolysis
≡Si–O—Si≡ + HOH
(iii)
[≡Si–O—Si≡]n
Polymerization
(iv)
Silica polymeric matrix

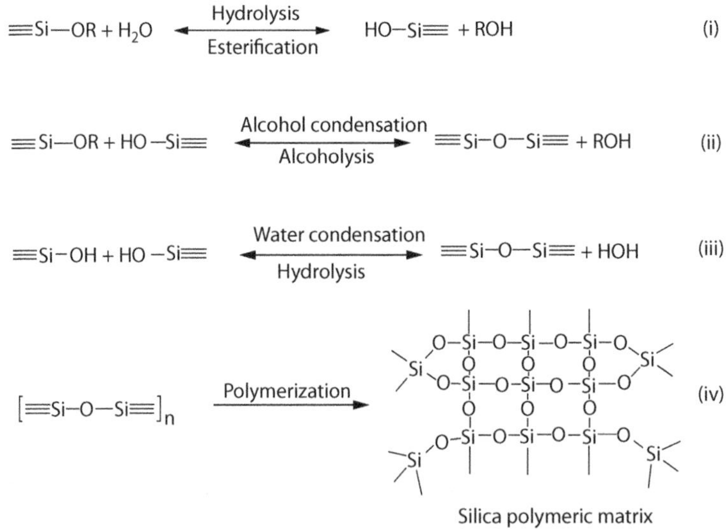

SCHEME 20.1 Chemical reactions in sol-gel synthesis. (Reproduced from Gupta, R. and Kumar, A., *Biomed. Mater.*, 3, 034005, 2008a, and references cited therein. With permission.)

in different configurations. Thin films can be produced on a piece of substrate by dip, spin, and spray coating. During the sol-gel transformation, the viscosity of the solution gradually increases as the sol becomes interconnected to form a rigid, porous network of gel. With further drying and heat treatment, the gel can be converted into dense ceramic or glass particles. During the drying process at ambient pressure, the solvent liquid is removed and substantial shrinkage occurs. The resulting material is known as a xerogel. When solvent removal occurs under supercritical conditions, the network does not shrink and a highly porous, low-density material known as an aerogel is produced. Heat treatment of a xerogel at elevated temperature produces viscous sintering (shrinkage of the xerogel due to a small amount of viscous flow) and effectively transforms the porous gel into a dense glass. As the viscosity of the sol is adjusted into a proper viscosity range, ceramic fibers can be drawn from the sol. Ultrafine and uniform ceramic powders are formed by precipitation, spray pyrolysis, or emulsion techniques (Diaz-Garcia and Badia Laino, 2005; Gupta and Chaudhury, 2007).

20.2.1 Merits and Demerits of Sol-Gel Method

Sol-gel-derived materials are inherently inhomogeneous in nature. Such heterogeneity is due to the variations in surface characteristics, viz., pore size, surface area and thickness uniformity (particularly in case of films), and local solvent composition, leading to the formation of a wide range of distinct physical and chemical environments in the dimensions of molecular scale. The synthetic (preparative) aging and drying conditions used for sol-gel processing determine the average physicochemical properties of the final material. These spatially varying properties influence the mechanical, optical, and other material characteristics. Better understanding of compositions vis-à-vis internal properties of these materials will thus provide a better way to control homogeneity.

Despite these limitations, the major advantages of sol-gel materials are: (a) compatible with many organic or inorganic reagents; (b) chemically, photo-chemically, and thermally stable as compared to organic polymer; (c) optically transparent and thus suitable for various spectroscopic-based analytical measurements; (d) cast as monoliths, coated as thin films on slides and fiber and ground into powder; (e) controllable surface area, average pore size, and its distribution and fractal dimensions which can be miniaturized to micron and even sub-micron (nano) size; (f) allow control of conductivity through the choice of the metal or metal alkoxide; (g) enhance the stability of the encapsulated molecules by virtue of the rigidity of the cage and most importantly prevent leaching of proteins due to the effective caging (Chaudhury et al., 2007).

20.2.2 Factors Affecting Sol-Gel Process

The dynamics of the sol-gel processes are dependent on various physical and chemical properties of the composition of the sol-gel, viz., R, type of catalyst, choice of precursors, pH, temperature, and solvent. The physicochemical properties of the internal environment are expected to undergo changes with the initial conditions as well as with storage (aging). The time courses of variation in these properties need to be understood and tailored for specific applications. In the following sections, factors affecting sol-gel process have been discussed.

20.2.2.1 Effect of Water-to-Alkoxide Molar Ratio

The hydrolysis reaction can be performed with R-values ranging from less than 1 to over 25 depending on the desired polysilicate product, e.g., fibers, bulk gels, or colloidal particles. An increased value of R is expected to promote the hydrolysis reaction. For small values of R, the alcohol-producing condensation reaction dominates (Scheme 20.1, Equation ii), whereas the water-producing condensation reaction becomes more important when R exceeds about 0.5. Higher values of R cause more complete hydrolysis of the monomers before significant condensation and also cause liquid–liquid immiscibility. However, alcohols produced as the by-product of the hydrolysis reaction and partial hydrolysis of the TEOS precursor lead to homogenization. Finally, because water is the by-product of the condensation reaction (Scheme 20.1, Equation iii), large values of R promote siloxane bond hydrolysis. The amount of water for hydrolysis has a dramatic influence on gelation time (t_g). Gel time decreased as the water content increased even though the sol became more dilute. This is because the silica network was more highly cross-linked when the value of R was high, evidenced by nuclear magnetic resonance (NMR) studies (Vega and Scherer, 1989). Thus, the condensation reaction was accelerated by excess water and the faster polymerization was started with reducing gelation time (Assink and Ray, 1988; Brinker and Scherer, 1990).

20.2.2.2 Effect of Catalyst

Hydrolysis is most rapid and complete when catalysts are used during sol-gel processing. Although mineral acids or ammonia are mostly used, other known catalysts are acetic acids, potassium hydroxide, amines, potassium fluoride, hydrogen fluoride, titanium alkoxide, vanadium alkoxide, and other metal oxides. Many authors reported that mineral acids are more effective catalysts than equivalent concentration of base. Sol-gel-derived silicon oxide networks, under acid-catalyzed conditions, yield primarily linear or randomly branched polymers which entangle and form additional branches resulting in gelation. On the other hand, silicon oxide networks derived under base-catalyzed conditions yield more highly branched clusters, which do not interpenetrate prior to gelation and thus behave as discrete clusters (Brinker and Scherer, 1990; Hench and West, 1990).

20.2.2.3 Effect of pH

The pH of sol plays a significant role in sol-gel processing as it affects hydrolysis, condensation, and gelation. Under acidic conditions, it is likely that an alkoxide group is protonated in a rapid first step (Scheme 20.2, Equation v) and electron density is withdrawn from silicon, making it more electrophilic and thus more susceptible to attack by water. Pohl and Osterholtz (1985) favored a transition state with significant bimolecular nucleophilic substitution reaction (S_N2)-type character. The water molecule attacks from the rear and acquires a partial positive charge. The positive charge of the protonated alkoxide is correspondingly reduced, making alcohol a better leaving group. The transition state decays by the displacement of alcohol accompanied by an inversion of the silicon tetrahedron. Under basic conditions, it is likely that water dissociates to produce nucleophilic hydroxyl anions in a rapid first step (Scheme 20.2, Equation vi). The hydroxyl anion then attacks the silicon atom. Iler (1979) and Keefer (1984) proposed an S_N2-Si mechanism in which OH– displaces OR– with inversion of the silicon tetrahedron. The condensation rate depends on pH of the sol. In silicate systems, the sticking probability is highest at intermediate pH, where the condensation rate is greatest.

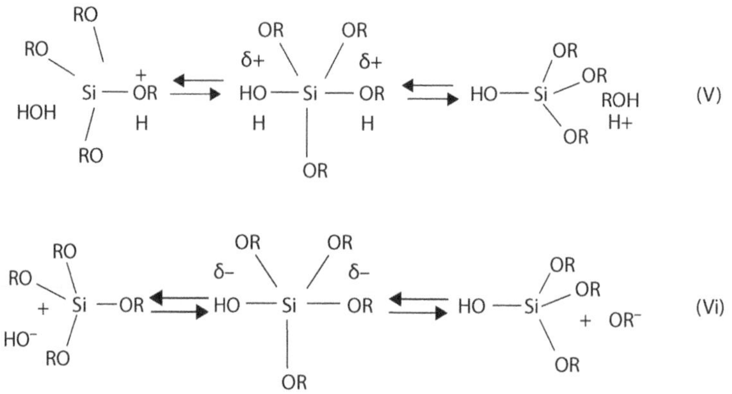

SCHEME 20.2 Mechanism of acid (Equation v) and base (Equation vi) catalyzed hydrolysis reaction.

The sticking probability is low near pH 2 and above about pH 10 where silanols tend to be deprotonated causing mutual repulsion of silicate particles. Gelation is the point when the condensing network has sufficient stiffness and yet is filled with solvent. The refractive index and pore volume are also very sensitive to pH. t_g is decreased by factors that increase condensation rate. Increase in the value of R, temperature, concentration of alkoxide, and decrease in the size of the alkoxy group reduce the t_g. At intermediate pH, gelation time is very low whereas at low pH, t_g is very high (Iler, 1979; Brinker and Scherer, 1990).

20.2.2.4 Solvent Effect

Traditionally, solvents are added to prevent liquid–liquid phase separation during the initial stages of the hydrolysis reaction and to control the concentration of silicate and water that influences the gelation kinetics. The availability of labile protons determines whether anions or cations are solvated more strongly through hydrogen bonding. Because hydrolysis is catalyzed either by hydroxyl or hydronium ions reduce the catalytic activity under basic or acidic conditions, respectively. Therefore, aprotic solvents that do not hydrogen bond to hydroxyl ions have the effect of making hydroxyl ions more nucleophilic, whereas protic solvents make hydronium ions more electrophilic. Hydrogen bonding may also influence the hydrolysis mechanism. The availability of labile protons also influences the extent of the reverse reactions, re-esterification or siloxane bond alcoholysis or hydrolysis (Scheme 20.1). Aprotic solvents do not participate in reverse reactions such as re-esterification or hydrolysis, because they lack sufficiently electrophilic protons and are unable to be deprotonated to form sufficiently strong nucleophiles (e.g., OH^- or OR^-). Alcohol is not simply a solvent. It can participate in esterification or alcoholysis reactions (Brinker and Scherer, 1990).

20.2.2.5 Effect of Solvent/Precursor Molar Ratio

Increased values of R generally promote hydrolysis. However, when R is increased while maintaining a constant solvent/silicate molar ratio, the silicate concentration is reduced. This in turn reduces the hydrolysis and condensation rates, causing an increase in the gel time. Increasing ethanol/precursor molar ratio at constant R also increases gel time due to decrease in viscosity of sols (Klein, 1988, Brinker and Scherer, 1990).

20.2.2.6 Effect of Salt

The addition of salt is well known to reduce gelation time due to change in the sol ionic strength, which reduces the surface charge and mutual repulsion of silica particles and hence increases the interparticle condensation reactions (Brinker and Scherer, 1990).

20.2.3 Sol-Gel-Derived Thin Films

Several analyte–matrix interactions in sol-gel-derived monoliths and glass particles are described in the literature. Sol-gel-derived thin films are desired because of the basic requirement of a short diffusion path for quick interaction and the detection of the analyte molecule. These thin films can be prepared by dip, spin, and spray-coating techniques. In case of dip coating, a high degree of thickness uniformity is achievable and can be controlled via the withdrawal speed. By comparison, thin film formation by spin coating causes a greater rate of solvent evaporation than dip coating and causing rapid changes in the physical properties of the sol-gel. The main factors that are important in the development of thin films are the uniformity and thickness of film, its adhesion to the substrate and resistance to cracking, designing of stable internal environment, and minimizing the potential of leaching of entrapped species. Ethanol is often used for homogenizing the immiscible water and alkoxide precursors in sol-gel processing and plays an important role in developing good optical-quality thin gel films. Sol-gel thin films have been prepared using sols diluted with alcohol and even in pure ethanol to decrease viscosity, enhance sol stability, and improve substrate wetting

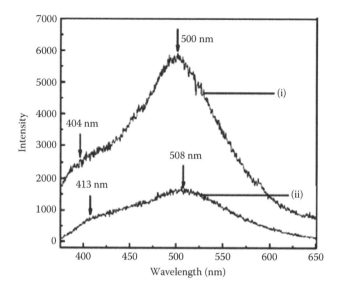

FIGURE 20.2 Emission spectra of H258 entrapped in (i) thin gel film, prepared at lower withdrawal speed (0.1 cm/min) using 45% ethanol concentration at constant water/TEOS ratio 4 at 35th day of observation and (ii) xerogel (1.5 year-old sample). (Reproduced from Gupta, R. and Chaudhury, N.K., *Biosens. Bioelectron.*, 22, 2387, 2007, and references cited therein. With permission.)

(Brinker et al., 1992). Gupta et al. (2005a) investigated the effect of ethanol concentration on internal environment of thin gel film prepared using different concentrations of ethanol from 15% to 60% at constant water-to-alkoxide molar ratio ($R = 4$). Fluorescent probe Hoechst 33258 (H258) entrapped in thin gel films prepared at 45% and 60% ethanol concentration showed dual emission bands at ~500 and 400 nm. This is because of the presence of mixed water and ethanol-like environment inside the pores of films. The emission characteristics of this molecule are sensitive toward solvent polarities. A higher concentration of ethanol led to a reduction in thickness and caused faster evaporation of entrapped solvent. Further, this caused the interaction between the probe molecule (H258) and silanol group as indicated by the blue band ~404 nm in thin gel films (Figure 20.2). Such a type of dual emission was also observed in bulk gel (xerogel) after 1.5 years storage, where aging as expected occurred very slowly (Figure 20.2). However, increasing ethanol concentration facilitates good coating but on the other hand it causes interaction between the sol-gel matrix and dopant. Thus, the film thickness should be optimum for biomolecules entrapment. Therefore, the conventional sol-gel procedures have been modified for the encapsulation of biological molecules using a buffer to bring the pH of the sol to a physiological range. A high concentration of acid and alcohol results in denaturation of most biological molecules (Ellerby et al., 1992). The thickness of the sol-gel-derived films is highly dependent on the gelation behavior of the sol, which depends upon the viscosity of the casting solution. The major factors affecting the rate of gelation of the hydrolyzed silane when mixed with buffer solution are: (a) ratio (R) and type of organosilane, (b) concentration and molecular weight of the polymer additives, and (c) type, concentration, and pH of buffer. The t_g increases with increase in amount of organosilane or polymer, buffer pH, and with decrease in buffer concentration. The gelation time also increases on going from TEOS to methyltriethoxysilane (MTES) to dimethyldimethoxysilane (DMDMS), suggesting that steric effect partially controls the gelation rate. The t_g decreases dramatically when phosphate buffer is used owing to phosphate-based catalysis of gelation. Porosity in thin gel films is also affected by sol pH (Goring and Brennan, 2002). Porous sol-gel thin films are a prerequisite for various biosensor applications; they can also be used as adsorbents, scaffolds for composite material synthesis, in separation technology, and molecular and biological engineering.

20.2.4 Characterization of Sol-Gel Matrix

The physicochemical nature of the local microenvironment within sol-gel-derived matrix is an important aspect in designing materials for sensing and bio-applications. Furthermore, the polarity, local microviscosity, interactions with pore walls, and preferential partitioning into a given phase experienced by the dopant will have an impact on the dynamics, stability, and accessibility of the interacting analytes.

A variety of analytical methods have been used to characterize the sol-gel materials. These include NMR (^{1}H, ^{29}Si, ^{17}O, and ^{13}C NMR), vibrational spectroscopy (Raman and infrared spectroscopy), small angle x-ray scattering (SAXS) and small angle neutron scattering (SANS), cryogenic gas adsorption analysis, thermogravimetry, and viscometry. However, these methods were applied for the bulk material and therefore reported the average properties surrounding the dopant molecules (Brinker and Scherer, 1990). Imaging ellipsometry has proven to be extremely valuable in determining thickness and refractive index profiles (Hurd and Brinker, 1988). Fourier transform infrared (FTIR) spectroscopy has been employed to monitor changes in solvent composition (Guglielmi and Zenezini, 1990). Brinker and Scherer (1985) used scanning electron microscopy (SEM), transmission electron microscopy (TEM), and nitrogen adsorption/desorption isotherm for the characterization of the physical structure of the desiccated gel. In recent years, fluorescence spectroscopy is the method of choice as it reports on the local microenvironment surrounding a probe molecule and facilitates the real-time characterization of biomolecular interactions. Several reports have appeared describing the use of fluorescence spectroscopy to characterize the internal environment of sol-gel-derived matrix and biomaterials. Steady-state and time-resolved fluorescence spectroscopy have been able to provide detailed information regarding the polarity, dynamics, and accessibility of the local environment within the sol-gel matrix, and how these factors change as a function of the preparation and aging protocols (Flora and Brennan, 2001).

In the literature, various studies have been reported related to the characterization of sol-gel matrix. Brinker and Scherer (1985) reviewed the mechanisms of gel formation in silicate systems derived from metal alkoxides. Kaufman and Avnir (1986) reported the use of photophysical probes for studying the sol-gel transition of silicon tetra-alkoxide undergoing hydrolysis and condensation–polymerization reactions. Kaufman et al. (1988) studied the detailed kinetics of water consumption during the early stages of the TMOS sol-gel polymerization process. Changes in water/silane molar ratio and in pH were found to affect markedly the kinetic behavior of water consumption. Matsui et al. (1989) observed significant shift (380–430 nm) in fluorescence emission maximum of 7-Azaindole (7-AI) during the gel-xerogel stages. Pouxveil et al. (1989) prepared aluminosilicate sols and gels doped with Pyranine (Py) by hydrolysis and polycondensation of an organometallic precursor. They showed that the water content of the medium surrounding the dye mainly controlled the ratio of the green and blue emission peaks, which reflected the relative degree of protonation of the excited Py molecules. Kinetic and structural information concerning the gel and its interaction with the dye were discussed. Narang et al. (1994a) provided detailed information on how the various stages of the sol-gel process, affect the photophysics of 6-propionyl-2-dimethylaminonaphthalene (PRODAN), an extremely solvent sensitive fluorescent probe, using steady-state emission anisotropy and time-resolved emission spectroscopy. Narang et al. (1994d) used static and dynamic fluorescence spectroscopy of Rhodamine 6G (R6G) in a sol-gel matrix to study the effect of aging time and hydrolysis pH on the local microviscosity. Dynamic anisotropy experiments showed that the R6G rotational reorientation dynamics were composed of two independent rotational correlation times. Lu et al. (1997) reported a sol-gel-based dip-coating method for the rapid synthesis of continuous mesoporous thin films on a solid substrate. They used fluorescence depolarization to monitor the evolution of the mesophase *in situ*, and see a progression through a lamellar to cubic to hexagonal structures. Klotz et al. (1999) discussed the problems related to the characterization of the porosity in the case of thin films and focused on the importance of the choice of the various synthesis parameters to tailor the porosity of the final material. Bonzagni et al. (2000) used steady-state and time-resolved fluorescence spectroscopy to determine the local microheterogeneity surrounding Pyrene molecules sequestered within TMOS-derived xerogel. Huang et al. (2000) used Py and PRODAN as *in situ* fluorescence probes to monitor the molecular mobility and chemical evolution during sol-gel silica thin film deposition by the dip-coating process. The internal environment of both bulk and thin gel films prepared from different sol compositions using various fluorescent molecules has been investigated as a function of long-term storage using fluorescence spectroscopy (Gupta et al., 2005a,b; Gupta and Chaudhury, 2009).

In addition, various studies have been reported to characterize the entrapped biological molecules within sol-gel matrix. These studies showed unambiguously that large-scale dynamics of biomolecules are strongly hindered in the glassy cage. Various physical forces, e.g., specific electrostatic interactions between silicate sites and protein surface residues and mechanical forces have been implicated to reduce flexibility of the entrapped proteins (Gupta and Chaudhury, 2007). Flora and Brennan (2001) studied the effect of storage and aging conditions on conformation and dynamics of protein human serum albumin (HSA), entrapped in TEOS-derived monoliths using time-resolved anisotropy decay measurements. Goring and Brennan (2002) characterized the organically modified silanes (ORMOSILS) precursors, viz., MTES, DMDMS, and polymer polyethylene glycol (PEG)-derived TEOS thin films having entrapped PRODAN or the protein HSA using steady-state emission measurement. In 2005, Brennan and coworkers characterized the silica monolith derived from TEOS along with biocompatible silane precursor, viz., diglycerylsilane (DGS) and containing covalently bound sugar moiety, viz., gluconamidylsilane (GLS) and showed that such derived matrix is much more biocompatible for protein entrapment than any conventional synthesized materials (Sui et al., 2005).

20.3 RECENT DEVELOPMENTS IN SOL-GEL PROCESSING

Although the utility of sol-gel matrices as hosts for organic and organometallic dopants was known for long, they are still attracting increased attention in basic research for designing appropriate sol compositions for the development of host matrix for biological applications. Conventional sol-gel procedures are usually unsuitable for the encapsulation of biomolecules because of high acidic condition and/or high concentration of alcohol, which lead to denaturation of biomolecules. Early in 1971, Johnson and Whateley reported their investigation on the use of silica gel for the immobilization of trypsin, an enzyme with esterase activity. They demonstrated that trypsin entrapped in silica gel was not removed by washing. They observed that the esterase activity of the trapped trypsin was 34% relative to solution whereas 10% loss in activity was observed when stored for 75 days at 4°C (Johnson and Whateley, 1971). In 1990, Braun and coworkers first reported the successful encapsulation of a purified enzyme alkaline phosphatase in a TMOS-derived sol-gel material. The immobilized enzyme exhibited 30% activity yield and improved thermal stability compared to solution. The bioactive glass was preserved in water for 2 months without losing activity (Braun et al., 1990). Perhaps the most significant development in this area has been reported by Ellerby et al. (1992) for a new mild procedure for protein encapsulation in sol-gel matrix without loss of reactivity or spectroscopic properties. The important features of this procedure were omission of alcohol and the raising of the pH of the precursor solution to physiological range after HCl-catalyzed hydrolysis of sol but before the addition of protein. In particular, these workers demonstrated a range of reversible spectroscopic reactions involving encapsulated metalloproteins such as copper-zinc superoxide dismutase (CuZnSOD) and established the viability of sol-gel-derived materials.

20.3.1 ORGANIC–INORGANIC HYBRID SOL-GEL MATERIALS

To date, most studies on sol-gel entrapped biomolecules have made use of the silane precursors TMOS and TEOS. Generally TEOS-based glasses are not likely to be amenable to practical applications, owing to long-term alterations in protein conformation and ligand binding; therefore, new sol-gel processing methods using different precursors, additives, and aging methods will be necessary to produce second-generation glasses for functional stabilization of biomolecules in native forms. Surface characteristics as well as uniformity in monoliths/thin films are one of the desirable criteria for sensing applications. During the drying phase, some of the larger pores are emptied while smaller pores remained wet by the solvent, creating large internal pressure gradients. This stress causes cracks in large monoliths and is also responsible for fractures in dry monolithic sensors upon immersion in water. Such cracks and fractures can be prevented by adding surface-active drying control chemical additives such as Triton-X and formamide to the sol-gel precursor solution. The incorporation of cationic surfactants such as cetyl pyridinium bromide has been proposed to prevent fractures of monoliths during gelation and on repeated wet dry cycles. These compounds form electrostatic bonds with deprotonated silanol groups, remain in the pores, and prevent drying fractures even after repeated immersion in aqueous solutions. Recently ORMOSILS, e.g., MTES, propyltrimethoxysilane (PTMS), DMDMS etc., have been employed in multifarious applications in industrial and medical fields and showed promising results in preserving the native activity of biomolecules. Table 20.1 lists some inorganic and most commonly used ORMOSILS precursors. The introduction of various functional groups such as amino, glycidoxy, epoxy, hydroxyl, etc., into alkoxide monomers leads to organically modified sol-gel glasses. ORMOSILS have several attractive features as compared to inorganic sol-gel and provides a versatile way to prepare modified sol-gel materials. The wettability of composite material can be tuned by a judicious choice of the ratio of hydrophilic to hydrophobic monomers. ORMOSILS and polymers are suitable for the retention of enzyme activity in sol-gel, evidenced from the studies on the use of polycationic polymers into ORMOSILS materials showing improved performance of flavoproteins. In addition, it was also reported that the incorporation of copolymers into silica-based glasses improved the activity of entrapped glucose oxidase (GOx) for amperometric detection of glucose. The biomolecules such as atrazine chlorohydrolase, lipase and HSA, entrapped in ORMOSILS showed improved performances including storage stability, excellent activity retention, etc. Due to these promising advantages, several enzymes have been successfully encapsulated into ORMOSILS and employed in design of biosensors. Further, the addition of polymers, viz., polydimethylsiloxane (PDMS), polyamides, polyacrylates, and PEG to regulate the inorganic condensation–polymerization process is also under investigation for improving the properties of sol-gel materials. Polyethers were also used in sol-gel processing mixtures to control pore size distribution. Addition of PEG to films improved the resistance of the films to cracking probably owing to greater hydration of the films during aging and hence a lower extent of hydration stress during rehydration. PEG doping also increased dynamics of biomolecule relative to undoped TMOS-derived composites. A large reduction in surface area was observed with PEG doping but no detectable change in pore size was reported. Despite these promising results, PEG depicted undesired alterations in encapsulated proteins. Tubio et al. (2004) showed that the quenching of the albumin tryptophan fluorescence by acrylamide in the presence of PEG was affected, because it separates the quencher molecule from the fluorophore, thus making the access of acrylamide to the tryptophan difficult. Dissociation of the phosphofructokinase tetrameric enzyme was reported in the presence of PEG (Reinhart, 1980) and altered the UV absorption spectrum of ribonuclease in 270–290 nm regions due to the modification of the tyrosine residue microenvironment (Poklar et al., 1999). PEG induced a displacement of the fluorescent probe 1-anilinonaphthalene-8-sulfonic acid (ANS) from its binding site in human albumin (Tubio et al., 2004). These results indicated compromised mobility and altered conformation of entrapped proteins. The inclusion of additives, viz., sorbitol and N-methylglycine (collectively referred to as osmolytes) during the immobilization of proteins into sol-gel-processed materials has been widely explored as a route to stabilize proteins against the denaturing stresses encountered upon entrapment. This has also increased thermal stability and biological activity of the encapsulated proteins by altering the hydration of the entrapped protein and increasing the pore size of the silica material, which improves substrate delivery, and thus activity as well as thermal stability. Thus, by the appropriate use of polymer dopants as well as the use of ORMOSILS, one can alter the ultimate

TABLE 20.1
List of Inorganic and Commonly Used ORMOSILS Precursors

S. No.	Inorganic Precursors	ORMOSILS
1.	C_3H_7O—Zr—OC_3H_7 with OC_3H_7 up and OC_3H_7 down Tertrapropyl - orthozirconate (TPOZ)	H_3CO—Si—OCH_3 with C_3H_7 up and OCH_3 down Propytrimethoxysilane (PTMS)
2.	C_2H_5O—Ti—OC_2H_5 with OC_2H_5 up and OC_2H_5 down Tetraethyl - orthotitanate (TEOT)	C_2H_5O—Si—OC_2H_5 with CH_3 up and OC_2H_5 down Methyltriethoxysilane (MTES)
3.	C_4H_9O—Al—OC_4H_9 with OC_4H_9 up Aluminium *tert* - butoxide	H_3CO—Si—OCH_3 with CH_3 up and CH_3 down Dimethyldimethoxysilane (DMDMS)
4.	H_3CO—Si—OCH_3 with OCH_3 up and OCH_3 down Tetramethyl - orthosilicate (TMOS)	C_2H_5O—Si—OC_2H_5 with CH_3 up and CH_3 down Dimethyldiethoxysilane (DMDES)
5.	C_2H_5O—Si—OC_2H_5 with OC_2H_5 up and OC_2H_5 down Tertraethyl - orthosilicate (TEOS)	(phenyl)—Si—OCH_3 with OCH_3 up and OCH_3 down Phenyltrimethoxy silane (PhTMOS)
6.	C_3H_7O—Si—OC_3H_7 with OC_3H_7 up and OC_3H_7 down Tetra - *n* - propoxysilane	(phenyl)—Si—OC_2H_5 with OC_2H_5 up and OC_2H_5 down Phenyltriethoxysilane (PhTEOS)
7.	C_4H_9O—Si—OC_4H_9 with OC_4H_9 up and OC_4H_9 down Tetrabutoxysilane	C_2H_5O—Si—OC_2H_5 with phenyl up and phenyl down Diphenyldiethoxysilane (DPDES)

TABLE 20.1 (continued)
List of Inorganic and Commonly Used ORMOSILS Precursors

S. No.	Inorganic Precursors	ORMOSILS
8.		$H_2C=CH-\overset{\overset{\displaystyle OC_2H_5}{\vert}}{\underset{\underset{\displaystyle OC_2H_5}{\vert}}{Si}}-CH_3$
		Methylvinyldiethoxysilane (MVDES)
9.		$H_3CO-\overset{\overset{\displaystyle OCH_3}{\vert}}{\underset{\underset{\displaystyle OCH_3}{\vert}}{Si}}-CH_2CH_2CH_2NH_2$
		3-Aminopropyltrimethoxysilane (APTMS)
10.		$H_3CO-\overset{\overset{\displaystyle OCH_3}{\vert}}{\underset{\underset{\displaystyle OCH_3}{\vert}}{Si}}-CH_2CH_2CH_2OCH_2$
		Glycidoxypropyltrimethoxysilane (GPTMS)
11.		$H_3CO-\overset{\overset{\displaystyle OCH_3}{\vert}}{\underset{\underset{\displaystyle OCH_3}{\vert}}{Si}}-H_2CH_2C$ ... $CH_2CH_2\cdot\overset{\overset{\displaystyle OCH_3}{\vert}}{\underset{\underset{\displaystyle OCH_3}{\vert}}{Si}}-OCH_3$
		Bis-(trimethoxysilylethyl)benzene (BTEB)

Source: Reprinted from Gupta, R. and Kumar, A., *Biotechnol. Adv.*, 26, 533, 2008b. With permission.

physicochemical properties of the material produced and may generate new customized platforms for bio-applications with improved analytical figures of merit (Tripathi et al., 2006; Gupta and Chaudhury, 2007).

20.3.2 BIOCOMPATIBLE SOL-GEL MATERIALS

Typical applications of sol-gel biomaterials include selective coatings for optical and electrochemical sensors and biosensors, stationary phases for affinity chromatography, immunoadsorbent and solid-phase extraction materials, controlled release agents, solid-phase biosynthesis, and unique matrices for biophysical studies (Podbielska and Ulatowska-Jarza, 2005). In recent years, a number of sol-gel-derived materials have been designed with the purpose of making the matrix more compatible with entrapped biological molecules for various biological applications. Gill and Ballesteros (1998) prepared polyglyceryl silicate (PGS) from a new class of precursor polyol silicates, polyol siloxanes, and glycerol for bioencapsulation under high biocompatibility and mild encapsulation conditions, which enabled the reproducible and efficient confinement of proteins and cells inside silica. The methodology was extended to metallosilicate, alkylsiloxane, functionalized siloxane, and composite sol-gels for fabrication of a physicochemical diverse range of biodoped polymers. The activities of hybrid materials were similar to those of the free biologicals in solution. In fact, the bioencapsulates performed better than those fabricated from TMOS, poly(methyl silicate), or alcohol-free poly(silicic acid) even when the latter were doped with glycerol. Jin and Brennan (2002) suggested that production of methanol or ethanol during sol-gel processing is detrimental to entrapped proteins and can lead to significant changes in the properties of the enzyme, including the Michaelis constant (K_m), catalytic constants (k_{cat}), and inhibition constant (K_I). To overcome ethanol effect, recently a number of new biocompatible silane precursors and processing methods have been reported that are based on glycerated silanes, sodium silicate, or aqueous-processing methods that involve removal of alcohol by-product by evaporation before the addition of protein. Besanger et al. (2003) have reported the development of the new silane precursor DGS, which is capable of maintaining entrapped enzymes in an active state for a significant amount of time due to the liberation of biocompatible reagent glycerol from DGS. In 2005, Brennan and coworkers reported that silica derived from biocompatible silane precursor, viz., DGS and containing covalently bound sugar moiety, viz., GLS is a much more biocompatible matrix for protein entrapment than any conventional synthesized materials. This was due to the release of protein-stabilizing compounds such as glycerol, upon hydrolysis of DGS (Sui et al., 2005). The same group reported first study on the successful immobilization of a fluorescence-signaling DNA aptamer within biocompatible sol-gel-derived materials, using sodium silicate and DGS precursors. Aptamers are single-stranded nucleic acids that are generated by *in vitro* selection. The high affinity of aptamers, their properties of precise molecular recognition, and the simplicity of *in vitro* selection make aptamers attractive as molecular receptors and sensing elements. Authors demonstrated that aptamers containing a complementary dabcyl-labeled nucleotide strand (QDNA) along with either a short complimentary strand bearing fluorescein (tripartite structure) or a directly bound fluorescein moiety (bipartite structure), remained intact upon entrapment within biocompatible sol-gel-derived monoliths and retained binding activity, structure-switching capabilities, and generated fluorescence signal, which is selective and sensitive to ATP concentration. Further, different properties of immobilized aptamers have been evaluated including response time, accessibility, and leaching. The properties of immobilized aptamers within sol-gel-derived monoliths were found to be similar to solution, with moderate leaching, only minor decreases in accessibility to ATP, and an expected reduction in response time (Rupcich et al., 2005a). Lin et al. (2007) prepared a miniaturized horseradish peroxidase (HRP)-entrapped bioreactor by one-step enzyme immobilization method using a biocompatible sol-gel processing method employing either DGS or sodium silicate as precursors and a covalently tethered sugar, N-(3-triethoxysilylpropyl) gluconamide as a silica modifier. Factors such as leaching, catalytic efficiency, and long-term stability were examined to assess the role of the precursor and modifier in influencing enzyme performance. The results showed that sodium-silicate-derived materials modified with covalently bound sugars at a level of 10 mol% were optically transparent and provided the highest catalytic turnover rate for entrapped HRP. The stability and reusability of the entrapped HRP was found to be satisfactory for at least 1 month in the GLS-doped sodium silicate materials, and the entrapped HRP was able to respond linearly to the presence of peroxide over the concentration range of 0–750 μM with a detection limit of 6 μM, demonstrating the potential of this material for the development of a reusable optical biosensor. These studies demonstrate that biocompatible sol-gel-derived materials have significant versatility for the entrapment of a range of biomolecules, extending the potential biotechnological applications of such materials.

20.3.3 Bioactive Sol-Gel Materials

Applications utilizing sol-gel as a porous material to encapsulate sensor molecules, enzymes, and many other compounds are most common; however, some potential applications of sol-gel-derived materials in biomedical applications are fast emerging. Biomedical applications require the design of new biomaterials and this can be achieved by merging sol-gel chemistry and biochemistry. The gel-derived materials are excellent model systems for studying and controlling biochemical interactions within constrained matrices with enhanced bioactivity because of their residual hydroxyl ions, micropores, and large specific surface. In all biomedical applications, the coating of the medical devices is an important issue. Materials used in medical devices should have appropriate structural and mechanical properties and ideally promote a healing response without causing severe bodily reactions. Medical device designers use various surface treatments such as coating that enhance or modify properties such as lubricity, hydrophilicity/hydrophobicity, functionality, and biocompatibility. Sol-gel technology offers an alternative technique for producing bioactive surfaces for various biomedical applications. Sol-gel thin film processing offers a number of advantages including low-temperature processing, ease of fabrication, and precise microstructural and chemical control. The sol-gel-derived film or layer not only provides a good degree of biocompatibility, but also a high specific surface area (which can be used as a carrier of adsorbed drugs) and an external surface whose rich chemistry allows easy functionalization by suitable biomolecules. Also, controlling the thickness and pore-size distribution of the silica coating provides a direct method to tailor the rate and duration of drug release. The rates of diffusion and release of a drug are correlated with the thickness and porosity of coated films that can be controlled via the withdrawal speed and sol compositions using the sol-gel dip-coating technique. Gao et al. (2005) evaluated the effects of the amount of channeling agents, the addition of colloidal silica, and the pH of the dissolution media on the release of the drug hydrochlorothiazide from compressed tablets. These tablets were spray coated using PDMS lattices with various PEG loadings as channeling agents. The rate of drug release was found to be constant in coated tablets containing up to 25% (w/w) PEG. Higher amounts of PEG resulted in nonlinear release patterns. The addition of colloidal silica decreased the rates of drug release. The pH of dissolution media affected the structures of the exposed PDMS films. SEM and density measurements showed that the films obtained after soaking in higher-pH media were more condensed, with corresponding changes in drug-release rates. Radin and Ducheyne (2007) described the synthesis of thin, resorbable, controlled-release bactericidal sol-gel films on a Ti-alloy substrate and determined the effect of processing parameters on its degradation and vancomycin release. Vancomycin is a potent antibiotic used in treating osteomyelitis. A close correlation between release and degradation rates suggested that film degradation is the main mechanism underlying the control of release and depends upon sol-gel processing parameters. The bactericidal properties of released vancomycin and the biocompatibility of the sol-gel films suggest great potential to prevent and treat bone infections in a clinical setting. The development of magnetic nanoparticles that can be used as drug delivery vectors remains a significant challenge for material scientists. Fernandez-Pacheco et al. (2004) described a simple and inexpensive method for the preparation of encapsulated magnetic nanoparticles consisting of a metallic iron core and an amorphous silica shell by using a modification of the arc-discharge method. The nanoparticles thus obtained present a much stronger magnetic response than any composite material produced up to now involving magnetic nanoparticles encapsulated in inorganic matrices, and the rich chemistry and easy functionalization of the silica outer surface make them promising materials for their application as magnetic carriers and can allow the binding of antibodies, proteins, medical drugs, or other biomolecules to the system. Energy-filtered transmission electron microscopy (EFTEM) of silica-coated iron nanoparticles is shown in Figure 20.3. Silica coating helps to make the particles biocompatible, preventing their aggregation and the degradation of the metallic core, and reducing the extent of clearance by the reticuloendothelial system. Beganskiene et al. (2007) prepared and characterized the modified sol-gel-derived silica coatings in which amino and methyl groups were introduced onto the colloidal silica. The coatings of colloidal silica (water contact angle 17°), polysiloxane sol (61°) methyl-modified sols (158° and 46°) with various wettability properties were tested for cell proliferation. Methyl-modified coating has proved to be the best substrate for cell proliferation. There is an increasing interest in the use of optical techniques for applications such as local treatment of tumors. Recently, much effort has been directed toward the development of novel methods including photodynamic therapy (PDT), which means the nonthermal destruction of tumors by the

FIGURE 20.3 Energy-filtered transmission electron microscopy (EFTEM) of silica-coated iron nanoparticles. (Reproduced from Fernandez-Pacheco, R. et al., *Nanotechnology*, 17, 1188, 2004. With permission.)

combined action of a chemical compound (photosensitizer) and low- or medium-energy laser radiation and interstitial laser-induced thermotherapy with laser energy, which is far lower than that used in ordinary laser surgery. Interstitial laser-induced thermotherapy is a quite new treatment modality designed for minimal invasive destruction of pathologic tissues in difficult-to-access environments (e.g., brain, liver). Fiber-optic laser applicators are used to perform interstitial therapy with laser light, where the applicator is inserted into the pathologic lesion and curing laser light is guided through the fiber. Silica-based sol-gel coatings are also used for the production of fiber-optic applicators for laser therapies as sol-gel-derived materials are optically transparent, frequently used for the construction of fiber-optic sensors, and allow optical measurements. These are also relatively safe and biocompatible for use within the human body. Sol-gel coatings on fiber cores depending on the value of R influence the light distribution and so it is possible to obtain various shapes of light beam emitted from the applicator (Gupta and Kumar, 2008a).

20.3.4 Molecular Imprinted Sol-Gel Materials

The concept of molecular imprinting is based on molecular interaction and has since then found applications in separation processes (chromatography, capillary electrophoresis, solid phase extraction (SPE), membrane separations), microreactors, immunoassays and antibody mimics, catalysis and artificial enzymes, biosensor recognition elements, and bio- and chemo-sensors. The field of molecularly imprinted polymers (MIPs) is extending with new applications such as recognition elements in intelligent drug delivery devices, in targeted drug delivery applications and in microfluidics devices with applications as analyte-sensing microvalves and microactuators. The molecular imprinting technique can be applied to different kinds of target molecules, ranging from small organic molecules (e.g., pharmaceuticals, pesticides, amino acids and peptides, nucleotide bases, steroids, and sugars) to polypeptides, high molecular proteins, and even whole cells. The principle underlying molecular imprinting is the assembly of a cross-linked polymer matrix around a template; when the template is removed, recognition sites are created, which are complementary to the template. The principle of molecular imprinting is shown in Figure 20.4. Generally, the fabrication of MIPs consists of three main steps: (a) prearrangement of the monomers around the target molecule, (b) polymerization in the presence of cross-linker, and (c) removal of the target molecule by extraction process. These MIPs can be stable in various critical chemical and physical conditions for a long time and reused without any alteration to the memory of the template. Usually MIPs have been prepared in the form of a macroporous monolith, then ground and sieved to the required

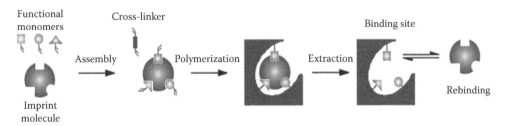

FIGURE 20.4 Schematic representation of molecular imprinting. (From Haupt, K., *Analyst*, 126, 747, 2001. With permission.)

particle dimensions. Recent improvements in the morphology of MIP particles have been achieved using a precipitation polymerization procedure that allows to obtain micro- or nanospheres with regular size and shape and particularly able to rebind effectively template molecule due to the high surface/volume ratio. The use of imprinted films in sensor development has attracted a lot of interest recently, because of the faster rate of diffusion through surface resulting in high sensitivity. In recent years, several groups have dealt with the application of imprinted films as recognition layers applied on various transduction systems, e.g., piezoelectric, amperometric, surface plasmon resonance (SPR), fluorimetric, and field effect transistors (FET) (Gupta and Kumar, 2008b).

In the literature, mostly MIPs were derived from organic polymers synthesized from vinyl or acrylic monomers by radical polymerization and using non-covalent interactions. These monomers can be basic, e.g., vinylpyridine or acidic, e.g., methacrylic acid, permanently charged, e.g., 3-acrylamidopropyltrimethylammonium chloride, hydrogen bonded, e.g., acrylamide, hydrophobic like styrene, or metal coordinating, etc. There are very limited reports where inorganic molecularly imprinted polymers (IMIPs) are being used. In 1949, Dickey observed that the adsorption of template methyl orange was more in imprinted silica gel as compared to non-imprinted gel. This publication appeared as the first documented demonstration of molecular imprinting in sol-gel matrix (Dickey, 1949). Pinel et al. (1997) used sol-gel chemistry to generate imprinted gel with (−) menthol as template. In contrast to Dickey's observation, there was no difference between imprinted and non-imprinted gel for the adsorption of template molecule. Hunnius et al. (1999) used sol-gel process to prepare new catalyst materials. Interestingly, they observed that the amorphous microporous oxide remembered the kinetic diameter of imprinted molecule (alcohol). His investigation provided another door for molecular imprinting to enter into inorganic oxides, particularly sol-gel materials. Silica-based materials are extremely rigid due to the high degree of cross-linking found in the (SiO_2) n network. This property is very important in the design and synthesis of imprinted materials, since both the size and shape of the cavities created by the template must be retained after the removal of the template. High thermal stability of sol-gel-derived material provides an easy way to remove imprint molecule using high temperature such as in calcinations method. In addition, sol-gel glasses are structurally porous and can be engineered to have extremely high surface area. These properties make silica sol-gel matrix an imprinting host. The selection and type of precursor play an important role in achieving selectivity in the resulting sol-gel MIPs. From a chemical and material standpoint, sol-gel-derived materials have a combination of properties which can hardly be achieved by other materials (Gupta and Kumar, 2008b).

20.3.4.1 Surface Sol-Gel Molecular Imprinting

In molecular imprinting, imprinted materials exhibit high affinity and selectivity but poor site accessibility to the target molecules because the template and functionality are totally embedded in the polymer matrices. Therefore, the kinetics of the sorption/desorption process is unfavorable and the mass transfer becomes slow. This problem can be overcome by using surface molecular imprinting, in which the imprinted materials with binding sites situated at the surface show many advantages including high selectivity, more accessible sites, fast mass transfer, and binding kinetics.

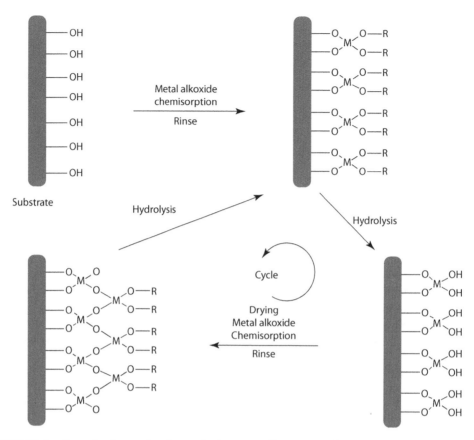

FIGURE 20.5 General scheme of surface sol-gel process. (Reproduced from Kunitake, T. and Lee, S.W., *Anal. Chim. Acta*, 504, 1, 2004. With permission.)

Traditionally, surface sol-gel process is used to prepare ultrathin metal oxide films. Since the sol-gel reactions proceed only on the surface, process is called surface sol-gel process. In this process, a solid substrate with hydroxyl groups on its surface is allowed to react with metal alkoxides in solution to form covalently bound surface monolayers of the metal alkoxide. The excessively adsorbed (physisorbed) alkoxide is removed by rinsing. The chemisorbed alkoxide monolayer is then hydrolyzed to give a new hydroxylated surface. The schematic of surface sol-gel process is shown in Figure 20.5. A very thin layer (~1 nm) of metal oxide layer can be deposited by using surface sol-gel process and can be repeated many times to give desired multilayers on the surface of support. Various guest molecules like organic, polymeric, biological, and metallic materials are readily adsorbable or imprinted using functionalized groups on metal oxide layer and removed by solvent washing. Thus, surface sol-gel process combined with molecular imprinting technique provides a new platform for various bio-applications (Gupta and Kumar, 2008b). Lee et al. (1998) demonstrated that the surface sol-gel process is superior as a means of molecular imprinting to the past imprinting techniques because in conventional sol-gel processes, time-consuming procedures of gel formation, pulverization, and the extraction process creates problems in creating imprinted cavities. Li et al. (2008) reported surface molecular imprinting in combination with sol-gel for protein recognition. The functional biopolymer chitosan (CS) as microsphere was chosen as polysaccharide core for the surface imprinting of bovine serum albumin (BSA) via covalent linkage. These microspheres were surrounded by (3-aminopropyl) trimethoxysilane (APTMS) and TEOS-derived hybrid sol-gel polymeric matrix in aqueous solution at room temperature (Figure 20.6). After template removal the protein-imprinted sol-gel surface exhibited a prevalent preference for the template

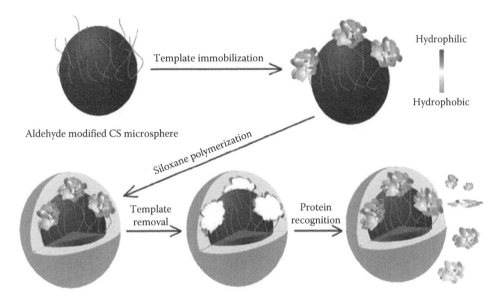

FIGURE 20.6 Schematic representation for synthesis of the protein-imprinted polymer on Chitosan microsphere using immobilized protein template. (Reproduced from Li, F. et al., *Talanta*, 74, 1247, 2008. With permission.)

protein in adsorption experiments, as compared with four contrastive proteins. The complementation in hydrophilicity/hydrophobicity was a major factor affecting the imprinting formation and template recognition. The grafting of imprinted layer through interfacial organic–inorganic hybridization improved the stability and reproducibility properties of the final material.

20.3.4.2 Bio-Imprinting

Detecting biomolecules (proteins, cells and microorganisms) in different matrices is becoming an increasingly important task in a variety of fields including bioprocess control, food technology, health care, and environmental analysis. Although molecular-imprinting technique is widely developed for imprinting organic molecules, this approach has not been adequately investigated for biomaterial applications. Very few reports are available on bio-imprinting in sol-gel. Zhang et al. (2006) described a novel method of combining sol-gel and self-assembly technology to prepare HSA-imprinted film on the surface of piezoelectric quartz crystal (PQC) Au-electrode modified with thioglycolic acid which was placed in a phenyltrimethoxysilane (PTMOS) and methyltrimethoxysilane (MTMOS)-derived sol. The effect of temperature, salts, and solvents on the performance of the sol-gel-imprinted film in air or in buffer was investigated. In this study, self-assembly sol-gel-imprinting technique proved to be an alternative method for the preparation of biomacromolecule-imprinted thin film. Dickert and Hayden (2002) presented the combination of mass-sensitive transducer with a surface-imprinting technique, as an ideal *in situ* analytical system for studying the interactions of biopolymers or even whole cells with surfaces. The regular patterns of molded polymer and sol-gel surfaces with yeast cells (*Saccharomyces cerevisiae*) as template were prepared that showed honeycomb-like structures as shown in Figure 20.7 and monitored cell concentration ranging from 10^4 to 10^9 cells/mL in flowing conditions of 10 mL/min. No unspecific adhesion of microorganisms was detected on non-imprinted polymers under flowing conditions. Lee et al. (2007) imprinted proteins (lysozyme or RNase A) in macroporous polysiloxane (silica) scaffolds using sol-gel processing. The quantity of surface-accessible protein, related to the number of potential binding sites was varied by changing the amount of protein loaded into the sol. Up to 62% of loaded protein was accessible. The imprinted scaffolds exhibited up to three times preference for binding their template molecules, even in the presence of a similar-sized competitor. The scaffolds

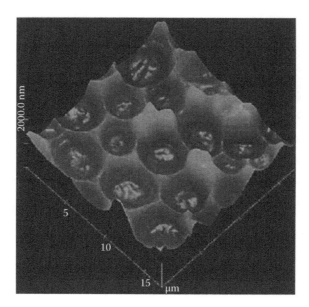

FIGURE 20.7 Sol-gel layer from titanium[IV] ethylate imprinted with *S. cerevisiae*. The coatings are extremely robust and scratch resistant. (Reproduced from Dickert, F.L. and Hayden, O., *Anal. Chem.*, 74, 1302, 2002. With permission.)

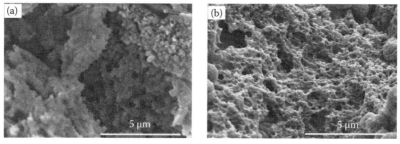

FIGURE 20.8 Higher magnification SEM images of protein (lysozyme)-imprinted scaffolds (a) before and (b) after exposure to protease to digest surface-accessible protein. (Reproduced from Lee, K. et al., *Acta Biomater.*, 3, 515, 2007. With permission.)

were highly textured and included micro-, meso-, and macropores. Figure 20.8a shows a higher magnification image of an imprinted scaffold, with noticeable protein embedded on the surface and Figure 20.8b shows that submicron-sized pores were revealed on the surface after the scaffold was exposed to protease solution to digest the surface-accessible protein.

20.3.4.3 Templated Xerogels

Sol-gel-derived xerogels are attractive for bioengineering applications because of the possibility of tailoring a xerogel's physicochemical properties by altering sol-gel processing conditions and exhibiting remarkable stability over time. Chambers group has prepared molecularly imprinted xerogels for the recognition of protein Ricin and investigated the protein–xerogel interactions using intrinsic fluorescence from the tryptophan residues within Ricin (Lulka et al., 2000). Despite the obvious attraction of luminescence-based detection, xerogels, and molecular imprinting, researchers have not developed a protein detection strategy that exploits the power of luminescence, the tunability of xerogels, and molecular imprinting. Tao et al. (2006) reported a new strategy for fabricating protein-responsive biosensor based on sol-gel-derived molecularly imprinted xerogels termed as "protein imprinted xerogels with integrated emission sites (PIXIES)," and described a methodology for rapidly producing and screening a wide variety of sol-gel-derived xerogel-based formulations and compared the analytical figures of merit for PIXIES to standard antibody-based assays (ELISA). The PIXIES platform is completely self-contained, and it achieves analyte recognition without a biorecognition element (e.g., antibody). In operation, the templated xerogel selectively recognizes the target analyte, the analyte binds to the template site, and binding causes a change in the physico-chemical properties within the template site that are sensed and reported by the luminescent probe molecule.

20.3.5 BIO-HYBRID NANOSTRUCTURED MATERIALS

Recent improvements in engineering at the nanoscale level have led to the development of a variety of novel nanostructured materials, viz., quantum dots, nanoshells, gold nanoparticles, paramagnetic nanoparticles, carbon nanotubes (CNTs), which after combining with biomolecules such as polysaccharides, polyesters, RNA and DNA, polypeptides, fibrous and globular proteins and enzymes, and inorganic substrates, such as silica and phyllosilicates, layered double hydroxides (LDHs), phosphates, and metal oxides result in bio-hybrid nanomaterials. These materials constitute an emerging interdisciplinary field in the frontier between life sciences, material sciences, and nanotechnology and are specially designed for exploring their potential applications in the clinical field for disease diagnostics and therapeutics. These bio-hybrid nanotools offer great potential for *in vivo* applications in the near future.

20.3.5.1 Sol-Gel-Derived Luminescent Superparamagnetic Nanoparticles

Superparamagnetic nanoparticles have been found to be very useful in biomedical applications such as in magnetic resonance imaging (MRI), targeted drug delivery, and magnetic separation. In bioanalysis, luminescence has been extensively utilized for detection and sensing. However, organic dyes have widely been used as signaling sources based on their emission properties but suffer from severe photobleaching during the detection process. In some cases, their low signaling intensity limits the achievable detection sensitivity. To solve these limitations, it is important to develop highly sensitive and photostable signaling materials. Great efforts have been made toward designing new labeling materials such as quantum dots, resonance light scattering particles, and dye-doped nanoparticles. Of these materials, the last type appears most promising. By encapsulating thousands of dye molecules into one protective nanoparticle, excellent photostability and enhanced signal density are to be expected. Silica turns out to be a very good material for a protective matrix on account of its proven biocompatibility and stability in most biosystems. Moreover, silica chemistry is well known, and standard chemistry protocols can be followed to conjugate various biomolecules to the silica surface, thus enabling silica-based particles to couple and label biotargets with selectivity and specificity (Gupta and Kumar, 2008a). Tan and coworkers have synthesized dye-doped silica nanoparticles with a reverse microemulsion technique and demonstrated their potential in biodetection and shown an increase of signal intensity by four orders of magnitude (Santra et al., 2001a,b; Zhao et al., 2003, 2004). Ma et al. (2006) combined the two useful functions, superparamagnetism and luminescence, along with an easily conjugated silica surface into multifunctional nanoarchitecture. The direct attachment of dye molecules to magnetic nanoparticles resulted in luminescence quenching. To avoid this problem, the magnetic core was surrounded by the first silica shell, and then the dye molecules were doped inside a second silica shell to concentrate the emission signal and enhance the photostability. The authors presented the sol-gel synthesis and detailed structural, magnetic, and optical characterizations of double shell structure. In recent years, the magnetic properties of nanometer-sized iron oxide powders such as $\alpha\text{-}Fe_2O_3$, $\gamma\text{-}Fe_2O_3$, and Fe_3O_4 have widely been studied for biomedical applications. These applications require small particle size, discrete and superparamagnetic iron oxide nanoparticles that can be successfully prepared by the sol-gel method due to lower annealing temperature. An et al. (2005) reported a novel synthesis of $\gamma\text{-}Fe_2O_3$ magnetic nanoparticles by a sol-gel method and characterized their superparamagnetic property for biomedical applications.

20.3.5.2 Sol-Gel/CNT Nanocomposites

The good electronic properties and electric conductivity of CNTs have shown various applications in electronics, scanning probe microscopy, in energy conversion and storage, reinforced composite materials, electromechanical actuators, and in any more areas. There are two main types of carbon nanotubes, viz., single-walled nanotubes (SWNTs), which consist of a single graphite sheet seamlessly wrapped in a cylindrical tube (Figure 20.9a through d) and multiwalled nanotubes (MWNTs), which comprise an array of such nanotubes that are concentrically nested like rings of a tree trunk (Figure 20.9e). The raw CNTs are electrochemically inert and before use have to be purified for removing impurities such as nanocrystal metal, amorphous carbon, and carbon nanoparticles that produce during synthesis. Thus, the pretreatment such as chemical oxidation or thermal treatment shortens the CNTs and leads to the partial oxidation and results in functional oxygenated groups at the open ends and defects along the sidewall of which some are electroactive. As a consequence, the electronic and structural properties of the nanotubes become modified. Many different functional groups like a series of amino acids, fluorescent probes, bioactive peptides, etc. can be placed at the tips and around the side walls of CNTs. Functionalized CNTs (f-CNTs) are highly soluble in water, which allows the formation of supramolecular complexes with biologically relevant macromolecules and substrates, based on electrostatic interactions. Thus, by surface modification solubility and properties of CNTs can be altered to a great extent and the resulting f-CNTs can be utilized for various biotechnological and biomedical applications. Functionalized CNTs are becoming relevant as a component of biosensing devices, in bioseparation and catalysis, in neuroscience research and

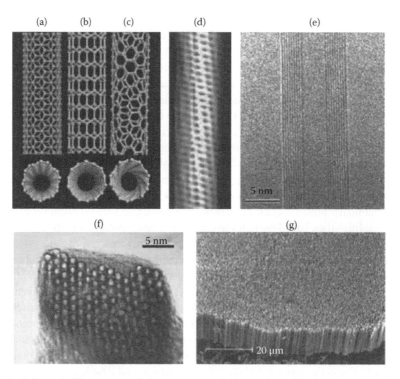

FIGURE 20.9 Schematic illustrations of the structures of (a) armchair, (b) zigzag, and (c) chiral SWNTs. Projections normal to the tube axis and perspective views along the tube axis are on the top and bottom, respectively. (d) Tunneling electron microscope image showing the helical structure of a 1.3 nm-diameter chiral SWNT. (e) TEM image of a MWNT containing a concentrically nested array of nine SWNTs. (f) TEM micrograph showing the lateral packing of 1.4 nm-diameter SWNTs in a bundle. (g) SEM image of an array of MWNTs grown as a nanotube forest. (Reproduced from Baughman, R.H. et al., *Science*, 297, 787, 2002. With permission.)

tissue engineering, as drug, vaccine, and gene delivery systems (Baughman et al., 2002; Gong et al., 2005; Bianco et al., 2005).

Nanocomposites have attracted great research and development interests. Sol-gel-derived ceramic carbon nanotubes nanocomposites can be prepared by doping the CNTs into a colloidal suspension of a sol solution that interconnect to form silicate gel matrix. Some kinds of siloxane polymer formed in the condensation and polycondensation reactions are hydrophobic depending upon the precursor used and can thus interact with the hydrophobic sidewall of the nanotubes through hydrophobic interactions. The non-covalent adsorption and growth of the silicate particle on the nanotubes were considered to effectively separate the aggregated nanotubes into single nanotubes or small bundles of nanotubes into several independent parts. The part shielded by the silicate particles are not electrochemically accessible and thereby cannot be used for electrode reactions. As a consequence, the intersection of CNT science with sol-gel chemistry not only provides a facile protocol for the preparation of CNT-based electrodes that are relatively useful for electrochemical studies, but also allows the prepared CNT-based electrodes to efficiently integrate the advantages both from the CNTs and from sol-gel electrochemistry. Furthermore, the CNT/sol-gel nanocomposite electrodes can be further exploited for the development of electrochemical biosensors by entrapping enzymes or proteins into the CNT/sol-gel nanocomposite because of the biocompatibility of the sol-gel matrix and facilitating the direct electron transfer between the redox protein and electrode surface. The direct electron transfer of enzymes and proteins such as cytochrome c (Cyt-c), GOx, catalase, HRP, myoglobin (Mb), and hemoglobin (Hb) in the presence of CNTs can be observed. CNTs also show electrocatalytic activities toward H_2O_2, NADH, ascorbic acid, dopamine, catechol, and homocysteine. The good catalytic activities of CNTs toward these molecules open their applications in amperometric sensors (Gong et al., 2005; Qian and Yang, 2006). Chen and Dong (2007) immobilized the HRP into the sol-gel-derived ceramic–carbon nanotube (SGCCN) nanocomposite film which provides a favorable microenvironment for HRP to perform direct electron transfer at glassy carbon electrode. The bioelectrocatalytic activity of HRP immobilized in the SGCCN film was superior to that immobilized in silica sol-gel film. Gavalas et al. (2004) reported a new type of composite material based on the combination of CNTs and aqueous sol-gel process using L-amino acid oxidase enzyme. The aqueous sol-gel process eliminates the generation of alcohol by-products and enables the gelation at neutral pH which allows the encapsulation of biomolecules into the CNT-containing composite without adverse effects on their activity, making them suitable for the development of stable biosensors. The surface of the CNT/sol-gel composite is shown in Figure 20.10. The average diameter of the nanotubes in the composite was determined to be between 200 and 300 nm (as compared to the 25–30 nm average diameter of the raw nanotubes), indicated the existence of the coating of sol-gel around the nanotubes.

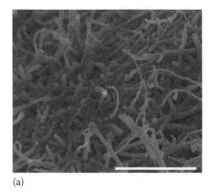
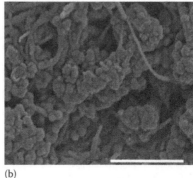

(a)   (b)

FIGURE 20.10 SEM pictures of the surface of the carbon nanotube sol-gel composite (a) without and (b) with 0.25 mg colloidal silica particles per mg of MWNT. Bar 5 μm. (Reproduced from Gavalas, V.G. et al., *Anal. Biochem.*, 329, 247, 2004. With permission.)

Engineering new bone tissue with cells and a synthetic extracellular matrix represents a new approach for the regeneration of mineralized tissues compared with the transplantation of bone. This requires a scaffold material upon which cells can attach, proliferate, and differentiate into a functionally and structurally appropriate tissue for the body location into which it will be placed. Bone tissue is considered as minerals and proteins and the minerals are mostly apatites such as hydroxyapatite (HAp) $[Ca_{10}(PO_4)_6(OH)_2]$, fluorapatite, and carbonate-apatite. In general, HAp is a main component of bone mineral and in some cases carbonate-apatite is a main hard tissue component as in dental enamel. HAp is widely accepted as a bioactive material and has excellent biocompatibility with hard tissues and high osteoconductivity, despite its low degradation rate, mechanical strength, and osteo-inductive potential. HAp has widely been used in tissue engineering, especially in bone and cartilage regeneration. Porosity in the HAp structure is important, since it allows for the growth of tissue and bone around a supporting framework and allows passage of nutrients. The porosity in the human bone structure varies widely with the type and function. Sol-gel technology offers an alternative technique for producing tailored porous bioactive and osteoconductive Si-substituted HAp, which is a highly promising material in the sense of bioactivity improvement (Gupta and Kumar, 2008a). Blanch et al. (2004) developed a methodology for improving HAp biocompatibility and bioactivity of HAp scaffolds appropriate for any specific biomedical application through controlling composition, impurity concentration, crystal size, and morphology using sol-gel process. CNTs possess excellent mechanical properties and can be used to increase the mechanical strength of HAp. Balani et al. (2007) used MWCNTs as reinforcement materials for imparting strength and toughness to brittle hydroxyapatite bioceramic coating. Plasma-spraying technique was applied for the uniform distribution of MWCNTs in HAp coating and further to address the issue of biocompatibility with living cells, human osteoblast hFOB 1.19 cells onto CNT reinforced HAp coating have been cultured. Unrestricted growth of human osteoblast hFOB 1.19 cells has been observed near CNT regions claiming assistance by CNT surfaces to promote cell growth and proliferation. Figure 20.11 shows the SEM image representing the growth of human osteoblast cells over CNT-HA nanocomposite.

20.3.5.3 Sol-Gel-Derived Gold Nanocomposites

Metal nanoparticles are attracting increasing attention in recent times due to their interesting size-dependent optical, electronic, and catalytic properties. Metal nanoparticles have been used in many electrochemical, electroanalytical, and bioelectrochemical applications owing to their extraordinary electrocatalytic activity. The possibility of tailoring metal nanoparticles, viz., of silver, copper, gold, and platinum with organic molecules has generated overwhelming interest in the design

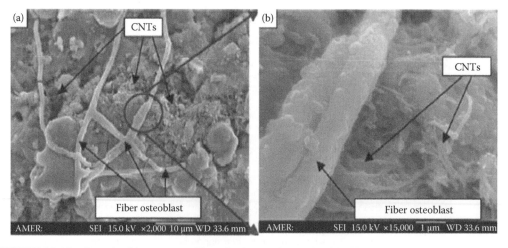

FIGURE 20.11 Growth of human osteoblasts is observed embedded in HAp matrix, and (b) cell growth alongside of CNTs *in vitro*. (Reproduced from Balani, K. et al., *Biomaterials*, 28, 618, 2007. With permission.)

of functionalized metal nanoparticles. Naked metal nanoparticles are unstable in organic solvents as they readily aggregate. Organic molecules with functional groups such as amines, thiols, isothiocynate, or silanes, which are often used to modify the metal nanoparticle surface, facilitate stabilization of sols (Thomas et al., 2002). Recently, sol-gel method has been used to prepare bulk and thin films doped with metal nanoparticles. Colloidal gold nanoparticles have been the subject of strong interest in areas of material science, biotechnology, and analytical chemistry due to their functions as molecular markers, diagnostic imaging, and catalysts, and as material of choice due to its well-established chemistry with organic ligands containing the thiol group (–SH) group. In particular, thiol-derivatized gold nanoclusters can be prepared in a one-step, two-phase reaction (Scheme 20.3). Gold colloid monolayer can be obtained by self-assembly on the surface of a thiol-containing electrode. The stability conferred by the attachment of thiols in self-assembled layer is remarkable, and materials prepared with this ligand (in solid form and in solution) show no sign of decomposition or loss of solubility even after several months of storage at room temperature (Fink et al., 1998). Gold nanoparticles can be strongly bound to the surface through covalent bonds to the polymer functional groups such as –CN, –NH$_2$, or –SH and can also be conjugated with biological macromolecules. Such composites are attractive for biological studies, particularly for developing sensors because of easy preparation, good biocompatibility, and bioactivity, and relatively large surface. The incorporation of metal nanoparticles into a variety of polymeric matrices to form nanocomposite films is attracting much attention particularly for sensor developments. The sol-gel processing offers a simple and versatile route to combine the favorable properties of organic–inorganic materials with biomolecules and metal nanoparticles and to prepare a three-dimensional silicate network. The gold nanoparticles are stabilized by the –NH$_2$ groups of the silicate sol-gel network derived from precursors containing –NH$_2$ group, viz., APTMS and (3-aminopropyl) triethoxysilane (APS). The –NH$_2$...Au$_{nps}$ affinity is apparently sufficient to increase the local concentration of the aminosilane molecules on the Au$_{nps}$ surface and to induce condensation and cross-linking of adjacent ethoxy groups to produce a three-dimensional silicate network, thus creating a stable silicate sol-gel network with stabilized Au$_{nps}$. The embedded Au$_{nps}$ can act as tiny conduction centers and it can facilitate the electron transfer events (Wu et al., 2005; Maduraiveeran and Ramaraj, 2007).

Further, the nanocomposites are quite useful for improving the sensitivity of conventional immunosensor based on sol-gel only. Wu et al. (2005) investigated the preparation and properties

Step I: phase transfer

$$AuCl_4^- \text{ (aq.)} \xrightarrow[\text{Toluene}]{\text{Tetra-}n\text{-octylammoniumbromide}} N(C_8H_{17})_4^+ AuCl_4^-$$

Step II: Reduction and stabilization with capping

$$N(C_8H_{17})_4^+ AuCl_4^- \xrightarrow[\text{NaBH}_4 \text{ (aq.)}]{C_8H_{17}SH} Au(C_8H_{17}SH)_n C_6H_5Me$$

Octanethiol stabilized gold
nanoparticles in toulene

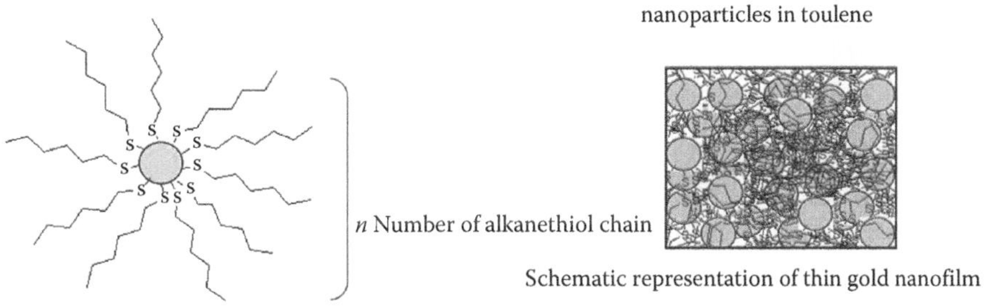

n Number of alkanethiol chain

Schematic representation of thin gold nanofilm

SCHEME 20.3 Chemical synthesis of alkanethiol stabilized gold nanoparticles.

of a novel capacitive immunosensor for the direct detection of the human IgG by self-assembling gold nanoparticles to the surface of the sol-gel-modified electrode. The gold electrode was initially modified with the (mercaptopropyl) triethoxysilane (MPTS) via the direct coupling of sol-gel and self-assembled technologies to introduce SH groups on the surface capable to capture gold nanoparticles from the solution, resulting in gold nanoparticle-derivatized interface. Then, the IgGAb was immobilized through strong electrostatic interaction by controlling the pH of antibody solution about its isoelectric point. When the antibody-immobilized gold electrode was immersed in a solution containing the specific antigen, the immuno-complexes formed resulted in the increase of the dielectric layer and induced a capacitance decrease directly proportional to the concentration of the target antigen. A schematic representation of the immobilization process of antibody onto the gold electrode and the antibody–antigen interaction is shown in Scheme 20.4. Tang et al. (2006) developed a novel potentiometric immunosensor for the detection of hepatitis B surface antigen by self-assembling gold nanoparticles to a thiol-containing sol-gel network derived from (3-mercaptopropyl) trimethoxysilane precursor over gold electrode. Yang et al. (2006) developed a novel electrochemical glucose biosensor derived from platinum nanoparticles–CNT–sol-gel nanocomposite. Here, the amine group containing sol-gel solution was selected to utilize the affinity of $-NH_2$ groups toward metal nanoparticles for the stabilization of the nanoparticles in solution. The resulting platinum nanoparticle-doped sol-gel solution was used as a binder for MWCNT and the immobilization of GOx enzyme for the fabrication of electrochemical biosensors. The combined electrocatalytic activity of such systems permits low-potential detection of hydrogen peroxide with remarkably improved sensitivity four times larger ($0.27\,\mu A/mM/cm^2$ at $0.1\,V$) than CNT-based biosensor.

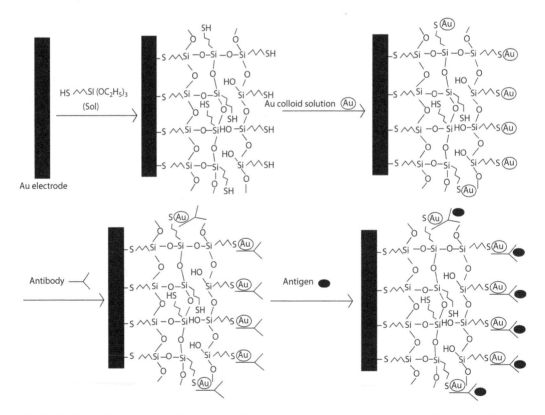

SCHEME 20.4 The schematic illustration of the process of the immobilization of antibody onto the gold electrode and interaction of antibody–antigen. (Reproduced from Wu, Z.S. et al., *Anal. Chim. Acta*, 528, 235, 2005. With permission.)

20.3.6 Sol-Gel-Based Microarrays

Over the past decade, advances in the fields of combinatorial chemistry, genomics, and proteomics have necessitated the development of a variety of high-throughput methodologies that allow multiple assays to be carried out in parallel. Among the most successful of these new screening tools are microarrays, a high-throughput technology that allows the detection of thousands of genes simultaneously. Parallelism, miniaturization, multiplexing, and automation are the main features of this technology. Such systems are suitable for applications as sensors and diagnostics, as screening devices for enzyme activity and inhibition, or as microreactors for biocatalytic synthesis and biotransformations. Biosensors have traditionally been designed to detect one analyte at a time in a sample. Multianalyte detection in a sample offers significant advantages. An ideal biosensor array should be entirely self-contained, rapid and simple to fabricate, and stable to avoid calibration and drift difficulties. Several research groups have developed biosensor arrays that can be used for the simultaneous detection of multiple analytes in a single sample (Liu et al., 2000; Cho and Bright, 2001, 2002a,b; Revzin et al., 2002; Moser et al., 2002; Biran and Walt, 2002; Christodoulides et al., 2002). Tsai and Doong (2005) developed an array-based enzymatic optical biosensor for the simultaneous determination of pH, urea, acetylcholine, and heavy metals. Bright and coworkers demonstrated for the first time the combination of sol-gel processing and pin-printing methods as a way to rapidly form ensembles of integrated, reusable, and stable biosensor arrays for the simultaneous detection of glucose and O_2 (Cho et al., 2002). Recent developments in genome sciences have led to the development of DNA microarray technology, a tool for the study of gene sequence, structure, and expression. DNA aptamers are single-stranded nucleic acids that are generated by *in vitro* selection and have several potential applications such as multianalyte biosensing, metabolite profiling, reporting enzyme activity, or affinity capture of specific analytes. In addition, protein microarrays to date provide applications such as high-throughput screening of antibody libraries and rapid evaluation of protein–protein interactions. These studies are becoming crucial for understanding cellular metabolism and opening new avenues for drug discovery. Brennan and coworkers have contributed significantly in the development of pin-printed protein microarrays and have demonstrated the usefulness of such arrays for kinetic studies, quantitative multianalyte biosensing, nanovolume inhibition assays, and in high-throughput drug-screening applications (Rupcich and Brennan, 2003; Rupcich et al., 2003, 2005b). Arenkov et al. (2000) developed a protein microchip composed of proteins covalently immobilized in polyacrylamide gel pads, suitable for immunoassays and enzyme activity measurements. Park and Clark (2002) developed versatile low-cost arrays of sol-gel-encapsulated enzymes (referred to as solzymes) patterned within multi-well PDMS films adhered to glass microscope slides, suitable for repeated assays of bioactivity or enzyme inhibition. However, despite recent advances in sol-gel processes for bioencapsulation, the fragility of the final gel structure, manifested by shrinkage of the gel, pore collapse, and/or poor adhesion to the substrate, remains a major problem for the miniaturization of sol-gel structures on microarrays or microfluidic chips.

20.4 SOL-GEL-BASED OPTICAL BIOSENSORS

Biosensors have potentially a large market, covering the areas of clinical diagnostics, process control, food technology, and military and environmental monitoring. The horizon for applications involving biosensors is continuously expanding. Two aspects that are probably the most problematic in developing a biosensor are (a) the incorporation of sensing molecules (bioselector) in a suitable matrix and (b) monitoring/quantitating the interactions between the sensing molecules and the analytes of interest. Various transducing elements, viz., electrochemical, optical, piezoelectrical, and calorimetrical have been utilized for monitoring such interactions over a long dynamic range. Recently, optical sensors or optrodes are gaining considerable interest in the detection of various analyte molecules. Optrodes have several advantages over conventional devices in terms of simplicity, miniaturization, low cost, disposability, freedom from electrical interference, high sensitivity, and remote sensing. Presently, sol-gel chemistry offers new and interesting possibilities for the promising encapsulation of heat-sensitive and fragile biomolecules (enzyme, protein, antibody and whole cells of plant, animal, and microbes) because sol-gel processing allows the formation of optical-quality porous glass at room temperature and provides a better way in comparison with adsorption, covalent binding, cross-linking, and encapsulation methods using organic polymeric supports for the entrapment of biological molecules. Although demonstrations of optical biosensor behavior employing sol-gel-derived materials are much fewer in number than for chemical sensors, recent developments in the sol-gel processing of biomaterials indicated that this is a challenging area of considerable potential for biosensor development (Mehrvar and Abdi, 2004; Kandimalla et al., 2006). The first optical biosensors built with sol-gel coatings were based on the affinity of hoemoglobin (Hb), myoglobin (Mb) or Cyt-c for carbon monoxide (CO) and nitric oxide (NO). The results reported clearly evidenced that sol-gels provide an ideal host matrix for metalloproteins with variable size, showing no adverse effects on their structure or activity (Blyth et al., 1995; Aylott et al., 1997).

20.4.1 Fiber-Optic Biosensors

The development of fiber-optic biosensors (FOBs) based on laser and modern low-cost optical fiber technology in recent years is increasing for clinical, pharmaceutical, industrial, and military applications. Excellent light delivery, long interaction length, low cost, and ability not only to excite the target molecules but also to capture the emitted light from the targets are the main points in favor of the use of optical fibers in biosensors. FOBs can be used in combination with different types of spectroscopic technique, e.g., absorption, fluorescence, phosphorescence, SPR. The detection principle of a large class of FOBs is based on the principle of total internal reflection fluorescence (TIRF). Optical fiber sensors based on doped sol-gel coatings are easy to prepare, since the sol-gel glass is highly compatible to silica or glass fibers. These sensors are typically inexpensive and provide the possibility of remote sensing and *in vivo* measurements. Through this technology, sensor arrays and sensors containing multiple reagents are easily developed. Additionally, since sol-gel glass is stable, inert, and nontoxic, these sensors can be applied to measurements in harsh environments, in medical diagnosis, and food industries. Besides these advantages, a wide variety of sensor designs are made possible. The development of optical biosensors based on sol-gel films is also becoming more relevant because of the fast diffusion of analytes from surface (Bosch et al., 2007; Chou et al., 2007; Jeronimo et al., 2007). Andreou and Clonis (2002) designed a portable fiber-optic cholinesterase biosensor for the detection and determination of pesticides carbaryl and dichlorvos as a three-layer sandwich. The enzyme cholinesterase was immobilized on the outer layer, consisting of hydrophilic modified polyvinylidenefluoride membrane and an intermediate sol-gel layer that incorporated bromocresol purple, deposited on an inner disk. Chou et al. (2007) described the application of a FOB to the real-time investigation of the interaction kinetics between fluorescein isothiocyanate (FITC)-conjugated monoclonal sheep antihuman C-reactive protein (CRP) antibody and CRP isoforms on the surface of optical fiber.

20.4.2 Protein and Enzyme Biosensors

The past decade has seen a significant increase in the number of reports describing the encapsulation of proteins and enzymes into inorganic silicate matrixes formed by the sol-gel processing of alkoxysilanes but in majority of cases, they retain some degree of the characteristic structure and biochemical functionality. However, these studies have concentrated on the entrapment of single protein or enzyme. A relatively new area of interest to molecular and cellular biologists is the study of protein–protein interactions, which is now accepted as an important mechanism in cellular signaling, growth, and differentiation (Avnir et al., 1994; Jin and Brennan, 2002).

Cyt-*c* is the most common electron transfer protein entrapped in sol-gel matrix, showed characteristic reactivity and chemical function (Dave et al., 1997; Shen and Kostic, 1997; Dunn et al., 1998). Edmiston et al. (1994) carried out fluorescence spectroscopic studies in order to monitor the possible denaturation of BSA labeled with acrylodan (Ac) entrapped in TMOS derived in sol-gel matrix when subjected to pH changes. Lundgren and Bright (1996) reported the first BSA-Ac-based biosensor for the quantification of ionic surfactant (cetyltrimethyl ammonium bromide). The biosensor consisted of an optical fiber, onto which Ac-labeled BSA was immobilized on TMOS-derived sol-gel coating.

Dave et al. (1994) designed several sol-gel based systems, based on encapsulation of enzymes and proteins that are capable of detecting biologically relevant analytes such as O_2, CO, NO, glucose, and oxalate. Amperometric glucose biosensors based on enzymes entrapped in sol-gel glasses have been one of the most investigated sensors. Wolfbeis et al. (2000) tested three different combinations of oxygen transducer, viz., the sandwich configuration, two-layer configuration, and powder configuration based on sol-gel immobilized GOx. In all cases, sorbitol was added to prepare more porous sol-gel thin film. The luminescent oxygen probe ruthenium (II) (4,7-diphenyl-1,10-phenanthroline)$_3$-(dodecylsulfate)$_2$ [Ru(dpp)] was entrapped in thin films for measuring the consumption of oxygen. The sandwich configuration provided the highest enzyme activity and the largest dynamic range (0.1–15 mM), but suffered from a distinct decrease in sensitivity upon prolonged use. The two-layer configuration had the fastest response time ($t_{90} = 50$ s), while the powder configuration provided the best operational lifetime. The storage stability of all configurations exceeded 4 months if stored at 4°C. Navas Diaz et al. (1998) introduced sol-gel HRP fiber-optic biosensor for hydrogen peroxide detection by chemiluminescence. Single use was the major drawback of this sensor. Aylott et al. (1997) encapsulated enzyme periplasmic nitrate reductase (Nap) extracted from the denitrifying bacterium *Thiosphaera pantotropha* within sol-gel matrix for the optical biosensing of nitrate ions. The reduction of nitrate by Nap resulted in a characteristic change in the UV/VIS absorption spectrum of the nitrate reductase. The nitrate biosensing system was fully reversible and was highly sensitive and selective to nitrate ions. The sol-gel matrix, did not affect the activity of the enzyme even after a storage period of up to 6 months and no leaching of the Nap from the sol-gel matrix was observed. Singh et al. (2007) covalently immobilized cholesterol oxidase (ChOx) and cholesterol esterase (ChEt) on to TEOS-derived sol-gel films and characterized these films using SEM, UV–VIS spectroscopic, FTIR spectroscopic, and amperometric techniques. The results of photometric measurements carried out on TEOS sol-gel/ChEt/ChOx films revealed thermal stability up to 55°C, response time as 180 s, linearity up to 780 mg/dL (12 mM), shelf life of 1 month, detection limit of 12 mg/dL, and sensitivity as 5.4×10^{-5} Abs./mg/dL. Doong and Shih (2006) fabricated a simple and novel titania sol-gel-derived optical biosensor coupled with carboxy seminaphthorhodamine-1-dextran (SNARF-1-dextran) as the fluorescent dye for the determination of glutamate in water and biological samples. The NADH-dependent glutamate dehydrogenase (GLDH) was trapped in titania sol-gel-derived matrix prepared by vapor deposition method and the surface morphology of immobilized spots was characterized by using SEM and atomic force microscopic (AFM). SEM and AFM images showed that the deposition of titania precursor at 27°C for 6.5 h was found to be suitable to form transparent titania sol-gel matrix to encapsulate GLDH and fluorescent probe. A dynamic range between 0.04 and 10 mM with the detection limit of 5.5 μM were observed. The responses to glutamate in biological samples also showed good performances, and the dynamic range and detection limit were 0.02–10 mM and 6.7 μM, respectively. In addition, the biosensor showed relatively high storage stability over more than 1 month. This study clearly demonstrated that simple vapor deposition method could be successfully used to form transparent titania sol-gel film for the fabrication of glutamate biosensors that are suitable for optical detection of glutamate in water and biological samples. The objectives, methods, and significant results of various studies involving the entrapment of proteins, enzymes, and antibodies in sol-gel matrix are shown in Tables 20.2 through 20.4, respectively.

TABLE 20.2
Encapsulated Proteins in Sol-Gel Matrix

S. No.	Objectives	Technique Used	Significant Results	References
1.	Entrapment of bovine copper zinc superoxide dismutase (CuZnSOD), Horse heart Cyt-c and Mb in TMOS-derived bulk	Absorption spectroscopy	Behavior of gel-encapsulated proteins were similar to solution	Ellerby et al. (1992)
2.	Dynamics of acrylodan (Ac) labeled BSA and HSA entrapped in TMOS-derived bulk	Steady-state and time-resolved fluorescence spectroscopy	(a) Acrylodan residue and proteins were able to undergo nanosecond motions within the biogels (b) Semiangle through which the BSA-Ac and HSA-Ac can process was same for a freshly formed biogel and native protein in buffer. Further it increased ~20° and 10° for BSA-Ac and HSA-Ac after 37 days of storage	Jordan et al. (1995)
3.	Detection and quantification of surfactant cetyltrimethyl ammonium bromide using BSA-Ac immobilized on silanized silica optical fiber	Steady-state fluorescence measurements	(a) Linear dynamic range extended from 5 to 60μM (b) t_{90} response precision (relative standard deviation) during 34 sensing cycle was 2.5% (c) On the lower side of optical fiber, biosensor performance decreased 38% after 25 days of storage	Lundgren and Bright (1996)
4.	Characterization of the influence of synthesis conditions (in presence of ethanol) on the Cyt-c stability and conformation within the TMOS-derived bulk and thin films	Absorption and impedance spectroscopy	Entrapment provided stabilization, functional activity as well as prevented denaturation of Cyt-c	Dave et al. (1997)
5.	Elucidation of structure, dynamics, distribution, and average environment of the entrapped Monellin (protein) labeled with N-acetyl tryptophanamide (NATA) in ultrathin monolith	Steady-state fluorescence and anisotropy measurements	Difference in kinetics and environment of proteins in glasses and in solutions	Zheng et al. (1997)

#				
6.	Interactions of Cyt-c and analyte in TMOS-derived bulk	Absorption spectroscopy	Entrapment provided stabilization and prevented denaturation and aggregation of Cyt-c	Dunn et al. (1998)
7.	Examination of the changes in the conformational motions of Cod III parvalbumin entrapped in TEOS-derived monolith with aging	Steady-state fluorescence spectroscopy	(a) The entrapped protein retained conformational flexibility similar to that observed in solution and remained accessible to analytes such as Ca^{2+} (b) Entrapment caused the apparent affinity constant for binding of Ca^{2+} (c) Fluorometric detection of Ca^{2+} could be done over a 600 µM range with a limit of detection of 3 µM and with no interference from divalent ions such as Mg^{2+}, Sr^{2+}, or Cd^{2+}	Flora and Brennan (1998)
8.	Entrapment and characterization of HSA, lipase, 7-Azaindole, and PRODAN in TEOS-ORMOSILS-derived bulk	^{29}Si, ^{13}C NMR, and fluorescence spectroscopy	(a) Improvement in function of entrapped HSA and lipase with increased ORMOSILS content (b) Suitable for encapsulation of lipophilic protein	Brennan et al. (1999)
9.	(a) Entrapment of an intact protein–peptide interaction, consisting of bovine calmodulin (bCaM) and melittin, into a TEOS-derived sol-gel bulk for drug screening	Steady-state and time-resolved fluorescence spectroscopy	Entrapped complex behaves similarly to the complex in solution and undergo reversible dissociation upon introduction of the denaturant guanidine hydrochloride	Flora et al. (2002)
10.	Determination of concentration distributions and conformations of BSA entrapped in sol-gels	Near-infrared multispectral imaging technique	(a) Inhomogeneity in distribution of BSA independent of its concentration within sol-gel matrix (b) No observable changes in the conformation of BSA at relatively high concentration (366 mg/mL) (c) Pronounced changes in the spectra of the BSA as a function of (sol-gel reaction) time, when the concentration of BSA was decreased to 220 mg/mL	Tran et al. (2004)

Source: Reprinted from Gupta, R. and Chaudhury, N.K., *Biosens. Bioelectron.*, 22, 2387, 2007, and references cited therein. With permission.

TABLE 20.3
Encapsulated Enzymes in Sol-Gel Matrix

S. No.	Objectives	Technique Used	Significant Results	References
1.	Entrapment of trypsin, GOx, and peroxidases in TMOS-derived xerogel disks	Absorption and fluorescence spectroscopy	(a) Xerogels prepared at pH <7 were practically devoid of acetylated trypsin activity, while at pH > 7 the activity increased with pH (b) Sol-gel optical glucose sensor was sensitive in the range between 0 and 100 mM of glucose (c) An efficient sol-gel immobilization of enzymes was expected at pH values above 8 and above the pI value of the enzyme	Braun et al. (1992)
2.	Entrapment of urease enzyme in spin coated TEOS-derived sol-gel thin films having sandwich configuration	Absorption spectroscopy	(a) Detection limit and response time was 0.5 mM and 10 s, respectively, for urea (b) The urease entrapped within films remained active (>95% of original activity) for at least 6 weeks if stored at 4°C	Narang et al. (1994b)
3.	Entrapment of GOx and peroxidase in spin coated TEOS-derived sol-gel thin films and comparison between immobilization methods, viz., physisorption, microencapsulation, and sandwich configuration	Amperometric and photometric method	(a) Sandwich configuration exhibited a fast response and high enzymes loading and stable for at least 2 months under ambient storage conditions (b) Response time and detection limit was 30 s and 0.2 mM, respectively, for sandwich configuration	Narang et al. (1994c)
4.	Entrapment of yeast alcohol dehydrogenase in TMOS-derived monoliths for sensing of alcohols and aldehydes	Fluorescence spectroscopy	(a) Aqueous propionaldehyde concentrations could be evaluated readily over a 0.1–10 mM range and those aqueous ethanol concentrations over a 10–1000 mM range (b) Limitation of this monolith approach involved the length of time to measure a response accurately as well as the potential for eventual diffusion of the soluble cofactor (NADH/NAD$^+$) into the analyte solution	Williams and Hupp (1998)
5.	Entrapment of GOx, lactate oxidase, and glycolate oxidase in TMOS-derived silica gel powder coated on glass slides	Absorption spectroscopy	(a) GOx retained most or all of its initial activity, while lactate oxidase and glycolate lost most of their activity (b) The half-life of GOx at 63°C increased upon immobilization 20-fold; the half-lives of lactate oxidase and of glycolate oxidase were not extended beyond those of the water-dissolved enzymes (c) Lactate oxidase was stabilized when electrostatically complexed with weak as well as strong base prior to immobilization, most of its activity retained and its half-life at 63°C increased 150-fold (d) Glycolate oxidase was not stabilized by weak base but stabilized by strong base prior to immobilization, its half-life at 60°C also increased 100-fold	Chen et al. (1998)
6.	Entrapment of HRP in TMOS-derived monoliths at extreme pH and temperature	Absorption spectroscopy	(a) Encapsulation in glass did not significantly alter the optical absorption spectrum of HRP	Cho and Han (1999)

#	Description	Method	Details	Reference
7.	Co-immobilization of cholesterol oxidase and HRP in TEOS-derived spin coated sol-gel thin films by physical adsorption, sandwich configuration, and microencapsulation	Spectrophotometric and electrochemical method	(b) Encapsulated enzyme was fully active even at pH 2. At high pH, both solubilized and encapsulated enzymes lost their catalytic activity (c) Encapsulation stabilized the enzyme at high temperatures (little change at 80°C)	Kumar et al. (2000)
8.	Optical biosensing of nitrite ions using cytochrome $cd1$ nitrite reductase encapsulated in TMOS-derived monolith	Absorption spectroscopy	(a) Response time of 10, 30, and 70 min for physically entrapped, physisorbed, and microencapsulated films, respectively, was observed for both spectrophotometric and electrochemical methods (b) The results of amperometric measurements undertaken on a sandwich configuration revealed a fast response time of 50 s and a lower limit of detection of 0.5 mM cholesterol (c) All enzyme sol-gel thin films were found to be stable for about 8 weeks at 25°C and 12 weeks at 4°C–5°C	Ferretti et al. (2000)
9.	Electrochemical entrapment of polyaniline (PANI) on to TEOS-derived films on indium tin oxide (ITO)-coated glass to immobilize lactate dehydrogenase (LDH) for sensing of lactate	Amperometric method	(a) No structural changes and retention of enzymatic activity of cytochrome $cd1$ nitrite reductase when encapsulated in a bulk sol-gel monolith (b) The detection of nitrite ions in the range of 0.075–1.250 mM was achieved, with a limit of detection of 0.075 mM (c) Sol-gel sandwich thin film structure enabled the determination of nitrite concentrations within ca. 5 min and stable for several months when the films were stored at 4°C	Chaubey et al. (2003)
10.	Entrapment of acetylcholinesterase (AChEs) enzyme in TMOS+MTMOS+PEG derived sol-gel on 7,7,8,8-tetracyanoquinodimethane (TCNQ) modified screen-printed electrode for the detection of carbamate insecticides	Amperometric method	(a) The amperometric response of the electrodes under optimum conditions of pH of the medium, substrate concentration, applied potential, and interference exhibited a linear relationship from 1 to 4 mM of lactate concentrations with a response time of about 60 s, a shelf life of about 8 weeks at 0°C–4°C (b) An attempt has been made to extend the linearity up to 10 mM for lactate by coating an external layer of polyvinyl chloride (PVC) over the sol-gel/PANI/LDH electrodes	Bucur et al. (2006)

Additional details for row 10:
(a) Wild and genetically engineered AChEs from *Drosophila melanogaster* (*Dm*) showed high sensitivity toward insecticides compared with cholinesterases from other sources
(b) The wild type and three mutant enzymes tested against three carbamate insecticides: carbaryl, carbofuran, and pirimicarb. The best limits of detection (LOD) obtained with the Y370A mutant for carbaryl (1×10^{-8} M), the E69W mutant for pirimicarb (2×10^{-8} M), and the I161V mutant for carbofuran (8×10^{-10} M)
(c) The introduction of mutations in the enzyme structure permits to obtain inhibition-based biosensors with a very good LOD

Source: Reprinted from Gupta, R. and Chaudhury, N.K. *Biosens. Bioelectron.,* 22, 2387, 2007, and references cited therein. With permission.

TABLE 20.4
Encapsulated Antibodies in Sol-Gel Matrix

S. No.	Objective	Technique Used	Significant Results	References
1.	Investigation of the aging and drying effect on the affinity constant of encapsulated polyclonal antifluorescein antibody in TMOS-derived monolith	Steady-state and time-resolved fluorescence measurements	(a) Encapsulation led to decreased in affinity in order of magnitude two, however, affinity constant (K_f) remains well over $10^7 \, M^{-1}$ (b) Storage time and conditions affected the affinity of the sol–gel encapsulated antifluorescein antibody	Wang et al. (1993)
2.	Entrapment of monoclonal anti-atrazine antibodies (Mabs) in TMOS + 10% PEG 400-derived sol-gel matrixes for the detection of atrazine, a widely used herbicide	ELISA method	(a) Leaching of the antibodies was found to be zero (b) Stability was tested under various storage conditions and was found to be 100% for at least 2 months at room temperature, compared with a drop of 40% in solution (c) The response time was found not to differ considerably from that obtained in solution	Bronshtein et al. (1997)
3.	Measurement of intrinsic fluorescence to probe the conformational flexibility and thermodynamic stability of a single tryptophan protein entrapped in TMOS-derived sol-gel monolith	Steady-state fluorescence and anisotropy measurements	(a) There were no significant improvements in either chemical or thermal stability when the protein was present in wet-aged monoliths. However, the long-term stability of the protein was improved sixfold when such monoliths were stored at 4°C (b) The steady-state fluorescence responses obtained during denaturation and the accessibility of native and denatured protein to quencher provided clear evidence that the entrapped protein had a smaller range of conformational motions compared to the protein in solution, and that the entrapped protein was not able to unfold completely	Zheng and Brennan (1998)
4.	Rotational reorientational dynamics of intact polyclonal anti-dansyl antibodies (Anti-DAN) labeled with dansyl-L-glycine (DAN) in TEOS-derived monolith	Steady-state and time-resolved fluorescence measurements	(a) No detectable rotational reorientation dynamics of Anti-DAN within an aged biogel on a timescale between ~100 ps and 250–270 ns (b) Equilibrium binding constant (K_b) that describes the anti-DAN association with its target hapten in a biogel was only fivefold less than the value for anti-DAN dissolved in aqueous buffer (c) K_b remained constant for at least 8 months at room temperature	Doody et al. (2000)

No.		Method		Reference
5.	Entrapment of monoclonal anti-TNT immunoglobulins in TMOS + 10% PEG 400 derived sol-gel column for the detection of trinitrotoluene (TNT)	ELISA method	(a) Binding was found to be highly reproducible, dose dependent, and only slightly (1.2–1.8-fold) lower than that in solution (b) No leaching from the matrix and were tolerant of absolute ethanol, acetone, and acetonitrile (c) Bound analytes could be easily eluted from the sol-gel matrix at high recoveries	Altstein et al. (2001)
6.	Comparison of ultrathin alumina sol-gel-derived films with SiO_2 sol-gel-derived films for codetermination of two liver fibrosis markers (hyaluronan, HA and laminin, LN) in mixed sample and human immunoglobulin (hIgG)	Capacitive immunoassay	(a) Compared with a SiO_2 matrix, alumina sol-gel-derived films are more suitable for capacitive immunoassay due to high specific area and less thickness (b) Alumina sol-gel-derived films showed reproducible linear responses to hIgG, LN and HA in the range of ~1–500, ~0.5–50, and ~1–50ng/mL, respectively (c) Immunosensor showed good selectivity for the antigens in mixed samples	Jiang et al. (2003)
7.	Preparation and characterization of TEOS-derived sol-gel immunosorbent doped with 2,4 dichlorophenoxyacetic acid (2,4-D)	Solid phase extraction method (SPE)	(a) A binding capacity of 130ng of 2,4-D methyl ester per mg of immobilized antibody, corresponding to 42% of the free antibody activity, was obtained with the best gels (b) For at least 8 weeks, the entrapped antibody retained more than 90% of its initial activity and after 14 weeks the remaining activity still was about 47% of the initial one	Vazquez-Lira et al. (2003)
8.	Entrapment of antifluorescein antibody in diglycerylsilane-derived sol-gel monolithic capillary column for immunoextraction	Laser-induced fluorescence detection	(a) Similar dissociation constants for fluorescein binding to the anti-fluorescein antibody in solution and in the meso/macroporous silica, indicating that the entrapped antibody retained its native conformation within such a matrix (b) The capillary-scale immunoaffinity columns operated at low backpressure using a syringe pump and capable of performing chromatographic separations, dependent on the presence of the antibody within the stationary phase. (c) Operated using in-line laser-induced fluorescence detection (d) Reusable columns even after exposure to 20% MeOH, with a loss of binding activity of 20% over five cycles	Hodgson et al. (2005)

Source: Reprinted from Gupta, R. and Chaudhury, N.K., *Biosens. Bioelectron.*, 22, 2387, 2007, and references cited therein. With permission.

20.4.3 IMMUNOSENSORS

In recent years, immunosensors developed for the diagnostic assay of biomolecules have attracted considerable interest. Electrochemical immunosensors are most widely used in affinity immunoassays and usually prepared by immobilizing various kinds of antibody or antigen on the electrode surface. The analytes are measured through the immunoreaction between the immobilized ligand and the analytes or labeled conjugate species. Enzymes have been extensively used as markers to improve the sensitivity of immunoassays by electrochemical amplification of the signals. Many enzymes such as HRP, alkaline phosphatase, laccase, and GOx have been used to label the antibody or antigen and to produce the electrochemically active species for amperometric immunosensor preparation. Correspondingly, some substrates such as H_2O_2 or O_2, naphthyl phosphate or p-aminophenol phosphate, O_2 or ferrocene and glucose or O_2 and immobilized HRP are needed in the test solutions for HRP, alkaline phosphatase, laccase, and GOx, respectively. Sometimes, when HRP is used as the marker, an additional substrate such as hydroquinone or o-aminophenol is also added to the detection solution as a mediator to transfer electrons between H_2O_2 and the enzyme (Du et al., 2003). Yang et al. (2002) immobilized antigentamicin antibody in mesoporous sol-gel biomaterials derived from TEOS and PEG for the development of an immunoaffinity column for the flow injection immunoassay of gentamicin. The immunoassay was based on the competition between gentamicin and FITC-labeled gentamicin for a limited number of encapsulated antibody binding sites. NaOH solution was used for the regeneration of encapsulated antibody binding sites after each measurement, which allowed the immunoreactor to be used for up to 20 times without any loss of reactivity.

20.4.4 MICROBIAL BIOSENSORS

A microbial biosensor consists of a transducer in conjunction with immobilized viable or nonviable microbial cells. Nonviable cells obtained after permeabilization or whole cells containing periplasmic enzymes have mostly been used as an economical substitute for enzymes. Viable cells make use of the respiratory and metabolic functions of the cell, the analyte to be monitored being either a substrate or an inhibitor of these processes (D'Souza, 2001). Carturan et al. (1989) reported the immobilization of yeast cells of about $50\,\mu m$ sizes in thin gel films of thickness of about $0.2\,\mu m$. The invertase activity of the yeast cells anchored to the sol-gel coated glass plates was measured 28% of that in solution. Jia et al. (2003) developed a novel type of biochemical oxygen demand (BOD) biosensor based on co-immobilization of microbial cells *Trichosporon cutaneum* and *Bacillus subtilis* in the sol-gel-derived composite material which was composed of silica and the grafting copolymer of poly(vinyl alcohol) and 4-vinyl pyridine for monitoring water. Good agreement was obtained between the results of the BOD sensor measurement and those obtained from conventional BOD_5 method for water samples. Jiang et al. (2006) prepared BOD sensing film embedded with an oxygen-sensitive Ru complex, by immobilizing three different kinds of microorganisms, *Bacillus licheniformis*, *Dietzia maris*, and *Marinobacter marinus* from seawater on a polyvinyl alcohol ORMOSILs for the BOD determination of seawater. The effects of temperature, pH, and sodium chloride concentration on the two microbial films were studied and after preconditioning, the BOD biosensor could steadily perform well up to 10 months. Nguyen-Ngoc and Tran-Minh (2007) developed a reagentless fluorescence optical biosensor based on vegetal cells, viz., *Chlorella vulgaris* entrapped in an inorganic translucent matrix produced from sol-gel technology for the determination of diuron as an anti-PSII herbicide.

20.5 BIO-APPLICATIONS OF SOL-GEL MATERIALS

More recently, new features of sol-gel technology have been developed for extending the potential of this technology for bio-applications. Various sol-gel materials derived from biocompatible organic–inorganic precursors and nanomaterials have appeared and extensively studied for bio-applications.

The development of more advanced novel sol-gel nanostructured hybrid materials particularly has become an attractive and current area of research for developing biosensors and opens up new possibilities in the field of molecular diagnosis and therapy.

20.5.1 Clinical Diagnostic Applications

The reliable and accurate information on the desired biochemical parameters is an essential prerequisite for effective health care, which requires accurate, fast, and inexpensive devices in medical diagnostics laboratories for the routine analysis of clinical samples like blood, serum, and urine. Therefore, the commercial development of biosensors for clinical diagnostic applications has attracted the greatest attention to provide miniaturized devices for quick measurements of clinical analytes. Biosensing devices are advantageous due to their specificity, sensitivity, fast response, small size, and convenience for self-testing at home and critical care at bed side in emergencies. A number of research papers and reviews are available in the literature showing the progress of developing biosensors for monitoring clinically important parameters, viz., blood glucose, cholesterol, CRP, lactate, urea, uric acid, creatinine, etc. Enzymes are well known as biological sensing materials in the development of biosensors due to their specificity. Moreover, the role of enzymes in clinical diagnosis has been known for several years. Since enzymes have poor stability in solutions, it is therefore needed to stabilize them by immobilization. In the immobilized phase, they gain excellent stability and can be reused. Numerous techniques of immobilization, viz., covalent linkage, physical adsorption, cross-linking, encapsulation, and entrapment have been known for stabilization of enzymes for the development of biosensing devices. The matrix or support to be chosen for immobilization depends on the nature of biomolecule and the method of immobilization. A number of matrices, such as polymeric films and membranes, gels, Langmuir Blogett films, carbon, graphite, diaphorase and conducting polymers, etc., have been used for the immobilization of biomolecules in the development of various types of biosensors (Malhotra and Chaubey, 2003).

Of all biosensors, the electrochemical glucose biosensor has been studied most. In the first glucose biosensor, the enzyme GOx was immobilized on electrodes. The design of glucose biosensors has been progressively improved over the years by the utilization of mediators, electroactive chemical compounds, and forming various polymeric matrices, especially sol-gel-derived hybrids and nanocomposite films. Chen et al. (2003) developed a new type of organically modified sol-gel/chitosan composite material for the construction of glucose biosensor. The incorporation of chitosan into TEOS-derived sol-gel matrix overcomes the brittleness of sol-gel matrix and also improves the long-term stability of the biosensor by providing good biocompatibility and stabilizing the microenvironment around the enzyme. Salimi et al. (2004) demonstrated immobilization of GOx enzyme within sol-gel-derived MWCNT films over graphite electrode. Compared with other types of biosensors for glucose detection, the modified biocomposite electrode exhibited excellent sensitivity and stability for the determination of glucose by the electrocatalytic oxidation of enzymatically liberated hydrogen peroxide at the reduced over potential. The implantable glucose biosensors are also under development. In order to develop a functional implantable sensor, all the different components of the sensor must display optimal performance, both in a biological and clinical way. Ideally, an outer interfacial membrane of an implantable glucose sensor should ensure appropriate transport of glucose to the sensor, while excluding proteins and other interference from the electrode. Furthermore, it should elicit minimal fibrous encapsulation with sufficient vascularization in order to warrant adequate glucose diffusion. Finally, to prevent denaturation of the enzyme GOx, which is the active component of the glucose sensor, stable membranes must be obtained without the need for high-temperature processing. In this contest, inorganic–organic-hybrid-derived coating can be advantageous to maintain the biocompatibility of sensor up to level of implantation. Gerritsen et al. (2000) investigated the biocompatibility of various sol-gel matrices derived from heparin, dextran sulfate, Nafion®, PEG, and polystyrene sulfonate *in vitro* in simulated body fluid (SBF), with cell culture using human dermal fibroblast and finally *in vivo* after subcutaneous implantation in rabbit in order to develop a coating for a subcutaneously implantable glucose sensor that elicits an optimal soft-tissue response. Different approaches have been proposed for developing sensitive, selective, reliable, and low-cost cholesterol sensors in several research papers (Kumar et al., 2000, 2006; Li et al., 2003; Tan et al., 2005; Singh et al., 2007) because of the significance of measuring this substrate in the clinical diagnosis of coronary heart diseases, arteriosclerosis, cerebral thrombosis, and miscellaneous other disorders.

Array biosensors are particularly well suited for clinical diagnostics applications. Tsai and Doong (2004) demonstrated the first use of a sol-gel-derived optical array biosensor for simultaneous analysis of multiple samples for the presence of multiple clinically important renal analytes with cross-interferences. Urease and creatinine deiminase were used to detect urea and creatinine, while GOx and uricase were co-immobilized with HRP to quantify glucose and uric acid in fetal calf serum (Figure 20.12). The reproducibility of array-to-array in 3 consecutive months was 5.4% ($n = 3$). Grant and Glass (1999) developed a novel fiber-optic biosensor for the detection of D dimer

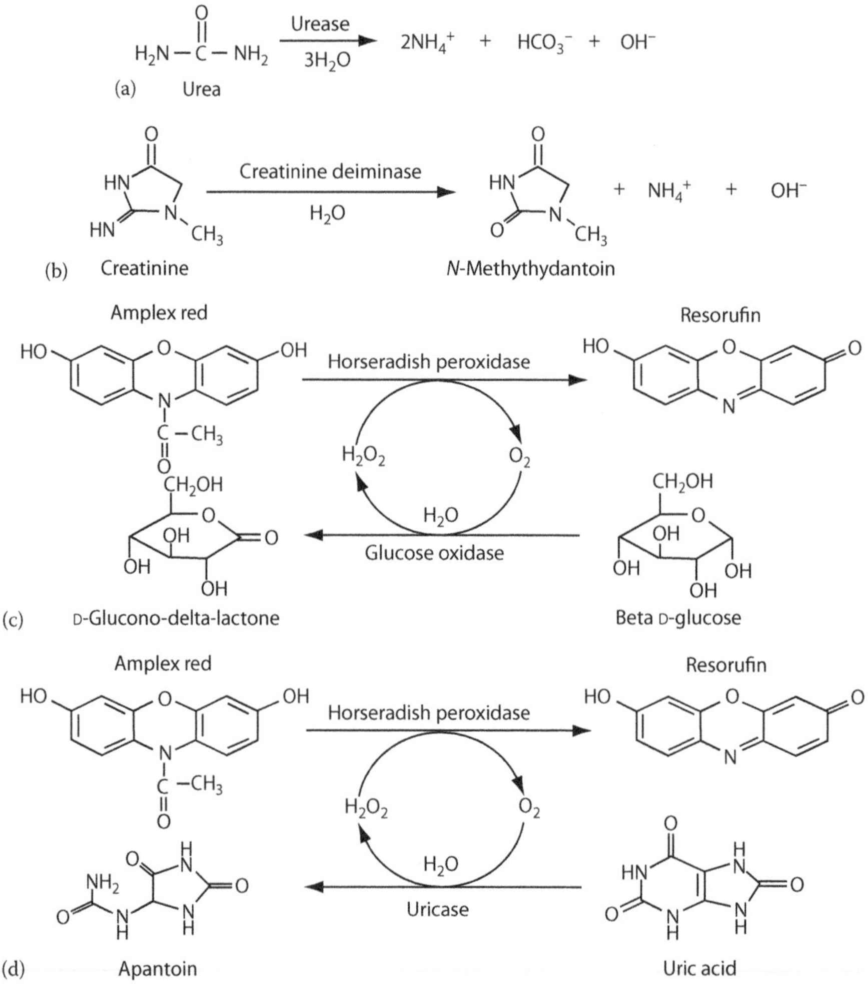

FIGURE 20.12 Detection principles of (a) urea, (b) creatinine, (c) glucose, and (d) uric acid. Urea and creatinine were detected by urease and creatinine deiminase, respectively. Glucose was detected by the combination of GOx and HRP and uric acid was detected by uricase and HRP. (Reproduced from Tsai, H.C. and Doong, R.A., *Anal. Biochem.*, 334, 183, 2004. With permission.)

antigens, which form from the dissolution of cross-linked fibrin clots. The presence of D dimer antigens above a threshold level is a clinical diagnostic used to determine the presence of such occlusions following a stroke. Fluorescein-labeled D dimer antibodies were immobilized on the tip of an optical fiber by dip coating from a silica sol-gel solution. When D dimer antigens combined with the antibodies, fluorescence intensity decreased. The response of the sensor was examined in phosphate-buffered saline (PBS), human plasma, and blood. Calibration plots for the sensor were obtained in the clinically significant D dimer concentration range from 0.54 to 6 µg/mL. A decrease in fluorescence intensity was observed in aged sol-gel-encapsulated tagged antibodies. The D dimer antibodies encapsulated in the sol-gel network was found to be viable for at least 4 weeks when stored at 4°C in PBS solution. Kwon et al. (2008) successfully applied the sol-gel-based protein microarray technology to a screening assay for hepatitis C virus (HCV) diagnosis with confirmatory test-level accuracy. The detection limit of assay was 1000 times more sensitive than that of the ELISA method.

20.5.2 BIOMEDICAL APPLICATIONS

Bioactive glasses and glass–ceramics have been attractive for several biomedical applications because of the nature of the spontaneous bond to living bone when implanted in a bony defect. Their flexibility can be increased by designing organic–inorganic hybrid materials through incorporating essential constituents for bioactivity (Si–O and calcium ions) with organic polymers. Organic–inorganic hybrid materials prepared by the sol-gel approach have rapidly become a fascinating new field of research in biomedical science. Ohtsuki et al. (2002) synthesized an organic–inorganic hybrid from 3-methacryloxypropyltrimethoxysilane (3-MPTS) and 2-hydroxyethylmethacrylate (HEMA), which formed the apatite layer when mixed with calcium chloride. Such a type of hybrid is expected to be a novel bone-repairing material with bioactivity as well as mechanical properties close to conventional poly(methyl methacrylate) (PMMA) bone cement. However, several studies have been reported in the literature for the coating of glucose sensors with various polymeric materials such as cellulose acetate, PEG, poly(vinyl chloride), polyurethane, and nafion, but their *in vivo* studies were not very promising. Bioactivity and compatibility of sol-gel-derived hybrids make them attractive for the coating of implantable biosensing devices *in vivo*, since they have been demonstrated to be highly compatible with proteins, enzymes, and other biomolecules. Kros et al. (2001) prepared a silica-based hybrid biocompatible coating, which can be used for future implantable glucose sensors by mixing TEOS as the main inorganic precursor with different organic molecules such as PEG, heparin, dextran sulfate, nafion, or polystyrene sulfonate. The toxicity of the coatings was examined *in vitro* using human dermal fibroblasts. All materials were found to be nontoxic and the cell proliferation rate of fibroblasts was found to be dependent on the additive. Glucose measurements using GOx-based sensors coated with the different hybrid films were performed both in buffered solutions containing BSA and in serum. Stable glucose responses were obtained for the coated sensors in both media. The dextran sulfate-derived sol-gel coating appeared to be most promising for future *in vivo* glucose measurements. The encapsulation of pancreatic islets for secreting insulin in sol-gel has also been reported. Pope et al. (1997) demonstrated the potential of silica-gel-encapsulated pancreatic islets of Langerhans by the measurement of insulin secretory response *in vitro* and blood sugar levels of diabetic mice *in vivo*. Peterson et al. (1998) developed a sol-gel-derived biocomposite material in the form of a capsule for the encapsulation of insulin-secreting murine islet cells as the first mammalian material in sol-gel using the drop-tower sphere generation and emulsion technique. Average pore sizes were 161 Å for drop-tower spheres and 105 Å for emulsion spheres. These capsules allowed the passage of insulin and cytokines but not the passage of antibodies. Implantation of encapsulated islets did not result in fibrosis of the capsule *in vivo*; and retrieval of capsules after 1 month *in vivo* documented continued insulin secretory capacity. Thus silica-based hybrid encapsulation provides a potentially useful alternative for the encapsulation of cells for transplantation or drug delivery.

20.5.2.1　Sol-Gel Hybrid Materials for Dental Applications

Inorganic–organic hybrid materials can be used as filling composites in dental applications. These composites feature tooth-like properties (appropriate hardness, elasticity, and thermal expansion behavior) and are easy to use by the dentist as they easily penetrate into the cavity and harden quickly under the effect of blue light. Moreover, these materials feature minimum shrinkage, are nontoxic, and sufficiently nontransparent to X-rays. Traditional plastic filling composites had long-term adhesion problems and a high degree of polymerization shrinkage resulting in marginal fissures. The dual character of the ORMOCERS® as inorganic–organic copolymers is the key for improving the properties of filling composites. The organic, reactive monomers are bound in the sol-gel process by the formation of an inorganic network. Thus, in the subsequent curing process, polymerization takes place with less shrinkage. Furthermore, abrasion resistance, in particular, is significantly enhanced by the existing inorganic Si–O–Si network. For example, in dental fillers, organic functionalities including ring-opening reactions, such as functionalized spyrosilanes, are commonly included in the hybrid network. Other systems are based on multiacrylate silanes, offering a high organic density. In addition, mechanical properties of the composite can be tuned through variation of the spacer between the silicon atom and the reactive functionality (Sanchez et al., 2005).

20.5.2.2　Sol-Gel-Derived Bioactive Glasses for Tissue Engineering

The concept of bioactive glasses was initially observed in 1971 for silicate glasses based on the system $SiO_2–CaO–Na_2O–P_2O_5$. Bioglass® is a trade name given to a series of such glass compositions. The relatively low silicon and high alkaline content lead to a rapid ion exchange in aqueous environments. This exchange generally leads to an increase in solution pH, which can be substantial for finely grained powders having high surface-to-volume ratios. The initially rapid release of sodium is accompanied by a somewhat slower release of other ion species, predominantly calcium and silica. Under certain conditions in solution, these ion species will precipitate onto the glass and onto other nearby surfaces to form calcium-containing mineral layers, or sometimes the outer glass surface itself can transform to hydroxy-carbonoapatite (HCA). The ability to build such a surface is sometimes referred to as a measure of the "bioactivity" of the glass. When implanted into the body, repair cells will colonize the bioactive surface, laying down new tissue on and in the glass. The reaction stages of bioglass are shown in Table 20.5. The activity is strongly dependent on the particle size and increases as particle size goes down. In a finely grained powder form, bioactive glasses have additionally demonstrated antimicrobial and anti-inflammatory properties. Other bioactive glass compositions, as well as other bioactive materials such as glass ceramics and calcium-phosphate-based ceramics (referred to generally as bioactive ceramics), have also been developed. Among the bioactive ceramics, bioactive glasses have the highest levels of bioactivity based on their rate of reaction and bone bonding. Bioactive glasses have many applications but these are primarily in the areas of bone repair and bone regeneration via tissue engineering. These glasses have been used successfully as bone-filling materials in orthopedic and dental surgery, but their poor mechanical strength limits their applications in load-bearing positions. However, many methods are reported to improve the mechanical strength of these bioactive glasses such as transformation of bioactive glasses into glass ceramics, fiber/particulate reinforced bioglass, using bioactive glass as a coating on a substrate, etc., but approaches to strengthen these materials decrease their bioactivity. Thermal treatments also affect the microstructure of bioactive glasses and hence their bioactivity (Gupta and Kumar, 2008a). Yurong and Lian (2005) observed that thermal treatment of gel-derived bioglasses results in phase separation and minor crystallization which leads to some changes in the microstructure of samples, such as the density and porosity that increase the bending strength and fracture toughness with some extent of decrease in bioactivity. The reason underlying this is the separation of silica-rich and phosphate-rich phases which causes a dramatic change in the bioactivity reactions at the material–biological fluid interface in consequence of the increase of viscosity of the phase-separated glass and the decrease of Si and Ca ions in solution. Balamurugan et al. (2006)

TABLE 20.5

Sequence of Interfacial Reactions Involved in Forming a Bond between Tissue and Bioactive Glass

Stages	Reactions
1.	Rapid exchange of Na^+ or Ca^{2+} with H^+ or H_3O^+ from solution: $\equiv Si-O-Na^+ + H^+ + OH^- \rightarrow \equiv Si-OH + Na^+ + OH^-$
2.	Loss of soluble silica in the form of $Si(OH)_4$ to the solution resulting from breakage of Si–O–Si bonds and formation of Si–OH at the glass/solution interface: $\equiv Si-O-Si \equiv + H_2O \rightarrow \equiv Si-OH + OH-Si \equiv$
3.	Condensation and repolymerization of a SiO_2-rich gel layer on the surface depleted in alkalis and alkaline-earth cations: $(RO)_3-Si-OH + HO-Si-(OR)_3 \rightarrow (RO)_3-Si-O-(OR)_3 + H_2O$
4.	Migration of Ca^{2+} and PO_4^{3-} groups to the surface through the SiO_2-rich layer forming a $CaO-P_2O_5$-rich film on top of the SiO_2-rich layer, followed by growth of the amorphous $CaO-P_2O_5$-rich film by incorporation of soluble calcium and phosphates from solution
5.	Crystallization of the amorphous $CaO-P_2O_5$ film by incorporation of OH^- or CO_3^{3-} anions from solution to form mixed hydroxyl-carbonate apatite layer
6.	Agglomeration and chemical bonding of biological moieties in the HCA layer
7.	Action of macrophages
8.	Attachment of mesenchymal stem cells
9.	Proliferation and differentiation of stem cells
10.	Generation of matrix
11.	Crystallization of matrix
12.	Proliferation of bones

Source: Reprinted from Gupta, R. and Kumar, A., *Biomed. Mater.*, 3, 034005, 2008a, and references cited therein. With permission. Copyright IOP.

prepared a bioactive gel-derived glass bulk 58S in the system $SiO_2-CaO-P_2O_5$ by using the sol-gel self-propagating method in order to realize the optimal matching between mechanical and biological properties. Xynos et al. (2000) investigated the concept of using bioactive substrates as templates for *in vitro* synthesis of bone tissue for transplantation by assessing the osteogenic potential of a melt-derived bioactive glass ceramic (Bioglass® 45S5) *in vitro*. Bioactive glass ceramic and bioinert (plastic) substrates were seeded with human primary osteoblasts and evaluated after 2, 6, and 12 days. Flow cytometric analysis of the cell cycle suggested that the bioactive glass-ceramic substrate induced osteoblast proliferation, as indicated by increased cell populations in both S (DNA synthesis) and G2/M (mitosis) phases of the cell cycle. SEM images of discrete bone nodules over the surface of the bioactive material, from day 6 onward, further supported this observation (Figure 20.13).

20.5.2.3 Sol-Gel-Derived Surface Modification of Metallic Implants

Metallic materials are an important class of implant materials because of their combination of strength and ductility as compared to polymers and ceramics. On the other hand, metallic materials are less corrosion resistant compared with the other two classes of implant materials. Corrosion reduces strength and causes premature failure of implants and may also impose harmful effects on the surrounding tissues. Stainless steels (SS), titanium alloys, and cobalt alloys are commonly used as biomaterials. Shape memory alloys (SMA) are a relatively new group of metallic biomaterials. The extraordinary properties of NiTi shape memory alloys are widely used in medical applications such as in orthodontics, cardiovascular, orthopedics, urology, etc., due to their unique shape memory effects, superelasticity, and good corrosion resistance. However, the high nickel content (about 50%) in NiTi has caused some concern about its safe use *in vivo* because Ni is allergenic and toxic

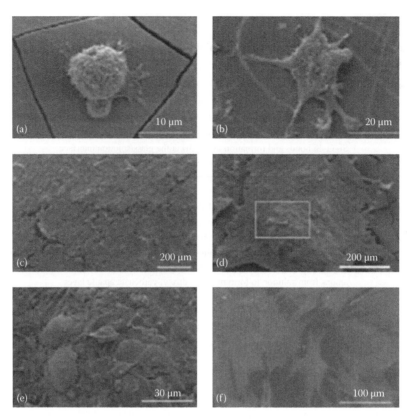

FIGURE 20.13   SEM images of osteoblasts cultured on Bioglass 45S5. (a) Two days after seeding, spherical osteoblasts contacted the substrate by means of numerous filopodia and fiber-like processes. (b) After 6 days of culture, cells were anchored to the substrate by multiple lamellipodia. (c) After 12 days of culture, cells appeared to be well spread and grew in multilayer fashion. (d) SEM image of a bone nodule present on Bioglass 45S5 on day 12. (e) Higher magnification of the central domed region of the bone nodule shown in (d). (f) After 6 days of culture, osteoblasts seeded on inert control substrate exhibited a smooth dorsal surface, adopted a flattened configuration, and formed confluent monolayers. (Reproduced from Xynos, I.D. et al., *Calcif. Tissue Int.*, 67, 321, 2000. With permission.)

when its concentration in the human body exceeds a certain level. In view of this, surface treatment of NiTi implants for improving corrosion resistance and hence reducing the amount of Ni released is necessary for biomedical implants. A number of surface treatment methods, namely chemical passivation, electropolishing, anodization, thermal oxidation, laser surface melting, nitriding, plasma ion implantation, and sol-gel-derived coating, have been reported. Out of these, the sol-gel route for surface modification of NiTi implants is of particular interest because of simple and inexpensive methodology, low temperature processing, and suitability for coating substrates of irregular shapes, such as implants. Conventional heat treatment for sol-gel-derived titania coatings is not desirable due to the higher temperature (400°C–500°C) employed because the thermomechanical properties of NiTi implants are highly sensitive to heat treatment. The sol-gel hydrothermal process is a potential low-temperature route for depositing oxide coating on NiTi implants and has been attempted by several authors (Gupta and Kumar, 2008a). Cheng et al. (2004) improved the corrosion resistance of NiTi implants via sol-gel dip coating with TiO_2, employing hydrothermal treatment to crystallize and densify the amorphous film, and to remove the organic residue. Electrochemical impedance spectroscopy (EIS) and polarization studies indicated a significantly larger increase in corrosion resistance compared with the coated samples dry heated at 500°C. Chiu et al. (2007) showed that steam crystallization is a feasible low-temperature treatment method for sol-gel-derived titania coating on NiTi in biomedical applications. The authors prepared dip-coated titania films via the sol-gel route using titanium butoxide [Ti(OC$_4$H$_9$)$_4$] as a precursor and further crystallized by treatment in steam at 105°C. Liu et al. (2003) prepared TiO$_2$ thin films on an NiTi surgical alloy by the sol-gel method. The electrochemical corrosion measurement indicated that the TiO$_2$ thin film was effective for improving the corrosion resistance of the NiTi alloy. *In vitro* blood compatibility of the film and the NiTi alloy was also evaluated by dynamic clotting time and blood platelet adhesion tests. The results showed that the NiTi alloy coated with the TiO$_2$ film had improved blood compatibility.

20.5.2.4 Biocidal Sol-Gel Coatings

Bacterial infection due to an implanted medical device such as prosthetic hip implants, central venous catheters, and urinary catheters, etc., is a potentially serious complication, typically leading to premature implant removal, which is costly, traumatic to the patient, and might be lethal. Despite various preventative methods such as sterilization, meticulous surgical procedure, and following proper infection control guidelines, invasive bacteria can be found at 90% of implantation sites immediately after surgery. The *Staphylococci* species including *Staphylococcus aureus* and *Staphylococcus epidermidis* are responsible for the majority of biofilms found on explanted orthopedic devices. Recently, new strategies are emerging for developing materials with antimicrobial activity that can be used for controlling and preventing microbial contamination of medical devices. The immobilization of an antimicrobial agent in a matrix capable of binding to different surfaces is an interesting way to develop such antibacterial materials. The sol-gel dip-coating process is an effective procedure for developing antimicrobial coatings as compared to other immobilization methods. Several studies have shown that silver or silver ions have broad-spectrum antibacterial activity against gram-positive and gram-negative strains, including antibiotic resistant strains (Gupta and Kumar, 2008a). Jeon et al. (2003) prepared silver-doped TEOS-derived silica thin films by the sol-gel method, showing an antibacterial effect against *Escherichia coli* and *Staphylococcus aureus*. Díaz-Flores et al. (2007) prepared antibacterial TEOS-derived sol-gel thin films and microcrystalline powders doped with Ag and Cu. Stobie et al. (2008) reported a potential solution to the problem of biofilm growth on short-term indwelling surfaces using low-temperature-processed silver-doped phenyltriethoxysilane (PhTEOS) sol-gel coating which reduced the formation of *Staphylococcus epidermidis* biofilms over a 10 day period. However, high temperature causes an increase in crystallinity of sol-gel coating and consequently a reduction in silver release kinetics. The antibacterial activity of Ag-doped PhTEOS coatings against Planktonic *S. epidermidis* is shown in Figure 20.14, where the release of silver ions caused an approximate 100% kill rate. Copello et al. (2006) immobilized the antimicrobial compound dodecyl-di(aminoethyl)-glycine in TEOS-derived xerogel films

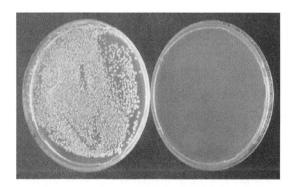

FIGURE 20.14 Antibacterial activity of Ag-doped PhTEOS coatings against planktonic *S. epidermidis* (CSF 41498) after 24 h. *Left*: Undoped coating. *Right*: Silver-doped coating. (Reproduced from Stobie, N. et al., *Biomaterials*, 29, 963, 2008. With permission.)

coated on a glass surface via the dip-coating technique. When antimicrobial-coated glasses were compared with antimicrobial-free coated glasses, the former showed greater than 99% reduction of colony forming units for *Escherichia coli*, *Pseudomonas aeruginosa*, and *Staphylococcus aureus*. Recently, a localized persistent concentration of nitric oxide (NO) in the vicinity of an invasive medical device may have proven to be a novel approach for reducing the implant-associated infection due to its short half-life (ranging from 1 s to a few minutes depending on the concentration of oxygen and the presence of NO scavengers such as oxyhemoglobin) in biological milieu. In the literature, it has been reported that NO gas destroys plated colonies of bacteria. A number of synthetic NO donors including nitrosothiols, nitrosamines, diazeniumdiolates, metal complexes, and organic nitrates/nitrites have been used to design polymer coatings capable of slowly releasing therapeutic levels of NO that are effective in reducing biofouling. Of these NO donor species, *N*-diazeniumdiolates have emerged as attractive candidates for designing more biocompatible coatings due to their ability to generate NO spontaneously under physiological conditions. When NO reacts with amines, a zwitterionic stabilized structure known as *N*-diazeniumdiolate is produced, which decomposes spontaneously in aqueous media to NO (Figure 20.15a). The rate of release of NO depends upon pH, temperature and/or the structure of the amine moiety. Sol-gel coatings capable of NO release have recently been shown to decrease bacterial adhesion (Gupta and Kumar, 2008a). Schoenfisch and coworkers (Nablo et al., 2001, 2005; Marxer et al., 2003; Nablo and Schoenfisch, 2003, 2005; Robbins et al., 2005) have reported significant work on the synthesis and characterization of sol-gel-derived materials (xerogels), where *N*-diazeniumdiolate NO donors were covalently bound to the xerogel backbone (Figure 20.15b). A range of inorganic–organic hybrid xerogels have been functionalized to release NO by incorporating diamine-containing organosilanes into a sol-gel matrix and their properties were tailored by varying the type and amount of alkyl- and aminosilane precursors and processing conditions (e.g., pH, catalyst, water content, and drying time and temperature). Several alkyl- and aminosilane precursors were evaluated for this purpose, including methyl-, ethyl-, and butyltrimethoxysilanes (MTMOS, ETMOS, and BTMOS, respectively), (aminoethylaminomethyl) phenethyltrimethoxysilane (AEMP3), *N*-(2-aminoethyl)-3-aminopropyltrimethoxysilane (AEAP3), *N*-(6-aminohexyl)aminopropyltrimethoxysilane (AHAP3), and *N*-[3-(trimethoxysilyl)-propyl]diethylenetriamine (DET3). Upon exposure to high pressures of NO (g), the diamine coordinates two molecules of NO to form a NO donor molecule known as a diazeniumdiolate. When the

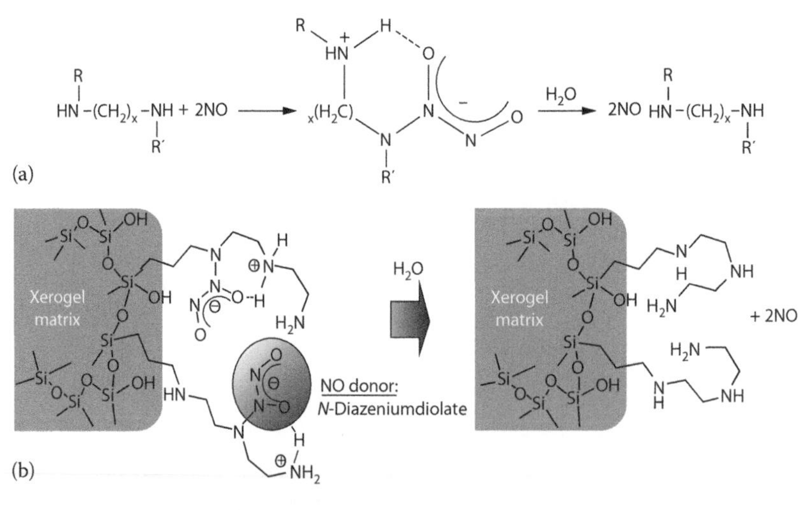

(a)

(b)

FIGURE 20.15 (a) Reaction of NO with amines to produce *N*-diazeniumdiolate NO donors followed by the subsequent generation of NO in the presence of water. (b) Schematic of NO generation from *N*-diazeniumdiolate-modified xerogel network occurring upon exposure to aqueous conditions. (Reproduced from Shin, J.H. and Schoenfisch, M.H., *Analyst*, 131, 609, 2006, and references cited therein. With permission.)

sol-gel is introduced into an aqueous environment, the diazeniumdiolate decomposes to NO and the diamine precursor. The local surface flux of NO generated from these xerogels significantly reduces the adhesion of *Pseudomonas aeruginosa* by up to 95%, demonstrating that NO release may represent a new class of antibacterial biomaterials. Nablo et al. (2005) coated medical-grade SS with a sol-gel film of 40% AHAP3 and 60% BTMOS. The bacterial adhesion resistance of NO-releasing coatings was evaluated *in vitro* by exposing bare steel, sol-gel, and NO-releasing sol-gel-coated steel to cell suspensions of *Pseudomonas aeruginosa*, *Staphylococcus aureus*, and *Staphylococcus epidermidis* at 25°C and 37°C. Cell adhesion to bare and sol-gel-coated steel was similar, while NO-releasing surfaces had significantly less bacterial adhesion for all species and temperatures investigated. The NO-releasing sol-gel materials for *in vivo* biosensor applications are also emerging and are in development. NO is a highly reactive radical and affects enzymatic activity. Shin et al. (2004) developed an NO-releasing hybrid sol-gel/polyurethane glucose biosensor by doping diazeniumdiolate-modified sol-gel particles in a polyurethane membrane on platinum electrodes, which was sandwiched with additional polyurethane membranes to reduce both enzyme inactivation by NO and sol-gel particle leaching. The response characteristics of the hybrid-NO-releasing glucose biosensor remained stable through 18 days, after which the line arrange decreased from 0 to 60 to 0–20 mM glucose, and the response time increased from less than 20 s to over 65 s. Oh et al. (2005) demonstrated the fabrication of a miniaturized needle-type glucose biosensor patterned with an *N*-diazeniumdiolate modified xerogel microarray for overcoming the reduced analyte permeability observed with xerogel films. Further studies are in progress for identifying methods to improve the duration of NO release and comprehensive *in vivo* biocompatibility testing of NO-releasing sol-gels.

20.5.3 Environmental Applications

Environmental pollution and damage on a global scale have drawn much attention for the development and implementation of more innovative technologies for water treatment and purifications. This is because of the presence of many microorganisms and naturally occurring organic and inorganic chemicals in water which exhibit carcinogenic, mutagenic, endocrine-disrupting, and toxic properties. All these water pollutants pose a serious health threat to both humans and aquatic life. The stimuli-responsive behavior of ORMOSILS-derived smart sol-gel materials can be useful for environmental applications by undergoing structural changes (in the form of swelling/shrinkage) at the molecular level to respond to changes in environmental variables like changes in temperature, pH, and salt concentrations, etc. In addition, development of new sol-gel nanocomposites can play an important role in environmental protection, e.g., nanocrystalline metal oxides are investigated as adsorbents for fuel desulfurization and as photocatalysts for the degradation of pollutants in air and water. Maduraiveeran and Ramaraj (2007) demonstrated the electrocatalytic activity of gold nanoparticles immobilized in APS-derived sol-gel network (Scheme 20.5) for the concurrent detection and determination of toxic chemicals, viz., hydrazine (N_2H_4), sulfite (SO_3^{2-}), and nitrite (NO^{2-}). These toxic chemicals are significantly important for the protection of human health because N_2H_4 is a neurotoxin and produces carcinogenic and mutagenic effects. The SO_3^{2-} ion is used as a preservative in beverage, food, and pharmaceutical products. The NO_2^- ion is an important precursor in the formation of nitrosamines, many of which are carcinogenic in nature.

20.5.3.1 Sol-Gel-Based Sensors for Water Pollutants

Transition metal ions can have profound biological effects in the environment and on human health even at extremely low concentrations, often as less than 1 mg/L. Therefore, it is important to detect and monitor levels of transition metals in low concentrations to assess the health risks and for environment monitoring. Cadmium (Cd), a major pollutant in some areas, is an example of a transition metal that can be selectively detected using fluorescence-based sensors. Anthryl tetra acid (ATA) has recently been reported to be a selective Cd^{2+} sensor. The Environment Protection Agency (EPA) standard for maximum Cd^{2+} concentration in drinking water is 0.005 mg/L. Long-term exposure of

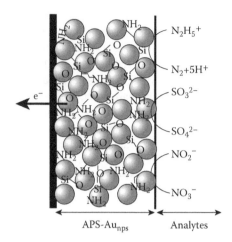

SCHEME 20.5 Schematic representation of gold nanoparticles embedded in silicate sol-gel matrix (APS–Au$_{nps}$) modified GC electrode and simultaneous electrocatalytic oxidation of N$_2$H$_4$, SO$_3^{2-}$, and NO$_2^-$. (Reproduced from Maduraiveeran, G. and Ramaraj, R., *Electrochem. Commun.*, 9, 2051, 2007. With permission.)

cadmium above this level can potentially cause kidney malfunctions. Trace quantities of copper (Cu) is essential intake for health, but excess amount (1.3 mg/L) has adverse effects. It causes gastrointestinal problems and kidney damage in case of long-term exposure. Recently, a selective Pyrene-based sensor for Cu^{2+} has been developed. Lead is also one of the major serious environmental and health hazards. Lead toxicity affects kidneys and reproductive systems. The FDA limit for lead in drinking water is 0.015 mg/L. Recently, a sensor for Pb^{2+} ions in the acetonitrile solutions has been reported. Organic reagents that change color upon interactions with metal ions are also reported to entrap in porous silica sol-gel matrices for the detection of various metal ions (Chaudhury et al., 2007). Zusman et al. (1990) developed the sensors for various metal ions such as Fe^{3+}, Al^{3+}, Co^{3+}, Ni^{2+}, Cu^{2+}, and Pb^{2+} as well as changes in pH by monitoring the characteristics color changes in the doped monolithic xerogel glass blocks. The response times were on the order of seconds and sensitivities were <10^{-7} mol/L. Further development resulted into disposable tube detectors for water analysis, which were based on doped sol-gel glasses (Kuselman et al., 1992; Kuselman and Lev, 1993). When a solution was sampled through a tube detector, a specific metal ion was complexed by the organic reagent, forming a colored section of the glass bed. The length of the colored section was related to the concentrations of the metal ions. A cyanide sensor has also been developed by encapsulating iron(III) porphyrin in a sol-gel-derived titanium carboxylate thin film (Dunuwila et al., 1994). A sensor for oxalate determination based on the use of oxalate oxidase encapsulated in sol-gel glass is also reported (Dave et al., 1994). Plaschke et al. (1996) immobilized porphyrin derivatives covalently bound to dextran in sol-gel-derived thin films to develop stable mercury sensor. The stability of the mercury sensor was found to be longer than 6 weeks. Flora and Brennan (1998) reported the development of a fluorometric detection of Ca^{2+} based on an induced change in the conformation of cod III parvalbumin entrapped within a sol-gel-processed glass. Many of these experimental demonstrated sensors could be very useful for detecting and monitoring the pollutants in drinking water.

20.5.3.2 Sol-Gel-Based Gas Sensor

Detection and monitoring of gases at low levels are often routine activities in laboratories and industries. Environmental monitoring of toxic gases has become very important due to an increasing level of pollutants as well as in the event of intentional accidents. In general, environmental monitoring system will require rugged systems. Sol-gel materials are very environment friendly and stable at high temperature and humidity. Sol-gel methods have been used in the detection of NO using cobalt tetrakis (5-sulfothienyl) porphyrin doped silica glass prepared by the sol-gel method (Eguchi et al., 1990). Blyth et al. (1995) demonstrated the feasibility of NO and CO sensing based on biomolecules such as Cyt-c, Mb, and Hb. The utility of various entrapped hemoproteins in sol-gel was also demonstrated for CO, NO, and O_2 sensing (Dave et al., 1994; Chung et al., 1995; Cox et al., 1996). Oxygen permeability of sol-gel coatings has been investigated by the selective oxygen quenching of phosphorescent probes (such as platinum octaethylporphine) doped in sol-gel glasses. It was also demonstrated that a series of sol-gel glass coatings with different degrees of oxygen permeability could be prepared by the sol-gel process using different compositions of precursors and different functional groups (methyl, phenyl, octyl, propyl, etc.). Therefore, fiber-optic oxygen sensors based on these sol-gel coatings can have a tunable optimum oxygen sensitivity range (Liu et al., 1992). Samuel et al. (1994), in an effort to solve reagent leaching problem, found that the polymerization of TMOS at high acidity and low water content resulted in non-leachable yet reactive sol-gel as demonstrated with oxygen sensing by fluorescence of entrapped Pyrene. MacCraith et al. (1992, 1993) developed oxygen sensor based on sol-gel glass doped with a ruthenium complex Ru^{II}-tris (2,2'-bipyridine) or R^{II}-tris (4,7-diphenyl-1,10-phenanthroline).

20.5.3.3 MI Sol-Gel-Based Detection Systems for Environmental Endocrine Disrupters

Endocrine disrupters constitute a wide group of environmental micropollutants that alter functions of the endocrine system and consequently cause adverse health effects in an intact organism or its progeny. Any chemical with the potential to interfere with the function of the endocrine systems like production, release, transport, metabolism, binding, action, or elimination of the natural hormones in the body are called endocrine-disrupting compounds (EDCs). The exposure to these compounds directly affects reproduction, developmental process, and thus evolutionary fitness. EDCs are both naturally occurring and synthetic, including female sex hormones, synthetic steroidal hormones, phytoestrogen, pharmaceutical products, industrial chemicals, polychlorinated compounds, phenols, polycyclic aromatic hydrocarbons (PAHs), pesticides, surfactants, compounds used in plastic industry, and heavy metals, which are entered in surface water by direct discharge of industrial, domestic, and municipal wastewater effluent, agricultural drains to streams and rivers, and overland flow after rainfall events. Many endocrine disruptors are persistent in the environment called persistent organic pollutants (POPs) and accumulate in fat; so the greatest exposures come from eating fatty foods and fish from contaminated water. The most potent endocrine disrupters are benzopyrene (carcinogenic PAHs), bisphenol-A (suspected chemical in plastic products), diethylstilbesterol (orally active synthetic nonsteroidal estrogen), atrazine (herbicide), 1,1-dichloro-2,2-bis(p-chlorophenyl) ethylene (metabolite of the insecticide DDT), and polychlorinated biphenyls (stable chemicals that may persist in the environment for many years). Although very low concentrations of these compounds are found in water, their widespread use and their capability to induce responses in animals at concentrations as low as ng/L or even pg/L levels have alerted scientists about the potentially dangerous consequences of their presence in the environment and on human health. Because of this, endocrine disrupters and their possible impact on human health is becoming an active area of research in environmental biotechnology. Several research activities have been started toward the better understanding of the endocrine disrupters issue, its potential effects on human health, and the development of a suitable detection system (biosensor) for the online monitoring of toxic micropollutants in the environment. Conventionally, SPE cartridges are used for isolation purposes. Imprinted polymers can also be used for isolation and screening of EDCs. Several studies have reported molecular imprinting of endocrine disrupters and used MIPs in column as online pretreatment device for their isolation and other structurally related chemicals with estrogenic activity in contaminated water coupled with analytical instruments like high performance liquid chromatography (HPLC), mass spectrometry (MS), and liquid chromatography. Lack of selectivity and low recovery of the analyte are the most common problems observed with

SPE cartridges as these problems can be overcome by coupling MIP with SPE (Gupta and Kumar, 2008–2009).

Sol-gel imprinting is also one of the new emerging fields in the area of molecular imprinting. Han et al. (2005) prepared a new molecularly imprinted amino-functionalized silica gel sorbent with binding sites situated at the surface for the template pentachlorophenol (PCP), by a surface-imprinting technique in combination with a sol-gel process, and applied it to online selective SPE coupled with HPLC for the determination of trace PCP in water samples. PCP is used as a general herbicide in agriculture and as an insecticide for termite control in the preservation of wood. Sol-gel molecular imprinting provides an easy approach for polymerizing sensitive layer of recognition element directly over mass-sensitive devices such as surface acoustic wave (SAW), quartz crystal microbalance (QCM), and SPR. However, sol-gel-based MIPs have found several analytical applications but still not much explored for endocrine disrupters. Thus, sol-gel molecular-imprinted polymer-based detection system can be effectively used for endocrine disrupters in near future.

20.5.4 Defense Applications

Recently, nanocomposites and molecularly imprinted materials are involved to a great extent for the development of a broad range of collectors and detectors for WMD for the purpose of homeland security and defense applications. Weapons of mass destruction include chemical warfare (CW) agents such as sarin, mustard gas, and VX, biological warfare agents such as anthrax and tetrodotoxin, and toxic industrial chemicals (TICs)—essentially any materials that can cause harm and casualties on a large scale. The features of collectors and detectors are typically a recognition site for the target WMD, and in the case of detectors, a transduction mode to translate the action of recognition to a detector for interpretation. A relatively popular approach for the detection and collection of WMD is through the immobilization of enzymes in a suitable polymeric matrix, e.g., acetylcholinesterase (AChE) enzyme, which recognizes organophosphate nerve agent poisoning and organophosphate hydrolase (OPH), which hydrolyses organophosphate compounds. From defense perspectives, sol-gel technology can be useful in a variety of ways, viz., for providing good immobilizing polymeric matrix for developing sol-gel UV pearls for protection from UV light, sol-gel-derived nanocomposites and molecularly imprinted materials for chemical and biological warfare agents and sol-gel-based nuclear radiation dosimeter, etc.

20.5.4.1 Sol-Gel UV-Pearls

Undesirable skin alterations and melanomas are created by sun-induced premature skin aging; therefore, use of efficient but biocompatible sunscreens are required. Nowadays, commercial sunscreens are directly applied to the skin and usually contain an extremely high amount of active ingredients. This can be detrimental to health when they are adsorbed by the skin and they are not photostable, thus generating free radicals that may cause damage to the DNA. The hybrid materials developed by sol-gel technology is the solution to some extent by encapsulating the active organic UV filters (80%, w/w of the final product) in silica microcapsules which reduces the contact of these potent chemicals with the skin and prevents damage from free radicals. Such silica microcapsules are marketed as UV-pearls by various companies for sunscreens and daily-wear cosmetics (Figure 20.16). Silica rainbow pearls containing organic dyes for cosmetic applications and silica active-pearls containing an effective acne medication such as benzoyl peroxide, which are as effective as antibiotics but do not causes bacterial resistance or stomach upset, are also developed by companies. Benzoyl peroxide in direct contact with skin provokes skin irritation, dryness, and hyper-pigmentation in many patients. Sol-gel active-pearls have undergone successful commercial development because embedding the benzoyl peroxide active ingredient in a silica shell prevents it from coming into contact with the epidermis while gradually delivering it into the follicular region

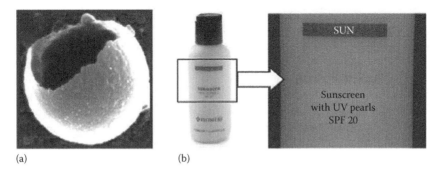

(a)   (b)

FIGURE 20.16 (a) Micrograph of a core–shell capsule for carrying functional contents. (b) Commercial Isomers Sunscreen SPF 20 with UV Pearls. (Reproduced from Sanchez, C. et al., *J. Mater. Chem.*, 15, 3559, 2005, and references cited therein. With permission from Taylor & Francis.)

where acne bacteria are found (Sanchez et al., 2005). Products of these types will be of immense importance for civilian and defense forces at high altitude.

20.5.4.2 Biomimetic Sol-Gel Sensor

Sol-gel process can be used in developing advanced materials for biomimetic sensors for countermeasures against chemical and biological warfare. The designing of nanoporous materials for biomimetic sensors is an important development in the area of management of bioterrorism. This was basically motivated by the mechanisms found in biological systems, for example, the designing of electronic nose, which is based on the natural sense of smell. An important application of these sensors is in detecting biological/chemical warfare agents such as anthrax, sarin, and mustard gas in the battlefield or in the environment. Generally, these sensors are expected to reproduce the human senses. In this pursuit, the critical parameters for designing the sensor are the detection time, concentration levels, and specificity of detecting agent/molecule. The choice of sensor material and its design are important steps that limit these parameters. Usually the sensor materials used are thin films of polymer or polymer–carbon black composite. The sensing mechanism is usually based on swelling of sensing materials by the analyte, which in turn depends upon the diffusion and adsorption characteristics of the materials. The diffusion depends upon the micro- and nanostructure of the polymer film and the adsorption depends upon the nature and extent of interactions between the polymer and the analyte. Sol-gel techniques are widely used to design nanoporous polymers and have been shown to be useful in designing materials with porosity and pore sizes between 0% and 80% and 1–50 nm respectively. Further, chemical grafting techniques can be used to modify the pore activity of these materials based on sulfonation and esterification reactions which impart hydrophilicity and bioactivity (Chaudhury et al., 2007).

20.5.4.3 Sol-Gel Based Radiation Dosimeters

The detection and monitoring of ionizing radiation in the environment due to accidental nuclear emergencies and/or in occupational exposure are of great importance for defense personnel and radiation workers in hospitals and nuclear industries. The importance of radiation monitoring has become all the more important after 9/11 terrorist attack. Although various nuclear instruments are available for monitoring radiation in the environment, these are not only expensive but also require technical support for both maintenance and operation. Moreover, it is known that a nuclear environment is hazardous also for instruments for routine operation. Another serious concern arises when the temperature rises during nuclear explosions; several types of monitoring equipment can start malfunctioning. However, it will not be practical to install available equipment in quick succession at several places in the event of nuclear emergencies or even monitoring of radiation level at different geographical locations. Because of all these reasons, suitable radiation monitoring techniques need to be developed.

Sol-gel method is emerging as a potential technology for the development of radiation dosimeters for several types of applications. Recently, few research groups have demonstrated the usefulness of sol-gel doped rare earth materials in radiation dosimetry. Luminescence properties of rare earth element, Eu^{3+} doped in sol-gel were found to be dramatically increased with ionizing radiation. The dose response appeared to be linear up to 400 Gy (Pedroza et al., 2002). Ferreira et al. (2001) reported a sol-gel-based new manufacturing process for thermoluminescent dosimeters. An interesting application of sol-gel combining with optical fiber has been reported as a medical radiation sensor for use in radiation therapy of cancer (Huston and Justus, 1998; Huston et al., 2002). Few commonly known porphyrin type molecules, viz., phthalocyaninies in sol-gel matrix have demonstrated radiation sensitivity up to 10^6 Gy. Some fluorescent molecules also tend to have dose-dependent radiation sensitivity. There have been few more reports on the possible use of optical fibers coated with transition elements again using sol-gel method for applications in radiation dosimetry in nuclear installation. Radiation hardness is another common phenomenon in optical fiber where the attenuation of visible light in certain wavelength range increases with absorbed dose. Mostly gamma radiation has been used for irradiation of sol-gel glassy matrix containing various probing molecules/elements and detection of absorbed dose has been carried out using either absorption or luminescence measurements. Both, absorption and fluorescence-based measuring methods are easy, simple, and can be translated into portable readout device (Chaudhury et al., 2007).

20.6 CONCLUSIONS AND FUTURE PROSPECTS

This chapter has presented an overview on the basic aspects of sol-gel process, its merits and demerits, factors affecting thin gel films, importance of the characterization of internal environment of sol-gel matrix, new developments in preparing sol-gel materials, viz., organic–inorganic sol-gel hybrid, biocompatible and bioactive, molecularly imprinted and nanocomposite materials for various clinical diagnostic, biomedical, defense, and environmental applications. However, the increasing number of publications reported in the literature reflects the unique features of sol-gel bioencapsulation, but some critical issues are still to be considered with respect to mass production and commercialization. A significant number of sensing devices reported exhibit one or more limitations that hinder their continuous use and some of them were not applied to real samples or practical situations. Many of the works referred, although describing optimization and, sometimes, showing preliminary results, present insufficient data concerning analytical performance. It is desirable to obtain sol-gel sensors with no reagent leaching and that can be used for a long period of time without changes in sensitivity and response time. Therefore, research should be emphasized on the critical issues concerning the sensor's performance, viz., microstructural stability, leaching, reversibility, response time, repeatability, sensitivity, and selectivity instead of simply demonstrating the sensing potential. In the recent years, rapid advances have been made in improving immobilization protocols and overcoming many hurdles of sol-gel technology. Incorporation of organic component in sol-gel, new class of biocompatible sol-gel precursor, and combining sol-gel process with molecularly imprinting and nanotechnology provide an enlarged range of possible applications of sol-gel technology. Microfabrication techniques such as biosensors array enable the production of new microsensors and bioelectronic devices. After decades of theoretical studies devoted to the full understanding of the sol-gel process, it is expected that the future years will provide a variety of new and improved sol-gel bio-applications.

ABBREVIATIONS

2,4-D	2,4 Dichlorophenoxyacetic acid
Ac	Acrylodan

AChE	Acetylcholinesterase
AEAP3	*N*-(2-Aminoethyl)-3-aminopropyltrimethoxysilane
AEMP3	(Aminoethylaminomethyl) phenethyltrimethoxysilane
AHAP3	*N*-(6-Aminohexyl)aminopropyltrimethoxysilane
ANS	1-Anilinonaphthalene-8-sulfonic acid
APS	(3-Aminopropyl) triethoxysilane
APTMS	(3-Aminopropyl) trimethoxysilane
ATA	Anthryl tetra acid
bCaM	Bovine calmodulin
BOD	Biochemical oxygen demand
BSA	Bovine serum albumin
BTEB	Bis-(trimethoxysilylethyl)benzene
BTMOS	Butyltrimethoxysilanes
ChEt	Cholesterol esterase
ChOx	Cholesterol oxidase
CNTs	Carbon nanotubes
CRP	C-reactive protein
CS	Chitosan
CuZnSOD	Copper zinc superoxide dismutase
Cyt-*c*	Cytochrome c
CW	Chemical warefare
DAN	Dansyl-L-glycine
DDT	Dichloro-diphenyl trichloroethane
DET3	*N*-[3-(Trimethoxysilyl)-propyl]diethylenetriamine
DMDES	Dimethyldiethoxysilane
DNA	Deoxyribonucleic acid
DPDES	Diphenyldiethoxysilane
EDCs	Endocrine-disrupting compounds
EFTEM	Energy-filtered transmission electron microscopy
EIS	Electrochemical impedance spectroscopy
ELISA	Enzyme-linked immunosorbent assay
EPA	Environment Protection Agency
ETMOS	Ethyltrimethoxysilane
f-CNTs	Functionalized carbon nanotubes
FDA	Food and Drug Administration
FET	Field effect transistors
FITC	Fluorescein isothiocyanate
FOBs	Fiber-optic biosensors
GLDH	Glutamate dehydrogenase
GPTMS	Gycidoxypropyltrimethoxysilane
Gy	Gray
HA	Hyaluronan
HAp	Hydroxyapatite
Hb	Hemoglobin
HCA	Hydroxy-carbonoapatite
HCV	Hepatitis C virus
HEMA	2-Hydroxyethylmethacrylate
hFOB	Human osteoblast
HPLC	High performance liquid chromatography
HRP	Horse radish peroxidase
IgG	Immunoglobulins

IgGAb	Immunoglobulins antibodies
IMIPs	Inorganic molecularly imprinted polymers
ITO	Indium tin oxide
K_I	Inhibition constant
K_f	Affinity constant
K_b	Binding constant
k_{cat}	Catalytic constant
K_m	Michaelis constant
LDH	Lactate dehydrogenase
LDHs	Layered double hydroxides
LOD	Limit of detection
LN	Laminin
Mb	Myoglobin
mg/dL	Miligram per deciliter
MIPs	Molecularly imprinted polymers
mL/min	Milliliter per minute
mM	Milimolar
MPTS	(Mercaptopropyl) triethoxysilane
MRI	Magnetic resonance imaging
MS	Mass spectrometry
MTMOS	Methyltrimethoxysilane
3-MPTS	3-Methacryloxypropyltrimethoxysilane
MVDES	Methylvinyldiethoxysilane
MWNTs	Multiwalled nanotubes
NADH	Reduced form of nicotinamide adenine dinucleotide
Nap	Periplasmic nitrate reductase
NATA	N-Acetyl tryptophanamide
ng/L	Nanogram per liter
NiTi	Nickel titanium
nm	Nanometer
OPH	Organophosphate hydrolase
PAHs	Polycyclic aromatic hydrocarbons
PANI	Polyaniline
PBS	Phosphate buffer saline
PCP	Pentachlorophenol
PDT	Photodynamic therapy
pg/L	Picogram per liter
PGS	Polyglyceryl silicate
PhTEOS	Phenyltriethoxysilane
PIXIES	Protein imprinted xerogels with integrated emission sites
PMMA	Poly(methyl methacrylate)
POPs	Persistent organic pollutants
PQC	Piezoelectric quartz crystal
PSII	Photosystem II
PTMS	Propyltrimethoxysilane
PTMOS	Phenyltrimethoxysilane
PVC	Polyvinyl chloride
QCM	Quartz crystal microbalance
RNA	Ribonucleic acid
SAW	Surface acoustic wave
SBF	Simulated body fluid

SGCCN	sol-gel-derived ceramic–carbon nanotube
SMAs	Shape memory alloys
SPE	Solid phase extraction
SPR	Surface plasmon resonance
SS	Stainless steel
SWNTs	Single-walled nanotubes
t_g	Gelation time
TCNQ	Tetracyanoquinodimethane
TEOT	Tetraethyl-orthotitanate (TEOT)
TICs	Toxic industrial chemicals
TIFR	Total internal reflection fluorescence
TNT	Trinitrotoluene
TPOZ	Tetrapropyl-orthozirconate
UV-VIS	Ultraviolet-visible
V	Volt
WMD	Weapons of mass destruction
Å	Angstrom
μm	Micrometer
$\mu A/mM/cm^2$	Microampere per milimolar per centimeter square
μM	Micromolar
μg/mL	Microgram per milliliter

REFERENCES

Altstein, M., Bronshtein, A., Glattstein, B., Zeichner, A., Tamiri, T., and Almog, J. 2001. Immunochemical approaches for purification and detection of TNT traces by antibodies entrapped in a sol-gel matrix. *Anal. Chem.* 73:2461–2467.

An, S.Y., Shim, I.B., and Kim, C.S. 2005. Easy synthesis and characterization of γ-Fe$_2$O$_3$ nanoparticles for biomedical applications. *J. Appl. Phys.* 97:10Q909–10Q910.

Andreou, V. and Clonis, Y. 2002. A portable fiber-optic pesticide biosensor based on immobilized cholinesterase and sol-gel entrapped bromocresol purple for in-field use. *Biosens. Bioelectron.* 17:61–69.

Arenkov, P., Kukhtin, A., Gemmell, A., Voloshchuk, S., Chupeeva, V., and Mirzabekov, A. 2000. Protein microchips: Use for immunoassay and enzymatic reactions. *Anal. Biochem.* 278:123–131.

Assink, R.A. and Ray, B.D. 1988. Sol-gel kinetics. III. Test of the statistical reaction model. *J. Non-Cryst. Solids* 99:359–370.

Avnir, D., Braun, S., Lev, O., and Ottolenghi, M. 1994. Enzymes and other proteins entrapped in sol-gel materials. *Chem. Mater.* 6:1605–1614.

Aylott, J.W., Richardson, D.J., and Russell, D.A. 1997. Optical biosensing of gaseous nitric oxide using spin-coated sol-gel thin films. *Chem. Mater.* 9:2261–2263.

Balamurugan, A., Sockalingum, G., Michel, J. et al. 2006. Synthesis and characterization of sol-gel derived bioactive glass for biomedical applications. *Mater. Lett.* 60:3752–3757.

Balani, K., Anderson, R., Laha, T., Andara, M., Tercero, J., Crumpler, E., and Agarwal, A. 2007. Plasma-sprayed carbon nanotube reinforced hydroxyapatite coatings and their interaction with human osteoblasts *in vitro. Biomaterials* 28:618–624.

Baughman, R.H., Zakhidov, A.A., and de Heer, W.A. 2002. Carbon nanotubes—The route toward applications. *Science* 297:787–792.

Beganskiene, A., Raudonis, R., Jokhadar, S.Z., Batista, U., and Kareival, A. 2007. Modified sol-gel coatings for biotechnological applications *Funct. Mater. Nanotechnol. J. Phys. Conf. Ser.* 93:012050.

Besanger, T.R., Chen, Y., Deisingh, A.K. et al. 2003. Screening of inhibitors using enzymes entrapped in sol-gel derived materials. *Anal. Chem.* 75:2382–2391.

Bianco, A., Kostarelos, K., Partidos, C.D., and Prato, M. 2005. Biomedical applications of functionalized carbon nanotubes. *Chem. Commun.* 571–577.

Biran, I. and Walt, D.R. 2002. Optical imaging fiber-based single live cell arrays: A high-density cell assay platform. *Anal. Chem.* 74:3046–3054.

Blanch, E., Garreta, E., Gasset, D. et al. 2004. Development of hydroxyapatite-based biomaterials scaffolds for hard tissue regeneration. *Eur. Cells Mater.* 7:1473–2262.

Blyth, D.J., Aylott, J.W., Richardson, D.J., and Russell, D.A. 1995. Sol-gel encapsulation of metalloproteins for the development of optical biosensors for nitrogen monoxide and carbon monoxide. *Analyst* 120:2725–2730.

Bonzagni, N.J., Baker, G.A., Pandey, S., Niemeyer, E.D., and Bright, F.V. 2000. On the origin of the heterogeneous emission from pyrene sequestered within tetramethylorthosilicate-based xerogels: A decay-associated spectra and O_2 quenching study. *J. Sol-Gel Sci. Technol.* 17:83–90.

Bosch, M.E., Ruiz Sánchez, A.J., Sánchez Rojas, F., and Ojeda, C.B. 2007. Recent development in optical fiber biosensors. *Sensors* 7:797–859.

Braun, S., Rappoport, S., Zusman, R., Avnir, D., and Ottolenghi, M. 1990. Biochemically active sol-gel glasses: The trapping of enzymes. *Mater. Lett.* 10:1–5.

Braun, S., Shetlzer, S., Rappoport, S., Avnir, D., and Ottolenghi, M. 1992. Biocatalysis by sol-gel entrapped enzymes. *J. Non-Cryst. Solids* 147–148:739–743.

Brennan, J.D., Hartman, J.S., Ilnicki, E.I., and Rakic, M. 1999. Fluorescence and NMR characterization and biomolecule entrapment studies of sol-gel-derived organic-inorganic composite materials formed by sonication of precursors. *Chem. Mater.* 11:1853–1864.

Brinker, C.J. and Scherer, G.W. 1985. Sol→gel→ glass: I. Gelation and gel structure. *J. Non-Cryst. Solids* 70:301–322.

Brinker, C.J. and Scherer, G.W. 1990. *Sol-gel Science: The Physics and Chemistry of Sol-Gel Processing.* New York: Academic.

Brinker, C.J., Hurd, A.J., Schunk, P.R., Frye, G.C., and Ashley, C.S. 1992. Review of sol–gel thin film formation. *J. Non-Cryst. Solids* 147–148:424–436.

Bronshtein, A., Aharonson, N., Avnir, D., Turniansky, A., and Altstein, M. 1997. Sol-gel matrixes doped with atrazine antibodies: Atrazine binding properties. *Chem. Mater.* 9:2632–2639.

Bucur, B., Fournier, D., Danet, A., and Marty, J.L. 2006. Biosensors based on highly sensitive acetylcholinesterases for enhanced carbamate insecticides detection. *Anal. Chim. Acta* 562:115–121.

Carturan, G., Campostrini, R., Dire, S., Scardi, V., and De Alteriis, E. 1989. Inorganic gels for immobilization of biocatalysts: Inclusion of invertase-active whole cells of yeast (*Saccharomyces cerevisiae*) into thin layers of silica gel deposited on glass sheets. *J. Mol. Catal.* 57:L13–L16.

Chaubey, A., Pande, K.K., and Malhotra, B.D. 2003. Application of polyaniline/sol-gel derived tetraethylorthosilicate films to an amperometric lactate biosensor. *Anal. Sci.* 19:1477–1480.

Chaudhury, N.K., Gupta, R., and Gulia, S. 2007. Sol-gel technology for sensor applications. *Defense Sci. J.* 57:241–253, and references cited therein.

Chen, H. and Dong, S. 2007. Direct electrochemistry and electrocatalysis of horseradish peroxidase immobilized in sol–gel-derived ceramic–carbon nanotube nanocomposite film. *Biosens. Bioelectron.* 22:1811–1815.

Chen, Q., Kenausis, G.L., and Heller, A. 1998. Stability of oxidases immobilized in silica gels. *J. Am. Chem. Soc.* 120:4582–4585.

Chen, X., Jia, J., and Dong, S. 2003. Organically modified sol-gel/chitosan composite based glucose biosensor. *Electroanalysis* 15:608–612.

Cheng, F.T., Shi, P., and Man, H.C. 2004. Anatase coating on NiTi via a low-temperature sol-gel route for improving corrosion resistance. *Scr. Mater.* 51:1041–1045.

Chiu, K.Y., Wong, M.H., Cheng, F.T., and Man, H.C. 2007. Characterization and corrosion studies of titania-coated NiTi prepared by sol-gel technique and steam crystallization *Appl. Surf. Sci.* 253:6762–6768.

Cho, E.J. and Bright, F.V. 2001. Optical sensor array and integrated light source (OSAILS). *Anal. Chem.* 73:3289–3293.

Cho, E.J. and Bright, F.V. 2002a. Integrated chemical sensor array platform based on a light emitting diode, xerogel-derived sensor elements, and high-speed pin printing. *Anal. Chim. Acta* 470:101–110.

Cho, E.J. and Bright, F.V. 2002b. Pin-printed chemical sensor arrays for simultaneous multi-analyte quantification. *Anal. Chem.* 74:1462–1466.

Cho, Y.W. and Han, S. 1999. Catalytic activities of glass-encapsulated horseradish peroxidase at extreme pHs and temperatures. *Bull. Korean Chem. Soc.* 20:1363–1364.

Cho, E.J., Tao, Z., Tehan, E.C., and Bright, F.V. 2002. Pin-printed biosensor arrays for simultaneous detection of glucose and O_2. *Anal. Chem.* 74:6177–6184.

Chou, C., Hsu, H.Y., Wu, H.T., and Tseng, K.Y. 2007. Fiber optic biosensor for the detection of C-reactive protein and the study of protein binding kinetics. *J. Biomed. Opt.* 12:024025–024129, and references cited therein.

Christodoulides, N., Tran, M., and Floriano, P.N. et al. 2002. A microchip-based multianalyte assay system for the assessment of cardiac risk. *Anal. Chem.* 74:3030–3036.

Chung, K.E., Lan, E.H., Davidson, M.S., Dunn, B.S., Valentine, J.S., and Zink, J.I. 1995. Measurement of dissolved oxygen in water using glass-encapsulated myoglobin. *Anal. Chem.* 67:1505–1509.

Copello, G.J., Teves, S., Degrossi, J., D'Aquino, M., Desimone, M.F., and Diaz, L.E. 2006. Antimicrobial activity on glass materials subject to disinfectant xerogel coating. *J. Ind. Microbiol. Biotechnol.* 33:343–348.

Cox, J.A., Alber, K.S., Tess, M.E., Cummings, T.E., and Gorski, W. 1996. Voltammetry in the absence of a solution phase with solids prepared by a sol-gel process as the electrolytes: Facilitation of an electrocatalytic anodic process in the presence of ammonia. *J. Electroanal. Chem.* 396:485–490.

Dave, B., Dunn, B., Valentine, J.S., and Zink, J.I. 1994. Sol-gel encapsulation methods for biosensors. *Anal. Chem.* 66:1120A–1127A.

Dave, B.C., Miller, J.M., Dunn, B., Valentine, J.S., and Zink, J.I. 1997. Encapsulation of proteins in bulk and thin film sol-gel matrixes. *J. Sol-gel Sci. Technol.* 8:629–634.

Diaz-Garcia, M.E. and Badia Laino, R. 2005. Molecular imprinting in sol-gel materials: Recent developments and applications. *Microchim. Acta* 149:19–36.

Diaz-Flores, L.L., Garnica-Romo, M.G., Gonzalez-Hernandez, J., Yanez-Limon, J.M., Vorobiev, P., and Vorobiev, Y.V. 2007. Formation of Ag–Cu nanoparticles in SiO_2 films by sol-gel process and their effect on the film properties *Phys. Status Solidi* C 4:2016–2020.

Dickert, F.L. and Hayden, O. 2002. Bioimprinting of polymers and sol–gel phases. Selective detection of yeasts with imprinted polymers. *Anal. Chem.* 74:1302–1306.

Dickey, F.H. 1949. The preparation of specific adsorbents. *Proc. Natl. Acad. Sci.* 35:227–229.

Doody, M.A., Baker, G.A., Pandey, S., and Bright, F.V. 2000. Affinity and mobility of polyclonal anti-dansyl antibodies sequestered within sol-gel-derived biogels. *Chem. Mater.* 12:1142–1147.

Doong, R. and Shih, H. 2006. Glutamate optical biosensor based on the immobilization of glutamate dehydrogenase in titanium dioxide sol–gel matrix. *Biosens. Bioelectron.* 22:185–191.

D'Souza, S.F. 2001. Microbial Biosensor. *Biosens. Bioelectron.* 16:337–353.

Du, D., Yan, F., Liu, S., and Ju, H. 2003. Immunological assay for carbohydrate antigen using an electrochemical immunosensor and antigen immobilization in titania sol–gel matrix. *J. Immunol. Methods* 283:67–75, and references cited therein.

Dunn, B., Miller, J.M., Dave, B.C., Valentine, J.S., and Zink, J.I. 1998. Strategies for encapsulating biomolecules in sol-gel matrices. *Acta. Mater.* 46:737–741.

Dunuwila, D.D., Torgerson, B.A., Chang, C.K., and Berglund, K.A. 1994. Sol-gel derived titanium carboxylate thin films for optical detection of analytes. *Anal. Chem.* 66:2739–2744.

Edmiston, P.L., Wambolt, C.L., Smith, M.K., and Saavedra, S.S. 1994. Spectroscopic characterization of albumin and myoglobin entrapped in bulk sol-gel glasses. *J. Colloid Interface Sci.* 163:395–406.

Eguchi, K., Hashiguchi, T., Sumiyoshi, K., and Arai, H. 1990. Optical detection of nitrogen monoxide by metal porphine dispersed in an amorphous silica matrix. *Sensors Actuators B* 1:154–157.

Ellerby, L., Nishida, C.R., Nishida, F. et al. 1992. Encapsulation of proteins in transparent porous silicate glasses prepared by the sol–gel method. *Science* 255:1113–1115.

Fernandez-Pacheco, R., Arruebo, M., Marquina, M., Ibarra, R., Arbiol, J., and Santamaria, J. 2004. Highly magnetic silica-coated iron nanoparticles prepared by the arc-discharge method. *Nanotechnology* 17:1188–1192.

Ferreira, M.P., De Faria, L.O., and Vasconcelos, W.L. 2001. A new manufacturing process to obtain thermoluminescent dosimeters using sol-gel method. *J. Sol-Gel Sci. Technol.* 21:173–176.

Ferretti, S., Lee, S.K., MacCraith, B.D., Oliva, A.G., Richardson, D.J., Russell, D.A., Sapsford, K.E., and Vidal, M. 2000. Optical biosensing of nitrite ions using cytochrome cd1 nitrite reductase encapsulated in a sol-gel matrix. *Analyst* 125:1993–1999.

Fink, J., Kiely, C.J., Bethell, D., and Schiffrin, D.J. 1998. Self-organization of nanosized gold particles. *Chem. Mater.* 10:922–926.

Flora, K. and Brennan, J.D. 1998. Fluorometric detection of Ca^{2+} based on an induced change in the conformation of sol-gel entrapped parvalbumin. *Anal. Chem.* 70:4505–4513.

Flora, K.K. and Brennan, J.D. 2001. Effect of matrix aging on the behavior of human serum albumin entrapped in a tetraethyl orthosilicate-derived glass. *Chem. Mater.* 13:4170–4179.

Flora, K.K., Keeling-Tucker, T., Hogue, C.W., and Brennan, J.D. 2002. Screening of antagonists based on induced dissociation of a calmodulin-melittin interaction entrapped in a sol-gel derived matrix. *Anal. Chim. Acta* 470:19–28.

Gao, Z., Nahrup, J.S., Mark, J.E., and Sakr, A. 2005. Poly(dimethylsiloxane) coatings for controlled drug release: III. Drug release profiles and swelling properties of the free-standing films *J. Appl. Polym. Sci.* 96:494–501.

Gavalas, V.G., Law, S.A., Ball, J.C., Andrews, R., and Bachasa, L.G. 2004. Carbon nanotube aqueous sol-gel composites: Enzyme-friendly platforms for the development of stable biosensors. *Anal. Biochem.* 329:247–252.

Gerritsen, M., Kros, A., Sprakel, V., Lutterman, J.A., Nolte, R.J.M., and Jansen, J.A. 2000. Biocompatibility evaluation of sol–gel coatings for subcutaneously implantable glucose sensors. *Biomaterials* 21:71–78.

Gill, I., Ballesteros, A. 1998. Encapsulation of biologicals within silicate, siloxane, and hybrid sol-gel polymers: An efficient and generic approach. *J. Am. Chem. Soc.* 120:8587–8598.

Gong, K., Yan, Y., Zhang, M., Su, L., Xiong, S., and Mao, L. 2005. Electrochemistry and electroanalytical applications of carbon nanotubes: A review. *Anal. Sci.* 21:1383–1393.

Goring, G.L.G. and Brennan, J.D. 2002. Fluorescence and physical characterization of sol-gel-derived nanocomposite films suitable for the entrapment of biomolecules. *J. Mater. Chem.* 12:3400–3406.

Grant, S.A. and Glass, R.S. 1999. Sol-gel based biosensor for use in stroke treatment. *IEEE Trans. Biomed. Eng.* 46:1207–1211.

Guglielmi, M. and Zenezini, S. 1990. The thickness of sol-gel silica coatings obtained by dipping. *J. Non-Cryst. Solids* 121:303.

Gupta, R. and Chaudhury, N.K. 2007. Entrapment of biomolecules in sol–gel matrix for applications in biosensors: Problems and future prospects. *Biosens. Bioelectron.* 22:2387–2399, and references cited therein.

Gupta, R. and Chaudhury, N.K. 2009. Probing internal environment of sol-gel bulk and thin films using multiple fluorescent probes. *J. Sol-Gel Sci. Technol.* 49:78–87.

Gupta, R. and Kumar, A. 2008–2009. Exogenous endocrine-disrupting compound. *Clin. Lab Int. N* 8:19–21.

Gupta, R. and Kumar, A. 2008a. Bioactive material for biomedical applications using sol-gel technology. *Biomed. Mater.* 3:034005, and references cited therein.

Gupta, R. and Kumar, A. 2008b. Molecular imprinting in sol-gel matrix. *Biotechnol. Adv.* 26:533–547.

Gupta, R., Mozumdar, S., and Chaudhury, N.K. 2005a. Effect of ethanol variation on the internal environment of sol–gel bulk and thin films with aging. *Biosens. Bioelectron.* 21:549–556.

Gupta, R., Mozumdar, S., and Chaudhury, N.K. 2005b. Fluorescence spectroscopic studies to characterize the internal environment of tetraethyl-orthosilicate derived sol–gel bulk and thin films with aging. *Biosens. Bioelectron.* 20:1358–1365.

Han, D.M., Fang, G.Z., and Yan, X.P. 2005. Preparation and evaluation of a molecularly imprinted sol-gel material for on-line solid-phase extraction coupled with high performance liquid chromatography for the determination of trace pentachlorophenol in water samples. *J Chromatogr. A* 1100:131–136.

Hench, L.L. and West, J.K. 1990. The sol-gel process. *Chem. Rev.* 90:33–72.

Haupt, K. 2001. Molecularly imprinted polymers in analytical chemistry. *Analyst* 126:747–756.

Hodgson, R.J., Brook, M.A., and Brennan, J.D. 2005. Capillary-scale monolithic immunoaffinity columns for immunoextraction with in-line laser-induced fluorescence detection. *Anal. Chem.* 77:4404–4412.

Huang, M.H., Soyez, H.M., Dunn, B.S., and Zink, J.I. 2000. In situ fluorescence probing of molecular mobility and chemical changes during formation of dip-coated sol-gel silica thin films. *Chem. Mater.* 12:231–235.

Hunnius, M., Rufinska, A., and Maier, W.F. 1999. Selective surface adsorption versus imprinting in amorphous microporous silicas. *Micropor. Mesopor. Mater.* 29:389–403.

Hurd, A.J. and Brinker, C.J. 1988. Optical sol-gel coatings: Ellipsometry of film formation. *J. Phys.* 49:1017–1025.

Huston, A.L. and Justus, B.L. 1998. *In Vivo Radiotherapy Dose Monitoring System.* Optical Science Division, Washington, DC.

Huston, A.L., Falkenstein, P.L., Justus, B.L., Ning, H., Miller, R.W., and Altenius, R. 2002. Multiple channel optical fiber radiation dosimeter for radiotherapy applications. *Proc. IEEE* 1:560–563.

Iler, R.K. 1979. *The Chemistry of Silica.* Wiley, New York.

Jeon, H.J., Yi, S.C., and Oh, S.G. 2003. Preparation and antibacterial effects of Ag–SiO$_2$ thin films by sol-gel method. *Biomaterials* 24:4921–4928.

Jeronimo, P.C.A., Araujo, A.N., and Montenegro, M.C.B.S.M. 2007. Optical sensors and biosensors based on sol–gel films. *Talanta* 72:13–27.

Jia, J., Tang, M., Chen, X., Qi, L., and Dong, S. 2003. Co-immobilized microbial biosensor for BOD estimation based on sol-gel derived composite material. *Biosens. Bioelecton.* 18:1023–1029.

Jiang, D., Tang, J., Liu, B., Yang, P., and Kong, J. 2003. Ultrathin alumina sol-gel-derived films: Allowing direct detection of the liver fibrosis markers by capacitance measurement. *Anal. Chem.* 75:4578–4584.

Jiang, Y., Xiao, L.L., Zhao, L., Chen, X., Wang, X., and Wong, K.Y. 2006. Optical biosensor for the determination of BOD in seawater. *Talanta* 70:97–103.

Jin, W. and Brennan, J.D. 2002. Properties and applications of proteins encapsulated within sol-gel derived materials. *Anal. Chim. Acta* 461:1–36.

Johnson, P. and Whateley, T.L. 1971. Use of polymerizing silica gel systems for the immobilization trypsin. *J. Colloid Interface Sci.* 37:557–563.

Jordan, J.D., Dunbar, R.A., and Bright, F.V. 1995. Dynamics of acrylodan-labeled bovine and human serum albumin entrapped in a sol-gel-derived biogel. *Anal. Chem.* 67:2436–2443.

Kandimalla, V.B., Tripathi, V.S., and Ju, H. 2006. Immobilization of biomolecules in sol-gels: Biological and analytical applications. *Crit. Rev. Anal. Chem.* 36:73–106.

Kaufman, V.R. and Avnir, D. 1986. Structural changes along the sol-gel-xerogel transition in silica as probed by pyrene excited-state emission. *Langmuir* 2:717–722.

Kaufman, V.R., Avnir, D., Pines-Rojanski, P., and Huppert, D. 1988. Water consumption during the early stages of the sol-gel tetramethylorthosilicate polymerization as probed by excited state proton transfer. *J. Non-Cryst. Solids* 99:379–386.

Keefer, K.D. 1984. *Better Ceramics Through Chemistry*, C.J. Brinker, D.E. Clark, D.R. Ulrich (Eds.), pp. 15–24. North-Holland, New York.

Klein, L.C. (Ed.) 1988. *Sol–Gel Technology for Thin Films, Fibers, Performs, Electronics and Specialty Shapes*. Noyes Publications, Park Ridge, NY.

Klotz, M., Ayral, A., Guizard, C., and Cot, L. 1999. Tailoring of the porosity in sol-gel derived silica thin layers. *Bull. Korean Chem. Soc.* 20:879–884.

Kros, A., Gerritsen, M., Sprakel, V.S.I., Sommerdijk, N.A.J.M., Jansen, J.A., and Nolte, R.J.M. 2001. Silica-based hybrid materials as biocompatible coatings for glucose sensors. *Sens. Actuators* 81:68–75.

Kumar, A., Malhotra, R., Malhotra, B.D., and Grover, S.K. 2000. Co-immobilization of cholesterol oxidase and horseradish peroxidase in a sol–gel film. *Anal. Chim. Acta* 414:43–50.

Kumar, A., Pandey, R.R., and Brantley, B. 2006. Tetraethylorthosilicate film modified with protein to fabricate cholesterol biosensor. *Talanta* 69:700–705.

Kunitake, T. and Lee, S.W. 2004. Molecular imprinting in ultrathin titania gel films via surface sol-gel process. *Anal Chim Acta* 504:1–6.

Kuselman, I. and Lev, O. 1993. Organically-doped sol–gel-based tube detectors: Determinations of. Iron (II) in aqueous solutions. *Talanta* 40:749–756.

Kuselman, I., Kuyavskaya, R.I., and Lev, O. 1992. Disposable tube detectors for water analysis. *Anal. Chim. Acta* 256:65–68.

Kwon, J.A., Lee, H., Lee, K.N. et al. 2008. High diagnostic accuracy of antigen microarray for sensitive detection of Hepatitis C virus infection. *Clin. Chem.* 54:424–428.

Lee, S., Ichinose, I., and Kunitake, T. 1998. Molecular imprinting of azobenzene carboxylic acid on a TiO_2 ultrathin film by the surface sol–gel process. *Langmuir* 14:2857–2863.

Lee, K., Itharaju, R.R., and Puleo, D.A. 2007. Protein-imprinted polysiloxane scaffolds. *Acta Biomater.* 3:515–522.

Li, F., Li, J., and Zhang, S.S. 2008. Molecularly imprinted polymer grafted on polysaccharide microsphere surface by the sol–gel process for protein recognition. *Talanta* 74:1247–1255.

Li, J., Peng, T., and Pengc, Y. 2003. A cholesterol biosensor based on entrapment of cholesterol oxidase in a silicic sol-gel matrix at a Prussian blue modified electrode. *Electroanalysis* 15:1031–1037.

Lin, T.Y., Wu, C.H., and Brennan, J.D. 2007. Entrapment of horseradish peroxidase in sugar-modified silica monoliths: Toward the development of a biocatalytic sensor. *Biosens. Bioelectron.* 22:1861–1867.

Liu, H.Y., Switalski, S.C., Coltrain, B.K., and Merkel, P.B. 1992. Oxygen permeability of sol-gel coatings. *Appl. Spectrosc.* 46:1266–1272.

Liu, X., Farmerier, W., Schuster, S., and Tan, W. 2000. Molecular beacons for DNA biosensors with micrometer to submicrometer dimensions. *Anal. Biochem.* 283:56–63.

Liu, J.X., Yang, D.Z., Shi, F., and Cai, Y.J. 2003. Sol-gel deposited TiO_2 film on NiTi surgical alloy for biocompatibility improvement. *Thin Solid Films* 429:225–230.

Livage, J., Henry, M., and Sanchez, C. 1988. Sol-gel chemistry of transition metal oxides. *Prog. Solid State Chem.* 18:259–341.

Lu, Y., Ganguli, R., Drewien, C.A. et al. 1997. Continuous formation of supported cubic and hexagonal mesoporous films by sol-gel dip-coating. *Nature* 389:364–368.

Lulka, M.F., Iqbal, S.S., Chambers, J.P. et al. 2000. Molecular imprinting of Ricin and its A and B chains to organic silanes: Fluorescence detection. *Mater. Sci. Eng. C: Biomim. Supramol. Syst.* C11:101–105.

Lundgren, J.S. and Bright, F.V. 1996. Biosensor for the nonspecific determination of ionic surfactants. *Anal. Chem.* 68:3377–3381.

Ma, D., Guan, J., Normandin, F. et al. 2006. Multifunctional nano-architecture for biomedical applications. *Chem. Mater.* 18:1920–1927.

MacCraith, B.D., McDonagh, C.M., O'Keeffe, G., Vos, J.G., O'Kelly, B., and McGilp, J.F. 1992. Evanescent-wave oxygen sensing using sol-gel-derived porous coatings. *Proc. SPIE* 1796:167–175.

MacCraith, B.D., McDonagh, C.M., O'Keeffe, G. et al. 1993. Fibre optic oxygen sensor based on fluorescence quenching of evanescent-wave excited ruthenium complexes in sol–gel derived porous coatings. *Analyst* 118:385–388.

Maduraiveeran, G. and Ramaraj, R. 2007. A facile electrochemical sensor designed from gold nanoparticles embedded in three-dimensional sol–gel network for concurrent detection of toxic chemicals. *Electrochem. Commun.* 9:2051–2055.

Malhotra, B.D. and Chaubey, A. 2003. Biosensors for clinical diagnostics industry. *Sens. Actuators B: Chem.* 91:117–127, and references cited there in.

Marxer, S.M., Rothrock, A.R., Nablo, B.J., Robbins, M.E., and Schoenfisch, M.H. 2003. Preparation of nitric oxide (NO)-releasing sol-gels. *Chem. Mater.* 15:4193–4199.

Matsui, K., Matsuzuka, T., and Fujita, H. 1989. Fluorescence spectra of 7-azaindole in the sol-gel-xerogel stages of silica. *J. Phys. Chem.* 93:4991–4994.

Mehrvar, M. and Abdi, M. 2004. Recent developments, characteristics, and potential applications of electrochemical biosensors. *Anal. Sci.* 20:1113–1126.

Moser, I., Jobst, G., and Urban, G.A. 2002. Biosensor arrays for simultaneous measurement of glucose, lactate, glutamate, and glutamine. *Biosens. Bioelectron.* 17:297–302.

Nablo, B.J. and Schoenfisch, M.H. 2003. Antibacterial properties of nitric oxide-releasing sol-gels. *J. Biomed. Mater. Res. A* 67:1276–1283.

Nablo, B.J. and Schoenfisch, M.H. 2005. *In vitro* cytotoxicity of nitric oxide-releasing sol-gel derived materials. *Biomaterials* 26:4405–4415.

Nablo, B.J., Chen, T.Y., and Schoenfisch, M.H. 2001. Sol-gel derived nitric oxide releasing materials that reduce bacterial adhesion. *J. Am. Chem. Soc.* 123:9712–9713.

Nablo, B.J., Prichard, H.L., Butler, R.D., Klitzman, B., and Schoenfisch, M.H. 2005. Inhibition of implant-associated infections via nitric oxide release. *Biomaterials* 26:6984–6990.

Narang, U., Jeffrey, J.D., Bright, F.V., and Prasad, P.N. 1994a. Probing the cytobactic region of PRODAN in tetramethylorthosilicate-derived sol-gels. *J. Phys. Chem.* 98:8101–8107.

Narang, U., Prasad, P.N., and Bright, F.V. 1994b. A novel protocol to entrap active urease in a tetraethoxysilane-derived sol-gel thin-film architecture. *Chem. Mater.* 6:1596–1598.

Narang, U., Prasad, P.N., Ramanathan, K., Kumar, N.D., Malhotra, B.D., Kamalasanan, M.N., Chandra, S., and Bright, F.V. 1994c. Glucose biosensor based on a sol-gel-derived platform. *Anal. Chem.* 66:3139–3144.

Narang, U., Wang, R., Prasad, P.N., and Bright, F.V. 1994d. Effects of aging on the dynamics of rhodamine 6G in tetramethyl orthosilicate-derived sol-gels. *J. Phys. Chem.* 98:17–22.

Navas Diaz, A., Ramos Peinado, M.C., and Torijas Minguez, M.C. 1998. Sol-gel horseradish peroxidase biosensor for hydrogen peroxide detection by chemiluminescence. *Anal. Chim. Acta* 363:221–227.

Nguyen-Ngoc, H. and Tran-Minh, C. 2007. Fluorescent biosensor using whole cells in an inorganic translucent matrix. *Anal. Chim. Acta* 583:161–165.

Oh, B.K., Robbins, M.E., Nablo, B.J., and Schoenfisch, M.H. 2005. Miniaturized glucose biosensor modified with a nitric oxide-releasing xerogel microarray. *Biosens. Bioelectron.* 21:749–757.

Ohtsuki, C., Miyazaki, T., and Tanihara, M. 2002. Development of bioactive organic–inorganic hybrid for bone substitutes. *Mater. Sci. Eng. C* 22:27–34.

Park, C.B. and Clark, D.S. 2002. Sol-gel encapsulated enzyme arrays for high-throughput screening of biocatalytic activity. *Biotechnol. Bioeng.* 78:229–235.

Pedroza, G., de Azevedo, W.M., Khoury, H.J., and da Silva, E.F. 2002. Gamma radiation detection using sol-gel doped with lanthanides ions. *Appl. Radiat. Isot.* 56:563–566.

Peterson, K.P., Peterson, C.M., and Pope, E.J.A. 1998. Silica sol-gel encapsulation of pancreatic islets. *Proc. Soc. Exp. Biol. Med.* 218:365–369.

Pinel, C., Loisil, A., and Gallezot, P. 1997. Preparation and utilization of molecularly imprinted silicas. *Adv. Mater.* 9:582–585.

Plaschke, M., Czolk, R., Reichert, J., and Ache, H.J. 1996. Stability improvement of optochemical sol–gel film sensors by immobilization of dye-labeled dextran. *Thin Solid Films* 279:233–235.

Podbielska, H. and Ulatowska-Jarza, A. 2005. Sol-gel technology for biomedical engineering. *Bull. Polish Acad. Sci. Technol. Sci.* 53:261–271.

Pohl, E.R. and Osterholtz, F.D. 1985. *Molecular Characterization of Composite Interfaces*, H. Ishida and G. Kumar (Eds.), p. 157. Plenum, New York.

Poklar, N., Petrovcic, N., Oblak, M., and Vesnaver, G. 1999. Thermodynamic stability of ribonuclease A in alkylurea solutions and preferential solvation changes accompanying its thermal denaturation: A calorimetric and spectroscopic study. *Protein Sci.* 8:832–840.

Pope, E.J.A., Peterson, C.M.J., and Braun, K. 1997. Bioartificial organs: Part I. Silica gel encapsulated pancreatic islets for the treatment of diabetes mellitus *J. Sol-Gel Sci. Technol.* 8:635–639.

Pouxveil, J.C., Dunn, B., and Zink, J.I. 1989. Fluorescence study of aluminosilicate sols and gels doped with hydroxy trisulfonated pyrene. *J. Phys. Chem.* 93:2134–2139.

Qian, L. and Yang, X. 2006. Composite film of carbon nanotubes and chitosan for preparation of amperometric hydrogen peroxide biosensor. *Talanta* 68:721–727.

Radin, S. and Ducheyne, P. 2007. Controlled release of vancomycin from thin sol-gel films on titanium alloy fracture plate material. *Biomaterials* 28:1721–1729.

Reinhart, G.D. 1980. Influence of polyethylene glycols on the kinetics of rat liver phosphofructokinase. *J. Biol. Chem.* 255:10576–10578.

Revzin, A.F., Sirkar, K., Simonian, A., and Pishko, M.V. 2002. Glucose, lactate, and pyruvate biosensor arrays based on redox polymer/oxidoreductase nanocomposite thin-films deposited on photolithographically patterned gold microelectrodes. *Sens. Actuators B: Chem. B* 81:359–368.

Robbins, M.E., Oh, B.K., Hopper, E.D., and Schoenfisch, M.H. 2005. Nitric oxide-releasing xerogel microarrays prepared with surface tailored poly (dimethylsiloxane) templates. *Chem. Mater.* 17:3288–3296.

Rupcich, N. and Brennan, J.D. 2003. Coupled enzyme reaction microarrays based on pin-printing of sol–gel derived biomaterials. *Anal. Chim. Acta* 500:3–12.

Rupcich, N., Goldstein, A., and Brennan, J.D. 2003. Optimization of sol-gel formulations and surface treatments for the development of pin-printed protein microarrays. *Chem. Mater.* 15:1803–1811.

Rupcich, N., Nutiu, R., Li, Y., and Brennan, J.D. 2005a. Entrapment of fluorescent signaling DNA aptamers in sol-gel derived silica. *Anal. Chem.* 77:4300–4307.

Rupcich, N., Green, J.R.A., and Brennan, J.D. 2005b. Nanovolume kinase inhibition assay using a sol-gel derived multicomponent microarray. *Anal. Chem.* 77:8013–8019.

Salimi, A., Compton, R.G., and Hallaj, R. 2004. Glucose biosensor prepared by glucose oxidase encapsulated sol-gel and carbon-nanotube-modified basal plane pyrolytic graphite electrode. *Anal. Biochem.* 333:49–56.

Samuel, J., Strinkovski, A., Shalom, S., Lieberman, K., Ottolenghi, M., Avnir, D., and Lewis, A. 1994. Miniaturization of organically doped sol-gel materials: A microns-size fluorescent pH sensor. *Mater. Lett.* 21:431–434.

Sanchez, C., Julian, B., Belleville, P., and Popall, M. 2005. Applications of hybrid organic–inorganic nanocomposites. *J. Mater. Chem.* 15:3559–3592, and references cited therein.

Santra, S., Wang, K., Tapec, R., and Tan, W. 2001a. Development of novel dye-doped silica nanoparticles for biomarker application. *J. Biomed. Opt.* 6:160–166.

Santra, S., Zhang, P., Wang, K., Tapec, R., and Tan, W. 2001b. Conjugation of biomolecules with luminophore-doped silica nanoparticles for photostable biomarkers. *Anal. Chem.* 73:4988–4993.

Shen, C. and Kostic, N.M. 1997. Kinetics of photoinduced electron-transfer reactions within sol-gel silica glass doped with zinc cytochrome c. Study of electrostatic effects in confined liquids. *J. Am. Chem. Soc.* 119:1304–1312.

Shin, J.H. and Schoenfisch, M.H. 2006. Improving the biocompatibility of *in vivo* sensors via nitric oxide release. *Analyst* 131:609–615, and references cited therein.

Shin, J.H., Marxer, S.M., and Schoenfisch, M.H. 2004. Nitric oxide-releasing sol-gel particle/polyurethane glucose biosensors. *Anal. Chem.* 76:4543–4549.

Singh, S., Singhal, R., and Malhotra, B.D. 2007. Immobilization of cholesterol esterase and cholesterol oxidase onto sol–gel films for application to cholesterol biosensor. *Anal. Chim. Acta* 582:335–343.

Stobie, N., Duffy, B., McCormack, D.E., Colreavy, J., Hidalgo, M., McHale, P., and Hinder, S.J. 2008. Prevention of *Staphylococcus* epidermidis biofilm formation using a low-temperature processed silver-doped phenyltriethoxysilane sol-gel coating. *Biomaterials* 29:963–969.

Sui, X., Cruz-Aguado, J.A., Chen, Y., Zhang, Z., Brook, M., and Brennan, J.D. 2005. Properties of human serum albumin entrapped in sol-gel-derived silica bearing covalently tethered sugars. *Chem. Mater.* 17:1174–1182.

Tan, X., Li, M., Cai, P., Luo, L., and Zou, X. 2005. An amperometric cholesterol biosensor based on multiwalled carbon nanotubes and organically modified sol-gel/chitosan hybrid composite film. *Anal. Biochem.* 337:111–120.

Tang, D., Yuan, R., Chai, Y., Zhong, X., Liu, Y., and Dai, J. 2006. Electrochemical detection of hepatitis B surface antigen using colloidal gold nanoparticles modified by a sol–gel network interface. *Clinical Biochem.* 39:309–314.

Tao, Z., Tehan, E.C., Bukowski, R.M. et al. 2006. Templated xerogels as platforms for biomolecule-less biomolecule sensors. *Anal. Chim. Acta* 564:59–65, and references cited therein.

Thomas, K.G., Zajicek, J., and Kamat, P.V. 2002. Surface binding properties of tetraoctylammonium bromide-capped gold nanoparticles. *Langmuir* 18:3722–3727.

Tran, C.D., Ilieva, D., and Challa, S. 2004. Inhomogeneity in distribution and conformation of bovine serum albumin in sol-gel: A closer look with a near infrared multispectral imaging technique. *J. Sol-Gel Sci. Technol.* 32:207–217.

Tripathi, V.S., Kandimalla, V.B., and Ju, H. 2006. Preparation of ormosil and its applications in the immobilizing biomolecules. *Sens. Actuators B* 114:1071–1082.

Tsai, H. and Doong, R. 2005. Simultaneous determination of pH, urea, acetylcholine and heavy metals using array-based enzymatic optical biosensor. *Biosens. Bioelectron.* 20:1796–1804.

Tsai, H.C. and Doong, R.A. 2004. Simultaneous determination of renal clinical analytes in serum using hydrolase- and oxidase-encapsulated optical array biosensors. *Anal. Biochem.* 334:183–192.

Tubio, G., Nerli, B., and Pico, G. 2004. Relationship between the protein surface hydrophobicity and its partitioning behaviour in aqueous two-phase systems of polyethyleneglycol-dextran. *J. Chromatogr. B* 799:293–301.

Vazquez-Lira, J.C., Camacho-Frias, E., Alvarez, A.P., and Vera-Avila, L.E. 2003. Preparation and characterization of a sol-gel immunosorbent doped with 2,4-D antibodies. *Chem. Mater.* 15:154–161.

Vega, A.J. and Scherer, G.W. 1989. Study of structural evolution of silica gel using proton and silicon-29 NMR. *J. Non-Cryst. Solids* 111:153–166.

Wang, R., Narang, U., Prasad, P.N., and Bright, F.V. 1993. Affinity of antifluorescein antibodies encapsulated within a transparent sol-gel glass. *Anal. Chem.* 65:2671–2675.

Williams, A.K. and Hupp, J.T. 1998. Sol-gel-encapsulated alcohol dehydrogenase as a versatile, environmentally stabilized sensor for alcohols and aldehydes. *J. Am. Chem. Soc.* 120:4366–4371.

Wolfbeis, O.S., Oehme, I., Papkovskaya, N., and Klimant, I. 2000. Sol-gel based glucose biosensors employing optical oxygen transducers and a method for compensating for variable oxygen background. *Biosens. Bioelectron.* 15:69–76.

Wu, Z.S., Li, J.S., Luo, M.H., Shen, G.L., and Yu, R.Q. 2005. A novel capacitive immunosensor based on gold colloid monolayers associated with a sol–gel matrix. *Anal. Chim. Acta* 528:235–242.

Xynos, I.D., Hukkanen, M.V.J., Batten, J.J., Buttery, L.D., Hench, L.L., and Polak, J.M. 2000. Bioglass t45S5 stimulates osteoblast turnover and enhances bone formation *in vitro*: Implications and applications for bone tissue engineering. *Calcif. Tissue Int.* 67:321–329.

Yang, H.H., Zhu, Q.Z., Qu, H.Y., Chen, X.L., Ding, M.T., and Xu, J.G. 2002. Flow injection fluorescence immunoassay for gentamicin using sol-gel-derived mesoporous biomaterial. *Anal. Biochem.* 308:71–76.

Yang, M., Yang, Y., Liu, Y., Shen, G., and Yu, R. 2006. Platinum nanoparticles-doped sol–gel/carbon nanotubes composite electrochemical sensors and biosensors. *Biosens. Bioelectron.* 21:1125–1131.

Yurong, C. and Lian, Z. 2005. Effect of thermal treatment on the microstructure and mechanical properties of gel-derived bioglasses. *Mater. Chem. Phys.* 94:283–287.

Zhang, Z., Long, Y., Nie, L., and Yao, S. 2006. Molecularly imprinted thin film self-assembled on piezoelectric quartz crystal surface by the sol–gel process for protein recognition. *Biosens. Bioelectron.* 21:1244–1251.

Zhao, X., Tapec-Dytioco, R., and Tan, W. 2003. Ultrasensitive DNA detection using highly fluorescent bioconjugated nanoparticles *J. Am. Chem. Soc.* 125:11474–11475.

Zhao, X., Hillard, L.R., Merchery, S.J. et al. 2004. A rapid bioassay for single bacterial cell quantitation using bioconjugated nanoparticles. *Proc. Natl. Acad. Sci. USA* 101:15027–15032.

Zheng, L. and Brennan, J.D. 1998. Measurement of intrinsic fluorescence to probe the conformational flexibility and thermodynamic stability of a single tryptophan protein entrapped in a sol-gel derived glass matrix. *Analyst* 123:1735–1744.

Zheng, L., Reid, W., and Brennan, J.D. 1997. Measurement of fluorescence from tryptophan to probe the environment and reaction kinetics within protein-doped sol-gel-derived glass monoliths. *Anal. Chem.* 69:3940–3949.

Zusman, R., Rottman, C., Ottolenghi, M., and Avnir, D. 1990. Doped sol–gel glasses as chemical sensors. *J. Non-Cryst. Solids* 122:107–109.

Index